HANDBOOK OF STRUCTURAL CERAMICS

Other McGraw-Hill Reference Books of Interest

Baumeister and Marks • MARKS' STANDARD HANDBOOK FOR MECHANICAL ENGINEERS
Bhushan and Gupta • HANDBOOK OF TRIBOLOGY
Brady and Clauser • MATERIALS HANDBOOK
Bralla • HANDBOOK OF PRODUCT DESIGN FOR MANUFACTURING
Brunner • HANDBOOK OF INCINERATION SYSTEMS
Corbitt • STANDARD HANDBOOK OF ENVIRONMENTAL ENGINEERING
Ehrich • HANDBOOK OF ROTORDYNAMICS
Elliot • STANDARD HANDBOOK OF POWERPLANT ENGINEERING
Freeman • STANDARD HANDBOOK OF HAZARDOUS WASTE TREATMENT AND DISPOSAL
Ganic and Hicks • THE MCGRAW-HILL HANDBOOK OF ESSENTIAL ENGINEERING INFORMATION AND DATA
Gieck • ENGINEERING FORMULAS
Grimm and Rosaler • HANDBOOK OF HVAC DESIGN
Harris • HANDBOOK OF NOISE CONTROL
Harris • SHOCK AND VIBRATION HANDBOOK
Hicks • STANDARD HANDBOOK OF ENGINEERING CALCULATIONS
Hodson • MAYNARD'S INDUSTRIAL ENGINEERING HANDBOOK
Jones • DIESEL PLANT OPERATIONS HANDBOOK
Juran and Gryna • JURAN'S QUALITY CONTROL HANDBOOK
Karassik et al. • PUMP HANDBOOK
Kurtz • HANDBOOK OF APPLIED MATHEMATICS FOR ENGINEERS AND SCIENTISTS
Parmley • STANDARD HANDBOOK OF FASTENING AND JOINING
Rohsenow, Hartnett, and Ganic • HANDBOOK OF HEAT TRANSFER APPLICATIONS
Rohsenow, Hartnett, and Ganic • HANDBOOK OF HEAT TRANSFER FUNDAMENTALS
Rosaler and Rice • STANDARD HANDBOOK OF PLANT ENGINEERING
Rothbart • MECHANICAL DESIGN AND SYSTEMS HANDBOOK
Schwartz • COMPOSITE MATERIALS HANDBOOK
Shigley and Mischke • STANDARD HANDBOOK OF MACHINE DESIGN
Townsend • DUDLEY'S GEAR HANDBOOK
Tuma • HANDBOOK OF NUMERICAL CALCULATIONS IN ENGINEERING
Tuma • ENGINEERING MATHEMATICS HANDBOOK
Wadsworth • HANDBOOK OF STATISTICAL METHODS FOR ENGINEERS AND SCIENTISTS
Young • ROARK'S FORMULAS FOR STRESS AND STRAIN

HANDBOOK OF STRUCTURAL CERAMICS

Mel M. Schwartz   Editor in Chief
*Sikorsky Aircraft Division
United Technologies Corporation*

McGraw-Hill, Inc.
New York   St. Louis   San Francisco   Auckland   Bogotá
Caracas   Lisbon   London   Madrid   Mexico   Milan
Montreal   New Delhi   Paris   San Juan   São Paulo
Singapore   Sydney   Tokyo   Toronto

To cara

Library of Congress Cataloging-in-Publication Data

Schwartz, Mel M.
   Handbook of structural ceramics / Mel Schwartz.
     p.   cm.
   Includes index.
   ISBN 0-07-055719-5
   1. Ceramics—Handbooks, manuals, etc. I. Title.
TP807.S343   1992
666—dc20   91-28316
                                                              CIP

Copyright © 1992 by McGraw-Hill, Inc. All rights reserved. Printed in the United States of America. Except as permitted under the United States Copyright Act of 1976, no part of this publication may be reproduced or distributed in any form or by any means, or stored in a data base or retrieval system, without the prior written permission of the publisher.

1 2 3 4 5 6 7 8 9 0   DOC/DOC   9 7 6 5 4 3 2 1

ISBN 0-07-055719-5

The sponsoring editor for this book was Robert W. Hauserman, the editing supervisor was Ingeborg M. Stochmal, and the production supervisor was Donald Schmidt. This book was set in Times Roman. It was composed by McGraw-Hill's Professional Book Group composition unit.

Printed and bound by R. R. Donnelley & Sons Company.

Information contained in this work has been obtained by McGraw-Hill, Inc., from sources believed to be reliable. However, neither McGraw-Hill nor its authors guarantees the accuracy or completeness of any information published herein, and neither McGraw-Hill nor its authors shall be responsible for any errors, omissions, or damages arising out of use of this information. This work is published with the understanding that McGraw-Hill and its authors are supplying information but are not attempting to render engineering or other professional services. If such services are required, the assistance of an appropriate professional should be sought.

CONTENTS

Preface ix

Chapter 1. General Introduction to Ceramic Materials   1.1

1.1. Introduction / *1.1*
1.2. Ceramics and Metals / *1.2*
1.3. Traditional and Advanced Ceramics / *1.8*
1.4. Generic Properties / *1.15*
1.5. Market Potential / *1.17*
1.6. Education and Ceramic Technology / *1.19*
    References / *1.20*
    Bibliography / *1.22*

Chapter 2. Mechanical Properties of Ceramics   2.1

2.1. Introduction / *2.1*
2.2. Ceramics Contrasted to Metals / *2.1*
2.3. Strength and Fracture Considerations / *2.4*
2.4. Ceramic Substitution—Design Considerations / *2.16*
2.5. Surface Effects / *2.24*
2.6. Composite Design Considerations / *2.26*
2.7. Nondestructive Evaluation / *2.29*
2.8. Conclusions / *2.30*
    References / *2.30*
    Bibliography / *2.33*

Chapter 3. Commercial Structural Ceramics   3.1

3.1. Introduction / *3.1*
3.2. Mechanical Testing of Ceramics / *3.3*
3.3. Standardization of Ceramics / *3.6*
3.4. Existing Handbooks, Databases, and Compilations / *3.8*
3.5. Commercial Structural Ceramics / *3.10*
3.6. Conclusions / *3.36*
    References / *3.39*
    Bibliography / *3.44*

Chapter 4. Powders—Forming, Processing, and Densification   4.1

4.1. Introduction / *4.1*
4.2. Ceramic Powders and Their Processes / *4.5*

4.3. Powders and Processing / *4.12*
4.4. Powders (Materials) / *4.41*
4.5. Reinforcements for Ceramics / *4.52*
4.6. Whiskers, Short Fibers, and Particulates / *4.60*
References / *4.77*
Bibliography / *4.87*

Chapter 5. Greenware Fabrication

5.1. Introduction / *5.1*
5.2. Ceramic Processes / *5.2*
5.3. Particle Packing / *5.6*
5.4. Rheology / *5.6*
5.5. Granulation / *5.6*
5.6. Forming Processes / *5.7*
5.7. Methods and Materials / *5.25*
References / *5.51*
Bibliography / *5.55*

Chapter 6. High-Temperature Processing and Consolidation

6.1. Introduction / *6.1*
6.2. Basics of High-Temperature Consolidation / *6.2*
6.3. Hot Pressing / *6.8*
6.4. Pressureless Sintering / *6.10*
6.5. Reaction Sintering / *6.13*
6.6. Overpressure Sintering / *6.16*
6.7. Hot Isostatic Pressing / *6.17*
6.8. Novel Methods / *6.26*
6.9. Other Processing Parameters / *6.36*
6.10. Conclusions / *6.37*
References / *6.37*
Bibliography / *6.42*

Chapter 7. Fabrication and Manufacturing Methods

7.1. Introduction / *7.1*
7.2. Manufacturing Methods / *7.1*
7.3. Machining with Ceramic Cutters / *7.38*
7.4. Joining of Structural Ceramics / *7.48*
7.5. Coatings / *7.84*
References / *7.94*
Bibliography / *7.105*

Chapter 8. Applications for Ceramic Materials and Processes

8.1. Introduction / *8.1*
8.2. Ceramics by Function and Material / *8.1*
8.3. Prognostication of Ceramic Applications / *8.5*
8.4. Markets for Structural Ceramics / *8.7*
8.5. Nondestructive Evaluation / *8.76*
References / *8.81*
Bibliography / *8.88*

Chapter 9. The Future of Ceramics

9.1. Introduction / *9.1*
9.2. Ceramics and Processes / *9.1*
9.3. Fabrication Processes / *9.10*
9.4. Nondestructive and Destructive Testing / *9.13*
9.5. Ceramic Applications / *9.14*
9.6. New Ceramic and Metal Materials / *9.17*
9.7. Reliability and Design / *9.38*
9.8. Future Trends / *9.40*
References / *9.41*
Bibliography / *9.48*

Index I.1

PREFACE

The hot tip of materials in the mid-1960s was "composites." This whispered word has changed, and the word for the decade is "ceramics." The industrial engineer understands the term as shorthand for a class of materials and a set of processes that may transform the next generation of industrial products just as surely as composites and thermoplastic plastics changed the present generation.

The gestation period for advanced structural ceramics is over; they are a reality and primed to explode. Some of the applications were never thought of previously. Ceramics are finding increased usage in many load-bearing applications, such as cutting tools, automotive-engine components, heat exchangers, wear parts, and aerospace. Other uses include gene-splicing equipment, office machinery, and replacement parts for the human body.

Unfortunately, the customer for ceramic components has had nowhere to turn for unbiased advice and guidance on the use of structural ceramics in engineering. The most experience is generally with the manufacturers, while specialists, who may know much about principles and the science, know little about the ins and outs of commercial products. The scientific literature is rather unstructured and unbalanced, weighted primarily in favor of the exotic or the unobtainable. The result has been trial and error and test.

This Handbook has been written with the aim to clear a straightforward path through the jungle of scientific and commercial literature, to provide processing techniques and procedures, classes of ceramics, current mechanical property data for design purposes, materials and their methods of manufacture, and subsequent fabrication and assembly methods, leading to the broad base of structural ceramic applications, including composites. This will answer most questions for the newly exposed and inexperienced user of ceramics as well as for the designer, ceramist, and materials and process engineer, who can then approach the use of ceramics in an appropriate way and will be in a better position to negotiate from a basis of some understanding and background knowledge.

In order for the forecasts expected of these twenty-first-century materials to be realized, ceramics must be refined to meet their statistical characterization, including: (1) designing for reliability; (2) improving the forming techniques to minimize the size and number of flaws in the finished piece; (3) improving or replacing the sintering process during which thermal conditions may create flaws; (4) reducing cost; (5) improving thermodynamic and mechanical stability in existing materials or with development of new materials; and (6) seeking nondestructive techniques that allow manufacturers to evaluate the finished part for flaw size and frequency.

Another development that remains is the emergence of effective new ways to toughen up ceramics in order to make them more competitive. One way is to work with ceramic-matrix composites. Ceramic–ceramic-reinforced composites are remarkable for their great hardness and heat resistance combined with thermal-shock resistance. Most important, their creep resistance, or ability to withstand high stress at elevated temperatures, is the best of any material yet produced. Since high-temperature creep of these materials occurs mainly by atomic diffusion at grain boundaries, their creep resistance is dependent on their grain size. The role of fiber reinforcements is thus to stiffen up the composite and prevent shape changes in components subject to high stress at elevated temperatures. Another approach is through transformation toughening; a third is by grain refinement.

All of these approaches are covered in this Handbook in order to show that the yet inadequate fracture toughness and the low damage tolerance of these brittle ceramic materials, which are both considered substantial obstacles to the use of these "perfectly ordered" ceramics in critical structural applications, are slowly being overcome.

Ceramists have long felt that other technical specialists, designers in particular, do not understand ceramics and often appear reluctant to learn about them. However, today ceramists both educate and interact with other technical disciplines and communities.

Second, finding ceramists with a high degree of mathematical cunning is difficult. Yet the real challenge facing today's ceramist is the need to learn about other related technologies and their economic factors, and then to take this information and, in collaboration with people from other disciplines, combine ceramics and other materials into components having an optimum balance between performance and total cost.

The key is to stiffen up materials, but the engineering is difficult because altering the composition of ceramics may also alter the directional properties of the material. Today we are moving toward engineering the right material for a job from the molecular level.

In the future the engineer and the ceramist may be able to participate in developing intelligent processing of materials (IPM). The IPM approach uses computerized "expert systems" to help make ceramic materials that have properties far surpassing those commonly used. IPM allows exceptional and reliable qualities to be built into materials during processing rather than by inspection. IPM technology differs from conventional automated processing that controls parameters such as temperature and pressure with preselected values. IPM allows for diversities in the incoming materials. It uses on-line measurements in real time to adjust a wide range of processing variables.

Many studies are indicating that U.S. manufacturers of advanced ceramics are facing a type of international competition today as never before. Every individual U.S. company is competing against foreign consortia organized on a country basis or even on a regional basis. In fact, the Japanese describe as "first" generation the traditional ceramics—brick, pottery, and so on—which are made from natural materials. Fine or high-tech ceramics are the "second" generation. They are trying to establish a "third" generation of ceramics with high reliability and toughness, with the aim of instituting an integrated technology system for these ceramics supported by performance estimation, testing, and simulation, and of designing technology for many applications. The United States must be able to compete in this billion dollar industry now and in the next decades.

Many of the obstacles can be surmounted by technologies now in the laboratories. Studies of synthetic raw materials for advanced ceramic composites are ongoing, with research to improve the performance of or to find substitutes for refractories requiring imported raw materials and to enable the development of advanced ceramic composites with high fracture toughness and temperature properties, which could replace metals requiring critical chrome, cobalt, manganese, or tantalum.

Other research efforts include studies of green-body and sintered microstructures, intergranular bond phases, and processing of high-purity raw materials, all directed at producing wear- and corrosion-resistant advanced ceramic materials.

I wish to acknowledge the able support and contributions of Drs. John B. Wachtman, Jr., and Dale Neisz, Center for Ceramics Research, Rutgers University, for Chapter 3, and Dr. Makuteswara Scrinivasan, consultant and formerly with the Carborundum Company and Union Carbide Corporation, for Chapters 2 and 6.

Mel M. Schwartz

ABOUT THE EDITOR

Mel M. Schwartz is an engineer with the Sikorsky Aircraft Division of United Technologies. Previously he was in charge of manufacturing research and development for the Aircraft Division of Rohr Industries and a manufacturing manager at Martin Marietta Corporation. He has taught welding metallurgy at San Diego City College and manufacturing processes and matrials at San Diego State University, and he has lectured at Yale University. Mr. Schwartz is a Fellow of SAMPE International and ASM International. He is a recipient of the SAMPE Lubin Award. He is the author or editor of several other books, including ASM International's *Ceramic Joining* and the second edition of McGraw-Hill's *Composite Materials Handbook*.

CHAPTER 1
GENERAL INTRODUCTION TO CERAMIC MATERIALS

1.1 INTRODUCTION

Ceramic materials technology and research and development in the United States are in a reasonably healthy state. However, several factors must be overcome for them to remain competitive into the future. Materials science and engineering are an enabling technology, highly leveraged in, and critical to, many areas of advanced technology. Our major trading partners in Asia and Europe are gaining on or exceeding our capabilities in the production of many materials and materials systems, ceramics being one of them.

This has come about through stepped-up investments in advanced ceramics materials technologies by other countries. As a result, the U.S. industry stands a greater risk of becoming "blindsided" by materials developments occurring elsewhere in the world. Other concerns include the substantial cutbacks in basic research by some materials industries, weaknesses throughout the U.S. materials R&D infrastructure in synthesis and processing research, our growing dependence on Europe and Asia for new developments in advanced-characterization instrumentation, and the low numbers of universities maintaining medium-sized facilities for ceramics engineering research.

"The opportunities are out there," said Dr. H. Kent Bowen of M.I.T. "Laboratory specimens have already been made that can fulfill the needs of the high-technology ceramics industry."

However, a lack of comprehensive government programs has hurt the U.S. effort, as well as a slow start by U.S. industry. Basic research has been satisfactory, but industry is too fragmented to take full advantage of it. There has not been enough cooperation and information exchange between powder producers, processors, manufacturers, machining companies, designers, and researchers.[1-3]

"Companies are so concerned about keeping information proprietary that one can't even buy ceramic powder to try and improve the processing of the material," according to Dr. D. Readey, Chairman of the Ceramic Engineering Department at Ohio State University.

Compounding the problem of lack of researchers is the unfamiliarity of product design engineers with the special properties of ceramics. "There are techniques today which permit design with brittle materials," says Dr. Readey. "If you have reasonably reliable data, you can design for a given stress. However, most materials engineers are completely ignorant of the ways to design with ceramics."

Some of the other negative perceptions of ceramic material usage include:

Very high raw-material purity requirements

Expensive, highly sophisticated processing

Inconsistent reliability or reproducibility

Insufficient standard reference materials

Lack of standard testing procedures

Poorly developed NDE procedures

Developments and plans for standard measurement methods and data gathering of fundamental material properties are ongoing programs. Therefore the high-performance ceramics based on carbides, nitrides, oxides, beryllides, and possibly aluminides are the future.[4-6]

1.2 CERAMICS AND METALS

The so-called advanced, or high-performance, or fine, or technical, or structural ceramics are materials such as silicon nitride (Si_3N_4), silicon carbide (SiC), zirconia (ZrO_2), boron carbide (B_4C), alumina (Al_2O_3), sialon (Si–Al–O–N), and beryllia (BeO). These materials have been developed to provide predictable and repeatable physical properties, allowing their use in applications such as seals in automobiles, electronic substrates, wear components, and high-temperature bearings.

Yet in the minds of many engineers, technical ceramics are perceived as still being in the laboratory. The fact is that not only did these materials enter the marketplace as replacements for other ceramics, they have displaced metals in several traditional areas.

The present definition of ceramics includes not only the traditional materials made by heating naturally occurring substances, but also the highly refined and synthesized materials engineered for modern chemical, mechanical, electrical, optical, and magnetic properties.

Most people think that a ceramic is brittle, high-melting, nonconducting (of both heat and electricity), and nonmagnetic. They also think that metals have opposite properties. These stereotyped viewpoints are not necessarily true for ceramics or metals. In fact, there is no clear-cut boundary between the two. Rather, there are intermediate compounds that have some aspects typical of ceramics and others typical of metals.

Ceramics and metals are two groups of materials whose main difference is that they each represent a different combination of physical properties. Because of their availability and predictability, most people are accustomed to working with, thinking in terms of, and designing with metals. This represents an almost subconscious acceptance of a typical metal's ductility and strength combined with electrical and thermal conductivity.

In addition, there are misconceptions about the properties of advanced ceramics based on experiences with traditional ceramics: "like a coffee cup, drop it, it breaks." Educating and making engineers aware of the increasing body of knowledge about these materials is the first step in viewing ceramics as legitimate materials for use in demanding applications.[7-9]

Perhaps the most significant difference between traditional metals and ceramics lies in their ductility. Owing to the highly localized covalent-ionic bonding typical of ceramics, their macroscopic properties are brittle in nature. Brittle materials behave elastically until they fail catastrophically, as opposed to ductile materials, which deform plastically after the elastic limit has been reached, but prior to failure.

The nature of a material is largely controlled by the type of bonding between its constituent atoms, which in turn is controlled by the electron configuration of the atoms. Elements with unfilled outermost electron shells interact with other atoms, such that electrons are shared or exchanged between these atoms to achieve full shells.

Pure metals consist of atoms of a single size and an electron configuration in a close-packed arrangement. The outer electrons are shared freely by all the atoms in the structure. This mutual sharing of electrons provides the bond force that holds the atoms together into a metal crystal. It also provides the basis for most of the properties that we associate with a metal: ductility, high electrical conductivity, thermal conductivity, and thermal expansion.

Atoms in a ceramic material are primarily held together by covalent or ionic bonding, or a combination of the two. In covalent bonding, electrons are shared, but only by two adjacent atoms. This results in a directional bond. Such bonds can produce strong, rigid three-dimensional structures (such as diamond, SiC, or Si_3N_4), fibrous chain structures (such as asbestos), and laminar structures (such as graphite, mica, and clays). Characteristics of covalent ceramics include high hardness, superior chemical inertness, no ductility, low thermal expansion, and low electrical conductivity.

Ionic bonding involves transfer of one or more electrons between adjacent atoms, producing oppositely charged ions, bonded by coulombic attraction. Examples of ceramics with a highly ionic bond include sodium chloride, calcium fluoride, and magnesium oxide. Al_2O_3, SiO_2, and ZrO_2 show both ionic and covalent characteristics. (See Fig. 1.1 and Table 1.1.)

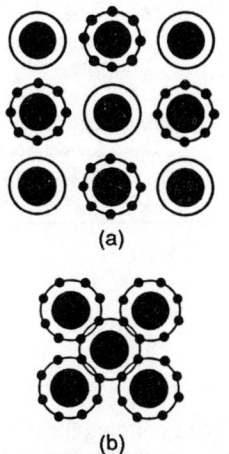

FIGURE 1.1 Bonding configurations in ceramic materials. Large black circles represent atoms, small ones, electrons. Rings surrounding atoms represent electron orbitals. (*a*) Ionic bond. (*b*) Very strong oriented covalent bond, consisting of shared electrons.

Ionic ceramics tend to form close-packed structures similar to those of pure metals, except that no free electrons are present and alternate atoms are of substantially different size. As a result, ionic ceramics have low ductility, high thermal expansion, and low electrical conductivity. Some have ionic conductivity at high temperatures because of diffusion of the charged ions through the structure. A good example is ZrO_2. At high temperatures, the negatively charged ions can move through the structure and carry an electric current. This material has been used in oxygen sensors to monitor the level of combustion in industrial processes.

TABLE 1.1 Properties of Some Advanced Ceramics

Type	Atomic bonding	Examples	Properties
Oxides	Ionic	Al_2O_3 (sapphire)	Hard-wearing, good creep properties
		Cr_2O_3	
		Fe_2O_3 (hematite)	
		MgO	
		ZrO_2 (PSZ)	
		$LiAl_2SiO_6$ (glass ceramic)	
Carbides	Less ionic interstitial compounds	ZrC	Very hard
		TiC	Very hard
		VC	Very hard
		NbC	High E moduli
		B_4C	High-temperature stability
	Covalent	SiC*	Poor creep properties
		WC	Used for cutting tools, abrasives, and dies
Nitrides	Covalent	BN (ambourite)	Low density
		Si_3N_4	High-temperature stability
		AlN	Very hard
		Sialon†	Good creep properties
		TiN	Used for cutting tools, gas turbine wheels, nozzles, and crucibles
Borides	Covalent	LaB_6	Excellent conductor
		ZrB_2	Used for electron microscope filaments, good creep properties

*SiC has properties more typical of nitrides.
†An alloy of Si, Al, O, and N.
Source: From Richardson.[3]

Ceramics Processing

In order to continue the enthusiastic swell that has been building for the potential of advanced ceramics as a unique class of engineering materials for certain high-performance applications, ongoing research is aimed at characterizing this independent class of materials, increasing the basic knowledge base, and developing the technology required to ensure the production of reliable, reproducible, and cost-competitive ceramic products.

Two specific areas receiving attention are inadequate fabrication techniques and the high sensitivity of ceramics to thermal and mechanical shock (inherent brittleness). Intense research worldwide is the focus of scientists, engineers, chemists, designers, and material specialists to solve these problems. While inadequate fabrication techniques can be solved in time as the technology matures, the problem of inherent brittleness cannot be avoided and ways must be found to manage it. (See Table 1.2.)

Ceramic materials are very sensitive to small flaws introduced during process-

TABLE 1.2 Characteristics of Advanced Ceramics

Advantages	
High melting point	Good dielectric properties
High stiffness	Thermal/electrical insulators
High hot strength	Semiconductor properties
High compressive strength	Ion-conductor properties
High hardness	Magnetic properties
Wear and corrosion resistance	Biocompatibility
Low density (lightweight)	Abundant raw materials

Limitations
Susceptible to thermal and mechanical shock (brittle)
Gaps in understanding and experience
Difficult to fabricate
Poor reproducibility
High cost

ing and in service. The multistage fabrication process for ceramics includes powder production, powder conditioning, shaping, and densification. Flaws, either physical (foreign inclusions or porosity) or chemical (impurities leading to second-phases, particularly liquids) can occur in each of these stages and cannot be corrected in subsequent processing stages.

The goal in powder processing is to achieve the highest degree of intimate mixing; in powder conditioning, to avoid the formation of hard agglomerates; in shaping, to avoid porosity; and in densification, to avoid the formation of glassy phases in grain boundaries. (See Table 1.3.)

Because the starting powder determines the ultimate quality of the ceramic part, there is a need to develop improved starting powders with high chemical purity and controlled particle sizes. Also needed are the tools necessary to characterize the physical and chemical properties of fine powders, as well as standard reference materials.

Wachtman[10] recently said: "Ceramics constitute an enabling technology, and under the right circumstances a ceramic material gives a performance enhancement which makes it a required component in achieving competitive performance. At the same time ceramics can have a high performance leverage. The amount of ceramic needed for a performance improvement is often only a small portion of an overall device or system." Therefore, to take advantage of these ceramic properties, the processing and design variables are critical. Three particular areas are receiving considerable attention:

1. Improved ceramic processing, to increase reliability by reducing structural defects and reducing costs
2. Improved design practice, to exploit their good properties while minimizing their limitations
3. Basic materials engineering principles, to design optimized compositions and microstructures for improved properties

Research and development activities to realize improvements in these areas include, according to Wachtman:[10]

TABLE 1.3 Potential Flaws during Manufacturing

Powder production
Unfavorable particle size, shape, and distribution
Off composition
Foreign inclusions
Hard agglomerates
Powder conditioning
Unfavorable agglomerate size distribution
Hard agglomerates
Varying agglomerate density distribution
Varying additive distribution
Insufficient binder
Organic fiber inclusions
Powder shaping (green compacts)
Porosity, voids, and cracks
Varying density
Nonuniform binder and additive distribution
Segregation
Residual binder
Organic inclusions
Densification
Porosity, micropore clusters, voids, and cracks
Nonuniform grain size and growth
Harmful grain-boundary phases (glassy phases)
Inclusions
Rough surface

Production of ceramics with finer grain sizes, extending even to sizes below one micron.

Fine grain size is important because the strength of ceramics is dependent on flaw size, which usually correlates with the size of the grains. Fine grain size material can give higher strength if the processing is carefully controlled. The lower the sintering temperature, the finer the grain size, and this leads to advantages in processing.

Ceramics are being systematically designed as multiphase composites. For example, a new family of ceramic composites composed of cordierite (magnesium aluminum silicate) and mullite (magnesium aluminate) is being developed for electronic substrate applications. The composite offers an advantage over the conventional alumina substrate in having a better thermal expansion match to silicon and a lower dielectric constant. The composite offers new processing challenges because a multicomponent suspension must be used and because sintering in two-phase systems presents problems of compatibility of deformation.

New chemical routes and better control of the chemistry of conventional processing routes is leading to better properties and better quality control.

The sol-gel process as well as other processes with distinct differences in detail, including the synthesis of very fine powders of uniform size by careful control of nucleation and growth, are discussed in Chaps. 4 through 6.

Another area receiving considerable attention is the use of computer modeling, which can permit the optimization of a set of properties rather than a single property. Other fields akin to processing as well as manufacturing technology (surface phenomena, joining, and tribological behavior) are being studied on a more scientific basis and are covered in Chaps. 7 and 8.

Material Processing

Advanced ceramic materials have attracted considerable attention because they present important new business opportunities and could improve national competitiveness significantly. Success in meeting these expectations, however, will depend on many conditions influencing the development and marketing of these advanced materials.

The displacement of outmoded materials by superior substitutes is a recurring theme in the history of materials development. Just as wood was displaced by bronze, which in turn was replaced by steel, so, too, are conventional steels (as well as other metals and glass) now being superseded by modern engineering polymers and structural ceramics. From this viewpoint, the introduction of advanced ceramic materials is evolutionary rather than revolutionary.

Nevertheless, distinct differences do exist between contemporary advanced ceramic materials and their forerunners. The most notable changes from past materials development were from new economic conditions and related information requirements. These changes include the rapid proliferation of new materials, the ability to serve small specialty markets, effects on manufacturing processes, and a much wider spectrum of scientific and industrial input.

Perhaps the most important feature of modern ceramic material development has been the accelerating rate at which new materials have been created and marketed.

A major cause of the dramatic growth in the number and variety of new ceramic materials has been the increasing ability of scientists to restructure materials at the molecular level. A seemingly infinite assortment of materials with different properties can be generated by reassembling their atomic structures. This capability is related to other factors that also have been instrumental in promoting a multiplicity of new and advanced structural ceramic materials, such as the birth of powerful computers and other sophisticated research tools and experimental techniques, the pressure of global competition, and the rise of worldwide information and communication systems for disseminating and exchanging scientific knowledge.

The rapid development of advanced materials has important implications. First, the large number and the diversity of new materials coming on stream will raise many questions about their effects on numerous issues. As the demand for data on materials increases, the ceramic industry is beginning to tax the capabilities of traditional materials information sources.

Another implication of rapid ceramic materials innovation is that the time-demand profile of future materials may be very brief, that is, the sequence of demand growth, maturation, and decline for new materials will be shorter than in the past since rapid technological improvements introduce so many superior competitors that substitution occurs earlier and more frequently. The shorter life span and greater turnover rate of future ceramic materials mean that the dynamics of materials development and marketing will be increasingly challenging.

The foregoing implications are significant to the government agencies and trade organizations that are expected to provide information on developments in the ceramic materials industry. Advanced ceramic materials offer advantages to so many manufacturing and industrial sectors that they have become significant to national competitiveness. First, improved ceramic materials and processes form the basis of all competitive manufacturing systems. Better ceramic materials can improve manufacturing competitiveness by raising the performance characteristics of products made from them and by reducing costs. Second, new ceramic materials that perform more efficiently provide manufacturers with opportunities to meet market demands with fewer resources; thus, overall resource scarcity and its attendant costs can be reduced. Third, new ceramic materials with improved capabilities are the means by which more demanding technological innovations can be converted into commercially feasible products.

New ceramic materials typically have been utilized first in the most technologically advanced and competitive sectors of industrial economies. Today those sectors are aerospace and automotive applications, biotechnology, communications, computerization, and electronics. In a sense, progress in materials science sets ultimate limits on the rates at which the technologically advanced sectors of industrialized economies can grow.[11-13] The significance of advanced ceramic materials to national competitiveness stems from more than their inherent performance and cost characteristics. New materials are modernizing contemporary U.S. manufacturing.[14,15]

1.3 TRADITIONAL AND ADVANCED CERAMICS

Traditional ceramics are largely made from natural raw materials, which are separated physically and reduced in size. The new ceramics generally require chemical conversion of raw materials into intermediate compounds, which lend themselves to purification and subsequent chemical conversion into the final form desired.[5] An extreme example of the production of pure oxide powder would involve refinement of pure metal from an oxide followed by reoxidation under controlled conditions. More typically for oxide powders, a soluble intermediate, such as carbonate, hydroxide, or a metal-organic compound, is formed, purified, and then transformed into an oxide. For the carbide and nitride ceramics, high-temperature reactions are generally used.[5]

Traditional ceramics are made using powders that have a distribution of particle sizes. For the new ceramics, one experimental line of work aims at producing very strong and reliable ceramics by working from very fine spherical, monodisperse powders (powders having a very narrow distribution of particle sizes). These monodisperse powders are made typically by techniques of colloidal chemistry for oxides and by controlled nucleation and growth in gas-phase reactions for the carbides and nitrides. However, most of the new ceramics are still made from powders with a broad distribution of particle sizes, but with smaller average particle sizes (less than 1 micrometer) than the traditional ceramics.[5]

Classes

Ceramics, one of the three major materials families, are crystalline compounds of metallic and nonmetallic elements. The ceramic family is large and varied, and includes such materials as refractories, glass, brick, cement and plaster, abrasives, sanitary ware, dinnerware, artware, porcelain enamel, ferroelectrics, ferrites, and dielectric insulators.[16,17] Also, intermetallic compounds, such as aluminides and beryllides, which are classified as metals, and cermets,[18] which are mixtures of metals and ceramics, are usually thought of as ceramic materials because of their physical characteristics which are similar to those of certain ceramics.

The following are the principal distinguishing characteristics of ceramics as a class:

1. They are crystalline materials, like metals, but because of fewer free electrons, they have little or essentially no electrical conductivity at room temperature.
2. They have high stability and, on average, higher melting points and greater chemical resistance than metals and organic materials.
3. They are generally the hardest of the engineering materials.
4. They are extremely stiff and rigid. Under mechanical stress, they have little or no yield and exhibit brittle fracture.

Today cement and concrete replace stone in most large structures. But cement, too, is a ceramic—a complicated but fascinating one. The understanding of its structure, and how it forms, is better now than it used to be, and has led to the development of special high-strength cement pastes which can compete with polymers and metals in certain applications.

The most exciting of all is the development, in the past 10 years, of a range of high-performance engineering ceramics. They have the potential to replace, and greatly improve on, metals in many very demanding applications. Cutting tools made of sialons or of dense alumina can cut faster and last longer than the best metal tools. Engineering ceramics are highly wear-resistant. They are used to clad the leading edges of agricultural machinery such as harrows, increasing the life by 10 times. They are inert and biocompatible, so that they are good for making artificial joints (where wear is a problem) and other implants. Modern body armor is made of plates of boron carbide or alumina, sewn into a fabric jacket.

There are mainly eight classes of materials that are of prime interest:

1. Glasses, all of them based on silica (SiO_2), with additions to reduce the melting point or give other special properties[19,20]
2. Traditional vitreous ceramics, or clay products, used in vast quantities for plates and cups, sanitary ware, tiles, bricks, and so forth[21]
3. New high-performance ceramics, now finding application for cutting tools, dies, engine parts, and wear-resistant parts
4. Cement and concrete, a complex ceramic with many phases and one of three essential bulk materials of civil engineering since approximately 1900 A.D.
5. Rocks and minerals, including ice
6. Ceramic composites, including fibers, particulates, and whiskers
7. Refractory groups, including oxides, carbides, borides, and nitrides
8. Miscellaneous

As with metals, the number of different ceramics is vast. But there is no need to remember them all. The generic ceramics discussed in the following sections embody the important features; others can be understood in terms of these. Although their properties differ widely, they all have one feature in common—they are intrinsically brittle. It is this brittleness that dictates the way in which they can be used.

They are, potentially or actually, cheap. Most ceramics are compounds of oxygen, carbon, or nitrogen with metals such as aluminum or silicon. All five are among the most plentiful and widespread elements in the earth's crust.

Glasses

Glasses are used in enormous quantities. As an engineering material, glass has a serious image problem. Broken bottles and cracked window panes are reminders that it is a brittle material. But this mindset is based almost entirely on a familiarity with the lowest-performance member of the glass family, soda-lime glass.[19]

Greatly superior glasses are available, some having properties that compare favorably with those of high-priced metals and other engineering materials. In addition, some properties of the more specialized glasses are not available in any other materials.

High-performance properties are readily produced by taking advantage of a relatively unrecognized property of glass—the ability to alter its behavior dramatically by changing its chemistry. Even when alterations are extreme, glass retains its ability to transmit light of a specific wavelength.[19]

Glass behaves much differently from common engineering materials, and some of its unique properties lead to its greatest advantages. For instance, unlike metals, glass does not have a specific melting point at which it exists together as a solid and a liquid. Rather, it gradually softens into a plastic and finally a liquid as the temperature increases. Glass can be formed and worked at temperatures just above the softening point. Two glasses of primary interest are the common window glass and the temperature-resistant borosilicate glasses. (See Table 1.4.)

Vitreous Ceramics

The potter has been a respected member of society since society developed. Pottery has survived the ravages of time better than any other product. The pottery of an era or civilization often gives the clearest picture of its state of development and its customs. Modern pottery, porcelain, tiles, and structural and refractory

TABLE 1.4 Generic Glasses

Glass	Typical composition, wt %	Typical uses
Soda-lime glass	$70SiO_2$, $10CaO$, $15Na_2O$	Windows, bottles, etc.; easily formed and shaped
Borosilicate glass	$80SiO_2$, $15B_2O_3$, $5Na_2O$	Pyrex; cooking and chemical glassware; high-temperature strength, low coefficient of expansion, good thermal shock resistance

Source: Adapted from Refs. 3, 9, 19.

bricks are made by processes which, though automated, differ very little from those of 2000 years ago. These included burnt clayware (500 B.C.), silicate glass (2500 B.C.), Roman cement ("pozzolana," burnt lime with volcanic ashes), whitewares of pottery and Chinese porcelain (9th century A.D.), lead Chinese crystal (12th century A.D.), heavy magnesite and chromite refractories in Staffordshire (18th century A.D.), and, finally, European porcelain (Limoges, Meissen, and Bohemia). All are made from clays, which are formed in the wet, plastic state and then dried and fired. After firing, they consist of crystalline phases (mostly silicates) held together by a glassy phase based, as always, on SiO_2. The glassy phase forms and melts when the clay is fired and spreads around the surface of the inert, but strong, crystalline phases, bonding them together. (See Table 1.5.)

Engineering Ceramics

The strength of a ceramic is largely determined by two characteristics, its toughness (K_{Ic}) and the size distribution of the microcracks it contains. A fully dense, high-strength ceramics class is now emerging which combines a higher K_{Ic} with a much narrower distribution of smaller microcracks, giving properties that make it competitive with metals, cermets, and even with diamond. (See Table 1.6.)

Cement and Concrete

Cement and concrete are used in construction on an enormous scale. Cement is a combination of lime (CaO), silica (SiO_2), and alumina (Al_2O_3), which sets when mixed with water. Concrete is sand and stones (aggregate) held together by a cement. (See Table 1.7.)

TABLE 1.5 Generic Vitreous Ceramics

Ceramic	Typical composition	Typical uses
Porcelain	Made from clays; hydrous	Electrical insulators
China	alumino-silicate such as	Artware, tableware, tiles
Pottery	$Al_2(Si_2O_5)(OH)_4$ mixed with	Artware, tableware
Brick	other inert minerals	Construction, refractory uses

Source: Adapted from Refs. 3, 9, 19.

TABLE 1.6 Generic Engineering Ceramics

Ceramic	Typical composition	Typical uses
Dense alumina	Al_2O_3	Cutting tools, dies; wear-
Silicon carbide, nitride	SiC, Si_3N_4	resistant surfaces, bearings,
Sialons	e.g., Si_2AlON_3	medical implants, engine
Cubic zirconia	ZrO_2 + 5 wt% MgO	and turbine parts, armor

Source: Adapted from Refs. 3, 9.

TABLE 1.7 Generic Cement and Concrete

Cement	Typical composition	Typical uses
Portland cement	$CaO + SiO_2 + Al_2O_3$	Cast facings, walkways, etc., component of concrete, general construction

Source: Adapted from Refs. 3, 9.

TABLE 1.8 Generic Natural Ceramics

Ceramic	Composition	Typical uses
Limestone (marble)	Largely $CaCO_3$	Building foundations, construction
Sandstone	Largely SiO_2	Building foundations, construction
Granite	Aluminum silicates	Building foundations, construction
Ice	H_2O	Arctic engineering

Source: Adapted from Refs, 3, 9.

Natural Ceramics

Stone is the oldest of all construction materials and the most durable. If used in a load-bearing capacity, it behaves like any other ceramic, and the criteria used in the design with stone are the same. One natural ceramic, however, is unique—ice. It forms on the earth's surface in enormous volumes: the Antarctic ice cap, something like 10^{13} square miles of pure ceramic. The mechanical properties are of primary importance in some major engineering problems, notably ice breaking and the construction of offshore oil rigs in the Arctic. (See Table 1.8.)

Ceramic Composites

The great stiffness and hardness of ceramics can sometimes be combined with the toughness of polymers or metals by making composites.[22] Glass- and carbon-fiber reinforced plastics are examples. The glass or carbon fibers stiffen the rather floppy polymer, but if a fiber fails, the crack runs out of the fiber and blunts in the ductile polymer without propagating across the entire section. Cermets are another example. They consist of particles of hard tungsten carbide bonded by metallic cobalt, much as gravel is bonded with tar to give a hard-wearing road surface (another ceramic composite). Bone is a natural ceramic composite, that is, particles of hydroxyapatite (the ceramic) bonded together by collagen (a polymer). Synthetic ceramic–ceramic composites (such as glass fibers in cement or silicon carbide fibers in silicon carbide) are now under development and will have important high-temperature applications in the next decade. (See Table 1.9.)

TABLE 1.9 Some Ceramic Composites

Ceramic composite	Components	Typical uses
Fiberglass	Glass-polymer	High-performance structures
CFRP	Carbon-polymer	High-performance structures
Cermet	Tungsten carbide–cobalt	Cutting tools, dies
Bone	Hydroxyapatite-collagen	Main structural material of animals
New ceramic composites	Alumina–silicon carbide	High-temperature and high-toughness applications

Source: Adapted from Refs. 3, 9, 18.

Refractory Groups

The most widely used common refractories are of the Al_2O_3 and the SiO_2 types. The compositions range from nearly pure SiO_2 through a wide range of Al_2O_3–SiO_2 combinations to nearly pure Al_2O_3.

Oxides.[23–25] The oxides can be divided into two groups, (1) single oxides that contain one metallic element and (2) mixed or complex oxides that contain two or more metallic elements. As a class, oxides are low in cost compared to other technical ceramics. Chapter 3 covers a variety of compositions and specific properties.

  Aluminum Oxide (Al_2O_3). Al_2O_3 is the most widely used oxide, chiefly because it is plentiful, relatively low in cost, and equal to or better than most oxides in mechanical properties. (See Chap. 3.)

  Beryllium Oxide (BeO). BeO is noted for its high thermal conductivity, which is about 10 times that of a dense Al_2O_3 at 499°C, three times that of steel, and second only to that of the high-conductivity metals (silver, gold, and copper). It also has high strength and good dielectric properties. However, it is costly and difficult to work with. Above 1649°C it reacts with water to form a volatile hydroxide. Also, because beryllia dust and particles are toxic, special handling precautions are required. The combination of strength, rigidity, and dimensional stability make BeO suitable for use in gyroscopes; and because of high thermal conductivity, it is widely used for transistors, resistors, and substrate cooling in electronic equipment.

  Magnesium Oxide (MgO). There are few commercially available dense MgO ceramics of the engineering type, primarily because MgO possesses few property advantages over Al_2O_3. On the other hand, MgO refractories are used extensively for their relatively high thermal conductivity, their refractoriness, and their resistance to basic slags. MgO ceramics tend therefore to be used in applications requiring resistance to corrosion at high temperatures.

  Zirconium Oxide (ZrO_2). There are several types of ZrO_2: a pure (monoclinic) oxide, a stabilized (cubic) form, and a number of variations such as Y_2O_3, MgO-stabilized ZrO_2, and nuclear grades. Ceramics with this type of strengthening can show advantages where certain properties are required. (See Table 1.10 and Chap. 3.)

  Thorium Oxide (ThO_2). ThO_2, the most chemically stable oxide ceramic, is only attacked by some earth alkali metals under some conditions. It has the highest melting point (3316°C) of the oxide ceramics. Like BeO, it is expensive, and as a result its use is limited. Some applications have been found in nuclear reactors.

TABLE 1.10 Typical Stabilizer Content of Various Forms of ZrO_2 Ceramic

Material type	Stabilizer type, mol %			
	CaO	MgO	Y_2O_2†	Mixed†
Fully stabilized*	10–15	12–16	6–9	Any appropriate combination
Partially stabilized	4–8	5–8	2–5	e.g., 2–5MgO plus 2–5CaO
Toughened, partially stabilized	5–9	7–9	2–4	Any appropriate combination
Tetragonal zirconia polycrystals	0–2	0–2	0–1	Any appropriate combination

*Lower levels of stabilizer require higher initial firing temperatures to achieve the cubic state and tend to be less stable in the temperature range 700–1000°C.

†Y_2O_3 is expensive, and is only used in high-value products when essential. The tendency is therefore to use MgO and CaO where possible. Y_2O_3-containing materials are said to destabilize less readily at high temperatures than CaO- or MgO-containing materials.

Source: Adapted from Refs. 3, 8, 9.

Titanium Oxide (TiO_2). Materials based primarily on TiO_2 itself have, in the main, been replaced by materials based on titanates, compounds of TiO_2, and other oxides, but silicate-bonded materials are still made in quantity for mechanical applications including thread guides.

Mixed Oxides. Except for zircon, the principal mixed oxides are composed of various combinations of MgO, Al_2O_3, and SiO_2. (See Chap. 3.)

Carbides. The carbide family contains materials with the highest melting points of all engineering materials. Unfortunately, because of their poor oxidation resistance, they cannot be used unprotected at high temperatures, except for SiC, which is useful at temperatures up to 1649°C.

Silicon Carbide (SiC). SiC has an attractive combination of high thermal conductivity, low thermal expansion, and low thermal shock. It is probably the most widely used carbide, and several forms of SiC, including self-bonded, also known as reaction-bonded or reaction-sintered, clay-bonded, and hot-pressed, are described in Chaps. 3 through 6.

Boron Carbide (B_4C). B_4C is best known for its extreme hardness and abrasion resistance. (See Chap. 3.)

Borides. The major materials in this group of refractory ceramics are borides of hafnium, tantalum, thorium, titanium, uranium, and zirconium. (See Chap. 3.)

Nitrides. Boron and silicon nitrides are the commercial materials in this group of refractory ceramics. (See Chap. 3.)

Silicon Nitride (Si_3N_4). One of the primary advantages of Si_3N_4 is its low coefficient of thermal expansion and hence its thermal shock resistance. Si_3N_4 components are available in two forms, reaction-bonded (RBSN) and hot-pressed (HPSN). (See Chaps. 3 through 5.)

Sialon (Si–Al–O–N) and Sintered Si_3N_4 (SSN). These products are essentially varieties of Si_3N_4. They have been developed in order to use sintering techniques instead of hot-pressing, which is expensive and limiting on shape, or reaction-bonding, which leaves the product with open porosity. Sialon ceramics derive their name from an acronym of their principal components, Si, Al, O, and N.[26] (See Chaps. 3, 5, 7, and 8.)

Boron Nitride (BN). BN is best known as the synthetic-diamond material Borazon. (See Chaps. 3, 7, and 8.)

Aluminum Nitride (AlN). AlN is another ceramic material typically synthesized from the elements, and fine powders are produced by milling.

Miscellaneous

Lanxide. Lanxides are formed in the reaction between a molten metal and oxygen in the air to some other vapor-phase oxidant. However, by controlling the molten metal's temperature and by adding traces of suitable dopant metals, a 1-inch-thick layer of a metal oxide composite can be grown on the liquid's surface.

Many types of metal may be used, but most work reported has been an alumina matrix reinforced by aluminum particles. Because the starting material is molten metal, careful control over composition is possible, as is extremely high purity. This results in properties far superior to those of conventional sintered ceramics. The high-temperature strength, for example, is much better than that of conventional Al_2O_3.

Chemically Bonded Ceramics (CBC). CBCs represent a new technology that produces high-performance ceramic parts and components at an attractively low cost. CBCs have 10 to 20 times the tensile strength of concrete. In terms of raw materials and energy costs, CBCs are generally 32 times cheaper than aluminum, 20 times cheaper than steel, and four to six times less expensive than plastic. CBC materials are mostly silicates, aluminates, and phosphates. Their low cost results from processing temperatures far below the 1093°C needed for producing conventional ceramics. Growth in these materials stems from their low cost, high performance, and the fact that the materials are neither toxic, flammable, nor petroleum-based. The material is used for plastic tooling, as well as for CBC armor, brake linings, roofing, wall panels, flooring, and electrical fixtures.

Intermetallics and Silicides. These materials, although technically metals because they are compounds of metals, are generally classified as ceramics. The three major classes are aluminides, beryllides, and silicides (silicon classed as a metal). They are hard and brittle in their polycrystalline form at room temperature, but they can be deformed plastically, like metals, at elevated temperatures. They are generally considered as having the greatest potential among ceramic materials for achieving low-temperature ductility.

Typically aluminides are nickel-aluminum compounds and titanium-aluminum compounds. Typical beryllides are those in which beryllium is compounded with either columbium, tantalum, or zirconium. Aluminides have the lowest melting point, beryllides the highest of the three major intermetallics. Silicides are metalloid compounds of silicon in combination with one or more metallic elements. They are closely related to the intermetallic compounds.

Glass Ceramics. Glass ceramics are a family of fine-grained crystalline materials made by a process of controlled crystallization from special glass compositions containing nucleating agents. They are sometimes referred to as devitrified ceramics or vitro ceramics.

1.4 GENERIC PROPERTIES

The technical ceramics now leaping to prominence are mostly polycrystalline, containing conglomerations of microcrystals. The extreme strength, chemical inertness, and heat and abrasion resistance of the technical ceramics stem largely from their structures. The strength of a given ceramic is strongly dependent on the size of its crystals grains. A material with large particles or with a random distribution of particle sizes is generally weaker than a ceramic with fine-grained or uniform-size particles. This is because large or random-size grains fit together with relatively large spaces in between, while smaller or more uniform grains can fit together more densely, with smaller spaces (Fig. 1.2).

The properties of ceramics and metals[27] result from a combination of the effects of atomic bonding and microstructure. The effects of bonding are primarily reflected in the intrinsic properties—chemical, physical, thermal, electrical, magnetic, and optical. Microstructure can also affect some of the intrinsic properties, but it has its major effect on the mechanical properties and the rate of chemical reaction. (See Chap. 3.)

Melting Temperature. Melting temperature is a function of the strength of the atomic bond. Multivalent ionic ceramics (BeO, Al_2O_3, ZrO_2) have much higher melting temperatures; covalent ceramics (TiC, HfC) have the highest melting temperatures.

Thermal Conductivity. Thermal conductivity is controlled by the amount of heat energy present, the nature of the heat carrier in the material, and the amount of heat dissipation. The primary ways to carry heat in ceramics are by lattice vibrations and radiation.

Thermal Expansion. The rate of thermal expansion of metals and ceramics is determined by the bond strength and the atomic structure. The greater the bond strength, the lower the expansion. Metals and ionic ceramics have close-packed atomic structures and a relatively high thermal expansion. On the other hand, covalent bonding is directional and produces structures having large open spaces. When a covalent ceramic is heated, a portion of the expansion can be absorbed by the open space within the structure or by bond-angle shifts, resulting in low expansion.

Toughness. Most ceramic metals are not tough. Once a crack in a ceramic starts, it propagates rapidly and results in fracture of the part. The problem of low toughness has been approached both from design and from materials aspects. The primary design approach has been to prestress the ceramic in compression by shrink-fit or lamination. This does not increase the toughness of the ceramic, but it does increase the resistance to crack initiation.

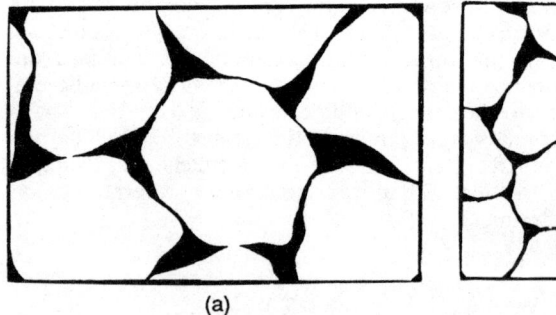
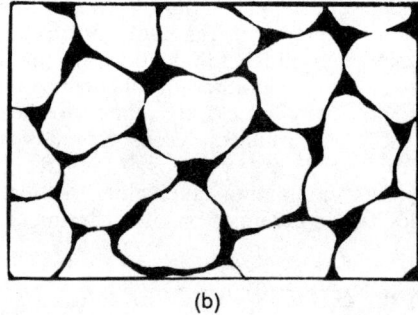

(a)   (b)

FIGURE 1.2 Larger spaces between crystal particles make coarse-grain ceramics (*a*) weaker than fine-grain ceramics (*b*) in which intercrystalline spaces are smaller.[27]

A broader method is to increase the toughness of the ceramic materials. Several techniques have been used, among them second-phase dispersion and fiber reinforcement. It is by second-phase dispersion that cermets such as WC–Co and TiC–Ni are made. The small additions of ductile cobalt and nickel allow redistribution of an applied load and minimize the stress concentration that normally causes crack initiation and propagation in the brittle ceramic.

Dual-ceramic particulate composites are being developed to satisfy applications where corrosion or oxidation of the composite metal phase is affected. The effective use of fiber reinforcement of glass, glass-ceramic, and ceramic matrices is in its infancy. More detailed discussions can be found in Chaps. 4, 8, and 9.

Ductility. Ductility is accommodated by dislocation movement along planes of atoms. For this to occur, a dislocation must (1) be present or be initiated easily, (2) have an activation energy below the fracture initiation energy for the material, and (3) have an unobstructed path for movement. These conditions are satisfied ideally in a pure metal having a close-packed structure.

Ionic-bonded ceramics have close-packed structures similar to the pure metals and thus have many potential slip planes. However, due to the opposite electrical charge of adjacent ions, each ion is stable only in a certain equilibrium position and coordination (number of nearest neighbors). A higher activation energy than is required for metals is necessary to move oppositely charged ions and cause slip. The situation is similar for covalent ceramics. The directionality of bonding will place atoms in equilibrium positions that require high activation energy for slip.

Ceramics fail in a brittle mode due to the presence of fabrication and structural flaws that result in stress concentration and fracture at a load well below the theoretical strength. Therefore ceramic components, because of their brittle nature, must be designed differently and more carefully than metals to avoid localized stress concentration resulting from impact, attachment, notches, thermal gradients, or other sources.

Mechanical Characteristics. A few ceramics have strengths above 173 MPa, but most have less. Ceramics are notable for the wide differences between their tensile and compressive strengths. (See Chap. 3.) As a class, ceramics are the most rigid of all materials. A majority of them are stiffer than most metals, and the modulus of elasticity in tension of a number of types runs as high as 340 to 442×10^2 MPa, compared with 197×10^2 MPa for steel.

1.5 MARKET POTENTIAL

The potential use of advanced structural ceramics may be grouped as near-term (5 years), intermediate term (by the year 2000), and long-term (2000 to 2025). The near-term market is estimated to be over $250 million.[8,28,29] It includes wear parts (seals, bearings, valves, nozzles), cutting tools, corrosion-resistant parts, and heat engine components (valve train components, thermal barrier coatings). Overcoming inadequate fabrication techniques and high sensitivity to thermal and mechanical shock will lead to a conservative estimated potential market of between $1 and $5 billion by the year 2000. Other disadvantages which have created major obstacles to structural applications for advanced ceramics include the present high cost and low reliability of manufactured parts (scrap rates as high as 25 to 50 percent). While experts see solutions to the problems in processing, reliability, and toughness within the next 10 years, large-scale commercialization will take longer. On the plus side, reports point to significant progress with the brittleness and reliability problems, achieved through better starting materials, improved processing, and the development of new composites.

Other studies[14,15] show that an estimated annual projected growth rate of 11 percent should occur between 1990 and 1995 for total U.S. shipments of advanced ceramics. The U.S. shipments could reach $10 billion in 2000, while Japanese studies predict their market as reaching $11.2 to $16.8 billion by the first decade of the next century.

In a worldwide survey conducted by CAMDEC (Ceramics Advanced Manufacturing Development & Engineering Center)[30] 46 percent of the respondents felt that advanced ceramic sales would be between $6 and $10 billion by the year 2000. In general, the anticipated markets, although large, were down somewhat from estimates in previous studies. Table 1.11 reflects another survey,[15,31] where the projected total value of advanced structural ceramic product shipments increases from $171 million in 1987 to $433 million in 1990, 22 percent from 1990 to 1995, and 18 percent from 1995 to 2000.

Demand for whisker-reinforced ceramic composites will exceed $385 million within 10 years. These ceramic composites are amalgams of advanced ceramics reinforced with micron-length whiskers, particulates, or fibers of SiC, for example. Whisker-reinforced ceramics are tough enough to be used in cutting tools, wear parts, heat exchangers, and heat engines. (See Chaps. 7 through 9.)

Between 1991 and 1996, the value of ceramic composite parts used in heat exchangers and commercial heat engines will easily exceed $200 million. Ceramic exchangers can operate at higher temperatures and in more corrosive environments than metal devices. Composite heat engine parts, such as diesels and turbines, will account for more than $150 million by 1996. Diesel engines that use composite "hot" sections could eliminate the need for a cooling system. Tough thermally stable composites that resist corrosion can be used to lengthen the life of turbines designed to burn heavily contaminated fuels, such as residual oil, coal, and sour gas.

By 1996, the U.S. Army could well be the largest buyer of ceramic composites. Key military uses will include diesel engines, turbine generators, armor, and gun-barrel liners.[32-34]

TABLE 1.11 U.S. Markets for Advanced Ceramics to the Year 2000 in 1988 Dollars

	1990		1995		2000	
	$M	%	$M	%	$M	%
Automotive, heat engines	81	19	310	27	820	31
Cutting tools	92	21	246	21	500	19
Wear parts and other industrial applications	150	35	320	28	720	27
Heat exchangers	15	3	50	4	100	4
Aerospace and defense-related applications	80	19	200	17	445	17
Bioceramics	15	3	34	3	60	2
Total	433	100	1160	100	2645	100
			1990–1995		1995–2000	
AAGR, %			22		18	

Source: Adapted from Ref. 15.

TABLE 1.12 Critical Barriers to Commercialization of Advanced Structural Ceramics

Barrier	Relative importance*
Increasing the understanding of ceramic characteristics by designers and users	4.1
Lack of applied R&D	3.9
Lack of developed markets for advanced ceramics	3.8
Lack of trained engineers and technicians	3.3
Lack of basic R&D in advanced ceramics	3.1
Lack of domestic high-quality raw materials	3.0

*Average respondents ranking: 1 (not critical) to 5 (very critical).
Source: From Cadotte et al.[30]

The development and application of advanced ceramic materials and products is also expected to help reduce to some extent U.S. dependence on foreign supplies of a variety of scarce or critical materials. Advanced ceramic materials, which are in abundant supply in the United States, do or will substitute in many applications for scarce materials such as tantalum, cobalt, and chromium. However, the extent of the reduction in U.S. demand for these scarce materials is expected to be relatively modest for the remainder of this century.

A prioritized ranking of the barriers to commercialization of advanced structural ceramics is shown in Table 1.12. The dominant concern has always been that of increasing the understanding of ceramic characteristics by designers and users. Table 1.12 also indicates a general level of satisfaction with the extent of basic R&D in advanced ceramics, but a need for increased applied research.

There are generally two problems with all predictions of advanced ceramics growth. First, almost all growth rates are taken from a currently very small base and then projected 10 to 15 years without adjusting for ultimate market saturation. Second, many of the market applications that might grow substantially in the future do not exist to any great extent today and still await the development of inexpensive raw materials, reliable fabrication techniques, and a total end-use system (such as advanced ceramic heat engines for automobiles) before initial commercialization can occur. The solution to these technical problems depends on the ability of industry and government to be technically innovative.

1.6 EDUCATION AND CERAMIC TECHNOLOGY

It is important that the upper management personnel in the key ceramic firms have the technical expertise to contribute and direct their personnel to understand the continuously changing and challenging structural ceramic technology. It is also important that they provide an adequate supply of skilled technical personnel to match the requirements of ceramic developments and research. This leads to an increased number of ceramists, ceramic engineers, chemists, material scientists, and other specialists who must be educated in the next decade to assist in developing ceramics to their full potential.

Currently, the lack of a science and engineering foundation for structural ceramics manufacturing is the dominant technical barrier preventing this technology from realizing its full potential for growth. Research on the analysis, design, and testing of ceramic materials and products is important, but it cannot contribute effectively to industrial competitiveness until it is linked to manufacturing. Manufacturing is the key that will unlock the economic leverage of products to be manufactured by industry.

The manufacturing of advanced structural ceramics is now based primarily on heuristics, or know-how. Remember that the performance characteristics of ceramics are determined largely by the microstructural arrangements that develop during the manufacturing process. Yet there is a shortage of basic information about how the performance of the material in the product is related to the manufacturing parameters. (See Chaps. 2 through 5.)

It is important to establish the fundamental basis in engineering science needed to support the development of manufacturing methods for ceramics and ceramic composites. Our technological objectives are to understand, simulate, and develop predictive models for the most important near-term processing methods, while building the required knowledge base. The ultimate goal is to integrate the ceramics design, materials, and production functions into a cost-effective system. The key interrelationships to be kept in mind relative to materials are properties, processing, and structure. They are discussed in Chaps. 2 through 6.

The long gestation period for advanced structural ceramics has passed, and reality is upon the users, manufacturers, and developers. Applications are developing rapidly, and many were never considered 5 years ago. The availability and commercialization of the new advanced structural ceramics will fundamentally change internal specification patterns in many existing industries, such as automobiles, manufacturing, energy, aerospace, bioceramics, and the defense-related fields.

Various types of the new advanced ceramics are now emerging from the laboratories and are undergoing field tests. Their manufacture is usually based on new findings obtained in connection with the advances in various fields of science and technology, although experience from the manufacture of traditional types of ceramics has also been utilized.[15] The materials are usually very sensitive to the maintenance of manufacturing conditions. This is why perfect control of the manufacturing processes, as well as of the properties of the final products, is vital.

Advanced structural ceramics promise to be among the most dynamic new high-technology markets of the 1990s and early 21st century. In many ways they are a structural engineer's dream material.

REFERENCES

1. E. J. Kubel, Jr., "Structural Ceramics: Materials of the Future," *AM&P Met. Prog.,* vol. 134, no. 2, pp. 25–33, Aug. 1988.
2. J. A. Spirakis, "Ceramics in a 'Metals World'," *AM&P Met. Prog.,* vol. 132, no. 3, pp. 48–51, Mar. 1987.
3. D. W. Richardson, "What Are Ceramics," *Chem. Eng.,* pp. 12–14; also, *Modern Ceramic Engineering; Properties, Processes and Use in Design,* Marcel Dekker, New York, 1982.
4. G. Fisher, "Refractory Uses—Practicality of High Technology Ceramics," *Ceram. Bull.,* vol. 66, pp. 1103–1108, 1987.
5. J. B. Wachtman and M. G. McLaren, "Advanced Ceramics: Structural Materials with a Hot Future," *Manuf. Eng.,* pp. 56–60, Feb. 1985.
6. T. L. Francis, "Advanced Ceramics in the United States," *Powder Met. Int.,* vol. 17, pp. 185–188, Apr. 1985.

7. H. Kent Bowen, "Advanced Ceramics," *Sci. Am.*, pp. 169–176, 1986.
8. "New Structural Materials Technologies: Opportunities for Use of Advanced Ceramics and Composites," Congress—Office of Techn. Assessment, Tech. Memo PB87-118253/KGC OTA TM-E-32, Sept. 1986, 88 pp.
9. K. Easterling, "Tomorrow's Materials," Inst. of Metals Gt. Britain, 1988, 109 pp.
10. J. B. Wachtman, "The Materials Effect in the Manufacturing Revolution: Emphasis on Advanced Ceramics," in M. V. Nevitt and N. D. Peterson (Eds.), *Advanced Materials in Manufacturing Revolution,* Argonne Natl. Lab., ANL89-3, Conf. 8806303, W31-109ENG38, 1989, pp. 52–63.
11. M. A. Steinberg, "Net Shape Technology in Aerospace Structures, V4, Fr1984-85; App: Future Composite Manufacturing Technology," Natl. Acad. of Science—Natl. Res. Counc., F49620-85C0107, Dec. 1986.
12. M. U. Islam, "Artificial Composites for High Temperature Applications: Review," Natl. Res. Counc. of Canada, DME007, NRC27323, Jan. 1987.
13. "Survey of Supply/Demand Relationship for Japanese Technical Information in United States: Field of Advanced Ceramics Research and Development," Commerce Dept., PB88-210943, Mar. 1988, 139 pp.
14. *Advanced Ceramic Materials,* Charles River Assoc., U.S. Dept. of Commerce, and Natl. Res. Counc., 1985, 651 pp.
15. "Advanced Structural Ceramics: Technologies, Economics and Market Opportunities," Business Comm. Co., Norwalk, Conn., GB-107, Dec. 1987, 354 pp.
16. W. L. Sheppard, Jr., "Ceramics in Chemical Service," 33d Biennial Rep. on Materials of Construction, *Chem. Eng.,* pp. 59–67, Oct. 24, 1988.
17. US–Japan Workshop on Dielectric and Piezoelectric Ceramics," KEIO University, Tokyo, Japan, July 1987.
18. P. S. Kislyj, "Cermets (Ceramic-Metallic Materials)," Army Foreign Science and Technol. Center, FSTC HT 0314-88, July 1988.
19. R. Caldwell, "Glass Is an Engineering Material," *Mach. Des.,* pp. 51–54, June 22, 1989.
20. S. Hampshire, "Preparation and Characterisation of Oxynitride Glasses and Glass-Ceramics," Natl. Inst. for Higher Education, Ireland, DAJA45-85C0050, May and Nov. 1987.
21. T. Kattamis, Univ. of Connecticut, Storrs, private communication.
22. "Ceramic Matrix Composites and Ceramic Fibers and Whiskers," Business Opportunity Rep., GB-110, Sept. 1988, 261 pp.
23. E. P. Rothman et al., "Potential of Ceramic Materials to Replace Cobalt, Chromium, Manganese, and Platinum in Critical Applications," Mater. Proc. Center, M.I.T., Cambridge, Mass., FR 1/84, Jan. 6, 1984, 292 pp.
24. "High-Technology Ceramics in Japan," Committee on the Status of High Technology Ceramics in Japan, NMAB-418, Natl. Acad. Press, Washington, D.C., 1984.
25. J. Jacobs and T. Kilduff, *Engineering Materials Technology,* Prentice-Hall, Englewood Cliffs, N.J., 1985, 656 pp.
26. "New Ceramic Looks to Future in Aerospace," *Ceram. Ind.,* pp. 36–37, Sept. 1984.
27. G. Graff, "Ceramics Take on Tough Tasks," *High Technol.,* pp. 68–73, Dec. 1983.
28. L. R. Johnson, A. P. S. Leotia, and L. G. Hill, "A Structural Ceramic Research Program: A Preliminary Economic Analysis," Energy and Environ. Sys. Div., Argonne Natl. Lab., ANL CNSV-38, Mar. 1983.
29. R. Wills, "Ceramics: Advances and Opportunities," Battelle Tech. Inputs to Planning, Battelle Columbus Labs., Columbus, Ohio, Rep. 21, 1980.
30. E. Cadotte, J. Brewer, and D. F. Craig, "Industry Survey on CAMDEC and the Barriers to Commercialization of Advanced Ceramics," *Ceram. Bull.,* vol. 66, pp. 1700–1701, Dec. 1987.

31. J. B. Wachtman, Jr. (Ed.), *Structural Ceramics,* Treatise on Materials Science and Technology Series, vol. 29, Academic Press, San Diego, Calif., 1989, 388 pp.
32. D. Lewis, III, "Research in Ceramics at the U.S. Naval Research Lab.," *Ceram. Bull.,* vol. 67, pp. 1349–1356, Aug. 1988.
33. A. P. Katz and R. J. Kerans, "Structural Ceramics Program at AFWAL Materials Lab," *Ceram. Bull.,* vol. 67, pp. 1360–1366, Aug. 1988.
34. J. W. McCauley et al., "Ceramics R&D at the U.S. Army Materials Technology Laboratory," Army Lab. Command, Watertown, Mass., Mater. Tech. Lab. MTL-TR-88-31, Mar. 1988.

BIBLIOGRAPHY

Boulet, J. A. M., "An Assessment of the State of the Art in Predicting the Failure of Ceramics," Univ. of Tennessee, ORNL/Sub-86-57598, DE08998, Mar. 1988, 26 pp.

Butler, E. P., "Transformation-Toughened Zirconia Ceramics," *Mater. Sci. Technol.,* vol. 1, pp. 417–432, June 1985.

"Ceramic Materials and Components for Engines," in V. J. Tennery (Ed.), *Proc. 3d Int. Sym. of the American Ceramics Society* (Las Vegas, Nev., Nov. 27–30, 1988), p. 1559.

Dickson, T., "Advanced Ceramics—Highlights of an Alabama Conf.," *Ind. Minor.,* pp. 79–81, Apr. 1985.

"Engineering Property Data on Selected Ceramics," vol. 1: "Nitrides," MCIC-HB-07, Aug. 1979, 107 pp.

"Engineering Property Data on Selected Ceramics," vol. 2: "Carbides," MCIC-HB-07, Aug. 1979, 135 pp.

"Engineering Property Data on Selected Ceramics," vol. 3: "Single Oxides," MCIC-HB-07, July 1981, 253 pp.

Lange, F. F., "Structural Ceramics: A Question of Fabrication Reliability," *J. Mater. Energy Syst., ASM,* vol. 6, no. 2, pp. 107–113, Sept. 1984.

"PSZ—A Breakthrough in Toughness," *Ceram. Ind.,* pp. 40–45, Apr. 1984.

Sheppard, L. M., "Sialon: Another Super Structural Ceramic," *AM&P,* vol. 2, pp. 35–39, Jan. 1986.

"Slow Growth for Ceramics," *Forecast '89, AM&P,* vol. 135, pp. 29–44, Jan. 1989.

CHAPTER 2
MECHANICAL PROPERTIES OF CERAMICS

2.1 INTRODUCTION

This chapter will address the general characteristics of high-performance ceramics, particularly with respect to their mechanical properties. As many of the materials are still evolving, with improvements occurring constantly, the reported properties may not represent the materials actually available. Nevertheless, this chapter will cover in detail the unique differences and similarities in the techniques used to determine the properties of ceramics as compared to metals and plastics. The inherent undesirable characteristic of ceramics is their brittleness, or low strain tolerance. In addition, they exhibit more variability in properties when compared to metals, thereby requiring a statistical estimation of service life. Also, their fracture toughness is very low, again when compared to metallic materials. Therefore a thorough understanding of the determination of the mechanical properties of high-performance ceramics is necessary for a judicial and reliable service application.

High-performance ceramics have very desirable use characteristics, including high temperature strength and oxidation and creep resistance, as well as corrosion and erosion resistance. They also exhibit resistance to crack propagation under compression, due to the fact that, in general, in ceramics, cracks close under compression. This is also the reason why ceramics exhibit high compression strengths, which may be an order of magnitude greater than their tensile strengths. Again, a good understanding of the methods of determining these properties is required for an educated selection of the ceramics suitable for a particular application.

2.2 CERAMICS CONTRASTED TO METALS

This section covers specific details with regard to the behavior of ceramics and metals pertaining to design and application. First, the general ceramic characteristics are described briefly.

Brittleness

The primary limitation to a wider use of ceramics is their brittleness. The brittleness is manifested by premature failure, either during preparation of the ceramics for use, such as in the course of machining or handling, when chipping at the corners and edges frequently occurs, or during service, when unexpected particle impact may lead to chipping. Often, the damage does not result in visible chipping, but in subsurface microcracks, which are not detectable by visual or other commonly used nondestructive methods. These microcracks can undergo growth due to stress corrosion to form a contiguous crack of Griffith critical length, leading to premature failure during service.

The propensity for chipping and brittle failure is governed primarily by the hardness and fracture toughness of the ceramic. Lawn and Marshall[1] have defined the brittleness index BI as

$$BI = \frac{H}{K_c} \quad (2.1)$$

where H is the hardness (resistance to deformation) and K_c is toughness (resistance to fracture). The ratio H/K_c reflects on the relative scales of deformation and fracture zones about a sharp contact site and thereby introduces a size effect into the competitive material responses.

To be of use in design, the brittleness index can be visualized in terms of a threshold load P^* above which fracture will initiate under contact loading conditions, leading to chipping. For sharp indenter, this critical load is defined as

$$P^* = \lambda_0 K_c \left(\frac{K_c}{H}\right)^3 \quad (2.2)$$

where $\lambda_0 = 1.6 \times 10^4$. The brittleness index, the threshold load to fracture, and the hardness and toughness values of some engineering materials are compared in Table 2.1.

It is realized that glass, a commonly used engineering material, is very brittle. Because of this reason extreme care is used in the packaging and handling of this material. This points to the fact that other engineering ceramics can also become useful in applications, provided that care is taken to build in this knowledge of

TABLE 2.1 Deformation and Fracture Properties of Engineering Materials

Material	Hardness, GPa	Toughness, MPa · m$^{1/2}$	Brittleness index, 1000 m$^{-1/2}$	Threshold load, N
Steel	5.0	50.0	0.1	800000.0
WC–Co	19.0	13.0	1.5	66.6
Sintered Si$_3$N$_4$	17.0	6.8	2.5	7.0
TiB$_2$	17.5	6.2	2.8	4.4
Sialon	14.0	5.2	2.7	4.3
Partially stabilized ZrO$_2$	12.7	4.8	2.6	4.1
Pyroceram 9606	7.0	2.9	2.4	3.3
Al$_2$O$_3$	18.3	5.9	3.1	3.2
NC 350 RBSi$_3$N$_4$	8.4	3.2	2.6	2.8
NC 132 HPSi$_3$N$_4$	16.5	5.1	3.2	2.4
AlN	13.0	3.3	3.9	0.9
NC 203 HPSiC	25.5	4.6	5.5	0.4
Sintered SiC	24.5	3.8	6.4	0.2
Glass	5.5	0.79	7.9	0.023

brittleness during application. In designing with ceramics, the threshold load [as defined by Eq. (2.2)] has the following implication: When the service load $P < P^*$, fracture does not occur in lieu of deformation, and therefore one should maximize hardness in design to limit the deformation. When the service load $P \gg P^*$, fracture occurs, and therefore one should maximize toughness in design to avoid fracture. When the service load $P = P^*$, both deformation and fracture control the mechanical behavior, and therefore one should maximize the ratio K_c/H in design. As will be seen later, within a material, or a material system, these properties are strong functions of density, and appropriate allowance should be made for this aspect in the design.

Low Strain Tolerance

The brittleness of ceramics is sometimes associated with their limited strain to failure. Since they are flaw-sensitive, catastrophic fracture precedes plastic deformation for polycrystalline ceramics. In addition, ceramics usually have high moduli. Therefore they have failure strains which are an order of magnitude or more lower than those for metals. This is illustrated in Fig. 2.1. The failure strains for typical ceramics based on their Young's moduli and tensile fracture strengths are tabulated in Table 2.2. This lack of strain tolerance is a critical barrier to a wider application of ceramics as engineering materials.

Variability of Properties

The final critical aspect that differentiates metallic materials from ceramics is the inherent and significant variability of the properties of ceramics. This variability

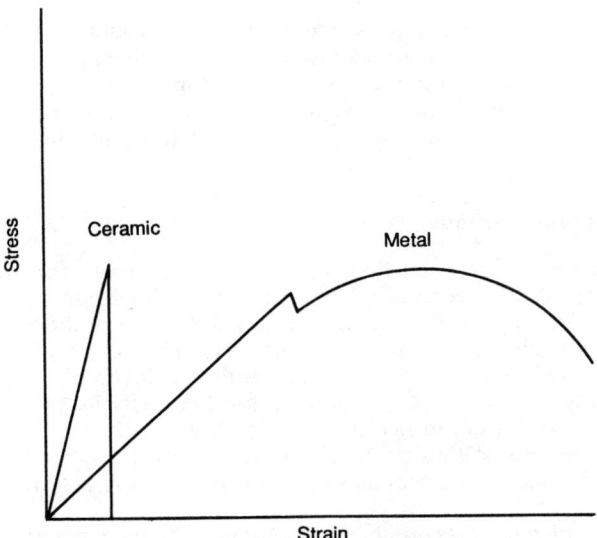

FIGURE 2.1 Schematic stress-strain diagram of typical polycrystalline ceramic compared with that of typical metal. Note absence of plastic deformation region for ceramic.

TABLE 2.2 Strain to Failure for Typical Engineering Ceramics

Material	Tensile strength, MPa	Young's modulus, GPa	Failure strain, %
Steel	550	205	5–10.0
SiC	240	410	0.06
Si_3N_4	675	310	0.22
Al_2O_3	240	345	0.07

is attributed to the differences in production methods. Commercial high-performance ceramics in general are not melt-produced, but are produced via powder consolidation and sintering. Greater potential is inherent in this process, as opposed to melt casting of metals, for a lack of homogeneity as well as for the introduction of flaws at various stages in processing. An existence of nonhomogeneity (of microstructure and chemistry) and process-induced flaws contribute to the variability in properties. Designing with single values for strength, for example, is thus not possible. One has to take into consideration the variability in strength by some statistical methodology. Therefore, probabilistic rather than deterministic design is practiced in engineering applications. This involves an understanding of the variability of properties such that reliability in service can be ensured. This particular factor can also be considered a barrier for a wider application of ceramics.

2.3 STRENGTH AND FRACTURE CONSIDERATIONS

The preceding discussions lead us into a detailed consideration of the strength characteristics and fracture methodology of ceramics which provoke catastrophic failure. The reasons for the variability in strength must be understood both as an inherent characteristic due to the existence of flaws of varying severity and as a characteristic arising from testing techniques related to sampling.

Strength-Testing Methods

Unlike with metals, where tensile tests are conducted routinely, various testing techniques are used for ceramics depending on the availability of material, test machine, and fixtures and the experience and preference of the individual or organization conducting and reporting on the tests. Tensile tests for ceramics are costly because of the diamond-machining operations involved in machining typical dog-bone tensile coupons. In addition, if not properly machined, the surface damage caused will result in inaccurate data. Also, because of low strain tolerance to bending, any slight eccentricity in loading will result in unusable data. Therefore techniques other than tension tests are routinely practiced in industry.

Flexure Test. Flexure tests are the most common routinely practiced method, as the machining of bars of rectangular cross section is easily accomplished. Also, no gripping of the specimen for support is needed, which permits easy adaptation to higher-temperature and hostile-environment testing. Here again, many methods are practiced, such as bending under three-point and four-point loading. Varying the inner spans to one-half or one-third of the outer span is commonly referred to as quarter-point and third-point loading, respectively. The flexure test, in general, samples critical surface flaws since the tensile stress is maximum at the outermost surface of the bend specimen. The configurations and relationships used to determine the strength are given in Fig. 2.2.

Biaxial Test. Biaxial strength tests are especially useful for testing small pucklike specimens, usually made during materials development. Several versions of this test are available, such as ball-on-ball, ring-on-ring, and ball-on-ring. Examples of some biaxial tests can be found in the literature. For example, equibiaxial tension can be achieved by concentric ring loading of a disk.[2] A schematic of the loading arrangement is given in Fig. 2.3. Another way of conducting a biaxial flexure test is the piston-on-three-balls technique.[3]

Alternatively, a ball-on-ring test arrangement can be used, where a disk specimen is supported on a circular ball-bearing race, as shown in Fig. 2.4.[4] Compared to the piston-on-three-balls case, where unequal radial stresses are present, here radial tensile and tangential stresses have predictable uniformity. Also, the support balls are free to rotate, thus permitting the testing of slightly warped specimens. The tangential and radial stresses are maximum and of equal value at the disk center directly below the loading ball, thus providing an equibiaxial tension test arrangement.

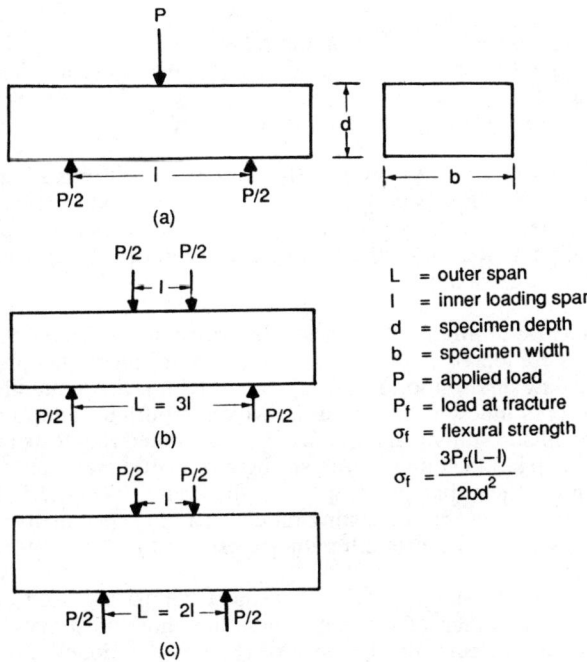

FIGURE 2.2 Schematic of flexural strength tests for ceramics. (*a*) Three-point bend. (*b*) Third-point loading, three-point bend. (*c*) Quarter-point loading, four-point bend.

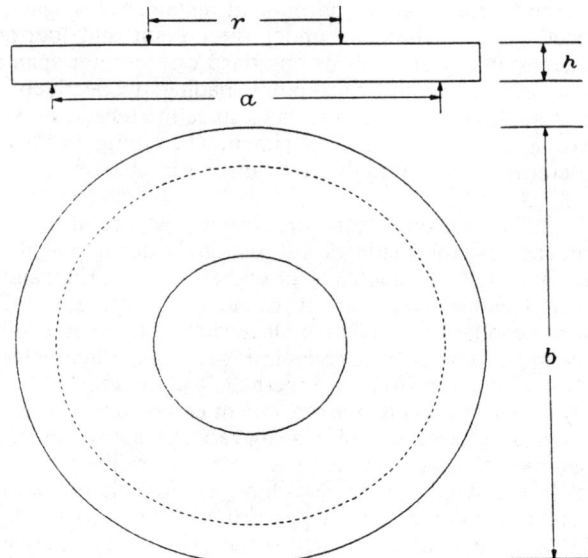

FIGURE 2.3 Schematic of equibiaxial tension loading arrangement. (*After Giovan and Sines.*[2])

Diametral Compression Test. The diametral compression test, or Brazil test, consists of loading a disk specimen on its thickness and across its diameter. The failure occurs along the diameter of the loading edges. This type of test is used infrequently because of the general nature of failure.[5]

C-Ring Test. C-ring tests are eminently suitable to test curved specimens without resorting to the expensive grinding required for flexure tests. Also, tubular ceramic specimens can be sliced into C-rings and loaded compressively. The fracture usually initiates along the outside surface of the C-ring sample, where the tensile stresses are maximum.[6]

Expanded Ring Test. Many of the tests described suffer from nonuniform stress distribution as compared to the uniform stress distribution encountered in tensile tests. As the flaw distribution in ceramics is random, this nonuniform stress distribution presents inherent data scatter, which is difficult and imprecise to analyze by probabilistic statistical methods. The expanded ring tests are closer to the uniaxial tensile test in estimating the strength of brittle materials.[7] The specimen configuration for the expanded ring test is illustrated in Fig. 2.5. As mentioned, the expanded ring is useful for testing tubular ceramic specimens after sectioning them into rings and gives strength values closer to the "true" tensile strength.

Tensile Test. The tensile strength is the most commonly used design property from mechanical and thermal loading standpoints, but unfortunately it is the most difficult to obtain for ceramics because of reasons mentioned. Nevertheless, the importance of obtaining this property is not lost in the ceramic community, and the most recent testing developments are geared toward this end. Those methods, which have simple objectives, such as performance simplicity, uncompli-

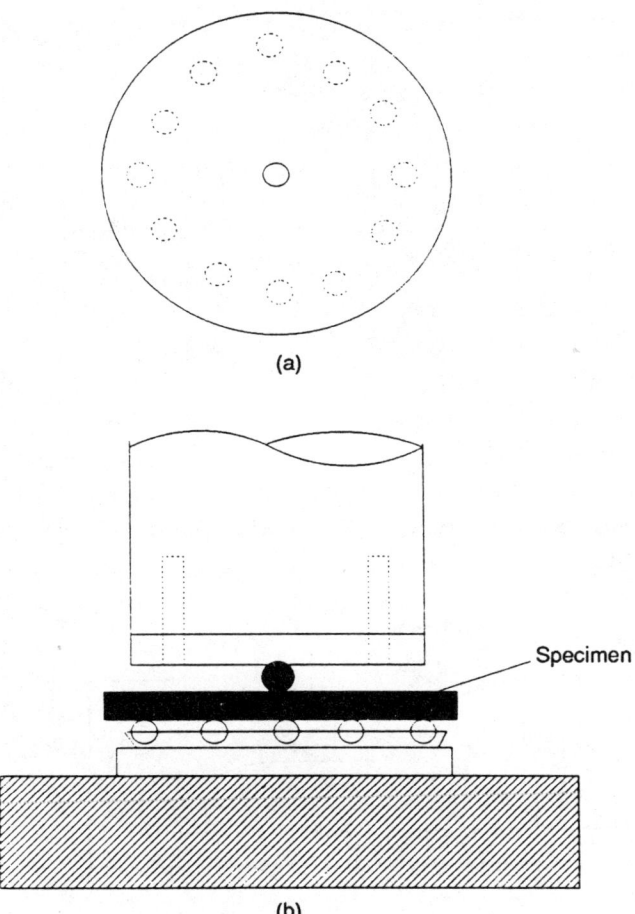

FIGURE 2.4 Schematic of ball-on-ring biaxial test. (*a*) Arrangement of loading and support balls. (*b*) Side view of test setup. (*After Rosenfield et al.*[4])

cated specimen geometry, suitability for ceramics and composites, and easy adaptability to elevated-temperature testing, while at the same time providing accurate uniaxial tensile properties, will be the most useful in the future.

There are several methods advocated in the literature, but none is practiced commercially. In a method advocated by Lange et al.[8] the so-called button-head tensile specimen is used. A more complicated load train assembly for high-temperature testing of ceramics was followed by Govila.[9] The geometry of the specimen is shown in Fig. 2.6.

In a novel approach, modeled after Kats,[10] Seshadri and Chia[11] have developed a simple technique which requires minimal machining for tensile test specimens or none at all. Flat ceramic specimens of rectangular cross section, with fanned ends, which can be pressed to shape easily by conventional dry pressing

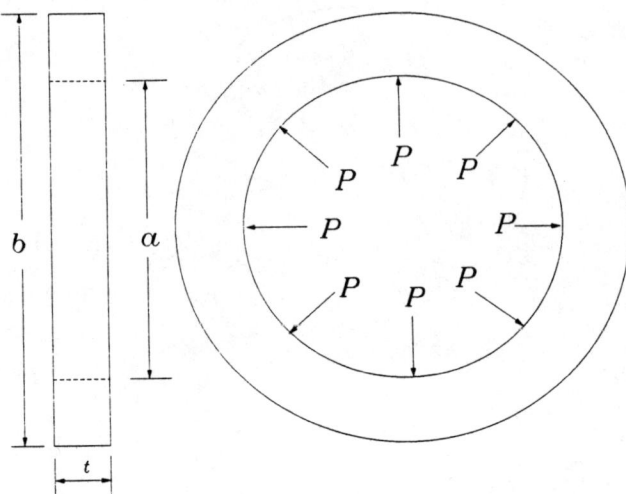

FIGURE 2.5 Schematic of specimen tested in expanded flexure.

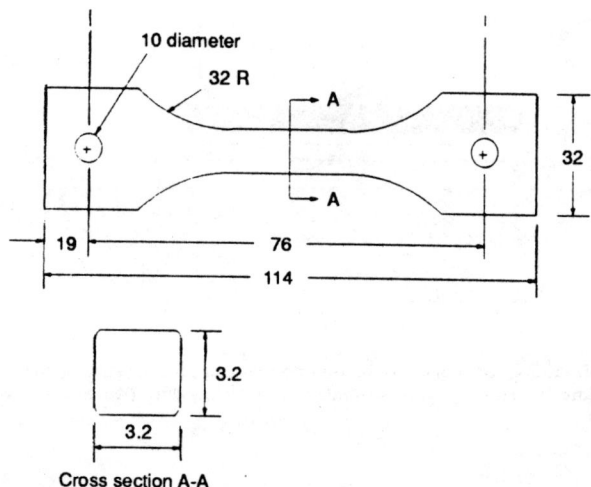

Cross section A-A

FIGURE 2.6 Schematic of tension specimen used by Govila.[9] All dimensions in millimeters.

or molded to shape by either slip casting or injection molding with simple tooling, can be used.

Compression Test. In general, ceramics exhibit compression strengths an order of magnitude greater than their tension strengths. This is because the imposed compressive loads result in crack closure promoting strengthening, as opposed to crack opening and growth promoting weakening, when tensile loads are imposed

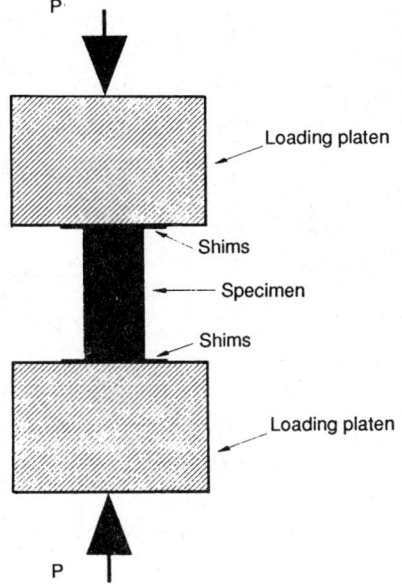

FIGURE 2.7 Schematic of compression test.

perpendicular to the crack plane. The compression tests are conducted on small rectangular or cylindrical specimens whose end surfaces are well machined to have absolutely parallel and flat surfaces. The use of compliant shims between the loading platen and the specimen is also practiced, as shown in Fig. 2.7.

Statistical Analysis

As mentioned previously, the variations in the strength properties of the ceramic necessitate the use of some form of statistical analysis in design. This section will be concerned with a commonly used analysis technique for handling the strength variations in ceramics.

Weibull Analysis. It is recognized that the existence of flaws in materials prevents them from achieving their theoretical maximum strength values. Further, the number of such flaws will depend on the total volume of the material. The greater the volume under stress, the greater the chance of finding the potential strength-limiting flaw. Hence, larger specimens yield lower strengths than smaller specimens, even though material strength, by itself, is invariant, fixed only by the atomic structure and the bond strength of the constituent atoms.

In brittle ceramics, experimental strength results show a definitive relationship between the probability of failure and failure strength.

As mentioned, the strength depends on the shape of the specimen and on the stress distribution dictated by the test methodology used. Nevertheless, the distribution of this property can be analyzed by a Weibull distribution function.

The advantage of using materials having high Weibull moduli, that is, less strength scatter, is readily realized in terms of tensile strengths reasonably close to bend strengths. Another important factor is the necessity to achieve maximum realizable Weibull moduli in products consistent with cost considerations since very high Weibull moduli do not necessarily translate into proportionately higher tensile strengths. These points are illustrated in Fig. 2.8.

It is apparent that as one moves from low-strength materials during the initial stages of new materials development to high-strength materials during advanced stages, the high Weibull modulus also needs to be achieved in order to realize a higher tensile strength, which is of importance in design and application. Also, a Weibull modulus greater than 20 does not result in a significant further increase in tensile strength.

Reliability and Safety-Factor Considerations. Most designers familiar with metals and alloys use a deterministic approach because the tensile strengths of the engineering metals and alloys are usually single-valued, with insignificant variations

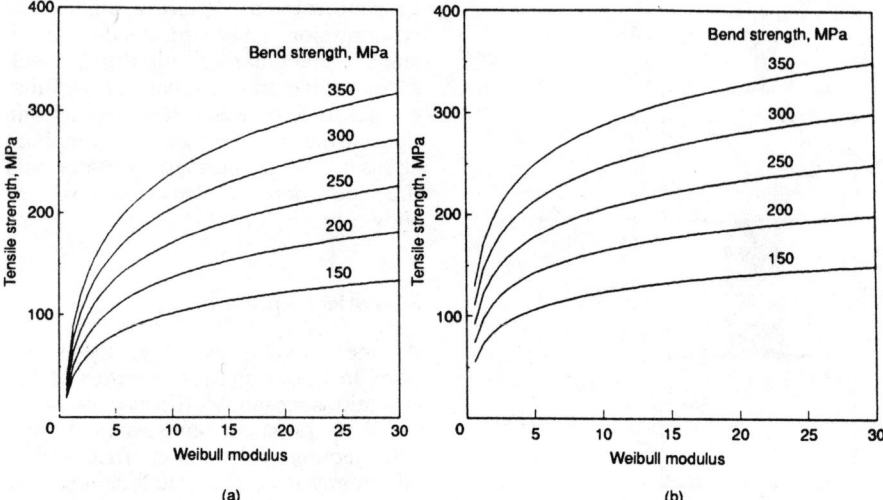

FIGURE 2.8 Effect of Weibull modulus on tensile strength estimate from bend strength data. (*a*) Volume-flaw based. (*b*) Surface-flaw based.

in a given population and composition. It is customary to use safety factors determined by experience and other fatigue data to arrive at a safe service stress at which reliability of the part can be guaranteed for the expected service life. With ceramics one must take into account the strength variability. However, probabilistic design approaches, as outlined previously, can be used to estimate safety factors.[12]

Fracture of Ceramics

The fracture of high-performance ceramics is understood to occur primarily because of tensile stresses at the cracks and cracklike flaws which are present during the manufacturing process, or are initiated and grow during service. The fracture mode can be of three types: primarily transgranular, primarily intergranular, or a combination of both. The type depends on the particular ceramic material. For example, alumina fractures primarily via intergranular mode, silicon carbide fractures by transgranular mode, and silicon nitrides fracture in mixed-mode fashion.

A considerable body of knowledge exists of the qualitative and microstructure-related issues with regard to failure origination, crack advance, and ultimate catastrophic failure in ceramic test bars tested in bending. Quantitative analyses are also abundant, correlating fracture stress with flaw properties and fracture toughness. However, such understanding is lacking in the area of component service failures, due to insufficient data and a lack of available detailed publications of ceramic-related service failures. However, the commonalities of the features on the fracture surface are worth noting.

Typical fracture origins are processing-related, such as powder agglomerates which were not sintered, agglomerates with ring cracks around them resulting from differential shrinking between matrix and the agglomerate, impurity inclusions, voids, delamination cracks, cracks, and large grains in a matrix of fine grains. Machining-related cracks can also contribute to premature failure, but these are difficult to detect by fractography. In general, processing-related flaws are the most detrimental, followed by machining cracks and microstructure-related flaws, such as large grains and agglomerates of pores. This information is represented graphically in Fig. 2.9 for sintered alpha silicon carbide.[13]

Fractography Considerations. In the failure analysis of a component in service, the objective is to estimate the failure stress *at the flaw* as determined by fractography in order to compare it with the service stress at the time of failure. This estimate generally gives information regarding stress concentration and enhancement at the flaw, which is important in further design iteration and the institution of appropriate nondestructive evaluation procedures.

Strength–Critical Flaw Analysis. Fractography is an important tool not only in the performance evaluations of ceramic components in service but also in research to improve processes that will minimize and ultimately eliminate critical flaws which severely inhibit reliability. For example, in the case of silicon nitride development it has been established by an earlier study[14] that the type of defect (chemistry), in addition to size, governs the fracture strength, as illustrated in Fig. 2.10.

It is seen that surface cracks are the most detrimental for a given size, emphasizing the need to improve the detectability of surface flaws in this material. The presence of tungsten carbide as a defect, possibly originating from milling operations during powder processing, is relatively harmless up to 600 μm. While this inclusion can be detected quite easily by simple X-ray radiography due to the

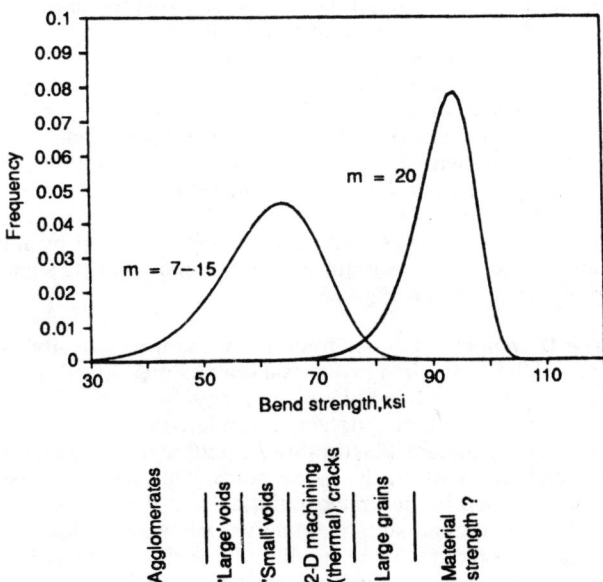

FIGURE 2.9 Effect of flaw type on strength distribution of sintered alpha silicon carbide. *m*—Weibull modulus.

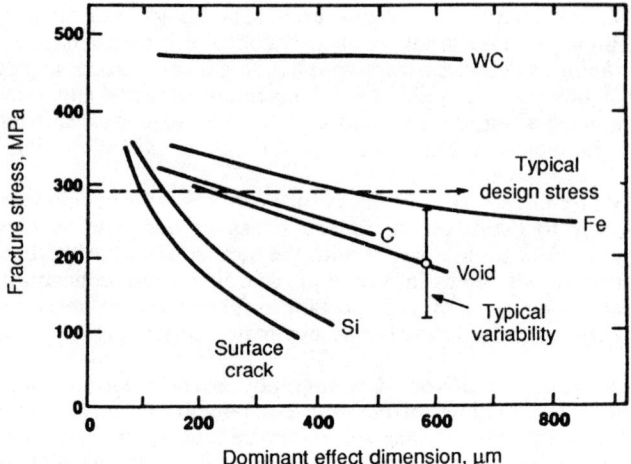

FIGURE 2.10 Effect of flaw size for several critical flaws in hot-pressed silicon nitride: (*After Evans.*[14])

vast differences in the densities of the inclusion and the matrix, the more detrimental critical defects such as voids, silicon inclusion, and surface cracks cannot be quite so readily detected by simple radiography.

A similar analysis has been reported for sintered silicon carbide[15] where the flaws were mainly processing-related pores, located mostly at the tensile surface of the bend specimen. In this work, the real flaws were simulated to be ellipsoids, and it was shown that at very small flaw sizes the expected dramatic increases in strength were not observed because of the nature of the possible cracks associated with the macroflaws as well as other stress-concentrating defects attached to the periphery of the main flaw. The findings are illustrated in Fig. 2.11.

The major unknown in understanding the strength–flaw size relationship is the effect of possible residual stresses in the material which are difficult to quantify, especially for ceramics with high elastic modulus. In addition, once the processing-related flaws are eliminated by systematic processing improvements and the proper machining practice is followed, thus eliminating machining-related cracks, the strength of polycrystalline ceramics is governed by intrinsic microstructure-related defects such as grain size and by nanodefects such as impurities within the grains, as shown in Fig. 2.9.

Classic Fracture Mechanics. Classic fracture mechanics is ideally suited for technical ceramics which behave in a brittle fashion. As mentioned, industrial ceramics are manufactured containing flaws, be they processing-related, handling-related (which includes machining defects), or microstructure-related. Because of the absence of any significant plastic flow, fracture occurring prior to ultimate strength as determined theoretically is the norm. This section considers some of the methods used to determine crack initiation and propagation resistance. No single method has been found suitable in all situations, even for a given ceramic. Variations in the data obtained by using different techniques on the same material are also common. In addition, different investigators, who used the same test technique and nominally the same procedure, have reported varying fracture toughness data on the same ceramic. Nevertheless, an attempt will be made to summarize the test techniques.

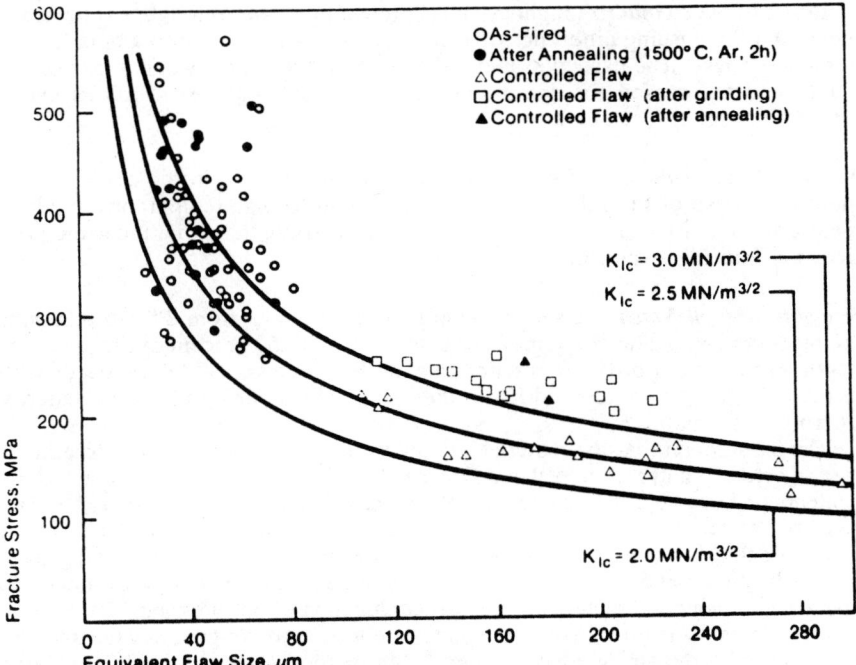

FIGURE 2.11 Effect of flaw size on fracture strength of sintered silicon carbide.

Linear Elastic Fracture Mechanics. The premise of linear elastic fracture mechanics is based on the catastrophic failure of a material from preexisting cracks or cracklike flaws during service stress. It is recognized that there exists a linear relationship, akin to linear elastic loading, between the applied stress and the applied stress intensity factor. At the instant the material fails, the applied stress and the stress intensity factor become the critical stress intensity factor K_{Ic}, which is referred to as *fracture toughness*. The fracture toughness of the material is a constant in the same sense as the strength. In other words, factors affecting strength, such as microstructure, residual stresses, heat treatment, grain chemistry, grain boundary, and triple grain junctions, can all contribute to fracture toughness variations for a given material. Furthermore, because of the intrinsic microstructural, chemical, and microstrain inhomogeneities arising from cooldown conditions after sintering, and residual stresses arising from finishing operations, there may exist localized fracture toughness variations within the same sample. It is therefore important to understand the different methods of determining the fracture toughness of ceramics.[16]

Fracture Toughness Determinations

This section examines in moderate detail the currently practiced fracture toughness determination methods, including specimen geometry and loading configuration. Many new methods are continuously being proposed, which are variations of some of the methods covered here. There exists some confusion in the ceramic literature between *macro* toughness measurement and *micro* toughness measurement, and the ensuing differences. However, we shall not get into a detailed discussion of these. The particular method chosen will depend on sample size availability and the reason for determining the fracture toughness, such as use at ambient or elevated temperature.

Notched Beam Method. This method has been widely used in the ceramic community because of the relative ease of adaptation to high-temperature environments, as well as its similarity to the bend test for strength so that the same practice and fixturing arrangement can be used.

Straight-Through Notch. Usually, wide notches, ranging from 0.08 to 0.50 mm, are machined with a diamond blade to a depth of 40 to 50 percent of the specimen depth. However, in order to practice this method, one should be knowledgeable of the effects of the notch width and the extent of machining damage created at the root of the notch.[17]

The assumption that the material exhibits isotropic properties is not true in the case of ceramic matrix composites and some monoliths that have significant additive contents or have been fabricated by hot pressing or extrusion processing,[18] which gives rise to texture in the microstructural features. In these instances, one has to determine three critical stress intensity factors for the three directions, paying attention to the direction of crack propagation with respect to the stress in relation to the major reinforcement direction relative to the direction of hot pressing.

Despite its difficulties in obtaining consistent fracture toughness values for the same ceramic, the single-edge notched beam method maintains wide popularity largely because of the ease of specimen preparation and the simplicity of testing.

Chevron Notch. The chevron method is also amenable to high-temperature tests. It is especially suitable if crack growth resistance behavior is encountered as it offers more chance for stable crack propagation so that analyses can be performed with regard to crack initiation as well as crack propagation.[19,20] Here a notch of triangle geometry is made at the center of the specimen with the height being at least 60 percent of the specimen depth.

Double-Cantilever Method. This method originated from a study in view of determining fracture surface energies by the cleavage technique.[21] In order to ensure that crack propagation occurs in a well-defined plane normal to the applied forces, Berry[22] suggested the procedure of machining grooves along the crack plane to contain the fracture. The mathematics of loading and the fracture have been variously analyzed[23,24] to determine the fracture toughness from the peak load for crack propagation and the specimen dimensions. A schematic of the specimen configuration is shown in Fig. 2.12.

Variations of this simple double-cantilever beam specimen include the so-called constant K specimens, in which the stress intensity factor K remains constant as the crack moves across the specimen length, such as the tapered double-cantilever beam specimen, the wedge-loaded tapered specimen, the double-torsion method, the constant-moment specimen, and the wedge-loaded specimen. The choice of a particular specimen will depend on the amount of specimen material available, the nature of the property determination required, that is, peak-load fracture toughness or complete crack propagation characterization, and also on the temperature and environments involved in the test. The reader is referred to any fracture mechanics text for the test configurations and equations used to determine K_{Ic}.

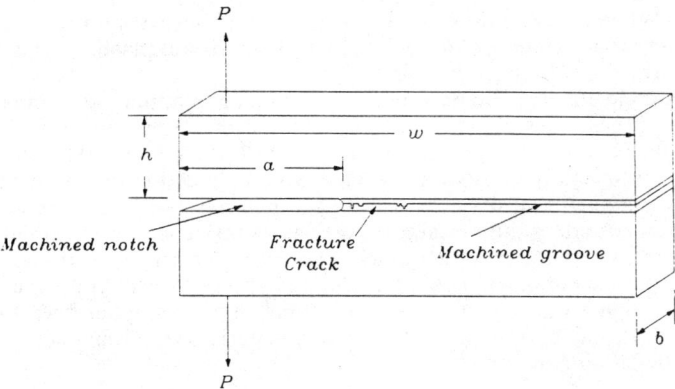

FIGURE 2.12 Double-cantilever beam specimen for fracture toughness measurements.

Surface-Flaw Bend Tests. The surface-flaw bend tests are easier to conduct and, perhaps, simulate well the behavior of naturally occurring cracks, cracklike processing defects, and machining-related surface cracks of the ceramics. The surface cracks can be introduced on the polished surfaces of the ceramic flexural bend bars by either Vicker's indentation or Knoop indentation. When a minimum of the threshold load for fracture is thus applied, radial cracks emanate from the corners of the indent in mutually perpendicular directions, the diameter of which depends on the fracture toughness and on the elastic-plastic response of the material. The Knoop indentation is the preferred method as one penny-shaped crack with the long axis as its diameter is introduced rather than the two orthogonal cracks introduced by Vicker's indentation. Assuming that no backward stresses are developed that will contribute to crack closure on removal of the load and that further stable crack growth due to stress- or corrosion-related phenomena does not occur, the fracture toughness K_c can easily be determined by measuring the stress to fracture the bend bar under flexure conditions.[25]

The advantage of this method for high-temperature fracture toughness determinations is obvious, provided that slow crack growth, due to environmental stress-assisted effects, does not occur during the test.[26]

Hertzian Fracture Method. When a spherical indenter is loaded onto a flat specimen surface, cracking occurs around the circumference of the indentation mark and extends into the bulk in the form of a cone in a completely elastic regime. Hertzian fracture testing, named after Hertz who first calculated the stress state for this configuration over a hundred years ago, refers to the study of this behavior. The stresses are predominantly compressive, except for a limited region near the surface and outside the contact region, where it is tensile. The maximum tensile stress occurs at the edge of the contact circle; it falls off rapidly with increasing radial distance. At a critical loading condition, a ring and cone crack develop in the brittle materials along the edge of the contact circle.

In practice, the critical load required to produce cracking exhibits considerable variation, leading to the application of statistical flaw theories to explain these effects.[27,28]

Indentation Method. Indentation techniques, especially Vicker's diamond pyramid tests, have long been recognized to be useful for surface characterization of materials. When sufficiently high loads are used, these tests produce specific crack patterns associated with the characteristic elastic-plastic indentation behavior in brittle materials.

Considerable efforts have been made in obtaining quantitative estimates of the fracture toughness of various ceramics from the measurement of the observed crack dimensions.[29-31] The relative ease of the test procedure and the economy in test specimen usage have resulted in extensive application of this technique in material development efforts. The significant material variables to be considered are the microstructure and surface conditions as influenced by residual stresses and phase transformations, which may occur near surfaces due to surface preparation in specific ceramics. Also, test variables to be considered include indentation load, impact velocity, and residence time. In general it has been found that this method is unable to distinguish between materials with toughnesses varying between 20 and 40 percent.

Critical Strain Energy Behavior

In cases where the ceramic exhibits resistance to fracture crack growth, resulting in an increase in failure strain, the critical stress intensity factor approach for fracture toughness is incomplete. The strain energy release approach is more pertinent, and the characterization of fracture data for fracture crack initiation and propagation gives a more realistic view for design, especially when probabilistic estimates are used. The critical strain energy release rate G_c is related to the critical stress intensity factor K_{Ic} through the relation

$$G_c = \frac{K_{Ic}}{1 - \nu^2} \tag{2.3}$$

where ν is Poisson's ratio.

2.4 CERAMIC SUBSTITUTION—DESIGN CONSIDERATIONS

The reader may already have sensed that in order to design with ceramics, special education is needed from the viewpoints of brittleness and variability of properties. It is by now well known that "one-to-one" substitution of a ceramic for a metal component is not advisable. Design comprises not only performance considerations, but also fabricability, inspectability, reliability levels, and statistical considerations. Aside from the technical considerations, cost might be the overriding factor, although total system benefit may result in some situations, such as under adiabatic engine conditions, where increased weight savings, in addition to low inertia and low friction without the use of coolants, is desirable. Total design requirements are given schematically in Fig. 2.13.

In general, the factors which must be evaluated can be broken down into two main groups: end-use considerations and fabrication considerations.[32] The end-use considerations are primarily performance-related and require the use of finite-element stress analysis or closed-form analytical stress solution, if available, in order to determine the safe operating stress level from ceramic properties. It is also imperative that the properties determined from the test coupons be translated into the component by testing test coupons cut from fabricated com-

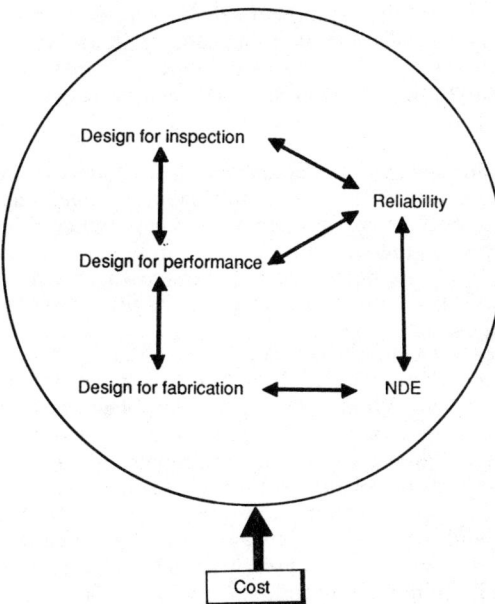

FIGURE 2.13 Design considerations in ceramics applications.

ponents intended for actual use. This is important because of processing differences, which may result in different types and locations of critical flaws in test coupons and components. Other factors that should be considered are the avoidance of impact loading, stress risers in the part, contact loading, and undercuts. Fabrication considerations address issues related to the ease of manufacture, inspection, and shipping in economic quantities.

There are several methods of ceramic fabrication available such as slip casting, cold dry pressing, cold isostatic pressing, extrusion, and injection molding, each with its own shape and size limitations. Dry pressing is well suited for mass production and automatic press. Closer dimensional tolerances can be maintained by this method. Cold isostatic pressing is suitable for producing large blanks of material with uniform green density, but is a slower method of production. Injection molding is generally suited for small components with a general thickness limitation imposed by the ability to remove the thermoplastic in the firing process. However, this may be the only economically viable method to produce complex-shaped gas turbines or other types of turbine components. Extrusion is eminently suitable for symmetrical cross sections such as thin-wall tubes. Tape casting is employed for producing parts with lateral dimensions which exceed the thickness by 25 times or more. The choice will depend, among other factors, on tooling requirements, part complexity, and volume required. We now consider some critical design aspects in detail.

Brittle Design for Component Fabrication

The foremost aspect is particle impact on brittle ceramics. As mentioned, the threshold load for crack initiation under pseudoimpact conditions is very low for these materials. Data relating to impact-related service damage for ceramics are very sparse, and theories dealing with this subject are of little relevance to actual use conditions. Also, generalizations of ceramic behavior based on fundamental properties have not yet been proven in actual service. It is therefore necessary, at this time, that a recommendation be made that impact loading be avoided, even if the impacting particles have low mass.

Sharp corners act as stress risers and can contribute to chipping or even catastrophic part failure well below the operating stress level. Ideally, a smooth radius should be provided. In small machined parts in which this is not practical, a chamfer greatly reduces the susceptibility to chipping or cracking. As illustrated schematically in Fig. 2.14, calculations have shown that the stress level is reduced from 296 to 221 MPa if a smooth radius is provided at the intersection of the hub and the backface.

Contact stresses are yet another aspect which needs close scrutiny. The relatively high elastic modulus and the low strain to failure translate into a lack of material flow under load. When two ceramic surfaces are in direct contact, the parts will only touch in a few "high spots." The load is then distributed over these highly localized areas, leading to extremely high Hertzian stresses. This problem can sometimes be alleviated by the use of a compliant layer of an appropriate material or a combination of different materials with relatively low, but varying elastic moduli between the ceramic surfaces. This compliant layer can then be elastically deformed, which greatly increases the surface area under load. Often, this layer is a thin metallic foil.

Compliant layers can also be used to accommodate the mismatch in the thermal expansion of a ceramic and the metal to which it is attached. In the case of the rotor, shown schematically in Fig. 2.15, direct contact of the metal and the ceramic resulted in a calculated stress of 470 MPa at the shaft. When a hypothetical compliant layer is incorporated, the stress is reduced to 210 MPa. However, the engineering aspects of putting the compliant layer into industrial practice need to be considered carefully in the selection of the material.

Another aspect to be considered is the possibility of trapping corrosive media in undercuts which might subsequently cause the ceramic to fail prematurely due to stress corrosion cracking and localized weakening of the component. Therefore, it is recommended that undercuts be avoided as much as possible in the design of the ceramic component.

As mentioned previously, the manufacturing method selected will depend on both cost considerations and manufacturability. As an example, Fig. 2.16 illustrates general considerations for ceramic mechanical face seals. The first two geometries shown can be made by an automatic dry-pressing operation, while the third part, shown with a grooved cross section, cannot be made by dry pressing. Injection molding can be considered as an option here, with appropriate tooling. It is noted that in Fig. 2.16 sharp corners must actually be chamfered or rounded, as appropriate, by known methods of finishing.

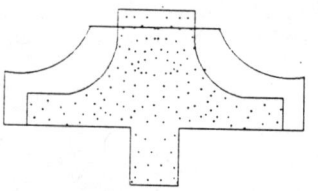
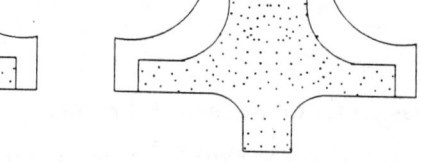

FIGURE 2.14 Schematic of a rotor.

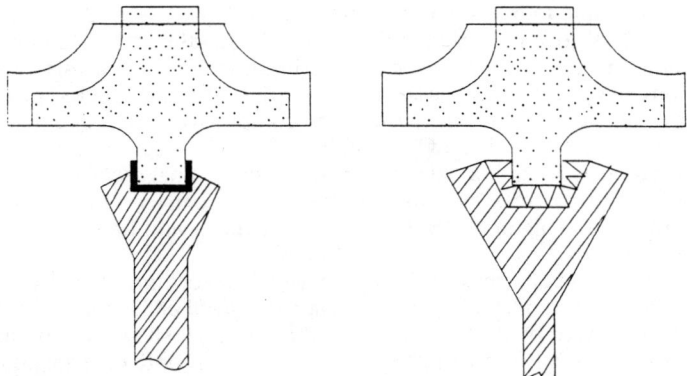

FIGURE 2.15 Use of compliant layer to minimize contact stresses in ceramic.

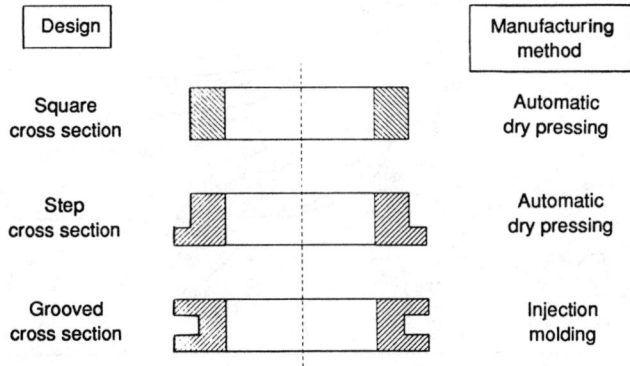

FIGURE 2.16 Design-manufacturing interaction for mechanical seals.

Finally, the design engineer must, for the most part, design parts that can be checked by simple inspection techniques. Reliability assurance and, therefore, part performance are directly related to the extent to which the parts can be completely inspected by simple inspection techniques. The maximum stress level in a component should be designed in areas which can be unambiguously inspected. The designer must be aware of the limitations that exist in nondestructive testing techniques; it is not absolutely assured that the component is defect-free simply because it passed inspection. The designer must design parts that can be inspected by existing techniques at low cost.

Analytical and Numerical Stress Analysis

As is well known, the modern design practice involves rigorous stress analysis of the component, taking into consideration actual use conditions. When closed-form analytical solutions exist, presumably under simple loading conditions, the stress map can be readily generated which may then be compared with the properties of the ceramic to be used in the application as discussed. Assuming that no slow crack growth occurs which will degrade the load-bearing capability, a survival probability range can be estimated. Shrink-fit stress analysis can be performed from analytical solutions for simple shrink-fit joints where needed.[33,34]

However, in cases where, because of the complexity of part geometry as well as the complexity of loading, which may include the superposition of thermal stresses in addition to multiaxial stress states, simple analytical solutions do not exist, finite-element stress analysis will need to be carried out to map out the stress distribution. This may involve both mechanical and thermal loads, either separately or jointly.[35]

A recent example of stress analysis in design has been provided by Katayama et al.[36,37] For a silicon nitride rotor based on a tip speed of 590 m/s, with a turbine inlet gas temperature of 900°C, Figs. 2.17 through 2.20 show the distribution of maximum principal tensile stress. The maximum stress, both centrifugal and thermal, occurs in the ring seal area of the neck of the back face. The finite-element

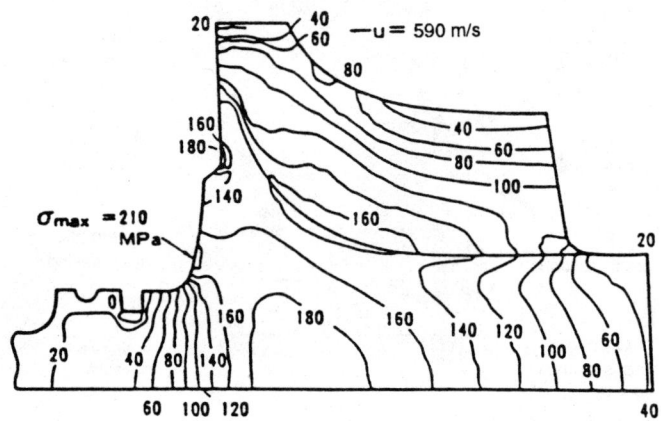

FIGURE 2.17 Centrifugal stress.

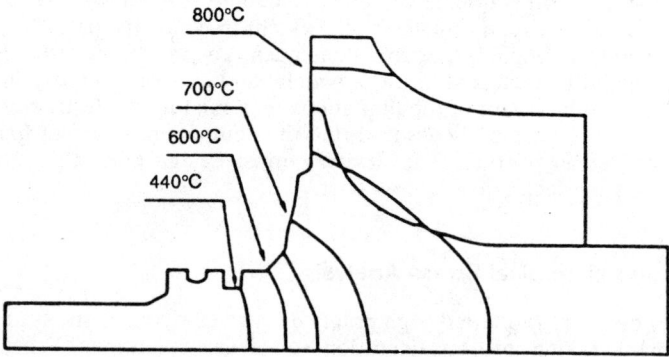

FIGURE 2.18 Temperature distribution.

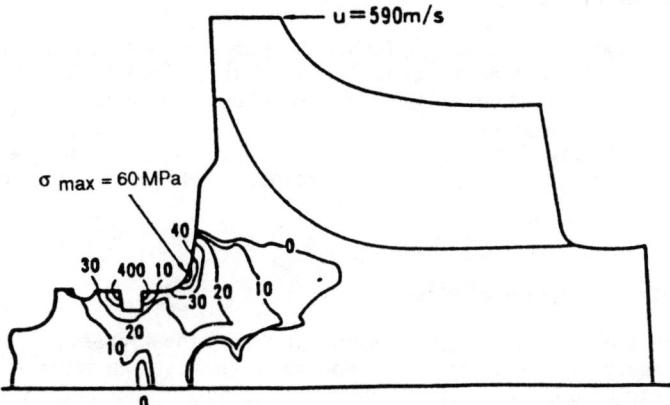

FIGURE 2.19 Thermal stress distribution.

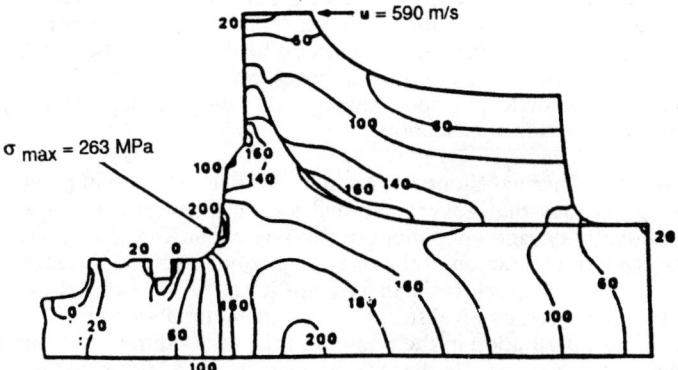

FIGURE 2.20 Combined stress distribution.

analysis design modifications were reported to have achieved a 20 percent reduction in stress compared to a ceramic rotor shaped like a conventional metal rotor, maintaining the same aerodynamic performance.

Probabilistic Design versus Deterministic Design

Once again the reader is reminded of the criticality of this aspect in design. Before published data can be used to design components, the designer might find it useful to check the data source to assure that the data were generated with an adequate statistical population of test samples. In addition, testing must be carried out using coupons cut from actual components to establish process reliability for performance targets. The design methodology should include the results of finite-element stress analysis together with those of statistical analysis to predict the probability of failure of the ceramic component.[38]

Multiaxial Service Stress

Rarely does one encounter unidirectional stresses in service. Rather, multiaxial stresses in addition to bend, torsional, and parasitic stresses are common. In addition, the time dependence of these stresses must also be considered. A thorough characterization of material behavior under such conditions is not possible in many cases. However, with specialized testing, some literature is available with respect to the multiaxial stress response of some high-performance ceramics.[39,40]

Stress Rate and Dynamic Effects

Practically all of the preceding discussions pertain to the strength property as determined under a constant stress rate condition. However, in actual service it is impossible to assure such a condition. Shock loading, either mild or severe, as well as loading, which varies over time, are commonplace. Unfortunately the literature data under these conditions are sparse for high-performance ceramics. In many instances, thermal shock, mechanical (vibrational) shock, and extraneous parasitic stresses, which arise from sources that are difficult to locate, may in fact result in premature failure of the component. Although the contributions from thermal shock can be estimated from theoretical considerations, those arising from mechanical shock and parasitic stresses can only be handled by prior design, application knowledge, and a history of product use. We shall consider some of these effects in this section.

Thermal Shock. Thermal shock of ceramics has been studied in great detail. The significant properties that govern thermal shock are strength, elastic modulus, thermal conductivity, and coefficient of thermal expansion. The severity of thermal shock is a strong function of the operating conditions and the geometry of the component. Thermal shock resistance is *not* an intrinsic material property. The effect of thermal stresses on different kinds of materials depends on the stress level, the stress distribution in the body, and the stress duration. Consequently it is impossible to define a single thermal stress resistance factor that is satisfactory for all situations.

The factor R' can be considered to be a material property since it depends on material parameters and can be described as a material resistance factor for thermal stress fracture. High fracture stress, low modulus of elasticity, low thermal expansion coefficient, and high thermal conductivity for a given material indicate good resistance to thermal stress failure.

Creep Effects. In general the creep strain of high-performance ceramics at elevated temperatures is several orders of magnitude less than that for metals at a given temperature and applied stress level. Provided that microstructural changes do not take place during creep, no substantial damage can be expected that will result in a reduction in service life. This is the case for very high purity ceramics such as sintered alpha silicon carbide. However, where large amounts of additives have been used in processing, such as in high-performance silicon nitride, microstructural changes are expected because of various chemical and physical changes that occur at the grain boundaries and triple grain junctions which are concentrated with glass, while the creep strain may be lower.[41,42]

Ceramic-Metal and Ceramic-Ceramic Joint Stresses

It is recognized that a ceramic has to be attached to another ceramic or a metal in practical applications. There are several ways of accomplishing this, such as by diffusion welding,[43] cosintering,[44] or suitable interlayers.[45,46] However, an impediment to the successful bonding of silicon nitrides and silicon carbides is their relative chemical inertness to several of the braze filler materials or metals. From a design viewpoint, excessive mismatches in Young's modulus, the thermal expansion coefficient, and Poisson's ratio can result in localized high stresses at the joint interfaces, resulting in either debonding or fracture within the ceramic.[47] Design considerations should include not only the length of the interface, but also corners where the stress situation may be more complex and may require finite-element analysis. For example, in the case of silicon nitride-Invar alloy joints, the highest residual stress perpendicular to the interface occurred near the corners in the rectangular bond face joint.[48] Of the variables that result in the tensile stress at the ceramic adjacent to the interface, the thermal expansion mismatch is the most significant.

For silicon carbide, assuming a tensile strength of 210 MPa, a host of materials with coefficients of thermal expansion (CTEs) of over $6.0 \times 10^{-6}/°C$ can be ruled out because of resulting tensile stresses greater than the strength of the ceramic, which will cause the ceramic to fail. For example, SiC cannot be successfully brazed to molybdenum (CTE = $5.5 \times 10^{-6}/°C$), but with an interlayer of tungsten (CTE = $4.7 \times 10^{-6}/°C$) joining can be accomplished. The ease of joining metals to Si_3N_4 reported by many investigators is due primarily to the lower elastic modulus (~0.6 of SiC) and CTE (~0.5 of SiC) of Si_3N_4. In order to circumvent these moduli and CTE mismatch limitations, many approaches have utilized interlayers with gradually varying properties. For example, silicon nitride has been joined successfully to steel by using an interlayer of kovar (Fe–29%Ni–15%Co); Si_3N_4, sialon, and SiC to WC–Co using an interlayer of Al–Si.

Once a successful joint has been accomplished, the lifetime of the joint will depend on the relative chemical inertness to the environment, which will include, among other factors, oxidation resistance and stress-assisted corrosion resistance. For example, a silicon nitride brazed with aluminum showed reduced strength in the presence of water.[49]

Uncertainties in Ceramic Lifetime Prediction

As mentioned previously, life prediction for ceramics is calculated from studies of slow crack growth. However, because of the statistical nature of the severity of the random flaws exposed to the stress-assisted corrosion environment, variations in failure times of an order of magnitude or more are encountered in practice. Several assumptions are made with respect to the variables in the fracture mechanics relation as to flaw growth in service, which are found to be not tenable under close scrutiny.[50]

In general, before the results of static fatigue or dynamic fatigue tests can be applied to any material to obtain slow crack growth data, appropriate failure mechanisms must be identified by extensive fractography in order to test the validity of the assumptions utilized in the failure prediction methodology. Also, it must be understood that the actual service conditions involve lower levels of stress and longer times to failure than the conditions evaluated in the tests. For materials resistant to slow crack growth, the errors in the data estimates from the test results are amplified considerably in the predicted lifetimes. One way to obtain usable engineering data is to construct confidence bands for the predicted lifetime. It may then become necessary to increase the number of specimens and to test at levels of stress closer to that of the service stress, in order to improve confidence in the predicted lifetimes.

2.5 SURFACE EFFECTS

It is usually realized that surface conditions can influence the properties of materials significantly. This effect is amplified in the case of ceramics, especially for those with high elastic modulus, because of the absence of plasticity. In addition to the effect of residual stresses at the surface, the rough geometric nature of the ceramic surfaces can also play an additional role in the wear and friction property. In general, reduced surface roughness produces significant decreases in friction under lubricated sliding conditions.[51]

Residual Stresses in Forming

Residual stresses can arise due to any of the ceramic manufacturing operations, including, but not limited to, constraints imposed during green forming such as isopressing, uniaxial pressing involving die wall friction, nonuniform and turbulent material flow during medium- and high-pressure injection molding, sintering, hot isostatic pressing, and subsequent machining operations. How the resulting strength can be affected by the presence of residual stresses was discussed in Sec. 2.3. In many instances, if special precautions are not followed in the high-temperature consolidation of, for example, closed-end components, then fractured components will result even though the residual stress effects are concentrated primarily at the surface.

Machining of Ceramics and Related Surface Effects

There have been several studies relating the effect of machining on the strength of both oxide and nonoxide ceramics. In general, finishing with finer-grit-sized grinding wheels result in higher strengths for the ceramic compared to coarse-grit wheels. For example, for a hot-pressed silicon carbide fabricated using 1500-mesh silicon carbide grains with 3 weight percent carbon and 0.4 weight percent boron as hot-pressing additives, strength "improvement" by as much as 20 percent was reported using a 600-grit diamond wheel as opposed to a 220-grit diamond wheel.[52] This observation was found to be valid for three different silicon carbide grain sizes in the matrix varying from 6 to 35 μm. Similar results were also reported for hot-pressed silicon nitride by Kawai et al.[53] Unfortunately, the excessive grain growth caused a large scatter in the strength data. Nevertheless, it was reported that the strength was invariant with a surface roughness of up to 1 μm, but decreased rapidly when the surface roughness increased from 1 to approximately 10 μm. It should also be realized that not only the final surface roughness, but grinding parameters such as wheel speed, down speed, cross speed, and table speed can also influence the extent of machining damage. However, the nature of surface damage is not well understood. For example, for a hot-pressed silicon nitride it was found[54] that there was a decrease in strength resulting from an increased downfeed rate. Strengths were the highest for materials machined at 0.03-mm downfeed rate.

Chemical Deposits from Ceramic Fabrication

Machining of the ceramic before use can, in many instances, eliminate furnacing deposits on the surface. Surface deposits on the sintered ceramic are quite common and result from the atmospheric sintering practice, which may include immersing the articles in a bed of ceramic grains. Also, because of chemical potential gradients within and surrounding the system being sintered, generation of ceramic products due to decomposition is possible. Although such deposits, especially when present in a nonhomogeneous manner, will impair the cosmetic value, it is quite possible that they may also be detrimental to the physical properties. For example, in the case of aluminum nitride, such deposits will result in localized electrical and thermal property changes, which are undesirable. In the case of sialon it has been well established that the nature of the surfaces resulting from fabrication has a significant effect on the room-temperature strength,[55] as shown in Fig. 2.21.

Relieving Surface Damage

The detrimental effects of surface damages in a ceramic can be removed, at least partially, in two ways. For machined and as-fired samples suitable heat treatment operation has proved to be very suitable. For machined samples, polishing can sometimes restore the original strength, relieving the residual stresses that result

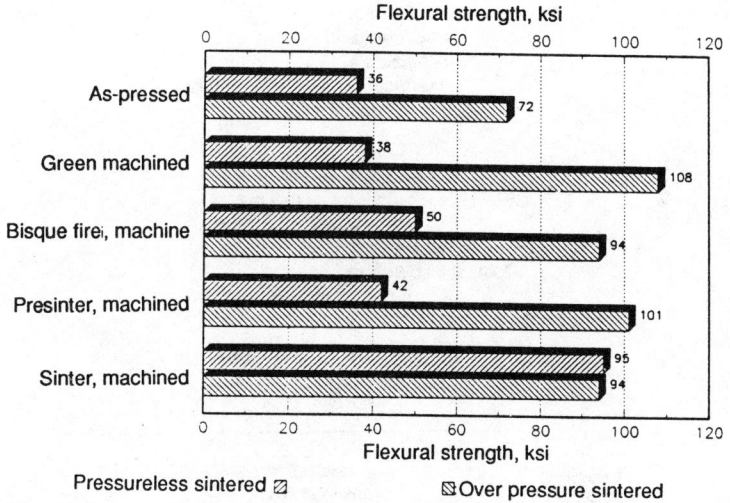

FIGURE 2.21 Strength of sialon ceramic as a function of heat treatment.

from improper machining practice. However, if chemical deposits are found in the surface, removal of the damage is much more difficult and the solution will depend on the chemical nature of the deposit, the substrate, and the property to be restored.

2.6 COMPOSITE DESIGN CONSIDERATIONS

Because of the complexity of the issues as well as the very early stages in the development of ceramic matrix composites, only a limited discussion of this subject is given in this section. The foremost challenges in composite design relate to system definition for intended use, reliable and reproducible fabrication practice, nondestructive evaluation, characterization, and establishment of methodology for reliability assurance. The design considerations vary depending on the type of the composite, namely, particulate reinforced, whisker or short-fiber reinforced, or continuous-fiber reinforced.

It should be realized that, in a majority of cases, a negligible increase of stiffness occurs due to the high moduli of the ceramic matrices compared to those of polymer and metal matrices. The reinforcements are utilized, with designed interface coatings which are amenable for separation under stress, to increase the relative fracture propagation resistance of the unreinforced matrix material. However, in engineering applications the most important property is the first matrix-cracking stress, which then determines the service stress that the material system can support without degradation due to subsequent environmental attack. This is illustrated in Fig. 2.22.

The prevailing deformation and fracture mechanisms have been identified in Fig. 2.22 at various stages of strain. Detailed mechanisms need to be understood for a specific system before application in order to differentiate the behavior from

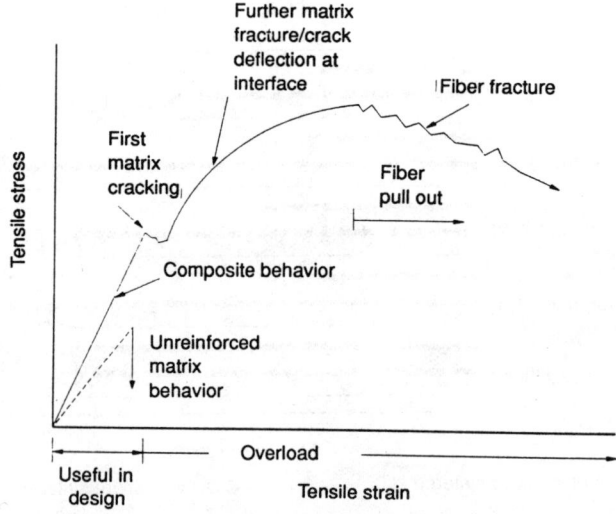

FIGURE 2.22 Stress-strain response of composite and relevance of increasing the first matrix-cracking stress in design.[56]

that of monolithic ceramics. Either the extent of the retention of the physical and chemical nature of matrix-reinforcement interface characteristics during service must be guaranteed, or the nature of change should be understood for lifetime predictions. Otherwise the strain response can become unreliable even in the so-called "graceful failure" regime, where toughening mechanisms may be understood but not relatively useful in practice. Some fundamental relationships of composite properties, based on the rule of mixtures, are briefly given here for the sake of completeness.[56,57]

Anisotropic Considerations

The nature of any composite is to exhibit properties that depend on the orientation. Even for the composite in which the reinforcement is randomly oriented, some minor inhomogeneous and localized orientation can be expected as a result of processing limitations. Many of the well-established theories and the practical understanding which exist for metal- and polymer-based composites are directly applicable to ceramic-matrix composites. (See, for example, Chamis.[57])

Strength. The strength of the oriented ceramic composite is, in general, higher in the direction of reinforcement orientation than perpendicular to it, as illustrated in Fig. 2.23. Figure 2.24 shows the stress-strain curves for a Si_3N_4–SiC system per Bhatt.[58]

Fracture Toughness. The mechanics of stress transfer across the reinforcement-matrix interface dictate that the fracture resistance behavior be different for different orientations of the composite. This aspect can be minimized in the randomly woven composites with superior crack propagation resistance in all

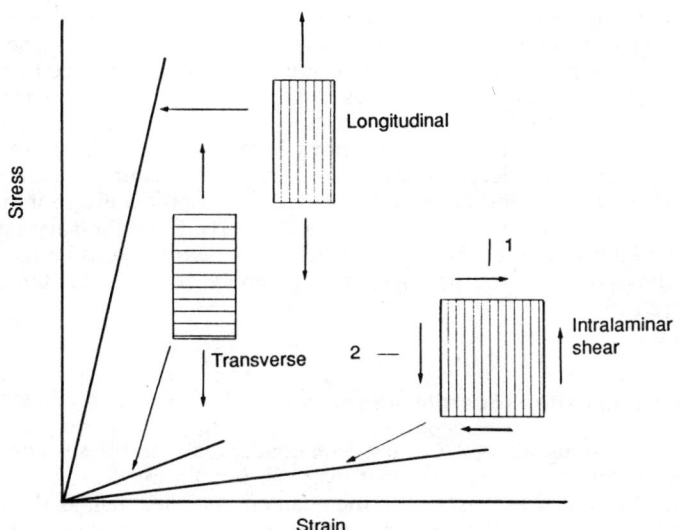

FIGURE 2.23 Stress-strain behavior of composite as a function of orientation.

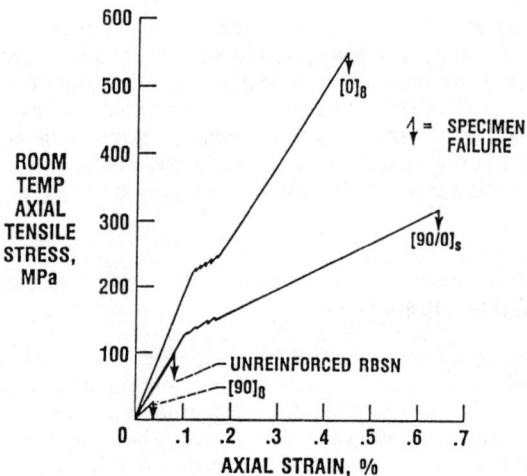

FIGURE 2.24 Stress-strain curves for reaction-bonded Si_3N_4 composite with continuous SiC-fiber reinforcement. (*From Bhatt.*[58])

directions. However, practical manufacturing considerations exist with respect to building a dense matrix around the woven reinforcement architecture.[59]

Interface Properties

As mentioned previously, the critical aspect, in both design and application, of the ceramic-matrix composite is the interface engineering. The interface must be both physically and chemically stable, but at the same time able to shear across the interface during stressing such that eventually the load is carried by the fibers which bridge the broken matrix surfaces from which the fibers have been pulled. The engineering of the interface involves considerations of chemical compatibility, application methods to assure uniformity in thickness and extent of bonding, and stability during service. Another challenge is establishing reliable property data on these interfaces by acceptable methods. It is premature to initiate a detailed discussion of these since many are still evolving. Preliminary indications are that the interface mechanical properties can be expected to be statistical in nature,[60] thus possibly requiring Weibull-type analyses and reliability considerations thereof.

Modeling Composite Properties—Behavior

Most of the modeling work so far has been concerned with the selection of reinforcement for toughness improvement over the unreinforced ceramic. The fundamental physical characteristics are the geometric shape, such as the effectiveness of two-dimensional platelets and three-dimensional particles, whiskers and short fibers, and long continuous fibers, and the mechanical and thermal characteristic mismatches, such as the mismatches in elastic moduli (E and G), Poisson's ratio, and thermal expansion coefficient.[61]

Design-Manufacture Interaction

As was established in the case of monolithic ceramics, active interaction among the designer and the manufacturer from the early concept stages is crucial for successful product demonstration, application, and end use. The constant mutual education in all aspects involving system definition, raw material considerations, manufacturing methods, in-process quality control aspects, properties definition, database creation for use in design, nondestructive evaluation, and final application is very critical to eventual success.

2.7 NONDESTRUCTIVE EVALUATION

The nondestructive evaluation of ceramic components assures the required reliability predicted by mechanical property database extrapolation. As mentioned previously, in-process quality control during powder preparation, green forming, processing-aids removal, and high-temperature consolidation and densification, machining, and subsequent heat treatment, as well as detailed postprocess inspection specifically related to component application, are important aspects which require good interaction between the user and the manufacturer.

Although details of in-process quality assurance procedures are proprietary, common aspects include powder qualification and checking the physical (morphological), chemical, and crystallographic nature of the constituent phases as well as chemical impurities in order to ensure proper flow properties during green compaction and sintering to produce active properties for subsequent high-temperature densification and grain growth. X-ray radiography, including microfocus facility and computed tomography, bulk wave ultrasonics, and, more recently, nuclear magnetic imaging, have all been developed successfully and used commercially with varying degrees of reliability in the making, shaping, and furnacing of ceramic bodies.[62]

The reliability of defect detection by using any of the nondestructive testing methods depends to a large extent on differences in the chemical nature of the flaw with respect to the matrix, its physical nature such as the density and definition of its geometry and shape, and finally its location and the ratio of its dimension to the thickness of the ceramic body. Many mathematical treatments are available in order to extrapolate defect properties from the nondestructive testing signals. However, the extent of severity of the flaw in the application cannot be determined unambiguously as yet. Therefore, the setting up of accept-reject criteria is currently based on experience and past performance history rather than on scientific foundation. The pitfalls of this practice are rather obvious. Components are falsely rejected or erroneously accepted, which results in product failures as well as added cost to the user.

Due to these reasons, in demanding applications such as gas turbine ceramic components, especially the rotor, it is prudent to proof test the components destructively at stress levels of between 105 and 120 percent of the service stress and use the survivors in the application. At the present time it is more costly to perform this proof test, especially at service temperatures and under environmental conditions, but this ensures the required levels of reliability better than any current nondestructive testing practice.

2.8 CONCLUSION

This chapter has given an overview of the philosophy and test methods used in the determination of, primarily, the mechanical properties of ceramics, although other (thermal and physical) properties are also of relevance in design. The key to the successful application of ceramics in engineering is an understanding of the mechanical response of ceramics as compared to metals, the statistical nature of the variation in properties, and the limited ductility. Within the last decade considerable advances have been made in establishing a better understanding of predictive models for reliability based on both a mechanical property database and nondestructive testing signals from test coupons as well as components.

The ceramics applications community has begun to appreciate the availability of a variety of different materials within the silicon carbide and silicon nitride families of nonoxide ceramics, as well as the zirconia- and alumina-based oxide ceramics with superior fracture resistance. This is similar to the availability of a variety of steels for specific uses. However, because of yet limited commercial use, the crucial questions of production methods, cost, and wider availability with alternate manufacturing sources remain to be answered.

Rapid progress toward commercialization begins with a successful demonstration of prototype components, which requires continuous designer-manufacturer interaction. The research is expected to provide within the next decade answers to many of the issues raised in this chapter, thus making these high-performance ceramics "now" products instead of products of the future.

REFERENCES

1. B. R. Lawn and D. B. Marshall, "Hardness, Toughness, and Brittleness: An Indentation Analysis," *J. Am. Ceram. Soc.*, vol. 62, pp. 347–350, 1979.
2. M. N. Giovan and G. Sines, "Biaxial and Uniaxial Data for Statistical Comparisons of a Ceramic's Strength," *J. Am. Ceram. Soc.*, vol. 62, pp. 510–515, 1979.
3. A. F. Kirstein and R. M. Wooley, *J. Res., Natl. Bur. Stand.*, JNBAA, vol. 71C, pp. 1–10, 1967.
4. A. R. Rosenfield et al., "Failure of Ceramics under Multiaxial Stresses," Battelle Columbus Labs., Rep., May 1979.
5. P. H. Conley, H. C. Chandan, and R. C. Bradt, "Dynamic Fatigue of Foamed Glass," in *Fracture Mechanics of Ceramics*, vol. 4, Plenum Press, New York, 1978, pp. 761–772.
6. M. K. Ferber, V. J. Tennery, and S. B. Waters, "Fracture Strength Characterization of Tubular Ceramic Materials Using a Simple C-Ring Geometry," *J. Mater. Sci.*, vol. 21, pp. 2628–2632, 1986.
7. S. G. Seshadri and M. Srinivasan, "Weibull Statistics for Expanded Ring Flexure Tests for Ceramics," *J. Mater. Sci.*, vol. 16, pp. 3052–3058, 1981.
8. F. F. Lange, E. S. Diaz, and C. A. Andersson, "Tensile Creep Testing of Improved Si_3N_4," *J. Am. Ceram. Soc.*, vol. 58, pp. 845–848, 1979.

9. R. K.Govila, "Uniaxial Tensile and Flexure Stress Rupture Strength of Hot-Pressed Si_3N_4," *J. Am. Ceram. Soc.,* vol. 65, pp. 15–21, 1982.
10. S. M. Kats, "Grip for Tensile Testing of Brittle Materials," *Ind. Lab.* (Engl. transl.), vol. 41, p. 1743, 1975.
11. S. G. Seshadri and K. Y. Chia, "Tensile Testing of Ceramics," *J. Am. Ceram. Soc.,* vol. 70, pp. C-242–C-244, 1987.
12. N. Hecht, Technology Development Contractors' Coordination Mtg., Dearborn, Mich., Oct. 1989.
13. M. Srinivasan and S. G. Seshadri, Carborundum Co., 1988, unpublished.
14. A. G. Evans, Rockwell Intl. Science Center, Rep., 1987.
15. S. G. Seshadri and M. Srinivasan, "Estimation of Fracture Toughness by Intrinsic Flaw Fractography for Sintered Alpha Silicon Carbide," *J. Am. Ceram. Soc.,* vol. 64, pp. C-69–C-71, 1981.
16. K. R. Selkregg et al., "Microstructural Characterization of Silicon Nitride Ceramics Processed by Pressureless Sintering, Overpressure Sintering, and Sinter/HIP," presented at the 14th Ann. Conf. on Composites and Advanced Ceramics, Am. Ceram. Soc., 1990.
17. M. Srinivasan and S. G. Seshadri, "Application of Single Edge Notched Beam and Indentation Techniques to Determine Fracture Toughness of Alpha Silicon Carbide," in S. W. Freiman and E. R. Fuller (Eds.), *Fracture Mechanics for Ceramics, Rocks, and Concrete,* ASTM Spec. Tech. Publ. 74S, 1981, pp. 46–68.
18. J. A. Salem, J. L. Shannon, Jr., and R. C. Bradt, "The Effect of Texture on the Crack Growth Resistance of Alumina," NASA TM–100250, 1987.
19. J. Nakayama, "Direct Measurement of Fracture Energies of Brittle Heterogeneous Materials," *J. Am. Ceram. Soc.,* vol. 48, pp. 583–587, 1965.
20. R. W. Davidge and G. Tappin, "The Effective Surface Energy of Brittle Materials," *J. Mater. Sci.,* vol. 3, pp. 165–173, 1968.
21. J. J. Gilman, "Direct Measurements of the Surface Energies of Crystals," *J. Appl. Phys.,* vol. 31, pp. 2208–2218, 1960.
22. J. P. Berry, "Determination of Fracture Surface Energies by the Cleavage Technique," *J. Appl. Phys.,* vol. 34, pp. 62–68, 1961.
23. P. P. Gillis and J. J. Gilman, "Double-Cantilever Cleavage Mode of Crack Propagation," *J. Appl. Phys.,* vol. 35, pp. 647–658, 1964.
24. S. M. Wiederhorn, A. M. Shorb, and R. L. Moses, "Critical Analysis of the Theory of the Double Cantiliver Method of Measuring Fracture Surface Energies," *J. Appl. Phys.,* vol. 39, pp. 1569–1572, 1968.
25. F. F. Lange, "Griffith Relation for Surface Cracks Placed in Bending," Westinghouse Electric Corp., Tech. Rep. 7, 1972.
26. J. J. Petrovic et al., "Controlled Surface Flaws in Hot-Pressed Si_3N_4," *J. Am. Ceram. Soc.,* vol. 58, pp. 113–116, 1975.
27. S. G. Seshadri and M. Srinivasan, "Hertzian Fracture Testing of Ceramics," in *Proc. Ceram. Soc.,* 1984.
28. H. Conrad, M. K. Keshavan, and G. A. Sargent, "Hertzian Fracture of Pyrex Glass under Quasi-Static Loading Conditions," *J. Mater. Sci.,* vol. 14, pp. 1473–1494, 1979.
29. K. Niihara, R. Morena, and D. P. H. Hasselman, "Evaluation of K_{Ic} of Brittle Solids by the Indentation Method with Low Crack-to-Indent Ratios," *J. Mater. Sci. Lett.,* vol. 1, pp. 13–16, 1982.
30. A. G. Evans and E. A. Charles, "Fracture Toughness Determination by Indentation," *J. Am. Ceram. Soc.,* vol. 59, pp. 371–372, 1976.
31. P. Chantikul et al., "A Critical Evaluation of Indentation Techniques for Measuring Fracture Toughness; II, Strength Method," *J. Am. Ceram. Soc.,* vol. 64, pp. 539–543, 1981.

32. R. S. Storm and M. Srinivasan, "Design Considerations for Fabrication of Sintered α-SiC Components," *Ceram. Eng. Sci. Proc.*, vol. 3, pp. 612–619, 1982.
33. S. Timoshenko, "Strength of Materials," 3d ed., Kreiger Publishing Co., New York, 1956, pt. 2, pp. 205–213.
34. E. J. Bunning, D. R. Claxton, and R. A. Giles, "Liners for Gun Tubes—A Feasibility Study," *Ceram. Eng. Sci. Proc.*, vol. 2, 1981.
35. J. Kidwell, L. Lindberg, and R. Morey, "ATTAP/AGT101—Year 2 Progress in Ceramic Technology Development," in *Proc. Automotive Technology Development Contractors' Coordination Mtg.*, Dearborn, Mich., Oct. 23–26, 1989.
36. K. Katayama et al., "Development of Nissan High Response Ceramic Turbocharger Rotor," presented at the West Coast Intl. Mtg., SAE Tech. Paper 861128, 1986.
37. K. Matoba et al., "The Development of Second Generation Ceramic Turbocharger Rotor—Further Improvements in Reliability," presented at the Intl. Congress and Exp., Detroit, Mich., SAE paper 880702, 1988.
38. G. G. Trantina and H. G. deLorenzi, "Design Methodology for Ceramic Structures," *Trans. ASME, J. Eng. Power*, pp. 1–8, 1977.
39. D. K. Shetty, A. R. Rosenfield, and W. H. Duckworth, "Biaxial Stress State Effects on Strengths of Ceramics Failing from Pores," in R. C. Bradt, A. G. Evans, D. P. H. Hasselman, and F. F. Lange (Eds.), *Fracture Mechanics*, Plenum Press, New York, 1983, pp. 531–542.
40. J. J. Petrovic and M. G. Stout, *J. Am. Ceram. Soc.*, vol. 64, p. 656, 1981.
41. A. G. Evans and A. Rana, "High Temperature Failure Mechanisms in Ceramics," *Acta Met.*, vol. 28, pp. 129–141, 1980.
42. R. L. Orr, O. D. Sherby, and J. E. Dorn, *Trans. ASM*, vol. 46, p. 113, 1954.
43. T. J. Moore, "Feasibility Study of the Welding of SiC," *Comm. Am. Ceram. Soc.*, vol. 68, pp. C-151–C-153, 1985.
44. M. Srinivasan, "Ceramics for Automotive Gas Turbines," 18th Summary Report, Automotive Technology Development Contractors Coordination Mtg., 1981, pp. 121–129.
45. K. Suganuma et al., "New Method for Solid-State Bonding between Ceramics and Metals," *Comm. Am. Ceram. Soc.*, pp. C-117–C-118, 1983.
46. R. C. Bill, "Thermal Shock-Resistant Ceramic Coatings on Metal," U.S. Army Material Development and Readiness Command, Rep. AVRADCOM–TR–79–28, 1983.
47. P. O. Cherreyron, N. J. Bylina, and J. G. Hanoosh, "Ceramic-to-Metal Bonding from a Fracture Mechanics Perspective," in R. C. Bradt, A. G. Evans, D. P. H. Hasselman, and F. F. Lange (Eds.), *Fracture Mechanics of Ceramics*, vol. 8, Plenum Press, New York, 1986, pp. 225–238.
48. K. Suganuma et al., "Influence of Shape and Size on Residual Stress in Ceramic/Metal Joining," *J. Mater. Sci.*, vol. 22, pp. 3561–3565, 1987.
49. K. Suganuma et al., "Stress Corrosion in Ceramic/Metal Joints," presented at the MRS Intl. Mtg. on Advanced Materials, Tokyo, Japan, 1988.
50. S. G. Seshadri and M. Srinivasan, "Evaluation of Slow Crack Growth Parameters for SiC Ceramics," *J. Mater. Sci.*, vol. 17, pp. 1297–1302, 1982.
51. J. Derby, J. MacBeth, and S. G. Seshadri, "Tribological Behaviour of Alpha Silicon Carbide Engine Components," in *Proc. Intl. Conf. on Wear*, Inst. of Mech. Eng., London, 1985, pp. 133–137.
52. D. C. Cranmer, R. E. Tressler, and R. C. Bradt, "Surface Finish Effects and the Strength-Grain Size Relation in SiC," *J. Am. Ceram. Soc.*, vol. 60, p. 230, 1977.
53. M. Kawai, H. Abe, and J. Nakayama, "The Effect of Surface Roughness on the Strength of Silicon Nitride," Asahi Glass Co., Japan, Res. Lab. Rep., 1985.
54. M. B. Thomas et al., "Effect of Machining Parameters on the Surface Finish and

Strength of Hot Pressed Silicon Nitride," presented at the Am. Ceram. Soc. Ann. Mtg., Chicago, Ill., 1986.
55. D. Campos-Loriz and M. Srinivasan, "Surface Effects on Sialon Strength," Carborundum Co., 1989, unpublished.
56. D. C. Larsen et al., "Test Methodology for Ceramic Fiber Composites; Results for Si/LAS, SiC/SiC, and C/SiC Composites," in *Proc. Joint NASA/DoD Conf.*, Cocoa Beach, Fla., NASA Conf. Publ. 2406, 1985, pp. 313–334.
57. C. C. Chamis, "Simplified Composite Micromechanics Equations for Mechanical, Thermal, and Moisture-Related Properties," in *Engineers Guide to Composite Materials*, ASM Intl., 1987.
58. R. K. Bhatt, "Strong, Tough SiC/RBSN Ceramic Composites," NASA Conf. Publ. 10025, *Hi Temp Rev.*, pp. 271–283, 1989.
59. F. K. Ko, "Preform Fiber Architecture for Ceramic-Matrix Composites," *Ceram. Bull.*, vol. 68, pp. 401–414, 1989.
60. A. G. Evans and D. B. Marshall, "The Mechanical Behavior of Ceramic Matrix Composites," *Acta Met.*, vol. 37, pp. 2567–2583, 1989.
61. J. Selsing, *J. Am. Ceram. Soc.*, vol. 44, p. 419, 1961.
62. See, for example, *Proc. Conf. on Nondestructive Testing of High-Performance Ceramics*, Boston, Mass., Aug. 25–27, 1967.

BIBLIOGRAPHY

Abe, H., "Mechanical Properties of Engineering Ceramics," *Ceram. Bull.*, vol. 64, pp. 1594–1596, Dec. 1985.

Anderson, R. M., "Testing Advanced Ceramics," *AM&P*, vol. 135, pp. 31–36, Mar. 1989.

Bradt, R. C., "Fracture Measurements of Refractories; Past, Present, and Future," *Ceram. Bull.*, vol. 67, pp. 1176–1178, July 1988.

Cao, H. C., et al., "High-Temperature Stress Corrosion Cracking in Ceramics," *J. Am. Ceram. Soc.*, vol. 70, pp. 257–264, 1987.

Chao, L. Y., D. Singh, and D. K. Shetty, "Effects of Subcritical Crack Growth on Fracture Toughness of Ceramics Assessed in Chevron-Notched Three-Point Bend Tests," ASME 88-GT-185, presented at the Netherlands Gas Turbine and Aeroengine Cong. and Exp., Amsterdam, June 5–9, 1988, 6 pp.

Clarke, D. R., and B. Schwartz, "Transformation Toughening of Glass Ceramics," *J. Mater. Res.*, vol. 2, pp. 801–804, Nov./Dec. 1987.

Cook, R. F., et al., "Crack Resistance by Interfacial Bridging: Its Role in Determining Strength Characteristics," *J. Mater. Res.*, vol. 2, pp. 345–356, May/June 1987.

Dalgleish, B. J., and E. B. Slamovich, "Creep Fracture of Ceramics," *AM&P*, vol. 2, pp. 30–35, Aug. 1985.

Datta, S. K., A. K. Mukhopadhyay, and D. Chakraborty, "Young's Modulus—Porosity Relationships for Si_3N_4 Ceramics—A Critical Evaluation," *Ceram. Bull.*, vol. 68, pp. 2098–2102, Dec. 1989.

Dauskardt, R. H., "Transformation Toughening and Its Implication for Crack Growth," presented at the ASM Intl. Ann. Mtg., Indianapolis, Ind., Oct. 2, 1989.

Dauskardt, R. H., W. Yu, and R. O. Ritchie, "Fatigue Crack Propagation in Transformation-Toughened Zirconia Ceramic," *J. Am. Ceram. Soc.*, vol. 70, pp. C-248–C-252, Oct. 1987.

Evans, A. G., "Fatigue in Ceramics," *Intl. J. Fracture*, vol. 16, pp. 485–498, Dec. 1980.

Evans, A. G., *Fracture in Ceramic Materials*, Noyes Publ., Park Ridge, N.J., 1984, 420 pp.

Evans, A. G., *Ceramic Containing Systems*, Noyes Publ., Park Ridge, N.J., 1986, 367 pp.

Evans, A. G., and B. J. Dalgleish, "Some Aspects of the High Temperature Performance of Ceramics and Ceramic Composites," in K. Iida and A. J. McEvily (Eds.), *Advanced Materials for Severe Service Applications,* Elsevier Applied Sci., London, 1987, pp. 91–118.

Evans, A. G., and B. J. Dalgleish, "Some Aspects of the High Temperature Performance of Ceramics and Ceramic Composites," in J. K. Tien and T. Caulfield (Eds.), *Superalloys, Supercomposites and Superceramics,* Material Science and Technology Series, Academic Press, Orlando, Fla., 1989, pp. 697–720.

Flom, Y., "Fracture Toughness of SiC/Al Metal Matrix Composite," NASA TM 100745, Aug. 1989, 18 pp.

Govila, R. K., P. Beardmore, and K. R. Kinsman, "Strength Characterization and Nature of Crack Propagation in Ceramic Materials," *Fractography Mater. Sci.,* pp. 225–245, Nov. 1979.

"Guide to Engineered Materials," *AM&P,* vol. 3, mid-June 1988.

Hartsock, D. L., and A. F. McLean, "What the Designer with Ceramics Needs," *Ceram. Bull.,* vol. 63, pp. 266–270, Feb. 1984.

Hecht, N. L., "High Temperature Tensile Strength and Tensile Stress Rupture Behavior of Structural Ceramics," presented at the 14th Ann. Conf. on Composites and Advanced Ceramics, Cocoa Beach, Fla., Jan. 14–17, 1990, Paper 43-C-90F.

Hoagland, R. G., and C. H. Henager, Jr., "Modeling Toughness Effects in Whisker-Reinforced Composites," presented at the ASM Intl. Ann. Mtg., Indianapolis, Ind., Oct. 2, 1989.

Katz, R. N., "Ceramics as Substitutes for Scarce Metals," AGARD CP356, AGARD Struct. & Mater. Panel, 57th Mtg., Vimeiro, Portugal, Oct. 9–14, 1983, pp. 14-1–14-8.

Lamon, J., "Ceramics Reliability; Statistical Analysis of Multiaxial Failure Using Weibull Approach and Multiaxial Elemental Strength Model," ASME 88-GT-147, June 1988, 9 pp.

Lange, F. F., "Microstructurally Developed Toughening Mechanisms in Ceramics," Rockwell Intl. Sci. Ctr., Contract N00014-77-C-0441,5 Tech. Rep., July 1978.

Larsen, D. C., et al., *Ceramic Materials for Advanced Heat Engines,* Noyes Publ., Park Ridge, N.J., 1985, 380 pp.

Lepisto, T. T., and T. A. Mantyla, "Room Temperature Fatigue Properties of PSZ-Ceramics," presented at the 14th Ann. Conf. of Composites and Advanced Ceramics, Cocoa Beach, Fla.; Jan. 14–17, 1990, Paper 78-C-90F.

Manderscheid, J. M., and J. P. Gyekenyesi, "Fracture Mechanics Concepts in Reliability Analysis of Monolithic Ceramics," NASA TM-100174, Aug. 1987, 16 pp.

Matsuo, Y., K. Kitakami, and S. Kimura, "Crack Size and Strength Distribution of Structural Ceramics after Non-Destructive Inspection," *J. Mater. Sci.,* vol. 22, pp. 2253–2256, 1987.

Michalske, T. A., and B. C. Bunker, "The Fracturing of Glass," *Sci. Amer.,* vol. 257, pp. 122–129, Dec. 1987.

Moorhead, A. J., and P. F. Becher, "Adaptation of the DCB Test for Determining Fracture Toughness of Brazed Joints in Ceramic Materials," *J. Mater. Sci.,* vol. 22, pp. 3297–3303, 1987.

Noguchi, K., et al., "Tensile Strength of Yttria-Stabilized Tetragonal Zirconia Polycrystals," *J. Am. Ceram. Soc.,* vol. 72, pp. 1305–1307, July 1989.

Osgood, C. C., "Fracture Mechanics Comes of Age, Parts I & II," *Mach. Des.,* pp. 123–127, 153–156, Feb. 12, 1987.

Page, R. A., "Study of High Temperature Failure Mechanisms in Ceramics," SWRI8578/4, AFOSR TR87-0738, F49620-85COO73, Apr. 1987.

Probst, H. E., "Substitution of Ceramics for High Temperature Alloys," NASA TM-78931, E-9674, 1978, 19 pp.

Quinn, G. D., "Flexure Strength of Advanced Ceramics—Round Robin Exercise," Army Lab. Command, MTL Rep. TR-89-62, July 1989.

Quinn, G. D., and F. Baratta, "Flexure Data—Can It Be Used for Ceramics Part Design?," *AM&P,* vol. 1, pp. 31–35, Dec. 1985.

Reuter, W. G., "Applicability of Fracture Mechanics to Lifetime Prediction of B&W Ceramic Heat Exchanger Tubes," EG&G Idaho, DOE/ID-10171 (DE87012038), Nov. 1986, 79 pp.

Rice, R. W., "Capabilities and Design Issues for Emerging Tough Ceramics," *Ceram. Bull.,* vol. 63, pp. 256–262, Feb. 1984.

Sheppard, L. M., "Tensile Testing of Ceramics," *AM&P,* vol. 131, pp. 11–15, May 1987.

Sheppard, L. M., "Microanalysis; Solving Microstructural Mysteries," *Ceram. Bull.,* vol. 68, pp. 1187–1195, June 1989.

Smith, P. C., "Making Ceramics Tougher," *Mater. Eng.,* vol. 104, pp. 25–28, Jan. 1987.

Srinivasan, M., and S. G. Seshadri, "Probabilistic Design and Reliability of Silicon Carbide Ceramics," *J. Mech. Des. (Trans. ASME),* vol. 104, pp. 635–642, July 1982.

Sutcu, M., "Weibull Statistics Applied to Fiber Failure in Ceramic Composites and Work of Fracture," *Acta Met.,* vol. 37, pp. 651–661, Feb. 1989.

Tighe, N. J., et al., "Application of Proof Tests to Silicon Nitride," NIST, Rep. NBSIR-77-1202(PB276652), Mar. 1977.

Tomizawa, H., and T. E. Fischer, "Friction and Wear of Silicon Nitride at 150°C to 800°C," ASLE 85-TC-4A-1, presented at the ASLE/ASME Tribology Conf., Atlanta, Ga., Oct. 8–10, 1985, 7 pp.

Usami, S., I. Takahashi, and T. Machida, "Static-Fatigue Limit of Ceramic Materials Containing Small Flaws," *Eng. Fracture Mech.,* vol. 25, pp. 483–495, 1986.

Wiederhorn, S. M., and B. J. Hockey, "Tensile Creep of Whisker Reinforced Silicon Nitride," presented at the 14th Ann. Conf. of Composites and Advanced Ceramics, Cocoa Beach, Fla., Jan. 14–17, 1980, Paper 2-CP-90F.

Zawada, L. P., and L. M. Butkus, "Room Temperature Tensile and Fatigue Properties of Silicon Carbide Fiber-Reinforced Aluminosilicate Glass," presented at the 14th Ann. Conf. of Composites and Advanced Ceramics, Cocoa Beach, Fla., Jan 14–17, 1990, Paper 98-C-90F.

Zipperian, D. C., "Preparation of Ceramic Materials for Surface Characterization," *Ceram. Bull.,* vol. 68, pp. 1196–1201, June 1989.

CHAPTER 3
COMMERCIAL STRUCTURAL CERAMICS

3.1 INTRODUCTION

This chapter is addressed primarily to the potential user of structural ceramics who may not be familiar with their processing and properties. The means of classifying structural ceramics are discussed, and the principal families are described. Attention is drawn to important and sometimes subtle aspects of microstructure that affect behavior. The principal aspects of the mechanical behavior of structural ceramics that affects their use are summarized, and the data needed to define such behavior are listed. The current situation regarding standards and standard test methods is briefly reviewed. The major compilations of mechanical property data on structural ceramics are identified. A summary table of manufacturers' data on commercially available structural ceramics is given. For each of the five major families of structural ceramics, the main types and their processing are described. Manufacturers' data are supplemented by selected data on similar research materials.

Commercial ceramic products are usually described by their chemical family (that is, their primary crystalline phase) and a specific manufacturer's designation. Thus, for example, silicon nitride NC132 refers to a predominantly silicon nitride (Si_3N_4) material made by the Norton Company under its designation of NC132. One might expect that property specifications would exist for a particular type of structural ceramic and that manufacturers would compete in supplying material to meet or exceed these specifications. Such a system of specifications is not yet in use and may never develop. An alternate approach, the adoption of sets of minimal acceptable property values by groups of users of ceramics, is another possible way in which standards for ceramics could develop. Structural ceramics are still under development, and neither the state of technical development nor the size of the market has yet caused the adoption of generally agreed-upon specifications or required property values. Work toward the standardization of advanced ceramics, including structural ceramics, is under way in several countries and is briefly outlined in this chapter. At present, however, commercial structural ceramics can best be discussed in terms of chemical families, with recognition of important groupings within each family, and with acknowledgement of the fact that properties within a grouping may vary significantly for materials made by different manufacturers. Before turning to a discussion of the various families of structural ceramics, it is important to recognize some general considerations that affect all such families.

1. The chemical family refers to the primary chemical constituent, but various ceramics of the same chemical family can vary considerably in their secondary chemical contents. For example, silicon nitride typically contains several percent of additives, such as Al_2O_3 and Y_2O_3, as well as SiO_2 from the surface of the Si_3N_4 powder. These additives form a liquid phase during heat treatment, producing a dense material as well as a second phase, or phases, at the grain boundaries of the primary phase. The grain-boundary phases often have a major effect on mechanical properties, especially at high temperatures.

2. Ceramic compounds often have metastable forms in addition to the stable form with the lowest free energy. Many commercial ceramics contain a portion which is in a metastable state. For example, the grain-boundary phase mentioned as a sintering aid often becomes a metastable glassy phase, which may crystallize during high-temperature service.

3. The mechanical behavior of ceramics intended for structural use is complex. This leads to difficulty in summarizing such behavior in tables and in determining a complete set of characterizing parameters. One must often work with incomplete data on a specific material and supplement them with data on the general behavior of similar materials. For design purposes, available data must often be supplemented with specific data needed for the application. This requires cooperation between the material supplier and the user to specify the material properties.

4. For a given chemical and phase composition, the mechanical properties of ceramics are strongly dependent on various aspects of microstructure. In particular, grain size and porosity affect strength. Fracture toughness is controlled by various mechanisms which are dependent on microstructure in complex ways. Much of the recent development work on structural ceramics is an attempt to discover microstructures which raise the strength and toughness, and to find processing routes for producing practical parts that have these optimum microstructures. Ideally, one would like to have a very thorough characterization of the chemical composition, phase composition, microstructure, and defects. Manufacturers of commercial ceramics can supply this type of data, but in varying degrees only, as the task of complete characterization of microstructure with respect to mechanical properties is difficult. The microstructural characteristics specified are often not sufficient in themselves to guarantee the mechanical properties listed. In effect, the user is relying on the manufacturer to maintain process control well enough that mechanical property data meet or exceed the manufacturer's claimed values. Various nondestructive evaluation techniques and proof testing are often used along with property data to better assure the performance of a component.

5. The nature of most conventional ceramic processing leads to difficulties in scaling up the size of parts and can result in a dependence of the strength on the size and shape of a part. This processing-related dependence occurs in addition to the dependence resulting from the statistical nature of strength in brittle materials. Strength data listed by manufacturers are generally taken on specimens cut from a larger shape typical of the manufacturer's usual production. Special orders for larger parts can sometimes result in different, and generally lower, strength values. These difficulties may be overcome by some new ceramic processing methods just being commercialized.[1]

6. The strength of ceramics is very often dependent on the state of the surface finish. Some ceramics are used in an as-fired state, and this is generally desirable

because machining is costly and can introduce flaws which may lower the strength. A ceramic part may, however, require machining of one or more surfaces to ensure the fit with other parts.

7. Properties measured by testing are affected by details of the test procedure as well as by all of the factors discussed.

In summary, manufacturers' data on the mechanical properties for a specific ceramic, rather than generic or research data, are essential for the use of commercial products. It may be necessary to supplement such data with estimates from generic data or even with a testing program, but manufacturers' data should be the starting point. These data are to be used with an understanding of the various factors outlined in this section and the realization that mechanical property values can vary if the circumstances of finishing and use do not correspond to the manufacturer's test conditions.

With regard to the data needed for design with structural ceramics, this chapter concentrates on mechanical property behavior. It is, however, important to note that stresses arising from thermal expansion and from the development of temperature gradients can be large and should be taken into account. Values of thermal expansion, thermal conductivity, and thermal diffusivity are therefore needed in addition to the mechanical properties discussed in this chapter.

3.2 MECHANICAL TESTING OF CERAMICS

Mechanical testing of ceramics may appear a mundane subject, but it is of critical importance. However, it will be discussed only briefly to indicate the most important points.

The mechanical behavior of ceramics generally falls into two temperature ranges, with quite different behavior under stress. Mechanical properties were discussed in Chap. 2. Some critical aspects of the mechanical behavior of ceramics are summarized here, and the consequences for the data needed to describe commercial ceramics adequately are indicated.

At room temperature and moderately elevated temperatures, ceramics are generally brittle. In normal mechanical testing they behave elastically at stresses up to the failure stress. They exhibit little or no plastic deformation, except under special loading conditions such as in hardness indents. The strength of a particular ceramic specimen is the result of its intrinsic resistance to crack growth and the size of the largest microcrack. The size of microcracks varies, and ceramics typically exhibit a scatter in strength about the average strength value. The strength distribution of a large number of nominally identical ceramic specimens is often characterized by a Weibull distribution, with a parameter m indicating the degree of variability of strength. In very slow mechanical testing, ceramics often exhibit slow crack growth. The strength of ceramics typically decreases with time under load.[2] This behavior is usually described empirically by the parameter n, which indicates how strongly the crack-growth velocity is dependent on the applied stress intensity. An entire philosophy and design methodology for load-bearing use of brittle ceramics has been built up based on this type of behavior[3] and is discussed in Chap. 2. This approach to mechanical design involves a detailed stress analysis and leads to a probability of failure of the entire ceramic part or assembly. In general, it would be desirable to know the average strength, the fracture toughness in rapid tests, the Weibull parameter m, and the crack-growth parameter n at all temperatures and pressures of intended use. Developing improved microstructures to give the highest possible values of these parameters is a central problem of structural ceramics. Most manufacturers give strength values and some give fracture toughness values, but few give values for Weibull or crack-growth parameters. The designer who wishes to use ceramics structurally may be forced to supplement data from manufacturers with research data on nominally similar ceramics to make an estimate of behavior under service conditions. This procedure should be used with caution because of the sensitivity of the behavior of ceramics to details of microstructure.

As the temperature is increased, a transition into a region of limited plasticity occurs. The onset of this transition depends on the ceramic involved and the details of its minor constituents, but typically occurs in the range of 900 to 1300°C. Increasing the temperature of onset is one of the major development goals of ceramics. This limited plasticity is typically confined to only a portion of the ceramic (such as the grain-boundary phases) and is geometrically constrained. This constraint leads to void formation and stress concentration. The general effect is a lowering of the strength, especially the long-term strength. Campbell et al.[2] distinguish between "rupture at high stress occurring by the extension of preexisting cracks" and "low-stress failures that occur by damage accumulation." Pioneering work on fatigue, deformation failure, and deformation maps was done by Quinn.[4,5] Low-stress failure takes a longer time under load and is usually preceded by appreciable creep. Research is in progress on safe-life prediction methods for ceramics used in this temperature range, but there is no generally accepted methodology. Indeed, although creep behavior is studied, there are no generally accepted parameters to characterize localized behavior in this temperature range. Instead some of the parameters used in the lower temperature range are typically measured into the lower portion of the limited-plasticity range. Thus strength is often measured in short-time tests up to a temperature where a substantial decline occurs.

In summary, knowledge of the following mechanical behavior data is desirable when considering a structural ceramic for engineering use:

1. Values of elastic moduli for all temperatures of intended use. In practice, manufacturers often give Young's modulus and Poisson's ratio at room temperature. Elastic moduli are not structure-sensitive so that the temperature dependence of the percent change in elastic moduli obtained from research specimens can be combined with the manufacturer's value to give good high-temperature estimates up to the temperature at which local plastic deformation begins.

2. Average strength and strength distribution as indicated by the Weibull statistical parameters. Ideally these data should be determined by tensile tests, and they should be available for each temperature of intended use. In practice, manufacturers usually give the average strength at room temperature as measured in bending. Some give an indication of the distribution by giving a maximum and a minimum value or a value of the Weibull m parameter. Different tests (such as four-point bending, three-point bending, C-ring, tensile testing) have different test "volumes" and may give different values for these parameters.

3. Values of fracture toughness as a function of crack length (the R-curve) for each temperature of intended use. In practice, some manufacturers give a fracture toughness value at room temperature but no crack-length dependence. Bulk toughness (such as by bending) can be different from microtoughness (as by Vickers indentation).

4. Curves of slow crack propagation velocity as a function of the stress intensity factor and environment for each combination of conditions under consideration for use. Such data are almost never given in manufacturers' sales literature.

5. Behavior under fatigue conditions, including static fatigue and dynamic fatigue, in the temperature range below which appreciable creep takes place. Such data are generally not available in sales literature.

6. Creep curves and stress rupture data at elevated temperatures. Almost no manufacturers give this type of data in their sales literature.

Thummler[6] has discussed a somewhat longer list of properties needed for design with ceramics, including behavior under multiaxial stresses and friction wear behavior. While these additional properties are needed for special situations, the aforementioned list is sufficient for most situations and constitutes more than is generally available for commercial ceramics.

It should be noted that manufacturers often have research data not included in their sales literature. They may be able to provide additional information for specific materials on an individual basis. We note again that one must be careful in combining research data with data on production material because of the sensitivity of some of the properties to microstructure, which can vary with component size and shape and with the details of the fabrication processes used.

A dominant aspect of research and development of structural ceramics for the last decade has been the effort to achieve higher toughness, primarily through transformation toughening with fine zirconia particles, toughening with fibers or whiskers, or toughening by growth of elongated grains. Evans[7] has summarized the history and given an overview of the theory of toughening. The primary impact so far on commercial ceramics has been the introduction of various toughened zirconias and whisker-toughened aluminas. Exciting research results on the toughening of silicon nitride by elongated grain growth are discussed later in this chapter. These suggest that a new generation of improved commercial silicon nitrides may be available in the future. Dramatically improved toughnesses have also been reported for fiber-reinforced ceramics, some of which are beginning to appear as commercial products.

Mechanical testing of ceramics has been reviewed by Anderson,[8] with special reference to the determination of fracture toughness. Sund and Nicholson have determined conditions for valid toughness measurements in chevron notch tests.[9] Mechanical test procedures for ceramic composites reinforced with ceramic fibers have been evaluated by Larson et al.[10] A central conclusion is that testing in flexure can be very misleading for composites because failure in flexure can occur by delamination. Tensile testing is also preferable for monolithic ceramics. Such testing should be performed at room temperature. In this way, a large volume of material may be exposed to a constant stress, so that the measured strength in the test is more representative of the conditions of service. At room temperature the principal effect concerns the statistics of failure. At high temperature there is also the question of creep, which should be measured separately in tension and in compression. A number of efforts have been made to develop accurate and practical means of tensile testing of ceramics.[11] Carroll et al.[12] have developed a relatively simple technique for tensile creep testing of ceramics which provides good alignment at minimal cost. Tensile tests usually require more complex and expensive specimens, as well as more complicated equipment, than bending tests, so that relatively few data measured in tension are available. Jakus and Wiederhorn[13] have investigated the use of bending tests in creep studies.

The surface condition can also affect measured strength. The effect of surface and near-surface damage on strength has long been recognized.[14] Recent work includes both experimental and theoretical studies of machining conditions to give maximum strength.[15,16]

3.3 STANDARDIZATION OF CERAMICS

Existing U.S. standards for ceramics were established largely under ASTM committees, which developed their own test methods. Some of these have been used for structural ceramics, but it is widely felt that standard test methods adapted to structural ceramics are needed. ASTM Committee C28 on Advanced Ceramics was recently established and is now in operation.[17] Test methods expected to become standards in 1991 include flexure testing at room temperature, elastic modulus by sonic procedures, and nondestructive evaluation processes for advanced ceramics. Test methods under development include flexure testing at elevated temperatures and the testing of nonmechanical properties such as density, porosity, and thermal expansion. In conjunction with the Versailles Project on Advanced Materials and Standards (VAMAS), the committee is developing a classification scheme for advanced ceramics, including structural ceramics.

Activities on standardizing advanced ceramics, including structural ceramics, in Europe, the United States, and Japan have been summarized by Padgett and are given in Table 3.1.[18] It is noteworthy how few standards in the mechanical properties area have been established despite ambitious plans. Work on standardizing test methods for ceramics is proceeding under the auspices of the Japan Institute for Standards (JIS) and appears to be most advanced.[18–20] Standards have been established for bending strength at room temperature (JIS R-1601), modulus of elasticity (JIS R-1602), bending strength at high temperatures (JIS R-1604), elastic modulus at high temperatures (JIS R-1605), tension testing at room and high temperatures (JIS R-1606), and fracture toughness (JIS R-1607). Other properties targeted for standardization in Japan as part of the program of the Japan Fine Ceramics Association (JFCA) are tensile strength, fracture toughness, high-temperature modulus of elasticity, Poisson's ratio, oxidation resistance, corrosion resistance, creep strength, thermal conductivity, specific heat, and compressive strength.

Under the aegis of the Japanese Ministry for International Trade and Industry (MITI) and the JFCA objectives were set for properties of ceramics for high-temperature structural use. Table 3.2 lists these objectives and Table 3.3 gives values achieved for silicon nitride and silicon carbide as reported by Ueda in 1988.[21] However, it is not clear that material available commercially, as opposed to special research samples, will consistently have properties of the caliber of those listed in Table 3.3.

An important international comparison of powder characterization,[22] microstructural characterization,[23] and strength data[24] on three structural ceramics has recently been completed under the auspices of the International Energy Agency. The responsible organizations in Germany, Sweden, and the United States tested three types of ceramics for strength. In the program were included a hot isostatically pressed silicon carbide made by Elektroschmelzwerke Kempten, a hot isostatically pressed silicon nitride made by ASEA CERAMA (now ABB CERAMA), and a sintered silicon nitride made by GTE WESGO. The resulting

TABLE 3.1 Standards Activities for Advanced Ceramics in Various Countries*,†

General area	Specific area	D	F	UK	US	J
Nomenclature				+	+	
General textural properties	Cracking			O	+	
	Density/porosity	O	*	O	+	
	Grain size	Δ		+	Δ	
	Surface texture	Δ	*	+		
Specimen preparation	Specimen shape		*			
	Surface preparation	Δ	*			
Mechanical properties, room temperature	Short-term strength	+	*	+	+	O
	Elastic properties	+	*	+	Δ	O
	Fracture toughness	+	*	Δ	Δ	+
	Fatigue	+				
	Static fatigue			Δ	Δ	
	Hardness	+		+	Δ	+
	Residual stress	Δ				
Mechanical properties, high temperature	Short-term strength, bend/tensile	Δ	*	Δ	Δ	O
	Elastic properties	Δ	*	Δ	Δ	O
	Impact					
	Fracture toughness	Δ	*			+
	Creep	Δ		Δ	Δ	+
	Fatigue	Δ				
	Low load deformation			O		
	Crack growth					
	Thermal shock resistance	O				
Physical properties, high temperature	Thermal expansion	O	*	O		Δ
	Thermal diffusivity			O		+
	Thermal conductivity	Δ				+
	Specific heat			+		+
	Permanent change in dimensions			Δ		
Electrical properties	Dielectric constant	O	*			
	Electrical resistance	O				
Corrosion	Oxidation			O		+
	Aqueous electrolytes			O		
Erosion						
Wear						
Adhesive joints			O			
Powders	Particle size		*		Δ	
	Specific surface		*		Δ	
	Compressibility	O	*		Δ	
	Rheology		*		Δ	
Processing	Green-body density	O	*		Δ	
	Sinterability		*		Δ	
Analysis	Structural		*		Δ	
	Nondestructive evaluation		*			
	Chemical		*			O
	Statistical		*			

*D—Germany; F—France; UK—United Kingdom; US—United States; J—Japan.
†O—completed standard; +—activity in progress; Δ—future activity; *—under consideration.
Source: After Padgett.[18]

TABLE 3.2 Objective Material Properties at Final Stage

Classification	Objective values
Materials usable at higher than 1400°C	Instant fracture ($\geq 1400°C$ in air): • Minimum guaranteed strength $\geq$ 400 MPa • Rate of rejection $\leq$ 20% • Weibull modulus (for reference) $m \geq 20$ Delayed fracture ($\geq 1400°C$ in air) by creep testing corresponding to 10,000-h holding: • Creep fracture strength $\geq$ 250 MPa Oxidation, corrosion, and wear resistance: • No degradation in strength and character after 200-h exposure in combustion-gas flow of $\geq$ 1400°C containing coal ash Fracture toughness for materials satisfactory with the above qualifications ≥ 8 MPa $\cdot$ m$^{1/2}$ at RT
Materials usable at higher than 1250°C	Instant fracture ($\geq 1250°C$ in air): • Minimum guaranteed strength $\geq$ 600 MPa • Rate of rejection $\leq$ 20% • Weibull modulus (for reference) $m \geq 20$ Delayed fracture ($\geq 1250°C$ in air) by creep testing corresponding to 10,000-h holding: • Creep fracture strength $\geq$ 250 MPa Oxidation, corrosion, and wear resistance: • No degradation in strength and character after 200-h exposure in combustion-gas flow of $\geq$ 1250°C containing coal ash Fracture toughness for materials satisfactory with the above qualifications ≥ 15 MPa $\cdot$ m$^{1/2}$ at RT

Source: From Japanese MITI-JFC program; after Ueda.[21]

values of the Weibull parameters σ_0 and m were 544 MPa and 7 for the ESK silicon carbide, and 682 MPa and 14 for the ASEA silicon nitride, and 695 MPa and 12 for the GTE silicon nitride.

3.4 EXISTING HANDBOOKS, DATABASES, AND COMPILATIONS

ASTM Committee E49 on Computerization of Material Property and various other committees and organizations are working toward unified designation systems to harmonize the various ways of designating materials which are now in use.[25] Meanwhile the different materials information sources must be considered and used separately.[26]

A comprehensive evaluation of the structural ceramics available in the early

TABLE 3.3 Properties of Materials for First-Stage Model (Fabricated in Second Phase)

Classification	Property*	Silicon nitride†	Silicon carbide†
High-strength material (1200°C)	Tensile strength, kgf/mm²	30 57.7	30 36.2
	Weibull modulus m	20 23.0	20 11.8
Corrosion-resistant material (1300°C)	Tensile strength, kgf/mm²	20 39.0	20 40.8
	Weibull modulus m	20 26.0	20 20.6
	Oxidation weight gain, mg/cm²	1 0.23	1 0.75
Wear-resistant material (800°C)	Tensile strength, kgf/mm²	50 77.7	50 49.2
	Weibull modulus m	22 20.9	22 11.8
	Specific wear rate, mm³/kg · mm	1×10^{-8} 5.0×10^{-8}	1×10^{-8} 0.2×10^{-8}

*Specific wear rate measured at room temperature; others measured at each specific temperature after 1000-h holding in air.
†Upper level—objective values; lower level—results.
Source: From Japanese MITI-JFC program; after Ueda.[21]

1980s was carried out under the sponsorship of the U.S. Air Force.[27] This report is especially valuable in presenting the temperature dependence of strength and the steady-state creep rate. The most extensive compilation of data on the engineering properties of ceramics was published in 1979.[26] Selected tables of more recent research results on structural ceramics are presented in *Ceramic Source 90.*[28] The first volume of a handbook produced by the National Physical Laboratory provides an introduction for the nonceramist; the second volume surveys data on high-alumina ceramics.[29] The most comprehensive and most nearly current information source is *Structural Ceramics,*[30] which presents numerous tables and graphs of properties.

The first release of a computerized database on structural ceramics is available from the NIST National Standard Data Reference Database Series.[31-34] This database* and a set of programs to access it are provided on a computer disk. It consists of a compilation of research data indicating the state of the technology and provides evaluated data on silicon carbides and silicon nitrides. In an analysis of the existing data on advanced ceramics, Munro et al.[33] conclude that "while measurements of certain properties, such as flexural strength, are plentiful, data on other properties, such as specific heat, thermal shock resistance, and creep crack growth, which are critical to the development of new applications of ceramics, are remarkably scarce."

A rather different database has been compiled by the Oak Ridge National Laboratory (ORNL).[35-37] It contains data generated by research programs on ceramics managed by ORNL and offers some indication of possible improvements that may occur in future commercial materials.

---

*The authors would like to thank R. G. Munro and E. F. Begley for allowing them to browse through a preliminary version of the NIST database.

Much of the research and development on structural ceramics in the last two decades has been supported by national programs. The Advanced Research Projects Agency of the U.S. Department of Defense initiated a ceramic gas turbine program in the early 1970s, and the Japanese MITI began work related to structural ceramics as early as 1974.[38] These programs have evolved through several stages and other federal agencies [U.S. Department of Energy (DOE) and NASA] are now the principal government supporters. These activities also led to the establishment of a recurring conference on ceramic materials and components for engines. The proceedings of these conferences are now a major source of information on structural ceramics.[39-41] These activities led to the founding of private associations in both Japan and the United States to promote the industry associated with advanced ceramics, including structural ceramics. These associations are the JFCA[42] and the U.S. Advanced Ceramics Association (USACA).[43]

Other major sources of information on structural ceramics are the American Ceramic Society's Annual Conference on Composites and Advanced Ceramic Materials,[44] and ASM International.[45]

3.5 COMMERCIAL STRUCTURAL CERAMICS

The structural ceramics of greatest significance fall into five families:

1. Alumina is the most widely used structural ceramic.
2. Silicon carbide is commonly used for its great hardness and good creep resistance at high temperatures.
3. Silicon nitride is a very promising ceramic material, which is undergoing perhaps the most intensive development program.
4. Sialon is a solid solution in the system Si–Al–O–N, which can be regarded as a derivative of Si_3N_4 by substitution of equal fractions of Al and O.
5. Zirconia has the advantage of a very high melting point plus a feature unique among the structural ceramics, that of exhibiting transformation toughening.

A structural ceramic is generally described as being either monolithic or composite. Any of the monolithic forms can be reinforced by particles, whiskers, or fibers, leading to a large family of possible composites. The distinction between monolithic and composite is not rigorous because even the monolithic forms usually contain a second phase used as a sintering aid, and so in a sense they are composites. The term composite is usually reserved for materials in which a second phase is introduced primarily for its strengthening or toughening effect rather than to assist processing.

The principal use of structural ceramics has been as monolithic materials. In addition, composites have shown promise, although costs and difficulties of processing have so far limited their commercial applications. These ceramic-ceramic composites (CCC) are of both the particulate-reinforced and the fiber-reinforced types. Work on ceramic-matrix composites (CMC) has been inspired by polymer-matrix (PMC) and metal-matrix composites (MMC). The fundamental theory for short-fiber and long-fiber composites is well developed,[46,47] and the prospects for CMCs becoming useful at very high temperatures have been considered systematically.[48,49] The CCCs are generally considered in three categories: particulate composites, whisker composites, and continuous-fiber composites.[50] The few types of CMCs that are commercially available will be included with the families of structural ceramics corresponding to the matrix of the composite.

Mechanical properties of commercial structural ceramics are listed in Table 3.4. No claim is made that this table is complete or current. It was constructed from sales literature supplied in 1989 by the companies listed. This table should not be used as a representation of currently available materials but only as a rough guide to the range of structural ceramics and their properties that are commercially available. Individual manufacturers should be contacted to obtain current data on their materials.

Alumina Ceramics

The term alumina is used in a precise sense to mean Al_2O_3. Sometimes it includes hydrates of alumina. Most commonly, alumina is used to include all ceramics in which the alpha phase (corundum) is predominant.

Alumina ceramics are often classified into high alumina (having more than 80 percent aluminum oxide) and porcelain (having less than 80 percent, but still being predominantly aluminum oxide).[29,51-53] High-alumina ceramics were first used in large quantity as spark plug insulators and as laboratory equipment. Subsequently they have found extensive use in mechanical devices and in electronics. More recently, high aluminas have been used as armor and in medical applications. The structural applications of alumina today probably exceed those of all other structural ceramics combined.

As formed by various chemical processes, aluminum oxide can occur in metastable forms, and Gitzen[51] reports that seven crystallographic phases have been found. However, all revert to the alpha phase on heating somewhat above 1000°C, so that only this phase of alumina need be considered in sintered ceramics.

Strength and other properties generally improve as the percentage of alumina is increased; however, costs also increase. This is for two reasons: (1) For lower percentages of alumina, less expensive raw materials can be used. (2) Processing is more difficult with higher-alumina materials. For these reasons, the best practical choice of an alumina is usually that possessing the lowest alumina content but having properties consistent with the desired application.

Mechanical Properties of Commercial Aluminas. The mechanical properties of aluminas are listed in Table 3.4. As generic figures, a flexural strength of about 300 MPa and a fracture toughness of 4 MPa · $m^{1/2}$ are typical of many commercial monolithic aluminas. However, the flexural strength values vary from as low as 150 MPa to over 500 MPa. Only a few manufacturers report Weibull m values, and those are generally low (8 to 10), although Pechiney gives an m value of 25 combined with a flexural strength of 300 MPa for one of its aluminas.

Data on commercial alumina-matrix ceramic composites are also included in Table 3.4. These fall into several categories.

First, alumina can be toughened by adding minor amounts of zirconia in a manner analogous to the transformation toughening of zirconia. The phenomenon of transformation toughening is discussed in the section on zirconia. The very high values of strength and toughness obtained in zirconia are not reached in transformation-toughened alumina, but substantial improvement over plain alumina is possible. Thus strength values as high as 830 MPa are reported and toughness values as high as 8.5 MPa · $m^{1/2}$, but not both in the same material. These transformation-toughened aluminas are inherently

TABLE 3.4 Mechanical Properties of Structural Ceramics as Listed by Their Manufacturers

Company	Specific material	Density, g/cm^3	Flexural strength, MPa	Weibull modulus m	Fracture toughness K_{Ic}, MPa·m$^{1/2}$	Young's modulus, GPa	Poisson's ratio
\multicolumn{8}{Mechanical properties of aluminas}
Coors	AD-85	3.41	296	—	3–4	221	0.22
Ceramics	AD-20	3.6	338	—	3–4	276	0.22
	AD-94	3.7	352	—	4–5	303	0.21
	AD-96	3.72	358	—	4–5	303	0.21
	AD-99.5	3.89	379	—	4–5	372	0.21
	AD-99.9	3.96	552	—	4–5	386	0.22
Feldmühle	RB92	3.99	300	8	4.9	420	0.22
	RK38	3.95	500	8	5.9	410	0.23
	RV679	3.9	300	6	5.4	390	0.23
	RN56	3.9	350	8	5.5	370	0.21
	RT405	3.9	280	6	4.2	350	0.23
	B40	3.82	300	8	4.2	360	0.22
	B600	3.8	290	8	4.9	350	0.23
	V35	3.79	330	9	4.0	350	0.23
	V38	3.75	330	9	4.0	350	0.23
	T201	3.77	280	7	4.0	370	0.23
Friedrichsfeld	F96	3.7–3.8	300	>10	—	350	0.22
	F99.7	3.9–3.95	350	>10	—	380	0.22
Greenleaf	94Al2O3	3.67	280	—	4.0	310	0.29
	97Al2O3	3.8	280	—	4.5	350	0.29
	99Al2O3	3.9	340	—	4.2	370	0.3
	GEM1	3.99	620	—	4.2	393	0.23
GTE	AL-500	3.67	345	—	—	—	—
	AL-600	3.72	365	—	—	—	—
	AL-300	3.76	296	—	—	—	—
	AL-995	3.86	310	—	—	—	—
Kyocera	A-56	3.9	529	—	—	382	—
	A-61	4.1	735	—	—	372	—
	A-150	4	313	—	—	382	—
Lodge	HM	3.68	350	—	6.1	300	0.21
Ceramics	FA	3.68	375	—	6.2	322	0.21
	FF	3.68	380	—	6.2	322	0.21
	FC	3.69	360	—	5.9	301	0.21
	FW	3.7	375	—	6.3	320	0.21
	EH	3.72	420	—	6	328	0.21
	DE	3.75	370	—	5.6	322	0.21
	CL	3.78	400	—	5.8	336	0.21
	AN	3.9	370	—	5.5	340	0.21
Matroc	Al998-051/061	3.96	400	—	5.7	380	—
	Al998-051/061	3.9	350	—	4.7	350	—
	Al998-001	3.85	350	—	3.8	—	—
	Al998-004	3.8	150	—	—	—	—
	Al993-024	3.88	250	—	4.7	—	—
Miyagawa Kasei	MA-1	3.95	441	—	—	0	—
	MA-2	3.6	294	—	—	0	—
NGK/NTK	Corundit-G	3.5	196	—	—	—	—
	Corundit-W	3.5	265	—	—	—	—
	Corundit-H	3.7	304	—	—	—	—

TABLE 3.4 Mechanical Properties of Structural Ceramics as Listed by Their Manufacturers (*Continued*)

Company	Specific material	Density, g/cm³	Flexural strength, MPa	Weibull modulus m	Fracture toughness K_{Ic}, MPa·m$^{1/2}$	Young's modulus, GPa	Poisson's ratio
	Mechanical properties of aluminas (*continued*)						
	Corundit-HC	3.7	304	—	—	—	—
	KP-33	3.6	294	—	—	—	—
	KP-92	3.6	294	—	—	—	—
	KP-95	3.7	340	—	—	280	—
	KP-990	3.9	490	—	—	370	—
	UHA-99	3.9	510	—	—	370	—
	P-6	3.6	294	—	—	—	—
	LP-18	3.5	275	—	—	—	—
	KP-99c	3.9	343	—	—	—	—
	Corundit-C	—	—	—	—	—	—
	Corundit-S	3.8	0	—	—	—	—
	G4	2.9	0	—	—	—	—
	AD-27	3.3	0	—	—	—	—
Nippon Steel	A101	3.90–3.98	392–490	—	3.0–4.0	0	—
	A111	3.95–3.98	411–539	—	3.5–4.5	382	—
	A120	3.60–3.80	343–411	—	3.0–4.0	0	—
	A130	3.95–4.00	490–588	—	4.0–4.5	0	—
Pechiney	AF920	3.6	300	20	4.0	240	0.27
	AF950	3.6	300	12	4.9	260	0.27
	AF970	3.7	300	—	4.6	280	0.27
	AF995	3.8–3.9	300	—	4.6	310	0.27
	AF997	3.8–3.9	300	25	4.6	310	0.27
	AF998	>3.9	380	—	6.0	310	0.27
Toshiba	AL-13	3.7	0	—	3–4	0	0.22
	AL-16	3.8	0	—	3–4	0	0.22
	AL-01	3.9	314	—	3–4	—	—
Toto	AL197	3.9	0	—	—	0	—
	AL190	3.9	0	—	—	0	—
	AL170	3.8	0	—	—	0	0.2
	AL920	3.6	0	—	—	0	—
	AL900	3.8	0	—	—	—	—
	AC291	3.8	0	—	—	0	—
	AC203	3.5	0	—	—	0	—
	AC270	3.4	0	—	—	0	—
W.R. Grace & Co.		3.45	293	—	—	248	—
		3.48	290	—	3.9	248	—
		3.57	320	—	—	293	—
		3.65	324	—	3.9	276	—
		3.70	341	—	4.5	318	—
		3.75	386	—	3.7	310	—
		3.90	345	—	4.3	331	—
	Mechanical properties of alumina-based composites						
Allied-Signal	AS-100	4.5	830	—	5.2	275	—
Coors Ceramics	Hot pressed	3.8	586	—	8	424	—
	Sintered	3.8	414	—	5	344	—
Feldmühle	SN60	4	450	10	5.5	390	0.23
	Sn80	4.12	600	10	5.8	380	0.24
	SH20	4.15	600	10	5.4	400	0.21
	SH1	4.3	600	10	5.2	400	0.21

TABLE 3.4 Mechanical Properties of Structural Ceramics as Listed by Their Manufacturers (*Continued*)

Company	Specific material	Density, g/cm³	Flexural strength, MPa	Weibull modulus m	Fracture toughness K_{Ic}, MPa·m$^{1/2}$	Young's modulus, GPa	Poisson's ratio
Mechanical properties of alumina-based composites (continued)							
Greenleaf	GEM2	4.26	760	—	4.0	395	0.22
	GEM4	3.78	350	—	3.1	355	0.22
	AB40	3.39	620	—	5.1	353	0.2
	AB50	3.28	620	—	4.5	377	0.2
	WG300	3.74	689	13	0.0	0	—
Dupont Lanxide	SiC/Al2O3/Al	2.9	477	—	19	200	0.29
	SiC/Al2O3/Al	3.4–3.5	450–500	—	7–7.5	310–330	0.25–0.29
Lodge Ceramics	AZ	4.1	500	—	8.5	280	0.25
Nippon Steel	A200	4.12–4.17	637–785	—	5.0–6.0	0	—
	A220	—	441–539	—	3.5–4.5	0	—
W.R. Grace & Co.	Transformation tough	4.15	585	—	5.5	0	—
Mechanical properties of aluminum titinates							
Feldmühle	TL10	3.35	27	≥20	—	18	0.2
	TL12	3.35	30	≥20	—	20	0.2
	TL20	3.65	65	≥20	—	30	0.2
NTK Spark Plug	TA-38	3.4	0	—	—	—	—
Mechanical properties of mullites							
NTK Spark Plug	AS	2.6	0	—	—	—	—
	AL	2.6	0	—	—	—	—
	G-1	2.9	0	—	—	—	—
Mechanical properties of magnesium oxide							
Feldmühle	MD10	3.48	180	—	—	260	—
Mechanical properties of multiphase ceramics							
NTK Spark Plug		2.4	0	—	—	—	—
Mechanical properties of silicon carbide							
Asahi Glass	C-600	3.12–3.17	588–637	—	5.1	402–422	0.14–0.16
	C-400	3.08–3.12	392–588	—	4.5	380–412	0.14–0.16
British Petroleum	SA	3.1	460	10	4.6	410	0.14
	KT	3.09	280	12	4.93	380	0.22
	KG	2.7	48	—	—	210	—
	ST	3.3	448	12	8	427	0.15
	Lucas 102	3.26	655	11	9.8	288	0.23
Ceradyne	Ceralloy 146	≥3.15	350–450	—	—	430	—
	Cerastar Rx	≥2.70	98–140	—	—	210	—
	Cerastar RB	≥3.10	378–406	—	—	366	—
Coors Ceramics	Sintered	3.1	552	—	3	400	—
	Reaction-bonded	3.1	462	—	4	393	0.19
	Nitride-bonded	2.6	48	—	—	152	—
Elektroschmelzwerk Kempten	EKasic D	≥3.1	410	10	3.2	410	0.17

TABLE 3.4 Mechanical Properties of Structural Ceramics as Listed by Their Manufacturers (*Continued*)

Company	Specific material	Density, g/cm³	Flexural strength, MPa	Weibull modulus m	Fracture toughness K_{Ic}, MPa·m$^{1/2}$	Young's modulus, GPa	Poisson's ratio
Mechanical properties of silicon carbide (*continued*)							
	EKasic HD post-HIP	≥3.17	430	12	3.2	430	0.16
	EKasic HD HIP	≥3.21	610	12	4	445	0.16
Feldmühle	CD100	3.06	350	8	3.8	400	0.16
	CS102	3.07	310	12	4.0	380	0.19
	CS107	3.04	360	15	4.0	390	0.19
Friedrichsfeld	SiC 198	3.1	350	>10	—	330	0.2
Ibiden	IBICERAM	3.08–3.15	0	—	5.6	0	0.13
Kyocera	SC-211	3.2	540	—	—	431	0.16
	SC-221	3	490	—	—	372	0.16
	SC-5500	2.2	39	—	—	—	—
Nippon Steel	C101	2.6	490–588	—	4	0	—
	CB107	4.5	392–490	—	4–4.5	0	—
	CB108	5	343–392	—	4	0	—
Norton	NC-203	3.3	800	—	3.9	435	
	NC-430	3.1	310	—	3.5	400	—
Toshiba	TSC-01	3.1	0	—	4.6	0	0.51?
Mechanical properties of silicon nitrides							
Allied-Signal	GN-10	3.31	779	10	6.5	306	—
	AS-440	3.22	862	11.7	6.47	310	—
Asahi Glass	N-600	3.10–3.15	588–687	—	5.3	284–304	0.24–0.25
	N-400	3.08–3.10	392–588	—	4.5	275–304	0.24–0.25
	N-600A	3.10–3.15	637–736	—	5.3	284–304	0.24–0.25
Ceradyne	Ceralloy 147	3.29	480–965	—	—	310	—
Coors Ceramics	Hot-pressed	3.31	906	—	6	311	—
	Reaction-bonded	2.5	338	—	3–4	179	—
Elektroschmelzwerk Kempten	EKasin R	2.3–2.8	150–300	10–20	2.0–3.5	80–200	0.22–0.25
	EKasin S	3.2–3.3	400–700	10–20	5.0–8.0	260–310	0.23–0.28
	EKasin D	3.16–3.35	600–850	15–25	6.0–8.5	290–325	0.26–0.28
Feldmühle	SL100	3.3	800	20	7.0	300	—
	SL200	3.25	750	15	6.0	280	—
	NH206	3.2	660	20	6.0	300	—
	NR115	2.4	220	15	—	150	—
Friedrichsfeld	HP 79	3.2	750	>20	—	320	0.26
GTE	SNW-1000	3.28	758	25	—	276	0.23
	SNW-2000	3.3	586	—	—	276	0.23
Kyocera	SN-220	3.2	590	—	—	294	0.28
	SN220M	3.2	676	—	6.3	294	0.28
	SN-235	3.2	980	—	7.0	304	0.28
	SN-251	3.4	650	—	6.9	310	0.27
	SN-733	3.2	980	—	—	304	0.27
NGK/NTK	EC-120	3.26	700	—	3.5	260	0.26
	EC-141	3.23	900	—	6	320	0.27

TABLE 3.4 Mechanical Properties of Structural Ceramics as Listed by Their Manufacturers (*Continued*)

Company	Specific material	Density, g/cm³	Flexural strength, MPa	Weibull modulus m	Fracture toughness K_{Ic}, MPa·m^{1/2}	Young's modulus, GPa	Poisson's ratio
	Mechanical properties of silicon nitrides (*continued*)						
	EC-152	3.26	1100	—	6.3	320	0.26
Nippon Steel	RBSN	2.6	0	—	—	0	—
	SN-NU	2.65–2.75	275–343	—	3–4	196–245	—
	SN-NU	2.70–2.80	313–373	—	3–4	49	—
Norton	NCX-34	3.3	310	—	6.6	310	—
	Noralide	3.2	990	—	5.4	310	—
Pechiney	Kernit 101	2.4	150	—	—	—	0.26
	Kernit 102	2.5	200	—	2.5	150	0.26
	Kernit 103	2.7	250	—	—	—	0.26
	Kersia 201	3.15	280	—	4.5	—	0.26
	Kersit 301	3.2	540	12	7.5	290	0.26
Shinagawa Refractories	SSN	3.15	0	—	—	265–304	—
	RBN	2.55	0	—	—	0	—
	RBS	2.65	0	—	—	0	—
Sumitomo Denko	NS	3.2–3.4	784–1070	—	—	0	0.25
	NT	3.6–3.9	834–980	—	—	0	0.25
Toshiba	TSN-01	3.18	0	5–6	—	0	0.027
	TSN-02	3.22	0	5–6	—	0	0.027
	TSN-03	3.22	883	6–7	—	0	0.027
	TSN-04	3.22	0	6–7	—	275	0.027
	TSN-05	3.2	0	5–6	—	275	0.027
	TSN-06	3.22	883	6–7	—	275	0.027
	TSN-07	3.22	0	5–6	—	275	0.027
	TSN-07	3.26	0	6–7	—	275	0.027
	TSN-08	3.21	0	6–7	—	0	—
	TSN-19	3.22	0	4–5	—	275	0.027
	Mechanical properties of titanium carbide						
Feldmühle	TC30	—	—	30	—	440	0.21
	Mechanical properties of zirconia						
Ceradyne	Ceralloy 408	≥5.79	1100–1700	—	—	210	—
Feldmühle	Zt35	5.74	350	25	6.0	210	0.3
	Zt40	5.74	520	25	8.1	210	0.3
	ZN100	5.98	950	20	10.5	210	0.3
Friedrichsfeld	FMZ	5.5–5.8	450	>20	—	180	0.23
Lodge Ceramics	ZT	5.9	1000	—	10	200	0.31
	SM	5.9	700	—	9	190	0.31
Kyocera	Z-201	5.9	980	—	—	—	0.21
Miyagawa Kasei	MZ-1	6.02	0	—	—	0	—
NGK/NTK	UTZ-10	5.8	750	—	7.3	230	—
	UTZ-20	4.9	1000	—	12.7	320	—
	UTZ-30	5.9	1000	—	9.2	220	—
Nilcra	MS	5.74	820	>30	8–12	205	0.31
	TS	5.73	716	>30	10–15	205	0.31
Nippon Steel	Z100	6.00–6.05	981–1275	—	10–12	0	—
	Z200	5.49–5.51	1079–1471	—	10–12	0	—

TABLE 3.4 Mechanical Properties of Structural Ceramics as Listed by Their Manufacturers (*Continued*)

Company	Specific material	Density, g/cm^3	Flexural strength, MPa	Weibull modulus m	Fracture toughness K_{Ic}, MPa · m$^{1/2}$	Young's modulus, GPa	Poisson's ratio
		Mechanical properties of zirconia (*continued*)					
Norton	YZ-110	6.05	1000	—	8.5	210	—
Matroc	ZR900-102	5.3	250	—	9.5	150	—
Pechiney	ZFCE	5.7	180	—	2.3	140	0.29
	ZFME	5.7	500	20	8.0	200	0.29
	ZFMES	5.4	180	—	2.3	140	0.29
	ZFYE	5.9	—	—	—	—	—
	ZFYT	6	920	15	10.0	220	0.29
Sumitomo Denko	RZ	6.0–6.1	1373–1765	—	—	0	0.32
Toshiba	TZR-01	6.05	0	—	>10	0	0.13
	TZR-02	6.09	0	—	>10	—	—
W.R. Grace & Co.	Tetragonal zirconia	6.10	827	—	11.0	0	—

cheaper than the transformation-toughened zirconias and are a promising family of materials for use near room temperature.

As for transformation-toughened zirconia, the strength of transformation-toughened alumina would be expected to drop with increasing temperature as the transformation-toughening effect is reduced by moving further away from the solidus temperature. This effect for fast-fracture experiments of Iio et al.[54] is shown in Fig. 3.1. The improved room-temperature strength of transformation-toughened alumina is also severely reduced by aging around 300°C. Iio et al. succeeded in making a TZP alumina which showed minimal reduction in room-temperature strength after aging at 300°C, but even this material showed a pronounced stress rupture effect around 300°C, as illustrated in Fig. 3.2. The pre-

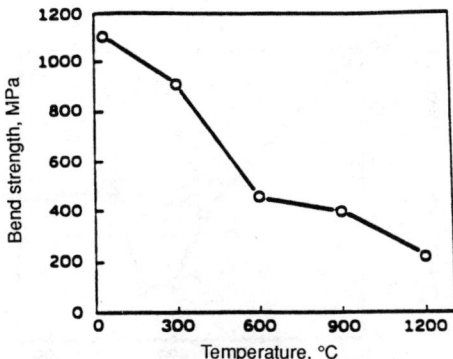

FIGURE 3.1 Effect of 300°C treatment on room temperature properties and temperature dependence of strength in fast-fracture experiments for alumina toughened with ZrO$_2$ stabilized with 2.5 mol % Y$_2$O$_3$.[54]

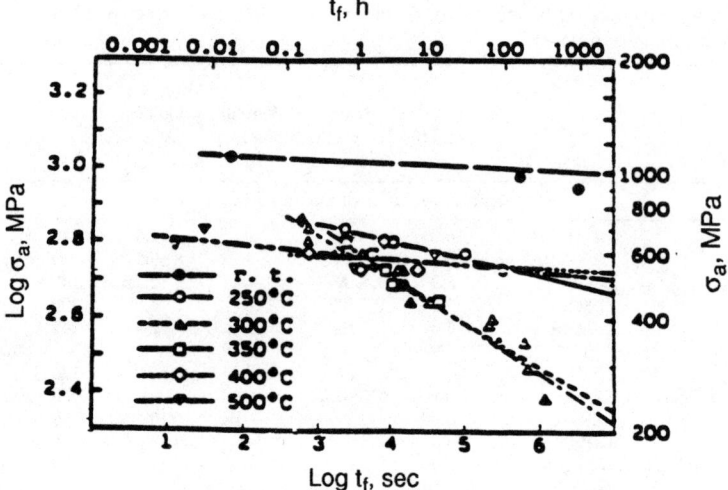

FIGURE 3.2 Stress rupture behavior at various temperatures for alumina toughened with ZrO_2 stabilized with 2.5 mol % Y_2O_3.[54]

dicted maximum stress for a 10,000-hour lifetime is given in Fig. 3.3 as a function of temperature, and the effect of temperature on toughness is shown in Fig. 3.4.

Second are the aluminas toughened with silicon carbide whiskers. When the material is made by hot pressing, strengths of 689 MPa combined with a toughness greater than 8 MPa · $m^{1/2}$ can be obtained. This material is especially interesting because the enhanced properties are retained up to about 1100°C in short-time tests.

Becher et al.[55] have experimentally investigated some of the theoretical predictions regarding the toughening effect of whiskers. They confirm that the toughness addition varies as the square root of the whisker radius, as shown in

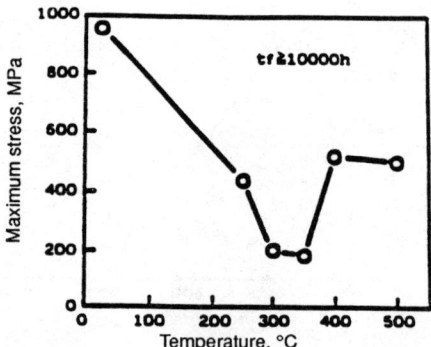

FIGURE 3.3 Predicted maximum applied stress for lifetime of 10,000 h at various temperatures for alumina toughened with ZrO_2 stabilized with 2.5 mol % Y_2O_3.[54]

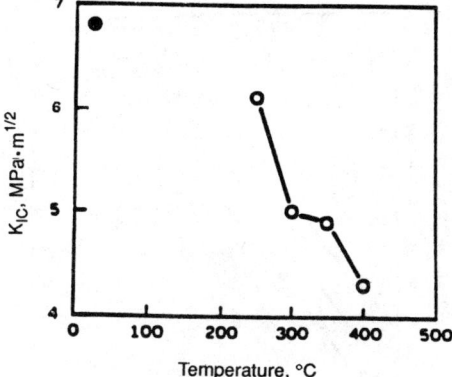

FIGURE 3.4 Fracture toughness decrease with temperature for alumina toughened with ZrO_2 stabilized with 2.5 mol % Y_2O_3.[54]

Fig. 3.5, and that the toughness increases with the square root of the volume fraction of whiskers, as illustrated in Fig. 3.6.

Campbell et al. discuss the mechanisms of toughening by whiskers.[56] They predict increases in fracture surface energy due to 20 vol % whiskers of about 80 J/m^2 in alumina (where the interface is under compression) and of about 30 J/m^2 in silicon nitride (where the interface is in tension).

Iio et al.[57] found that the strength of alumina reinforced with silicon carbide whiskers reaches a maximum of 700 MPa at 30 wt % whiskers, but that the toughness continues to increase to 8 MPa · $m^{1/2}$ at 40 wt %.

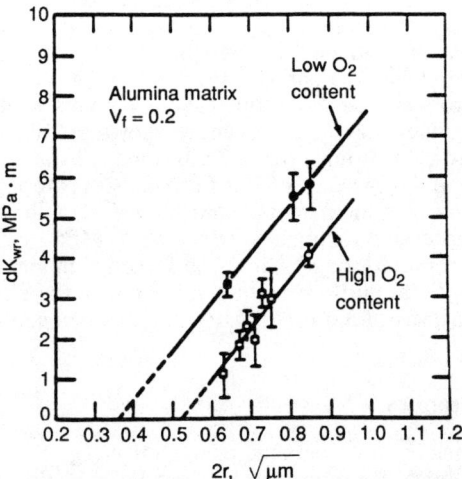

FIGURE 3.5 Toughness increase of SiC whisker-reinforced alumina as a function of whisker diameter.[55]

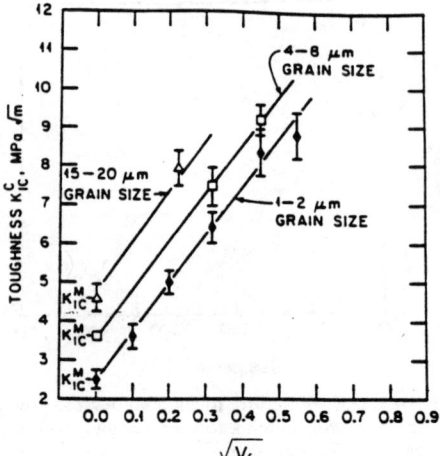

FIGURE 3.6 Toughness of SiC whisker-reinforced alumina as a function of volume fraction of whiskers.[55]

There is some doubt about the degree of property retention for long times at high temperatures in air. Luthra and Park[58] found that the mass gain due to the oxidation in air of alumina reinforced with SiC whiskers was larger than expected for free SiC whiskers, so that property deterioration may be rapid.

Whisker reinforcement of alumina does reduce high-temperature creep rates. Lin and Becher[59] found a reduction in creep rates of about two orders of magnitude in the bending of Al_2O_3 + 20% SiC_w in the temperature range of 1200 to 1300°C. Arellano-Lopez et al.[60] found a somewhat smaller reduction in creep rates in compression; the creep resistance increased with increasing whisker concentration.

Third are aluminas strengthened or hardened by the addition of second-phase particles other than zirconia. The particles used include titania, titanium carbide, silicon nitride, and boron carbide. Substantial improvements in strength and modest improvements in toughness can be achieved, as shown in Tables 3.5 and 3.6.

Fourth are second-phase toughened aluminas made from liquid aluminum by a special oxidation reaction process of the Lanxide Corporation. These generally have some remaining aluminum metal, and so are actually a three-component composite. Very impressive toughness values of 29 MPa · $m^{1/2}$ and strengths of 997 MPa have been reported for SiC fiber-reinforced alumina or aluminum.[61] For alumina with SiC particles and residual aluminum from the directed metal oxidation process, toughnesses of 5.6 to 7.5 MPa · $m^{1/2}$ have been reported.[62]

Silicon Carbide Ceramics

Silicon carbide is a widely used ceramic characterized by high hardness and good strength retention to high temperatures. It has good oxidation resistance due to a protective SiO_2 surface layer. Its thermal shock resistance is considered good for a ceramic due to its relatively low thermal expansion and high thermal conductivity.

Silicon carbide exists in a cubic form termed beta silicon carbide and in a variety of noncubic forms, which are collectively termed alpha silicon carbide. Chemical conversion from polymer precursors generally leads to the beta form, while carbothermal reduction at high temperature produces the alpha form. The alpha form is the more stable at high temperatures, but the beta form is sufficiently stable that it can be sintered.[38,63]

Silicon carbide is produced most commonly by the Acheson process, in which current is passed through a mixture of silica sand and carbon to drive the reaction $SiO_2 + 3C \rightarrow SiC + 2CO$. When carried out at 2600°C or higher, the alpha form results, but the beta form can be made by operating around 1500 to 1600°C. The resulting mass is crushed, milled, leached, and washed to produce silicon carbide powder.

Silicon carbide can also be produced from polymer precursors and by vapor-phase reactions. (See Chaps. 4 and 5.) These processes have been used to make powder, but are especially important in the production of fibers and whiskers of silicon carbide.

Silicon carbides in solid form can be grouped into three categories according to the processing used: reaction-sintered, hot-pressed, and sintered.[63]

Reaction sintering refers to the reaction between carbon and silicon to form silicon carbide. A mixture of silicon carbide powder and carbon (or a carbonaceous precursor) is heated in contact with molten silicon. The result is a nearly complete conversion to silicon carbide, although most materials made by this process contain an excess of carbon or silicon.

Hot pressing of silicon carbide is done at 2000°C or higher and produces a dense and very hard material. A sintering additive such as alumina is used to promote liquid-phase sintering. A process for sintering beta silicon carbide by using carbon and boron as additives was discovered in 1973. The process was later extended to include alpha silicon carbide.[64]

Mechanical Properties of Silicon Carbide. The mechanical properties of silicon carbides are summarized in Table 3.4. Mechanical properties of silicon carbides studied in the DOE-NASA programs and summarized in the ORNL database are listed in Tables 3.5 and 3.6.[37]

Hecht et al.[65] summarized reported property ranges for three generic types of silicon carbide and also carried out a comprehensive study of selected commercial ceramics, including several silicon carbides. Tables 3.7 through 3.9 give the characteristics of the commercial silicon carbides chosen for study and present a range of mechanical and other physical properties measured in the study. Tables 3.8 and 3.9 are especially valuable in giving Weibull moduli and short-term strengths at high temperatures.

Data on strength and steady-state creep rates were collected by Larsen et al.[27] for many silicon carbide materials. These data must be used with caution because of the possibility of improvements in properties since 1983, but they give a good perspective on high-temperature properties. Figure 3.7 shows that the strength of sintered silicon carbide holds up well to temperatures as high as 1500°C. In contrast, the strength of either siliconized or hot-pressed silicon carbide drops off at high temperature. Larsen and Adams[27] reflect the steady-state creep rate as a function of stress for several temperatures and silicon carbides.

Ghosh et al.[66] report the temperature dependence of the modulus of rupture and the toughness of alpha silicon carbide to 1400°C. It is noteworthy that there is no indication of a decrease of either property with increasing temperature in short-time tests.

TABLE 3.5 Background Information for Silicon Nitrides and Silicon Carbides

Material	Material class	Batch code	Fabricator	Primary matrix	Additives	Information source	Process
Hexoloy SA	ASIC	SOHIO/UDRI1	SOHIO	α-SiC		BM11/87, p.70	Pressureless sintered
Hexoloy SA*	ASIC	SOHIO/UDRI1	SOHIO	α-SiC		ORNL/TM-10705, p.285+	Sintered SiC
SSC	SSIC	GTELSSC1	GTE	SiC		BM11/87, p.89	
AC8	HPSN	GTE/UBE-E10	GTE	Si_3N_4	8.3 wt % CeO, 1.5 AlO	BM4/88, p.94	
AM4	HPSN	GTE/UBE-E10	GTE	Si_3N_4	4.4 wt % MgO, 1.5 AlO	BM4/88, p.94	
AY6	HPSN	GTE/UBE-E3&10	GTE	Si_3N_4	6 wt % Y_2O_3, 1.5 AlO	BM4/88, p.96	Hot-pressed 100 min at 1725°C at 34.5 MPa
AY6	HPSN	GTE/UBE-E03	GTE	Si_3N_4	6 wt % Y_2O_3, 1.5 AlO	BM4/88, p.96	Hot-pressed 100 min at 1725°C at 34.5 MPa
AY6	HPSN	GTE/UBE-E10	GTE	Si_3N_4	6 wt % Y_2O_3, 1.5 AlO	BM4/88, p.96	Hot-pressed 100 min at 1725°C at 34.5 MPa
AY6*	HPSN	GTE/UBE-E10	GTE	Si_3N_4	6 wt % Y_2O_3, 1.5 AlO	BM4/88, p.94	
AY6H	HPSN	GTELAY6H1	GTE	Si_3N_4	6 wt % Y_2O_3, 1.5 AlO	BM11/87, p.89	HIPed Si_3N_4
AY6H + AlN coat	SN	GTELAY6H1+COAT1	GTE	Si_3N_4 + AlO coat	6 wt % Y_2O_3	BM12/88, p.115	
EC-152	SN	NGK/UDRI	NGK	Si_3N_4		BM10/88, p.91	
GTE PY6	HPSN	GTEL/UDRI1	GTE	Si_3N_4		BM11/87, p.70	
GTE PY6*	HPSN	GTEL/UDRI1	GTE	Si_3N_4	6 wt % Y_2O_3	ORNL/TM-10705, p.285+	
NC-132	HPSN	BCD-SA9/NTRW	Norton/TRW Ceramics	Si_3N_4		ORNL/TM-10705, p.195+	Injection-molded and HIPed
NC-132	SN	NORTON1/NC132	Norton/TRW	Si_3N_4		BM12/88, p.151	
NT-154	SN	NORTON/UDRI	Norton/TRW	Si_3N_4		BM10/88, p.90	
NT-154	SN	NORTON/ORNL1	Norton/TRW	Si_3N_4		BM12/88, p.19	

RBSN	RBSN	GTELRBSN1	GTE	Si_3N_4		BM11/87, p.89	Reaction-bonded Si_3N_4
RBSN+AlN coat	SN	GTELRBSN1+COAT1	GTE	Si_3N_4 + AlO coat		BM12/88, p.115	
Si_3N_4	SN	NORTON-3205	Norton	Si_3N_4		BM11/87, p.21	
Si_3N_4	SN	ACC/CODE2	Garrett Processing	Si_3N_4		ORNL/TM-10469, p.52	
Si_3N_4	SN	NORTON/BASE	Norton	Si_3N_4	4 wt % Y_2O_3	BM1/88, p.162	
Si_3N_4	SN	NORTON/3X015	Norton/TRW	Si_3N_4	4 wt % Y_2O_3	ORNL/TM-10838, p.55	
SiN+ZrO/CaO+MgO	SNZROCAOMGO	ROCKWL-1/30ZRO	Rockwell	Si_3N_4		ORNL/TM-10308, p.40	
SiN+ZrO/MgO+MgO	SNZROCAOMGO	ROCKWL-1/30ZRO	Rockwell	Si_3N_4		ORNL/TM-10308, p.40	
SiN+ZrO/YO+AlO	SNZROYOALO	ROCKWL-1/30ZRO	Rockwell	Si_3N_4		ORNL/TM-10308, p.40	
SiN+ZrO/YO+AlO	SNZROYOALO	ROCKWL-1/20ZRO	Rockwell	Si_3N_4		ORNL/TM-10308, p.40	
SN-252	SN	KYOCERA/UDRI	Kyocera	Si_3N_4		BM11/87, p.21	
SN-4	SN	HOWMET/UDRI1	Howmet	Si_3N_4		BM10/88, p.90	
XL144	HPSN	NORTON/UDRI1	Norton	Si_3N_4		BM11/87, p.70	
XL144*	HPSN	NORTON/UDRI1	Norton	Si_3N_4		ORNL/TM-10705, p.285+	Hot pressed

*Second analysis for same material.
Source: From ORNL database; after Keyes.[37]

TABLE 3.6 Room-Temperature Properties of Silicon Nitrides and Silicon Carbides

Material	Batch code	Density, g/cm^3	Young's modulus, GPa	Modulus of rupture, MPa	Hardness test	Hardness value	Thermal expansion	K_{Ic}, MPa·m$^{1/2}$	Weibull modulus m	Information source
Hexology SA	SOHIO/UDRI1	3.17	427	331	Vicker	2539 kg/mm^2	5.4	2.0	8.9	BM11/87, p.70
Hexology SA*	SOHIO/UDRI1	3.17	300	325	Vicker	2740 kg/mm^2	4.5	4.0	—	ORNL/TM-10705, p.285+
SSC	GTELSSC1	—	—	401	Knoop, 1 kg	2579 kg/mm^2	—	2.2	7.6	BM11/87, p.89
AC8	GTE/UBE-E10	3.30	284	879	—	—	—	5.1	—	MB4/88, p.94
AM4	GTE/UBE-E10	3.12	251	706	—	—	—	3.6	—	BM4/88, p.94
AY6	GTE/UBE-E3&10	3.25	—	932	—	—	—	4.7	—	BM4/88, p.96
AY6	GTE/UBE-E03	3.25	—	995	—	—	—	4.1	—	BM4/88, p.96
AY6	GTE/UBE-10	3.25	—	845	—	—	—	4.7	—	BM4/88, p.96
AY6*	GTE/UBE-E10	3.25	293	983	—	—	—	5.0	—	BM4/88, p.94
AY6H	GTELAY6H1	—	297	822	Knoop, 1 kg	1348 kg/mm^2	—	3.4	6.8	BM11/87, p.89
AY6H+AlN coat	GTELAY6H1+COAT1	—	—	—	—	—	—	—	—	BM12/88, p.115
EC-152	NGK/UDRI	—	—	—	—	—	—	—	—	BM10/88, p.91
GTE PY6	GTEL/UDRI1	3.24	293	641	Vicker	1458 kg/mm^2	3.4	3.0	10.8	BM11/87, p.70
GTE PY6*	GTEL/UDRI1	3.24	267	800	—	—	3.5	4.3	—	ORNL/TM-10705, p.285+
NC-132	BCD-SA9/NTRW	—	—	—	—	—	—	—	—	ORNL/TM-10705, p.195+
NC-132	NORTON1/NC132	—	—	—	—	—	—	—	—	BM12/88, p.151
NT-154	NORTON/UDRI	3.23	—	311	—	—	—	—	—	BM10/88, p.90
NT-154	NORTON/ORNL1	—	—	—	—	—	—	—	—	BM12/88, p.19
RBSN	GTELRBSN1	—	—	273	Knoop, 1 kg	1004 kg/mm^2	—	3.5	7.6	BM11/87, p.89
RBSN+AlN coat	GTELRBSN+COAT1	—	—	—	—	—	—	—	—	BM12/88, p.115

Material	Product			Test				Reference
Si₃N₄	NORTON-3205	—	—	—	—	—	—	BM11/87, p.21
Si₃N₄	ACC/CODE2	—	—	—	—	—	—	ORNL/TM-10469, p.52
Si₃N₄	NORTON/BASE	—	—	—	—	4.6	—	BM1/88, p.162
Si₃N₄	NORTON/3X015	3.24	662	Vicker, 10 kg	16.0 GPa	4.2	—	ORNL/TM-10838, p.55
SiN+ZrO/CaO+MgO	ROCKWL-1/30ZRO	—	—	—	—	—	—	ORNL/TM-10308, p.40
SiN+ZrO/MgO+MgO	ROCKWL-1/30ZRO	—	—	—	—	—	—	ORNL/TM-10308, p.40
SiN+ZrO/YO+AlO	ROCKWL-1/30ZRO	—	—	—	—	—	—	ORNL/TM-10308, p.40
SiN+ZrO/YO+AlO	ROCKWL-1/20ZRO	—	—	—	—	—	—	ORNL/TM-10308, p.40
SN-252	KYOCERA/UDRI	—	—	—	—	—	—	BM11/87, p.21
SN-4	HOWMET/UDRI1	3.23	204	—	—	—	—	BM10/88, p.90
XL144	NORTON/UDRI1	3.23	314	Vicker	1421 kg/mm²	3.4	3.1	BM11/87, p.70
XL144*	NORTON/UDRI1	3.23	382	Vicker	1600 kg/mm²	3.5	4.3	ORNL/TM-10705, p.285+

*Second analysis for same material.
Source: From ORNL database; after Keyes.[37]

TABLE 3.7 Summary of Reported Property Values for Selected Commercial Silicon Nitrides and Silicon Carbides

Material type	Density, g/cm³	Coefficient of thermal expansion at 20–1000°C, 10⁻⁶/°C	Young's modulus, GPa		Fracture toughness K_{Ic}, MPa·m$^{1/2}$	Flexure strength, MPa		Tensile strength, MPa	
			20°C	1400°C		20°C	1400°C	20°C	1400°C
Hot-pressed Si₃N₄ (HPSN)	3.10	3.3	300	175	4.8*	750	300†	375	150
Range	3.07–3.37	3–3.9	250–325	175–250	2.8–6.6	450–1100	0–600	—	—
Sintered Si₃N₄ (SSN)	3.13	3.5	245	—	4.3	415	70	—	—
Range	2.8–3.4	—	195–315	—	3.0–5.6	275–840	0–700	—	140
Reaction-sintered Si₃N₄ (RSSN)	2.4	2.9	175	155	3.6	200	250	170	
Range	2.0–2.8	2.5–3.1	100–220	120–200	—	50–300	0–400	70–210	35–150
Hot-pressed SiC (HPSC)	3.23	4.55	440	380	3.9	500	300	200	
Range	3.2–3.3	4.3–5.4	430–450	—	3.0–4.0	300–800	175–575	—	—
Sintered SiC (SSC)	3.1	4.5	395	372	4.0	375	380	—	—
Range	3.0–3.2	4.4–4.8	375–420	300–400	2.5–6.5	275–535	240–450	—	—
Reaction-sintered SiC (RSSC)	3.0	4.33	360	275	—	310	190	77	—
Range	2.9–3.1	4.3–4.4	350–375	200–320	—	175–450	70–450	—	—

*K_{Ic} increases at temperatures above 1000°C.
†Above 1200°C dynamic fatigue shows slow crack growth.

TABLE 3.8 Flexural Strength Measurements Made in Phase II

Material	Supplier	Machine temperature, °C	Flexural crosshead speed, cm/s	Strength, MPa	Standard deviation, MPa	Weibull modulus m	Fracture origin
α-SiC	Sohio-Hexoloy	25	0.0064	331	41	9	Surface flaws
		1300	0.0064	494	90	8	Surface flaws
		1300	0.00004	500	77	8	Surface and edge flaws
		1450	0.0064	500	52	10	Surface and edge flaws
		1450	0.00004	455	59	9	Surface and edge flaws
PY6	GTE	25	0.0064	641	69	11	Surface flaws
		1450	0.0064	393	69	8	Surface, edge, and subsurface flaws (mostly inclusions)‡
		1450	0.00004	241	36	7	Flaws and inclusions‡ at or below surface, possible SCG
XL144, batch 1*	Norton/TRW	25	0.0064	538	97	6	Surface and edge flaws
		1300	0.0064	604	44	13	Flaws and inclusions¶
		1300	0.00004	551	93	9	At or below surface
		1450	0.0064	453	53	12	Surface and edge flaws, subsurface inclusions¶
		1450	0.00004	213	12	6	At or below surface, SCG
XL144, batch 2†	Norton/TRW	25	0.0064	787	41	22	—
		1450	0.0064	537	76	7	—

*Specimens machined from a hot-pressed billet at the University of Dayton.
†Specimens prepared at Norton/TRW.
‡Inclusions: Fe, Mo, Cr, Cu. ¶Inclusions: Fe, Cr.

TABLE 3.9 Flexural Strength Measurements Made in Phase III

Material	Supplier	Machine temperature, °C	Flexural crosshead speed, cm/s	Strength, MPa	Standard deviation, MPa	Weibull modulus m	Fracture origin
α-SiC	Sohio-Hexoloy	25	0.004	402	52	9	Surface flaws
		1000	0.004	416	30	18	Surface flaws, pits
		1000	0.00004	412	37	12	Surface flaws, pits, and inclusions
		1200	0.004	435	53	9	Surface flaws, pits
		1200	0.00004	441	6	10	Surface flaws, pits
		1400	0.004	436	48	11	Surface and edge flaws, pits
		1400	0.00004	390	36	15	Surface and edge flaws, pits
SN-4	Howmet	25	0.004	977	129	12	Surface flaws
		1000	0.004	744	91	—	Surface flaws
		1000	0.00004	514	47	—	Surface flaws, possible SCG
		1200	0.004	402	40	—	Surface flaws
		1200	0.00004	312	15	—	Surface flaws, possible SCG
		1400	0.004	190	—	—	Severe oxidation and blistering
SN-252	Kyocera	1000	0.004	526	31	20	At surface and edge due to large whiskers
		1000	0.00004	466	26	20	
		1200	0.004	478	43	13	
		1200	0.00004	485	35	20	
		1400	0.004	459	29	19	
		1400	0.00004	498	29	21	Severe plastic deformation prior to fracture
CVD SiC	CVD	25	0.0064	554	120	—	—
		1400	0.0064	556	28	—	—
NT-154	Norton/TRW	25	0.004	907	79	15	Surface and edge flaws
		1400	0.004	610	30	23	Surface and edge flaws

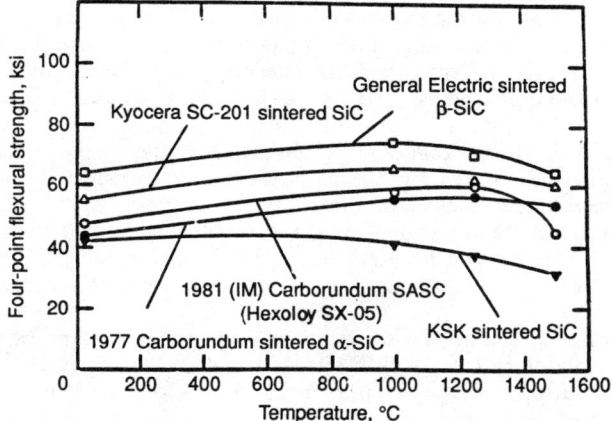

FIGURE 3.7 Flexural strength of sintered silicon carbide materials as a function of temperature.[27]

Carroll et al.[67] show that the creep of ceramics in tension is usually different from that in compression and that creep in bending is a combination of the two. Tests by Carborundum show the creep in tension and in compression for KX01 reaction-bonded silicon carbide and that the rupture time correlates well with the minimum (steady-state) creep rate for this material. They were able to calculate the rupture time in bending as a function of the applied stress and also found good agreement.[67] Carroll and Tressler[68] have determined a threshold stress for creep damage in tension of 132 MPa for siliconized (reaction-sintered) silicon carbide at 1100°C.

Silicon Nitride Ceramics

Silicon nitride is one of the strongest structural ceramics. It has good oxidation resistance due to the presence of a protective SiO_2 layer. Its thermal shock resistance is considered reasonably good for a ceramic due to its low thermal expansion, high strength, and relatively low elastic modulus.[69]

Monolithic silicon nitride is classified into categories by the processing technique used to form the solid, because the technique used produces characteristic microstructures and the resulting properties. These processing categories are:

1. Reaction bonding
2. Sintering and gas-pressure-assisted sintering
3. Hot pressing and hot isostatic pressing

To understand the reason for these categories and the limitations on possible processing routes, it is necessary to consider the crystal chemistry and phase stability, including temperature and gas-pressure effects.

Pure silicon nitride (Si_3N_4) occurs in two crystallographic forms, alpha and beta. Both are hexagonal. The alpha form is made up of alternate layers of the beta form and its mirror image.

The stability of the polymorphs of silicon nitride is complex because "silicon nitride" often contains some amount of oxygen. Generally speaking, powders of silicon nitride are predominantly in the alpha form. Solid specimens are predominantly in the beta form. This transformation of alpha to beta is thought to occur by solution of the alpha phase in the grain-boundary liquid and precipitation of the more stable beta form during densification by sintering or hot pressing. The generally accepted view is that beta silicon nitride is the thermodynamically stable form at all temperatures and the alpha version is formed only under special conditions when kinetic factors predominate.[70]

Silicon nitride is stable against oxygen only at very low partial pressures of oxygen (10^{-21} atm oxygen in 1 atm nitrogen).[70] In air, silicon nitride rapidly forms a coating of SiO_2 on the surface. This forms a protective coating against further oxidation.

Reaction-bonded silicon nitride is made by forming a shape of pure silicon powder and reacting this body with nitrogen at high temperature. The resulting material suffers from having open porosity in the 15 to 20 percent range, but has the advantage of being free of sintering additives which could degrade high-temperature mechanical properties. Silicon nitride fabricated by this process has essentially zero shrinkage during processing, which is a major advantage for achieving good dimensional tolerances. The room-temperature strength of materials fabricated by this process is substantially lower than that of sintered or hot-pressed materials due to this residual porosity.[71]

Silicon nitride powder compacts can be sintered to full density under 1 atm of nitrogen using combinations of rare-earth oxides with aluminum oxide as sintering aids. These sintering aids combine with the SiO_2 layer on the surface of the silicon nitride powder particles to form a eutectic liquid at the sintering temperature, and the material densifies by reactive liquid sintering.

If a nitrogen overpressure is used to prevent the dissociation of silicon nitride, a higher sintering temperature can be used as fewer additives are required to achieve a fully dense material which has better high-temperature properties than materials sintered at lower temperatures. However, the fabrication cost is increased. Nitrogen overpressures for this process are generally 10 MPa or less.

Pressure can also be applied after the powder compact has been sintered to the point where all the remaining pores are isolated from the surface. This process reduces the amount of additives required and enhances the properties of the material. This pressure has been applied by a separate hot isostatic pressing step, generally at gas pressures of 200 MPa or below in cold-wall autoclaves that contain high temperature furnaces. Silicon nitride can also be sintered in a single temperature cycle, in which the material is first sintered to the closed-pore stage under low pressure and then a higher pressure is applied to achieve final densification. Pressures up to 200 MPa can be applied if the sintering is carried out in a hot isostatic press. If the process is carried out in a pressurized sintering furnace, the applied pressure is usually 10 MPa or less.

Hot-pressed silicon nitride is usually made with the use of additives similar to those used for sintering, which allow full density to be achieved through reactive liquid-phase sintering. The resulting silicon nitride material can have highly desirable properties, which are still being improved as the processing is refined to produce microstructures with elongated grains. The shapes which can be formed are limited and the process cost is relatively high. As with sintering, the additives leave second phases, which can soften at high temperatures and degrade mechanical properties. However, the applied pressure allows the use of a lower percentage of sintering aids, and high-temperature properties are generally better than those for sintered material.

Hot isostatic pressing can be used with the aid of a cladding. A coating of glass particles can be used which remains porous during the first stages of heating and allows removal of gas from binder burnout. At intermediate temperatures the glass melts and forms a viscous cladding, which remains gastight at the working temperature and allows the application of gas pressure. Since the applied pressures are higher than for hot pressing, the higher-temperature properties of silicon nitride prepared by this process are generally superior to those of sintered or hot-pressed materials.

Reaction sintering and sintering or hot isostatic pressing can also be combined by starting with a silicon powder compact containing the oxide sintering aids used for sintering silicon nitride powder compacts.[72] For this process the powder compact is first nitrided as with reaction-bonded silicon nitride. This gives a silicon nitride compact containing sintering aids that is 70 to 80 percent dense. This material can then be fully densified by sintering or hot isostatic pressing. The two principal advantages over silicon nitride powder compacts are lower-cost starting powders and lower shrinkage. The lower shrinkage makes it easier to maintain dimensional tolerances.

A similar process which starts with a mixture of silicon and silicon nitride powders containing sintering aids has also been reported.[73]

Mechanical Properties of Silicon Nitrides. The mechanical properties of commercial silicon nitrides are listed in Table 3.4. Mechanical properties of silicon nitrides studied in the DOE-NASA programs and summarized in the ORNL database are listed in Tables 3.5 and 3.6.[37]

Data on strength and steady-state creep rates that have been collected by Larsen et al.[27] for many silicon nitride materials must be used with caution as discussed for silicon carbide. The temperature dependence of the strength of reaction-sintered silicon nitride is shown in Fig. 3.8. The room-temperature strength is relatively low, but it is retained to high temperatures. The strengths of some typical sintered silicon nitride materials have also been compiled by Larsen et al.[27] The room-temperature strengths are considerably higher than for reaction-bonded silicon nitride, but they drop off at high temperatures. Similar behavior is shown for hot-pressed silicon nitride.[27] Steady-state flexural creep

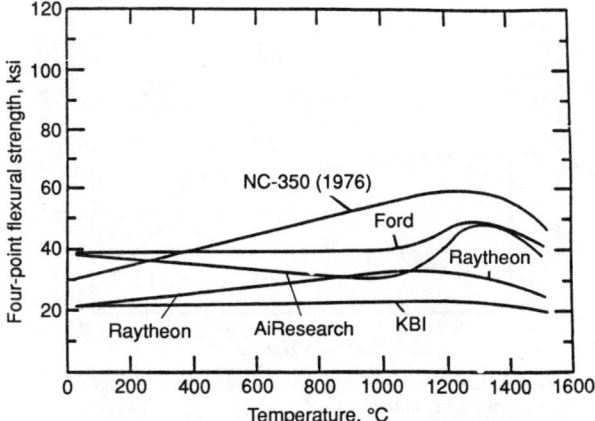

FIGURE 3.8 Flexural strength of reaction-sintered silicon nitrides.[27]

rates for various reaction-sintered silicon nitrides as well as for various hot-pressed silicon nitrides have also been reported by Larsen et al.[27]

Hecht et al.[65] summarized reported property ranges for three generic types of silicon nitride and also carried out a comprehensive study of selected commercial ceramics, including several silicon nitrides. Tables 3.7 through 3.9 list the characteristics of the commercial silicon nitrides chosen.

Dutta et al.[74] have summarized data from the literature on the effect of porosity on Young's modulus and strength and examined the utility of various equations to represent this effect.

Efforts to improve the properties of silicon nitride have taken the direction of improving the microstructure or reinforcing the silicon nitride matrix by particles or whiskers.[37] Two main themes have been the improvement of high-temperature creep resistance through engineering grain boundaries to contain higher-melting-point glassy phases (or ideally crystalline phases) and the production of elongated grains to increase toughness. In unreinforced silicon nitride the grain morphology has been recognized as controlling the toughness, as summarized by Matsuhiro and Tomonori[75]; well-elongated beta silicon nitride grains give high toughness. Yeh et al.[76] have reported substantial improvement in both room-temperature and high-temperature strength by careful microstructure control, as shown in Fig. 3.9. Not only is the room-temperature strength of their GN-10 material considerably better than that of the baseline material, but the strength of the GN-10 material remains higher at all temperatures despite the drop-off at high temperature associated with grain-boundary phase softening. Tanaka et al. hot-pressed specimens without any sintering additive.[77,78] They apparently obtained liquid-phase sintering through the trace of silica present in the starting powder. The strength was found to hold up better in short-time tests at temperatures up to 1400°C. Li and Yamanis[79] made silicon nitride material with a fracture toughness of 10.6 MPa · $m^{1/2}$ by gas-pressure sintering, and this material exhibited R-curve behavior reflecting a dependence of toughness on crack length.

Masuda et al.[80] found that sintered silicon nitride undergoes cyclic fatigue. They showed that the strength drops to about 40 percent of its initial value in 10^{10} cycles in bending. The stress dependence of cyclic fatigue in bending was also shown in the form of failure probability plots.[80] Nikkila and Mantyla[81] found that

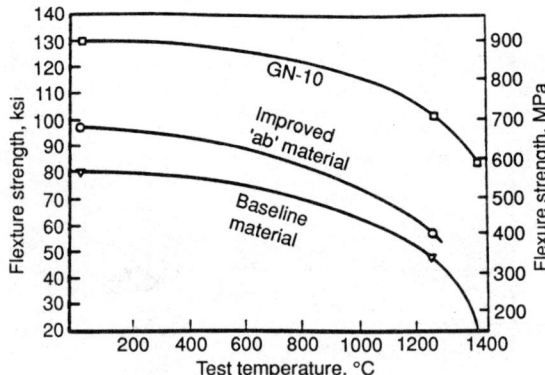

FIGURE 3.9 Four-point flexural strength of silicon nitride as a function of temperature.[76]

long dynamic fatigue lives in bending can be expected at stress levels lower than two-thirds of the short-time bend strength. For higher stresses they found that fatigue is affected by the compressive stress portion of the cycle, so that an estimate of dynamic fatigue life based on crack propagation in tension could give a dangerous overestimate of fatigue life. Rawlins et al.[82] have studied dynamic and static fatigue of silicon nitride at 1000°C and found a slow crack-growth mechanism of failure. Tajima et al.[83] studied static, cyclic, and dynamic fatigue in gas-pressure sintered silicon nitride and concluded that, at room temperature, all results are explainable in terms of slow crack growth, which is unaffected by cycling. The material is highly resistant to dynamic fatigue up to 900°C. At 1000°C slow crack growth does occur under cyclic loading. Van Der Biest et al.[84] found that hot-pressed silicon nitride with 9 wt % Y_2O_3 had a crystalline silicon-yttrium-oxynitride intergranular phase and had excellent creep and static fatigue resistance compared to other silicon nitrides. Under inert conditions the static fatigue limit is about 300 MPa at 1400°C. Xu et al.[85] have improved high-temperature strength and toughness up to 1400°C by using Y_2O_3 plus La_2O_3 as a sintering additive. Pasto et al.[86] studied creep in bending of hot-pressed silicon nitride with 6 wt % Y_2O_3 and 1.5 wt % Al_2O_3, and constructed the failure mechanism map shown in Fig. 3.10. This map is particularly interesting as it indicates the need to allow for creep in designing with this material at any appreciable stress above 1000°C.

Kodama et al.[87,88] have studied the toughening of silicon nitride by the simultaneous addition of silicon carbide whiskers and silicon carbide particles, and found a significant improvement over the use of either additive alone. For 20 vol % whiskers and 10 vol % 40-μm particles, the toughness was 10.5 MPa · $m^{1/2}$ and the flexural strength was 550 MPa. The retention of properties to high temperatures is shown in Figs. 3.11 and 3.12.

Baldoni and Buljan[89] studied creep in four-point bending of hot-pressed silicon nitride with SiC dispersoids. Large (8-μm) particles increased the creep rate compared to the monolithic material, but whisker additions increased the resistance to creep.

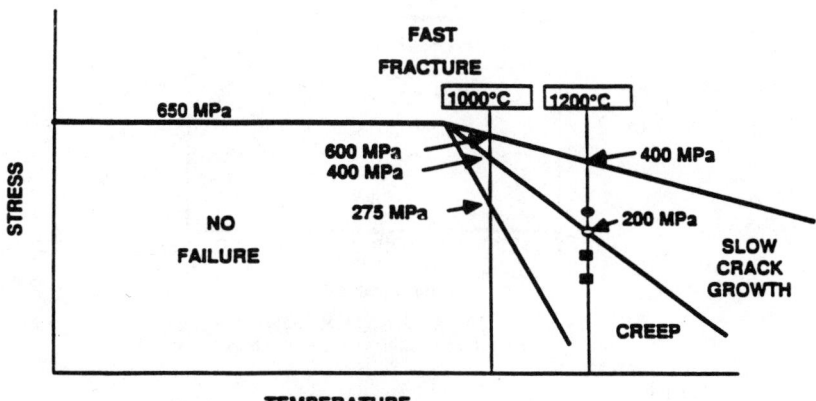

FIGURE 3.10 Schematic failure mechanism map for hot-pressed silicon nitride with 6 wt % Y_2O_3 and 1.5 wt % Al_2O_3.[86]

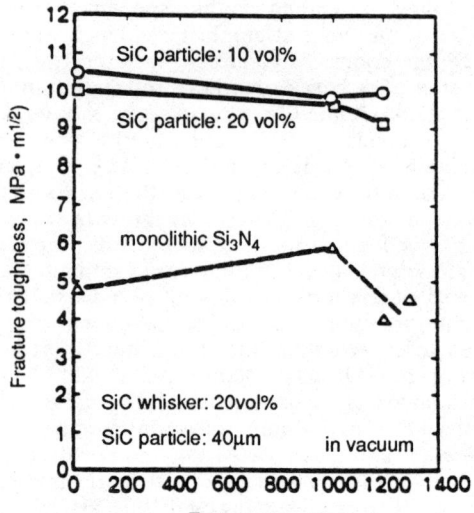

FIGURE 3.11 Flexural toughness at high temperature of silicon nitride reinforced with both whiskers and particles.[88]

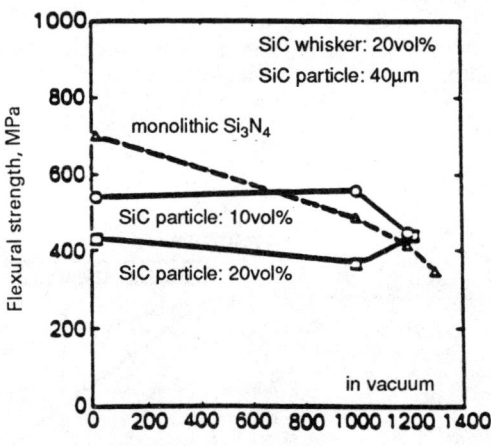

FIGURE 3.12 Flexural strength at high temperature of silicon nitride reinforced with both whiskers and particles.[88]

Beta Sialon Ceramics

An extensive family of ceramics can be made by a coupled substitution of equal atomic percentages of (Al + O) for (Si + N) in the beta-silicon nitride structure. These are generally termed sialons. Beta sialon thus forms the family of compositions indicated by $Si_{3-x}Al_xO_xN_{4-x}$, with $0 \leq x \leq 2$ to remain within the solid solution range.[70]

Ekstrom and Olsson[90] have made dense single-phase beta sialon ceramics at 1700°C and 200 MPa using the glass-encapsulated hot isostatic pressing technique.

Yabuta et al.[91] have succeeded in making a beta sialon without the rare-earth sintering aid by adding aluminum from aluminum isopropoxide solution to silicon nitride followed by gas-pressure sintering and hot isostatic pressing. The resulting material had a three-point bending strength of 800 MPa, which it retained to 1300°C in short-time tests.

Tiegs[92] has made a sialon-matrix composite reinforced with 20 vol % SiC whiskers that has a strength of 600 MPa and a fracture toughness of 7 MPa · $m^{1/2}$.

Zirconia Ceramics

Zirconia ceramics have attracted much attention since the discovery that materials which are very strong and tough at room temperature can be made by control of the phases present. Understanding of the phase transitions is fundamental to understanding the properties of zirconia ceramics.

In pure zirconia the monoclinic phase is stable below 1170°C, the tetragonal phase is stable from 1170 to 2370°C, and the cubic phase is stable above 2370°C. The actual transformation in coarse unconstrained powder occurs between 700 and 900°C on cooling and between 1050 and 1200°C on heating. As the powder size decreases, the relative contribution to the total free energy from the surface free energy increases, leading to the result that the tetragonal phase is more stable in finer particles and persists to lower temperatures.

Certain additives (such as MgO, CaO, Y_2O_3, and Ce_2O_3) form a cubic-phase solid solution at high temperatures. The amount of additive can be chosen so that upon cooling, conversion to the tetragonal phase and subsequently to the monoclinic phase still occurs, but at lower temperatures.

The excellent room-temperature strength and toughness properties of properly made zirconia ceramics arise from retaining all, or a portion, of the material in the tetragonal form at room temperature.

Zirconia undergoes a sudden and significant contraction on heating through the monoclinic to tetragonal transformation and a corresponding expansion on cooling. Associated with this volume change is a pressure dependence of the transformation temperature. A pressure of 40 kbar shifts the transformation temperature from 1170 to 0°C.

Cannon[93] classifies transformation-toughened zirconia into two categories: partially stabilized zirconia (PSZ) and tetragonal zirconia in polycrystalline form (TZP). PSZ contains enough stabilizing additives (usually MgO) to allow sintering in the cubic phase. It is subsequently annealed at a lower temperature in the cubic plus tetragonal phase field to form small precipitates of tetragonal zirconia in a cubic matrix. When a crack forms in the material, the high stress at the crack tip transforms the precipitates near the crack tip to the high-volume monoclinic phase because of the stress or pressure dependence of the tetragonal to monoclinic transformation. The volume increase associated with this transformation reduces the stress at the crack tip and results in high strength and toughness. Thus the phase transformation has the same effect as plastic deformation at the tip of a crack in a metal. Cannon references a typical commercial PSZ material as containing about 8.5 percent MgO and having a phase composition of 58 percent cubic, 37 percent tetragonal, and 5 percent monoclinic.

TZP is nearly 100 percent tetragonal at room temperature. Strengths as high as 2400 MPa have been reported in this material. The toughening and strengthening mechanism in these materials is similar to that for PSZ, but there is disagreement on the mechanisms for retaining nearly 100 percent tetragonal phase at room temperature. However, it is clear that a grain size of less than about 1 μm is required. Yttria (2.0 to 3.5 mol %) is the most common stabilizing agent, although CeO_2 and other rare-earth oxides have been used alone or in combinations.

Both strength and toughness of PSZ depend on the microstructure, which, in turn, depends on the thermal history as well as the chemical composition. A further complexity is that the measured toughness depends on the crack length, increasing with increasing crack length to a constant value for relatively large cracks (the so-called R-curve behavior).

Mechanical Properties of Zirconias. The mechanical properties of commercial zirconias are listed in Table 3.4. Mechanical properties of zirconias studied in the DOE-NASA programs and summarized in the ORNL database are listed in Table 3.10.[34] As shown in Fig. 3.13, the tensile strength of stabilized zirconia falls rapidly with increasing temperature. This is to be expected from the nature of the toughening effect. Tetragonal zirconia has a very high bending strength value of 1400 MPa and a toughness of 6 MPa · $m^{1/2}$ or more, but both quantities decrease sharply with increasing temperature.[94]

As discussed for alumina transformation-toughened with zirconia, transformation-toughened zirconia itself is subject to deterioration at slightly elevated temperatures if moisture is present.[95]

Liu and Brinkman[96] have studied fast fracture and cyclic fatigue of MgO–partly stabilized zirconia using a special self-aligning tensile testing system. They found that true tensile strength was considerably lower than the strength determined in bending toughness.

3.6 CONCLUSIONS

Structural ceramics of good quality are available from a wide variety of manufacturers. Data from sales literature are generally not sufficiently extensive when designing for long-term or high-temperature service. Design can be carried out by supplementing manufacturers' data with research results on similar materials, but care must be taken to understand the dependence of properties on microstructure. Progress continues at such a rapid pace that still better commercial materials are likely to become available. Effective standardization remains an elusive goal.

TABLE 3.10 Background and General Material Information for Partially Stabilized Zirconias

Material	Material class	Batch code	Supplier	Process	Vintage	Matrix	Stabilizer	Density, g/cm³	Modulus of elasticity, GPa	Modulus of rupture, MPa	Hardness test	Hardness, GPa	Weibull modulus m
1985	ZIRCONIA TTZ	HIT1985/MTL	Hitachi	Hot-pressed	1985	ZrO_2	2 mol % Y_2O_3	6.038	213	1169	Knoop	12.4	3.6
1986H	ZIRCONIA TTZ	KOR1986H/MTL	Koransha	HIPed	1986	ZrO_2	3 mol % Y_2O_3	6.045	214	1261	—	—	8.8
1986S	ZIRCONIA TTZ	KOR1986S/MTL	Koransha	Sintered	1985	ZrO_2	3 mol % Y_2O_3	5.966	210	640	Knoop	10.8	9.5
AC-SENSOR	ZIRCONIA PSZ	AMTL-A/ACS82	AC Sparkplug		1982	ZrO_2	Y_2O_3	5.670	213	—	Vickers	10.71	10.2
APC-TTZ	ZIRCONIA TTZ	AMTL-K/APCK	Amer. Feldmuehle		?	ZrO_2	MgO	0.000	215	—	—	—	—
CERAD-PSZ	ZIRCONIA PSZ	AMTL-F/CERAD82	Ceradyne	Hot-pressed	1982	ZrO_2	Y_2O_3	5.600	180	—	—	—	—
COORS-TZP	ZIRCONIA TZP	AMTL-I/COORS84	Coors Porcelain		1984	ZrO_2	Y_2O_3	5.940	211	—	Vickers	12.08	4.5
COORS-ZDM	ZIRCONIA TTZ	AMTL-G/COORS81	Coors Porcelain		1981	ZrO_2	MgO	5.290	149	—	—	—	21.4
COORS-ZDM	ZIRCONIA TTZ	AMTL-H/COORS81	Coors Porcelain		1983	ZrO_2	MgO	5.650	199	—	Vickers	8.99	4.2
COORS-ZDY	ZIRCONIA PSZ	AMTL-J/COORS81	Coors Porcelain		1981	ZrO_2	Y_2O_3	5.530	189	—	—	—	16.0
CZ-203	ZIRCONIA TTZ	CERAM-CZ203/MTL	Ceramatec	Sintered	1987	ZrO_2	12 mol % MgO			—	—	—	—
MS-TTZ	ZIRCONIA TTZ	AMTL-E/NILSEN82	Nilsen, USA		1982	ZrO_2	MgO	5.650	208	—	Vickers	9.69	CeO_2 13.4
NGK-TZP	ZIRCONIA TZP	AMTL-M/NGKM	NGK Sparkplug		?	ZrO_2	Y_2O_3	0.000	198	—	Vickers	12.82	—
NGK-TZP	ZIRCONIA TZP	AMTL-N/NGKN	NGK Sparkplug		?	ZrO_2	Y_2O_3	5.770	198	—	Vickers	11.42	13.5

TABLE 3.10 Background and General Material Information for Partially Stabilized Zirconias (*Continued*)

Material	Material class	Batch code	Supplier	Process	Vintage	Matrix	Stabilizer	Density, g/cm³	Modulus of elasticity, GPa	Modulus of rupture, MPa	Hardness test	Hardness, GPa	Weibull modulus m
NILSEN-TTZ	ZIRCONIA TTZ	AMTL-C/NILSENC	Nilsen, USA		?	ZrO_2	MgO	0.000	227	—	Vickers	9.78	—
NRL-TZP	ZIRCONIA TZP	AMTL-B/NRL82	Naval Research Lab		1982	ZrO_2	Y_2O_3	5.770	202	—	Vickers	10.93	—
TASZIC	ZIRCONIA TTZ	TOSTASZIC/MTL	Toshiba Ceramics	Sintered	1985	ZrO_2	2–3 mol % Y_2O_3	5.880	200	633	Knoop	10.1	6.2
TOR-TZPHP	ZIRCONIA TZP	AMTL-O/TORAY83	Toray	Hot-pressed	1983	ZrO_2	Y_2O_3	5.950	215	—	Vickers	12.55	—
TOR-TZPSIN	ZIRCONIA TZP	AMTL-R/TORAY83	Toray	Sintered	1983	ZrO_2	Y_2O_3	5.900	213	—	Vickers	11.34	—
TOSH-TZP	ZIRCONIA TZP	AMTL-P/TOSHIBA83	Toshiba Ceramics		1983	ZrO_2	Y_2O_3	5.930	209	—	Vickers	11.56	10.2
TS-TTZ	ZIRCONIA TTZ	AMTL-D/NILSEN82	Nilsen, USA		1982	ZrO_2	MgO	5.660	227	—	Vickers	9.12	14.1
TZP-110	ZIRCONIA TZP	ACTZP110/MTL	AC Sparkplug	Sintered	1985	ZrO_2	2.6 mol % Y_2O_3	5.835	204	753	Knoop	11.1	12.2
Z-191	ZIRCONIA TZP	AMTL-O/NGK84	NGK Sparkplug	Sintered	1984	ZrO_2	Y_2O_3	5.770	—	—	—	—	13.5
Z-191	ZIRCONIA TZP	NGKZ191/MTL	NGK-Locke	Sintered	1985	ZrO_2	3 mol % Y_2O_3	5.869	208	873	Knoop	10.9	15.2
Z-201	ZIRCONIA TTZ	KYOZ201/MTL	Kyocera	Sintered	1985	ZrO_2	2.8 mol % Y_2O_3	5.853	201	745	Knoop	10.5	8.8
Z-701	ZIRCONIA TTZ	KYOZ701/MTL	Kyocera	HIPed	1988	ZrO_2	Y_2O_3	—	—	—	—	—	—
ZIRCOA2120	ZIRCONIA TTZ	AMTL-S/CGW	Corning Glass Works		1982	ZrO_2	MgO	5.580	194	—	Vickers	9.0	7.7
ZT-35	ZIRCONIA PSZ	AMTL-L/AFC82	Amer. Feldmuehle		1982	ZrO_2	MgO	5.510	215	—	Vickers	9.32	5.9

Source: From ORNL database; after Booker.[36]

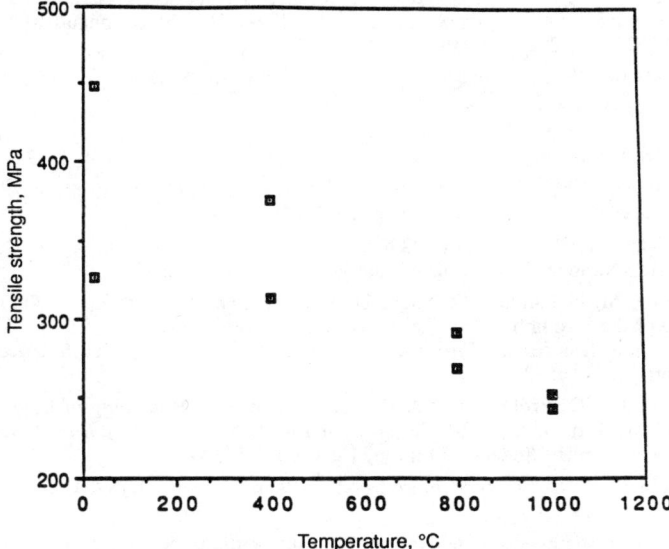

FIGURE 3.13 Tensile strength of MgO partially stabilized zirconia as a function of temperature.[36]

REFERENCES

1. M. S. Newkirk, H. D. Lesher, D. R. White, C. R. Kennedy, A. W. Urquhart, and T. D. Claar, "Preparation of Lanxide® Ceramic Matrix Composites: Matrix Formation by the Directed Oxidation of Molten Metals," *Ceram. Eng. Sci. Proc.*, vol. 8, pp. 879–885, 1987.
2. G. H. Campbell, B. J. Dalgleish, and A. G. Evans, "Brittle-to-Ductile Transition in Silicon Carbides," *J. Am. Ceram. Soc.*, vol. 72, pp. 1402–1408, 1989.
3. A. F. McLean and D. L. Hartsock, "Design with Structural Ceramics," in J. B. Wachtman, Jr. (Ed.), *Structural Ceramics,* vol. 29 of *Treatise on Materials Science and Technology,* Academic Press, Orlando, Fla., 1989, pp. 27–97.
4. G. D. Quinn, "Static Fatigue in High-Performance Ceramics," ASTM Spec. Tech. Testing Publ. 844, 1984.
5. G. D. Quinn, "Fracture Mechanism Maps for Advanced Structural Ceramics; Part 1: Methodology and Hot-Pressed Silicon Nitride Results," *J. Mater. Sci.*, vol. 25, pp. 4361–4376, 1990.
6. F. Thummler, "Engineering Ceramics," in G. de With, R. A. Terpstra, and R. Metselaar (Eds.), *Euro-Ceramics,* vol. 3, Elsevier Applied Science, London, 1989, pp. 3.81–3.99.
7. A. G. Evans, "Perspective on the Development of High-Toughness Ceramics," *J. Am. Ceram. Soc.*, vol. 73, pp. 187–206, 1990.
8. R. M. Anderson, "Testing Advanced Ceramics," *Adv. Mater. Process.*, pp. 31–36, Mar. 1989.
9. J. Sung and P. S. Nicholson, "Valid K_{IC} Determination via In-Test Subcritical Precracking of Chevron-Notched Bend Bars," *J. Am. Ceram. Soc.*, vol. 72, pp. 1033–1036, 1989.

10. D. C. Larson, S. L. Stuchly, and J. W. Adams, "Evaluation of Ceramics and Ceramic Composites for Turbine Applications," Air Force Wright Aeronautical Labs., Rep. AFWAL-TR-88-4202, Dec. 1988.
11. L. C. Mejia, "High Temperature Tensile Testing of Advanced Ceramics," *Ceram. Eng. Sci. Proc.,* vol. 10, pp. 668–681, 1989.
12. D. F. Carroll, S. M. Wiederhorn, and D. E. Roberts, "Technique for Tensile Creep Testing of Ceramics," *J. Am. Ceram. Soc.,* vol. 72, pp. 1610–1614, 1989.
13. K. Jakus and S. M. Wiederhorn, "Creep Deformation of Ceramics in Four-Point Bending," *J. Am. Ceram. Soc.,* vol. 71, pp. 832–836, 1988.
14. R. W. Rice and B. J. Hockey, "The Science of Ceramic Machining and Surface Finishing II," National Bureau of Standards, Spec. Publ. 562, 1979.
15. A. Gallee, M. Nakamura, E. Nagy, D. McGarry, and S. Peteves, "Influence of Machining of the Strength of Hot-Pressed Silicon Nitride," in G. de With, R. A. Terpstra, and R. Metselaar (Eds.), *Euro-Ceramics,* vol. 3, Elsevier Applied Science, London, 1989, pp. 3.538–3.542.
16. E. Lujbrink, "Control of Surface Damage in Abrasive Machining of Engineering Ceramics," in G. de With, R. A. Terpstra, and R. Metselaar (Eds.), *Euro-Ceramics,* vol. 3, Elsevier Applied Science, London, 1989, pp. 3.543–3.547.
17. S. J. Schneider, "Advanced Ceramics—What's in a Name?," *ASTM Stand. News,* pp. 28–30, Oct. 1989.
18. G. C. Padgett, "Review of the National and International Standardization of Advanced Ceramics," in G. de With, R. A. Terpstra, and R. Metselaar (Eds.), *Euro-Ceramics,* vol. 3, Elsevier Applied Science, London, 1989, pp. 3.63–3.70.
19. S. Schneider, Jr., "Advanced Ceramics Standards Development," in G. de With, R. A. Terpstra, and R. Metselaar (Eds.), *Euro-Ceramics,* vol. 3, Elsevier Applied Science, London, 1989, pp. 3.71–3.75.
20. "FC Annual Report for Overseas Readers," Japan Fine Ceramics Association, no. 5, 1987.
21. M. Ueda, "The Studies Participated by Industrial Companies and Principal Results until the Second Phase of the Project," presented during the visit of the 5th Overseas Technology Team of Japan to the United States in October 1988; publ. in Ref. 23, pp. II-9–II-10.
22. "Characterization of Ceramic Powders: Data and Analyses—Final Report," Natl. Inst. of Standards and Technology, Mar. 1990.
23. "Characterization of Sintered Silicon Nitride and Silicon Carbide Structural Ceramics," KemaNord Industrikemi, Oct. 1989.
24. "Flexural Analysis of Flexure Strength Data," University of Karlsruhe, Germany, June 1989.
25. J. Rumble, Natl. Inst. of Standards and Technology, Chairman of ASTM E29, private communication.
26. *Engineering Property Data on Selected Ceramics,* vol. 1: *Nitrides*; vol. 2: *Carbides*; vol. 3: *Single Oxides,* Metals and Ceramics Information Center, Battelle Columbus Div., Rep. MCIC-HB-07, Aug. 1979; reprinted July 1987; available from American Ceramic Soc., Westerville, Ohio.
27. D. C. Larsen and J. W. Adams, "Property Screening and Evaluation of Ceramic Turbine Materials," Air Force Materials Lab., Rep. AFWAL-TR-83-4141, Apr. 1984.
28. *Ceramic Source 90,* annual source book, American Ceramic Soc., Westerville, Ohio, 1990.
29. R. Morrel, *Handbook of Properties of Technical and Engineering Ceramics,* pt. 1: "An Introduction for the Engineer and Designer," 1985, reprinted 1989; pt. 2: "Data Reviews, Section I, High-Alumina Ceramics," 1987; Her Majesty's Stationary Office, London.

30. B. Wachtman, Jr. (Ed.), *Structural Ceramics,* vol. 29 of *Treatise on Materials Science and Technology,* Academic Press, Orlando, Fla., 1989.
31. R. G. Munro and C. R. Hubbard, "Property Database for Gas-Fired Applications of Ceramics," *Ceram. Bull.,* vol. 68, pp. 2084–2090, 1989.
32. R. G. Munro, F. Y. Hwang, and C. R. Hubbard, "The Structural Ceramics Database: Technical Foundations," *J. Res. Natl. Inst. Stand. Tech.,* vol. 94, pp. 37–47, 1989.
33. R. G. Munro, E. F. Begley, and T. L. Baker, "Strengths and Deficiencies in Published Advanced Ceramics Data," *Ceram. Bull.,* 1991.
34. R. G. Munro and E. F. Begley, "NIST Structural Ceramics Database," preliminary version, 1991.
35. M. K. Booker, "Ceramic Technology for Advanced Heat Engines Program Data Base: A Summary Report," Oak Ridge National Lab., Rep. ORNL/M-462, undated.
36. B. L. P. Booker, "Ceramic Technology for Advanced Heat Engines Program Data Base: September 1988 Summary Report," Oak Ridge National Lab., Rep. ORNL/M-755, Mar. 1988.
37. B. L. P. Keyes, "Ceramic Technology for Advanced Heat Engines Program Data Base: September 1989 Summary Report," Oak Ridge National Lab., June 1989.
38. J. B. Wachtman, Jr., R. C. Bradt, R. F. Davis, R. Raj, D. W. Richerson, and N. J. Tighe, "Japanese Structural Ceramics Research and Development," Science Applications International Corp., July 1989.
39. S. Somiya, E. Kanai, and K. Ando (Eds.), *Proc. 1st Intl. Symp. on Ceramic Components for Engines,* KTK Sci. Publ., Tokyo, Japan, 1983.
40. W. Bunk and H. Hausner (Eds.), *Proc. 2d Intl. Symp. on Ceramic Materials and Components for Engines,* Verlag Deut. Keram. Ges., Germany, 1986.
41. V. J. Tennery (Ed.), *Proc. 3d Intl. Symp. on Ceramic Materials and Components for Engines,* American Ceramic Soc., Westerville, Ohio, 1989.
42. Japan Fine Ceramics Association, 22-13 Toranomon 1-chome, Minato-ku, Tokyo 105, Japan.
43. U.S. Advanced Ceramics Association, 1440 New York Ave., N. W., Washington, D.C. 20005.
44. R. E. Barks (Ed.), *Proc. 13th Ann. Conf. on Composites and Advanced Ceramic Materials, Ceram. Eng. Sci. Proc.,* July–Aug., Sept.–Oct. 1989.
45. S. J. Schneider, Natl. Inst. of Standards and Technology, private communication.
46. T. W. Chou and A. Kelly, "Mechanical Properties of Composites," *Ann. Rev. Mater. Sci.,* vol. 10, pp. 229–259, 1980.
47. R. L. Lehman, "Ceramic Matrix Fiber Composites," in J. B. Wachtman, Jr. (Ed.), *Structural Ceramics,* vol. 29 of *Treatise on Materials Science and Technology,* Academic Press, Orlando, Fla., 1989, pp. 229–291.
48. W. B. Hillig, "Prospects for Ultra-High-Temperature Ceramic Composites," General Electric Co., Rep. 85CRD152, Aug. 1985.
49. "High Temperature Materials for Advanced Technological Applications," Natl. Materials Advisory Board, Rep. NMAB-450, June 1989.
50. R. A. Bradley, D. E. Clark, D. C. Larsen, and J. O. Stiegler, "Whisker- and Fiber-Toughened Ceramics," ASM, Materials Park, Ohio, 1988.
51. W. H. Gitzen, "Alumina as a Ceramic Material," American Ceramic Soc., Westerville, Ohio, 1970.
52. E. Dorre and H. Hubner, *Alumina,* Springer-Verlag, New York, 1984.
53. L. D. Hart (Ed.), *Alumina Chemicals: Science and Technology Handbook,* American Ceramic Soc., Westerville, Ohio, 1990.
54. S. Iio, M. Watanabe, and Y. Matsuo, "Static Fatigue of TZP-Al_2O_3 Composite," *Ceram. Eng. Sci. Proc.,* vol. 10, pp. 1374–1382, 1989.

55. P. F. Becher, C. H. Hsueh, P. Angelini, and T. N. Tiegs, "Toughening Behavior in Whisker-Reinforced Ceramic Matrix Composites," *J. Am. Ceram. Soc.,* vol. 71, pp. 1050–1061, 1988.

56. G. H. Campbell, M. Ruhle, B. J. Dalgleish, and A. G. Evans, "Whisker Toughening: A Comparison between Aluminum Oxide and Silicon Nitride Toughened with Silicon Carbide," *J. Am. Ceram. Soc.,* vol. 73, pp. 521–530, 1990.

57. S. Iio, M. Watanabe, M. Matsubara, and Y. Matsuo, "Mechanical Properties of Alumina/Silicon Carbide Whisker Composites," *J. Am. Ceram. Soc.,* vol. 72, pp. 1880–1884, 1989.

58. K. L. Luthra and H. D. Park, "Oxidation of Silicon Carbide-Reinforced Oxide-Matrix Composites at 1375 to 1575°C," *J. Am. Ceram. Soc.,* vol. 73, pp. 1014–1023, 1990.

59. H. T. Lin and P. F. Becher, "Creep Behavior of a SiC-Whisker-Reinforced Alumina," *J. Am. Ceram. Soc.,* vol. 73, pp. 1378–1381, 1990.

60. A. R. de Arellano-Lopez, F. L. Cumbrera, A. Dominguez-Rodiguez, K. C. Goretta, and J. L. Routbort, "Compressive Creep of SiC-Whisker-Reinforced Al_2O_3," *J. Am. Ceram. Soc.,* vol. 73, pp. 1297–1300, 1990.

61. P. Barron-Antolin, G. H. Schiroky, and C. A. Andersson, "Properties of Fiber-Reinforced Matrix Composites," *Ceram. Eng. Sci. Proc.,* vol. 9, pp. 759–766, 1988.

62. C. A. Andersson and M. K. Aghajanian, "The Fracture Toughening Mechanism of Ceramic Composites Containing Adherent Ductile Metal Phases," *Ceram. Eng. Sci. Proc.,* vol. 9, pp. 621–626, 1988.

63. M. Srinivasan, "The Silicon Carbide Family of Structural Ceramics," in J. B. Wachtman, Jr. (Ed.), *Structural Ceramics,* vol. 29 of *Treatise on Materials Science and Technology,* Academic Press, Orlando, Fla., 1989, pp. 99–159.

64. J. A. Coppola, L. N. Hailey, and C. H. McMurty, "Sintered Silicon Carbide Body," U.S. Patent 4,312,954, Jan. 26, 1982.

65. N. L. Hecht, D. E. McCullum, and G. A. Graves, "Investigation of Selected Si_3N_4 and SiC Ceramics," in *Ceramic Materials and Components for Engines,* American Ceramic Soc., Westerville, Ohio, 1989, pp. 806–816.

66. A. Ghosh, M. G. Jenkins, K. W. White, A. S. Kobayashi, and R. C. Bradt, "Elevated-Temperature Fracture Resistance of a Sintered Alpha-Silicon Carbide," *J. Am. Ceram. Soc.,* vol. 72, pp. 242–247, 1989.

67. D. F. Carroll, T. J. Chuang, and S. M. Wiederhorn, "A Comparison of Creep Rupture Behavior in Tension and Bending," *Ceram. Eng. Sci. Proc.,* vol. 9, pp. 635–642, 1988.

68. D. F. Carroll and R. E. Tressler, "Effect of Creep Damage on the Tensile Creep Behavior of a Siliconized Silicon Carbide," *J. Am. Ceram. Soc.,* vol. 72, pp. 49–53, 1989.

69. M. L. Torti, "The Silicon Nitride Family of Structural Ceramics," in J. B. Wachtman, Jr. (Ed.), *Structural Ceramics,* vol. 29 of *Treatise on Materials Science and Technology,* Academic Press, Orlando, Fla., 1989, pp. 161–194.

70. J. Weiss, "Silicon Nitride Ceramics: Composition, Fabrication Parameters, and Properties," *Ann. Rev. Mater. Sci.,* vol. 11, pp. 381–389, 1981.

71. J. S. Haggerty, A. Lightfoot, J. E. Ritter, P. A. Gennari, and S. V. Nair, "Oxidation and Fracture Strength of High-Purity Reaction-Bonded Silicon Nitride," *J. Am. Ceram. Soc.,* vol. 72, pp. 1675–1679, 1989.

72. J. A. Mangles and G. J. Tennenhouse, "Densification of Reaction-Bonded Silicon Nitride," *Bull. Am. Ceram. Soc.,* vol. 55, pp. 1216–1218, 1980.

73. B. Nyberg, L. K. L. Falk, R. Pompe, and R. Carlsson, "Some Features of Nitrided Pressureless Sintered (NPS) Silicon Nitride Materials Made by Modified Preparation Routes," in J. B. Wachtman, Jr. (Ed.), *Structural Ceramics,* vol. 29 of *Treatise on Materials Science and Technology,* Academic Press, Orlando, Fla., 1989, pp. 155–163.

74. S. K. Dutta, A. K. Mukhopadhyay, and D. Chakraborty, "Assessment of Strength by

Young's Modulus and Porosity: A Critical Evaluation," *J. Am. Ceram. Soc.,* vol. 71, pp. 942–947, 1988.
75. K. Matsuhiro and T. Tomonori, "The Effect of Grain Size on the Toughness of Sintered Si_3N_4," *Ceram. Eng. Sci. Proc.,* vol. 10, pp. 807–816, 1989.
76. H. Yeh, H. Fang, and K. Teng, "Process Improvement for Si_3N_4 for Heat Engine Applications," *Ceram. Eng. Sci. Proc.,* vol. 9, pp. 1333–1342, 1988.
77. I. Tanaka, G. Pezzotti, R. Okamote, and Y. Miyamoto, "Dense Silicon Nitride without Additives: Sintering and High Temperature Behaviors," *Ceram. Eng. Sci. Proc.,* vol. 10, pp. 817–822, 1989.
78. I. Tanaka, G. Pezzotti, R. Okamote, and Y. Miyamoto, "Hot Isostatic Press Sintering and Properties of Silicon Nitride without Additives," *J. Am. Ceram. Soc.,* vol. 72, pp. 1656–1660, 1989.
79. C. W. Li and J. Yamanis, "Super-Tough Silicon Nitride with R-Curve Behavior," *Ceram. Eng. Sci. Proc.,* vol. 10, pp. 632–645, 1989.
80. M. Masuda, T. Some, M. Matsui, and I. Oda, "Cyclic Fatigue of Sintered Si_3N_4," *Ceram. Eng. Sci. Proc.,* vol. 9, pp. 1371–1382, 1988.
81. A. P. Nikkila and T. A. Mantyla, "Cyclic Fatigue of Silicon Nitrides," *Ceram. Eng. Sci. Proc.,* vol. 10, pp. 646–656, 1989.
82. M. H. Rawlins, T. A. Nolan, L. F. Allard, and V. J. Tennery, "Dynamic and Static Fatigue Behavior of Sintered Silicon Nitrides: II, Microstructure and Failure Analysis," *J. Am. Ceram. Soc.,* vol. 72, pp. 1338–1342, 1989.
83. Y. Tajima, K. Urashima, M. Watanabe, and Y. Matsuo, "Static, Cyclic and Dynamic Fatigue Behavior of Silicon Nitride," in *Ceramic Materials and Components for Engines,* American Ceramic Soc., Westerville, Ohio, 1989, pp. 719–728.
84. O. Van Der Biest, C. Weber, and L. A. Garguet, "Role of Oxidation in Creep and High Temperature Failure of Silicon Nitride," in *Ceramic Materials and Components for Engines,* American Ceramic Soc., Westerville, Ohio, 1989, pp. 729–738.
85. Y. R. Xu, T. S. Yen, and X. R. Fu, "Grain Boundary Tailoring of High Performance Nitride Ceramics and Their Creep Property Studies," in *Ceramic Materials and Components for Engines,* American Ceramic Soc., Westerville, Ohio, 1989, pp. 739–750.
86. A. E. Pasto, W. C. Van Schalkwyk, and F. M. Mahoney, "Creep Behavior of Yttria- and Alumina-Doped Silicon Nitride," in *Ceramic Materials and Components for Engines,* American Ceramic Soc., Westerville, Ohio, 1989, pp. 776–785.
87. H. Kodama, T. Suzuki, H. Sakamoto, and T. Miyoshi, "Toughening of Silicon Nitride Matrix Composites by the Addition of Both Silicon Carbide Whiskers and Silicon Carbide Particles," *J. Am. Ceram. Soc.,* vol. 73, pp. 678–683, 1990.
88. H. Kodama and T. Miyoshi, "Fabrication and Properties of Si_3N_4 Composites Reinforced by SiC Whiskers and Particles," *Ceram. Eng. Sci. Proc.,* vol. 10, pp. 1072–1082, 1989.
89. J. G. Baldoni and S. T. Buljan, "Creep and Crack Growth Resistance of Silicon Nitride Composites," *Ceramic Materials and Components for Engines,* American Ceramic Soc., Westerville, Ohio, 1989, pp. 786–795.
90. T. Ekstrom and P. O. Olsson, "Beta-Sialon Ceramics Prepared at 1700°C by Hot Isostatic Pressing," *J. Am. Ceram. Soc.,* vol. 72, pp. 1722–1724, 1989.
91. K. Yabuta, H. Nishio, H. Okamoto, S. Umebayashi, and K. Kishi, "High Temperature Bending Strength and Oxidation Resistance of Beta-Sialon Prepared with Minimal Grain Boundary Phase," in *Ceramic Materials and Components for Engines,* American Ceramic Soc., Westerville, Ohio, 1989, pp. 622–630.
92. T. Tiegs, "SiC Whisker-Reinforced Sialon Composites: Effect of Sintering Aid Content," *Ceram. Eng. Sci. Proc.,* vol. 10, pp. 1101–1107, 1989.
93. W. R. Cannon, "Transformation Toughened Ceramics for Structural Applications," in

J. B. Wachtman, Jr. (Ed.), *Structural Ceramics,* vol. 29 of *Treatise on Materials Science and Technology,* Academic Press, Orlando, Fla., 1989, pp. 195–228.
94. J. Sung and P. S. Nicholson, "Residual-Stress-Severity Relationships in Ceramics," *J. Am. Ceram. Soc.,* vol. 73, pp. 1318–1322, 1990.
95. T. T. Lepisto, and T. A. Mantyla, "A Model for Structural Degradation of Y-TZP Ceramics in Humid Atmosphere," *Ceram. Eng. Sci. Proc.,* vol. 10, pp. 658–667, 1989.
96. K. C. Liu and C. R. Brinkman, "Exploratory High-Temperature Tensile and Cyclic Fatigue Characterization of Commercial MgO-PSZ," in *Ceramic Materials and Components for Engines,* American Ceramic Soc., Westerville, Ohio, 1989, pp. 841–855.

BIBLIOGRAPHY

Andriyevskiy, R. A., "Silicon Nitride and Materials Based Upon It," AFSC, Rep. FTD ID (RS) T 1192-86, Apr. 1987.

Bellosi, A., P. Vincenzini, and G. N. Babin, "Stability of Si_3N_4–Al_2O_3–ZrO_2 Composites in Oxygen Environments," *J. Mater. Sci.,* vol. 23, pp. 2348–2354, July 1988.

"Ceramic Application and Design," *Ceram. Ind.,* vol. 130, no. 2, pp. S1–S18, Feb. 1988.

Freiman, S. W., "Stress Corrosion of Ceramic Materials," Natl. Bureau of Standards, Rep. AR 10/86–9/87, N00014-87F0007, Jan. 1988.

Miyoshi, K., "Adhesion, Friction and Micromechanical Properties of Ceramics," NASA, TM 100782, 1988, 24 pp.

Takatori, K., and O. Kamigaito, "Mechanical Properties of Hot-Pressed Alumina-Sialon Composites," *J. Mater. Sci. Lett.,* vol. 7, pp. 1024–1026, Oct. 1988.

CHAPTER 4
POWDERS—FORMING, PROCESSING, AND DENSIFICATION

4.1 INTRODUCTION

The major advance in fabrication methods during the past decade is in the area of processing prior to firing. The dramatic advances in powder quality in the direction of both the ultrafine and the monodisperse have brought about improvements in reliability and in processing convenience (such as lower firing temperatures). When combined with forming methods, striking enhancement of properties can occur.

The advances have indeed raised the question of the extent to which perfection should be sought in powders. Generally, the advances achieved by progress in powder refinement have underlined the benefits to be gained. However, since minor variations in packing density are predicted to become more extreme as densification proceeds, perfection is to be seen as a metastable condition. There are strong grounds therefore for using stabilizing features in the powder together with whatever quality of packing has been achieved. Boundary pinning additives can be seen to play such a role, though it would be rewarding to seek other methods.

A further issue concerns the merits of various powder methods in comparison with more direct alternatives. Although increased skill in powder handling has permitted fuller exploitation of the driving force present in the high surface area of fine powders, an alternative is to use direct applied pressure, such as hot isostatic pressing. Cost considerations will be significant in deciding this question, but it is worth noting again that the two techniques are not mutually exclusive. If progress can continue at a rate to match that of the last decade, then by 1996 users of advanced ceramics will be faced with an impressive array of ceramic opportunities.

Processes range from sol-gel to hydrothermal methods, rapid solidification technology (RST), and dynamic compaction (DCT). They make ideal processing combinations by which new amorphous materials (such as metallic glasses and complex ceramic compositions) can be formed as powders and subsequently densified into monolithic product shapes without undesirable crystallization. Other techniques include lasers, plasma, and self-propagating high-temperature synthesis (SHS). All of these are reviewed in this chapter.

Ceramic Problems

Ceramics pose their own set of problems. Those processed in the traditional manner are highly inconsistent, due primarily to ceramic powders that consist of relatively large, random-sized particles. There have also been severe problems caused by an uneven distribution of particles in the ceramic slurry and with the organic binders that hold the particles together during molding.

The random size and shape of particles result in irregular clusters in the ceramic slurry. These clusters prevent dense packing of the powders in the unfired ceramic, which results in a product with porous sections. This is responsible for the varied strengths, dimensions, surface finishes, and electrical properties of ceramics.

Some ceramic processes have approached the random-particle problem by producing their own submicrometer ceramic powders consisting of particles that are nearly uniform in shape and size and much smaller than particles in commercially available powders. These particles pack together in a tight, consistent, predictable matrix. The finished product has few weak spots, its surface is smooth, and the process is repeatable.

However, Lange[1] quotes Dr. John Wachtman of Rutgers University: "characterization of ceramic powders is central and crucial to ceramic processing and part reliability." Rejection rates of ceramic parts can constitute up to 75 percent of total manufacturing costs. He further quotes: "Better characterization of starting materials is needed for advanced ceramics, in terms of their uniformity, purity, and particle size. Powder characterization can play a key role in reducing costs by providing in-process control from the starting raw material through forming and sintering processes, thereby considerably improving both reliability and reproducibility."[1]

Finally he stated that improved nondestructive evaluation methods are needed for presintered (green) ceramics along with a means to correlate green-state characterization with the properties and performance of sintered parts. Especially important are techniques to determine the distribution of binders and sintering aids. Engineering reliability is a matter of processing reliability. Most forming methods are generally unacceptable for ceramics. Some advanced ceramics, for example Si_3N_4 and SiC, decompose prior to melting. Glass-ceramic processing, a special melt-forming method that takes advantage of newtonian rheology to form shapes and crystallization after solidification to control microstructure, produces nonequilibrium phase assemblages and is limited to glass-forming chemistries. Columnar grain growth and uneconomical deposition rates are disadvantages for chemistries that can be shaped by vapor condensation methods. Liquid precursor methods, such as sol-gel processing, suffer from large volume changes during fluid removal, pyrolysis, or densification, which limits this method to shaping small bodies, that is, particles, thin films, and fibers. Most advanced ceramics are formed as powder compacts made dense by heat treatment. Although powder processing is a many-bodied problem prone to heterogeneities and nonuniform phase distributions, it is the most efficient method to form ceramics.

Basics of Powders

Powder processing involves four basic steps: (1) powder manufacture, (2) powder preparation for consolidation, (3) consolidation to an engineering shape, and (4) densification or microstructural development that eliminates void space and produces the microstructure that optimizes properties. Each step has the potential for introducing a detrimental heterogeneity which either persists during further processing or develops into a new heterogeneity during densification and microstructural development.[1]

As strength is sensitive to stress concentrators, microstructural heterogeneities that are stress concentrators can be best observed by fracture and examination of fracture origins. Many microstructural heterogeneities stem from the powder itself. Agglomerates are a major heterogeneity in powders. The attractive interparticle forces responsible for free particle agglomeration include van der Waals and capillary forces.

Outside stresses cannot always be avoided, and they dramatically show the great liability of ceramics: brittleness. The rigid orientation of molecules within a crystalline structure, coupled with a strong ionic bonding within each crystal, gives ceramics their desirable properties, especially a lack of electrical conductivity. However, when crystal is joined to crystal in the firing process, the joining is initiated by a physical rather than a molecular attraction. Each crystal has its own bound water, which is driven off in the firing process; this creates a surface-absorption adherence. But differences in crystal size, impurities in the materials being joined, and minor differences in crystal formation produce nonuniform boundaries between crystals. When the crystals undergo physical stress, the intercrystalline bonding does not hold, and boundary separations, called microcracks, are formed. Continued stress causes the microcracks to join, and eventually results in fracture of the ceramic. Most of the ceramics used for technical purposes are constructed from fine-particle powders. Users must choose the powders, know that they bind together, and know how these two variables will affect the microstructural properties of the final ceramic. For most purposes, the ideal microstructure is uniform, dense, and has as few inclusions (foreign particles) and air pockets as possible. Thus the key step is to understand the particle-binding process that creates such a ceramic. In thermodynamic terms, each particle has a certain amount of free energy related to its surface area, an amount that increases as the particle decreases in size. We know from the second law of thermodynamics that reactions tend to occur that decrease the amount of free energy in the system. Therefore two particles in contact tend to diffuse together as a way of reducing their individual surface free energies. For example, two particles of identical size, 200 μm in diameter, in contact for 1 h at 1200°C interpenetrate by one-half micrometer to several micrometers, depending on the physical and chemical conditions. But if a particle 200 μm in size is in contact with a 5-μm particle, the smaller one will diffuse and disappear into its larger neighbor. Hence, to arrive at the ideal ceramic microstructure, the best powders to use will be small, pure, and uniform in size.

Agglomerates. Agglomerates limit densification[2,3] and are common to all powder processing routes. They can be classified as either soft, that is, particles held together by van der Waals forces which can be broken apart with surfactants, or hard, that is, partially sintered groups of particles which require attrition to dismember. All dry powders contain soft agglomerates. Hard agglomerates are common to powder manufacturing routes that involve heating to decompose an oxide precursor. Some processing routes (spray drying) purposely form large agglomerates to produce a flowable "powder" for dry pressing.

Agglomerates with different bulk densities can persist during powder consolidation to form cracklike voids during densification because of their different shrinkage rates relative to the surrounding powder compact.[4] The cracklike voids can be eliminated, or reduced in size, by two methods. First, the cracklike voids present in the near theoretically dense (>98%) bodies can be closed by hot isostatic processing. The second approach is more fundamental to processing science. It involves dispersing the powder or powders in a liquid containing a surfactant that eliminates the soft agglomerates and consolidating the powder from the slurry state to form the desired engineering shape, which is dried and sintered.[5]

Organic and Inorganic Inclusions. Powders contain organic and inorganic inclusions introduced by both the manufacturer and the processor during powder preparation for consolidation. Organic inclusions can be lint or hair, for example. Some of these heterogeneities are introduced when powders are milled to reduce the size of hard agglomerates or when they are exposed to the environment. Organic inclusions disappear during densification and leave irregular shaped voids when they burn out during sintering. These irregular shaped voids are found at fracture origins. Inorganic inclusions can react with the powder during densification or produce microcracks during either cooling from the densification temperature or subsequent stressing.

One method for eliminating such voids is simply to burn out the organic matter at a low temperature, cool to room temperature, isothermal-press the powder compact, and then sinter. Isothermal pressing after burnout and before sintering can eliminate the void produced by the organic inclusions and further increase the average strength of transformation-toughened ZrO_2 from 1050 to 1300 MPa.

Control of Grain Size. Bimodal and large-grain-size microstructures must be avoided to obtain high strengths. Large single grains within a fine grain matrix are common fracture origins. It is also well known that the average strength is inversely proportional to the average grain size. Similar to the case of inclusions discussed, localized stresses arise within and around grains due to thermal expansion and elastic anisotropy. The conditions for microcracking in single-phase polycrystalline ceramics are similar to those of inclusions, that is, a critical grain size exists for spontaneous microcracking during cooling. Control of grain growth is therefore required for mechanical integrity (Fig. 4.1).

It is widely accepted that control of grain growth can be achieved with the addition of a chemically compatible second phase. Two-phase composites can be designed to avoid microcracking. It should be noted that when colloid routes[1] are used, a uniform phase distribution can be achieved.

It can be concluded that the mechanical reliability of a ceramic can be increased by controlling grain growth with a dispersed second phase. This approach can be optimized with strict controls on phase distribution. A cost-effective approach, however, is the use of colloidal methods for powder treatment and consolidation. Colloidal methods (powder dispersion, sedimentation, and slurry consolidation) effectively reduce the size of agglomerates and inclusions. They also uniformly distribute second phases needed to control abnormal grain growth or to produce residual surface compressive stresses through environmental reactions. Finally if reliable ceramics are to be produced, methodologies must be developed to ensure, with a high probability, that heterogeneities will be eliminated from powders and that they will not be reintroduced in subsequent processing steps.

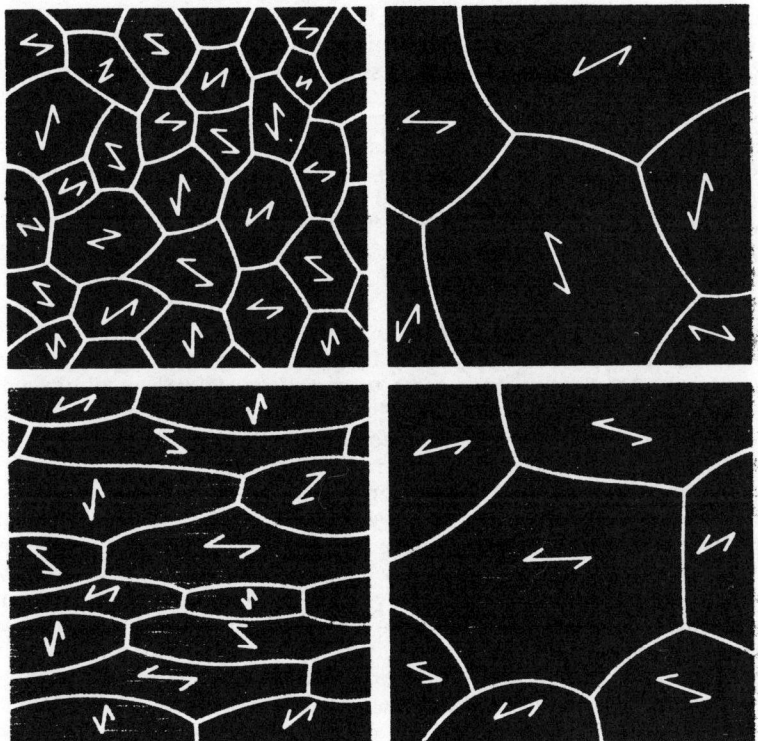

FIGURE 4.1 Size, shape, and orientation of grains within a ceramic determine significant properties.

4.2 CERAMIC POWDERS AND THEIR PROCESSES

Powders

The properties of the starting powder largely dictate the subsequent stages in ceramic processing and the quality of the final product. It is generally considered that the following properties of a powder are important:

1. *Size:* Small particle sizes (submicrometer) favor more rapid densification and sintering at lower temperature.
2. *Size distribution:* A uniform particle size is thought to avoid the growth of large grains which reduce strength.
3. *Shape:* Angular and acicular particles are difficult to pack as uniform green bodies. Inhomogeneity in packing will result in corresponding flaws in the fired component.
4. *State of aggregation:* Aggregates of small particles sinter more rapidly than the surrounding particles, thus opening up a flaw.

5. *Purity:* The chemical purity and stoichiometry of ceramic powders can profoundly influence the sintering and densification process. A target is to produce powders of high purity to which selected additives may be introduced to aid in sintering and densification.

There are paradoxes in these specifications. For example, the tendency for powders to aggregate increases as size is reduced. Aggregates may vary in strength so that some are easily dispersed in subsequent process steps while others persist even under conditions of high shear. This alone may be responsible for the surprising fact that a ceramic processor can select nominally similar powders from different manufacturers, treat them in the same way, and produce entirely different results.

The recognition of the vital role of powder quality has attracted much attention to sophisticated methods for powder manufacture by both wet chemistry and gas-phase reactions. The aqueous route to ceramic powders is well exemplified by the Bayer process for alumina and the sulfate process for titania,[6] and this type of route (hydrolysis, hydroxide precipitation, and calcination) is being upgraded as knowledge of the complex chemistry involved grows. This is reflected in recent refinements in zirconia powders. There is much interest in the controlled hydrolysis of metal alkoxides to produce uniform particles,[7-9] a benefit being high purity (99.99%). Much work remains to be done to scale up such methods, and one disadvantage is that hydrolysis reactions seldom produce the appropriate phase, most often yielding hydrated phases which require calcination for best performance. Calcination is likely to cause the powder to sinter, producing aggregates and negating the advantage of this route to synthetic powders. Under different conditions of hydrolysis, alkoxides give gels rather than particles, and after casting, these can be dried and then fired to a ceramic. The apparent advantage of this method is that it avoids starting with a powder at all, but there are serious disadvantages. The gel can shrink more than tenfold on drying and firing, and therefore one must be careful to avoid cracking.[10] In producing fibers and coatings by the sol-gel method shrinkage can be more easily accommodated.[11-13] One advantage of this method, however, is seen when it is used to produce a mixture of, for instance, the oxides of two or more metals. This will be discussed in a later section.

Another promising route to synthetic ceramic powders is gas-phase reactions in a hot zone provided by a plasma or laser.[14] Gases such as SiH can be reacted with CH_4 or NH_3 to produce SiC and Si_3N_4. Anhydrous phases are produced directly (an advantage over hydrolysis methods), but the powders can be so fine that they present problems in collection and processing. Nevertheless, gas-phase reactions are in commercial use to produce ceramic particles of closely controlled size, shape, and purity.

Also, there is interest being shown in using new techniques for synthesizing improved powders with these processes:

Solution techniques

Vapor-phase techniques

Salt decomposition

Each of these has advantages and disadvantages. Solution techniques can provide high-purity materials with a wide range and close control of composition. Also, trace materials can be added easily and homogeneously. These techniques include solvent vaporization, precipitation filtration, solvent extraction filtration (such as sol-gel for solvent extraction), and solvent combustion. Vapor-phase techniques can produce very fine unaggregated powders which can be spherical. Other techniques include vapor condensation (such as aerosol), vapor decomposition, vapor-vapor reaction (such as plasma and laser), and vapor-solid reactions.[15] Specifically, oxides can be synthesized via plasma, aerosol, flame hydrolysis, sol-gel, or hydrothermal processes.

The aerosol process produces spherical powder and has the capability of producing mixed oxides. Flame hydrolysis produces oxide powders of extremely high purity in a very flexible and uniform manner, while sol-gel can yield powders which are quite pure without requiring high temperatures or pressures. Finally, the hydrothermal process can form powders with well-controlled size and shape by precipitation and sol-gel solution.

Other approaches in addition to these include organometallic, aqueous precipitation, thermite techniques, metal nitridation, and chemical vapor deposition (CVD). There are many differences in the types of materials that each route can produce, and no one route is satisfactory for all oxide and nonoxide powders. In addition, the physical and chemical properties of the powders that result can be dramatically different.

Before discussing the individual synthesis routes, one common hurdle faced by all the methods is the economics of the process on a production scale. At the present market levels, all of the advanced ceramic materials of interest are used in very limited quantities. Because of the small volumes, none of these synthetic routes can take advantage of the economies of scale. This economy of scale favors the routes with less expensive raw materials and greater dependence on capital.

Some routes, such as the organometallic precursors and sol-gels, utilize high-purity but usually expensive raw materials and equipment; therefore they are less likely to be economical at the larger volumes. Of course, as market demand increases, the cost of the organic raw materials can also benefit from a reduction in cost due to economies of scale. For any of the synthesis routes the volumes must be increased significantly before costs can be reduced. Conversely, the costs of the ceramic powders must be decreased to obtain market interest in the material at large volumes. The dilemma facing the industry must be overcome by forward-looking consumers and forward-pricing suppliers.

Processes

Organometallic Synthesis. The synthesis of ceramic powders and near-net-shaped parts is possible from the family of materials called organometallics. The fundamental characteristic of these materials is that they contain combinations of carbon and metal atoms bonded together, which can be produced in a variety of chemistries and stoichiometries. Organometallics can be considered one of the sol-gels; they allow the user great flexibility in producing precursors in conventional chemical reactor schemes. The ability of the chemist to control the particle size and molecular chemistry by using the organometallic processes gives more degrees of freedom in producing unique parts. In particular, very-low-temperature densification is possible, which provides the user with the opportunity to produce unique grain structures in ceramic materials. Control of chemistry on the microstructural level is at a maximum when using the organometallic techniques available.

If the improvement in performance levels using organometallics is only incremental, the new method in most cases will not be accepted. The improvements must be revolutionary or actually "enabling" before the organometallic methods will be accepted as the norm. The increased costs are a result of the high cost of the raw materials and the capital equipment necessary for the production and processing into ceramics.

Organometallic production of ceramics offers the potential for excellent chemistry control. This is particularly true when compared to processes using raw materials that are extracted from natural resources and may carry impurities along with them which cannot be removed by conventional processing. It has been the experience of researchers in organometallic processes, however, that there remains some difficulty in producing ceramics without excess carbon or nitrogen, or both, in the composition. The excess carbon or nitrogen produces phases in the bulk material that detract from the performance, most often at high temperatures, of the resultant ceramic pieces when compared to the "pure-phase" materials produced from more conventional routes.

For the reasons of high cost, sensitive handling, and limited impurity control, the organometallic processes will have significant difficulty in making inroads into the structural ceramics marketplace of the future. A small niche currently being exploited is the production of SiC/Si_3N_4 fibers, whereby organometallic polymer precursors are utilized to fabricate fibers that are then pyrolyzed into the ceramic compositions. These fibers prove expensive. However, for initial applications and study the high cost can be tolerated.

Sol-Gel. Solution-gelation (sol-gel) ceramic production processes have been studied by many academicians and industrial scientists for many years. The route is similar to the organometallic route, but sol-gel materials may be either organic or inorganic in nature. The two routes have many of the advantages and disadvantages in common. One additional problem to both routes, which was not mentioned earlier, is that the amount of ceramic material in the starting precursor is quite low. This results in dramatic shrinkages during the final ceramic firing, and leads to significant difficulties in producing stress-free, net-shaped, and near-net-shaped (NNS) parts. For this reason, large-cross-section parts at full density are difficult to achieve.

It is likely that the sol-gel route will find applications in very specialized bulk uses, such as for radomes and windows,[16] in thin films for structural and electronic uses, and as "ceramifiable" binders for ceramic powders. One exception is an abrasive application, in which the large shrinkages of the sol-gel materials result in appropriately sized grit when the ceramic stress-cracks.

Laser Synthesis. The long search for useful applications of laser technology in chemistry has resulted in the production of research quantities of high-quality, unique-morphology ceramic powders. The laser provides the energy for chemically converting the gases that are injected into the laser beam into extremely fine, well-controlled powders. Ceramic parts produced from laser-synthesized powders in laboratory equipment have demonstrated properties that rival the best powders produced by more conventional means.

The ability to scale the process into realistic volume levels at realistic costs and the availability of low-cost gases of the proper chemistry (that are able to couple with the laser energy supplied) limit the future usefulness of this synthetic technique. The application of lasers to a variety of materials is limited by the availability of suitable precursors. Although efforts to prove that the technique is commercially viable exist, it remains difficult to see where it can compete with other processes currently available or being developed. However, this process remains a critical tool for researchers who are using it to develop new materials that can then be mimicked by more cost-effective means.

Plasma Synthesis. Although similar to the laser method, the use of plasma energy to synthesize ceramic powders has progressed somewhat further toward commercialization. The major reason for this progress is the scalability of the plasma process. Large-capacity synthesis chambers can be built that approach those necessary for economical production, and the flexibility of the process is broader than that of the laser process.

The major roadblocks for the success of plasma powders are the ability to produce agglomerate-free materials and the ability to handle and fabricate the very fine powders into the required green shapes. Agglomeration of the powders presents a problem for the fabricator because lower-density green parts are formed. The low density results from the significant amount of entrapped porosity in the agglomerates that cannot be sintered out even with high temperatures. Unless improvements of a revolutionary nature are shown in the final product, it is unlikely that the plasma process will be the process of choice for producing bulk powders for ceramic markets. The plasma process can produce thin films and coatings that may be quite useful.

Hydrothermal or Aqueous Processes. The synthesis of oxide-based materials by either aqueous or hydrothermal processes has been of interest because of several factors, most importantly lower costs. The aqueous process produces precursor materials such as hydroxide, nitrate, or carbonate forms that require further heat treatment to produce the useful ceramic phases. These processes use somewhat elevated temperatures but only atmospheric pressures.

Using the precipitation of anhydrous crystalline powders from aqueous solutions at temperatures above their boiling points, powders made by this method feature controlled particle morphology and composition. The powders are also highly crystalline and anhydrous, and calcination and milling are therefore not required. Since the powders are reactive to sintering, densification temperatures which are lower than those employed with conventional powders can be used. Other advantages include small particle size and distribution, high purity, and homogeneity. A variety of oxides has been made—Fe_2O_3, ZrO_2, and Al_2O_3—for magnetic and electronic applications. A major advantage of hydrothermal processing is its ability to produce high-quality products from impure feedstocks. With the technique, waste materials can be converted into marketable coproducts. The hydrothermal syntheses utilize both high temperatures and higher than atmospheric pressures to produce oxide powders directly without the need for further heat treatment to produce the desired ceramic phases.

Oxide materials produced by either of these methods offer great economic incentives for the fabricator. While they are less expensive than the organometallic materials, however, it is unlikely that bulk alumina can be produced more effectively by these methods. The difficulty facing the commercialization of these technologies is the development of markets of sufficient size to allow the economies of scale to be utilized. The development of the families of oxide materials depends on breakthrough technologies in bringing the toughness of the oxide into the range of the improved SiC and Si_3N_4 materials.

Carbothermal Synthesis. The production of nonoxide powders by the carbothermal reaction can be accomplished in a variety of ways. In all cases, the reaction is one between carbon and a metal oxide. In some cases, the reaction with nitrogen is necessary to produce the nitride materials. Techniques within the carbothermal family include electric arc, thermite, and self-propagating, high-temperature synthesis. The processes can produce large-particle materials that require attrition for proper ceramic sizes, or they can produce very small particles directly in a variety of sizes. The modified electric-arc process produces large "logs" of widely variable ceramic materials that need both selection and attrition to obtain the purity and sizes needed for advanced ceramic applications.

The thermite process takes advantage of the reaction of metals such as aluminum or magnesium with carbon and oxides to produce carbides. The limitation of this process is the range of materials that can be produced, the cost and purity of the starting metals, and the need for final separation of the product from the starting materials. The self-propagating, high-temperature synthesis processes take advantage of the thermite approach but attempt to produce final shapes using the energy from the synthesis in the densification.

Some companies have recently developed unique reactor approaches that produce very-high-purity powders with very tightly controlled particle-size distributions. These processes are cost-effective at high volumes because of the low cost of the starting materials and the elimination of high-cost attrition equipment with its added impurity contributions. Carbides, nitrides, and borides are some of the families that can be synthesized with slight modifications to the reactor design or conditions. The carbothermal reduction of alumina has been used to form AlN and AlON powders. For nitrides (such as Si_3N_4 and AlN) a gas reaction called thermal decomposition is often used. Several Japanese firms use this reaction, which combines a chloride (such as $SiCl_4$) and ammonia to produce an intermediate imide and ammonium chloride.[16]

Metal Nitridation. Direct reaction at high temperatures of metals with atmospheres such as nitrogen are a viable approach to nonoxide powders of reasonable, but not top, quality. The powders suffer from the normal impurities in the metal and the inability to control many of the parameters that control particle-size growth and morphology (Table 4.1).

Chemical and Physical Vapor Deposition. As a technique for powder preparation, it is unlikely that chemical vapor deposition (CVD) or physical vapor deposition (PVD) holds any promise for strictly powders. These processes are being developed to produce ceramic-ceramic composites by filtration of a ceramic material into substrates and carbon fiber preforms. Their major drawback is the slow speed of deposition, which results in relatively high fabrication costs.

Nondestructive Evaluation. A most critical key area which traditionally has not been investigated thoroughly as related to powders and powder fabrication is nondestructive evaluation (NDE). This is a most crucial process, which must be performed throughout the manufacturing of ceramic powders, compacts, products, and applications.

NDE includes techniques for determining the flaw size, orientation, and distribution, which are the primary factors controlling the strength of ceramic materials. It is therefore important to detect these flaws and characterize them without destroying the part. Techniques being evaluated and developed for the detection of flaws include microradiography and low-frequency acoustic methods for unfired parts and high-frequency ultrasonics, acoustic microscopy, radiography, holographic interferometry, and infrared scanning for sintered parts.[17,18]

TABLE 4.1 Comparison of Powder Processes

Material	Process	Specific surface area, m²/g	Particle size, μm
Si_3N_4	Nitridation of Si	20	0.1–1
Si_3N_4	Reduction of silica	5	0.2–1.2
Si_3N_4	Vapor-phase reaction	4	0.2–4
Si_3N_4	Thermal decomposition of $Si(NH)_2$	12–15	0.1–0.3
MgO (UBE Ind.) (depends on grade)	Vapor-phase reaction	7.8–129	0.05–0.2
ZrO_2 (Toya Soda) (wide range of grades)	Hydrolysis of metal salts	8–23	0.3–0.5
Acetate-derived oxides (ZnO, NiO, Mn_3O_4, $NiMn_2O_4$)	Evaporative decomposition of solutions	6–56	
Nitrate-derived (ZnO, NiO, Fe_2O_3, $ZnFe_2O_3$)	Same	6–11.5	1–20 (aggregates)
Yttria	Emulsion precipitation		0.6
PLZT	Precipitation/spray drying		0.1–3.0
TiB_2, TiC, TiN, SiC	Combustion synthesis	<25 (TiB_2)	1–10
TiB_2	Carbothermal reduction		5–10
SiC	Same	20	0.1
AlN	Thermal decomposition		4.2
SiC	Plasma synthesis	60–130	10–30 nm
Al_2O_3, ZrO_2, MgO	Same		2–150 nm
SiC	Laser synthesis		19–65 nm
B_4C	Same		10–100 nm
Lead titanate	Hydrothermal synthesis	30–100	10–100 nm
Li_2ZrO_4	Hydrolysis of alkoxides	Up to 190	
$BaTiO_3$	Hydrothermal synthesis	6–12	0.08–0.17
Yttria	Homogeneous precipitation		0.4
Lanthana, Neodymia, Ceria	Same		1–6
MgO–ZrO_2, MgO, and Y_2O_3–ZrO_2	Coprecipitation		20 nm
$CaZrO_3$	Evaporative decomposition	35	0.1–0.5
Nb_2O_5	Controlled hydrolysis of $Nb(OC_2H_5)$	5	0.5–0.6
Carbides, borides	Reductive dehalogenation of halides		1–5

Source: From Millberg.[16]

4.3 POWDERS AND PROCESSING

Although most of the basic raw materials for ceramics occur abundantly in nature, they must be extensively refined or processed before they can be used to fabricate structures. The entire silicon group of silicon-based ceramics (other than silica) does not occur naturally. Compositions of silicon carbide, silicon nitride, and sialon must all be fabricated from gases or other ingredients. Even minerals that occur naturally, such as bauxite, from which alumina is made, and zircon sands, from which zirconia is derived, must be process-controlled before use to control purity, particle size and distribution, and heterogeneity.[19]

The crucial importance of powder preparation has been recognized in recent years. Particle sizes and size distributions are critical in advanced ceramics to produce uniform green (unfired) densities so that consolidation can occur to produce a fully dense, sintered, ceramic part.[20]

One of the major ceramic processing barriers is powder production and characterization. Problems that arise involve optimizing powder processes in order to obtain consistent and well-characterized powder morphology and chemistry to minimize the presence of critical flaws that could lead to failure in service. The need to minimize flaws in the starting powders is particularly acute in the case of load-bearing structural ceramics in heat engines and in the case of gas sensors where performance is sensitive to surface properties. Better characterization of both bulk and surface chemistries is needed to obtain better in-process control and reliability in the final product. Better analytical techniques are needed to obtain accurate quantitative and spatial distribution information on elements such as oxygen, nitrogen, carbon, and boron. The chemical characterization of starting powders is important because of its critical effect on subsequent processing steps and final properties.

Ceramics have no available stored energy-absorption mechanisms other than the creation of surface area (crack growth). Physical flaws such as pores or porous regions, foreign particles, large grains, or local regions of nonuniform density lead to the catastrophic failure of ceramics. In addition, different thermal expansions of inhomogeneities lead to thermal stresses when parts cool from sintering temperatures, and different elastic moduli lead to nonuniform stress distribution under working stresses. The effects of both conditions are additive and produce stress concentrations in the vicinity of flaws.

The most common source of processing flaws in ceramics are pores, either single, in limited groups, or in various cluster sizes. Typically, single pores are several times larger than the material's grain size; however, pores or groups of pores smaller than the grain size can also be sources of failure in larger-grain structures. One common cause of irregularly shaped pores is the formation of a pore or series of pores around an agglomerate during part fabrication. Another major cause of porosity is incomplete powder compaction due to local concentrations of organic matter, which is burned out during firing.

Chemical-related flaws are associated with minor second phases (impurities) in grain boundaries, which dissolve small amounts of the major phase at high temperatures and spread along the boundaries. These glassy phases (containing SiO_2) soften at elevated temperatures, enhancing creep, which produces cavitation eventually leading to flaws due to cavity coalescence. Since many of these second phases are purposely added to enhance sintering, sintering aids must be carefully selected to avoid or minimize the degradation of high-temperature properties. (See Chap. 5.)

Continued research has led to methods of eliminating physical flaws from ceramic powders. Agglomerates, clusters of fine ceramic particles formed during powder conditioning and held together by electrostatic forces, are retained in the green compact. Agglomerates can be removed by dispersion in liquid (weakly bonded clusters) or by sedimentation from the liquid (strongly bonded clusters). Agglomerate formation in very fine powders also can be minimized by modifying the surface chemistry of individual particles to make them repel each other. This process results in uniform particle dispersion in the slurry stage and uniform characteristics in the finished part.

Inclusions, either organic (such as dust or lint) or inorganic (such as iron particles), picked up during powder production or conditioning also lead to harmful flaws. Organics result in pores after they burn out upon firing, while inorganic matter can act as stress raisers. The former can be removed by prefiring green parts and repressing, and the latter by sedimentation. Porosity, while being the most difficult flaw to remove, can be minimized by consolidating to near theoretical density, such as by hot isostatic pressing.

Chemical flaws can be controlled by using additives that produce refractory secondary phases, which have little effect on final properties. For example, both Y_2O_3 and MgO additives used for hot-pressed Si_3N_4 form liquid silicates; however, Y_2O_3 silicates are more refractory than MgO silicates.[21]

Researchers have recently developed a nondestructive method for detecting flaws in compacted ceramic powders. The ultrasonic sensor method could become an important tool for ceramics producers in that it would allow them to fully automate the inspection of compacted powders while the material is in the mold. This will help ceramics producers screen out defective parts prior to costly processing steps. The system provides information on the uniformity and density of materials at almost any stage of compaction. Since testing can be performed with the part in its mold prior to the firing or sintering stage, the ultrasonic system offers a way to control the quality of ceramic powders without having to handle the fragile materials while they are in their green or unfired state. This method was developed as a fallout of work applying nondestructive evaluation to materials for in-line monitoring and process control during manufacture.

Another major barrier facing the structural ceramic industry today is cost-effectiveness. This can be related to the technical capability to produce reliable and durable ceramics. Reliability can be defined as consistency in properties, while durability is related to corrosion, wear, and environmental stability under in-service conditions. Both factors can be controlled through advanced processing and intelligent process control, which requires models, test methods, and sensors capable of real-time measurements. As a result, the underlying scientific theme is to develop models to describe the processing-structure-property relationships of monolithics (those ceramics cast as a single piece) and composites.[22] Even though the identification and measurement of key powder properties for reliable manufacturing of ceramics are difficult but crucial requirements, they must be performed for the eventual successful process model development for both monolithic and composite ceramics.

Raw-Material Selection

Ceramic raw materials are generally more readily available and less expensive than the critical materials used in specialty steels and superalloys. The differences between traditional and high-technology ceramics are due in part to the raw materials used. Clays and many of the nonclay minerals are associated with traditional ceramics and are not generally used in newer advanced ceramics. However, some materials such as Al_2O_3, ZrO_2, and SiC have both traditional and high-tech applications. The ability of these materials to fulfill both roles is a function of powder preparation and manufacturing techniques.[23,24]

Powder Production. Ceramic powders are processed from naturally occurring minerals, by-products of mining operations, or chemical reactions. Powders such as Al_2O_3 SiO_2, MgO, CaO, and ZrO_2 are derived from naturally occurring minerals or sands, which are heat-treated, leached, or reacted to remove water and impurities.

Powders are mixed, blended, granulated, crushed, and synthesized. Crushing is generally done in a ball mill and may be done wet or dry. The former is more effective since it keeps the particles together and prevents the suspension of fine particles in air. The ground particles can then be mixed with a variety of additives, including dopants, binders, plasticizers, deflocculents, lubricants for mold release, and wetting agents. Dopants can control grain growth or help achieve higher final densities in ceramics. The use of dopants, although providing many beneficial results, can also have a detrimental influence on the material properties. Segregation of dopants at the grain boundaries can weaken the final part.

Powder Processes. Chemical solution techniques provide a relatively convenient means for achieving powders of high purity and fine size. First a suitable liquid solution containing the cations of interest is prepared and analyzed. A solid particulate phase may be formed by precipitation, solvent evaporation, or solvent extraction. Segregation is minimized by combining the ions in a precipitate or gel phase or by extracting the solvent in a few milliseconds from a microscopic drop. The solid phase is usually a salt that can be decomposed without melting by calcination at a relatively low temperature.

Precipitation techniques have been widely investigated for preparing submicrometer-size high-purity oxide powders. This technique is used commercially to produce Al_2O_3 with a purity exceeding 99.995 percent, and these techniques have been used as commercial catalysts.

An alternative procedure (solvent evaporation and extraction techniques) has been used to prepare special powders and to disperse the solution containing the ions of interest into microscopic volumes and then remove the solvent as a vapor, forming a salt. Spray drying has been used to produce and dry salt particles 10 to 20 μm in diameter. The fast drying that occurs in a few milliseconds reduces segregation. This technique has been reported to give good results for the preparation of Mg-stabilized beta alumina from mixed nitrates.[16] Another approach has been to absorb the chemical solution into a microporous organic material such as cellulose. The loaded fiber is first pyrolyzed and then calcined in a controlled atmosphere. The porous agglomerates are easily comminuted. This technique has been used to produce colloidal sized particles or fibers of a wide variety of oxides and carbides.

Silicon nitride powder may be formed by reacting silicon tetrachloride and ammonia in a plasma below 1000°C, while the thermal decomposition of $(CH_3)_2SiCl_2$ and CH_3SiH_5 vapor has also been used to produce high-purity SiC. Vapor-phase techniques produce submicrometer-size, well-dispersed particles entrained in large volumes of gas.

Synthesizing processes[24-28] include solution techniques, vapor-phase techniques, and salt deposition. Each has advantages and disadvantages. Solution techniques can provide high-purity materials with a wide range and close control of composition. These include solvent vaporization, precipitation filtration, solvent extraction filtration (such as sol-gel for solvent extraction), solvent combustion, and colloid science. Vapor-phase techniques can provide very fine unaggregated powders which can be spherical. These techniques include vapor condensation (such as aerosol), vapor decomposition, vapor-vapor reaction (such as plasma and laser), and vapor-solid reactions.[29-31]

Specifically, oxides can be synthesized via plasma, aerosol, flame hydrolysis, hydrothermal processes, or sol-gel. However, the particles made via plasma may be too fine, the particle-size control (both mean and distribution) may be difficult, and the process is not completely understood. The aerosol process can give narrow distributions, produce spherical powder, and has the capability of producing mixed oxides. Flame hydrolysis produces oxide powders of extremely high purity in a very flexible and uniform manner. The hydrothermal process can form powders with well-controlled size and shape by precipitation from solution. The process tends to be rather expensive since high temperature, pressures, and acidic environments are usually involved. Sol-gel can yield powders which are quite pure without requiring high temperatures or pressures.[32-35] Instead, process control is obtained through control of the surface chemistry of polymerization reactions. Most sol-gel activity in the past has concentrated on producing fibers, coating, or monoliths instead of powder; however, this has changed.

Sol-gel methods can produce ceramics and glasses of higher purity than is possible with conventional methods. Given the control over purity in sol-gel processes, it follows that it is also possible to produce materials of controlled composition.

Finally, organometallic compounds have been dissolved in alcoholic solutions, hydrolyzed, and polymerized to form gels. The excess water and unreacted organics are then removed by vacuum or thermal treatment. The gels are porous, and on heating to temperatures below the glass transition, most of the pores, if not all, are eliminated. The dense glasses from the gels apparently have properties identical to the melt-formed glasses.[36] Glass ceramics have been made this way, and therefore a method exists for using the sol-gel approach for the fabrication of composites with a glass or glass-ceramics matrix at much lower temperatures.

The borides, carbides, and nitrides can be synthesized via plasma, laser, carbothermic reactions, molten salts, or from polymers.[37-40] The laser has potential low capital cost, high energy efficiency, and produces high-purity materials with good control of particle-size distribution. The drawbacks of a laser process are that it is limited to those materials which can be produced from reactants that couple with the laser.

The other processes for synthesizing powders for low-cost ceramic composites include an exothermic process and explosive compacting techniques.[37] The former can produce composites of the oxide-carbide type made by reducing oxides in the presence of carbon. The promise of a fast, low-cost process has been the main attraction and has been called self-propagating high-temperature synthesis, combustion synthesis, or thermite reactions. These reactions have produced Al_2O_3/TiC, Al_2O_3/SiC, MgO/SiC, and some other oxide-carbide compositions.[41,42]

Vapor Phase. To improve particle-size control and morphology, researchers are exploring more advanced vapor-phase reactions, using an alternative heat source to drive the reaction. Nitride and oxide powders can be formed in a vapor-phase reaction under a high-temperature gas generated by the plasma. Rapid quenching of the gases produces ultrafine, ultrapure powders, requiring no milling or grinding (Table 4.1).

The plasma can be produced by either a dc arc or radio-frequency (RF) inductive coupling. The latter requires no electrodes and has the ability to inject reactants axially. However, the agglomeration and formation of a graphite coating are problems that must be eliminated. Although dc arc plasma has electrodes which can corrode, it nonetheless offers higher thermal efficiency with lower capital investment, higher plasma temperatures, and greater stability (Table 4.2).[29]

The dissociation temperature of the starting elemental powder is an important factor in plasma synthesis; carbides are much easier to process than nitrides. Ultrafine plasma-synthesized nitride and carbide powders tend to be extremely air-sensitive. Further, the accompanying higher surface area can cause greater pyrophoricity, but it also improves sintering behavior.

Some companies use vapor-phase pyrolysis techniques to produce submicrometer sinterable Si_3N_4 powders with high alpha-phase content and a narrow particle-size distribution by reacting tris (dimethylamino) silane in an inductively heated Al_2O_3 tube. The vapor of the silane is introduced into the reactor under its own vapor pressure or by a carrier gas, H_2. After pyrolysis the powders then require a mild treatment to remove excess carbon. The powders are crystallized at 1400 to 1600°C. The final product has particle sizes ranging from 0.1 to 1.0 μm, with a mean of 0.25 μm.

Plasma synthesis, on the other hand, is usually a single-step continuous process, with a wide variety of reactants available. Both RF induction and arc plasmas can be used to synthesize nitride powders. In general Si_3N_4 has been produced with 5- to 30-nm particles containing both α and β phases, or in amorphous form (Table 4.3).[29] Not only has plasma synthesis been used to produce Si_3N_4 and SiC, but it has successfully produced TiN, which is not subject to NH_4Cl contamination like Si_3N_4 since it is formed directly from N_2.

U.S.S.R., U.S., and Japanese researchers have all reported plasma synthesization of AlN, SiC, and Si_3N_4,[29,43-45] while U.S. researchers believe that a glow-discharge technique can be used to produce any material for which volatile reactants are available, including ceramics and ceramic precursors.[46] There are advantages and disadvantages to the glow-discharge powder synthesis method (Table 4.4), and a question remains with regard to the extent of agglomeration after calcination.

Just as spray drying, which will be discussed later, can be considered to be a vapor-phase processing method, there are also thermal plasmas, such as dc arc jets and electrode induction,[47] the glow discharges or cold plasmas,[46,47] and aerosol decomposition,[47] also known as spray pyrolysis, spray roasting, or evaporative decomposition. Spray drying relies on the mechanical generation of droplets containing precursors dissolved in a solvent followed by thermal decomposition of the precursors in the particles to form the product.

Aerosol decomposition allows the production of chemically homogeneous multicomponent powders with controlled morphology and particle-size distribution. A carrier gas is passed through an aerosol generator containing precursors dissolved in a suitable solvent, resulting in fine droplets of the solution. The droplets are sent through a heated-wall reactor where the solvent evaporates and the precursors decompose to form the product. A variety of oxide materials have been produced, including Al_2O_3, Y_2O_3/ZrO_2, and ceramic superconductors.

A novel plasma technique whose antecedents go back nearly a century[48] has been developed. This technique—the reactive electrode submerged arc (RESA)

TABLE 4.2 Powder Synthesis Comparison

Process	Advantages	Disadvantages	Compositions
Carbothermal reduction	Process can be automated, some control of chemistry	Can be somewhat energy and capital cost-intensive; usually requires milling, which can produce impurities; large scaleup may be difficult; expensive raw materials	TiB_2, SiC, other nonoxides
Solid-solid, solid-gas, combustion synthesis	Reaction is usually self-propagating (requiring no external heat source) and is fast (within seconds)	Exothermic, volatile reactions; sometimes low density or low yields; densification may require high pressures; addition of dopants may be required	TiB_2, TiC, Si_3N_4, other nonoxides, composites
Vapor-phase synthesis	High purity; no aggregation; ease of preparation; narrow size distribution; versatility; homogeneity	Limited chemistry, ternary compounds difficult; low yields; reactant gases expensive	Oxides, nonoxides (nitrides, carbides), metals, binary compounds
Laser synthesis	Wide range of composition; short reaction times; uniform heating rates; improved process control; uniform size distribution; minimum agglomeration	Volatile reactants; expensive equipment; powder yields can be low; contamination may be a problem for certain reactions	Refractory materials (nitrides, borides, silicides, carbides), transition metal compounds, oxides with other emission lives
Plasma synthesis	Highly efficient, simple, continuous process; homogeneous mixtures; high surface areas; oxides have wide range of starting materials available; very fast quench rates	Requires high power (10^3 kW); large capital and operating costs; higher surface areas cause greater pyrophoricity; health hazards due to inhaling; carbides, nitrides are sensitive; low powder yields; some agglomeration, nonreproducibility	Oxides, carbides, nitrides, mixtures (SiO + AlN, SiC + SiC, Al + AlN)

Source: From Sheppard.[29]

TABLE 4.3 Plasma Synthesis of Ceramic Powders

Compound	Starting materials	Plasma type
Carbides		
SiC	CH_3SiCl_3	RF/Arc
SiC	$SiO_x + CH_4$	Arc
SiC	$SiH_4 + CH_4$	RF
WC	$W + C/W + CH_4$	Arc
WC	$W_3O + CH_4$	Arc
TiC	$TiCl_4 + CH_4 + H_2$	Arc
TaC	$Ta + CH_4$	RF
TaC	$TaCl_5 + CH_4 + H_2$	Arc
B_4C	$BCl_3 + CH_4 + H_2$	RF
Nitrides		
Si_3N_4	$SiCl_4 + NH_3 + H_2$	RF/Arc
Si_3N_4	$SiH_4 + NH_3$	RF
Si_3N_4	$Si + N_2/NH_3$	RF/Arc
AlN	$AlNH_3$	RF
TiN	$TiCl_4 + N_2 + H_2$	RF
TiN	$Ti + N_2$	RF
ZrN	$Zr + N_2$	RF/Arc
TaN	$Ta + N_2$	Arc
MgN	$Mg + N_2$	Arc
NbN	$Nb + N_2$	Arc
VN	$V + N_2$	Arc
HfN	$HfCl_4 + N_2 + H_2$	RF
BN	$BCl_3 + N_2 + H_2$	RF
Oxides		
Al_2O_3	$Al/AlCl_3 + O_2$	RF/Arc
Al_2O_3/Cr_2O_3	Al halide + O_2 + CrO_2Cl_2	RF
SiO_2	$SiCl_4 + O_2$	RF
SiO_2/Al_2O_3	$Si + Al + O_2$	RF
TiO_2, TiO_2/Cr_2O_3	$TiCl_4 + O_2 + CrO_2Cl_2$	RF
ZnO, Sb_2O_3, BaO, SiO_2, MgO	Oxides	Arc
MgO	$Mg(NO_3)_2$(aq)	RF
ZrO_2, ZrO_2/Al_2O_3	$Zr(NO_3)_2$(aq) + $Al(NO_3)_3$(aq)	RF
ZrO_2/SiO_2	$Zr(NO_3)_2$(aq) + silicone oil	RF

Source: From Sheppard.[29]

method—is based on a liquid approach. With this technique, metal electrodes are submerged in a dielectric fluid. An arc is struck, causing localized vaporization of the electrode and a reaction between the metal and the fluid vapor. The apparatus is shown in Fig. 4.2. The vapor and solid mixture quenches to extremely fine submicrometer spherical particles. The particles produced include Al_2O_3, ZrO_2, TiN, or Si_3N_4 suspended as a solution.

The use of laser-heated gas-phase reactors for ceramic powder synthesis has

TABLE 4.4 Advantages and Disadvantages of Glow-Discharge Technique

Advantages	Disadvantages
High purity and relative freedom from oxygen contamination under vacuum conditions	Further calcination needed prior to consolidating bulk bodies to shape
Powder stoichiometry variation and control facilitated in low-temperature reactors	Long reactor residence time
Novel metastable materials synthesis potential in nonequilibrium plasmas	
Submicrometer powders	
Simplified powder handling	

Source: After Anderson et al.[47]

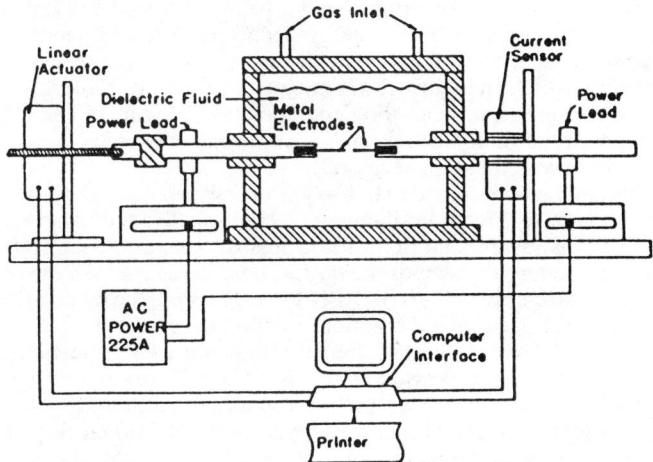

FIGURE 4.2 Schematic of microprocessor-controlled RESA apparatus for fine-powder preparation.[29]

advantages, including production flexibility, synthesis of submicrometer nonoxide powders, and relatively high purity. Disadvantages of the approach include relatively small primary particle sizes (which limits green density), high raw-material costs, and process complexity and uncertainty. Scaleup is difficult as a result of the small body of experience which exists about this type of processing. Using the direct coupling of laser radiation to gas-phase molecules, laser-driven reactions may have certain advantages over plasma synthesis and other gas-phase reactions. The laser energy produces a small, well-defined reaction zone. Using both high temperatures and short reaction times, the technique results in precision control of the powders' composition, size, and size distribution.[49]

The Massachusetts Institute of Technology has pioneered research in laser synthesis of powders. They have used a CO_2 laser beam that intersects the reactant gas stream orthogonally to produce Si powders that densify to 95 percent of theoretical density. These could be nitrided to yield reaction-bonded Si_3N_4 with 50- to 100-μm grain sizes. The fully nitrided material is harder, with a higher modulus of rupture (MOR), when compared to conventional materials. In addition, M.I.T. researchers[29] have improved the laser synthesis process to boost SiC yields to over 95 percent by weight. Such powders as Si_3N_4, boron, B_4C, Al_2O_3, and TiB_2 have been made. Similarly, French researchers, using M.I.T.'s orthogonal crossflow configuration, have synthesized SiC powders with yields greater than 99 percent from SiH_4–C_2H_2 mixtures, using a continuous-wave 600-W CO_2 laser.[29]

Other firms have produced B_4C powders from CO_2 laser driven pyrolysis of BCl_3–H_2–CH_4 (or C_2H_4) mixtures. The density was found to be greater than 98 percent theoretical; microhardness and elastic modulus were higher than for conventional B_4C; and fracture toughness and transverse rupture strength were similar to that of dense B_4C.[29]

Laser synthesis and the dissociation of chloromethylsilanes by ultraviolet excimer laser radiation have been shown to be promising methods of synthesizing fine ceramic powders. A potential for efficient usage of laser energy and high conversion of the feedstock to product have been demonstrated consistently in experiments. The mechanism of the energy-absorption process seems to be rapid decomposition of a fraction of the feedstock by the laser photons, generating new chemical species, which then absorb the majority of the laser energy.[50] Powders are formed by condensation from the decomposition products. The powder produced is dependent on the feedstock, being amorphous in some cases, with β-SiC being observed in the remainder (Fig. 4.3).

Reaction-bonded silicon nitride (RBSN) is usually made from commercial silicon powder about 5 to 15 μm in diameter.[51] Historically, the room-temperature strength of RBSN is lower than that of hot-pressed silicon nitride (HPSN) since the strength is limited by the size of the largest pore present.[51] On the other hand, high-temperature strength values of RBSN can exceed those of HPSN, since sintering aids are not used in the reaction-bonding process.[51,52]

High-strength RBSN has been produced by nitriding extremely fine (0.3-μm-diameter) high-purity silicon powder synthesized from laser-heated SiH_4. The 76 percent dense RBSN samples have resulted in average strength values that are over 75 percent greater than for samples made from traditional Si powder, with toughness and hardness about 10 percent greater. The specific strength and

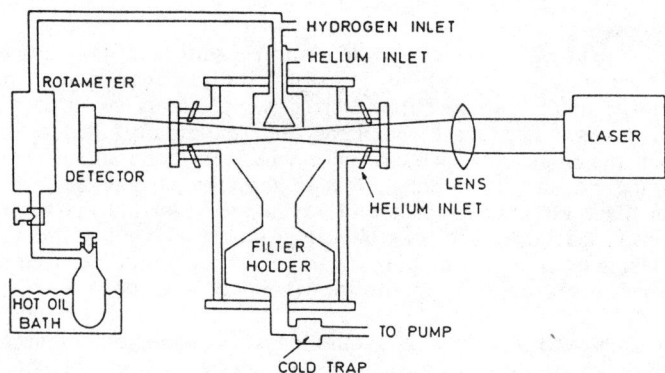

FIGURE 4.3 Powder production apparatus.[50]

toughness of this SiH_4-derived RBSN are superior to that of α-HPSN. Oxidation resistance is also significantly better than for conventional RBSN. The most important aspect of these results[53–55] reveals largely unexplored opportunities for other processes and materials that are capable of achieving highly perfect microstructures in bodies that need not be fully dense.

Recent interest in high-temperature structural ceramics has led to the development of chemical vapor-phase synthesis techniques for the formation of a variety of high-quality oxidic and nonoxidic ceramic powders. With regard to the gas-phase reaction process, chemical vapor precipitation (CVP)[56] of submicrometer Si_3N_4 powders has been produced by using thermal and laser activation.[57] Characteristic of the laser-driven gas-phase synthesis is a well-defined reaction zone which facilitates a considerable degree of control over composition, size, and size distribution of the product powders. In the laser-CVP of Si, SiC, and Si_3N_4, SiH_4 is invariably used as the Si reactant,[56,57] although in the traditional thermally activated CVP of powders and thin films of Si_3N_4, chlorinated silanes, that is, SiH_2Cl_2 and $SiCl_4$, are frequently used.

From the network described[58] it is clear that Si and Si_3N_4 can be made from SiH_2Cl_2, and Si_3N_4 from Si_3N_4 with NH_3, using laser-chemical vapor precipitation (L-CVP) with a sensitizer. The particle sizes are similar to those reported using SiH_4 and NH_3 as reactants. The synthesis of Si and Si_3N_4 powder by L-CVP in a continuous-flow reactor using chlorinated SiH_4 reactants is a new step in ceramic powder synthesis.

Other vapor-phase methods briefly discussed earlier include the following:

1. Ceramic powders produced by intraparticle reaction of droplets generated by cooling a hot vapor precursor, such as a metal alkoxide. An advantage of this approach is its ability to produce nearly monodisperse particles of a single material or of coated particles. Disadvantages of the process include high raw-material costs. Scaleup is difficult because of the relatively small technical knowledge base for the process, and volatile precursors are required.[47]

2. Heated-wall reactors are a versatile method for ceramic powder generation. Gaseous reactants are introduced into a heated reactor where reactions take place to produce condensable species which nucleate to form particles. Advantages of the process include simplicity (only a gas-handling system, furnace, and powder-collection system are required) and versatility (a variety of oxide and nonoxide powders can be produced). Disadvantages include the requirement of volatile precursors, and the fact that only synthesis of single-component systems has been demonstrated.[47]

3. Spray drying can be considered to be a vapor-phase processing method and is applied on a commercial scale for ceramic powder production. However, spray drying typically results in the production of hard agglomerates with sizes greater than 1 μm. Control of chemical homogeneity depends on the solution-processing steps undertaken before spray drying, and control of the spray-drying process variables primarily affects only the agglomerate size and size distribution.

Spray drying is the process by which a fluid feed material is transformed into a dry powder by spraying the feed into a hot drying medium.[59] When making ceramic powders, the feed material is usually a water-based suspension, often called a slurry, and the drying medium is heated air. The slurry is pumped to an atomizer located in the drying chamber and broken into a large number of droplets which quickly achieve a spherical shape because of surface tension effects. Properly done, spray drying is an economical and continuous operation that produces powder of uniform and repeatable character. The majority of spray-dried ceramic powders are generated from water-based slurries.

In summary, spray drying is capable of producing a granulated, highly flowable powder of uniform bulk density and controlled moisture content. The slurry character is the major factor controlling the bulk density of the granules. Higher solids content will result in higher granule densities. A thorough characterization of the feed material and the powder is required for process optimization.[60]

Carbothermal. TiB and SiC powders have been produced using carbothermal processes. By blending TiO_2, C, and 98.5 percent pure B_4C powder together before agglomerating, mixing, and then drying, the agglomerated blend is charged into a continuous furnace and the product output is TiB_2. Another manufacturer plans to use a similar process and apply the technology to making B_4C powder.[29] β-SiC powder also can be made via carbothermal reduction.

Nitrides, such as Si_3N_4 and AlN, can be produced via thermal decomposition. A Japanese firm reacts silicon tetrachloride ($SiCl_4$) with ammonia (NH_3) at lower temperatures to give silicon di-imide and ammonium chloride. An intermediate compound, Si_2N_3H, is then heated to 1500°C, which crystallizes into α-Si_3N_4 powder. The powder consists of agglomerates, which are milled into a finer powder of equiaxed particles. Another firm nitrides highly pure aluminum metal under carefully controlled conditions in a special furnace to produce AlN.[29]

The Tokyo Institute of Technology[61] has developed a carboreduction method using arc image heating for preparing 1-μm-diameter TiN and TiC powders from starting materials of titania and graphite. By using an arc-imaging furnace, the heating times and temperatures are reduced significantly. For instance, single-phase TiN was synthesized within 30 s, while in an argon atmosphere, single-phase TiC was prepared. In Yugoslavia[61] researchers used carbothermal reduction for synthesizing SiC powders from carbon-black-doped silica gel. This gel was derived from colloid silica.

Included in the carbothermal processes is self-propagating high-temperature synthesis (SHS). This process of making nonoxide materials (such as carbides, nitrides, borides, chalcogenides, hydrides, intermetallic compounds, and sulfides) is by combustion synthesis. This self-propagating high-temperature synthesis, originally developed in the U.S.S.R., usually involves exothermic reactions above 2500°C. The combustion process itself can be stable or unstable and does not require a furnace. The high temperatures remove any volatile contaminants by vaporization (Fig. 4.4).

Combustion synthesis can involve several different types of reactions. For example, an oxidation reduction, or thermite, can produce multiphase compositions such as cermets. Two starting metallic elements—titanium and boron, for example—can be combined to make TiB_2.[62] There are also combinations of these two types, as well as reactions requiring chemical activators. Preheating is required when the heat of reaction is insufficient for liquid-phase diffusion. The exothermic reaction can occur simultaneously throughout the whole material or may be propagating, where a synthesis wave passes through the material. For example, the reaction of titanium and carbon powders to form TiC releases 737 cal/g, with a calculated adiabatic temperature (no heat moved in or out of the system) of 3240°C. Strongly exothermic reactions like this have been used as sources of energy in many kinds of practical applications. For example, "thermite" reactions have been used to weld metals, as in the welding of cracks in railroad rails. During the past decade the work of a group of Soviet scientists[63-65] has led to the

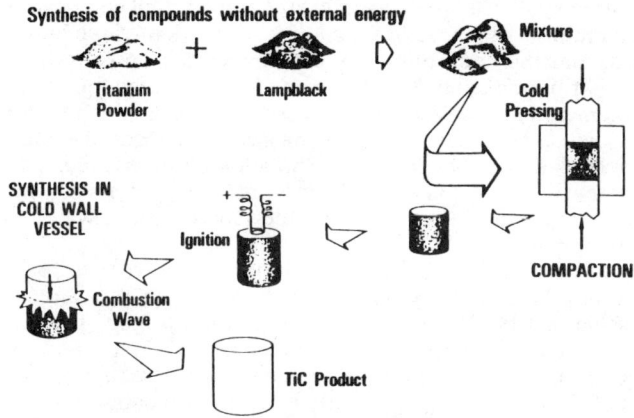

FIGURE 4.4 Self-propagating high-temperature synthesis.[67]

development of a process for the synthesis of refractory materials. To illustrate the process, we refer to the reaction between transition metals and carbon powders. Cold-pressed compacts of these mixed powders are ignited at one end in an argon atmosphere or in a vacuum by means of hot metal wires, pulsed lasers, electric arcs, or high-intensity lamps (Table 4.5). Upon ignition of the compact, a combustion wave rapidly (several seconds) propagates through the mass, converting the reactants to metal carbides. Besides carbides, β-sialon, and microcomposites, $SiC + Al_2O_3$ and $TiC + Al_2O_3$ have been formed.

As a class of refractory materials, the nitrides, such as AlN, TiN, ZrN, HfN, TaN, BN, and Si_3N_4, have all been formed by the SHS process. The process for synthesizing nitrides proceeds in a fairly predictable manner. The metal powders are cold-pressed into porous compacts. Subsequently these compacts are ignited in nitrogen gas or liquid nitrogen by a resistance-heated tungsten wire. The self-propagating combustion wave quickly transforms the metal powder into porous solid nitrides.

TABLE 4.5 Typical SHS Reactions

Simple binary compounds*:
  $Ti(s) + C(s) \rightarrow TiC(s)$
  $2Ti(s) + N_2(lg) \rightarrow 2TiN(s)$

Compositions more complex than binaries:
  $Ti(s) + 0.7C(s) + 0.3N(l) + Ti(CN)(s)$

Cermet compositions:
  $wTi(s) + xB(s) + yC(s) + zCu(s) \rightarrow lTiB_2(s) + mTiC(s) + nCu(s)$

Thermitelike reactions:
  $3TiO_2(s) + 3C(s) + 4Al(s) \rightarrow 3TiC(s) + 2Al_2O_3(s)$

Reactions with chemical activators:
  $3Ti(s) + NaN_3(s) \rightarrow 3TiN + Na(g)$

*g—gas, l—liquid, s—solid.
Source: From Frankhouser.[67]

In combustion synthesis, both particle size and starting composition affect the reaction temperature. The heat of reaction is a function of the amount of liquid phase formed, and this fact ultimately determines the porosity of the product. Particle-size distribution controls initiation and rates of reaction, while increasing the starting-element particle size can decrease the propagation rate for certain reactions. The final particle size, as well as phase percentages, are functions of the starting particle size of either reactant. The addition of dopants can reduce porosity and increase density, and the addition of inert materials, which have the intended final composition, can decrease the reaction rate, wave velocity, and combustion temperature.[16,66]

Since the final product is often in a porous powder form, subsequent densification may be necessary. Such methods include pressureless sintering, uniaxial pressing, isostatic pressing, casting, and hot rolling. The reactor capacities available vary from 2.5 to 30 liters, and a block of three 16-liter reactors can average 90 kg/h of product in continuous production. Production at the pilot facility in the U.S.S.R.[67,68] for TiC recently reached a level of more than 1000 tons per year.

During the course of the Soviet development of SHS technology, various techniques have evolved to control the synthesis reactions. Of four such innovations, one is used to slow down and three to intensify the reactions. A brief description follows.

1. Kinetic braking is used to reduce SHS reaction intensity. This is accomplished by inclusion of the previously reacted product within the mixture of reactants (sometimes in portions as large as 65 percent of the total). This process variation has been used in the production of AlN and TiB_2.
2. Thermal explosion is used to ignite some SHS reactions when a higher-reaction temperature is desired. The mass of reactants is simply preheated (usually to 149 to 319°C) until it self-ignites (and continues therefrom by self-propagation).
3. Chemical furnace is a means of identifying some SHS reactions, especially when a high mass element (such as W) is the metal reactant. A blanket of material that attains a more intense reaction than the desired product surrounds it and serves as a reaction booster.
4. Chemical activators are added to various reactant mixtures to intensify the reactions. One example is the use of an oxide and metallic reducing agent. This technique has been used in Soviet SHS technology to produce cast (and dense) ceramic shapes.

Finally, reaction hot pressing (RHP) is a process that utilizes SHS reactions in situ in a uniaxial hot press to form densified ceramic products. RHP takes advantage of the favorable thermodynamics of SHS reactions to rapidly form and densify product phases (Fig. 4.5). This process offers the potential for the formation of unique phase assemblages and cost-effective conditions for the production of standard materials.[69–71] Among the variety of RHP materials produced are TiB_2–Al_2O_3, TiB_2–Al_2O_3–Si_3N_4, and Al_2O_3–TiC. A number of these have been prototyped and tested in various applications, including metal cutting inserts, pump seals, and bearings. One RHP product, TiB_2–Al, has been successfully developed for aluminum electrorefining applications.

Hydrothermal. Techniques somewhat intermediate between the solid-state reaction of mixed-oxide particles and the special chemical techniques may offer a

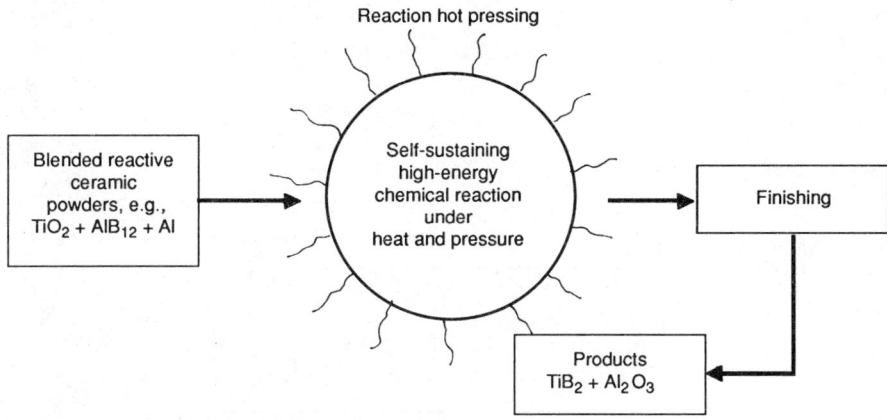

FIGURE 4.5 Schematic of reaction hot-pressing process.[69]

compromise between added expense and improved homogeneity. Mixtures of oxide powders or gels may form crystalline mixed-hydroxide compounds when heated 180 to 700°C in the presence of steam having a pressure of 1 to 100 MPa. A mixed-oxide compound of fine grain size is formed on calcining to a relatively low temperature. This technique, called hydrothermal synthesis, has been used to produce a variety of mixed-oxide compounds.[72]

Decomposition of a salt commonly produces a relatively fine reactive product. Silicon nitride and silicon aluminum oxynitride have been produced by heating discrete oxide particles in a nitrogen atmosphere at high temperature.

ZrO_2 and Y-PSZ powders have been prepared by the hydrothermal method.[73] Among the hydrothermal precipitation, hydrothermal crystallization, hydrothermal hydrolysis, and hydrothermal oxidation methods, hydrothermal precipitation is an excellent method for producing Y-PSZ powder. The content of tetragonal ZrO_2 was 75 percent, relative density was over 99 percent, and the bending strength was 920 to 1050 MPa.[74] Figure 4.6 shows the process flow in producing powder by the hydrothermal homogeneous precipitation method.

The hydrothermal crystallization technique can be used alone or in combination with supercritical drying. For instance, researchers have prepared pure and stabilized aluminum/titanium oxide powders by coprecipitation and hydrothermal crystallization of aluminum and titanium salts. Aluminum titanate was then formed by reaction sintering.[61]

A joint project was undertaken by scientists from Iceland, Sweden, Finland, and Denmark, which involved the hydrothermal synthesis of yttria-stabilized zirconia powders. Although such a process can produce particle sizes ranging from 10 to 20 nm, such sizes are often too small for certain forming processes, such as hot isostatic pressing. To minimize this problem, hydrothermally produced powders have been dried supercritically. In this drying method, the liquid and precipitate are heated in an autoclave to a temperature and pressure above the precipitate's critical point. Pressure is then reduced to atmospheric pressure and the precipitates produced in aqueous solutions are washed and slurried in methanol (Fig. 4.7).[8]

Polymer Pyrolysis. Researchers around the world are working on a variety of techniques, processes, and methods, the goal being to develop a process that

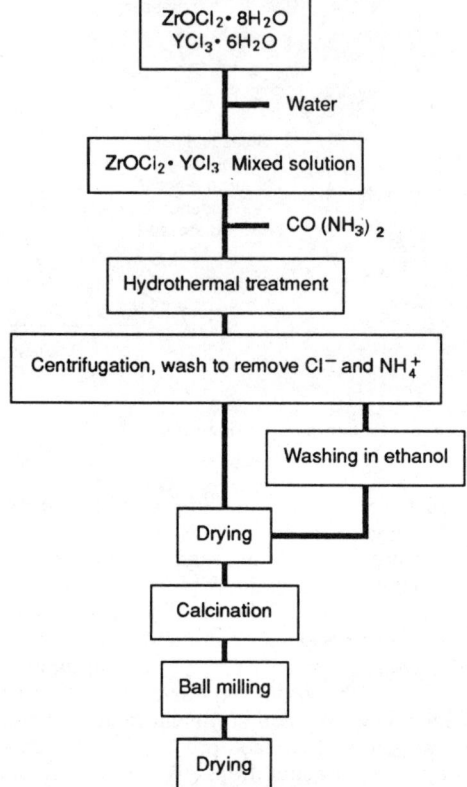

FIGURE 4.6 Process flowchart of hydrothermal homogeneous precipitation method.[73,74]

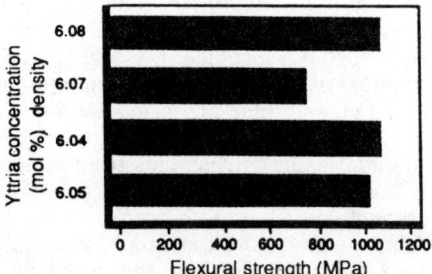

FIGURE 4.7 Hydrothermal process has produced materials with good flexure strength.[61]

enables the manufacture of ceramics more efficiently than conventional processes. The need for soluble or fusible precursors whose pyrolysis will give the desired ceramic material has led to a new area of macromolecular science, that of preceramic polymer chemistry.[75–77] Ceramic powders by themselves are difficult to form into bulk bodies of complex shape. Preceramic polymers could, in principle, serve in such applications, either as the sole material from which the shaped body is fabricated, as a matrix in a shaped composite, or as a binder for the ceramic powder from which the shaped body is to be made. In either case, pyrolysis of the green body would then convert the polymer ceramic material, hopefully of the desired composition.

Ceramic fibers of diverse chemical compositions are sought for application in the production of metal-, ceramic-, glass-, and polymer-matrix composites.[77] The presence of such ceramic fibers in a matrix, provided they have the right length-to-diameter ratio and are distributed uniformly throughout the matrix, can result in very considerable increases in the strength (that is, fracture toughness) of the resulting material. To prepare such ceramic fibers, a suitable polymeric precursor is needed, one which can be spun by melt-spinning, dry-spinning, or wet-spinning techniques into fibers which then can be pyrolyzed (with or without a prior cure step).

In order to have a useful preceramic polymer, considerations of structure and reactivity are of paramount importance. Not every inorganic or organometallic polymer will be a useful preceramic polymer. Although preceramic polymers are potentially "high value" products if the desired properties result from their use, the more generally useful and practical systems will be those based on commercially available, relatively inexpensive starting monomers. Preferably, the polymer synthesis should involve simple, easily effected chemistry which proceeds in high yield. The preceramic polymer itself should be liquid, or, if solid, it should be fusible or soluble in at least some organic solvents, that is, it should be processable. Its pyrolysis should provide a high yield of ceramic residue, and the pyrolysis volatiles preferably should be nonhazardous and nontoxic.[77] The key to one method is the use of polysilazane precursors that allow researchers to handle ceramics more like pliable plastics than solid materials. Called amorphous covalent ceramics, amorphous materials are achieved by heating a precursor solution instead of by cooling a melted phase, the usual method. The amorphous stage can be final, or it can be the intermediate toward a polycrystalline or crystalline state.

Polysilazane precursors have been produced in Si_3N_4 and WC coatings in only 20 min, instead of the usual several hours, and at temperatures of 450 to 800°C instead of the typical 1700°C. Although a high content of polysilazane (38 percent) produces a high green density (62 percent of the theoretical value), this density does not change after pyrolysis. However, the material has minimal shrinkage during sintering and has mechanical properties similar to those of conventionally produced Si_3N_4. Strength has been found to be dependent on the heating temperature in a nitrogen atmosphere. The process starts with a liquid phase of organometallic polymers such as polycarbosilane or aminoborazine. Organic fractions are heated away and the result is an inorganic polymer of amorphous carbide or nitride. The cooled, solidified solution stays amorphous and is highly covalent. When the material is heated to its final state, it maintains thermal stability to 1200°C and even higher. SiC, Si_3N_4, oxycarbide, molecular composite carbides of silicon and titanium, and silicon and aluminum have been produced by this technique.

Another group of scientists have synthesized polyaminoalane precursors by anodic dissolution of metallic aluminum in an electrolyte. The latter contains a primary amine, acetonitrile, as solvent and a tetraalkylammonium salt as supporting electrolyte. The solution is then heated to 750 to 1100°C, which polymerizes the aminoalane compounds and removes the solvent and excess amine. Heating under vacuum produces a syrupy fluid, which is finally converted to a polymeric powder. Calcination of this powder under ammonia up to 750°C leads to completely amorphous products. At 850°C, crystalline aluminum nitride is obtained.

These "molecular and polymeric" (organometallic precursor) systems,[78,79] through variations in precursor preparation and conversion chemistry, provide an opportunity to (1) control the product purity and microstructure (better and more consistent properties), (2) fabricate difficult-to-obtain shapes such as fibers, and (3) allow the possibility of lower temperature processing. The use of such precursors in the preparation of covalent nonoxide solid solutions is a new and unexplored area. The methodology used involves the copyrolysis of molecularly mixed precursors to SiC and AlN. This methodology, when applied to systems such as SiC–AlN, may provide the ability to control the chemistry at the molecular level and produce materials with high homogeneity and purity. The chief source of AlN precursors for this work has been from the reaction of various trialkylaluminum compounds with ammonia.[79,80]

Markedly different chemistry has been observed for pyrolyses of polycarbosilanes and mixed polycarbosilane-AlN precursors under N_2 and NH_3 atmospheres at 350 to 1000°C. Ammonia plays a key role in removing organic groups from SiC precursors to produce Si_3N_4. Homogeneous SiC–AlN solid solutions can be formed via organometallic precursors, and this system shows great promise in providing routes to both SiC–AlN solid solutions and high-purity Si_3N_4 from common organometallic precursors.[79]

Potentially, silicon oxynitride can be just as good a ceramic as silicon nitride. The major advantage of Si_2ON_2 prepared from a hybrid polymer over the conventional methods is that stoichiometric Si_2ON_2 can be obtained at low temperatures with high purity and without minor phases. The hybrid polymer can be prepared by readily available (and relatively cheap) polysiloxane and polysilazane, and its composition can be easily modified by varying the mixing ratio of these two precursor oligomers.[80,81] High ceramic-conversion yield has made these polymers ideal for ceramic-ceramic composite fabrication, and it was found that Si_2ON_2 and carbon fibers are potentially good combinations for ceramic-ceramic composites. The absence of strong bonding between carbon fibers and Si_2ON_2 matrix should provide greater toughness for these composites. Through the reaction of hybrid polymers with aluminum hydroxides, polymer precursor systems can be extended from the Si_2ON_2 to the Si–Al–O–N system.[81]

Organometallic/metal organic chemicals have been used to synthesize molecular composites of SiC–SiO_2, SiC–Al_2O_3, and SiC–TiC. This is a potential method for producing mixed oxide and nonoxide composite powders at low temperature with molecular-level homogeneity, using the combination of crosslinked polysilastyrene (PSS) and metal alkoxides. First, the technique should provide high-purity composite powders which are sinterable at lower temperatures, that is, those compositions difficult to sinter otherwise.[82,83] Second, the test results[82] suggest that the concept of mixing two precursors of a composite at a molecular level may be applied to many other systems as well. Testing of the mechanical properties of these mixed composites will determine whether advantages in physical properties are achieved by this molecular chemistry processing method.

The pyrolysis of polymers has attracted more and more attention as a potential route to nonoxide ceramics, such as SiC and Si_3N_4. The method is now to be extended to hexagonal boron nitride (hBN) ceramics. hBN is structurally similar to graphite, but has better resistance to high temperatures, electricity, and chemical attack. Potential applications include high-temperature crucibles, industrial abrasives, and electronic packing materials. Researchers have concentrated on short- and long-chain compounds based on aminoboranes, borazenes, and borazanes. They prepared several preceramic oligomeric gels by cross-linking starting compounds and swelled and consolidated gels, foams, and films, which were then pyrolized into amorphous and microcrystalline hBN.

Cost is the instigating factor behind the extensive efforts made in investigating pyrolysis methods. hBN powders are usually made by high-temperature reaction of boric acid and nitrogen-containing bases, then sintered at high temperature into finished parts. Both steps are energy-intensive. hBN parts can also be made by CVD, which tends to be expensive and is limited to the production of low-density foams and fibers. Pyrolysis, on the other hand, is more convenient and versatile.

Sol-Gel. Sol-gel techniques, such as precipitation of hydroxides from salt solutions (chlorides, nitrates, sulfates) by the addition of ammonia, or the hydrolysis of metal alkoxides, have been in development for 15 years. The target has been to find a powder preparation method that offers both excellent chemical homogeneity and small particle size. The worldwide goal of all sol-gel work has been and remains ultrahomogeneity or nanoheterogeneity.[84-88]

Gel processes have two common characteristics: (1) they are used to convert metal salt solutions into ceramic solids, and (2) the gel structure and the gel properties are of controlling importance to the preparation and properties of the solid ceramic products. While the basic particles are colloidal in size, the gel form of the colloidal particles can resist shearing forces. A gel may be described as a rigid or solid form of a colloid.

A sol is a colloidal state in which the solid matter is composed either of small particles or of macromolecules. A gel differs from the sol in that it is characterized by a chemical and structural three-dimensional network. This network is formed and the connectedness established at a point called the gel point. The interconnection may be realized by forming bridges between the colloidal particles of a sol, or by developing cross-links between smaller polymeric units in solution. The gels obtained in the two cases are termed colloidal and polymeric gels,[89] respectively.

Processes designated as sol-gel and gel-sphere processes are subdivisions of the more general division of gel processes. When a stable sol (a colloidal dispersion in a liquid medium) is prepared and followed by a gelation step, the overall process may be called a sol-gel process. The term sol-gel has been used incorrectly for processes that do not involve a sol. The term gel-sphere refers to the formation of the gel from liquid drops, which may be solutions or sols, to give solid spheres that resemble tiny beads. Some gel processes are either sol-gel or the gel-sphere, some are both types, and some are neither.

A general outline of sol-gel processes includes (1) the chemistry of precursors; (2) the resultant dispersed sol, or a solution; (3) gelation or precipitation to a shape which may range from powders to monoliths and includes coatings and fibers; (4) the drying stage; and (5) densification, and in some cases crystallization to a final ceramic product (Fig. 4.8). The sol-gel process begins with a reactive precursor material that is converted into a final product by chemical and thermal means. The precursor is prepared in a colloidal suspension or solution (sol) that undergoes a gelling stage (gel) and a final fusion.[86]

Sol-gel processes can proceed from chemical precursors of various forms. The precursors most often used are (1) the metal salts and (2) the metal alkoxides. The

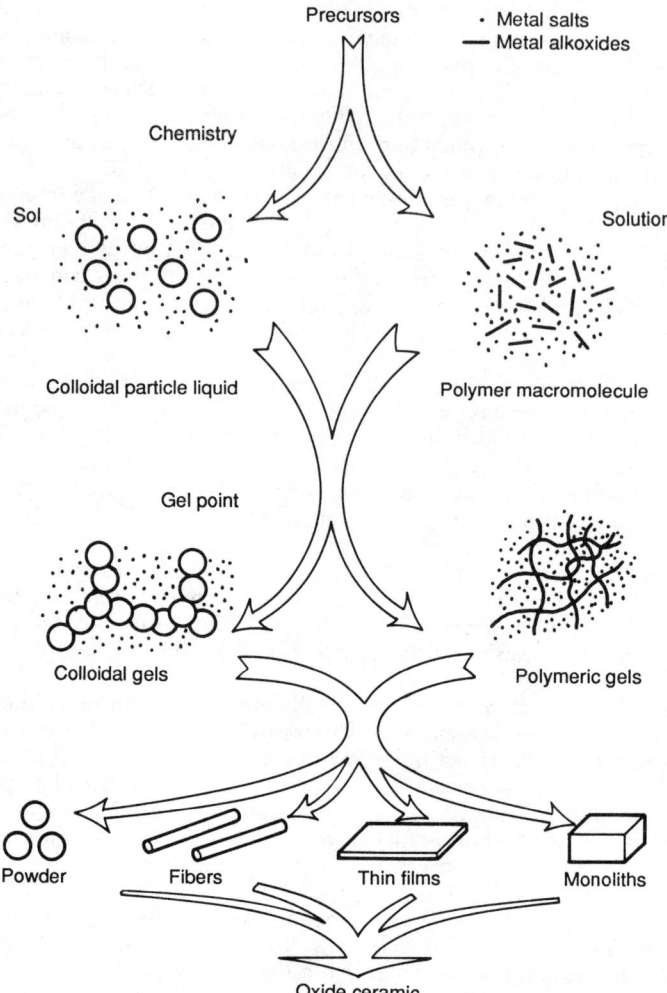

FIGURE 4.8 Map of sol-gel process routes.[87,89]

metal salts are generally cheaper. Their chemistry can lead to various states of polymerization, depending on the conditions. This complexity is increased further by the anions that constitute impurities, which must be eliminated or which can affect the processing and properties of the resulting materials. The metal alkoxides are generally more expensive than metal salts, but they allow one to obtain high-purity materials; and they have a chemistry which is also very rich.[36,89,90]

The characteristics of the dried gels identify the gel processes and their advantages as compared with other conventional ceramic processes. The dried gels are porous solids of high surface area. The gel processes start with solutions of high purity that can be easily mixed to give homogeneous compositions.[85]

The gel processes can be grouped and described in terms of the reactions that change sols into gels (Fig. 4.9). One classification is known as external gelations, with reactions that involve mass transfer to a second phase or fluid, as compared with internal gelations without mass transfer.[16] Internal gelation is more difficult to control, but the mass transfer during external gelation can result in nonhomogeneous or shell-like structures and can also limit the gelation rates.

The six gelation processes shown schematically in Fig. 4.9 are described in greater detail in Table 4.6. The starting metal-salt solutions, the solid gel with low concentrations of other solutes, and the ceramic product are common to all processes. The preparation of feeds and the gelation procedures show large differences. Four of the processes include washing operations that generate liquid wastes for disposal and other similar problems, while two do not require washing. The thermal treatments are critical and differ greatly, depending on the gel properties before treatment and the desired product properties.[85]

The most widely studied sol-gel processes in recent years are internal gelation methods (type III in Table 4.6), in which metal alkoxides are hydrolyzed—chemically decomposed by splitting chemical bonds and adding the elements of water. In these processes, the organic metal salts are dissolved in an organic solvent such as ethanol.[16,35,78,91] Type IV is an external gelation method[35,91] where water is removed from aqueous sols and put into an immiscible alcohol, and gel spheres are formed.[78,85,91]

Sol-gel processes take advantage of the high degree of homogeneity that can be achieved by mixing on the colloidal scale and the excellent bonding and sintering properties of colloids resulting from their very high specific surface energy. The feasibility of sol-gel processes is used for synthesizing powders of tetragonal zirconia–alumina composites and coating whiskers to control their interface properties to matrix phases. These are just two examples which demonstrate the application of sol-gel processing.

In other work it has been demonstrated that up to 14 vol % unstabilized tetragonal ZrO_2 can be retained in an alumina matrix by using a sol-gel system in which the boehmite (AlOOH) to α-alumina transformation is seeded with α-alumina. This result is attributed to the homogeneous mixing of two sol systems and the development of a uniform, aggregate-free, fine-grain α-alumina after transformation of the boehmite gel to α-alumina.[92]

To improve the transformation toughening of sialon, mullite, and alumina and, therefore, enhance the high-temperature strength and fracture toughness of these transformation-toughened ceramics, an interesting approach was used utilizing sol-gel processing.[40] The approach was to use a ZrO_2–HfO_2 solid solution in place of ZrO_2 as the toughening addition to these materials to preserve the transformation-toughening effect at temperatures up to 1370°C.

To obtain the ZrO_2–HfO_2 powder, zirconia and hafnium oxychlorides were dissolved together to form a gel. This gel was washed, dried, and sintered to produce the powder, combined with the matrix material, cold-pressed, and sintered at 1620°C for 0.5 to 3 h.

Tests show that the K_{Ic} increases significantly with an increase of the retained tetragonal fraction of total ZrO_2 and HfO_2(ZH). An alumina matrix with 15 vol % ZH had a retained fraction of 0.36 and a K_{Ic} of 4.58.[40] In Japan[61] a sol-gel method is being used to produce alumina sheets for multilayer ceramic capacitors up to 100 μm thick and having a flexural strength of 5300 kg/cm^2. After firing, the sheets reach almost full density and show no porosity, and microwave drying is used to control the shrinkage before firing.

Other researchers at Nishi Tokyo University are using sol-gel to homoge-

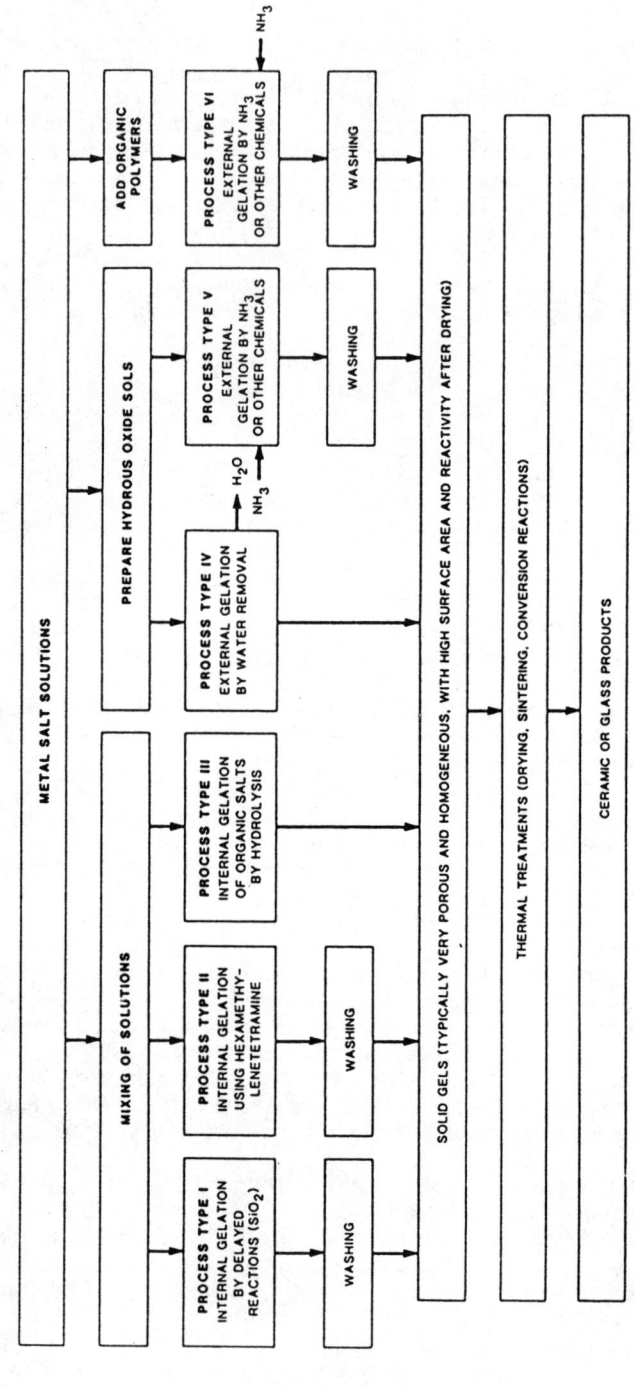

FIGURE 4.9 Types of gel processes.[85]

TABLE 4.6 Characteristics of Gel Processes

Characteristic	I	II	III	IV	V	VI
Type of gelation	Internal	Internal	Internal	External	External	External
Feed to gelation	Sodium silicate plus acidic solution	Metal salt solution plus HMTA* and urea	Metal alkoxides in organic liquids	Aqueous sols (hydrous oxides)	Aqueous sols (hydrous oxides)	Aqueous solution with organic polymers
Gelation mechanism	Slow formation of silicate gels	Reactions of HMTA	Hydrolysis of metal alkoxides	Removal of H_2O	Reaction with NH_4OH	Reaction with NH_4OH
Difficult or troublesome process steps	Control of gelation	Washing; disposal of liquid wastes	Preparation (or cost) of feed; drying	Preparation of sol	Preparation of sol; washing	Washing; thermal decomposition of organic polymer
Typical example and product composition	Sodium silicate mixed with $Al_2(SO_4)_3$–H_2SO_4 solution; SiO_2–Al_2O_3	$UO_2(NO_3)_2$–HMTA–urea solution drops into hot trichloroethylene; UO_3 or UO_2 spheres	$Si(OC_2H_5)_4$ in ethyl alcohol plus H_2O; SiO_2 glasses	ThO_2 sol drops in 2-ethyl-1-hexanol; ThO_2 spheres	ThO_2 sol drops in NH_3 gas over NH_4OH solution; ThO_2 spheres	Solution drops of $UO_2(NO_3)_2$–polyvinyl alcohol with NH_3 over NH_4OH solution; UO_2 spheres

*HMTA—hexamethylenetetramine [$(CH_2)_6N_4$] hydrolyzes in acidic solutions as follows: $(CH_2)_6N_4 + 6H_2O + 4H^+ \rightleftarrows 4NH_4^+ + 6HCHO$.
Source: From Haas.[85]

neously disperse zirconia in mullite to enhance fracture toughness. Room-temperature flexural strength and fracture toughness values were 500 MPa and 4.3 MPa · m$^{1/2}$, respectively. These properties were retained up to 1000°C and then decreased to 350 MPa and 2.5 MPa · m$^{1/2}$ at 1300°C. Similar work at Nagoya University[61] has involved the preparation of zirconia-oxide composites with a surface-modification technique. Dispersed oxide particles are coated with zirconium alkoxide by preferential hydrolysis. The technique has been applied to the zirconia-mullite system with similar results, as well as in numerous laboratories in several countries.[93-98]

A novel sol-gel ceramic composite process combines two toughening methods—zirconium oxide phase transformation and silicon carbide whisker reinforcement. The resulting composite has high oxidation resistance, modulus, and strength, and exceptionally good static and dynamic toughness. ZrO_2 is also produced as a sol-gel, and in a second novel processing twist, it is applied as a thin film onto the surface of alumina particles. This creates a fine-grained phase transformation boundary interface that absorbs shocks. The SiC reinforcing whiskers added to the composite act as a shock deflection barrier. The resulting sol-gel can be cast to shape, injection-molded, or dried and the powders consolidated by conventional means.

The applications described as well as future gel processes always require empirical testing. Whenever a new application is first considered, it may seem logical to use existing procedures and equipment with as few changes as possible. This conservative approach, however, may be unwise because other process conditions and equipment may be more suitable for an application.

The first step toward the selection of a gel process for a particular application is to clearly define the product properties needed and their relative order of importance. Unless some of the important product properties are derived from the unique properties of gels, the use of a gel process is probably not justified. Other ceramic powder preparation processes, such as precipitation or thermal decomposition, are probably more economical or practical if they can provide all of the product properties needed. Among the product property requirements that may justify the use of a gel process are the following:

Good ceramic sinterability (to near theoretical density at relatively low temperatures)

High chemical reactivity or surface area

Mixed composition

Homogeneity of composition or structure

Control of shape (spheres, film)

Control of microstructure by gelation, aging, and calcination conditions

Control of impurities or chemical compositions

Control of particle size, such as sphere diameter

The suitability ratings of the gel processes are given in Table 4.7. The ranking of product properties needed should be compared with the suitability of gel processes and product compositions.[85] Good ceramic sinterability and high gel-surface areas are common to all six types of processes listed, and low impurities and homogeneous compositions or structures are usually possible.

A ceramic abrasive grinding wheel that can remove metal 50 percent faster than standard wheels has enabled workers to increase part production fivefold.

TABLE 4.7 Suitability Ratings of Gel Processes

	Internal, silicate gels I	Internal, using HMTA* and urea II	Internal, hydrolysis of metal alkoxides III	External, removal of H$_2$O from sol IV	External, sol + NH$_3$ V	External, NH$_3$ + solution with organic polymers VI
Good ceramic sinterability	Fair	Excellent	Excellent	Excellent	Good	Fair
High surface area	Excellent	Excellent	Excellent	Good	Good	Excellent
Mixed compositions	Fair	Good	Excellent	Fair	Fair	Good
Homogeneous compositions or structure	Good	Excellent	Good	Excellent	Good	Fair
Spherical shape	Good	Excellent	Poor	Excellent	Good	Good
Formation of films	Poor	Poor	Excellent	Fair	Poor	Poor
Control of microstructure	Fair	Excellent	Excellent	Good	Fair	Poor
Low impurities	Fair	Good	Excellent	Good	Good	Good
Control of particle size	Fair	Good	Fair	Good	Good	Fair
Economical (feed cost, process cost)	Excellent	Fair	Poor	Good	Good	Excellent
Technical practicality or difficulty	Excellent	Good	Fair	Fair	Good	Fair
Reported product compositions	SiO$_2$, SiO$_2$–Al$_2$O$_3$, and other silicates	UO$_3$, UO$_3$–PuO$_2$, ThO$_2$–UO$_3$, ZrO$_2$, UO$_3$–C, Al$_2$O$_3$, and PuO$_2$–C	SiO$_2$, SiC, Al$_2$O$_3$, TiO$_2$, ZrO$_2$, Ta$_2$O$_5$, etc., and mixed oxides	ThO$_2$, UO$_2$, UO$_2$–PuO$_2$, rare-earth oxides, ZrO$_2$, and ThO$_2$–C	ThO$_2$, ThO$_2$–UO$_3$, and UO$_2$–PuO$_2$	ThO$_2$, UO$_3$, UO$_3$–PuO$_2$, and UO$_3$–C

*HMTA—hexamethylenetetramine [(CH$_2$)$_6$N$_4$].
Source: From Haas.[85]

The new grinding-wheel material was produced by sol-gel processing, and it is believed that the wheel life of the ceramic material is much higher than for standard wheels. This wet chemistry solution versus the traditional dry ceramic processes is the first innovation in glass abrasive materials in more than 40 years. It should be noted that the strength of the material and the fine particle shape are expected to give it an advantage over standard grinding materials.

Furthermore, researchers are producing dense abrasive grit of composite nonoxide ceramics by a combination of sol-gel and reaction-sintering techniques.[99,100] The achievement opens the way to a wider choice of small abrasive grits suitable for the high temperatures of grinding. The researchers have produced composites of titanium nitride dispersed in a matrix of either aluminum nitride or γ-aluminum oxynitride. Starting with an alumina sol, obtained by dispersing boehmite powder into water, they seed the sol by milling a portion of it with aluminum oxide to create α-Al_2O_3 crystallites. Next, they disperse carbon black and titanium oxide into the sol and gel, dry, crush, screen, and calcine the result. The gel is reaction-sintered at 1900°C in an N_2 atmosphere. Density of the finished grit is >90 percent. The titanium nitride in the matrix forms by carbothermal reduction of the TiO_2. Whether the matrix around the titanium nitride becomes aluminum nitride or γ-aluminum oxynitride depends on the carbon content of the sol.

Special Processes. There are several other processes which are either laboratory curiosities, experimental, or not completely accepted as a full-blown process for producing ceramic powders.

1. *Chemical vapor deposition (CVD):* CVD is a process which was originally developed as a means of coating a substrate by an elevated-temperature gas-phase reaction. For producing SiC, the most common method involves thermal decomposition of methyltrichlorosilane (CH_3SiCl_3) to SiC at a temperature of about 1000°C, and SiC vapor deposits as a solid onto the substrate. This principle has been applied successfully to building up a matrix material through "chemical vapor infiltration" of a porous preform, with the result that a composite material is produced. Appropriate manipulation of temperature and pressure gradients is the key to attaining a product of reasonable density within a reasonable time period.[101,102] In the CVD process, a controlled flow of hydrogen mixes with heated methyltrichlorosilane (MTS) and enters the reaction chamber through a temperature-controlled injector. In this chamber, the following reaction occurs:

$$CH_3SiCl_3 + H_2 \rightarrow SiC + 3HCl + H_2$$

At the proper combination of (1) reactant composition, (2) reactant flow rate, (3) pressure, (4) flow pattern, (5) substrate type, and (6) temperature, it is possible to deposit a smooth, strong, leaktight layer of SiC.[103] The CVD process can produce high-purity SiC with very fine grain size. A simplified diagram of a CVD reactor used for the deposition of SiC is shown in Fig. 4.10. This particular arrangement shows a graphite susceptor heated by RF induction, although resistance heating can also be used. The reactor can be modified to allow deposition to occur on the outside or the inside of the heated surface.

The properties of CVD materials can be varied to some extent by controlling the CVD process parameters, by incorporating dopants or a second phase in the material, or by performing a postdeposition treatment. The large-scale CVD process presents unique engineering and technical issues, which are quite different from those of the thin-film deposition. A recirculating type of flow pattern is cru-

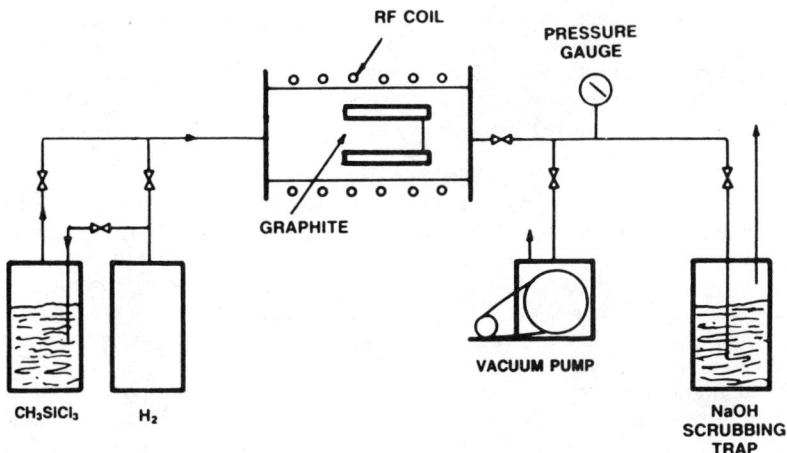

FIGURE 4.10 Simplified diagram of chemical vapor deposition SiC reactor.[103]

cial to the successful operation of the CVD process. There is a need to develop a fundamental understanding of phenomena such as the bowing and nodular growth to further improve properties of the CVD materials.[104]

An outgrowth of CVD may be OMCVD, a system similar to conventional CVD, but using organometallic precursors instead of pure reagents. It has the advantages of operating at lower temperatures (300°C) and producing higher-purity materials.[105,106]

2. *Functionally gradient processing:* It is hoped that in several years ceramic materials will have flexural strengths of 700 MPa at 1400°C and a fracture toughness of 30 MPa · m$^{1/2}$ for aerospace applications. At present materials with these properties are not available, and it still may take several years to obtain toughnesses even on the order of 10 MPa · m$^{1/2}$.[106]

Functionally gradient materials (FGM) are ceramics that continuously vary in composition through the thickness of the material (Fig. 4.11). The materials have a structure consisting of a surface layer of refractory ceramics with a metal phase continuously increasing within the ceramic matrix, eventually forming a metal base. Heterogeneous distributions of fibers, whiskers, and micropores also are possible. Such materials are expected to have high-temperature resistance with low internal thermal stress, good fracture toughness, and enhanced metal-bonding capabilities.[106]

A number of processes have been investigated, including plasma vapor deposition (PVD), plasma spray, P/M,

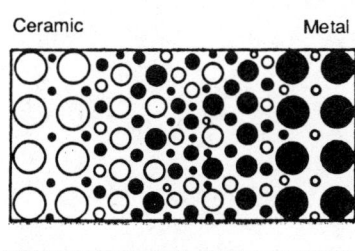
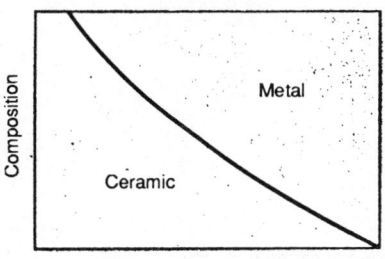

FIGURE 4.11 Schematic of FGM concept.[106]

and high-pressure combustion sintering (HPCS). For example, HPCS combines an exothermic reaction involving several elements with pressure to form and densify a variety of materials. The technique produces a variety of ceramic-base, multiphase components using ceramic-ceramic and ceramic-metal systems (Table 4.8). A variation developed from the HPCS method utilizes a reactant which is encapsulated in glass and ignited by the exothermic heat of the surrounding ignition agent. TiB_2 + Ti (10 wt %) has been densified to over 98 percent of theoretical density, and $TiC_{0.8}$ and TiC_2 + Ni (10 wt %) have been densified to 100 percent density at 100-MPa argon gas pressure. Metal canning also is possible and requires no preheating, a step necessary to soften the glass capsule.[106]

In another development program[106] SiC/C was produced by continuously controlling the composition of the CVD process gas made up of silicon chloride (H_2 carrier) and propane. Voids that appear in the SiC/C layer are said to have a beneficial effect on stresses within the material during thermal cycling.

TiB_2/Cu gradient material has also been produced. The composite, a configuration of intermediate ceramic-metal layers that link TiB_2 to pure Cu, is the result of SHS.

Gradient materials are especially well-suited to high-temperature applications. The Space Shuttle, for example, uses 10-cm-thick tiles to enable it to withstand 1300°C, but gradient materials with an integral metal backing could be used to reduce tile thickness to 1 cm. Finally, the process, when fully available, has the mass-production potential of near-net-shape components. In addition, microstructures can be engineered to produce ceramic composites since the synthesis reaction occurs simultaneously with densification. Various metals also can be added to achieve full densification as well as to obtain new microstructures. These materials are expected to be widely used in functional and biomedical applications, in addition to mechanical and structural applications.

3. *Exothermic filtration combustion:* By utilizing a self-propagating exothermic filtration combustion process, researchers claim they have a faster method,

TABLE 4.8 Some Monoliths and Composites FGM-Produced by HPCS

Monoliths	
Carbides	TiC, SiC
Borides	TiB_2, TiB, ZrB_2, NbB_2
Silicides	$MoSi_2$, $TiSi_2$
Intermetallics	TiN, TiAl, NiAl, CoAl
Composites	
2Ti + 2B + C = TiC + TiB_2	
3Ti + B_4C = TiC + $2TiB_2$	
Ti + Si + 2C = TiC + SiC	
$3TiO_2$ + 4Al + 3C = 3TiC + $2Al_2O_3$	
TiO_2 + Zr + C = TiC + ZrO_2	
$3SiO_2$ + 4Al + 3C = 3SiC + $2Al_2O_3$	
SiO_2 + 2Mg + C = SiC + 2MgO	
$2B_2O_3$ + 4Al + C = B_4C + $2Al_2O_3$	
WO_3 + 2Al + C = WC + Al_2O_3	

Source: From Miyamoto.[106]

use less energy, and produce more homogeneous ceramics than by previous methods. AlN has been one of the materials under investigation and was produced by the process.[107] Aluminum powder is ignited by a small energy impulse from a tungsten lighter in an N_2 atmosphere. As the N_2 filters through the reaction front, it combines to form a nitride. Under low pressure, the reaction produces ceramic powder. At higher pressures it yields fully densified net-shaped parts. In addition to nitrides, similar exothermic reactions have been used to produce titanium diboride, which is made by reacting titanium powder with amorphous boron powder. SiC and HfC are also possibilities.

4. *Gas-phase processing (flame):* The generation of fine, ultrapure, uniform, spherical, loosely agglomerated particles is currently being evaluated through the use of a continuous self-sustaining flame reaction route.[108] Direct synthesis of these ceramic powders from gas-phase reactants is a promising route for achieving the desired properties. The ideal gas-phase process would cause rapid-particle-forming reactions to occur in a region of controlled-reagent concentrations, thus minimizing nucleation and growth times (hence yielding small particles with a narrow size distribution). These conditions can be realized by providing a sharp thermal gradient in a flowing reactant mixture. This has been done using lasers, RF plasma heating, heated flow tube reactors, and now with the flame process. An additional advantage of such gas-phase methods over other methods, such as CVD or solid-state synthesis, is the avoidance of contact of the reactants or products with walls, where contamination often occurs.

Flames are widely used as synthesis reactors for producing fine particles such as carbon black, titanium dioxide, and silica. The flame system for synthesizing ceramic powders shares the advantages of other gas-phase synthesis methods, but avoids some of their disadvantages. One important advantage is simplicity, which leads to direct scalability and low process equipment costs, an advantage over laser-based methods. Compared to plasma heating, flame synthesis is a milder process with a much better defined temperature, pressure, and composition history. Compared to heated flow-tube reactors, flame synthesis provides a much more compact reaction zone, higher throughput, more uniform conditions, and it eliminates wall interactions. This new flame synthesis process for producing fine high-purity Si_3N_4 powders has been demonstrated experimentally.[108] Mixtures of SiH_4 and N_2H_4 or SiH_4, NH_3, and N_2H_4 were found to support stable flames over a wide range of pressures and reagent ratios.

5. *Cryochemical processing:* The principle of this process is the rapid freezing of aqueous solutions of salts followed by removing the water as ice, either by extraction or by sublimation. Powders and monolithic ceramics can be made. In making powders, the solution is first converted to uniformly sized droplets (40 to 200 μm), the droplets are frozen (cryocrystallization), immersed in acetone at 40°C to extract the water, then decomposed by heating. The basic principle here of freezing monosized sprayed droplets, then removing the solvent to produce a uniform mixture of fine particles, has been around for some time. However, the developers in the U.S.S.R. claim that the powders can be formed (with little or no binders) into highly dense materials with some unusual properties. For example, they have made a sodium β-alumina so dense and unreactive that it can be boiled for a long time in concentrated H_2SO_4 without being hydrolyzed. Variations from this process include cryoimpregnation, which involves impregnating a substrate or covering its surface with a salt solution, rapidly freezing, then removing the water by sublimation. Aggregation is thus prevented, and a composite with a homogeneous distribution of the introduced material is formed. It could be

a useful way to achieve a uniform distribution of a minor constituent in the pores of a substrate or on its surface, for example, to distribute a sintering aid uniformly in a powder compact or to put a catalyst on a catalyst support.

6. *Attrition or turbomilling:* Turbomilling consists of intense agitation of a milling medium, the material to be milled, and a suspending liquid. Turbomilling is recognized as a process that can produce ultrafine high-purity ceramic powders without contamination through high-shear and high-speed wet grinding of a fine fraction of material by a coarse fraction. The machine (turbomill) consists of a cylindrical container, a cagelike stator composed of vertical bars attached to rings at the top and bottom, a cagelike rotor composed of vertical bars fixed to an upper disk attached to a drive shaft, and a frame that holds the motor and machine components (Fig. 4.12).

In general, the turbomilling process has been much more efficient than other processes such as ball milling and vibration milling. Tests have shown that attrition milling in silicon nitride mills is a fast way of producing fine silicon nitride powders with minimum pickup of impurity. Consequently, because of its high rate of reduction of particle size and low rate of absorption of oxygen, attrition milling is preferred over vibratory and ball milling.[110,111]

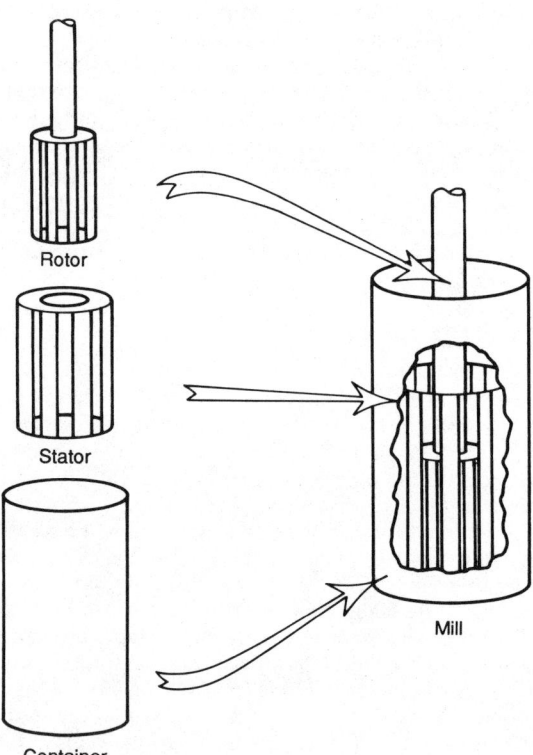

FIGURE 4.12 Turbomill used at U.S. Bureau of Mines.[109]

4.4 POWDERS (MATERIALS)

Most of the structural ceramics under consideration are actually families of ceramic materials, in much the same way as steels and brasses are alloy families. As in the case of an alloy family, a ceramic material family has certain attributes in common. For example, in the case of silicon nitrides, some of the attributes held in common are strongly covalent bonding, dissociation rather than melting at high temperatures (>1900°C), low coefficients of thermal expansion (which lead to the excellent thermal shock behavior of these materials), and excellent oxidation resistance due to the formation of a protective SiO_2 coating. These similarities aside, each individual member of the silicon nitride family is a unique material made by a unique process. These materials have very different properties as a consequence of particular processing. In addition, the ease and cost of producing useful engineering shapes differ widely from one processing method to another.

Silicon Carbide

Silicon carbide is in itself a compound produced by encasing a graphite electrode in a mixture of sand and coke and heating to approximately 2125°C. Crystalline silicon carbide is then ground to submicrometer particles. The reaction of silica sand with carbon produces SiC and CO,

$$SiO_2 + 3C = SiC + 2CO$$

The carbon source is generally coke. Salt and sawdust are added to the reaction, salt to create chlorides to purify the system, and sawdust to aid in gas circulation. Low levels of alumina (Al_2O_3; <0.25 percent) and iron oxide (Fe_2O_3; <0.1 percent) must be maintained to control the undesirable formation of graphite and silicon metal.[23,112] This Acheson process for manufacturing SiC powder is normally carried out in a refractory furnace trough. An outline of the overall silicon-based ceramic production processes is given in Fig. 4.13.

There are several types of silicon carbide. Of these, only two are actually composed almost entirely of SiC and, therefore, offer the high-temperature performance and chemical inertness that this material can possess. These two, known as pressureless sintered, or self-sintered, silicon carbide, are made without silicon sintering aids to avoid a liquid phase and low-temperature limits of reaction-bonded materials. These materials have no free silicon or graphite, thus allowing use at temperatures to 1371°C in oxidizing environments, with excellent resistance to strong acids and caustic solutions. The two types of SiC differ in the initial raw materials and processes used. One method uses α-phase SiC powder, compacted and sintered without pressure. The nearly 100 percent α-phase composition of the resulting product results in slightly better abrasion resistance than other types of SiC. The other method starts with a β-phase powder that converts to at least 80 percent α-phase during the pressureless sintering process. This procedure results in a more continuous microstructure than the sintered α-material as the individual particles lose their shape and grow into larger crystals during sintering.[113]

Both methods have the potential to produce good materials, with virtually no difference in the properties obtainable. However, because the α-phase method is less sensitive to processing variables, it is easier to produce in complex shapes such as

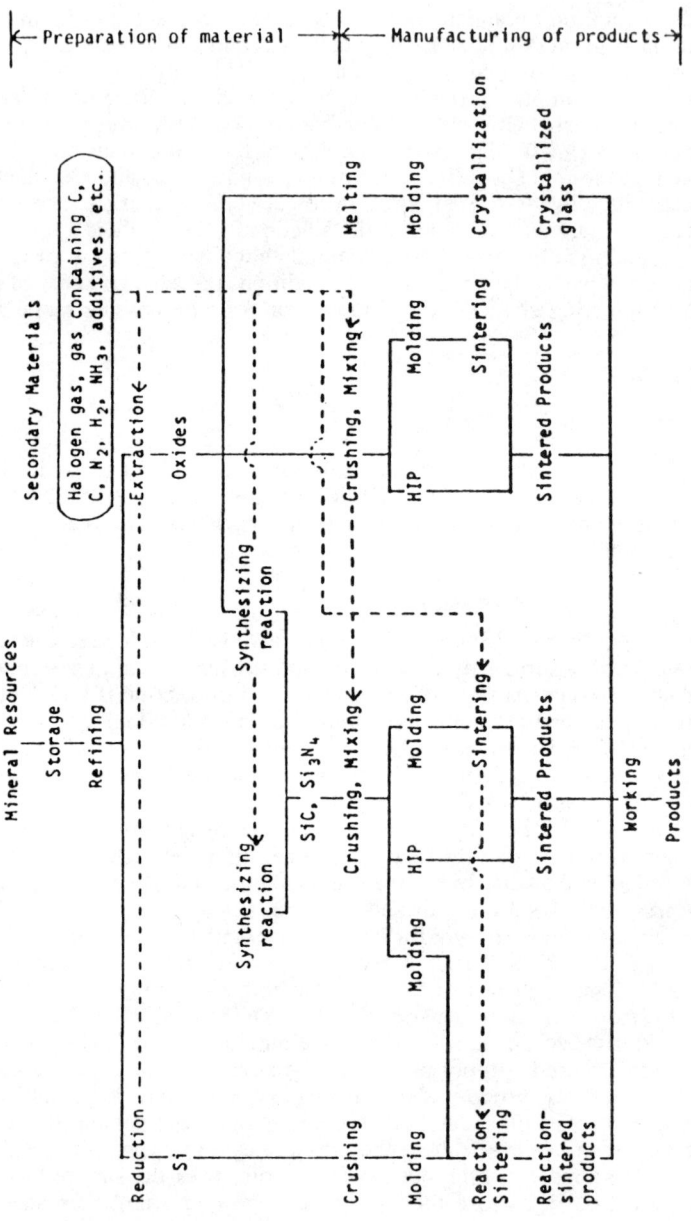

FIGURE 4.13 Outline of production processes for silicon-based ceramics.[23]

turbine components. The α-phase powders, which are widely used as abrasives, are also more readily available and less expensive than β-phase powders.

A new low-temperature process to make precursors for SiC has been developed.[114] The chemical reactions take place at 130°C, which compares with 1500°C for conventional processes. At such high temperatures, reactions are difficult to control and form unwanted by-products, which lower the reliability and durability of finished ceramic parts.

The powders are made by reacting two or more metal halides with an alkali metal at 130°C. The alkali binds to the halides, permitting the other components to react. In a typical reaction, silicon and carbon tetrachloride are reacted with sodium to form a silicon precursor. While final crystallization takes place at high temperatures, purity is still high because the initial reaction took place at low temperatures. The process also works with titanium diboride, titanium carbide, and boron carbide.

One other approach is to use polycarbosilane (organosilicon polymer), which can be pyrolyzed to form β-SiC through an amorphous state. The fine-grained SiC ceramics show similar flexural strength to that of the ordinary hot-pressed SiC.[115]

Silicon Nitride

Silicon nitride exists in two phases, α and β, both of which have hexagonal crystal structures. The α-phase has a unit cell approximately twice as large as that of the β-phase. The silicon nitride–based ceramics include hot-pressed, reaction-bonded, sintered, and hot isostatically pressed materials, as well as the β-sialons, a solid solution of Al_2O_3 and other metal oxides in the β-Si_3N_4 lattice.[116,117]

Si_3N_4 is most commonly produced by the direct nitriding method shown in Fig. 4.14a. The reaction occurs at 1450°C and must be carefully controlled to generate high-quality material. Fig. 4.14b and c illustrates other Si_3N_4 production techniques. The silica-reduction method is currently in use, while the vapor-phase synthesizing method is in the experimental stage.[23]

Si_3N_4 components are available in two forms: reaction-bonded and hot-pressed. As with all ceramics, there is a profound variation between grades of Si_3N_4, and the suitability of Si_3N_4 for a particular application cannot be judged from tests on a single material. It is critical to use materials that have been made with good-quality control procedures and a good understanding of the influence of processing parameters on results. Manufacturers of quality ceramics stress that many potential ceramics applications fail because the developer tests a generic Si_3N_4 (or other ceramic) without regard for the differences between suppliers.

According to one supplier, good structural Si_3N_4 parts can be made only with powder that has high α-phase content, a consistent and controllable composition, high purity, uniform distribution of any impurities that might be present, small particle size with narrow size distribution, equiaxed crystallites, and good dispersability. Good Si_3N_4 powders contain at least a 90:10 ratio of α- to β-phase, required because the sintering mechanism of Si_3N_4 involves a transformation from the α-phase to the β-phase. Without the high α-phase content, the sintered material will not develop the interlocked-grain structure required for high strength.

High purity is important, but absolute purity is not. Sintering of Si_3N_4 can take place at reasonable temperatures only because of the presence of small amounts of impurities that are liquid at sintering temperatures. One of the best sintering

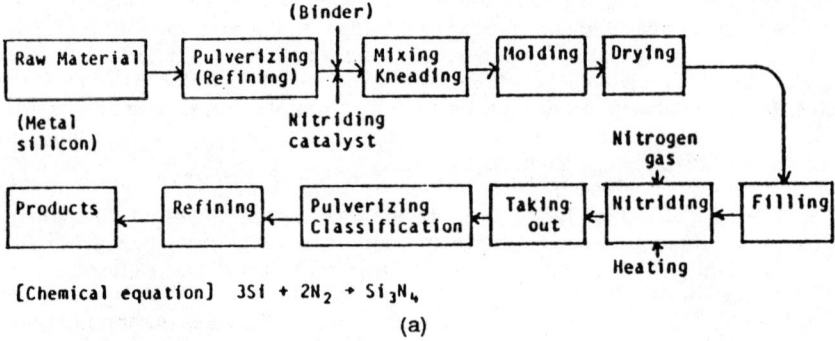

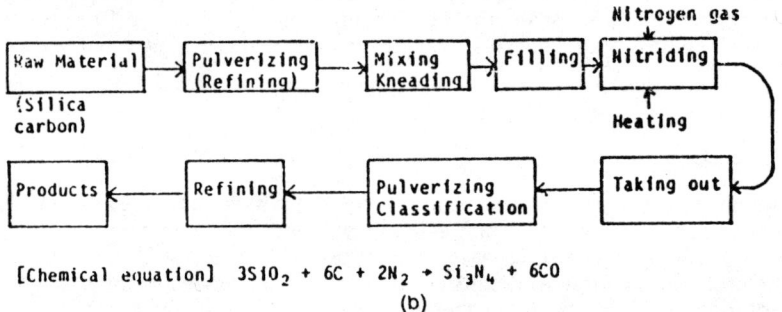

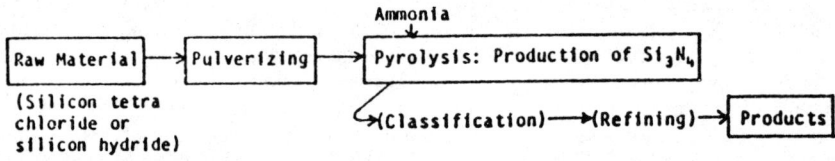

FIGURE 4.14 Production of Si_3N_4.[23] (a) Direct nitriding. (b) Silica reduction. (c) Vapor-phase synthesizing.

aids is 6 wt % yttria, which promotes the needed diffusion of Si_3N_4 while causing a minimum amount of detrimental glass formation.

Iron contamination is especially detrimental to Si_3N_4, reacting with Si_3N_4 during sintering to form a liquid iron silicide and nitrogen gas. The resulting metallic particles, deposited in large pores created by the nitrogen, typically cut the strength of the material in half. Typical commercial Si_3N_4 powders contain on the order of 100 ppm of iron; those made by reaction of ammonia with chlorosilanes (such as $SiCl_4$) have iron contents below 10 ppm. Calcium, nickel, and chromium are also harmful contaminants. Ideal particle sizes range from about 0.2 to about 20 μm.[118]

Currently, Si_3N_4 powders are manufactured by four main techniques: (1) nitridation of metallic silicon ($3Si + 2N_2 \rightarrow Si_3N_4$); (2) gas-phase ammonolysis of silicon tetrachloride ($3SiCl_4 + 4 NH_3 \rightarrow Si_3N_4 + 12 HCl$); (3) carbothermic reduction of silicon dioxide ($3SiO_2 + 6C + 2N_2 \rightarrow Si_3N_4 + 6CO$); and (4) thermal decomposition of silicon di-imide [$3Si(NH)_2 \rightarrow Si_3N_4 + 2NH_3$].[119] All four routes use inorganic reagents and synthetic techniques designed to minimize contamination with carbon, oxygen, and various metallic impurities (see Fig. 4.14).[120,121] Recently the pyrolysis of cross-linked polycarbosilanes, polysilanes, and polysilazanes in an NH_3 atmosphere which yielded amorphous, low-carbon Si_3N_4 powders was examined.[119]

Aluminum Nitride

Ultrafine extremely pure aluminum nitride powder has now become available in commercial quantities and should make the replacement of highly toxic beryllia substrates with nontoxic AlN economically attractive. In addition the AlN powders are expected to substitute heavily for boron nitride powders in electrical insulating applications due to the higher heat dissipation factor of AlN along with its superior dielectric properties. The relative ease with which AlN can be fabricated (cold-pressing and sintering) will allow the production of complex shapes from this material for unusual applications in furnace fixturing, source holders for ion implantation systems, insulators for plasma jet furnaces, and many other areas. Other materials such as BN must be hot-pressed and machined, which add to the expense of fabrication for those powders. In dispersions of silicone oils and resins, these powders increase the thermal conductivity significantly and decrease the thermal expansion of the resulting composite. This allows engineers to tailor their materials to the end-use requirements. Typical purities (metals basis) of AlN powders are in excess of 99.7 percent.[122,123]

Japanese researchers have developed a method to make ultrafine AlN particles with mean diameters of 0.013 μm. They use a modified arc system to create an argon-hydrogen plasma that blasts 91-μm aluminum powder and nitrogen into the ultrafine particles. The powders are 10 times more thermally conductive than alumina, and may replace it as a substrate in denser, hotter microelectronics that require more heat bleed off.

Boron Nitride

Two recently developed inexpensive low-temperature processes for making high-purity BN powders have been announced.[124] One method reacts boron halides and nitrogen compounds at −75°C to 200°C. The other mixes boron and nitrogen compounds in an NH_3 atmosphere held at 750°C. It is expected that these new methods will be useful in fabricating BN components for nuclear energy control rods, microelectronic parts, and metal processing equipment. Using CVD,[125] the codeposition of a BN matrix containing AlN was developed.[125] BN and AlN single-phase coatings were prepared by CVD, as well as BN + AlN composite coatings. The phases were either amorphous or crystalline, depending on the deposition temperature and reagent concentration. Figure 4.15 illustrates the system used, and Fig. 4.16 represents the matrix composite, where one type

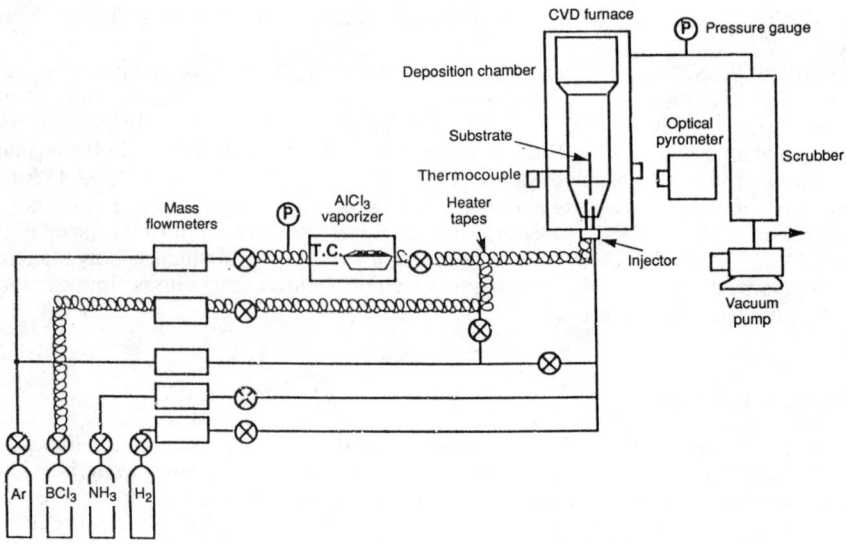

FIGURE 4.15 Schematic diagram of CVD system used to prepare BN + AlN composites.[125]

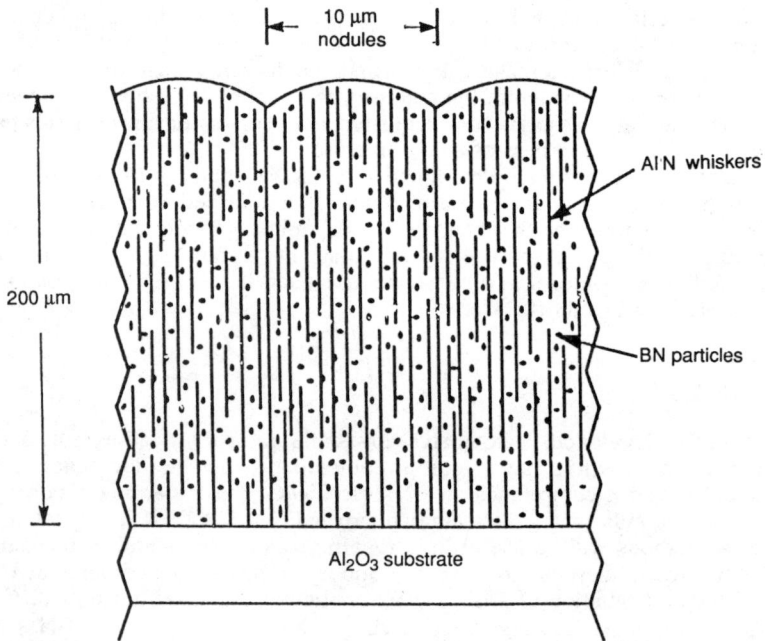

FIGURE 4.16 Ideal representation of composite containing turbostatic BN and anisotropic AlN.[125]

of deposit consisted of a turbostatic BN matrix containing oriented single-crystal whiskers of AlN. It should be emphasized that more R&D work is needed with regard to an understanding of the kinetic behavior and growth mechanisms of CVD-produced BN, AlN, and, most importantly, BN + AlN.

Titanium Diboride

In the past, a major deterrent to the full exploitation of titanium diboride has been the unavailability of large-scale powder production capability. This major problem has been solved through the development of a high-temperature continuous-flow furnace, a modification of the well-known cement kiln, designed to operate up to 2400°C in inert and reducing atmospheres.

In operation, a mixture of the finely powdered reactants is fed continuously into a furnace, where it reaches the reaction temperature in a few seconds. The reaction product is rapidly discharged from the furnace as an incandescent stream into a water-cooled quench chamber, where it is cooled to below red heat in less than a minute.

As produced, the powder consists of loosely bonded agglomerates of monocrystalline grains. By controlling reactant chemistry and furnace parameters, the powder can be tailored to meet specific requirements. For example, by reducing the temperature and increasing the feed rate, it is possible to produce a very fine monocrystalline agglomerate. Conversely, for applications in which a coarsely crystalline product is desirable, an increase in the reaction temperature and an increased residence time can be used to obtain the necessary material. Of interest, the products still consist of agglomerates of monocrystalline grain, but the degree of bonding is increased. After gentle deglomeration and milling, these titanium diboride powders have been sintered to over 90 percent of theoretical density without the use of sintering aids. These powders, because of their fine particle sizes, are expected to find extensive use in the electrolytic production of aluminum.

Titanium boride has a unique combination of properties. It is wet by molten aluminum, but it is not appreciably attacked or dissolved by such molten aluminum. Other potential applications for this material are in wear parts and in high-performance ballistic armor, as well as in cutting tools, bearings, grinding compounds, and metallization boats.

Other approaches in producing fine high-purity titanium diboride powders are low-temperature routes, which include using the exothermic homogeneous reaction between gaseous titanium trichloride and boron trichloride at 1200°C. There is also a multistep heterogeneous reaction between titanium metal sponge and gaseous boron trichloride at an even lower temperature (650 to 750°C). The formation of TiB_2 powder by the second method is particularly surprising because one would expect to end up with titanium boride coated titanium sponge. It is believed that since some of the intermediate reactions are endothermic (absorb heat) and others are exothermic (evolve heat), the material breaks down (gaseous intermediates are formed) and the net result is a very fine pure powder. Of particular importance, from the standpoint of reducing production costs, is that when powders formed by this "low-temperature" reaction were compacted by hot pressing at ~1500°C, products of very near 100 percent of theoretical density were fabricated (96 percent). This is a considerably lower temperature than required for sintering powders produced by other methods (2000 to 2200°C).

Alumina

Although alumina (Al_2O_3) is a naturally occurring mineral [gibbsite, $Al(OH)_3$], a number of processing steps must be carried out before it can be used in high-technology ceramic applications. The calcination of $Al(OH)_3$ in turn produces the hexagonal α-Al_2O_3, or corundum, by the Bayer process.[126,127] Minerals such as bauxite are pulverized, melted with soda, separated, and calcined (Table 4.9). Another method of Al_2O_3 production is alkoxide hydrolysis, in which Al metal is synthesized to aluminum alkoxide, then to an aluminum hydroxide and calcined to produce Al_2O_3 and water. The alumina produced is available as alpha aluminum trihydrate [α-$Al(OH)_3$], calcined alumina, and a variety of lower grades.

The traditional alumina ceramics discussed previously are not suitable for high-temperature structural applications. However, within the last decade a family of toughened aluminas has been developed that have suitable properties to be considered for such applications. Alumina has been toughened by the addition of ZrO_2 dispersoids, zirconia-toughened alumina (ZTA), and by the incorporation of SiC whiskers, SiC-whisker-toughened alumina, to be discussed later in this chapter.[126,128]

ZTA consists of a polycrystalline α-Al_2O_3 matrix with a dispersion of ZrO_2 particles. The ZrO_2 can be either tetragonal or monoclinic, or both, and is almost without exception located in the alumina grain boundaries. However, a ZTA containing intragranular t-ZrO_2 (tetragonal) particles and intergranular m-ZrO_2 (monoclinic) particles has been produced by sol-gel methods.[129,130] ZTAs are toughened by transformation toughening when t-ZrO_2 particles are present. These materials also exhibit a great deal of toughening due to microcracking when m-ZrO_2 is present. When both phases (tetragonal and monoclinic) of ZrO_2 are present, both transformation and microcrack toughening are active.

Most commercial ZTAs contain both phases due to the difficulty in controlling the zirconia particle-size distribution. In general ZTAs containing primarily m-ZrO_2 will have excellent toughness and thermal shock resistance, but will be weak. Conversely, ZTAs containing primarily t-ZrO_2 will have excellent strength, but will be moderately tough and have poor thermal shock resistance. Depending on the application, the percentage of each phase can be controlled by heat treatment and doping. Although ZTAs containing a large volume fraction of dispersed zirconia have been produced, it appears that the optimal amount is 15 vol % dispersed phase.[131] Most ZTAs are produced by mechanical mixing of ZrO_2 and Al_2O_3 powder and subsequent densification. However, advanced powder processing appears to hold a great deal of promise for this family of ceramics. Densification can be accomplished by sintering in the 1500 to 1550°C range or by hot pressing or hot isostatic pressing.

Sialon ($Si_3Al_3O_3N_5$) derives its name from its basic chemical formula and can be produced in several grades. By varying the degree of substitution of aluminum and oxygen atoms for silicon and nitrogen, respectively, in the β-silicon-nitride lattice, properties can be tailored to specific applications. During sintering, crystalline Si_3N_4 in the α modification transforms to the β modification. If aluminum and oxygen are present in the liquid phase, some amount will be incorporated into the β grains during the transformation, forming a solid solution of Si_3N_4 according to the formula $Si_{6-z}Al_zO_zN_{8-z}$, which is referred to as sialon. On cooling, the liquid becomes a glassy grain-boundary phase which degrades mechanical properties, particularly at elevated temperatures.[132-135] Finally, it should be noted that sialons are solid solutions of metal oxides in the β-Si_3N_4 crystal structure.

β-sialon ceramics are commonly given by their general formula, presented in

TABLE 4.9 Alumina Production Techniques

Al(OH)$_3$ calcining method (Bayer process)

Raw material (bauxite) → Pulverization → Caustic soda melting → Separation of aqueous solution of residue → Deposition of alumina (↑ Water) → Washing → Burning → Products

Chemical equations:

$Al_2O_3 + 2NaOH + 2NaAlO_2 + H_2O$
$NaAl_2O_2 + 2H_2O + NaOH + Al(OH)_3$
$2Al(OH)_3 + Al_2O_3 + 3H_2O$

Alkoxide hydrolysis method

Raw material (metal aluminum) → Synthesis of alkoxide → Hydrolysis (aluminum hydroxide) (↑ Water) → Burning → Refining → Products

Chemical equations:

$Al + 3ROH + Al(OR)_3 + 3/2H_2$ (synthesis of alkoxide)
$Al(OR)_3 + 3H_2O + Al(OH)_3 + 3ROH$ (hydrolysis)
$2Al(OH)_3 + Al_2O_3 + 3H_2O$ (burning)

Source: From Rothman et al.[23]

the preceding paragraph, where $0 < z \leq 4.2$ at normal pressure. However, if the basic powder is Si_3N_4, the sialons are named $Si_{3-x}Al_xO_xN_{4-x}$. Thus $x = z/2$ as compared with the first formula.

Commercial sialon ceramics contain about 4 to 10 percent Y_2O_3 and have an Al_2O_3:AlN ratio in excess of unity, which permits complete pressureless densification by liquid-phase sintering.[136] Sialons are synthesized by reacting together silicon nitride, silica, alumina, and aluminum nitride. The Lucas Cookson Syalon Company has developed a method to produce a series of sialon materials. One has a high-α-phase Si_3N_4 powder, which is mixed with alumina, a polytype phase, and yttrium oxide. The yttria acts as a liquid-phase sintering aid (Y–Si–Al–O–N), and the stable polytype phase eliminates the hydrolyzation of AlN, which otherwise prevents fabrication via aqueous solutions, as in slip casting.[137]

Other approaches for β-sialon powder processing include the metal alkoxide process, which has produced an average particle size of 0.6 μm and flexural strengths of 900 MPa at room temperature and of 600 MPa at 1200°C.[138]

Zirconia

Zirconia (ZrO_2) powders are produced as natural by-products in copper/phosphate/magnetite and baddeleyite mining. They can also be produced from zircon ($ZrSiO_4$) sands or chemical hydrolysis. The hydrolysis of a mixture of $ZrOCl_2$ and YCl_3 precipitates a mixed hydroxide. Subsequent stages involve calcination in the region of 850 to 950°C to obtain an optimum dispersion of yttrium with a uniform particle size; milling to obtain a dispersed powder; and spray drying with or without the presence of binders to obtain a weakly agglomerated free-flowing powder. The resultant powder has densities greater than 99 percent theoretical, which are achievable at sintering temperatures as low as 1350°C.[139] Other companies are producing fully or partially stabilized ZrO_2 powders with crystallite sizes <30 nm. Such ultrafine powders are produced by a coprecipitation method from chloride or alkoxide and densify to theoretical density at ~1400°C.[140]

Another approach starts with a 10:1 blend of 1-μm Zr powder and 0.1-μm C powders. To that is added 5 to 10 vol % of CaO or Y_2O_3 as stabilizers. The powder subsequently is processed by heating to 1500°C, depressurized in a vacuum furnace, oxidized at 800°C, and a resulting white ZrO_2 powder (2- to 3-μm diameter) is produced. ZrO_2 has three stable allotropes. The cubic form is stable above 2370°C, the tetragonal form between 2370 and 1170°C, and the monoclinic form below 1170°C. The transformation from tetragonal to monoclinic form with decreasing temperature at approximately 1170°C is quite disruptive and renders pure ZrO_2 useless as a high-temperature structural ceramic. This disruption is caused by a 6.5 percent volume expansion upon transformation from tetragonal to monoclinic form. The addition of certain stabilizing oxides (such as MgO, CaO, or Y_2O_3) will suppress the disruptive phase transformation of ZrO_2. These additions in small amounts result in a two-phase material, that is, partially stabilized zirconia (PSZ) consisting of a cubic matrix and tetragonal (t) or monoclinic (m) precipitates, or both, depending on thermal history.[29,126,141–147]

PSZ has been in limited use for decades. Without stabilization, ZrO_2 rapidly transforms from a monoclinic to a tetragonal form and back again upon heating and cooling. The addition of a sufficient amount of a stabilizing oxide, such as 10 percent MgO, keeps it in the cubic phase. Partial stabilization yields a cubic phase with inclusions of a transforming phase. A major breakthrough for the material came some 15 years ago, when it was discovered that if the particles of the transforming phase were kept below 1 μm in size, then the strength and toughness were dramatically improved. The single-phase cubic material results from a high-temperature solution treatment. Treatment at lower temperatures results in the two-phase cubic-tetragonal material. Extended treatment at lower temperatures results in overaging and the cubic-monoclinic material.

Unlike the silicon nitrides and carbides in which the particular mode of processing determines the properties and microstructure, the different members of the zirconia family are best classified by their microstructures, which are, in general, processing-independent, that is, starting with different types of powders and utilizing different consolidation techniques, identical final microstructures can be obtained by manipulating the thermal history. This ability makes zirconia ceramics unique among structural ceramics. The classifications by microstructure are conventional PSZ, fine-grained PSZ, tetragonal zirconia polycrystal (TZP), fine-grained monoclinic, overaged conventional, and single-crystal PSZ. These microstructures are shown schematically in Fig. 4.17. Conventional PSZ ceramics are stabilized by the addition of MgO, CaO, Y_2O_3, or rare-earth oxides. They are usually sintered in the cubic solid solution field (1600 to 1900°C).

TZPs are a fine-grained single-phase material stabilized by Y_2O_3 (Y-TZP) or rare-earth oxides. The constraint imposed by grains on each other allows the retention of the tetragonal phase. Sintering takes place in the tetragonal field (1300 to 1500°C). Y-TZP ceramics are presently the toughest and strongest ceramics and are most likely the toughest and strongest of all polycrystalline ceramics yet developed.[148]

Future

A clear sense of important advances in ceramic raw materials has taken place and continues. Also arising is a sense of market development problems, which may inhibit commercialization or even the optimization of properties through re-

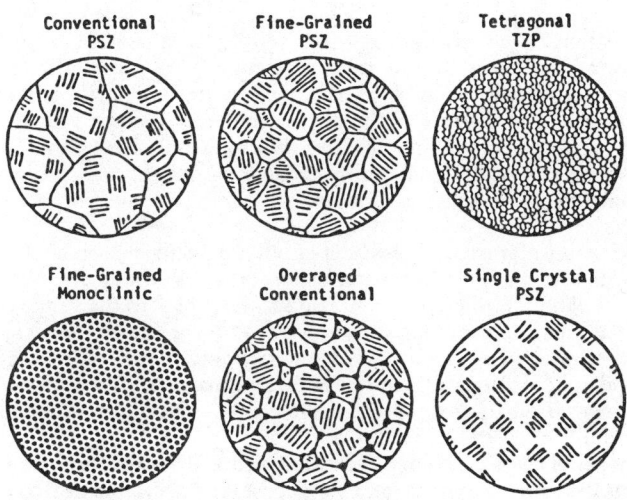

FIGURE 4.17 Microstructures of zirconias.[126]

search. One recurring theme is the need for cooperation between raw-material producers and the users of powders.

Raw-material producers need to understand the behavior of their powders in subsequent processing and should work closely with processors. The latter, however, consider some of the crucial details of their processing to be proprietary. Nevertheless, an optimistic spirit must prevail, and further improvements in available ceramic raw materials are to be expected. The ideal powder does not yet exist, although there have been significant advances.[149] Such a powder will have a high surface area to allow facile sintering, yet will not be strongly aggregated or contaminated with large particles, and foreign materials will be absent.

4.5 REINFORCEMENTS FOR CERAMICS

Advances in processing and design engineering are improving the reliability of ceramics. New processing methods are making more uniform parts with fewer of the flaws that lead to failure. Advanced analytical methods using X rays and ultrasonics are now able to spot dangerous voids before they cause trouble, and a computer-aided design technique called finite-element analysis (FEA) is helping engineers gauge a part's probability of failure. In addition, ceramists, chemists, and material researchers continue their work evaluating the various principal compounds that possess particular physical and chemical properties as well as the manufacturing route. This usually involves processing to a temperature designed to suit the material. In many cases other elements such as fibers, whiskers, and particulates are added to obtain desired properties. The engineering properties of the final product depend not only on the chemistry, but also on the result of the fabrication process. The applications for a particular product depend on these engineering properties, so it is necessary to understand the manufacturing process and the dependence of properties on it in order to assess a material's potential.

Using the aforementioned ideas and concepts, reinforcements were initially made with glass matrices. This led to new materials such as SiC/LAS, which is SiC fibers in lithium aluminosilicate, a glass matrix composite of glasses based on oxynitride systems such as Si–Y–Al–O–N. SiC reinforcement of glass matrices lends a new connotation to the term "strength" in materials. Comparing a carbon-carbon composite to a SiC/LAS glass matrix material shows that both fracture toughness and high strength can be attained. This has been demonstrated by attempts to shatter either material by driving a nail through it. The superior toughness of the SiC/LAS-glass composite prevents the nail from penetrating while not shattering, while the carbon-carbon composite allows penetration without shattering.

There exists an almost limitless availability of matrices or continuous-phase materials for use as part of these composites. This allows engineering of specific properties through material selection. The same applies to the range of reinforcement materials or discontinuous-phase materials; but these are somewhat more restricted. In general, the composites using ceramics can be classified by matrix type, with the specific matrix material and reinforcement material and their configurations being used to designate a system. By convention, the reinforcement material is noted first and the matrix second.

Four general classes of composites are of interest to ceramists: organic matrix, glass matrix, metal matrix, and polycrystalline ceramic matrix. Composite systems, by their definition, extend beyond the ceramic areas of interest, but for material composites these four classes define the majority.[150] Regardless of the materials combined as composites, several factors are influential in producing successful reinforcements. First, the modulus of elasticity of the reinforcement materials should be at least twice that of the matrix in order to produce an effective load transfer to the reinforcing phase. Second, thermal expansion properties of the fiber must be close enough to the matrix to limit prestressing of the fiber during matrix processing. This is particularly crucial when an anisoptropic reinforcement arrangement is produced. Third, chemical reactivity between the fiber and the matrix must be limited, while good mechanical bonding is attained, to prevent fiber degradation and allow greater physical property differences (thermal expansion) between the fiber and the matrix. In fact, Air Force Major Steve Wax stated at the 1984 Metal and Ceramic Matrix Processing Conference that "the fiber needs to be poorly bonded to the matrix. If the ceramic is tightly bonded to the fiber, then a crack would slide through it, which makes it brittle. If you have an interface between the fiber and the matrix, it will stop the crack. In that way ceramics can be made so they don't shatter catastrophically. We must have ceramics that have noncatastrophic behavior before we will ever use them in manned systems."

Why Ceramic Composites?

The argument for utilizing fiber reinforcement in ceramic materials differs from that applied to traditional resin (and glass)-based systems. In the latter the basic objective is to use the strength and stiffness properties of the fibers, the matrix being there purely to transfer load to the fibers, to prevent fiber-fiber contact and subsequent degradation, and to protect the fibers from environmental attack. Ceramics (except glass-glass ceramics), however, already have high strength and stiffness, and the purpose of the fibers in this case is to provide a mechanism for toughness, for example, to act as "crack" stoppers, add some defect tolerance, and prevent catastrophic failure. Excluding particulate-reinforced systems, two basic types of ceramic composites exist: (1) those containing whiskers or short fibers arranged in a random or two-dimensional fashion, and (2) those with long continuous fibers generally aligned in a single or small number of directions. The behavior of the two classes of materials has been reviewed in the literature.[150-159]

Potential Problem Areas

Although composite ceramics offer the promise of pseudoductility, a number of critical problem areas must first be addressed. The major areas of concern are the availability of suitable fibers and an associated fiber-coating technology, cost-effective fabrication routes, and, most importantly, innovative composite design and stress methodology.

The availability of fibers with high strength and stiffness, low density, small diameters, and thermal stability, both during fabrication and under operating conditions and hostile environments, is seen to be crucial.[160] More refractories with higher temperature capability as well as greater impact resistance are properties that will have to characterize the next generation of fibers in order for the revolution in ceramic composites to continue.

Furthermore, fiber coatings are likely to be required, first to limit the fiber-matrix chemical interaction and second to act as mechanically weak boundaries for toughening, that is, to control the fiber-matrix interface in order to obtain the desired mechanical, thermal, and physical properties. Also, the long-term compatibility of these coatings with both fibers and matrices will be important in maintaining as-fabricated composite properties.

Since short-fiber–whisker composites are typically fabricated via hot pressing,[161,162] this is easier than producing continuous-fiber composites. Continuously reinforced ceramic composites can be prepared via solid, liquid, or gaseous routes. In the solid (powder) route, hot pressing is preferred, and high-density materials result. Fiber damage is, however, likely to occur. This problem can be overcome by liquid or gaseous impregnation, but different limitations now arise because of the need for multiple impregnations with liquid precursors and long infiltration times with gaseous precursors.

Fabrication techniques such as chemical vapor infiltration (CVI), sol-gel, and polymer infiltration discussed earlier still require development and refinement to ensure that adequate reliability, reproducibility, and surface finish can be achieved in components of complex geometry.

In order to utilize the full potential of composite ceramics, a radically new approach to component design methodology is required. The introduction of new modes of material behavior will necessitate a change from the earlier empirical approach to material development and component life, to one based on an understanding of the physical phenomena involved. This will require the development of sophisticated modeling techniques, innovative component design, predictive component performance techniques, improved process monitoring, and nondestructive material inspection (NDE). In addition, mechanical test methods, joining techniques, and damage tolerance criteria will also have to be developed.

Inorganic Fibers

Inorganic fibers consisting of mixtures of alumina and silica have been made for some time, and because of their high melting points, these fibers are mainly used for insulation applications involving high temperatures, and in composites. If the silica content is high, these fibers can be produced by melt spinning using methods similar to those used in making glass fibers. With the advent of space travel, however, high-performance fibers with even higher temperature resistance than aluminosilicate fibers were required.

The most widely used inorganic fiber is glass. Most organic fibers consist of long molecules that are generally oriented in the direction of the fiber axis; these fibers are normally at least partly crystalline. In contrast, the long molecules of glass fibers, although generally oriented axially like organic fibers, are not crystalline but amorphous. Most of the molecules in inorganic fibers other than glass are at least partly crystalline as well as oriented, and some of these fibers consist of single crystals.

The inherent strength of a fiber increases as the fiber diameter is decreased. The relationship between fiber tensile strength and fiber diameter has been shown by Anderegg in work with glass fibers.[163] A sevenfold decrease in fiber diameter (19 to 2.5 µm) showed an 8 percent increase in ultimate tensile strength and the modulus increased as well. The generally accepted explanation for these phenomena is that there are fewer imperfections and faults as the fiber diameter is reduced, with perhaps the fewest imperfections occurring in the single-crystal whiskers.

A review of the literature on the composition of inorganic fibers shows that the molecules involved contain elements fairly closely grouped in the periodic tables. A list of compounds used to make inorganic fibers includes silica, alumina, aluminum silicate, titania, zirconia, boron, boron carbide, boron nitride, graphite, silicon boride, silicon carbide, and silicon nitride.

Inorganic fibers have been prepared by many methods, including (1) rod drawing as practiced in making wires, (2) passage through an orifice as used in melt-spun fibers, (3) vapor deposition as used in vapor-plating boron on a tungsten core, (4) crystal growth from a melt solution, or by the vapor-liquid-solid technique used in making whiskers, and (5) chemical reactions such as that for making SiC fibers by the pyrolysis of polycarbosilane precursor fibers.

Metal Oxide Fibers. There is much incentive to make aluminosilicate fibers with a high Al_2O_3 content or 100 percent Al_2O_3 fibers because of the higher temperature resistance and higher moduli of these fibers, which makes them attractive for use in composites. However, the high melting point and low viscosity of molten Al_2O_3 preclude melt spinning, so that other methods of forming fibers have been developed. One of the first such methods involved dispersing aluminum salts in a precursor fiber such as rayon, heating to burn off the organic material, and finally sintering to produce the alumina fiber.[164] Two other methods developed for high Al_2O_3 aluminosilicate fibers have advanced to commercial production, namely, the slurry and solution processes. The slurry process involves extruding slurries of alumina or alumina hydrate particles, and drying and heating the resulting fibers to produce a polycrystalline filament.[165] The solution process[164,165] involves extruding and drawing into fibers a viscous solution of aluminum compounds, followed by drying and heating the resulting fiber at 1000 to 1200°C to burn off any organic material.[34,166]

As the temperature is increased, not only do the crystal forms change, but the crystal size increases and the porosity decreases.[165] With these changes there is also a decrease in tensile strength and shrinkage at high temperatures, as well as increases in modulus, in crystallinity, and in stability. Table 4.10 shows most of these changes for alumina fibers containing approximately 4 percent silica. Other novel methods include the unidirectional freezing of an alumina hydrogel, which is then thawed.[167] Alumina fibers have also been prepared as whiskers by the vapor-phase method.

The properties of Al_2O_3 fibers lead to their use in high-performance composites. A 20-μm-diameter 99 percent α-alumina fiber had a tensile strength of 1400

TABLE 4.10 Property Changes during Processing of Alumina Fibers Containing ~4% Silica, Major Phase

	Eta	Eta/ Gamma	Gamma	Gamma/ delta	Delta	Delta/ Theta	Theta/ mullite	Alpha/ mullite
Approximate crystal size, Å	60				500		1000	2000+
Crystallinity, %	50	62	68	77	79	86	97	100
α-alumina, %					7	16	20–50	100
Pore volume, mm³/g	200		187	121	73	46	0	0
Shrinkage at 1400°C, 1 h	18	17	14	8	6.5	3.5	0.5	0
Tensile strength for 3.5-μm-diameter fiber, MPa	1800				1500			500

Source: From Birchall et al.[165]

MPa when measured at a gauge length of 10 mm.[164] The same fiber when coated with silica to heal any surface flaws had a tensile strength of 1900 MPa. This high tensile strength and modulus in alumina composites is retained after being heated in air at 1000°C.[164]

Polycrystalline α-alumina fibers have been incorporated into aluminum, magnesium, lead, zinc, and copper to make composites in the form of sheets, plates, tubes, rods, and beams. Wetting of α-alumina fibers by aluminum is difficult; the problem is alleviated, however, if 3 percent lithium is added to the aluminum.[164] A composite containing 30 vol % of randomly oriented staple alumina fiber in an aluminum alloy has been shown to have a tensile strength tenfold that of the unreinforced alloy at automobile engine temperatures.[164]

Another metal oxide fiber used in high-temperature applications is high-purity silicon dioxide fiber. One of the first methods of preparing this fiber was by forcing 6- to 7-mm-diameter silica rods formed from molten high-purity silica through a graphite guide and over a vertically oriented oxyhydrogen burner operating at 1800°C. A later method of producing silicon dioxide fiber was by melting a glass containing 75 wt % of SiO_2 and 25 wt % of Na_2O in a typical glass furnace at 1100°C, spinning the fiber from an orifice at the furnace bottom, passing the fiber in front of a gas flame while stretching it, and finally subjecting the fiber twice to an acid-leaching process to remove the Na_2O.

Other methods include fibers consisting of greater than 99.6 percent silica produced by spinning concentrated sodium silicate solutions, followed by conversion to fibrous polymeric silicic acid in an acid bath and subsequent dehydration to silica.[167] Finally, another method of making high-purity silicon dioxide fiber takes a polymer of silicon alkoxide and is spun by a sol-gel process. Heating the fiber to 1000°C produces a 99.999 percent pure quartz fiber, which is resistant to temperatures up to 1000°C.[151,159,166,168]

Other Oxide Fibers. Other oxide fibers that have been produced include calcia-stabilized zirconia fiber and an alumina-silica boria fiber. To produce the zirconium oxide fiber stabilized with calcium oxide, an aqueous mixture of zirconium oxychloride, zirconium acetate, and calcium oxide with a suitable water-soluble polymer such as polyvinyl alcohol is extruded and the resulting fiber heated and then sintered. The alumina-silica-boria fibers are produced by extruding a mixture of an aqueous solution of basic aluminum acetate with an aqueous dispersion of colloidal silica and dimethyl formamide, followed by heating first at 870°C, which converts the fibers to metallic oxides, and finally by heating at 1000°C.[169] Many additional oxide fibers, including alumina-chromia, titania, titania-silica, thoria, thoria-silicate, zirconia, zirconia-alumina, zirconia-yttria, zirconium silicate, and calcia-chromia, are described in the literature,[170] as well as methods for preparing whiskers of alumina, zirconia, thoria, magnesium oxide, zinc oxide, tungsten oxide, and many others.[171]

Boron-Based Fibers. The excellent mechanical properties and low density of boron fibers make them attractive for use in fiber-reinforced composites, especially in lightweight construction. Early attempts were made to produce a continuous boron fiber. However, the fiber was so brittle that it had little utility.[170] Later work succeeded in producing a fiber with greater utility by coating boron onto a tungsten core using a vapor deposition process. With the advent of commercial carbon fibers, vapor deposition of boron on a carbon core produced a low-density, high-strength fiber. Whether the core fiber on which the boron is coated is tungsten or carbon, the method for vapor-depositing the boron is similar.[172]

Boron fibers are manufactured worldwide today on both tungsten and carbon cores, with a range of diameters from 50 to 315 μm and with thermostable coatings of silicon and boron carbides.

Boron Carbide Fibers. Boron carbide fibers can be prepared by the same vapor deposition process used to prepare boron fibers, using mixtures of boron trichloride and methane or carboranes, which give a boron carbide mantle rather than a boron mantle over a tungsten core filament. Boron carbide fibers can also be prepared by heating carbon fibers in the presence of a vaporized boron halide.[170]

Boron Nitride Fibers. Boron nitride fibers with high strength and high modulus have been prepared by nitriding boric oxide fibers followed by hot stretching. They can also be prepared by nitriding boron fibers made by vapor deposition of boron on a tungsten core. Boron nitride fibers have been made into composites with a boron nitride matrix for use in electrical and electronic applications requiring a material that acts as both an electrical insulator and a thermal conductor. Boron nitride fibers have been used in aluminum composites because they are one of the few materials easily wet by molten aluminum.[169] Recently boron nitride fiber mats have been produced for use as electric cell separators in a lithium sulfide battery.[170]

Silicon-Based Fibers Other than Silica. Silicon carbide fiber is one of the most important fibers for high-temperature use. It has high strength and modulus and will withstand temperatures even under oxidizing conditions up to 1800°C, although the fibers show some deterioration in tensile strength and modulus properties at temperatures above 1200°C. It has advantages over carbon fibers for some uses, having greater resistance to oxidation at high temperatures, superior compressive strength, and greater electrical resistance.

There are two commercial processes for making continuous silicon carbide fibers: (1) by coating silicon carbide on either a tungsten or a carbon filament by vapor deposition to produce a large filament (100 to 150 μm in diameter), or (2) by melt spinning an organic polymer containing silicon atoms as a precursor fiber followed by heating at an elevated temperature to produce a small filament (10 to 30 μm in diameter). Fibers from the two processes differ considerably from each other, but both are used commercially.

Silicon carbide whiskers as small as 7 μm in diameter can be made by a number of different processes. Although these whiskers have the disadvantage in some applications of not being in continuous-filament form, they can be made with higher tensile strength and modulus values than continuous silicon carbide filaments.[151,172–174]

Vapor-Deposited Silicon Carbide Fibers. Carbon- or tungsten-cored fibers coated with silicon carbide are produced in the temperature range of 1200 to 1400°C by a chemical vapor deposition process in the same type of reaction chamber as the one used for the production of boron fibers.[172,173]

Polycarbosilane-Precursor Silicon Carbide Fibers. In this approach to making silicon carbide fibers, a polycarbosilane is synthesized and melt spun at 350°C. The resulting precursor fiber is heated in a vacuum or in an inert gas at 1000°C to produce a continuous filament containing β-SiC crystallites and having a high tensile strength and Young's modulus. This process can produce a small-diameter fiber compared to the large-diameter fibers produced by the vapor deposition process.[175–178]

Silicon Carbide–Carbon Fibers. In a recent work, Hasegawa and Okamura[179] describe the formation of a silicon carbide–carbon fiber by heat treating a precursor fiber spun from a mixture of a polycarbosilane and pitch. The polycarbosilane was made from dimethylsilane. The silicon carbide–carbon fibers were made by melt spinning the precursor mixture and heating the resulting fiber in a vacuum at temperatures up to 1400°C.

Silicon Carbide–Nitride Fibers. Silicon carbide–nitride fibers have been prepared by heating at high temperatures a precursor fiber made by melt spinning a polycarbosilazane. The melt-spun precursor fiber had a tensile strength of 1.2 MPa, a modulus of 187 MPa, and was resistant to oxidation at temperatures up to 1200°C.[180–182]

Silicon Nitride Fibers. Two approaches have been taken to produce reinforced Si_3N_4 ceramic fibers. One has been to incorporate modest amounts of short fibers and produce a composite which should to a great extent reflect the properties of the monolith, but with considerably improved toughness. The other has been, in analogy with the development of conventional fiber composites, to utilize the high strength and stiffness of the fiber by producing composites with high volume fractions of continuous fibers in which the matrix is of secondary importance to strength, but imparts its elevated temperature characteristics. However, so far developments of such Si_3N_4-based composites have only been successful in producing composites with improved properties for low-temperature applications.[19,23,78,151,183–185]

In order for Si_3N_4 fibers and a ceramic or metallic matrix to be combined to form a successful composite, the following criteria must be satisfied:

Thermal expansion mismatch between fiber and matrix minimized

Relative elastic moduli of the two components maximized, the aim being that the fibers bear a greater proportion of the load than the matrix

Chemical compatibility of the two components and their external environment maximized under the temperatures of fabrication and use

Oxidation resistance and thermal stability maximized

Fiber reinforcement with Si_3N_4 has been utilized to obtain both strengthening and toughening in other brittle matrix systems. So far work with Si_3N_4-based materials has met with limited success. Other approaches include the pyrolyzing of pitch into SiC, subsequently cross-linking the SiC–Si_3N_4 fibers by pyrolyzing polycarbosilazane resins. In both processes, fibers can be easily spun into preforms, then pyrolyzed into stiff ceramics.[186] The SiC–Si_3N_4 process, which is experimental, utilizes extruded polycarbosilazanes which are spun into preforms and pyrolyzed at 1000°C for 3 h. Heating converts the resin into a cross-linked fibrous mixture of SiC and Si_3N_4.

So far, fibers of 25 to 30 μm have been extruded. Up to 9.15×10^4 cm of fiber has been drawn on a takeup wheel in 3 min. Researchers increased the amount of cross-linking by heating the fiber to 50°C in 100 percent relative humidity for several days. During that period, strength and heat resistance rose and, surprisingly, brittleness dropped. Softening, which defines the fibers' maximum use temperature, reached 800°C after 9 days.[187] Other gel approaches have utilized polycyclomethylsilazanes (PCMS, APCMS)[188] to produce Si_3N_4 fibers; however, this work is very preliminary.

A continuous Si_3N_4 fiber, which is claimed to be equal in performance to top-quality carbon fiber and can compete with SiC fibers as well, called Nicalon fiber, has been developed by a Japanese company.[189] The Nicalon fibers (40%Si, 50%C, 8%O, 2%N) have tensile strength of 2413 to 3790 MPa and moduli ranging from 190 to 410 GPa. While the Si_3N_4 fibers are still experimental, their potential looks promising since they are simpler and less expensive to fabricate than SiC. Continuous filaments of 5- to 10-μm diameter have been produced with a typical 10-μm-diameter fiber having a tensile strength of 2447 MPa, a modulus >200 GPa, and has withstood continuous use at 1200°C.[190] The raw material is a by-product of the production of silane, which contributes to the processing route being cheaper than that used to produce an equivalent SiC fiber.

In relation to glass fibers, the Si_3N_4 continuous fiber, which has been also produced in the United States as HPZ, has a two and one-half to three times greater tensile modulus, up to 2.45×10^5 MPa, whereas HPZ has a tensile strength measured at 3312 MPa, which is also higher than that of glass. HPZ is said to have outstanding compression strength, which increases its strength-to-weight properties. This, in turn, broadens the material's design flexibility within a part compared to glass fiber and, depending on the resin system, to carbon fiber as well.

Zirconium Oxide Fibers. High-strength ZrO_2–Y_2O_3 fibers have been produced from a metastable acetate-based precursor. The key requirement for high-strength fibers is a uniform fine-grained microstructure. The addition of 1.5 to 5 mol % Y_2O_3 to ZrO_2 achieves this result by inhibiting grain growth during sintering.[1,143,145] Moreover, if the grain size is kept below ~0.5 μm, the tetragonal crystal structure is retained at room temperature.

In achieving strengths of 1.5 to 2.6 GPa, solutions of zirconium acetate, as a fiber precursor, and yttrium nitrate were used to produce the fibers. The two necessary ingredients to obtain high-strength fibers were demonstrated here: a uniform fine-grained microstructure (which requires a critical amount of additive or second phase to control grain growth) and small-diameter fibers (to avoid cracking during shrinkage). These concepts are equally applicable to other techniques of fiber or fabric forming (such as soaking of cellulose with solutions, followed by burnout).[191]

ZrO_2 staple precut fiber has become a ZrO_2 composite diaphragm for use in electrolysis cells. The ZrO_2 cloth is woven from fiber in the above application.[192] ZrO_2 textiles are used as electrode separators, diaphragms, and gaskets in electrochemical devices and have excellent stability in high-temperature electrolytes.

Conclusions and Recommendations

Fiber diameters classify the various fiber morphologies which include whiskers (<1 μm), staple (1 to 10 μm), continuous multifilament yarn (5 to 25 μm), and continuous monofilament (>100 μm). The varieties of primary fiber available and discussed previously are given in Table 4.11. These include the aggregate forms such as wool or rigid preforms (whiskers, staple), to be discussed later in this chapter, and the yarns and wovens (continuous fibers). Some of these fibers and their properties are listed in Table 4.12.

The variety of fibers have been processed by innumerable techniques, simple in

TABLE 4.11 Fiber Compositions and Types Available

Chemical type	Whiskers	Staple	Yarns	Monofilament
Carbon	—	—	Yes	—
Silicon carbide	Yes	—	Yes	Yes
Alumina-silica	—	Yes	Yes	—
Boron	—	—	—	Yes
Zirconia	—	Yes	—	—
Silicon nitride	Yes	—	—	—

Source: From Stacey.[192]

some cases and ingenious in others. For example, one engine manufacturer is assembling gas turbine parts by using a tape which contains the fibers plus glass powder in an organic binder. The tape is cut up and stacked just like any other composite tape to shape the required part, then heated to about 982°C. The heat removes the binder, and the glass particles soften and flow to form the completed part.

Using this relatively simple procedure to produce fibers and, subsequently, parts, as an example, we must consider various approaches[193] to optimize commercially viable fiber fabrication from the viewpoints of cost and prototype nature of some continuous fibers:

Fibers are coated with ceramic or glass powder, laid up in the desired orientation, and hot-pressed.

Fibers or woven cloth are laid up and then infiltrated by CVD to bond the fibers together and fill in a portion of the pores.

Fibers are woven into a three-dimensional preform and then infiltrated by CVD.

A fiber preform is infiltrated with a ceramic-yielding organic precursor and then heat-treated to yield a ceramic layer on the fibers. This process is repeated until the pores are minimized.[194-198]

4.6 WHISKERS, SHORT FIBERS, AND PARTICULATES

Whiskers or microcrystalline fibers, with their small diameters and minimal flaws or imperfections, have very high strength and modulus values. Many materials including metal oxides, carbides, halides, and nitrides have been produced as whiskers.[199] The metal oxides include Al_2O_3, MgO, $MgO-Al_2O_3$, Fe_2O_3, BeO, MoO, NiO, Cr_2O_3, and ZnO. Whiskers grow by two different mechanisms: tip growth and basal growth. In tip growth, high enough temperatures are used so that the vapor pressure of the whisker material is sufficiently high to allow the material to be transported as vapor to the tip of the crystal, where it is condensed. Tip growth occurs in the formation of SiC whiskers as well as Si_3N_4, Al_2O_3, and AlN whiskers.[199] In basal growth, the whisker material migrates to the base of the whisker and extrudes the whisker from the substrate.

TABLE 4.12 Typical Properties of Various Fibers

Fiber type	Manufacturer	Diameter, μm	Density, mg/m^3	Modulus, GN/m^2	Tensile strength, MN/m^2	Specific modulus E/ρ	Specific strength
B(W) monofilament	Avco	140	2.4	390	4000	162	1660
SiC(C) monofilament	Avco	140	3.0	420	4000	140	1330
SiC yarn (Nicalon)	Nippon Carbon	15	2.6	185	2700	71	1040
SiTiCO yarn (Tyranno)	Ube Industries	10–13	2.4	220	>2500	92	>1040
SiC whiskers (Silar)	Arco Metals-Silag	0.6	3.2	700	7000	220	2200
SiC whiskers (Tokomax)	Tokai Carbon	0.1–0.5	3.19	400–700	3000–14,000	170	1000–4000
Si$_3$N$_4$ whiskers (SNW no. 1)	Tateho Chemical	0.5–2.0	3.18	380	14,000	120	4000
Carbon yarn (Celion 1000)	Celanese	7.0	1.75	230	3550	130	2030
Carbon yarn (Celion G50)	Celanese	6.6	1.78	360	2500	200	1400
Carbon yarn (Celion GY70)	Celanese	8.4	1.96	520	1860	265	950
24%SiO$_2$/14%B$_2$O$_3$ Al$_2$O$_3$ (Nextel 312)	3M	11	2.7	152	1720	56	640
28%SiO$_2$/2%B$_2$O$_3$ Al$_2$O$_3$ (Nextel 440)	3M	11	3.1	220	1720	71	550
15%SiO$_2$/Al$_2$O$_3$ yarn	Sumitomo	17	3.2	200	1500	62	470
Al$_2$O$_3$ yarn (FP)	du Pont	20	3.9	380	>1400	97	>360
5%SiO$_2$/Al$_2$O$_3$ staple (Saffil RF)	ICI plc	1–5	3.3	300	2000	90	600
5%SiO$_2$/Al$_2$O$_3$ staple (Saffil HA)	ICI plc	1–5	3.4	>300	1500	>90	440
50%SiO$_2$/Al$_2$O$_3$ staple (Fiberfrax)	Sohio/Carborundum	1–7	2.73	105	1000	38	360
Mullite staple (Fibermax)	Sohio/Carborundum	1–7	3.0	150	850	50	280

Source: After Stacey.[192]

Fracture Toughness and Strengthening

The mechanisms responsible for whisker toughening include crack deflection and both whisker bridging and whisker pullout within a zone immediately behind the crack tip. Compared to some continuous-fiber-reinforced ceramics, in which extensive fiber pullout occurs, whisker pullout is more limited. This is, in part, a result of the short whisker lengths, typically <50 μm, and hence small pullout lengths as compared to those (tens of millimeters) that can be obtained in continuous-fiber-reinforced ceramics. Increases in toughness by extensive fiber or whisker pullout require minimization of the shear strength of the fiber-matrix interface. Fiber-whisker bridging requires at least modest interfacial strength to transfer load to the fiber and high fiber-whisker tensile strength to sustain the applied stress within the wake of the crack tip.[200]

In order that a particular whisker and matrix will combine to form a successful composite, three areas must be considered: (1) the difference between the coefficients of thermal expansion of the fiber (CTE_f) and the matrix (CTE_m), (2) the elastic moduli E of matrix and fiber, and (3) the chemical compatibility between fiber and matrix at processing temperatures. If $CTE_f > CTE_m$, the matrix is under radial tension and tangential compression and the fibers are under tension. Precompression of the matrix by the fibers (such as SiC whiskers in cordierite, mullite, and Si_3N_4) could result in strengthening. If $CTE_f < CTE_m$, the matrix is under tangential tension and the fiber is under compression. This applies to SiC whiskers in Al_2O_3, ZrO_2, and B_4C matrices. In these cases, radial tensile stresses are introduced across the whisker-matrix bond, which can influence the whisker-matrix interfacial shear strength.[201]

In principle, whiskers should confer superior properties because of their higher aspect ratio (length divided by diameter). However, whiskers are brittle and tend to break up into shorter lengths during processing. This reduces their reinforcement efficiency, and makes the much higher cost of whisker reinforcement hard to justify. Development of improved processing techniques could produce whisker-reinforced composites with mechanical properties superior to those made from particulates.

A disadvantage of using whisker reinforcement is that whiskers tend to become oriented by some processes, such as rolling and extrusion, producing composites with different properties in different directions (anisotropy). Anisotropy can be a desirable property, but it is a disadvantage if it cannot be controlled precisely in the manufacture of the material. It is also more difficult to pack whiskers than particulates, and thus it is possible to obtain higher reinforcement:matrix ratios (fiber volume fraction) with particulates. Higher reinforcement percentages lead to better mechanical properties such as higher strength.

In spite of the potential disadvantages of whisker reinforcement, the utilization of whiskers and particulates as toughening constituents appears more near term and, to some extent, economically more appealing due to simpler processing. The concepts of composite toughening by particulate and short-fiber dispersoids are listed in Table 4.13. In advanced heat engines, where lifetimes of 10,000 hours are a goal, microcracking leads to a reduction in the component strength and the elastic modulus.

Phase-transformation toughening also has limited promise for Si_3N_4-based composites. Careful consideration of the various additives that could toughen a Si_3N_4 matrix by this means has indicated that many would combine chemically with the SiO_2-based second phase during consolidation. However, some of them may be applicable in RBSN.

TABLE 4.13 Potential Toughening Mechanisms for Ceramic-Matrix Composites with Particulate or Whisker Dispersoids

Mechanism	Description
Microcracking	Crack interaction with residual strain field around dispersoid to create process zone ahead of crack tip
Phase transformation	Dispersoid volumetric change by phase transformation creating a process zone to shield crack from applied stress
Crack impediment	Crack blunting by dispersoid or line tension impeding crack propagation
Crack deflection	Surface toughening and crack tilting and twisting during propagation around dispersoid caused by thermal expansion mismatch or elastic modulus mismatch stresses
Crack bridging	Pullout and crack bridging by whisker dispersoids

Source: From Buljan et al.[202]

Crack impediment by the ductile dispersoids may be another somewhat less attractive option. It has been shown that in a metal dispersoid–brittle matrix composite, the advancing crack front may be blunted when it encounters the ductile dispersoid obstacle, which can deform plastically. The plastic deformation absorbs sufficient energy to stop crack propagation and therefore toughens the ceramic composite. A line tension effect can occur, which is analogous to dislocation motion past obstacles, according to Buljan et al.[202]

The predominant toughening mechanism that has so far been explored in composite development for applications in severe environments has been crack deflection and bridging by crack interaction with the dispersed, hard refractory particulates or whiskers. The deflection mechanism originates in the presence of a stress field surrounding the dispersoid along the matrix-dispersoid interface, which is caused by thermal expansion or elastic modulus mismatch. The residual strain state in the Si_3N_4 matrix surrounding such dispersoids is, after densification, radial tension and tangential (hoop) compression, according to Buljan et al.[202]

Toughening by crack deflection has been postulated to be independent of dispersoid size but dependent on dispersoid volume, irrespective of the sign of thermal expansion mismatch stresses. However, this has only been demonstrated in glass-matrix systems,[202] which are free of matrix grain boundaries.

The development of composites requires careful consideration of the dispersoid shape as well as its properties. It has been hypothesized that crack deflection, which produces tilting and twisting of the advancing crack front, is an effective toughening mechanism and that the degree of toughening is dependent on the dispersoid shape. On a theoretical basis, rod-shaped particles are predicted to be more effective toughening agents than disk-shaped particles, which are more effective than spheres. Toughness increases of up to four times have been predicted when rod-shaped dispersoids are used.

Whisker Fabrication

Manufacturing processes attempt to produce materials that have the reinforcement placed in a somewhat controlled fashion, which usually means even distribution either completely throughout or to designated regions of the component. Two principal methods have been used:[203]

1. Use of liquid metals either to infiltrate fiber bundles and preforms or to mix with particles
2. Solid-state methods, which mix metallic powders with the reinforcement and then either hot-press and extrude or roll the mixture to a product form

On some occasions, a liquid phase will form in the part during the hot pressing referred to in method 2. Preforms or intermediate products (such as sheet or wire) produced under liquid metal infiltration techniques (method 1) may be bonded or deformed to shape by a number of solid-state processes, such as diffusion bonding. More complete representations of the possible production routes for composites containing continuous fibers and for those with discontinuous reinforcement are given in Figs. 4.18 and 4.19, respectively, and discussed in Chap. 5.

Prior to discussing the fabrication of discontinuous composites (see Chaps. 5 and 9), seven of the most important methods of producing whiskers for these composites are described[170]:

1. *Vapor phase—evaporation and condensation:* This method of crystal growth involves evaporation of the fiber source material, mass transport through the vapor phase, and condensation at the growth site under conditions of low supersaturation. Examples of whiskers grown in this manner are aluminum oxide, zinc oxide, zinc sulfide, cadmium sulfide, and tungsten oxide.

2. *Vapor phase—chemical reaction:* In this method, crystal growth involves a chemical reaction to produce the fiber-forming material in the vapor phase, with deposition of the product on a substrate. Whiskers grown in this way include Al_2O_3 by reaction of aluminum vapor, hydrogen, and water vapor, SiC by reacting a silicon-containing material, a halogenated substance, and carbon, and B_4C whiskers from the reaction of B_2O_3 vapors with a hydrocarbon gas.

3. *Growth from solution:* This method involves the growth of crystals from saturated or supersaturated solutions. Examples include TiO_2 whiskers, α-MnO_2 whiskers, and Zn_2SiO_4 whiskers.

4. *Growth from melt solutions:* This method involves the growth of crystals from melt solutions using solvents or fluxes. Examples include β-Si_3N_4 and TiO_2 whiskers.

5. *Growth directly from melts:* In this method whiskers are grown directly from melts rather than from melt solutions. Examples include the production of continuous Al_2O_3 and other metal oxide filaments by a technique in which the filaments are drawn from a molten zone at one end of a feeding rod.

6. *Vapor-liquid-solid (VLS) method:* This method involves growth of a whisker at the tip site of the growing fiber with precipitation at the solid-liquid interface from a solution supersaturated with material supplied from the vapor. Examples include SiC whiskers, B whiskers, and α-Al_2O_3 whiskers.

7. *Growth from gels:* This method is useful for materials which cannot readily be made into whiskers by other methods. It involves the use of a gel such as silica gel to hold a whisker in position while it is being grown.

SiC Whiskers. Silicon carbide whiskers have been produced by several methods. Most of these, including the VLS process, involve combining Si and C at high

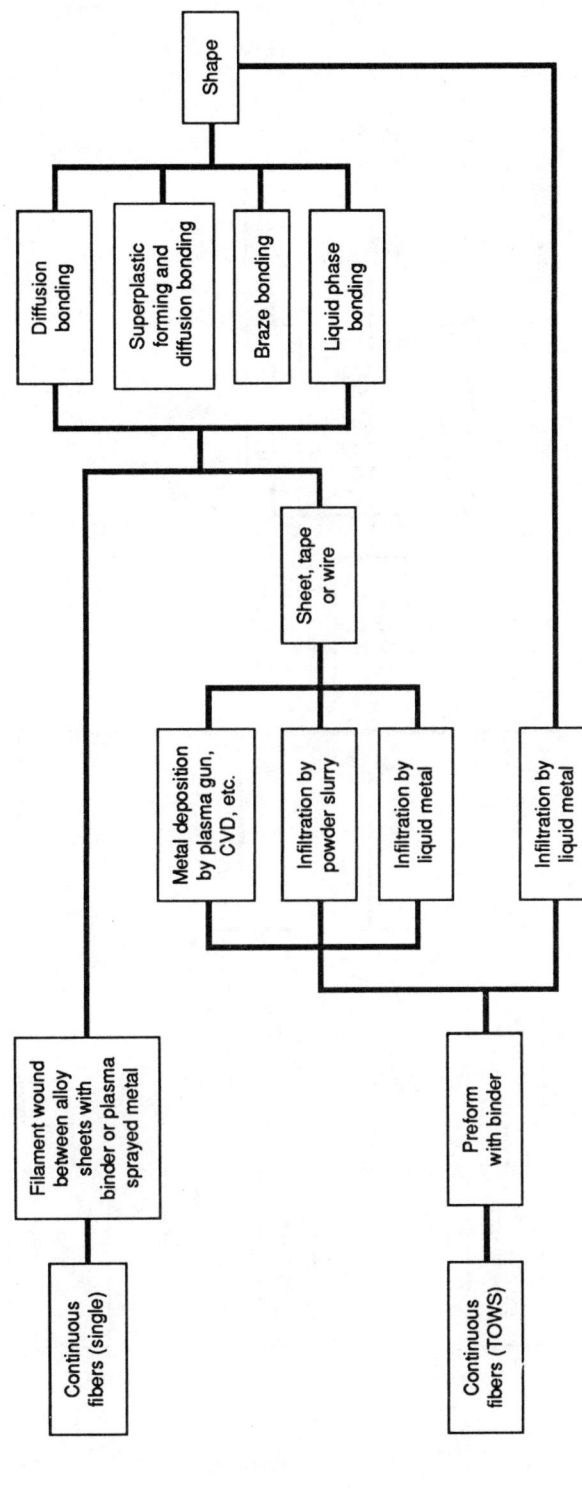

FIGURE 4.18 Process routes for production of continuous-fiber-reinforced composites.[203]

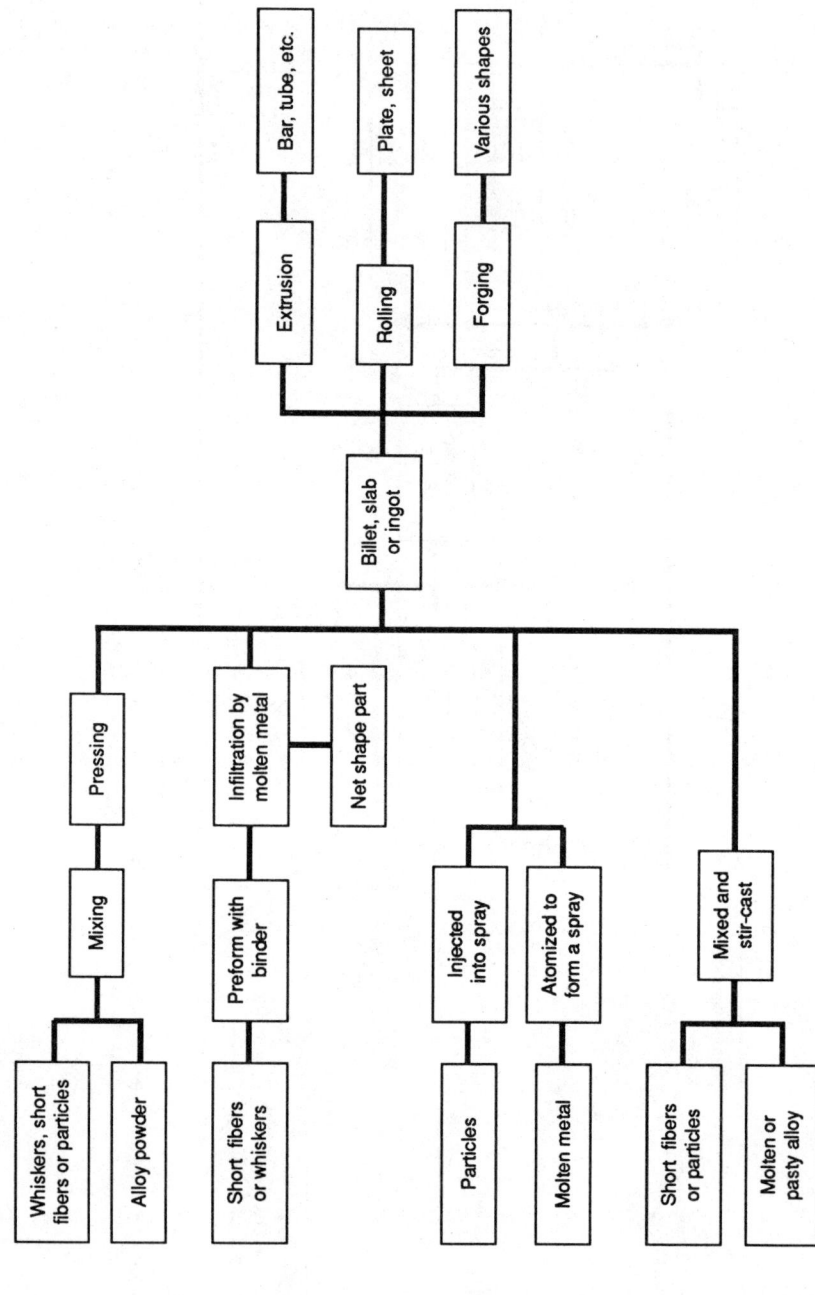

FIGURE 4.19 Process routes for production of discontinuous-fiber or particle composites.[203]

temperatures (1200 to 1500°C) under conditions that lead to the growth of high-purity single crystals. One of the early processes for making SiC whiskers involved heating 1 part by volume of SiO with 3 parts of CO in an inert or reducing carrier gas at 1300 to 1500°C. The SiC whiskers that formed on the colder parts of the reactor tube had diameters of 4 to 6 μm and lengths of 50 mm. A process for SiC whiskers has been developed in Japan[205] specifically for fiber-reinforced ceramic applications. The new whisker has a diameter of 1 μm compared with the 0.4- to 0.5-μm diameters of whiskers produced for fiber-reinforced plastics and metals, and also possesses a surface coating to enhance various properties. The whiskers are currently under evaluation and tests. Another group of Japanese researchers have looked for faster, cheaper ways to make SiC whisker reinforcements. The process starts when powders of fine metallic Si and another material that releases C are mixed and heated to 900 to 1300°C in an argon-oxygen atmosphere. The oxygen reacts with the C to form CO, and with the Si to form SiO_2. These eventually react to create SiC and O_2. Growth rates are outstanding, 2 to 3 mm in 10 min. Whisker sizes peak at 10 mm and 0.05 μm.[206]

SiC whiskers have been made from the SiO_2 and C present in rice hulls. Rice hulls contain between 15 and 20 wt % SiO_2. Rice hulls are first heated to 700 to 900°C in an oxygen-free atmosphere to remove volatiles; the residue contains approximately equal weights of SiO_2 and free carbon. SiC is formed by heating the residue in a stream of N_2 or NH_3 gas to temperatures between 1500 and 1600°C for about 1 h. The reaction proceeds via the formation of SiO, and the CO formed has to be flushed away by the flowing gas atmosphere in order to maintain the reaction. When the reaction is complete, the residue is heated to 800°C in air to remove the remaining free carbon. The resulting yield of SiC is about 10 percent of the original rice hulls. Small proportions of SiO_2 and Si_3N_4 are also present. Most of the SiC is present in the form of particles, only about 10 percent being in the form of whisker crystals. These have diameters of 0.1 to 1 μm and lengths of about 50 μm.

β-SiC whiskers have been grown by the H_2 reduction of methylchlorosilane on carbon substrates at 1500°C (Table 4.14), while others have produced SiC whiskers with combinations of chlorosilanes, carbon monoxide, and methane; reactions of dimethyldichlorosilane with phenylmethyldichlorosilane; heating a mixture of SiO_2, C, and NaF at 1200 to 1400°C; and reacting Si powder and propylene at 1300°C in a flowing H_2 atmosphere containing a few percent of hydrogen sulfide.

A liquid catalyst is used in the VLS process, which is used in making β-SiC whiskers. Here V stands for vapor feed gases, L for liquid catalyst, and S for solid crystalline whisker growth. Crystal growth occurs by the transport of material in vapor form to the surface of a molten droplet of the catalyst. The feed material is then deposited from the molten droplet to form a single-crystal whisker that increases in length as material is transported from the vapor through the liquid to the solid crystalline phase. Hence the term VLS process. A substrate is initially seeded with particles of a suitable catalyst—iron alloys can be used for SiC—and under suitable conditions crystal growth occurs.

In a production process, a graphite reactor is used, the temperature is raised to 1400°C, the catalyst particles melt and form liquid balls, and porous insulation brick impregnated with a fine powder mixture of silica and carbon generate the required silicon monoxide. 10 percent CO is added to the reactor feed gases to control the reaction, and other feed gases include 1 percent CH_4, 80 percent H_2, and 9 percent N_2. The role of the liquid catalyst is to form a liquid solution interface into which carbon and silicon atoms are extracted from the vapor phase. When this solution of

TABLE 4.14 Mechanical and Physical Properties of Reinforcement Components for Silicon Carbide

Property	Fibers				Whiskers		
	Carbon (HM)	PCS-SiC*	CVD-SiC†	Al_2O_3	α-SiC	β-SiC	α-Si_3N_4
Filament diameter, μm	6.5–7.5 (pan) 10–14 (pitch)	9–15	100–140	15–25	5–20	0.05–1.5	0.1–1.6
Filament length, mm	Endless	Endless	Endless	1.000	2–20	0.02–0.2	0.02–0.2
Tensile strength, GPa	2.5–3.5 (1.4–2.3 coated)	1.9–3.0	2.5–3.7	1.2–1.4	8–21	3–21	3–14
Young's modulus, GPa	320–390	180–200	380–420	340–400	450–490	400–490 (700)	385
Density, g/cm³	1.8–1.9	2.55–2.58	3.4–3.5	3.90–3.95	3.21	3.18–3.19	3.18
Thermal expansion coefficient, 10^{-6}/K							
Parallel	−0.5–0.2	3.1	4.2–4.5	7.0–8.6	4.4–4.6	4.4–4.7	2.4–2.5
Perpendicular	10–15	3.1	4.2–5.2	7.0–8.6	4.4–4.6	4.1–4.7	2.4–2.5

*PCS—polycarbosilane-derived fibers.
†Chemically vapor-deposited fibers.
Source: After Fitzer and Gadow.[204]

carbon and silicon in the catalyst is supersaturated, solid SiC is precipitated. As precipitation continues, the whiskers grow, lifting the catalyst balls. A uniform-sized catalyst particle will produce a uniform-sized whisker.

The form of the SiC whiskers produced can be controlled by the composition of the gas in the reactor. Under appropriate conditions, whisker crystals about 6 μm in diameter and 5 to 100 mm long have been manufactured.[207] However, on the average SiC whiskers produced in reactors have been 4 to 6 μm in diameter, approximately 10 mm in length, with an average tensile strength of 8.4 GPa and an elastic modulus of 581 GPa. These values are 3.5 and 3.2 times higher, respectively, than those of continuous-filament polycrystalline Nicalon SiC fibers made from polycarbosilane.

Some laboratories have aimed at VLS processing in unconventional ways to form reinforcing staple yarns from long whiskers. While a yarn would not be as strong as a fiber tow, it would let designers orient the whiskers to optimize properties in one dimension. The laboratories' focus has been on longer whiskers that can be wound into yarns. They have grown whiskers up to 102 mm long and need only 6.4- to 25.4-mm length for stable yarns. The whiskers are too stiff to go through an ordinary textile spinning process, so researchers are trying to wind them together with adhesive, form an oriented form, then burn out the adhesive.

A second approach to producing VLS whiskers starts with an external source of Si ($SiCl_4$). The process cannot yet be controlled to produce more than research-sized yields; however, it has great potential. First, a more homogeneous nutrient mix is obtained by bubbling through $SiCl_4$, since SiO is a heavy gas that does not mix completely. Second, there is better control of the Si concentration and whisker growth can be speeded up.

If one is looking for a single process that can yield multiple whiskers, a newly developed vapor-solid (VS) system is available for the carbothermal reduction of silicon dioxide and an auxiliary halide bath to yield silicon carbide, silicon nitride, and sialon whiskers. SiO_2 is dissolved in a halide bath to form a molten fluorosilicate bath and SiO gas. The gases are then reacted with nitrogen or carbon monoxide to form the various whiskers.

Researchers can switch between whisker types by controlling phase stability in the Si–C–N–O system as they adjust temperature, pressure, and carbon activity. The Si_3N_4 phase is stable at lower temperatures and carbon activity, while SiC grows more stable as temperature and carbon activity rise. SiC and Si_3N_4 whiskers grow in the <210> and <110> directions. Sialon whiskers contain uniformly distributed aluminum. Sizes vary with the whisker type. SiC, for example, ranges from 0.75 to 1 μm in diameter and up to 10 mm in length. Grown in the <110> direction, it is smooth; in the <111> direction, very rough.

SiC Whiskers and Matrices. Putting SiC whiskers into a Si_3N_4 matrix is not easy as reaction temperatures usually turn SiC into Si_3N_4. However, if starting powders have less than 100 ppm impurities, they can be nitrided to completion in 10 min at 1250°C, that is, 200°C cooler and two orders of magnitude faster than other processes. Dr. J. Haggerty of M.I.T. claims that this system will work with considerably larger parts, and since it is a highly exothermic reaction, energy costs are low. Also he has found that the matrix has five to ten times better oxidation resistance up to 1400°C than HPSN.[208]

Other successful work[209] has shown that SiC-whisker-reinforced Si_3N_4 composites can be fully densified by pressureless sintering at 1800 to 1850°C and subsequent hot isostatic pressing or by gas-pressure sintering at distinctly higher temperatures.

Several investigations have recently shown that whisker-reinforced ceramics present improved mechanical properties when compared with monolithic ceramics. Flexural strength and fracture toughness have been investigated thoroughly, and improvement in these properties increases with fiber loading[201,210] and raw-material purity. The improvement is explained by different mechanisms, assuming that nonelastic energy is absorbed in the composite material. As examples of these mechanisms, whisker pullout, crack deflection, and crack branching are mentioned. In a study[211] fully dense SiC-whisker-Al_2O_3 composites were manufactured by hot isostatic pressing, and the test results indicated an increased strength and fracture toughness with fully linear-elastic behavior. Heat treatment in an argon atmosphere further increased strength. It was also found that a reduction of the stress intensity promotes higher strength and fracture toughness since the whisker, which has the higher Young's modulus and good bonding, spreads the load within the matrix, and also enhances the ability of whisker reinforcement to suppress grain growth in the matrix.[211]

A process under development uses injection-molding methods[212] for making complex preforms of particulates, whiskers, or fibers and has been extended to make high-density green moldings of particulates. The particulates can vary in size from about 3 μm to 6.4 mm and can be loaded in the green body to 75 percent or more of theoretical density. Ceramics include SiC, BC, and AlN while the matrices include aluminum, magnesium, and titanium. When the process is used to make injection-molded preforms of fibers or chopped whiskers, it achieves loadings of 40 percent or more. The fibers can have aspect ratios as high as 100:1 and can be maintained in random orientation. Whether loading whiskers, fibers, or particulates, the process can produce net-shaped parts accurate to +0.1 percent for particulate-reinforced parts and to +0.5 percent for fiber- or whisker-reinforced composites.

The process consists of the whiskers, fibers, or particulates being mixed with thermoplastic binders and surfactants to ensure complete wetting and dispersion of the reinforcements in the process. The mix is injected into a mold, solidified, and removed as a shape strong enough to be handled easily. The binder is removed, leaving the preform still strong enough for normal handling, and the part is put into another mold to receive a metal or ceramic matrix. Molding is done at low pressures, so much larger shapes are possible than normally associated with injection molding. The metal matrix can be introduced by squeeze casting or other pressure or vacuum-assisted techniques.

The process has been used to make a whisker-reinforced curved hexagonal mirror blank 152 mm in diameter and only 6.4 mm thick, and in the future lie components several centimeters long and up to 50.8 or 76 mm thick. Supertough whisker-reinforced ceramics for cutting tools have become a reality. Exhibiting fracture toughness of 8 MPa · $m^{1/2}$ and three-point flexure strength of 1600 MPa, this material is based on Al_2O_3 and reinforced with 25 percent SiC whiskers and 10 percent PSZ. It is processed by being mixed for 48 h and then sintered at 1500°C. Mullite continues to contend for many high-temperature ceramic uses, and it is relatively inexpensive. Researchers have improved properties through zirconium oxide stabilization and SiC whisker reinforcement of mullite.

As a general rule, 30 percent loadings of either ZrO_2 or whiskers have the same effect on the room temperature flexural strength and fracture toughness: both boost strength 50 percent and double toughness. When mullite + ZrO_2 is reinforced with SiC, it grows 50 percent stronger and 25 to 50 percent tougher than unreinforced mullite + ZrO_2. The thermal shock resistance of mullite, measured by quenching progressively hotter samples in water until they microcrack, rose to 381°C for both SiC–mullite and Sic–mullite + ZrO_2 from 277°C for plain mullite. The mullite composite was hot-pressed from a matrix made by the alkoxide process, while ZrO_2 stabilization was adjusted by varying the amount of Y_2O_3 and MgO additions.

Using SiC-whisker reinforcement in glasses and glass ceramics produces a significant increase in strength toughness, and moduli over the unreinforced materials. Whisker-reinforced glass ceramics also have been sintered to full density with Weibull moduli greater than 20 at room temperature. The maximum use temperature of these materials, which also have the advantage of near-net-shaped processing, is 1200°C.[213,214]

Some companies[213] have gone one step further and combined SiC whiskers with SiC continuous fibers and an alkaline-earth aluminosilicate-based glass. These hybrid composites have shown substantial improvements in mechanical performance compared with fiber reinforcements alone, in terms of elastic limit and transverse and interlaminar shear strengths. These properties were increased by two to four orders of magnitude.

By combining Nicalon SiC fiber and 10 percent SiC whiskers and the glass matrix the following values were obtained:

Hybrid elastic limit 650 MPa, from 330 MPa for SiC fibers

Transverse strength 134 MPa, from 47 MPa

Interlaminar shear strength 51 MPa, from 12 MPa for SiC fibers

Finally there are single-crystal α-SiC platelets, which have several advantages over whiskers. They do not agglomerate like whiskers, so they are easier to disperse, nor do platelets retard sintering or interfere with densification, which makes them ideal for pressureless sintering. Platelets are grown to size by a diffusion process that uses growth dopants to promote surface diffusion above 2100 to 2200°C. The platelets are grown as α-phase, which withstands higher temperatures than β-phase whiskers. The final product is a smooth, perfect hexagon, unlike other processes in which the whiskers appear to have been ground. Mechanical studies must be performed to learn how to process the platelets into a composite.[215,216]

Si_3N_4 Whiskers. Si_3N_4 whiskers, with their superb heat resistance, may one day be the reinforcement of choice for high-strength high-temperature ceramic composites. Scientists have not yet given Si_3N_4 whiskers the same scrutiny brought to the more common SiC whiskers, but this is changing. Researchers are trying to determine how processing affects whisker crystallinity, extent of polymorphic transformation of the Si_3N_4, and how to control whisker segregation during processing.

Researchers in Japan claim that they have produced Si_3N_4 whiskers by treating rice husks in HNO_3 before firing the product in a tubular carbon vessel under reduced pressure. They find this to be a cheap way to make whiskers which are virtually free of defects. The whiskers produced are heat-treated at 1400 to 1450°C in a N_2 atmosphere and are reported to be 0.1 to 0.5 μm in diameter, 300 to 1800 μm long. The scientists say that concentrations of defects, mostly due to α-type crystals, are low.

Another group of researchers is using powder metallurgy techniques to make a superplastic composite based on β-Si_3N_4 whiskers and aluminum powder. The aluminum is 2124, a type of superduralumin containing 5 percent copper and particles 44 μm or smaller. The reinforcing β-Si_3N_4 whiskers are 10 to 20 μm long and 0.5 to 1.5 μm in diameter. The process involves mixing the aluminum powder with 20 vol % whiskers using an ultrasonic cleaning apparatus. After mixing, the composition is press-sintered under vacuum and then recompressed in atmosphere to produce a billet 400 mm in diameter. The billet is hot-extruded into a 6-mm-diameter rod, which is then heat-treated to yield the composite material.

To grow β-Si_3N_4 whiskers, just modify a process already used to yield cubic SiC and α-Si_3N_4 fibers. To get α-Si_3N_4, one produces a reaction between C and SiO_2 at 1450°C to generate silicon monoxide, which is then reacted with a mixture of NH_3, H_2, CO, or N_2. To get β-Si_3N_4, raise the temperature and flow rates and use a simple tubular reactor. The result is a white lightly packed layer of β-Si_3N_4 up to 6 mm thick, which could be useful in thermal barrier tiles. The single-crystal β-Si_3N_4 whiskers are about 1 μm in diameter and 5 cm long.

Si_3N_4 Whisker Fabrication. Two methods have been developed to use Si_3N_4 whiskers in a metal-matrix composite (MMC) or particulate-reinforced MMC. One method is to injection mold ceramic whiskers and short fibers into preforms. These preforms can economically produce complex shapes from such desirable reinforcements as SiC and Si_3N_4 whiskers and Saffil Al_2O_3 fibers. Subsequently the preforms can be made into MMC by squeeze casting and other pressure and vacuum processes, or be used to reinforce sections of monolithic metal structures.

Injection molding high-aspect-ratio (length to diameter) whiskers has produced uniform agglomerate-free preforms from fibers and whiskers whose aspect ratios range from 20:1 to 100:1. Usually whiskers with a high aspect ratio suppress injectability. To injection mold whiskers, one mixes whiskers or fibers with surfactants to aid wet-out and thermoplastic binders. (One can also add ceramic or metal particulates to the mixture and still maintain the random orientation of the fibers.) After molding, the organic phase is burned off. The resulting preforms can be made in any shape, are as strong as chalk, and require no special handling. The process allows one to make larger parts at lower pressures than commonly associated with injection molding of plastics.

A second technique uses an aspiration method for making preforms. For this method, it is important to disperse whiskers homogeneously in a solvent. If there are some agglomerates of whiskers in the solution, these will remain in the composite.[217–219]

The squeeze-casting process is useful for the fabrication of composites. The advantage of squeeze casting can be summarized as follows:

1. High pressure is effective to improve wetting between ceramic fibers (whiskers) and liquid metals.
2. This process is suitable for near-net-shaped production of composites.
3. This process is suitable for the production of casting which has composites in the required part.
4. Equipment is not expensive.
5. Production operation is easy.

In one version of the squeeze-casting process, relatively low die temperatures (200 to 300°C) are used to prolong die life in industrial applications. As liquid metals solidify quickly and the reaction time is short, degradation of whiskers is minimal. Since SiC is harder than Al_2O_3 or Si_3N_4, for example, SiC whiskers are superior to other whiskers in this respect. However, the strength of Si_3N_4 whiskers are almost the same as that of SiC whiskers (Table 4.15). To fabricate high-strength whisker-reinforced metals by squeeze casting, the following points need to be considered:

1. Whiskers are thin and long enough for strengthening.

TABLE 4.15 Properties of Si_3N_4 Whisker Compared with SiC Whisker

Whisker	Diameter, μm	Length, μm	Density, g/cm^3	Tensile strength, GPa	Elastic modulus, GPa	Hardness (Vickers)
Si_3N_4	0.1–1.6	5–200	3.18	13.8	379	1800*
SiC	0.05–1.0	50–200	3.19	2.1–13.8	551	2800*

*Sintered material.
Source: From Nishida et al.[219]

2. Whiskers are made of hard material.
3. Whiskers are dispersed in the matrix as homogeneously as possible.
4. Bonding between whisker surface and matrix is sufficiently strong.
5. Binder for forming preforms is unfavorable, because the binder might affect the bonding at the whisker-matrix interface, and agglomerates of the binder might cause initiation of a crack.
6. Nonfibrous inclusions must be removed.
7. During forming of preforms, nonuniform stress must be removed as it causes inhomogeneous distribution of whiskers in the matrix.

Al_2O_3 Whiskers. Single-crystal α-Al_2O_3 whiskers have been found to be easier to disperse than SiC whiskers, so higher reinforcement loadings can be achieved.[215] The diameters of aluminum oxide whiskers (4 to 7 μm versus ≈0.5 μm for SiC) make them less prone to agglomeration and therefore easier to disperse. (Their aspect ratio is 10 to 13.) Their larger size also makes them less likely to float in the air and become a health problem, a potential threat to those who work with SiC whiskers.

A new Al_2O_3-based ceramic whisker, called Alborex, for use as a reinforcing material in fiber-reinforced plastics and metals has been developed.[220] Using the flux method, with aluminum sulfate and aluminum borate as precursor materials and a potassium sulfate fluxing agent to synthesize the Al_2O_3 whisker, a mass production method was developed. The whisker has a composition of $9Al_2O_3 \cdot 2B_2O_3$. It consists of a very fine, white, needle-shaped crystal filament having an average thickness of 0.5 to 1.0 μm and an average filament length of 10 to 30 μm. It has a melting point of 1440°C, excellent thermal resistance, and its acidity is generally neutral. Its tensile elasticity is 40 tons/mm^2, close to that of a SiC whisker, and its strength is greater than that of a potassium hexatitanate ($K_2Ti_6O_{13}$) whisker. Since its Al_2O_3 content is 86 percent, it has great stability with respect to metals such as aluminum.

Miscellaneous Ceramic Whiskers. Whiskers of TiB_2 for reinforcing aluminum have many advantages over SiC. TiB_2 is more thermodynamically stable than SiC and less reactive. SiC cannot be used in aluminum above its 650°C melting point, but TiB_2 is stable and wets out all the way up to 900°C. It can also be a stable reinforcement for Ti, Si_3N_4, and SiC. The TiB_2 whiskers are grown from gaseous precursors in an easy-to-control process (a variant of the VLS process).

NiO, MgO, and MgO–NiO whiskers have been grown from their molten salts by a variant of the VLS process (used to grow SiC whiskers). Crystals up to 5 mm long and from 1 μm to 10 μm in diameter have been grown. In the process molten chlorides are combined with one or more alkali-earth chlorides in water vapor over a refractory substrate (Al_2O_3, fused SiO_2, Pt). It takes from 1 to 4 h at 800 to 1000°C.

A new process has been developed to produce primarily synthesized sialon material consisting of mainly silicon oxynitride (SiO_2–N_2) and aluminum oxynitride (Al–O–N) in solid solution. The resulting material is a homogeneous Si–Al–O–N whisker-powder blend. The process involves a specially designed combustion-synthesis technique that is said to yield at least 20 wt % whisker in the composite. Such a composite is expected to have enhanced fracture toughness and improved creep properties compared with conventional sialons. These benefits are in addition to already good sialon properties, including high strength, wear resistance, low coefficient of expansion, and high resiliency in corrosive environments.

Particulates

Most frequently there are two types of discontinuous reinforcement: whiskers and particulates. The most common types of particulates are Al_2O_3, BC, SiC, TiC, and WC. Whiskers generally cost more than particulates. For instance, SiC whiskers cost $95 per pound, whereas SiC particulate costs $3 per pound. In terms of tailorability, a very important advantage in MMC applications, particulate reinforcement offers various desirable properties. BC and SiC, for instance, are widely used, inexpensive commercial abrasives that can offer good wear resistance as well as high specific stiffness. TiC offers a high melting point and chemical inertness, which are desirable properties for processing and stability in use. WC has high strength and hardness at high temperature.

In composites, a general rule is that mechanical properties such as strength and stiffness tend to increase as the reinforcement length increases. Particulates can be considered to be the limit of short fibers. Particulate-reinforced composites are isotropic, that is, they have the same mechanical properties in all directions.[146, 221–225]

A novel use of Al_2O_3 or ZrO_2 particles has been investigated and a process developed. It starts with a three-dimensional felt or woven fabric shaped into a preform, which is typically a cylinder of felted carbon; but nylon, rayon, and other polymers also work. Submicrometer powdered particles are packed into the preform network by pressure filtration. The particle size has to be small enough to flow into the spaces around the woven or random felted fibers. The particles can be infiltrated into the preform by any of several methods, but the usual way is to flow them around the fibers in a slurry under pressure ranging from 1 to 15 MPa or higher, depending on how fast results are wanted. This is done in a metal die. The packed preform goes into a furnace to pyrolyze the fibers at a temperature of about 600°C for carbon and about 400°C for nylon. Pyrolysis leaves pour channels in the ceramic where the fibers were burned away. Molten metal—usually aluminum—is then squeeze-cast into the pour channels and solidified under pressure. About 170 MPa of ram pressure does the job for aluminum reinforcing in ZrO_2. The process is applicable to a wide variety of fabric materials and architectures and many preform shapes, as well as to interesting combinations of ceramics and metals.

Molecular-Scale Reinforcements

A second-phase addition, typically in the 10- to 20-vol % range, and on a molecular scale, can also reduce a ceramic's brittleness. The second phase can consist of particle phase transformation toughening, where the particles inhibit crack propagation by crack-tip deflection and crack-tip stress redistribution. The reinforcement must be both thermodynamically and chemically stable to follow retention of improved mechanical properties to higher temperatures without loss of physical property advantages inherent in the matrix material. The transformation of tetragonal ZrO_2 to the monoclinic phase, by the addition of Y_2O_3 or other stabilizers, is a typical example. Such additions increase the fracture toughness of ZrO_2 ceramics to about 20 MPa $\cdot$ m$^{1/2}$, which is five times that of conventional ceramics. However, the transformation toughening is dependent on the type and amount of solute (HfO_2, CeO_2, or Y_2O_3), the stability of the tetragonal phase, precipitate size, and grain size. To achieve even greater toughness, whisker reinforcements can be added to the toughened zirconias.

This technique, called dual toughening, has been demonstrated with SiC whiskers. The zirconias can also be used as reinforcements themselves, and can replace transformation-toughened Si_3N_4 with dispersions of partially stabilized Y_2O_3–CaO–ZrO_2, called particulate composites. Initial results look promising with strengths and toughnesses of up to 1200 MPa and 13.8 MPa $\cdot$ m$^{1/2}$, respectively.[221] The aforementioned materials belong to a class of composites called molecular or particulate composites. These composites can be obtained by conventional powder mixing, from coprecipitated or alkoxide-derived powders, or from rapidly solidified powders. Investigators have examined Al_2O_3–ZrO_2 eutectic materials, partially stabilized ZrO_2 containing Y_2O_3, CeO_2, or other stabilizers, and Al_2O_3:Cr_2O_3–ZrO_2:HfO_2.

By controlling the tetragonal-to-monoclinic ratio of the dispersed ZrO_2 particles, toughness can be increased. Toughnesses greater than 10 MN/m$^{3/2}$ have been obtained with 40 percent tetragonal content. Suspension processing also can be used to make particulate composites. For example, TiB_2–TZP (tetragonal ZrO_2 polycrystals) has been made in this way. Hot pressing is used for complete densification.

Other ceramics matrices can be reinforced with particles. Scientists have found that α-SiC, reinforced with TiB_2 particles, retains mechanical properties at high temperatures. The mixture, combined with binders, can be shaped using conventional ceramic processing methods such as injection molding, extrusion, isopressing, casting, and green machining. Other particulate composites may require special densification techniques. Al_2O_3–TiC particulate composites usually require additional hot pressing in addition to conventional sintering techniques to achieve full density. Glass matrices also can be reinforced with discrete particles for improved strength and toughness. Sometimes called dispersion hardening, the particulate additions can increase strength and toughness by stopping crack growth, having a higher modulus than the matrix, and inhibiting grain growth or the metallization of glass at high temperatures. For instance, Al_2O_3 and glass can prereact to produce particulate-glass matrix composites containing various borosilicate glasses. Fracture strengths are greater than 70 MPa.

Researchers are looking at sol-gel methods to achieve uniform particle distribution on a molecular scale in the glass matrices. They mix a polysilane SiC precursor with a silica gel derived from tetramethoxysilane (TMOS). Cogellation and in-situ cross-linking of polysilane then occur. Compared to the unreinforced SiO_2, the composites had a threefold increase in hardness and twofold increase in fracture toughness.

Other molecular composites using sol-gel techniques include SiC–SiO_2, SiC–Al_2O_3, and SiC–TiC. Polysilastyrene and alkoxide precursors for SiC and SiO_2, Al_2O_3 and TiC, respectively, are mixed in the presence of a catalyst.[16] Another approach for making molecular composites is the synthesizing of Al_2O_3–TiC and

Al_2O_3–SiC composites by carbothermal reduction of oxides (TiO_2 or SiO_2) in the presence of aluminum, which is a thermite type of reaction. The composite powders are sintered in argon and a 35 percent increase in fracture toughness over conventional SiC has been achieved.[221]

Foams, Spheres, and Other Reinforcements

Tiny spheres, as small as 150 μm in diameter, can be made at the rate of 400 per second. In this system a microsphere generator is used, whereby molten ceramics emerge from a nozzle and drop through a tube where solidification occurs. Many applications exist for ceramic and glass microspheres. Among the myriad applications are low-density fillers for metals and ceramics, fluidized-bed heat exchangers, solid fuel for rockets, catalyst holders, fusion-power fuel containers, and shock-wave dampers.

The challenges of ceramics manufacturing continue to fuel the fire for innovative approaches for reinforcement fabrication. Scientists have designed a rapid liquid-forming process to produce paper-thin ceramic "aerospheres" between 2 and 4 mm in diameter. The process relies on a proprietary coaxial extrusion nozzle with an inner aperture that releases a gas, while an outer ring extrudes a ceramic-powder slurry like bubble blowing. The spheres could provide long-lasting insulation for industrial furnaces or act as emissions-absorption pellets in catalytic converters for automobiles.

Other scientists are developing a new family of ceramics that can be shaped as easily as plastics. So-called amorphous covalent ceramics are noncrystalline versions of familiar ceramics such as Si_3N_4 or WC. They are fired at temperatures of 427 and 760°C as compared to the usual 1649°C. Firing times are also much shorter—20 min versus several hours. The amorphous materials are not quite as hard and heat-resistant as their crystalline counterparts. They will not stand up to temperatures higher than those at which they are fired. However, because they start as liquid polymers (as do ordinary plastics), they can be injection-molded or sprayed onto surfaces before being fired. The technology could speed manufacturing of ceramic objects with complex shapes and simplify the production of ceramic thin films.

Lightweight open-celled ceramic foams are now available and are described as exceptionally rigid and heat-resistant. The foams are being applied to heat exchangers, aerospace structures, filters, and porous electrodes. Materials include Si_3N_4, SiC, and Al_2O_3, which can be fabricated into sandwich and impregnated forms.

Like monolithic foam and cellular-ceramic foam, composite foams are light in weight and have pores that can be left open for filtering or closed for barrier purposes. Cellular ceramics are typically 10 to 20 percent dense and have pores ranging down to micrometer sizes. Foams can be less than 5 percent dense and usually have from 10 to 200 pores per inch, with the composites reaching the higher pore counts and lower densities. Monolithic foams are all ceramic. They are usually made by dipping a precursor matrix in a slurry or slip, drying the result, and then sintering it to volatilize the precursor away. Composites consist of a matrix—usually carbon—and an infiltrate or deposit, with the matrix serving only as a locator for the added material. In composite refractory foams the matrix remains in the finished product. Cellular ceramics and monolithic foams have been on the market for several years; composite foams are still in development.

To make composite refractory foams, researchers start by pyrolyzing an organic precursor to get a reticulated carbon structure. They then either deposit or infiltrate a thin film on the carbon ligaments. Next they heat the structure to deposition temperature and pull reaction gases through it at a uniform rate. The first layer of the deposit forms at nucleation sites on carbon until the surface is coated. Then the refractory film continues to grow on the crystal faces of the deposit. Any refractory material suitable for chemical-vapor techniques can be used, as can oxides, nitrides, and borides.

Materials that have been used as coatings include Hf, Cb, Pt, Rh, Si, Ta, Ti, W, and Zr, depending on the features wanted. Cb on Rh gives a lightweight structure resistant to oxidation at high temperatures, while structures of B, BC, and SiC have high stiffness-to-weight ratios. Researchers can either leave the surface pores open for filtering or closed for barrier purposes. They can also create a density gradient through the cross section to blend, say, a high-density surface for strength with a low-thermal-conductivity interior section for insulation. Parts can be made to near net shape by machining the lattice before further processing. An advantage of composite refractory foams is that they are tailorable to a wide range of needs within their limits of structural strength. For their weight they can have high tensile, flexural, and compressive strength in various loading directions. They can have high toughness and thermal shock resistance compared to monolithic ceramics, while they are low in thermal conductivity because of their long heat paths and low density.

A procedure has been developed for the preparation of small-diameter, hollow ceramic fibers. The process enables the preparation of fibers from a variety of compositions. The low bulk density of the hollow fibers makes them attractive for thermal insulation or in ceramic fiber–polymer matrix composites for certain electronic applications. Preparation of the hollow ceramic fibers starts with activated carbon fibers. These fibers are prepared from polyacrylonitrile fibers by established techniques. The typical procedure involves treatment of the carbon fiber with humidified CO_2 at elevated temperatures. Reaction of the carbon with hot water vapor results in the formation of CO_2 and pitting of the fiber surface. This mild oxidation process is known to create surface porosity on the fibers. Prolonged exposure results in localized destruction of the carbon fiber and breakage. The creation of surface porosity is a key feature that allows the formation of the hollow ceramic fibers. The fibers have outer diameters of 2 to 4 µm and aspect ratios greater than 100:1. Examples of fibers prepared include TiO_2, $BaTiO_3$, ZrO_2, $CaO-ZrO_2$, and $Y_2O_3-ZrO_2$.[226]

REFERENCES

1. F. F. Lange, "Powder Processing Science and Technology for Increased Reliability," *J. Am. Ceram. Soc.,* vol. 72, pp. 3–15, 1989.
2. W. H. Rhodes, "Agglomerate and Particle Size Effects on Sintering Yttria-Stabilized Zirconia," *J. Am. Ceram. Soc.,* vol. 64, pp. 19–22, 1981.
3. F. F. Lange, "Sinterability of Agglomerated Powders," *J. Am. Ceram. Soc.,* vol. 67, pp. 83–89, 1984.
4. F. Lange and M. Metcalf, "Processing-Related Fracture Origins: II, Agglomerate Motion and Cracklike Internal Surfaces Caused by Differential Sintering," *J. Am. Ceram. Soc.,* vol. 66, pp. 398–406, 1983.
5. F. F. Lange, "Structural Ceramics: A Question of Reliability," *J. Mater. Energy Sys., ASM,* vol. 6, pp. 107–113, 1984.
6. J. McDermott, "Ceramics," *Technology,* pp. 19–30, Nov.–Dec. 1981.

7. T. G. Langon, *J. Met.,* vol. 8, pp. 12–16, 1980.
8. B. Fegley and E. A. Barringer, in *Proc. Materials Research Symp.,* vol. 32, 1984, pp. 187–197.
9. W. B. Rhodes, *J. Am. Ceram. Soc.,* vol. 64, pp. 19–22, 1981.
10. K. S. Mazdiyasni, *Ceram. Int.,* vol. 8, p. 42, 1982.
11. J. L. Woodhead and D. L. Segal, *Chem. Br.,* vol. 20, pp. 310–313, 1984.
12. J. D. Birchnall, W. Watt, and B. J. Perov (Eds.), *Handbook of Composites,* vol. 1, Elsevier, Amsterdam, 1985, pp. 115–154.
13. H. Dislich, *Angew. Chem.,* vol. 12, pp. 428–438, 1971.
14. S. C. Danforth and J. S. Haggerty, *Ceram. Eng. Sci. Proc.,* vol. 2, pp. 466–479, 1981.
15. D. W. Johnson, Jr., "Non-Conventional Powder Preparation Techniques," presented at the 6th Ann. Northern Ohio Section, American Ceramic Soc. Symp. on Modern Trends in Ceramic Processing; Theory and Practice, 1984.
16. L. S. Millberg, "The Synthesis of Ceramic Powders," *J. Met.,* pp. 9–13, Aug. 1987.
17. D. S. Kupperman, "NDE of Green and Sintered Structural Ceramics," presented at the Japan–U.S. Seminar on Fundamental Ceramics, Univ. of Washington, Aug. 15, 1984.
18. S. J. Klima, G. Y. Baaklini, and P. B. Aoel, "Nondestructive Evaluation of Structural Ceramics," NASA TM–88978, 1987, 23 pp.
19. "New Structural Materials Technologies; Opportunities for Use of Advanced Ceramics and Composites," Congress-OTA, Tech. Memo TM–E–32, 1986, 88 pp.
20. L. M. Sheppard, "Advances in Processing of Ceramics," *Ceram. Bull.,* vol. 67, pp. 1644–1653, 1988.
21. E. J. Kubel, Jr., "Structural Ceramics; Materials of the Future," *AM&P,* vol. 132, pp. 25–33, Aug. 1988.
22. S. M. Hsu, "Ceramics; Technical Activities 1988," Inst. for Materials Science and Eng., Nat. Inst. of Standards and Technology (IMSE), Gaithersburg, Md., Ceramics Div., NISTIR-88/3840, Feb. 1988, 92 pp.
23. E. P. Rothman, G. B. Kenny, and H. K. Bowen, "Potential of Ceramic Materials to Replace Cobalt, Chromium, Manganese and Platinum in Critical Applications," Materials Processing Center, M.I.T. Industrial Liaison Prog. Rep. 9–11–85, M.I.T., Cambridge, Mass., Final Rep. Contract 333–6530.2, OTA, Jan. 6, 1984, 284 pp.
24. J. S. Reed, *Introduction to the Principles of Ceramic Processing,* Wiley, New York, 1986, 486 pp.
25. *Advanced Structural Ceramics,* BCC Inc., Norwalk, Conn., Dec. 1987, 355 pp.
26. "Trends in Plastics, Ceramics, and Technology," *Met. Prog.,* vol. 125, pp. 89–90, 92, Jan. 1984.
27. *High-Tech Mater. Alert, Tech. Insights,* Englewood, N.J., vol. 14, pp. 1–2, Oct. 1987.
28. L. M. Sheppard, "Ceramics at the Cutting Edge," *AM&P,* vol. 131, pp. 73–79, Aug. 1987.
29. L. M. Sheppard, "Vapor-Phase Synthesis of Ceramics," *AM&P,* vol. 131, pp. 53–58, Apr. 1987.
30. "Chemistry Seen as Key to Successful Ceramics," *Mater. Eng.,* pp. 46–47, Aug. 1982.
31. K. Sawano, J. S. Haggerty, and H. K. Bowen, "Formation of SiC Powder from Laser-Heated Vapour Phase Reactions," *J. Ceram. Soc. Jpn.,* vol. 95, pp. 64–69, 1987.
32. *Sol-Gel Production of High Performance Ceramics and Glasses,* Int. Conf., Gorham Adv. Mater. Inst., Marco Is., Fla., Dec. 1989, 100 pp.

33. "Chemically Derived Ceramics and Glasses-Sol Gel Processing," BCC Inc., Norwalk, Conn., GB-114, Sept. 1988, 350 pp.
34. L. C. Klein, *Sol-Gel Technology for Thin Films, Fibers, Preforms, Electronics, and Specialty Shapes,* Noyes Publ., Park Ridge, N.J., Dec. 1987, 398 pp.
35. C. F. Lewis, "Premium Ceramics from Sol-Gel," *Mater. Eng.,* pp. 49–53, Apr. 1987.
36. J. D. Mackenzie, "New Materials for Spacecraft Stability and Damping—A Feasibility Study," UCLA, Final Rep., AFOSR-TR-86-0308, Dec. 1985, 51 pp.
37. L. M. Sheppard, "Powders that Explode into Materials," *AM&P,* vol. 2, pp. 25–32, Feb. 1986.
38. J. B. Holt, "Exothermic Process Yields Refractory Nitride Materials," *IR&D,* pp. 88–91, Apr. 1983.
39. O. Yamada, Y. Miyamoto, and M. Koizumi, "High Pressure Self-Combustion Sintering of TiC," *J. Am. Ceram. Soc.,* vol. 70, pp. C206–C208, Sept. 1987.
40. L. M. Sheppard, "New-Generation Ceramics," *AM&P,* vol. 1, pp. 39–42, Sept. 1985.
41. Z. A. Munir, "Synthesis of High Temperature Materials by Self-Propagating Combustion Methods," *Ceram. Bull.,* vol. 67, pp. 342–349, Feb. 1988.
42. R. R. Irving, "SHS Ceramic Process R&D Heightening," *Metalwork. News,* pp. 18–20, Mar. 6, 1989.
43. D. S. Phillips and G. J. Vogt, "Plasma Synthesis of Ceramic Powders," *MRS Bull.,* vol. 12, no. 7, p. 54, 1987.
44. E. P. Rothman and H. K. Bowen, "New and Old Ceramic Processes; Manufacturing Costs," Ceramics Processing Res. Lab., Rep. 63, M.I.T. Industrial Liaison Prog. Rep. 4-14-88, M.I.T., Cambridge, Mass., June 1986, 29 pp.
45. *High-Tech Mater. Alert,* vol. 5, p. 9, Apr. 1988.
46. "World Report on Advanced Ceramics," *Tech. Insights,* vol. 1, p. 6, Sept. 1989.
47. H. Anderson, T. T. Kodas, and D. M. Smith, "Vapor-Phase Processing of Powders; Plasma Synthesis and Aerosol Decomposition," *Ceram. Bull.,* vol. 68, pp. 996–1000, May 1989.
48. A. Kumar and R. Roy, "Reactive-Electrode Submerged-Arc Process for Producing Fine Non-Oxide Powders," *J. Am. Ceram. Soc.,* vol. 72, pp. 354–356, Feb. 1989.
49. G. L. Messing et al., "Laser Processing of Ceramics," Pennsylvania State Univ., Final Rep. 6/1/85–11/30/85, DOE/ER/10942—T2, DE86011954, May 1986.
50. J. A. O'Neill et al., "Production of Fine Ceramic Powders from Chloromethylsilanes Using Pulsed Excimer Radiation," *J. Am. Ceram. Soc.,* vol. 72, pp. 1130–1135, July 1989.
51. A. J. Moulson, "Review; Reaction-Bonded Silicon Nitride; Its Formation and Properties," *J. Mater. Sci.,* vol. 14, pp. 1017–1051, 1979.
52. M. L. Torti et al., "Silicon Nitride and Silicon Carbide for High Temperature Engineering Applications," *Proc. Br. Ceram. Soc.,* vol. 22, pp. 129–147, 1973.
53. S. C. Danforth and J. S. Haggerty, "Mechanical Properties of Sintered and Nitrided Laser-Synthesized Silicon Powder," *J. Am. Ceram. Soc.,* vol. 66, pp. C-58–C-59, 1983.
54. J. E. Ritter et al., "High-Strength Reaction-Bonded Silicon Nitride," *Adv. Ceram. Mater.,* vol. 3, pp. 415–417, July 1988.
55. "Ceramics," *New Mater. Jpn.,* vol. 6, pp. 2–3, July 1989.
56. R. A. Bauer et al., "Laser-Chemical Vapor Synthesis of Submicron and Silicon Nitride Powders from Halogenated Silanes," *Ceram. Sci. Eng. Proc.,* vol. 9, pp. 949–956, 1988.
57. E. J. Jaquemijns, P. J. van der Put, and J. Schoonman, "Vapour Phase Synthesis of

Ultrafine Silicon Nitride Powders," *High Temp.—High Pressures,* vol. 20, pp. 31–34, 1988.
58. R. A. Bauer et al., "Laser-Chemical Vapor Precipitation of Submicrometer Silicon and Silicon Nitride Powders from Chlorinated Silanes," *J. Am. Ceram. Soc.,* vol. 72, pp. 1301–1304, July 1989.
59. K. Masters, *Spray Drying,* 4th ed., Wiley, New York, 1985.
60. S. J. Lukasiewicz, "Spray-Drying Ceramic Powders," *J. Am. Ceram. Soc.,* vol. 72, pp. 617–624, April 1989.
61. L. M. Sheppard, "International Trends in Powder Technology," *Ceram. Bull.,* vol. 68, pp. 978–985, May 1989.
62. V. A. Lavrenko et al., "High-Temperature Oxidation of Composites Based on Titanium Diboride," AFSC, Ohio, FTD ID(RS)T 0357-88, June 1988.
63. A. G. Merzhanov and I. P. Borovinskaya, "Self-Propagating High Temperature Synthesis of Refractory Inorganic Compounds," *Dokl. Akad. Nauk SSR,* vol. 204, p. 366, 1972.
64. A. G. Merzhanov, A. K. Filonenko, and I. P. Borovinskaya, "New Phenomena in Combustion of Condensed Systems," *Dokl. Akad. Nauk SSR,* vol. 208, p. 892, 1973.
65. A. G. Merzhanov and I. P. Borovinskaya, "A New Class of Combustion Processes," *Combustion Sci. Technol.,* vol. 10, p. 195, 1975.
66. Kiser Research, Inc., Memo. Briefing on Soviet Solid Flame Ceramics Processing, Washington, D.C., Apr. 18, 1989.
67. W. L. Frankhouser, "Advanced Processing of Ceramic Compounds," SPC, U.S. Dept. of Defense Rep., DARPA, May, 1986, and Noyes Data Corp., 1987, 188 pp.
68. R. R. Irving, "New Soviet Ceramics Technology Coming to U.S.," *Metalwork. News,* p. 28, May 8, 1989.
69. D. S. Weiss, "Reaction Hot Pressing; Ceramic Processing by Self-Propagating High-Temperature Synthesis," SME EM89-121, Feb. 1989, 13 pp.
70. J. D. Walton, Jr., and N. E. Poulos, "Cermets from Thermite Reactions," *J. Am. Ceram. Soc.,* vol. 42, pp. 40–49, 1959.
71. Z. A. Munhir, "Synthesis of High Temperature Materials by Self-Propagating Combustion Methods," *Ceram. Bull.,* vol. 67, pp. 342–349, 1988.
72. S. Komarneni et al., *Adv. Ceram. Mater.,* vol. 1, pp. 87–92, 1986.
73. J. A. Pask and A. G. Evans, *Ceramic Microstructures '86—Role of Interfaces,* vol. 21 of *Materials Science Research,* Plenum Press, New York, 1987, 947 pp.
74. S. Somiya et al., "Microstructure Development of Hydrothermal Powders and Ceramics," in *Ceramic Micro. '86,* Plenum Press, New York, 1987, pp. 465–472.
75. K. J. Wynne and R. W. Rice, *Ann. Rev. Mater. Sci.,* vol. 14, p. 297, 1984.
76. R. W. Rice, *Am. Ceram. Soc. Bull.,* vol. 62, p. 889, 1983.
77. D. Seyferth, "Synthesis of Some Organosilicon Polymers and Their Pyrolytic Conversion to Ceramics," Dept. of Chem., M.I.T. Industrial Liaison Prog. Rep. 2-17-89, M.I.T., Cambridge, Mass., Feb. 1989, 33 pp.
78. J. D. Mackenzie and D. R. Ulrich, *Ultrastructure Processing of Advanced Ceramics,* Wiley, New York, June 1987, 1010 pp.
79. M. L. J. Hackney et al., "Organometallic Precursors to $Al_wSi_xN_yC_z$ Ceramics," in *Ultrast. Proc. of Adv. Ceramics,* Wiley, New York, 1987, pt. 1, Chap. 6, pp. 99–111.
80. K. Niihara et al., "Nanostructure and Mechanical Properties of SiC Consolidated Using Organosilicon Precursors," in *Ultrast. Proc. of Adv. Ceramics,* Wiley, New York, 1987, pt. 4, Chap. 40, pp. 547–556.
81. Y. -F. Yu and T. -I. Mah, "Silicon Oxynitride and Si–Al–O–N Ceramics from

Organosilicon Polymers," in *Ultrast. Proc. of Adv. Ceramics,* Wiley, New York, 1987, pt. 6, Chap. 59, pp. 773–782.

82. B. I. Lee and L. L. Hench, "Molecular Composites of SiC/SiO$_2$, SiC/Al$_2$O$_3$, and SiC/TiC," *Ceram. Bull.,* vol. 66, pp. 1482–1485, Oct. 1987.

83. I. Kimura et al., "Preparation of SiC–Al$_2$O$_3$ Composite Powder," *J. Mater. Sci. Lett.,* vol. 6, pp. 1359–1360, Nov. 1987.

84. "Ceramic Research at the Powder Metallurgy Laboratory of the Max-Planck-Institut in Stuttgart," *PMI,* vol. 20, pp. 26–28, Aug. 1988.

85. P. A. Haas, "Gel Processes for Preparing Ceramics and Glasses," *Chem. Eng. Prog.,* pp. 44–52, Apr. 1989.

86. "Sol-Gel Production of High-Performance Ceramics and Glasses," Gorham Adv. Mater. Inst., Multi-Client Study, Mar. 15, 1988.

87. T. Woignier, "Al$_2$O$_3$–TiO$_3$ and Al$_2$TiO$_5$ Ceramic Materials," Aerospatiale Aquitaine, St. Medard en Jalles, France, SNIAS 881–430–122, 1988, 16 pp.

88. R. Roy, "Exploitation of the Sol-Gel Route in Processing of Ceramics and Composites," Pennsylvania State Univ., Mater. Res. Lab., AFOSR TR 87–1193, Contract F49620-85C0069, July 1987, 103 pp.

89. A. C. Pierre, D. R. Uhlmann, and A. Hordonneau, "Ceramic Composites Made by Sol-Gel Processing," Dept. of Mater. Sci. and Eng., M.I.T. Liaison Prog. Rep. 1–35–86, M.I.T., Cambridge, Mass., 1986, 6 pp.

90. D. R. Uhlmann, B. J. J. Zelinski, and G. E. Wnek, "The Ceramist as Chemist—Opportunities for New Materials," Dept. of Mater. Sci. and Eng., M.I.T. Industrial Liaison Prog. Rep. 10-9-84, M.I.T., Cambridge, Mass., 1984, 12 pp.

91. W. D. Bond and P. F. Becher, "Synthesis of Alumina-Zirconia Powders by Sol-Gel Processing," ORNL, Rep. Conf-870254-2, Contract AC05-840R21400, Feb. 23, 1987, 18 pp.

92. G. L. Messing and M. Kumagai, "Low-Temperature Sintering of Seeded Sol-Gel-Derived, ZrO$_2$-Toughened Al$_2$O$_3$ Composites," *J. Am. Ceram. Soc.,* vol. 72, pp. 40–44, Jan. 1989.

93. D. J. Green, R. H. J. Hannink, and M. V. Swain, *Transformation Toughening of Ceramics,* CRC Press, Boca Raton, Fla., 1988, 256 pp.

94. H. G. Sowman, "Alumina-Boria-Silica Ceramic Fibers from the Sol-Gel Process," Noyes Publ., Park Ridge, N.J., 1987, Chap. 8, pp. 162–183.

95. S. Sakka, "Fibers from the Sol-Gel Process," Noyes Publ., Park Ridge, N.J., 1987, Chap. 7, pp. 140–161.

96. C. J. Brinker, A. J. Hurd, and K. J. Ward, "Fundamentals of Sol-Gel Thin Film Formation," Sandia Natl. Labs., Albuquerque, N.M., Rep. Sand-87-0402C, Contract ACO4-76DP00789, Feb. 23, 1987, 23 pp.

97. B. D. Fabes et al., "Stronger Glass Via Sol Gel Coatings," Dept. of Mater. Sci. and Eng., M.I.T. Industrial Liaison Prog. Rep. 9–34–85, M.I.T., Cambridge, Mass., 1985, 10 pp.

98. L. Montanaro and B. Guilhot, "Preparation of Microspheres from an Alumina-Zirconia Sol," *Ceram. Bull.,* vol. 68, pp. 1017–1020, May 1989.

99. J. P. Mathers and T. E. Forester, "Sol-Gel Preparation of Non-Oxide Abrasives," *Ceram. Bull.,* vol. 68, pp. 1330–1336, July 1989.

100. C. J. Brinker et al., *Proc. Better Ceramics Through Chemistry III Symp.,* vol. 21, Mater. Res. Soc., Pittsburgh, Penn., AFOSR TR 89-0444, AFOSR88-0145, Final Rep. 4/88–3/89, Aug. 1988.

101. T. M. Maccagno, "Processing of Advanced Ceramics which Have Potential for Use in Gas Turbine Aero Engines," Natl. Res. Council of Canada, NAE-AN-58, NRC 30057, Feb. 1989, 37 pp.

102. R. F. Bunshah, "Two Vapor Coating Processes Cover the Field of Applications," *R&D,* pp. 73–78, Aug. 1989.
103. W. E. Cole et al., "Research and Development of a CVD Ceramic Composite Heat Exchanger for Industrial Waste Heat Recovery," U.S. Dept of Energy, Tegogen, Inc., DOE/ID/12544-3 (DE89020622), Contract DE-AC07-84ID12544, May 1988, 191 pp.
104. J. S. Golla and R. L. Taylor, "Monolithic Material Fabrication by Chemical Vapour Deposition," *J. Mater. Sci.,* vol. 23, pp. 4331–4339, Dec. 1988.
105. *High-Tech Mater. Alert,* vol. 5, p. 2, Feb. 1988.
106. Y. Miyamoto, "Ceramics Reach for the Sky and Beyond," *AM&P,* vol. 4, p. 3, Mar. 1988.
107. *High-Tech Mater. Alert,* vol. 3, p. 2, Apr. 1986.
108. D. B. Olson, "New Process for Synthesis of Silicon Nitride Powders for Advanced Ceramics," Aerochem Res. Lab., Aerochem TP469, ARO21828.1MSS, Contract DAAG29-84COO23, Nov. 1987, 20 pp.
109. J. L. Hoyer, "Ultrafine Ceramic Powders Produced by Turbomilling," *Ceram. Bull.,* vol. 67, pp. 1663–1668, Oct. 1988.
110. D. E. Wittmer, "Alternative Processing Through Turbomilling," *Ceram. Bull.,* vol. 67, pp. 1670–1672, Oct. 1988.
111. T. P. Herbell, M. R. Freedman, and J. D. Kiser, "Particle Size Reduction of Si_3N_4 Powder with Si_3N_4 Milling Hardware," NASA TM-86864, N86-24839, in *Proc. 10th Ann. Conf. on Composites and Adv. Ceramic Materials,* Cocoa Beach, Fla., Jan. 19–26, 1986, 19 pp.
112. S. Mito, "The Ceramics Revolution; An Industrial Army to Bring About the Fourth Industrial Revolution," PHP Res. Labs., Kyoto, Japan, FSTC-HT-0215-85, AD-B103793, transl. Leo Kanner Assoc., Redwood City, Calif., Mar. 1986, 196 pp.
113. Y. Ikuhara, H. Kurishita, and H. T. Toshinaga, "Grain Boundary Structure and Mechanical Properties of Covalent-Bonded Ceramics," in *Interfaces in Polymer, Ceramic & Metal Matrix Composites,* Elsevier Sci. Pub., London, 1988, pp. 673–684.
114. W. Lee et al., "Chemical Vapor Deposition of Silicon Carbide Using a Novel Organometallic Precursor," Dept of Chem. and Mater. Eng., Rensselaer Polytechnic Inst., DCTR5, Contract NOOO14-85KO632, May 15, 1989, 7 pp.
115. H. Kodama and T. Miyoshi, "Preparation of Fine-Grained SiC Ceramics from Pyrolyzed Polycarbosilane," *Adv. Ceram. Mater.,* vol. 3, pp. 177–179, Feb. 1988.
116. G. A. Gogotsi et al., "The Effect of the Composition on the Mechanical Properties of a Material Based on Silicon Nitride," *Poroshkovaia Metallurgia, AN USSR,* pp. 93–96, Feb. 1980.
117. D. S. Yan and T. S. Yen, "High Performance Nitride Materials Research in the Chinese Academy of Sciences," *Mater. Res. Bull.,* vol. 22, pp. 1249–1257, Sept. 1987.
118. C. F. Lewis, "Silicon Nitride; The Rock-Solid Performer," *Mater. Eng.,* pp. 30–33, May 1989.
119. G. T. Burns and G. Chandra, "Pyrolysis of Preceramic Polymers in Ammonia; Preparation of Silicon Nitride Powders," *J. Am. Ceram. Soc.,* vol. 172, pp. 333–337, Feb. 1989.
120. G. Ziegler, J. Heinrich, and G. Wötting, "Relationships between Processing, Microstructure and Properties of Dense and Reaction-Bonded Silicon Nitride," *J. Mater. Sci.,* vol. 22, pp. 3041–3086, Sept. 1987.
121. B. Durham, M. Murtha, and G. Burnet, "Si_3N_4 by the Carbothermal Ammonolysis of Silica," *Adv. Ceram. Mater.,* vol. 3, pp. 45–48, Jan. 1988.
122. S. Y. Kuo, A. V. Virkar, and W. Rafaniello, "Modulated Structures in SiC-AlN Ceramics," *J. Am. Ceram. Soc.,* vol. 70, pp. C125–C128, June 1987.

123. *Ceram. Ind. Jpn.,* vol. 4, pp. 1–15, May 1988.
124. "High Purity Boron Nitride Powder Synthesis," NTIS PB 86-217940/WMS, 1986.
125. W. J. Lackey et al., "Ultrafine Microstructure Composites Prepared by Chemical Vapor Deposition," Georgia Inst. of Technology, Contract NOOO14-87-K-0036, Ann. Rep. A-4699-2, Jan.–Dec. 1988, 38 pp.
126. G. L. Leatherman and R. N. Katz, "Structural Ceramics; Processing and Properties," in J. Tien and T. Caulfield (Eds.), *Superalloys, Supercomposites, Superceramics,* Materials Science and Technology Series, Academic Press, Orlando, Fla., 1989, 755 pp.
127. J. C. Southern, "Ceramic Aluminas; Manufacturing, Use and Characteristics," SME, EM 89–118, *Adv. Ceram. '89,* Philadelphia, Penn., Feb. 20–22, 1989, 23 pp.
128. *Ceramic Whiskers, Fibers and Ceramic Matrix Composites,* BCC Inc., Norwalk, Conn., GB-110, Oct. 1988, 261 pp.
129. P. Becher, in S. Somiya (Ed.), *Zirconia Ceramics I,* Uchida Rokakuho Publ., Tokyo, 1983, pp. 151–159.
130. F. Erdogan and P. F. Joseph, "Toughening of Ceramics through Crack Bridging by Ductile Particles," *J. Am. Ceram. Soc.,* vol. 72, pp. 262–270, Feb. 1989.
131. N. Claussen, in N. Claussen, M. Ruhle, and A. Heuer (Eds.), *Advances in Ceramics,* American Ceramic Soc., Westerville, Ohio, 1984, pp. 325–351.
132. D. D. Messier, "The Alpha/Beta Silicon Nitride Phase Transformation," AMMRC Rep. AMMRC-TR-77-15, May 1977.
133. L. J. Gauckler and G. Petzow, "Representation of Multi-Component Silicon Nitride Systems," in F. L. Riley (Ed.), *Nitrogen Ceramics,* Noordhoff, The Netherlands, 1977, pp. 41–62.
134. F. F. Lange, "Phase Relations in the System Si_3N_4-SiO_2-MgO and Their Interrelation with Strength and Oxidation," *J. Am. Ceram. Soc.,* vol. 61, pp. 53–56, 1978.
135. D. A. Bonnell, T. Y. Tien, and M. Ruhle, "Controlled Crystallization of the Amorphous Phase in Silicon Nitride Ceramics," *J. Am. Ceram. Soc.,* vol. 70, pp. 460–465, July 1987.
136. T. Ekström and P. Olsson, "β-Sialon Ceramics Prepared at 1700°C by Hot Isostatic Pressing," *J. Am. Ceram. Soc.,* vol. 72, pp. 1722–1724, Sept. 1989.
137. L. M. Sheppard, "SIALON—Another Super Structural Ceramic," *AM&P,* vol. 2, pp. 35–39, Jan. 1986.
138. *New Mater.—Japan (Ceramics),* p. 5, Aug. 1989.
139. T. M. Allen, I. Birkby, and R. Stevens, "Nature and Effects of Defects Introduced during Fabrication of Zirconia Engineering Ceramics," *Powder Metall.,* vol. 31, pp. 23–27, Jan. 1988.
140. A. Kato, "Study on Powder Preparation in Japan," *Ceram. Bull.,* vol. 66, pp. 647–650, Apr. 1987.
141. A. Kato, K. Inoue, and Y. Katatae, "Sintering Behavior of Yttria-Stabilized Zirconia (YSZ) Powders Prepared by Homogeneous Precipitation," *Mater. Res. Bull.,* vol. 22, pp. 1275–1281, Sept. 1987.
142. S. K. Chan, "Polymorphic Transformations of Zirconia," Argonne Natl. Lab., Ill., Conf. 870843-11, W31-109, ENG38, 1987, 27 pp.
143. D. B. Marshall, M. R. James, and J. R. Porter, "Structural and Mechanical Property Changes in Toughened Magnesia-Partially-Stabilized Zirconia at Low Temperatures," *J. Am. Ceram. Soc.,* vol. 72, pp. 218–227, Feb. 1989.
144. G. Jingkun, "Brittleness and Toughening of Ceramics," AF Systems Command, FTD ID(RS)T 0419–88, July 1988.
145. D. B. Marshall, "Transformation Toughening of Ceramics," RI., SC5444 AR, AFOSR TR87–1854, AR2 9/86–8/87, Contract F49620–85-C-0143, Oct. 1987, 116 pp.

146. W. F. Kladnig, "Fracture Behavior of Duplex Al_2O_3-ZrO_2 Ceramics," *Mater. Chem. Phys.*, vol. 18, pp. 181–191, June 1987.

147. A. H. Heuer, "Transformation Toughening in ZrO-Containing Ceramics," *J. Am. Ceram. Soc.*, vol. 70, pp. 689–698, Oct. 1987.

148. J. J. Swab, "Properties of Yttria-Tetragonal Zirconia Polycrystal (Y-TZP) Materials after Long-Term Exposure to Elevated Temperatures," Army Lab. Command, MTL TR-89-21, FR3/89, Mar. 1989.

149. W. H. Rhodes and S. Natansohn, "Powders for Advanced Structural Ceramics," *Ceram. Bull.*, vol. 68, pp. 1804–1812, Oct. 1989.

150. R. W. Davidge, "Review of Composites—Behavior of Materials," *Composites*, vol. 18, pp. 92–98, Apr. 1987.

151. *Proc. Metal and Ceramic Matrix Composite Processing Conf.*, vol 2, Metals and Ceramic Information Center, Battelle Columbus Labs., Columbus, Ohio, DLA900-83-C-1744, Nov. 13–15, 1984, 284 pp.

152. B. Budiansky, "Small-Scale Crack Bridging and Fracture Toughness of Particulate-Reinforced Ceramics," Harvard Univ., Mech. 104, Contract NOOO14-86-K-0753, June 1987, 33 pp.

153. J. V. Milewski, "Efficient Use of Whiskers in the Reinforcement of Ceramics," *Adv. Ceram. Mater.*, vol. 1, p. 36, 1986.

154. V. C. Li, "Fiber Reinforced Structural Ceramics for Construction," M.I.T., Cambridge, Mass., ARO-24621 4-EG-UIE, Contract DAALO3-86-G-0197, Jan. 1988, 22 pp.

155. R. P. Vozzola, "Fracture Toughness Testing of Ceramic Matrix Composite," Thesis, Air Force Inst. of Technology, AFIT/GAE/AA/87 D-24, WPAFB School of Eng., Dec. 1987, 95 pp.

156. T. G. Nieh, C. M. McNally, and J. Wadsworth, "Superplasticity in Intermetallic Alloys and Ceramics," *J.Met.*, pp. 31–35, Sept. 1989.

157. S. W. Freiman et al., "Mechanical Property Enhancement in Ceramic Matrix Composites," Natl. Bureau of Standards (IMSE), Gaithersburg, Md., Ceramics Div., NBSIR 88-3798, Apr. 1988, 60 pp.

158. H. Icikawa et al., "Mechanical and Electrical Properties of SiC Fibers (NICALON) and their Composites," presented at the 1st Japan Intl. SAMPE Symp. and Exhibition, Chiba, Japan, Nov. 29, 1989, 12 pp.

159. K. M. Prewo, "Evaluation of Advanced Fibers for the Reinforcement of Glass and Ceramic Matrix Composites," presented at the 1st Japan Intl. SAMPE Symp. and Exhibition, Chiba, Japan, Nov. 29, 1989, 15 pp.

160. T. Mah et al., "Recent Developments in Fiber Reinforced High Temperature Ceramic Composite," *Am. Ceram. Soc. Bull.*, vol. 66, pp. 304–308, 317, Feb. 1987.

161. J. Cornie et al., "Processing of Metal and Ceramic Matrix Composites," *Am. Ceram. Soc. Bull.*, vol. 65, pp. 293–304, Feb. 1986.

162. L. J. Schioler and J. J. Stiglich, Jr., "Ceramic Matrix Composites; Literature Review," *Am. Ceram. Soc. Bull.*, vol. 65, pp. 289–292, Feb. 1986.

163. F. O. Anderegg, *Ind. Eng. Chem.*, vol. 31, p. 290, 1939.

164. J. D. Birchall, "Inorganic Fibers," in M. B. Bever (Ed.), *Encyclopedia of Mater. Sci. and Eng.*, Pergamon Press, Oxford, 1986, pp. 2333–2335.

165. J. D. Birchall, J. A. A. Bradbury, and J. Dinwoodie, "Alumina Fibers," in W. Watt and B. V. Perov (Eds.), *Handbook of Composites*, North-Holland, Amsterdam, 1985, pp. 115–155.

166. T. Tsuchiya, "Synthesis of Al_2O_3-SiO_2 Glasses from Sol-Gel Process and Properties," presented at the 1st Japan Intl. SAMPE Symp. and Exhibition, Chiba, Japan, Nov. 28, 1989, 10 pp.

167. T. Maki and S. Sakka, "Preparation of Porous Alumina Fibres by Unidirectional Freezing of Gel," *J. Mater. Sci. Lett.,* vol. 5, pp. 28–30, 1986.
168. J. Tretzel et al., "Production, Processing, and Application of Enka Silica Fibers," 57th Ann. TRI Conf., Charlotte, N.C., Apr. 21–22, 1987.
169. W. C. Miller, in M. Grayson (Ed.), *Encyclopedia of Textiles, Fibers and Nonwoven Fabrics,* Wiley, New York, 1984, pp. 443–450.
170. P. Brocke, H. Schurmans, and J. Verhoest, *Inorganic Fibers and Composite Materials; A Survey of Recent Developments,* Pergamon Press, Oxford, 1984.
171. E. S. Chin, "Structures and Properties of Magnesium Base Composites," Army Lab. Command, Mater. Tech. Lab., Watertown, Mass., MTL-TR-88-41, Dec. 1988, 9 pp.
172. M. M. Schwartz, *Composite Materials Handbook,* McGraw-Hill, New York, 1983, 600 pp.
173. J. O. Carlsonn, "Silicon Carbide Fibers," in M. B. Bever (Ed.), *Encyclopedia of Mater. Sci. and Eng.,* Pergamon Press, Oxford, 1986, pp. 4406–4408.
174. Y. Yiming et al., "Preparation and Properties of Porous Silicon Carbide Fiber," presented at the 1st Japan Intl. SAMPE Symp. and Exhibition, Chiba, Japan, Nov. 29, 1989, 10 pp.
175. S. Yajima et al., "Synthesis of Continuous Silicon Carbide Fiber with High Tensile Strength and High Young's Modulus," *J. Mater. Sci.,* vol. 13, pp. 2569–2576, 1978.
176. Y. Hasegawa and K. Okamura, "Synthesis of Continuous Silicon Carbide Fibre; Part 3: Pyrolysis Process of Polycarbosilane and Structure of the Products," *J. Mater. Sci.,* vol. 18, pp. 3633–3648, 1983.
177. H. Ichikawa et al., "Synthesis of Continuous Silicon Carbide Fibre; Part 5: Factors Affecting Stability of Polycarbosilane to Oxidation," *J. Mater. Sci.,* vol. 21, pp. 4352–4358, 1986.
178. S. Yajima, "Silicon Carbide Fibres," in W. Watt and B. V. Perov (Eds.), *Handbook of Composites,* North-Holland, Amsterdam, 1985, pp. 201–232.
179. Y. Hasegawa and K. Okamura, "Synthesis of Continuous Silicon Carbide Fibres," *Yogyo Kyokai Shi,* vol. 95, pp. 99–103, 1987.
180. B. G. Penn et al., "Preparation of Silicon Carbide Nitride Fibers by the Pyrolysis of Polycarbosilazane Precursors," *Polym. Eng. Sci.,* vol. 26, pp. 1191–1194, Sept. 1986.
181. T. Tani and S. Wada, "SiC Matrix Composites Reinforced with ZrB_2 Particulate Internally Synthesized by the Carbothermal Reaction of ZrO_2," presented at the 1st Japan Intl. SAMPE Symp. and Exhibition, Chiba, Japan, Nov. 29, 1989, 18 pp.
182. R. T. Bhatt, "Effects of Fabrication Conditions on Properties of SiC Fiber Reinforced Reaction-Bonded Silicon Nitride Matrix Composites (SiC/RBSN)," NASA, E 3169, TM 88814, Jan. 22, 1986, 18 pp.
183. D. Seyferth, "The Preparation of Silicon-Containing Ceramics by Organosilicon Polymer Pyrolysis," M.I.T., Cambridge, Mass., TR-25, Contract NOOO14-82-K-0322, ONR, June 25, 1988, 32 pp.
184. F. K. Ko, "Preform Fiber Architecture for Ceramic-Matrix Composites," *Ceram. Bull.,* vol. 68, pp. 401–414, Feb. 1989.
185. S. T. Buljan and V. K. Sarin, "Silicon Nitride-Based Composites," *Composites,* vol. 18, pp. 99–106, Apr. 1987.
186. J. Lipowitz, H. A. Freeman, and R. T. Chen, "Composition and Structure of Ceramic Fibers Prepared from Polymer Precursors," *Adv. Ceram. Mater.,* vol. 2, pp. 121–128, Feb. 1987.
187. *High-Tech Mater. Alert,* vol. 4, p. 3, Feb. 1987.
188. Y. D. Blum, "Polymer Precursors to Silicon Nitride and Other Ceramics; Synthesis and Applications," SRI Intl. D89-1315, Jan. 1989, 22 pp.

189. *New Mater.—Japan,* vol. 5, pp. 6–7, July 1988.
190. T. Isoda et al., "High Purity Continuous Silicon Nitride Fibers," presented at the 1st Japan Intl. SAMPE Symp. and Exhibition, Chiba, Japan, Nov. 29, 1989, 15 pp.
191. D. B. Marshall, F. F. Lange, and P. D. Morgan, "High-Strength Zirconia Fibers," *J. Am. Ceram. Soc.,* vol. 70, pp. C-187–C-188, Aug. 1987.
192. M. H. Stacey, "Production and Characterization of Fibers for Metal Matrix Composites," *Mater. Sci. Technol.,* vol. 14, pp. 227–230, Mar. 1988.
193. D. Ulrich and R. A. Wagner, "A New Process for Final Densification of Ceramics," ARPA Order 5172, AFOSR Contract F49620-85-C-0053, Final Rep. AFOSR-TR-88-1008, 2/85–5/88, 62 pp.
194. A. R. Bunsell, "Development of the Fine Ceramic Fibers for High Temperature Composites," *Mater. Forum,* vol. 11, pp. 78–84, Jan. 1988.
195. H. Fukunaga, "Processing Aspects of Squeeze Casting for Short Fiber Reinforced Metal Matrix Composites," *Adv. Mater. Manuf. Proc.,* vol. 3, pp. 669–687, Apr. 1988.
196. A. N. Scoville, P. Reagan, and F. N. Huffman, "Evaluation of SiC Matrix Composites for High Temperature Applications," *Adv. Mater. Manuf. Proc.,* vol. 3, pp. 643–668, Apr. 1988.
197. W. H. Atwell et al., "Advanced Ceramics Based on Polymer Processing," Dow Corning Corp., vol. II: Comp. Tech., AFWAL-TR-85-4099, Contract F33615-83-C-5006, Final Rep. 2/8/84 to 2/7/85, Dec. 1987, 622 pp.
198. A. P. Katz and R. J. Kerans, "Structural Ceramics Program at AFWAL Materials Lab.," *Ceram. Bull.,* vol. 67, pp. 1360–1366, Aug. 1988.
199. J. V. Milewski, "Whiskers," in M. B. Bever (Ed.), *Encyclopedia of Mater. Sci. and Eng.,* Pergamon Press, Oxford, 1986, pp. 5344–5346.
200. P. F. Becher, C. H. Hsueh, and P. Angelini, "Toughening Behavior in Whisker-Reinforced Ceramic Matrix Composites," *J. Am. Ceram. Soc.,* vol. 71, pp. 1050–1061, Dec. 1988.
201. G. C. Wei and P. F. Becher, "Development of SiC-Whisker-Reinforced Ceramics," *Am. Ceram. Soc. Bull.,* vol. 64, pp. 298–304, Feb. 1985.
202. S. T. Buljan et al., "Dispersoid-Toughened Silicon Nitride Composites," GTE, ORNL, Contract DE-AC05-84OR21400, ORNL/Sub85-22011/1 Final Rep., Sept. 1988, 63 pp.
203. S. J. Harris, "Cast Metal Matrix Composites," *Mater. Sci. Technol.,* vol. 4, pp. 231–239, Mar. 1988.
204. E. Fitzer and R. Gadow, "Fiber-Reinforced Silicon Carbide," *Am. Ceram. Soc. Bull.,* vol. 65, pp. 326–335, Feb. 1985.
205. D. Broussard and F. Pochet, "Fibre Reinforced Ceramic Matrix Composites in Japan," *Ind. Ceram.,* no. 839, pp. 442–444, 1989.
206. *High-Tech Mater. Alert,* vol. 5, p. 9, Apr. 1988.
207. J. G. Morley, "High-Performance Fibre Composites," in *Reinforcing Fibers,* Academic Press, Orlando, Fla., 1987, Chap. 2, pp. 21–88.
208. *High-Tech Mater. Alert,* vol. 5, p. 1, Mar. 1988.
209. K. G. Nickel et al., "Thermodynamic Calculations for the Formation of Si/C-Whisker-Reinforced Si_3N_4 Ceramics," *Adv. Ceram. Mater.,* vol. 3, pp. 557–562, Nov. 1988.
210. T. N. Tiegs and P. F. Becher, "Sintered Al_2O_3-SiC-Whisker Composites," *Am. Ceram. Soc. Bull.,* vol. 66, pp. 339–342, Feb. 1987.
211. L. Björk and L. A. G. Hermansson, "Hot Isostatically Pressed Alumina-Silicon Carbide-Whisker Composites," *J. Am. Ceram. Soc.,* vol. 72, pp. 1436–1438, Aug. 1989.
212. *World Report on Advanced Ceramics,* vol. 1, pp. 5–6, Sept. 1989.

213. L. M. Sheppard, "Toughening Glass with Fiber Reinforcements," *Ceram. Bull.,* vol. 67, pp. 1779–1782, Nov. 1988.
214. R. Chaim, L. Baum, and D. G. Brandon, "Mechanical Properties and Microstructure of Whisker-Reinforced Alumina–30 vol % Glass Matrix Composite," *J. Am. Ceram. Soc.,* vol. 72, pp. 1636–1642, Sept. 1989.
215. *High-Tech Mater. Alert,* vol. 5, p. 10, June 1988.
216. J. Lankford, "Ceramics Research at the Southwest Research Institute," *Ceram. Bull.,* vol. 68, pp. 1418–1423, Aug. 1989.
217. Y. Nishida and H. Matsubara, "Effect of High Pressure on the Solidification of Al and Al-25.4 wt % Si Alloy," *Z. Metallk.,* vol. 71, pp. 189–194, 1980.
218. Y. Nishida et al., "Fundamental Study on the Squeeze Casting," *Bull. Jpn. Inst. Met.,* vol. 19, pp. 895–902, 1980.
219. Y. Nishida et al., "Si_3N_4 Whisker Reinforced Aluminum Alloys Fabricated by Squeeze Casting," *Bull. Jpn. Inst. Met.,* vol. 27, pp. 429–438, 1988.
220. "Alumina-Based Whiskers for Fiber-Reinforced Plastics and Metals," *New Tech. Japan,* vol. 17, pp. 19–20, Aug. 1989.
221. L. S. Millberg, "The Search for 'Ductile Ceramics'," *J. Met,.* vol. 39, pp. 10–13, Nov. 1987.
222. S. V. Prasad and P. K. Rohatgi, "Tribological Properties of Al Alloy Particle Composites," *J. Met.,* vol. 39, pp. 22–26, Nov. 1987.
223. S. Y. Oh, J. A. Cornie, and K. C. Russell, "Particulate Wetting and Metal:Ceramic Interface Phenomena," Dept. of Mater. Sci. and Eng., M.I.T. Ind. Liaison Prog. Rep. 6-44-87, July 1987, 55 pp.
224. *High-Tech Ceram. News,* vol. 1, Aug. 1989, 10 pp.
225. S. V. Kamat, J. P. Hirth, and R. Mehrabian, "Mechanical Properties of Particulate-Reinforced Aluminum Matrix Composites," *Acta Metall.,* vol. 37, pp. 2395–2402, Sept. 1989.
226. R. J. Card, "Preparation of Hollow Ceramic Fibers," *Adv. Ceram. Mater.,* vol. 3, pp. 29–31, Jan. 1988.

BIBLIOGRAPHY

Adair, J. H., "An Overview of Ceramic Powder Processing," presented at the ASM Intl. Ann. Mtg., Indianapolis, Ind., Oct. 2, 1989.

Advanced Materials, vol. 11, Jan. 8, 1990, 53 pp.

Ayral, A., et al., "Zirconia by the Gel Route," in G. De With, R. A. Terpstra, and R. Metselaar (Eds.), *Euro-Ceramics,* 1989, vol. 1, pp. 1268–1274.

Benedetti, A., et al., "Structural Properties of Ultrafine Zirconia Powders Obtained by Precipitation Methods," *J. Mater. Sci.,* vol. 25, pp. 1473–1478, 1990.

Bunsell, A. R., *Fibre Reinforcements for Composite Materials,* Composite Materials Series, vol. 2, Elsevier, 1989, 479 pp.

Burkland, C. V., et al., "CVI-Processed Ceramic Matrix Composites," presented at the Aeromat '90 Conf. and Exposition, Long Beach, Calif., May 23, 1990.

"Ceramic Matrix Composites: High Temperature Effects" (Jan. 72–Sept. 89), PB89-871412, NTIS, Springfield, Va., Sept. 1989, 38 pp.

Chu, C. Y., and J. P. Singh, "Mechanical Properties and Microstructure of SiN Whisker-Reinforced Si_3N_4 Matrix Composites," Paper 19-C-90F, presented at the 14th Ann. Conf. on Comp. and Adv. Ceram., Cocoa Beach, Fla., Jan. 14–17, 1990.

Claussen, N., et al., "Development and Testing of Short Fibre and Whisker Reinforced Ceramic Composite Materials," *Keram. Z.,* vol. 41, pp. 484–487, 1989.

Davis, R. F., "Plasma-Assisted Chemical Vapor Deposition of Ceramic Films and Coatings," in I. A. Aksay, G. L. McVay, and D. R. Ulrich (Eds.), *Proc. Sci. of Adv. Ceram., Mater. Res. Soc. Symp.*, San Diego, Calif., Apr. 27–28, 1989, vol. 155, pp. 213–225.

DeRenzo, D. J., *Ceramic Raw Materials*, Noyes Publ., Park Ridge, N.J., 1987, 890 pp.

Doremus, R. H., "Ceramic Powders from Organic Polymers and Gels," presented at the ASM Intl. Ann. Mtg., Indianapolis, Ind., Oct. 2, 1989.

Elschenbroich, C., and A. Salzer, *Organometallics, A Concise Introduction*, VCH Publ., New York, 1989, 490 pp.

Flinn, B. O., M. Ruhle, and A. G. Evans, "Toughening in Composites of Al_2O_3 Reinforced with Al," *Acta Metall.*, vol. 137, pp. 3001–3006, Nov. 1989.

Foulds, W., "SCS-6 Monofilament Reinforced Silicon Nitride Ceramic Composites," presented at the Aeromat '90 Conf. and Exposition, Long Beach, Calif., May 23, 1990.

"Fourteenth Ann. Conf. on Comp. and Adv. Ceram., Jan. 14–17, 1990," *Ceram. Bull.*, vol. 68, pp. 1970–1981, Nov. 1989.

Garnier, J. E., S. Keck, and A. S. Fareed, "Structural Ceramic and Metal Matrix Composites Fabricated by the Directed Metal Oxidation and Pressureless Metal Infiltration Processes," presented at the Aeromat '90 Conf. and Exposition, Long Beach, Calif., May 23, 1990.

Geoghegan, P. J., "Ceramic Matrix Composites by Chemical Vapor Infiltration," presented at the Aeromat '90 Conf. and Exposition, Long Beach, Calif., May 23, 1990.

Gulden, T. D., J. E. Sheehan, and C. L. Thompson, "Fabrication and Mechanical Behavior of Mullite Fiber Reinforced Amorphous Silicon Nitride Matrix Composites Using Chemical Vapor Infiltration," presented at the 14th Ann. Conf. on Comp. Mater. and Structures, Cocoa Beach, Fla., Jan. 17–19, 1990.

Holmes, J. W., "Elevated Temperature Mechanical Behavior of SiC Fiber-Reinforced SiN Composites," presented at the Aeromat '90 Conf. and Exhibition, Long Beach, Calif., May 23, 1990.

Johnson, H. H., "Microfabrication of Ceramic Fibers and Grids," presented at the ASM Intl. Ann. Mtg., Indianapolis, Ind., Oct. 2, 1989.

Koczak, M. J., et al., "Inorganic Composite Materials in Japan; Status and Trends," ONRFEM7, Nov. 1989, 53 pp.

Konsztowicz, K. J., and S. G. Whiteway, "Toughening in ZTA Composites with Non-Transforming Zirconia," Paper 76-C-90F, presented at the 14th Ann. Conf. on Comp. and Adv. Ceram., Cocoa Beach, Fla., Jan. 14–17, 1990.

Kools, F., and O. Fiquet, "Wet Pressing for Forming Advanced Ceramics," in G. De With, R. A. Terpstra, and R. Metselaar (Eds.), *Euro-Ceramics*, 1989, vol. 1, pp. 1253–1257.

Lee, R. R., and W. C. Wei, "Fabrication, Microstructure and Properties of SiC-AlN Ceramic Alloys," Paper 50-C-90F, presented at the 14th Ann. Conf. on Comp. and Adv. Ceram., Cocoa Beach, Fla., Jan. 14–17, 1990.

Lightfoot, A., C. Ker, and J. S. Haggerty, "Properties of RBSN and RBSN-SiC Composites," Paper 30-C-90F, presented at the 14th Ann. Conf. on Comp. and Adv. Ceram., Cocoa Beach, Fla., Jan. 14–17, 1990.

Lowden, R. W., "Chemical Vapor Infiltration Processing of Ceramic Matrix Composites," presented at Ceramtec '90, Dearborn, Mich., June 12–13, 1990.

Mangels, J. A., "Powder Preparation and Fabrication of Ceramics for High Technology Applications," presented at the ASM Intl. Ann. Mtg., Indianapolis, Ind., Oct. 2, 1989.

Mathers, J. P., T. E. Forester, and W. P. Wood, "Sol-Gel Preparation of Non-Oxide Abrasives," *Ceram. Bull.*, vol. 68, pp. 1330–1336, July 1989.

McEntire, B. J., "Processing of Powders for Gas Turbine Applications," presented at the ASM Intl. Ann. Mtg., Indianapolis, Ind., Oct. 2, 1989.

Messier, D. R., R. P. Gleismer, and R. E. Rich, "Y-Si-Al-O-N Glass Fibers," U.S. Army Lab. Command, MTL, TR-89-44, May 1989, 11 pp.

Mohr, D. L., P. Desai, and T. L. Starr, "Production of Silicon Nitride/Silicon Carbide Fibrous Composites using Polysilazanes as Pre-Ceramic Binders," presented at the 14th Ann. Conf. on Comp. and Adv. Ceram., Cocoa Beach, Fla., Jan. 14–17, 1990.

Nagelberg, A. S., "The Effect of Processing Parameters on the Growth Rate and Microstructure of Al_2O_3/Metal Matrix Composites," in I. A. Aksay, G. L. McVay, and D. R. Ulrich (Eds.), *Proc. Sci. of Adv. Ceram., Mater. Res. Soc. Symp.*, San Diego, Calif., Apr. 27–28, 1989, vol. 155, pp. 275–282.

Ramachandran, S., "Modified Alumina Fibres," E. I. Dupont De Nemours, UK Patent Appl. GB2219791A, Dec. 20, 1989, priority date Sept. 24, 1987, US100760.

Reymer, A. P. S., "Dry-Pressing of Ceramic Powders," in G. De With, R. A. Terpstra, and R. Metselaar (Eds.), *Euro-Ceram.*, vol. 1, pp. 1.253–1.257, 1989.

Rice, R. W., "Toughening in Ceramic Particulate and Whisker Composites," Paper 15-C-90F, presented at the 14th Ann. Conf. on Comp. and Adv. Ceram., Cocoa Beach, Fla., Jan. 14–17, 1990.

Sata, N., "Functionally Gradient Material Developed by SHS Process," *New Tech. Jpn.*, vol. 18, p. 17, 1990.

Sawyer, L. C., et al., "Strength, Structure, and Fracture Properties of Ceramic Fibers Produced from Polymeric Precursors, I, Base-Line Studies," *J. Am. Ceram. Soc.*, vol. 70, pp. 798–810, Nov. 1987.

Schwab, S. T., D. L. Davidson, and R. C. Graef, "Use of Preceramic Polymers in the Fabrication of SiC Fiber Reinforced Si_3N_4 Composites," presented at the 14th Ann. Conf. on Comp. Mater. and Structures, Cocoa Beach, Fla., Jan. 17–19, 1990.

Scoville, A. N., and P. Reagan, "Processing and Characterization of SiC Components Fabricated via Chemical Vapor Composite (CVS) Deposition Process," presented at the 14th Ann. Conf. on Comp. Mater. and Structures, Cocoa Beach, Fla., Jan. 17–19, 1990.

Sherman, A. J., and R. H. Tuffias, "Advanced Ceramic Reinforcements by Chemical Vapor Deposition (CVD)," Paper 83-C-90F, presented at the 14th Ann. Conf. on Comp. and Adv. Ceram., Cocoa Beach, Fla., Jan. 14–17, 1990.

Sim, M., A. Morrone, and D. E. Clark, "Processing and Microstructure of Y-TZP/Al_2O_3 Fibers," Paper 15-CP-90F, presented at the 14th Ann. Conf. on Comp. and Adv. Ceram., Cocoa Beach, Fla., Jan. 14–17, 1990.

Singh, R. P., and R. D. Doherty, "Synthesis of Titanium Nitride Powders under Glow Discharge Plasma," *Mater. Lett.*, vol. 9, pp. 87–89, Jan. 1990.

Solomah, A. C., et al., "Mechanical Properties, Thermal Shock Resistance, and Thermal Stability of Zirconia-Toughened Alumina-10% Silicon Carbide Whisker Ceramic Matrix Composite," *J. Am. Ceram. Soc.*, vol. 73, pp. 740–743, 1990.

Stambaugh, E. P., "Hydrothermal Processing—An Emerging Technology," *Mater. Des.*, vol. 10, pp. 175–185, 1989.

Suganuma, K., and K. Niihara, "Fibre Reinforcement Control," *Ceram. Jpn.*, vol. 24, pp. 937–944, 1989.

"Third Intl. Conf. on Ceram. Powder Processing Science, Feb. 4–7, 1990," *Ceram. Bull.*, vol. 68, pp. 1907–1913, Nov. 1989.

Veltri, R., et al., "Sol-Gel Derived Matrix Composites," *Powder Met. Inter.*, vol. 21, pp. 18–20, Dec. 1989.

Wai, C. M., and S. G. Hutchison, "A Thermodynamic Study of the Carbothermic Reduction of Alumina in Plasma," *Metall. Trans.*, vol. 21b, pp. 406–408, Apr. 1990.

Wills, R., M. Pascucci, and F. Jelinek, "Ceramic-Ceramic Composites—A State-of-the-Art Report," MCIC-86-51, Metals and Ceramics Inform. Center, Battelle-Columbus Div., Columbus, Ohio, Jan. 1986, 56 pp.

Yoshimatsu, H., et al., "Preparation of ZrO_2-Al_2O_3 Porous Ceramics from Spray-Dried Powder of Zr-Al Metallo-Organic Compound," *J. Ceram. Soc. Jpn.*, vol. 97, pp. 1365–1371, 1989.

CHAPTER 5
GREENWARE FABRICATION

5.1 INTRODUCTION

Since reproducibility and reliability are the major obstacles to the widespread use of advanced ceramic materials, these concerns must be addressed by a careful examination of the various ceramic manufacturing processes discussed in this chapter—greenware fabrication.[1-5] The development of ceramic greenware formulations requires a thorough understanding of the critical process parameters. These parameters include powder characteristics (particle size and distribution, surface area and chemistry), polymer properties (binders, dispersants, and lubricants), solvent composition, solids loading, and processing conditions. Controlling these variables is critical to the production of all high-quality ceramic greenware.

Greenware processes include tape casting, dry pressing, injection molding, extrusion, slip casting, and melting and casting.

The technology for metal-matrix composite (MMC) and ceramic-matrix composite (CMC) processing is reaching production in some cases, while in others it is still in the prototype and experimental stages.[3] It is essential that the manufacturing method ensure good bonding between matrix and reinforcement, and that no undesirable matrix-fiber interface reactions result.

MMC production processes can be divided into primary and secondary processing methods, though these categories are not as distinct as with monolithic metals. Primary processes are those used first to form the material and can be broken down into combining and consolidation operations. Secondary processes can be either shaping or joining operations. Table 5.1 lists different manufacturing methods and notes the types of operations that are included in each method.

Continuous Reinforcement

There are primary and secondary processes involved in the manufacture of MMCs with continuous reinforcement. The basic methods of combining and consolidating MMCs include liquid-metal infiltration, modified casting processes, and deposition methods such as plasma spraying. Hot pressing consolidates and shapes MMCs. Diffusion bonding consolidates, shapes, and joins MMCs. Chapters 6, 7, and 9 give more details on consolidation, MMCs, and CMCs.

Discontinuous Reinforcement

Primary and secondary processes are also involved in the manufacture of MMCs with discontinuous reinforcement (Table 5.1). The most common methods for producing particulate- and whisker-reinforced MMCs are powder processing

TABLE 5.1 Manufacturing Methods for Metal-Matrix Composites

	Combines	Consolidates	Shapes	Joins
Primary methods				
Casting (squeeze casting, compocasting, gravity casting, low-pressure casting)	x	x	x	
Diffusion bonding		x	x	x
Liquid infiltration (gravity, inert gas pressure, vacuum infiltration)	x	x		
Deposition (chemical coating, plasma spraying, chemical vapor deposition, physical vapor deposition, electrochemical plating)	x	x		
Powder processing (hot pressing, ball-mill mixing, vacuum pressing, extrusion, rolling)	x	x	x	
Secondary methods				
Shaping (forging, extruding, rolling, bending, shearing, spinning, machining)		x	x	
Machining (turning, boring, drilling, milling, sawing, grinding, routing, electrical discharge machining, chemical milling, electrochemical milling)			x	
Forming (press brake, superplastic, creep forming)		x	x	x
Bonding (adhesive, diffusion)				x
Fastening				x
Soldering, brazing, welding				x

Source: From Ref. 4.

techniques. Other processes for discontinuously reinforced MMCs are liquid-metal infiltration and casting.

At the present time there is no one MMC manufacturing method that holds great promise for reducing costs, although there seems to be some agreement that for particulate-reinforced MMCs, liquid-metal infiltration and powder metallurgy techniques are likely candidates for future development.

One of the major advantages of particulate- and whisker-reinforced MMCs is that most of the conventional metalworking processes can be used with minor modifications (see Table 5.1). A wide variety of joining methods are applicable, including mechanical fastening techniques, metallurgical methods used with monolithic metals, and adhesive bonding, as used for polymer-matrix composites.[5] (See Chap. 7.)

5.2 CERAMIC PROCESSES

Ceramic processing commonly begins with one or more ceramic materials, one or more liquids, and one or more special additives called processing aids. The starting materials or the batched system may be beneficiated chemically and physically using operations such as crushing, milling, washing, chemical dissolving, settling, flotation, magnetic separation, dispersion, mixing, classification, deairing, filtration, and spray drying. The forming technique used will depend on the consistency of the system (that is, slurry, paste, plastic body, or granular material) and will produce a particular unfired shape with a particular composition and microstructure. Drying removes some or all of the residual processing liquids. Additional operations may include green machining, surface grinding, surface smoothing and cleaning, and the application of surface coatings such as electronic materials or glaze. The finished material is then commonly heat-treated to produce a sintered microstructure. The sintered product may be a single component or a multicomponent composite structure.

Before scientific insights of ceramic processing had been gained, the properties of the product were often correlated with changes in a processing operation to identify the more important superficial variables. Viewed as a science, ceramic processing is the sequence of operations that purposefully and systematically change the chemical and physical aspects of a structure, which we call the characteristics of the system.

As a result, raw materials are now more beneficiated, more consistent, and often much simpler in composition. Modern instrumentation for analyzing ceramics materials and systems is more automated and precise, and with microcomputer accessories, quantitative data are obtained quickly and presented in a convenient format. Computer-controlled, closed raw-material handling systems improve the precision, efficiency, employee health, and safety in batching and mixing particulate materials. The processing pressure and temperature are more precisely monitored and controlled. In the factory, on-line monitoring and control of the production processes is practiced in some industries during some stages of processing.

The principles of processing science can provide insights into fundamental causes of behavior, procedures for modifying and controlling materials and processes, and avenues for improving manufacturing productivity. The role of ceramic processing science in ceramic manufacturing will surely increase in the years ahead.[6]

Additives for Processing

In processing ceramic materials, several processing additives must be incorporated in a batch to produce the flow behavior and properties requisite for forming. These additives may be categorized conveniently as follows:

1. Liquid or solvent medium
2. Surfactant (wetting agent)
3. Deflocculant
4. Coagulant
5. Flocculant or binder
6. Plasticizer
7. Lubricant
8. Bactericide or fungicide

Although most of these substances are added in relatively small amounts and some may be eliminated in a later stage of processing and do not appear in the final product, from a processing perspective they are essential materials. The selection and the control of these additives are often the key to successful processing or the development of an improved process.[6]

Liquids and Wetting Agents

Liquids are used in ceramic processing to wet the ceramic particles to provide a viscous medium between them, and to dissolve salts, compounds, and polymeric substances in the system. The admixed liquid changes the state of dispersion of the particles and alters the mechanical consistency.

A surfactant is a substance added to reduce the surface tension of the liquid or the interfacial tension between the surface of the particle and the liquid to improve wetting and dispersion.

Water is the major liquid used in ceramic processing, and it is often refined to improve its purity or consistency. It is polar and has a high surface tension. Nonaqueous liquid systems are less polar in nature and are good solvents for relatively nonpolar substances; their solutions provide greater ranges in properties. The liquid system used must dissolve additives, yet permit their adsorption at interfaces to assist in dispersion.

Deflocculants and Coagulants

Deflocculants and coagulants are essential additives that modify the interparticle forces, the agglomerate structure, and the consistency of the processing system. Particle charging and steric hindrance may produce the deflocculation of particles in polar liquids, whereas steric hindrance produces deflocculation in nonpolar liquids. Deflocculants producing particle charging may be either simple or polymer electrolytes; the type and concentration must be controlled to control interparticle repulsion. Coagulation is produced by crowding counterions in the double layer, sometimes called overdeflocculation or by adding counterions of a higher valence; both compress the double layer and reduce the repulsion between particles. The role of deflocculants and coagulants in real systems is sometimes difficult to analyze because of aging caused by the effects of soluble impurities and time-dependent dissolving and adsorption processes.[6]

Flocculants, Binders, and Bonds

Polymer molecules and coagulated colloidal particles that are adsorbed and bridge between ceramic particles may provide interparticle flocculation, and are commonly called flocculants or binders. An adsorbed flocculant may improve the wetting of the particles (wetting agent), increase the apparent viscosity significantly (thickener), retard the settling rate of the particles (suspension aid), alter the dependence of the apparent viscosity on the flow rate or temperature (rheological aid). In more condensed systems, such as granules for pressing, extrusion bodies, cast bodies, and unfired ceramic coatings, the flocculant may improve the plasticity (body plasticizer), reduce the liquid migration rate (liquid retention agent), alter the liquid requirement (consistency aid), or improve the green strength (binder). In ceramic processing these additives are more commonly referred to as binders rather than flocculants, and the term binder will be used here as a general term for these additives.

Other types of binders are waxes, resins, gels, low-temperature reaction bonds, and hydraulic cements. A film, a wax, a resin, or an additive that polymerizes into a gel may bind the particles together. Additives that react with the ceramic particles and polymerize or crystallize are called reactive bonds. Hydraulic cements are inorganic bonds that react with water and form very fine needlelike hydrated calcium silicate, or calcium aluminate minerals which coat the particles and form a bonded network.

Decomposition of the binder during firing commonly produces gas and pores, which must be eliminated during sintering to obtain a dense ceramic. For both technical and economic reasons, the binder must be conveniently dispersed and admixed into the system, and it must be sufficiently time-stable during subsequent processing. For example, different types of binders are needed for each process. In tape casting it is necessary for the binder to be a film former as it comes out of solution. The recommended binder in dry pressing is a "soft" polymer, which usually has a low average molecular weight.

The selection of binders for producing high-quality green bodies with high yields must be balanced with the ease of binder removal. Knowledge of the binder burnout characteristics of the chosen polymer is critical to the final production steps. Polymers decompose mainly following a free-radical mechanism, chain scission, or sublimation. In all cases volatiles are produced, which must be removed from the structure without disrupting the compact. In addition, polymers decompose differently when mixed with a ceramic as compared to their pristine decomposition. It is believed that acidic or basic sites on the ceramic powder have a catalytic effect on the decomposition. Other parameters affecting burnout include furnace size, the atmosphere, and the loading of the furnace.

Plasticizers, Foaming and Antifoaming Agents, Lubricants, and Preservatives

The processing engineer often uses small amounts of other types of additions, in addition to deflocculants, coagulants, and binders, to develop a satisfactory processing system. A plasticizer is added to modify the viscoelastic properties of a condensed binder-phase film on the particles. Binders used in ceramic processing must be plasticized to produce a moldable composition. A water-soluble binder is plasticized by adsorbed moisture, and the relative humidity must be controlled to control the plasticizing effect. An organic liquid with a lower vapor pressure than water is commonly used as the primary plasticizer. The plasticized binder is of lower strength, but is more deformable and resistant to failure on impact. The strength of the binder may increase when the plasticizer is lost during drying.

The tendency for bubbles to persist in a slurry may be reduced by adding a small amount of antifoaming agent; a foaming agent may increase the stability of gas bubbles. Special surfactants may be added to promote foaming, or to reduce the stability of bubbles and eliminate a surface foam.

A lubricant is a surfactant that is strongly adsorbed and especially effective in reducing the coefficient of friction of the surfaces of ceramic particles and metal dies. A molecular boundary lubricant must be capable of being strongly adsorbed, but of such a molecular size and structure that a smoother low-friction surface is produced. Finally, a preservative is used when enzymatic degradation of a binder must be controlled.

5.3 PARTICLE PACKING

Particle packing is controlled by the particle size and shape. (For example, coarser particles pack better than fine particles.) Most of the packing models assume spherical particles. While most ceramic particles are nodular, the theoretical models still apply. In bimodal powder distributions, the packing of ceramic powders improves when the coarse-to-fine ratio is 6:1 or 7:1. For alumina, optimum packing is obtained when the weight fraction of the fine is about 20 percent. Agglomeration prevents ceramic powders from packing densely. The degree of agglomeration also impacts slip viscosity. The greater the level of agglomeration, the higher the viscosity. Agglomeration is also detrimental to the texture of tape and can generate voids in spray-dried powder.

For example, a monosize system with a packing density of about 62 percent can be increased above 75 percent by adding a specific proportion of a finer size that packs efficiently in the interstices among the coarse. This type of sizing is used to fabricate relatively dense coarse-grained systems.

5.4 RHEOLOGY

Rheology is the science of deformation and flow. Knowledge of rheological behavior is essential in designing or selecting equipment for storing, pumping, transporting, milling, mixing, atomizing, and forming a ceramic system. Rheological measurements are an integral part of the research and development of slurry systems, and rheological tests are used in programs for monitoring and controlling the consistency and behavior of slurries for casting, spray drying, or glazing.

The rheology of suspensions and slurries is a complex topic. Physical properties and interactions between liquid, colloid, and particle components can vary widely. Particles may range from granular sizes to colloids. Added electrolytes and polymers may change interparticle forces and the state of dispersion significantly. The interparticle spacing depends on the concentration of the particles (solids loading) the state of dispersion, and the particle packing.

5.5 GRANULATION

Powders for pressing, calcining, and melting operations are commonly granulated. Compaction and extrusion processes are used to produce relatively large granules of higher density than the bulk powder. Spray granulation is used to produce granules that are satisfactory for use as a pressing powder for some fine-grained technical ceramics. The process must be controlled well to produce granules approximately spherical and reproducible in size and density. Granules produced by spray drying are commonly more homogeneous. The formulation of the precursor slurry and drying process must be properly designed to avoid the formation of granules with large internal pores and irregular shapes.

Spray drying is used widely for preparing granulated pressing feed from powders of alumina, carbides, and nitrides. Batch mixing in a slurry form and atomizing facilitate mixing and the achievement of mixedness. When properly controlled, spray drying produces nearly spherical, relatively dense granules larger than 20 μm in size, which flow and compact well in pressing operations.

5.6 FORMING PROCESSES

Forming transforms the unconsolidated system of feed material into a coherent, consolidated body having a particular geometry and microstructure. The selection of a ceramic-forming operation for a particular product is very dependent on the size, shape, and dimensional tolerances of the product, the requisite microstructure characteristics, the levels of reproducibility required, and capital investment and productivity considerations. Other considerations include the surface character after forming, die or mold requirements, energy requirements, and safety.

The mechanics of forming are very dependent on the consistency and rheological response of the material. Coarse granular materials and granulated powders are commonly formed by uniaxial dry pressing in a hard die or isostatic pressing in a flexible mold. Feed materials having a plastic consistency are formed by extrusion, pressure molding, injection molding, and jiggering operations. In conventional plastic forming, the degree of saturation of the feed material increases when the material is compressed; feed material used in ordinary injection molding is saturated with a thermoplastic or thermosetting polymer resin.

Warm molding is a plastic forming method, but in contrast to injection molding, a duroplastic binder or plasticizer is used. The process is similar to dry pressing in that the cold mixture is fed into a preheated mold. The mixture may also be squeezed into the mold by a piston. In this case, the process is similar to injection molding. There are many variants of the two basic processes, in terms of the method of feeding the material, tooling, and the combination of duro- and thermoplastic material used.

Warm molding offers some advantages over injection molding. The appearance of flow textures is strongly reduced, and since very high pressures can be used, good densities can be achieved. This avoids shrinkage and ensures good strength in the product. Furthermore, burnout is easier and more rapid since the danger of deformation due to softening of the plastic is avoided.

Pressing

Pressing is the simultaneous compaction and shaping of a powder or granular material confined in a rigid die or a flexible mold. Powder feed for industrial pressing is in the form of controlled granules containing processing additives produced by spray drying or spray granulation. Coarse granular compositions containing a binder are commonly in the form of a poorly flowing, semicohesive mass. For reasons of productivity and the ability to produce parts ranging widely in size and shape to close tolerances with essentially no drying shrinkage, pressing is the most widely practiced forming process. Products produced by pressing include a wide variety of fine-grained technical aluminas, including chip carriers and spark plugs, engineering ceramics such as cutting tools and refractory sensors, and grinding wheels.

Pressing by means of punches in hardened metal dies, commonly called dry pressing, is generally used for pressing parts thicker than 0.5 mm and parts with surface relief in the pressing direction. Isostatic pressing in flexible rubber molds, commonly called isopressing,[7] is used for producing shapes with relief in two or three dimensions, shapes with one elongated dimension such as rods and tubes, and very massive products with thick cross sections. Pressed products are also produced by a combination of metal die and isostatic pressing and by roll pressing.[6]

Dry Pressing. Dry pressing is a frequently used process for shaping ceramics. Without doubt it contributes to net-shape processing, particularly if no cutting is required before or after sintering. Usually it is used for simple shapes, such as gas turbine shrouds, which can be produced by axial pressing between two dies. The process can only be used when the quality of the material produced is adequate for the intended application.

The process is usually performed on automatic or semiautomatic presses; it is a relatively fast operation, enabling long runs of components to be made. It is usual to make the pressed shape as close to the final required shape as possible, with due allowance for firing shrinkage. Surprisingly complex shapes can be made, limited only by the cost of the mold or die and the ingenuity of the die maker. In fact the cost of the die is usually a significant factor in determining the cost of the component, and complex dies would not normally be used, except for long runs, as upwards of 5000 parts. For shorter runs, or for shapes that cannot be molded, it is common practice to press a blank and then machine it to the required shape prior to firing. The pressed blank is usually strong enough to withstand gentle handling, and hand shaping is not uncommon, particularly for difficult edges or profiles and for the removal of flash.[7] The process produces rapid and uniform compaction, and good part-to-part geometric reproducibility is achieved (Fig. 5.1).[8]

There are, however, limitations to dry pressing. Only flat or axisymmetric shapes can be formed, and the length-to-diameter ratio cannot exceed 3:1.

In dry pressing, the moisture content is generally below 4 percent, and lubricants such as stearic acid or wax are usually added to the powders. Finally, it

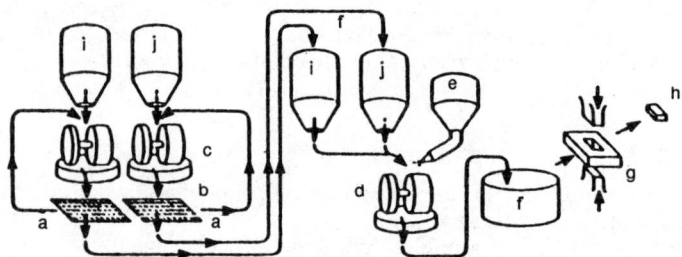

FIGURE 5.1 Simplified flow diagram of dry-pressing operation. a—oversize; b—screen; c—grinder; d—wet mixing; e—water; f—storage; g—dry pressing; h—formed wire; i, j—granules of powder.[8]

should be noted that dry pressing is essentially a plastic-forming process, similar to a forging operation.

Die Pressing. Die pressing is a unique process, with the die cavity filled with a metered quantity of powder feedstock. Although it is not strictly a dry-shaping method, the water content being typically around 10 percent and the pressed shapes requiring further drying before firing, the shaping technology is the same as for completely dry pressing of materials not containing clay. In die pressing it is usual to add an organic binder to the mix during the wet-mixing stages. The mix is then dried and granulated (often spray-dried) to form a free-flowing powder.

Die pressing has some limitations. Attention has to be paid to the ability to fill the mold with powder feedstock, and to transmitting pressure to all parts of the mold cavity so that the pressed shape has uniform density and no weak edges. Variations in pressed, or green, density lead to variations in fired density, and uneven firing shrinkage leads to distortion. For these reasons there are limitations on the ratio of the length in the direction of pressing to the diameter or wall thickness of various parts of the shape. Some variation of properties from point to point can be expected. Discussion with manufacturers on these aspects can be fruitful, and may lead to the use of a pressed blank followed by machining to ensure good properties in critical areas. An example might be an edge requiring a specified surface finish and hardness for wear resistance.[7]

Isostatic Pressing. Isostatic pressing can produce parts with tightly controlled dimensions and shapes and excellent properties when combined with contour grinding. Therefore the isostatic process leads to better density and thus better mechanical properties, but normally it does not lead to final shape.

The isostatic pressing technique, in which pressure is applied hydrostatically, is used to some extent in overcoming the problems of uneven density found in die pressing. There are two main versions of this process. In wet-bag pressing a rubber mold bag, sometimes with an inserted pin or mandrel to produce a shaped bore, is filled with powder and placed in a chamber of pressure-transmitting fluid, which is then pressurized. The applied pressure on the powder mass is more or less hydrostatic, and compaction is uniform (on the mandrel if this is included). After decompression, molds are removed and the part is ejected (Fig. 5.2). Dry-bag pressing is a variant of the dry-pressing process and is used for the production of standardized shapes in which there is no contact between operator and pressurizing fluid. The rubber tooling set forms part of the pressurizing system in the press, and the shape is normally pressed on a suitable mandrel, which facilitates its removal from the tooling set after compaction. Pressure is applied radially by means of the pressurized liquid medium between a flexible mold and a rigid shell (Fig. 5.3). Dry-bag pressing is used for the manufacture of insulators for vehicle spark plugs because the process is capable of being run automatically and continuously to produce very large numbers of pieces, as well as for grinding media and hollow tubes. Wet-bag processing by comparison is very slow and tends to be used primarily for the production of large blanks of pressed powder for subsequent machining, or for small runs of large pieces. The pressed, or green, density tends to be lower than in die pressing, but is more homogeneous.

Advantages of the isostatic pressing process are the potential for large part size, a capability for thick and variable walls, green-density gradients are minimized (density is uniform), and there is no length-to-diameter ratio limitation.

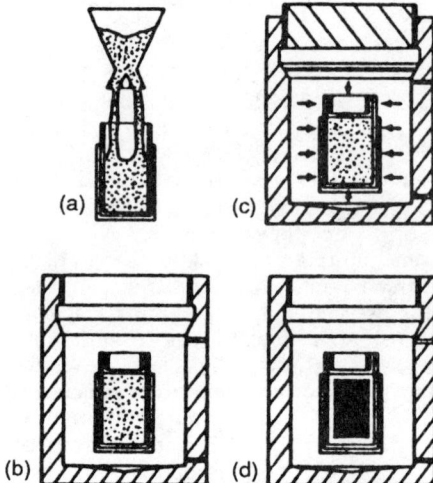

FIGURE 5.2 Wet-bag isopressing steps. (*a*) Filling. (*b*) Loading. (*c*) Pressing. (*d*) Decompression prior to part removal. (*Courtesy of Loomis Products Co., Levittown, Penn.*)

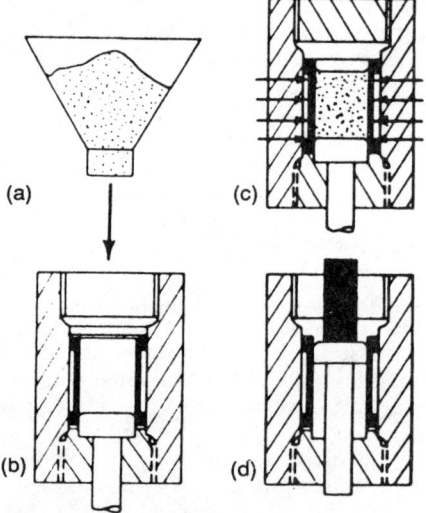

FIGURE 5.3 Dry isopressing steps. (*a*) Loading. (*b*) Opened die. (*c*) Pressing. (*d*) Ejection after decompression. (*Courtesy of Loomis Products Co., Levittown, Penn.*)

Limitations include a limited complexity, sensitivity to powder control, and a need for machining.[4,6–10]

There is also isostatic dry pressing which provides the opportunity to produce almost any complex shape by mechanically machining a pressed, or if necessary a presintered, green part. This is a costly method, but it is justified for the manufacture of prototypes and other small-quantity products such as pistons for diesel engines.

In addition to the benefits which accrue from the isostatic nature of the applied forces, isostatic pressing has also important advantages over other ceramic forming methods. For nonoxide ceramics it allows compaction of dry (nonhydrated) powders with suitable binders into large components, and it is the only practicable method of making large tubes and hollow shapes. Tooling design allows multiphase heterogeneous bodies to be made with little change to the processing parameters over those for a similar monolithic body. Green compaction of ceramic shapes containing a high-volume fraction of whiskers or fibers for ceramic reinforcing is extremely difficult by any pressing technique other than an isostatic process. Against all these advantages, however, must be set the disadvantage that isostatic pressing is a batch technique. Considerable effort has gone into developments of automation and robotic handling, but success has been limited. Dry-bag isostatic pressing of small components has been semiautomated, but for large pieces isostatic pressing remains a labor-intensive process.

It must be noted that hot pressing and hot isostatic pressing of engineering ceramics at temperatures up to 1800°C are widely used in attempts to combine green forming and firing. These are covered in Chap. 6.

Other Methods. Two other press processes include combination pressing, which uses a combination of dry pressing and dry-bag isopressing and is primarily used for dinnerware, and roll pressing. (See Reed[6] for more details on these processes.

Extrusion

Plastic clay bodies have traditionally been formed by plastic pressing, jiggering, and extrusion. These processes are also used to a lesser extent for forming ceramic bodies plasticized with an organic binder.

Wet pressing refers to the pressing of plastic granules in a rigid die. Pressures are lower than for dry pressing, and the die components are heated or spray-lubricated to facilitate ejection.

In jiggering,[6] a section of deaired, extruded body is compressed and sheared between the surfaces of a lubricated template or roller tool and a rotating permeable mold.

Extrusion is the shaping of the cross section of a cohesive plastic material by forcing it through a rigid die. Stiff plastic forming or extrusion is a well-established technique in the oxide ceramic industries.

Products formed by extrusion include hollow furnace tubes, honeycomb catalyst supports, transparent alumina tubes for lamps, electrical porcelain insulators and graphite electrodes ranging up to more than 1 ton in size, and flat substrates.

Warm and cold extrusion can be performed by mixing sufficient amounts of binders and plasticizer materials to, say, sialon powder to enhance flow characteristics.

Typical components for extruding are thermocouple tubes made of reaction-bonded silicon nitride and of Si_3N_4-bonded silicon carbide. Also rings of the same material have been made for some time by extruding and cutting to the required length.

The choice of raw materials is a critical factor in successful extrusion and in the related process of injection molding.[11] The raw-material parameters are the particle-size distribution and particle shape of the ceramic and the characteristics of the organic vehicle used to confer plasticity on the mix. It has been shown that a broad distribution of particle sizes allows the solid content of an extrudable mix to be increased without increasing the viscosity of the system. (It should also be noted that similar conclusions apply to slip casting, and that there is a similarity with the requirements for achieving high green density in pressed powders.[11] The effect of particle shape on the characteristics of the mix is similar to the classical arguments concerning the packing density of powders; as the particles become more angular, packing efficiency decreases.

The organic systems used for extrusion and injection molding are complex materials which are required to fulfill several criteria. These may be identified according to their function as either a binder or a plasticizer. The binder should:

1. Confer fluidity on the powder
2. Wet the powder
3. Confer green strength on the compact
4. Leave a low residue after burnout

The plasticizer must be compatible with the binder and have low volatility to ensure retention during processing, which is necessary for consistency of powder volume loading.

A great advantage of the extrusion process is the possibility of incorporating ceramic fibers to produce composite materials and to use the process to give a preferred orientation in these materials. For more complex shapes extrusion is the most suitable and can yield high production rates.

The stages in extrusion are:

1. Material feeding
2. Consolidation and flow of feed material in barrel
3. Flow through a tapered die or orifice
4. Flow through the finishing tube of constant or nearly constant cross section
5. Ejection

Piston and auger extruders are normally used.

Plastic Transfer Pressing and Jiggering

In ancient times, bricks and pottery were formed by pressing a section of plastic material in a mold. Later, hollowware was formed by pressing one's hands against material supported on a rotating wheel, which is called jiggering.

Parameters in plastic pressing and jiggering include the thickness and diameter of the feed, thickness-reduction ratio, material flow rate, surface features, the motion of the contouring tool and its lubrication, imposed vacuum, pressure, temperature, and the flow properties of the body. Fabrication begins as the feed material is cut and transferred to the cavity of the mold. On lowering the forming tool, the body makes contact with the mold and then flows against the mold and laterally in a plane between the tool and the mold.

Defects in jiggered and pressed plastic pieces include those observed in extruded material. Plastic transfer pressing is commonly used for forming large shapes with detailed surface relief and relatively deep nonsymmetrical shapes. Jiggering is widely used for forming circular and elliptical shapes with relatively thin walls (less than 5 mm) and deeper circular shapes.

Injection Molding

Injection molding is probably the most interesting method for mass-producing small and medium-sized parts having complex geometries. This method, which is well known for producing plastic parts, was adopted for oxide ceramics in the midthirties. It is a process that is complementary to all others in that it is a method of making complex irregular shapes that would defy simple uniaxial pressing and would be difficult to machine from blanks. In this process,[6] the ceramic powder mass is mixed with sufficient thermoplastic organic binder in heated mixers to enable the mixture to be injected under pressure and at raised temperature into a cooler mold cavity in which it solidifies. The mixture is granulated and fed into an extruder where it is pressed by a piston or screw system through the injection chamber and into a cooled mold. Here it hardens, and after a very short time, the green part can be removed.

The green parts are tested in an oxidizing atmosphere, to drive off the organic binder, and then finally fired (Fig. 5.4). Some of the processing parameters that must be closely controlled in injection molding to produce acceptable parts[13] are powder,[14] organic binder,[15] removal of organic binder, components,[15-18] tool (mold), strand thickness, material and mold temperature, and filling pressure and velocity.

The key issues related to injection molding are the removal of the high organic content in such a manner as to prevent disruption of the ceramic body and also to keep the resulting voids small enough so that their elimination will be obtained during the sintering process.[19-21]

Most green parts out of the mold are at final tolerances, but 10 to 20 wt % of plastic binder is needed to thicken the material and must burn out during firing. Allowing the escape of combustion products currently limits designers to cross-section thicknesses of less than 19 mm. But alternate binders such as water-soluble resins or other aqueous-based systems may give more leeway. After the

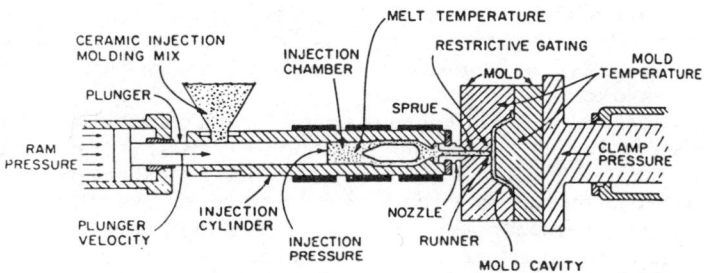

FIGURE 5.4 Schematic diagram of plunger-type injection-molding machine.[12]

resin stiffens in a heated mold, firing could readily drive off the remaining water. Another way may be, prior to firing, to flush out the part with gases or solvents that combine with the plastic binder as the temperature rises. Work has been done with liquid carbon dioxide, for example.

Injection molds used for ceramics are not merely mating halves, as is common for plastic. Molds used to form ceramic parts such as turbine wheels for automobile turbochargers and for aircraft auxiliary power units work like camera shutters. In a camera, flat leaves overlap, extend, and rotate inward to close the aperture. The leaves of a mold are made in three dimensions, so that when they come together, an inner cavity has the desired shape.

The main advantages to the process are good part-to-part geometric reproducibility and accuracy in calculating and controlling green density[19] and determining sintering shrinkages. Limitations include that tooling is expensive and binder removal often creates body defects in thick sections and is time-consuming. Also, binders create residual green-body stresses.[20]

The key to the future of injection molding includes:

Understanding the rheology of highly loaded suspensions of ultrafine particles

Developing binder systems that will result in improved binder-removal behavior

Developing a basic understanding of the binder-removal process

Developing improved binder-removal techniques especially for thick and thick/thin cross sections

Developing new models and incorporating existing models of the injection-molding process into a computer-aided engineering system for ceramic molding, utilizing developed information

Minimizing the stress state of the part in the tool and during binder removal[21]

Casting

Casting processes are used to produce a self-supporting shape called a cast from a specially formulated slurry. The yield strength of the cast may be increased by the partial removal of the liquid, which concentrates the solids, or by the gelation, polymerization, or crystallization of the matrix phase. An aqueous slurry containing fine clay has been traditionally called a slip, and conventional casting is commonly called slip casting.

Slip casting has been known for producing clay-based products for more than two centuries. It is based on the fact that when electrically charged particles are held in suspension in a liquid of suitable composition, an unusually low volume of liquid is required to make the solid powder mass fluid. Since this method is also suitable for ceramic powders other than clay, it is of interest for producing complicated components from nonoxide ceramics, and has been successful with relatively low mold costs.

In slip casting, the slurry is poured or pumped into a permeable mold having a particular shape. Capillary suction and filtration concentrate the solids into a cast adjacent to the wall of the mold. The motivating force for the separation of the liquid may also be pressure applied to the slurry (pressure casting), a vacuum applied to the mold (vacuum-assisted casting), or centrifugal pressure.

Therefore, there exist several variations of the basic slip-casting process. In pressure and vacuum casting, the slip is shaped in the mold under pressure or vacuum. In centrifugal casting, the porous mold is rotated, and in thixotropic casting, chemical agents are added to promote curing and to reduce the amount of water required. In tape casting, a controlled film of slurry forms on a substrate. Evaporation of the liquid from the film during controlled drying transforms the film of slurry into a flexible, rubbery tape or sheet.

Slip Casting. The slip, which is a suspension of the powder in water or organic liquids, is prepared with suitable additives for rheology controls—frequently one of the critical trade secrets. This slip is poured into a porous mold which absorbs the liquid—mostly water—and the ceramic body is allowed to set. Slip casting requires the slip to have the correct pH to keep the particles in suspension, and the correct particle-size distribution to give strength to the cast shape when it is dry. Since no pressure is applied in the process, slip-cast shapes may have a lower bulk density (green or unfired density) than mechanically shaped components, and hence rather higher firing shrinkage. Bubbles in the slip can leave voids in the casting.

The slip must be chemically stable with the liquid medium and rheologically stable over the life of the slip and the duration of the casting. Furthermore, the slip chemistry controls the rheology and the rheology controls both the green density and the casting rate of the slip.

After removal from the mold the component is dried and then usually forwarded to the sintering process. Figures 5.5 and 5.6 show schematically the steps in the slip casting of thin-walled and solid ceramic shapes.[9,22]

Good properties of the slip itself do not guarantee a good casting process. It must be possible to cast the slip without streaks or bubbles, but also the mold must allow the water to be withdrawn evenly through its porous walls, thus allowing the part to set uniformly. Withdrawal of liquid from the slip into the mold lowers the liquid content adjacent to the wall until a rigid layer of the body is built up. The rate of buildup decreases with time as the mold wall becomes saturated with liquid and the thickening component becomes less permeable.

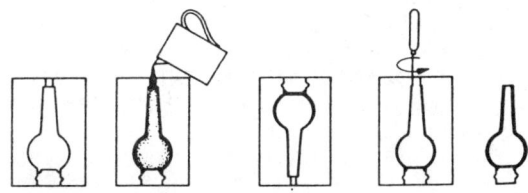

FIGURE 5.5 Slip casting a thin-walled ceramic shape.[22]

FIGURE 5.6 Steps in solid casting of ceramic slip.[22]

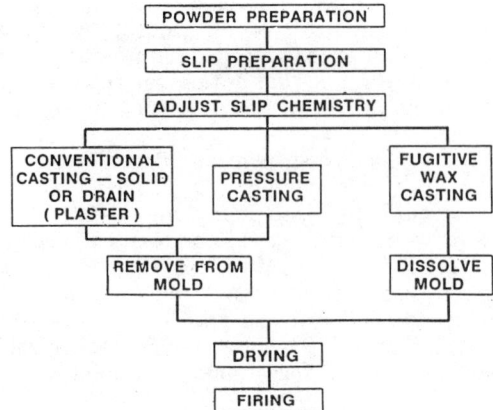

FIGURE 5.7 Slip-casting process flow.[21]

The mold must be designed such that the shrinkage, which occurs because of the removal of the liquid, is allowed to take place freely. Finally, the part must have sufficient strength that it can be handled and can be removed from the mold without sticking. The typical steps in the process flow of slip casting ceramics are illustrated in Fig. 5.7.

Slip casting was, and to some extent still is, an art. While it is used more commonly for producing sanitary ware, it is also being used for producing such parts as combustor baffles, scroll bodies, and connecting ducts. In general, it is used for hollow shapes with a uniform wall thickness of 2.5 to 12.7 mm, as variable walls are difficult to produce. The process does not lend itself to high-volume production.

In the wet-shaping method of slip casting, where the shaping depends on the rheological properties of the wet mass, it is often possible to combine two or more parts produced by the primary shaping method to make a composite component by joining them, while still wet, with a small amount of ceramic slip as an adhesive and then blending the joint to eliminate external evidence. An example of this is the closure of one end of an extruded tube by hand working. In critical conditions, such joints tend to be sources of weakness, and manufacture in one piece is to be preferred.

In many cases, because of the relatively low cost of tooling for slip casting, the process may be economically viable for making prototypes and for small batch production. When larger quantities are required, the most economic route would normally be by die or isostatic pressing, but this can lead to changes in properties.

Slip casting offers a simple process that requires little capital equipment and has low tooling costs. It is ideal for a laboratory or pilot production process and can produce small or large components to near net shape and in a variety of geometries from simple to complex.[21]

The future for the slip-casting process depends on[21]:

Understanding the rheology of highly loaded suspensions of ultrafine particles

Developing alternative slip-casting media (nonaqueous systems)

Developing a better understanding to predict and control green density

Developing techniques to pressure-cast complex-shaped articles

GREENWARE FABRICATION

Developing new mold and mold-removal techniques for the fugitive-wax slip-casting process

Developing techniques to incorporate pressure casting and fugitive-wax slip casting

Improving the technology for drying

Since slip casting is a batch process, it is not easily adapted to continuous operation. There is, however, one technique, tape casting, which has been developed to satisfy demand for simple flat, thin ceramic tiles for insulating substrates in the electronic industry.

Tape Casting. Doctor-blade processing, or tape casting, is a specialized method of producing thin, flat, uniform strip which can subsequently be cut into rectangular tiles. A thick slip, often using an organic vehicle for quick drying, is cast on a belt to a specified thickness and allowed to dry before being cut up. The thickness of the tape is a function of the height of the blade, the viscosity of the slurry, the speed of the carrier film, and the drying shrinkage. The casting speed is dependent on the thickness of the tape, the evaporation rate, and the length of the machine, all of which control the drying time (Fig. 5.8).

Tape defects include cracks, camber, local regions of low density, and surface defects consisting of unacceptable average roughness and large surface pores. Cracks are caused by a differential shrinkage. Low-density regions are caused by retarded sintering in regions containing powder agglomerates. Surface pores larger than 5 μm are undesirable, as they may interrupt metallization lines; pores are caused by bubbles in the slurry, segregated organic, and pullouts on separating the tape from the carrier.

The process has been applied mainly to oxides and is normally carried out by replacing the single slip-casting mold by a continuous, absorbent belt onto which the slip is poured. The belt provides the initial drying and then passes through a hot dryer before the green tile is cut and removed from the belt for firing. The entire process can be automated and is now the subject of considerable development effort for processing of electronic ceramics.[23]

The method is used almost exclusively for the production of thin electronic substrates where the control of surface finish and flatness is of overriding

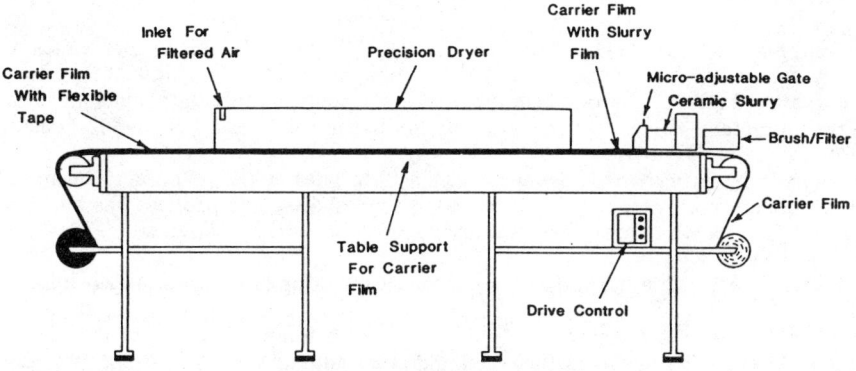

FIGURE 5.8 Schematic of continuous tape-casting equipment.[6]

importance.[6,7] These include multilayer ceramic electronic packaging, multilayer titanate capacitors, piezoelectric devices, thick- and thin-film insulators, ferrite memories, and catalyst supports. In producing electronic packaging, metal circuit patterns and film resistors are commonly printed on the green sheet and cofired. Multilayer packaging is produced by stacking printed sheets, laminating these under pressure, punching for size and shape, inserting metal via interconnections, and sintering.

Vacuum Casting. Vacuum casting is widely used for forming very porous refractory insulation having a complex shape. The slurry typically contains partially deflocculated ceramic powder and chopped refractory fiber. The fiber increases the viscosity and liquid requirement as well as the porosity of the cast. When the desired wall thickness has been cast, the preform is withdrawn from the slurry and removed for drying and a chemical set.[6]

Pressure Casting. Pressure casting has been investigated as a forming technique for complex refractory shapes, and mechanized, relatively automatic systems are now used in some industries. In pressure casting, the mold serves as a filter, and the casting time is controlled by regulation of the external pressure. The molds need not be dried, and the humidity of the workplace is lower. In studies of pressure casting versus slip casting[24] it has been shown that due to the reduced water content of the cast, the drying shrinkage of about 3 to 3.5 percent has been reduced to 1 to 1.5 percent, and increased shrinkage anisotropy has been found. The casting time of 1 to 2 h has been accelerated to less than 20 min. As a result, improved productivity and uniformity have been found, that is, there is less time for rheological and chemical changes to occur.

Disadvantages of pressure casting include problems with conventional mold materials and the greater capital cost.

Thixotropic Casting. Casting slurries containing a gelling, reactive, or hydraulic binder are used to produce a variety of refractory shapes and molds for metal casting, to repair refractory structures, and to produce concrete structural materials. For dense materials, the mold or cavity is usually filled with a very concentrated slurry of minimal liquid content and mechanical vibration is often used to induce flow and eliminate air pockets. The yield strength and viscosity of the cast are increased by the initial "set," and the ultimate strength is increased when the binder system hardens.[6]

Electrophoretic and Fugitive-Wax Slip Casting. These two casting processes are very rarely used. The first includes the electrophoretic deposition of particles onto a conducting substrate as a means for forming thin tubular shapes and coatings on metals. Cross sections must be limited in thickness, because the casting rate is relatively slow.[25]

The latter process was developed by a U.S. automobile manufacturer and allows complex-shaped articles to be cast to net shape. The process has also produced articles which could not otherwise be produced by slip casting.

Compocasting. Compocasting is a process for making discontinuously reinforced metal-matrix composites. The reinforcement is added to a semisolid matrix alloy and the mixture is agitated vigorously. The agitation repeatedly breaks the alumina skin formed on the surface of liquid aluminum and brings it in intimate contact with the reinforcement. The reinforcement is also trapped by the primary solid phase and prevented from agglomerating. When sufficient mixing is produced, the composite slurry is cast. This method has been referred to as SS (semisolid-semisolid), which is the state of the matrix during the two manufacturing steps of mixing and casting.

Two variations of the compocasting process that have been developed are SL (semisolid-liquid) and LL (liquid-liquid). Figure 5.9 shows the basic compocasting process and its variations. Composites have been made using all three methods. For example, in the production of 2024 Al, reinforced with 3- and 12-mm-long FP-alumina fibers, squeeze casting has been used as a complementary casting technique. The SS route produced castings with uniform fiber distribution, but casting was difficult due to the high viscosity of the slurry. The SL route gave good fiber distribution and casting was easy. The LL route gave castings with nonuniform fiber distribution, but a large amount of fibers could be added without any problem. Addition of alumina fibers to 2024 Al increased its modulus of elasticity considerably.

Finally discontinuously reinforced metal has been produced by compocasting and its two variations, but the metallurgical quality must be controlled closely to get good strength values.[26]

Dynamic Compaction (Shock-Wave Processing)

Dynamic compacting of advanced ceramic powders with explosives is a relatively new process. Early exploratory work of the 1960s demonstrated that the shock process produced extraordinary conditions which are potentially useful for materials technology. Furthermore, continued activity provided specific quantitative information on the details of the shock process and how the process can be used in ceramic powder technology.[27–29] The considerable data demonstrate that

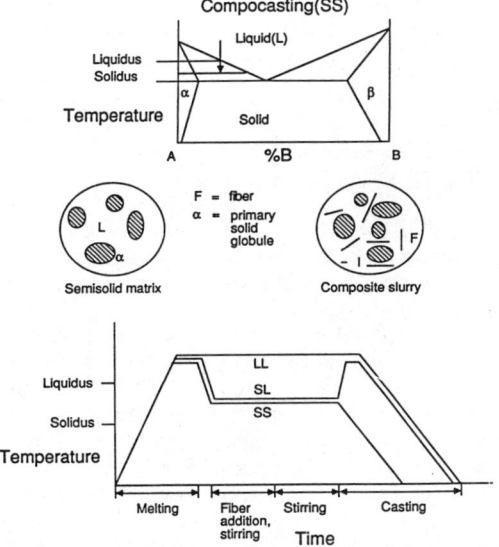

FIGURE 5.9 Compocasting and its variations.[26]

the shock process produces a unique combination of excitations which are not achieved in other processes, and herein lies the ceramic materials impact.

Nevertheless it should be recognized that in spite of progress in understanding the shock compression process in powders, our knowledge is still quite limited. The very uniqueness of the shock process and the unusual nature of the shock-treated materials provide a major impediment to the development of commercial processes and a major challenge to research activities.[27]

Dynamic compaction is a processing technique with considerable market potential for manufacturing a number of advanced ceramics. The following key factors support this view[30]:

1. Product purity
    a. Little contamination from processing environments
    b. Little reaction among components in composites
2. Operational aspects
    a. Rapid processing rate
    b. Excellent reproducibility
    c. Minimal limitations on size scaleup and aspect ratio
    d. Processing at considerably lower temperatures than other methods
3. Cost aspects
    a. Low capital investment
    b. Single process step to final shape

The goal of researchers is to densify refractory ceramic products in a single-step processing operation and produce high-density final shapes in oxides, carbides, and nitrides.

Perhaps the greatest competitive advantage of dynamic compaction can be realized in fabricating either sizes that are too large or shapes that are not suitable for hot compaction on presses. Conversely, overriding competition is anticipated in situations where small pieces can be cold-pressed at high rates and batch-sintered or where shapes are extremely complex, especially with sharp changes in section thickness and surface contour (such as those often formed by injection molding).

The special features gained by shock loading are:

1. High pressures and temperatures generated over a short period of time
2. Deformation and fracture of shocked material
3. Promotion of metastable microstructures

The main benefits of shock loading ceramic powders are the high degree of possible densification, the increased rate of chemical reactivity, and a predisposition to sintering. The latter feature is of particular interest for nonoxide ceramics, which are normally difficult to sinter.[31]

In the current state of development there is evidence that the high pressures and high temperatures generated in shock loading lead to higher-density monoliths produced from powder, due to localized surface melting of the powders. Second, the explosive loading can lead to eliminating sintering and densification additives and decreasing sintering temperatures due to fracture of the powder into fine particles. Finally the process has yielded crackfree densified products in the following shapes: flat platelets and other rectilinear forms, solid right circular cylinders, and tubes.

Monolithic ceramic compounds that have been compacted include Al_2O_3, AlN, B_4C, sialons, SiC, TiB_2, and TiC. The compaction of microcomposites (such as Al_2O_3 in an AlN matrix) and a macrocomposite (a steel mesh in AlN) also has been demonstrated.[30] The major benefit cited for shock loading of powders is a significant increase in density after subsequent sintering. For a given sintering time the density increases by 31 percent at 1600°C and by 25 percent at 1700°C over that of unshocked alumina. This density increase is achieved concurrently with a two- to tenfold reduction in grain size, which should also improve mechanical properties.[32–34]

Other benefits from shocking SiC ceramics appear to be in the areas of consolidation and sintering enhancement. The main benefit of shocking Si_3N_4 is the reduction in the amount of sintering aids required, particularly MgO and Y_2O_3. These additions reduce creep resistance and oxidation resistance. Early results have shown a 97 percent dense TiB_2 in a 25.4-mm-high block while AlN has been produced in a 90 percent dense 76.2-mm-long plate and 97 percent dense 50.8-mm-outer-diameter tube.

The potential application of ceramic materials[31,35] as lightweight armor is most often considered by armament specialists. The ceramic products that are required for armor are usually particles or platelets, and in some limited applications cast products may be provided. The armor shapes that would be most amenable to fabrication by the dynamic compaction process are platelets, cylinders, or simple variations of those configurations (Table 5.2).

The production of composite armors, both micro and macro types, appears to be

TABLE 5.2 Potential Applications for Dynamic Compaction Techniques

Application areas	Critical products
Armaments and relevant materials	Lightweight armor tiles Ceramic gun barrels or lines Duplex (metal-ceramic) kinetic-energy penetrators Composite (metal-ceramic) periscope tubes and deep-submergence vessels
Ceramic engines and components	Structural engine components (e.g., housings, liners, shrouds, shields, baffles, insulators, combustion chambers) Spark plugs Turbine blades Roller bearings
Electronic and optical hardware	Capacitors and semiconductors (stacked layers) Magnets (e.g., $SmCo_5$ and metallic glasses) Integrated optics (fiber-optic waveguides on ceramic substrates)
Other mechanical and structural components	Nonmetallic heat exchangers and heat pipes Ceramic cutting tools and drilling crowns Ceramic nozzles, pump seals, and valves (for use in liquid-metal service, turbines, weldments)
Rapid solidification technology	Turbine blades, nozzles, and magnets Superconductors Microelectronics

especially well suited for dynamic compaction. A major advantage in the fabrication of macrocomposite armors is elimination of undesirable chemical reactions.

A second potential application in the armaments category is the production of kinetic energy (KE) penetrators with a composite structure. Emphasis in materials selections, especially for medium and large calibers, has shifted from very hard metals to fracture-tough, heavy metals in monolithic designs. However, the optimum penetrator may be one that combines the very hard, strong material with the heavier, tougher material, rather than one that comprises both of those requirements in a single material. Dynamic compaction could be used, for example, to densify a ceramic compound as an insert or series of inserts inside a metal sleeve.[30] The final product would have a relatively tough metal shell fitted around one or more slugs of a relatively hard, strong ceramic.

A third potential application for dynamic compaction in the armaments category is the production of ceramic gun barrels or internal liners for metallic barrels. Some ceramic compounds (such as titanium diboride) are much more resistant to high-velocity gas erosion than metals. Since metal erosion is a major problem that limits the performance capability and service life of metal gun barrels, a full ceramic substitution or an inner surface for metal barrels is a rational solution to mitigate those problems.

Another military hardware item that has the tubular shape of a gun barrel but a greater length (approximately 40 to 50 ft) is the periscope tube used in submarines. The compacted product could have inner and outer metal sheaths enclosing an annular ceramic core, as shown in Fig. 5.10. Neither the tubular shape nor the length pose a problem for dynamic compaction technology. If the ceramic material for the design concept shown in Fig. 5.10 were titanium diboride, the overall weight reduction in comparison to a stainless-steel tube would be one-third (~500 lb).

Once applications such as composite gun barrels and periscope tubes have been mastered, another potential application to consider would be fabrication of ceramic or glass-ceramic (such as Pyroceram) cylindrical housings for deep-submergence structures. As the depth of submergence is increased, advanced ceramics are expected to be preferred for fulfilling the severe requirements implicit in this application.

Cryogenic Processing

New research is taking some of the black art out of the process of producing ceramic shapes. The National Institute of Standards and Technology (NIST) has shown that the ingredients for carbide and boride ceramics can be partly reacted at a low temperature, −130°C, to "freeze" the mixture's chemistry while composition and purity are under precise control. The resulting compound is then fused at a high temperature. Moreover, says Joseph J. Ritter, an NIST researcher, this technique opens the door to new ceramics, since some chemical reactions can be achieved only at relatively low temperatures.

Drying

Products formed by casting, paste processing, and plastic forming must be dried in a controlled manner to remove the interstitial liquid phase prior to firing in a furnace. Drying is the removal of liquid from a porous material by means of its transport and evaporation into a surrounding unsaturated gas or, in some cases, a

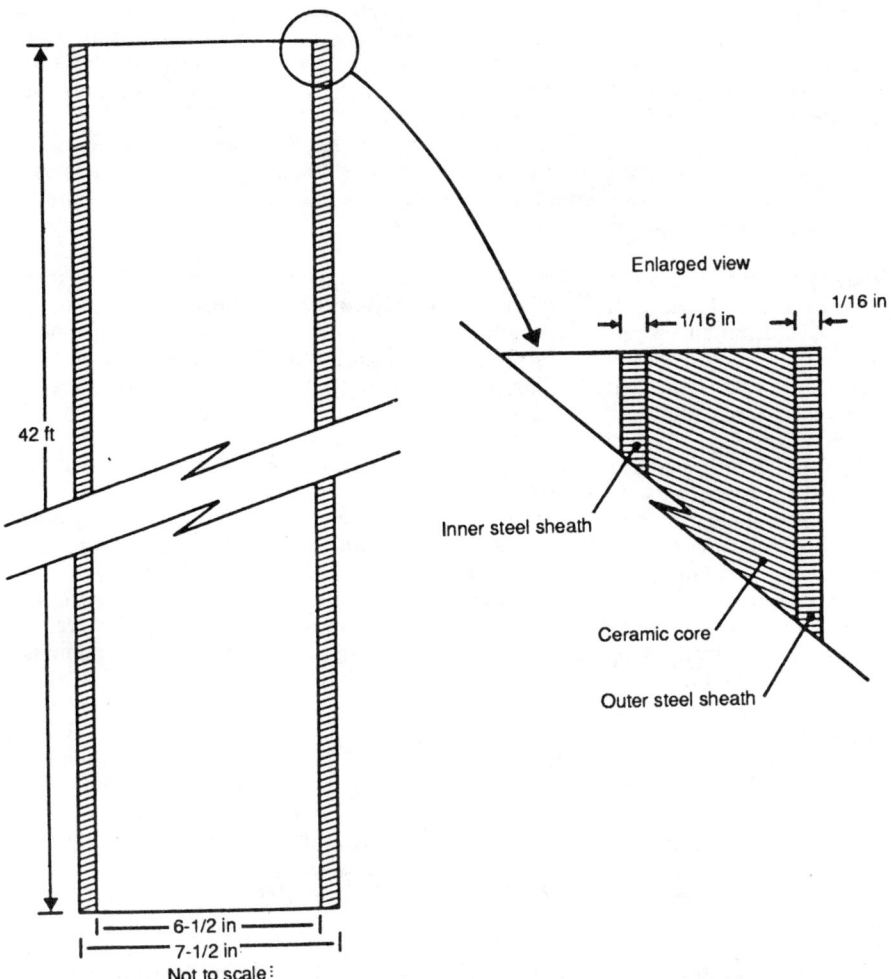

FIGURE 5.10 Concept for composite dynamically compacted periscope tube.[30]

desiccating liquid. It is an important operation prior to firing in processing bulk raw materials and products shaped by plastic forming and casting.

The evaporation of processing liquids is relatively energy-intensive, and drying efficiency is always an important consideration. Drying must be carefully controlled, because stresses produced by differential shrinkage or gas pressure may cause defects in the product.

Freshly cast plaster molds and very large wet-processed shapes are commonly dried slowly by open-air drying. Shaped ceramic products requiring drying and working molds are usually dried in a controlled manner in fabricated metal dryers.[6]

Drying involves the transport of energy into the product; liquid is transported through pores to the meniscus, where evaporation occurs, and by vapor transport through pores. In a drying system, heat energy must be brought to the surface of the product, and vapors must be carried away. The mechanisms of evaporation and mass and thermal transport must be considered before deciding which mode of the drying process is to be considered and selected.[6] The drying rate must be considered and depends on the temperature of the liquid in the body and the temperature, humidity, and flow rate of the drying air. After initial heating, the product dries at a constant rate during which shrinkage commonly occurs. When the drying rate is very fast or nonuniform, the constant-rate period is relatively short, and the differential shrinkage can cause cracks.

Mode of Drying. Conduction and convection drying are commonly used for drying ceramic products. Convection drying is used to heat the product and remove vapors, and the air circulation may be designed to facilitate the drying of a particular size, setting, or shape. Infrared drying can be used to reduce the drying time, while vacuum-assisted drying reduces the partial pressure of vapors. There are several other modes of drying.

Microwave Drying. Microwaves are generally reflected by electrical conductors, transmitted by electrical insulators, and adsorbed by dielectrics. Liquid water behaves like a dielectric when subjected to a microwave field. When drying ceramic insulators, the microwaves are preferentially adsorbed by the water, and the product temperature during drying may never exceed 50°C, that is, the high surface temperatures in conventional drying are avoided. Potential uses for microwave drying are in the processing of temperature-sensitive products, more rapid drying, drying products of large cross sections, and drying products containing colloidal materials such as gels.

Slurry Drying. Slurried raw materials supported on a metal belt and tape-cast films are dried continuously and relatively rapidly. In drying a mineral slurry on a metal belt, heat is supplied by conduction through the belt and by the hot drying air, which may exceed 400°C. In tape casting, the thin slurry film cannot be dried too rapidly, because the drying is one-directional and the liquid permeability of the tape is very low. The air flow and temperature are carefully controlled to maximize the constant-rate period and minimize liquid concentration gradients until shrinkage has ceased.

Other Drying Methods. For every product with a finite drying shrinkage there is some critical drying rate that will cause defects in the product. When drying thick products or a product of very low permeability K_p, drying must be conducted in such a way that the liquid concentration gradients are not large. This is controlled-humidity drying. By heating the product in humid air, the surface evaporation of the liquid is arrested and also the viscosity is reduced prior to heating.

Supercritical drying may be used to minimize the effects of the surface tension of the liquid during drying. In supercritical drying, the product is heated in an autoclave until the liquid becomes a supercritical fluid. Freeze drying can be used for drying products where liquid formation and product heating must be avoided.

Spray drying is relatively efficient compared to the convection drying of formed products, because the material is well dispersed in the drying medium, the diffusion path is shorter, and the high specific surface area contributes to a higher rate of evaporation per unit mass of product.

5.7 METHODS AND MATERIALS

Dry Pressing

In studying and comparing the dry-pressing process versus injection molding, it appears that they consist largely of the same working steps (Table 5.3). During dry pressing, in the case of geometrically simple-shaped parts with large dimensional tolerances, the pieces often do not need any mechanical aftertreatment. In this case the production costs are comparatively low.

For small, simple parts, axial dry pressing is unquestionably the most economical process. Consequently this process is most widely spread. In most cases isostatic dry pressing and injection molding produce qualitatively equal or even better products, but they are much more expensive.

In large-volume thick-walled parts, axial pressing becomes a problem. The degassing problems that arise during compression and the density gradients developed in the shaped part often make it necessary to change the forming process. Isostatic or quasi-isostatic pressing processes are then advantageous. Unfortunately, product quality and production costs are interrelated. Purely isostatic compacting can only be attained at relatively high production costs. However, quasi-isostatic compression can be achieved much more economically, such as with rubber dies. Stress arising during sintering may result in cracks and structural defects in the shaped part. Despite these limitations isostatic or quasi-isostatic pressing processes are frequently used.

It is immediately obvious that parts with geometrically complicated shapes cannot be pressed axially. With axially or isostatically operating presses, only crude preliminarily shaped parts can be produced, from which the actual shape must then be developed through extensive mechanical treatment. This path is occasionally chosen today, although, because of high costs, it is very disadvantageous.

Due to the problems mentioned, the injection-molding process is an interesting alternative with the following superior features:

1. No density gradient in the part
2. Narrow dimensional tolerances
3. Geometrically complicated structures realizable
4. Low surface roughness when fine powders are used
5. Fine-grained structure in the sintered parts
6. Economically advantageous with large numbers of pieces and complex geometry

TABLE 5.3 Comparison of Dry-Pressing and Injection-Molding Processing

Processing steps	Dry pressing	Injection molding
Starting powder	×	×
Mixing with binder		×
Molding	×	×
Binder removal		×
Sintering	×	×
Mechanical treatment		

Source: After Lange and Müller.[16]

Dry-pressing techniques have been applied to slide bearings from siliconized SiC. The production and fabrication technology for seal rings and slide bearings has gone in two directions. Where production quantities are less than about 5000 parts, dry pressing and automatic machining are used. Above 5000 parts, dry pressing to shape is the more economical solution.

Over 100,000 of these components have been produced with close as-fired tolerances and with negligible deviation in essential properties.[36] The overall yield on the production line has been between 85 and 90 percent.

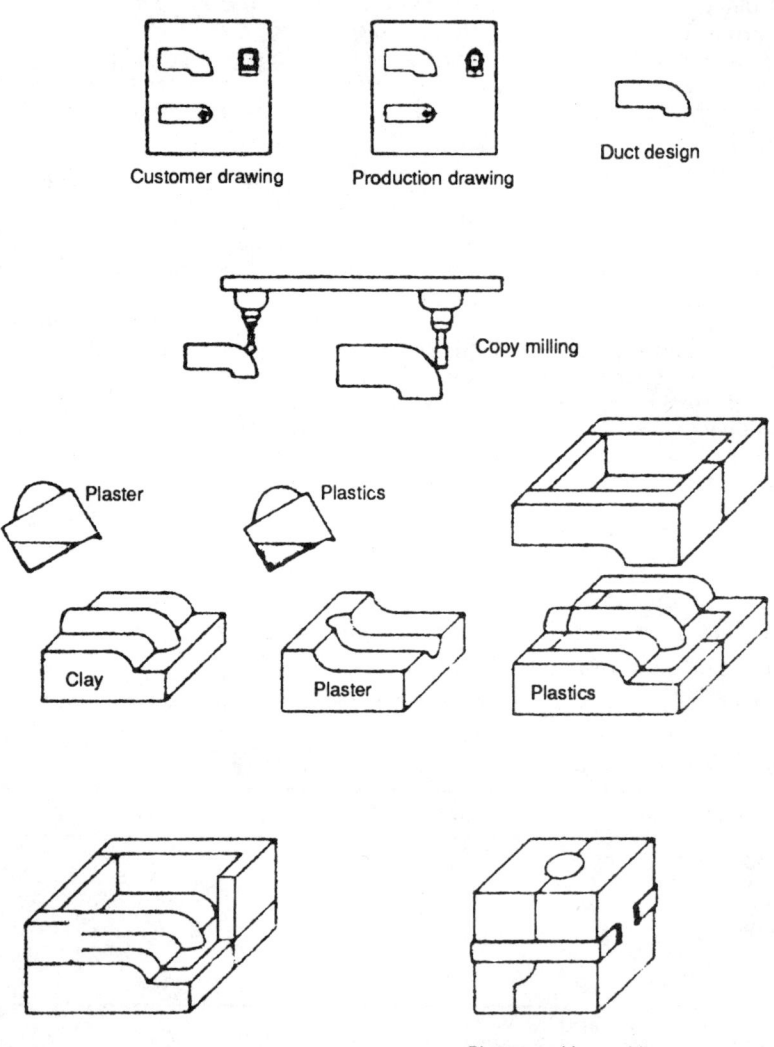

FIGURE 5.11 Axial and isostatic dry-pressing process.[13]

Isostatic and Slurry Pressing

Figure 5.11 depicts the axial and isostatic dry-pressing process used in fabricating SiC components. Reaction-bonded Si_3N_4 pistons for diesel engines have also been fabricated successfully by isostatic dry pressing. Table 5.4 lists a variety of shaping methods used on SiC and Si_3N_4.

A recently completed study compared the slurry-pressing and dry-pressing characteristics of a Si_3N_4 powder. The objective was not only to produce a stronger, more reliable slurry-pressed Si_3N_4 monolithic material, but also to determine whether slurry pressing is a possible way to compact tough, reliable, homogeneous Si_3N_4 composites with whisker or particulate additions.

By combining colloidal techniques with an innovative slurry-pressing technique that avoids agglomeration, the incidence of critical flaws induced by pro-

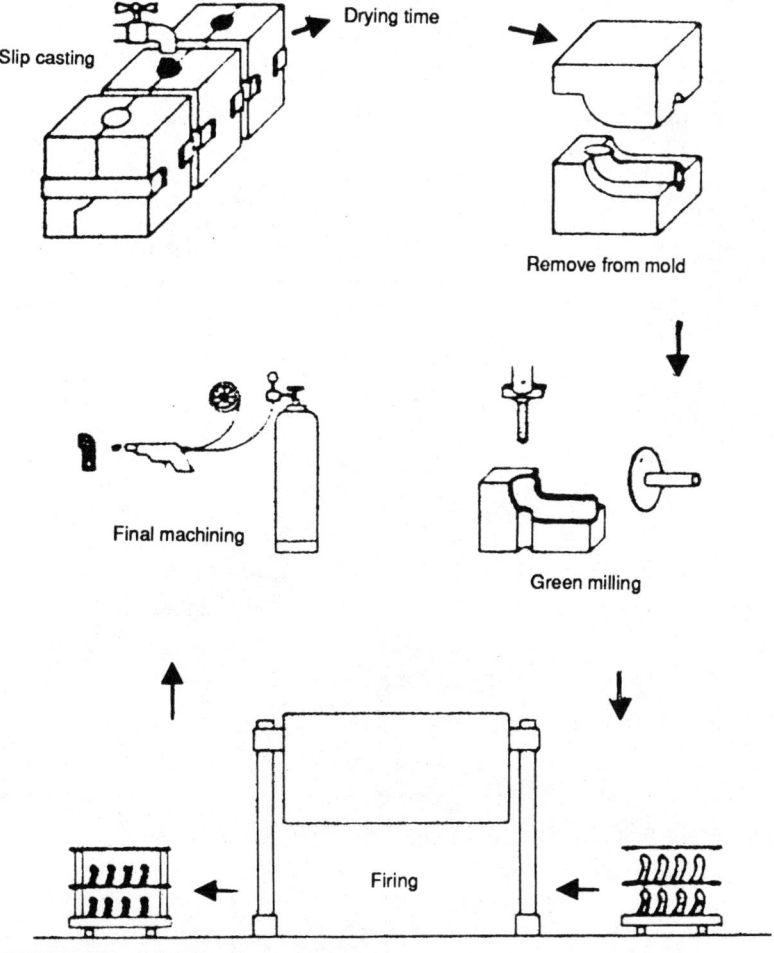

FIGURE 5.11 (*Continued*)

TABLE 5.4 Shaping Methods for Nonoxide Ceramics

Product*	Dry pressing		Injection molding	Warm molding	Ex- truding	Slip- casting	Hot pressing
	Axial	Isostatic					
N, reaction-bonded	x	x	x	x	x	x	
N, hot-pressed							x
C, reaction-bonded	x	x	x	x	x	x	
C, silicon-infiltrated	x	x	x	x	x	x	
C, recrystallized	x	x				x	
C, sintered	x	x				x	
C, hot-pressed							x

*N—silicon nitride; C—silicon carbide.
Source: From Gugel.[13]

cessing could be reduced.[36,37] A slurry-pressing die was developed by the researchers (Fig. 5.12).[36–39]

On pressing, the liquid is pushed out of the holes in the base plate and, by vacuum, drawn through the hole in the plunger. Filter paper on the porous stainless-steel plates between which the slurry is pressed traps the powder as the liquid is expelled from the slurry. Preformed bulk filter paper that swells during pressing prevents leakage between plunger and die and between die and base plate. Though currently used only to form flat 51.8-mm-diameter disks, the die concept is amenable to forming moderately complex ceramic shapes, including whisker-reinforced shapes.

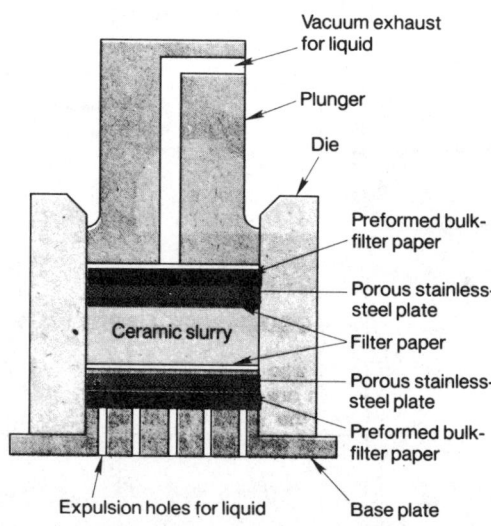

FIGURE 5.12 Slurry-pressing die for forming ceramics.

The average flexural strength for dry-pressed samples, which was 235.6 MPa with a critical flaw size of about 100, was compared with that of aqueous-slurry-pressed samples of 440.8 MPa and a critical flaw size of about 25. Compared with as-sintered dry-pressed silicon carbide, the as-sintered slurry-pressed material has demonstrated 87 percent greater strength. The strength improvement is attributed to reduced porosity—both in the green, or unsintered, condition and after sintering—as well as to smaller critical flaws. For similar reasons, as-sintered Si_3N_4 combined with 6.4 wt % Y_2O_3 and 5.8 wt % SiO_2 (sintering additives) proved 10 to 15 percent stronger.

At a laboratory in England,[40] a slurry impregnation method was used to make continuous SiC-fiber-reinforced glass (Pyrex) ceramics. The parts were pressed at 6 to 10 MPa at 1000°C. Densified composites in the shape of flat, curved, or grooved plates are possible.

Extrusion

Recent studies [41] have indicated that sedimentation volume, torque, and extrusion pressure can be used as effective tools for aiding in ceramic processing of alumina. By the use of these techniques cellulose ethers have demonstrated a variety of properties to control the rheology of ceramic systems by acting as deflocculants and binders. One should be cautious in transferring these results to other ceramic systems, since an effective interaction of cellulosics with ceramic powder may strongly depend on other intermediates or additives used in the total processing system.[42-44]

Injection Molding

One of the keys to successfully manufacturing a ceramic part is to select a fabrication method that is simple and inexpensive to use, yet able to handle complex shapes. This is the case with injection molding (Fig. 5.13), which was selected as the forming method in the development of high-strength, high-reliability SiC parts with complex shapes suitable for use in advanced heat engines.[45] The process is capable of forming complex parts adaptable for mass production on an economically sound basis. Weibull characteristic strengths in excess of 550 MPa with Weibull moduli to about 16 were realized, and this met the program goals.

Other government-sponsored programs [Advanced Turbine Technology Applications Project (ATTAP)] are component-fabrication-oriented.[46] Major emphasis has been placed on developing high-volume, near-net-shape fabrication technology by injection molding of turbine stators, high-pressure injection molding of rotors and vanes, and injection molding of transition ducts.[47-49]

Injection molding for producing stators looks promising because this part has a small cross section, complex shape, and minimal binder removal problems. Many variables can be controlled simultaneously. The latter is necessary for optimum production.

In the injection molding of rotors, strength values of 700 MPa have been attained, some of which is due to residual levels of carbon from the starting SiC powders.

In looking at injection molding for transition ducts, it has the potential for producing high volumes at low costs, is reproducible, and can produce near-net shapes. However, it is still difficult to control the dewaxing (or binder burnout)

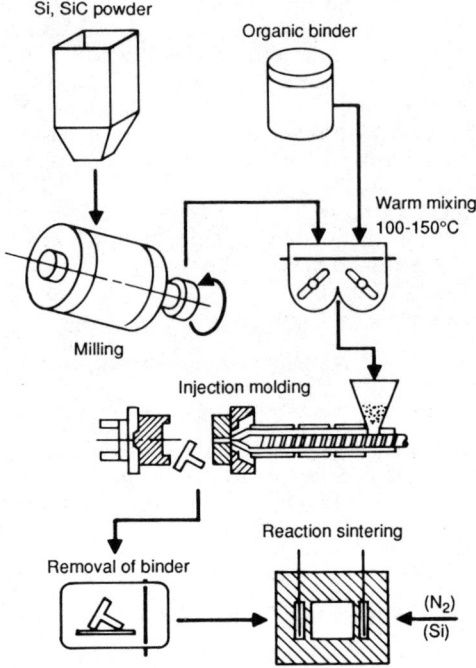

FIGURE 5.13 Injection-molding process.[13]

step of the process, which could cause significant problems when scaling up the process.[49]

Application of "preliminary" tooling has certain advantages whereby fewer experimental dies are required, tool delivery times are reduced, and flow modeling is simplified. In addition, material requirements can be significantly reduced and, perhaps most important of all, the tooling can easily be scaled up for production (Fig. 5.14).

There are other processing developments being carried on to improve injection molding. One way is to use an aqueous-based molding system, although this approach has limits with regard to large cross sections, binder-removal problems, and a high die cost at low volumes. Another approach would be to mix a gel-based organic material with water and a ceramic powder; controlling the moisture content could be a limiting factor.[49]

Another novel process being explored is a low-pressure binderless injection-molding technique. This process has the potential for improved dimensional control and thus reduced machining requirements and higher yields. Low-pressure molding also allows the use of simpler lower-cost tooling, and fabrication cycle times can be reduced as well because there is no binder to remove.[49–54]

Injection molding of Si_3N_4 has been demonstrated on nozzles for welding and for oil burners. The production technology used allows molding to shape with very little subsequent cleaning. Special shapes sometimes need to be finished by machining. This production technique has been automated with a mechanized removal of the feed head.

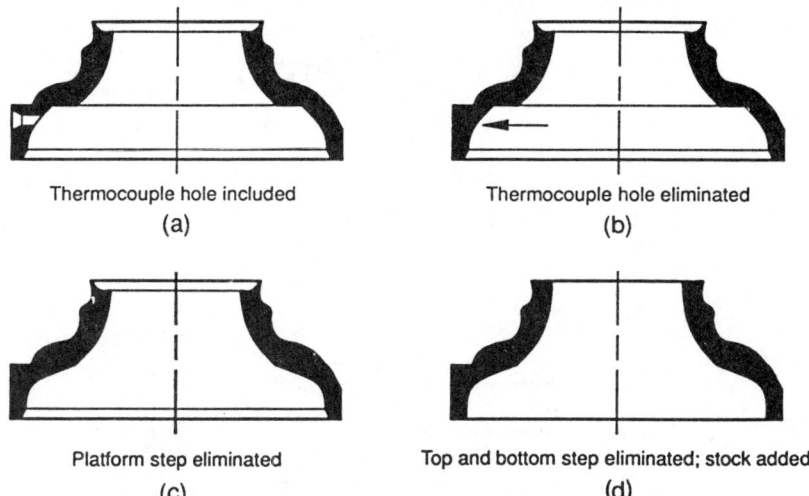

FIGURE 5.14 Mold design for injection molding of transition ducts has undergone several changes to eliminate (a) knit lines, (b) cracks, and (c) molding turbulence and density gradients, and to provide (d) adequate machining stock.[47]

In addition, axial turbine blades, gasifiers, and powder turbine vanes have been injection-molded from Si_3N_4. Small-cross-section parts such as those described previously have been routinely injection-molded and sintered to greater than 98 percent theoretical density (3.26 g/cm^3) with excellent dimensional control and with state-of-the-art properties.[17,52] New processing techniques have overcome the problem of internal and external cracking during binder removal of injection-molded Si_3N_4 parts. By utilizing proprietary processing,[17] which included HIP, results have shown that complex large-cross-section Si_3N_4 parts can be fabricated by injection molding and sintering or HIP.

Injection molding was selected as the process to demonstrate near-net-shape fabrication technology capable of producing ceramic matrix composite automotive engine components. The system selected was a Si_3N_4 + 30 percent SiC whisker material.[53]

Batches of Si_3N_4 + 30 percent SiC whiskers for injection molding were prepared by two process routes, preblend and compound blend. In the preblend case, dispersed whiskers were first added to Si_3N_4, and this material was then mixed with the binders. The alternate approach was to mix whiskers, Si_3N_4, and the binder using the high-shear mixing action of the compounder to blend the components. Microstructural examination of as-molded and HIP-processed parts (compounder-blended powder) showed a large number of Si_3N_4 agglomerates which were not broken down during compounding. Samples fabricated using preblended powders were homogeneous. Based on these observations, preblending was selected as the powder preparation route of choice (Fig. 5.15).[53]

Improvements in strength and fracture toughness of Si_3N_4 have been achieved through the composite approach. Si_3N_4-based composites containing up to 30 vol % of SiC whiskers could be injection-molded and densified by HIP to cover 99 percent theoretical density.

The room-temperature fracture toughness was increased by 40 percent, with a

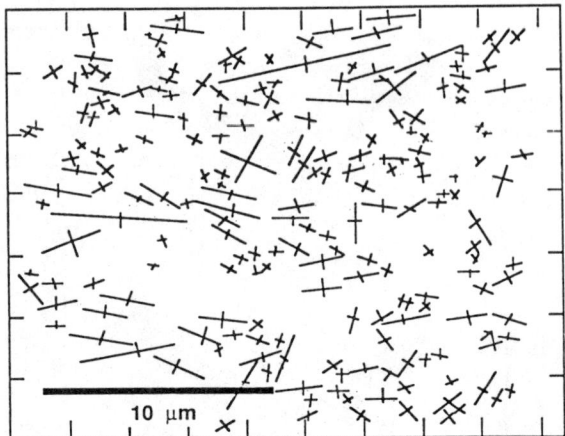

FIGURE 5.15 Digitized SiC whiskers in material processed by injection-molding and HIP, showing orientation and length and diameter of each measured whisker.[53]

concomitant 25 percent strength increase by the addition of 30 vol % SiC whiskers. At 35 vol % of whisker additions, an increase in fracture toughness of 70 percent was observed. Due to intergranular fracture, the degree of fracture-toughness increase was found to depend on the matrix grain size, dispersoid content, and its size relative to the matrix. The acicular nature of Si_3N_4 grains reduces the effectiveness of particulate dispersoids as toughening agents. Whisker additions offer considerably higher potential in fracture toughness and strength enhancement.[56–58]

Considering the results obtained for the Si_3N_4–SiC_w system, it appears reasonable to expect further improvements in fracture toughness through an increase of the amount and aspect ratio of dispersed phase. This approach, in either case, is expected to increase difficulties in processing. Matrix microstructure plays an important role in determining the composite properties, suggesting that further development of these composites through dispersoid additions must be concurrent with additional matrix microstructure tailoring.[56,57]

Injection molding of ceramic components provides flexibility in metal casting and forming. A wide variety of complex oxide ceramic shapes can be produced by the injection-molding process. All sizes and shapes of ceramics from 0.4 mm thick to lengths exceeding 500 mm are produceable by a proprietary process. Products, for example, include ceramic cores used in the lost wax casting process to make hollow shapes. This is accomplished by casting metal around a premolded ceramic core and, after solidification, dissolving the core to form the desired hollow metal shapes. More complex ceramic pieces and molds as well as near-net-shape metal parts of more difficult configurations can be produced (Figs. 5.16 through 5.20).[59] These parts do not require the stringent fracture-toughness and fatigue requirements of the various engine components discussed in the preceding sections.

The application of low-pressure ceramic injection molding has been used to make a variety of small, intricate parts of alumina, magnesia, zirconia, forsterite, steatite, and cordierite, and in the future the process will be extended to nonoxide ceramics.

FIGURE 5.16 Injection-molded ceramic filter used in casting superalloys. Piece is 140 mm in diameter by 70 mm high and slots are 12 mm wide.[59]

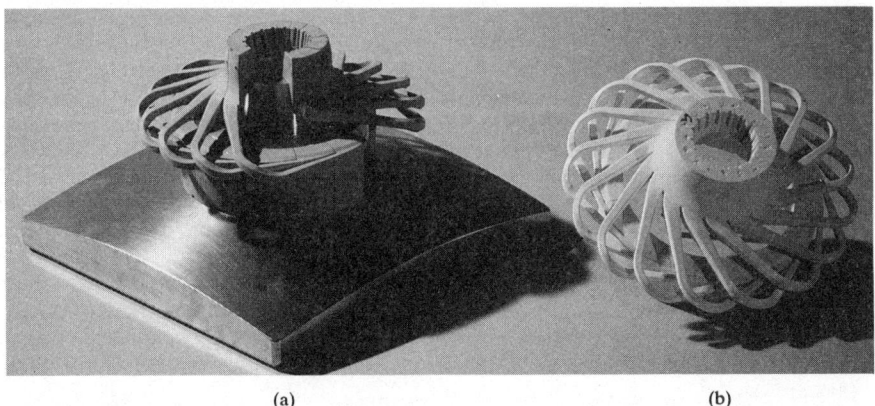

(a)   (b)

FIGURE 5.17 Elaborate ceramic assembly. (*a*) Fixtured assembly of 17 identical unfired pieces. (*b*) Fired ceramic core (~140-mm diameter).[59]

Any shape that can be made by high-pressure injection molding also can be made using pressures as low as 690 kPa. At this pressure, parts can be made with walls (as thin as 0.39 mm), complex shapes, holes as small as 0.25 mm in diameter, and with internal and external threads.

Low-pressure injection molding typically requires a binder content as high as 10 to 20 wt %, compared to 5 to 8 wt % for high-pressure molding. Technology has been developed for removing the wax binder overnight from the molded part without deforming it. The proprietary "dewax" method allows parts to be transferred to the furnace and fired in a single step without support systems other than

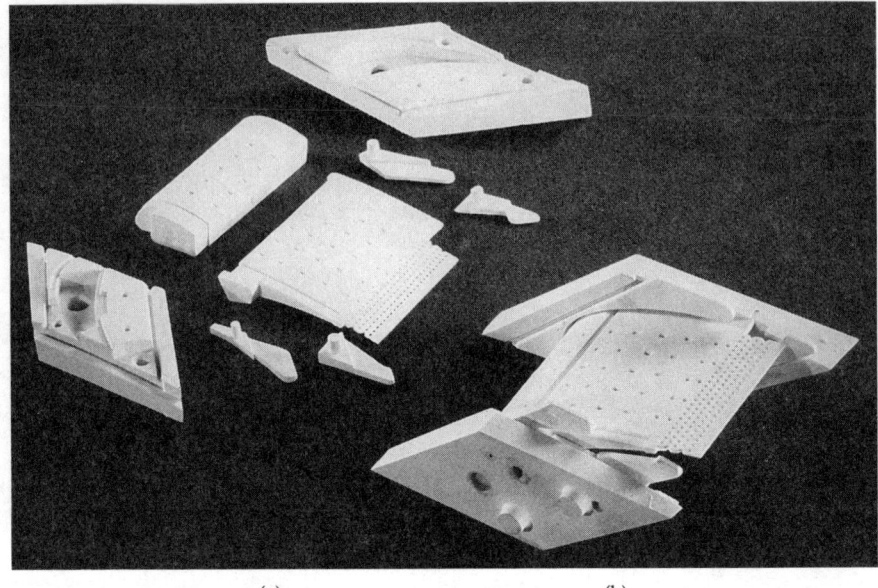

(a)   (b)

FIGURE 5.18 Ceramic assembly for production of air-cooled turbine vane. (*a*) Eight components. (*b*) Ceramic assembly (part length is ~150 mm).[59]

the shelf they stand on. Aluminum dies used in this process are easier to machine than the steel required for high-pressure processes, thereby making small-lot runs of 50 to 500 pieces feasible. The process involves mixing the particles with wax binder and a surfactant, then stirring the mixture, which is heated above 85°C. The resultant highly viscous fluid can be injected using only air pressure. Because the aluminum die is an efficient heat conductor, the injected material freezes rapidly. Parts with small cross sections (up to 1.9 mm) can be removed from the die in 10 to 15 s. Particle-size distribution is critical. For full density, the initial grain size should be 1 to 2 μm.

New techniques have cut the time for degreasing injection-molded zirconia to prevent defects from 5 days for a 10-mm molding to 1 h by optimizing the binders that are used.

Researchers mix their binder from six resin types that have good flow properties and which decompose and vaporize at temperatures below 400°C. The six are a polypropylene resin, a polystyrene resin, an acrylic resin, paraffin, stearic acid, and a plasticizer. In the right mixture the combination requires a binder-to-ceramic ratio of only 46.9 percent instead of the more usual ratio of 50 percent or more. Zirconia fired with the new binder has an average flexural strength of 61.9 kg/mm², which is nearly double that of equivalent alumina.

For ceramic injection-molding procedures which use wax binders in the production of powder-based parts, melt wicking is commonly used to debind the components prior to sintering. Because debinding is often a time-consuming procedure, the influence of such process variables as powder size, part height, green density, and temperature have been investigated to reduce the amount of time required for debinding by melt wicking.

Therefore it is apparent that small changes in the processing variables in melt wicking exert great influence on the resulting debinding rate. Understanding

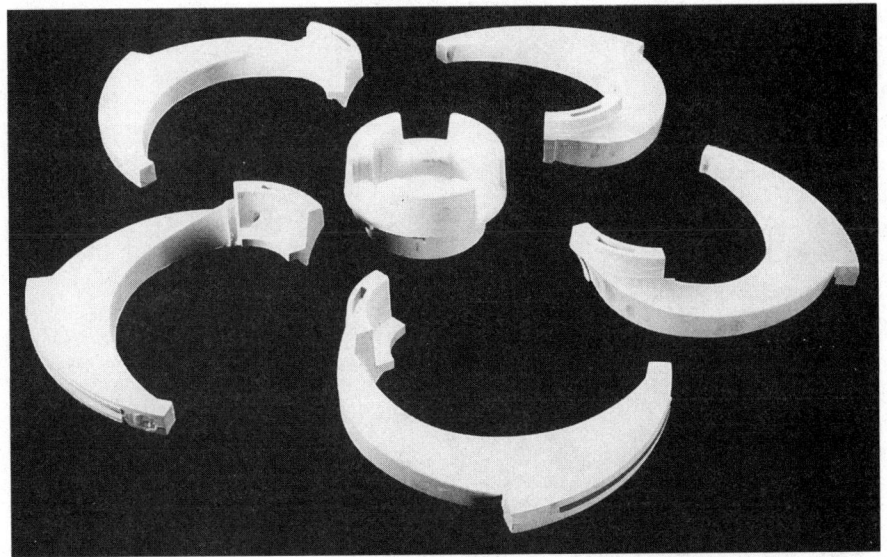

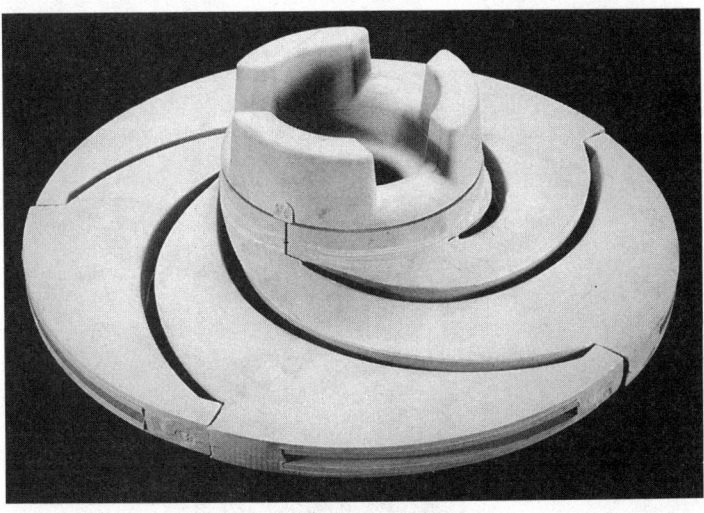

(b)

FIGURE 5.19 Ceramic core for pump impeller. (a) Five vanes and other components. (b) Assembled core (~240-mm diameter).[59]

these effects through modeling and experimentation enables producers to optimize debinding conditions and reduce processing time, a significant limitation in ceramic injection molding.[56,60]

Finally the combination of SiC and Si_3N_4 ceramic materials in the manufacturing of turbocharger rotors is one of the newest applications of injection molding. Combinations of sintered silicon carbide, sintered silicon nitride, and

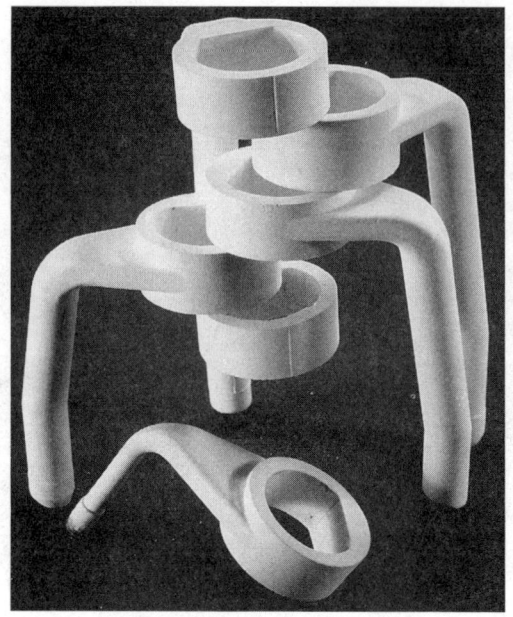

(a)

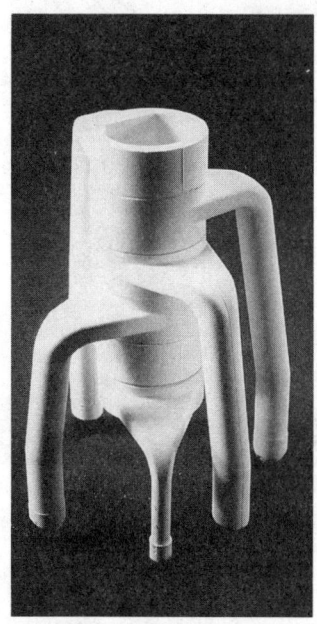

(b)

FIGURE 5.20 Ceramic part made by injection molding and firing "arms" individually and then assembling them on a specially designed rod. (*a*) Five "arms." (*b*) Finished ceramic piece (~80 mm long).[59]

sintered reaction-bonded silicon nitride appear capable of replacing metallic turbochargers, which are heavier in weight. The ceramic replacement is capable of withstanding extremely high temperatures while also meeting high mechanical resistance up to temperatures above 1100°C, high resistance to oxidation and corrosion, and thermal shock resistance.

SiC can be distinguished not only by its good resistance to oxidation and corrosion, but also because it retains its room-temperature strength to above 1400°C. Si_3N_4 has very good strength at room temperature, which, however, decreases at temperatures above 1000°C.[57,58]

Slip Casting

With the introduction of ceramic turbocharger rotors into automobiles a significant step has been taken with advanced ceramics, which represents a challenge to overcome. The factors currently limiting their use include:

1. Problems in manufacturing reliable, reproducible rotors (including ceramic-metal joining technology to bond the ceramic to a metal shaft)
2. Lack of mass-production facilities
3. Difficulties associated with penetrating an existing market
4. High costs

Ceramic turbochargers could be cost-competitive with metallic rotors. Because of the performance benefits associated with the ceramic rotor (such as lighter weight, which corresponds to less transient lag), a 40 percent price premium could be expected in the initial phases of introduction.[61]

Injection-molding as well as slip-casting techniques have been developed to form turbocharger rotors. Both techniques allow the manufacture of intricate shapes, though the densities achieved, and thus the final properties, vary with the solvents and binders used in the manufacturing process. Differences between the processes include (1) powder preparation (additions of binders, deflocculants, and water for slip casting versus wax, polystyrene, and a surfactant for injection molding), (2) equipment and labor requirements (slip casting uses low-cost molds and is labor-intensive; injection molding requires expensive equipment and tooling), and (3) the predensification firing phase (in slip casting a small amount of water and organics is removed, while injection molding entails a complex and extremely sensitive binder-removal process).

The estimated processing costs for Si_3N_4 turbocharger rotors suggest an ~15 percent advantage of slip casting over injection molding if the same quality product and the same yields are assumed. The only real conclusion one would make is that these estimates are so close that there is no a priori advantage in one technique over the other.

Slip casting has been utilized with new formulations of Si_3N_4, TiN, and BN, which can be cast into simple and complex shapes in open and two-piece plaster molds. These ready-to-use water-based formulations have produced low-stress parts, and therefore slips are an economical alternative to isostatic-pressed Si_3N_4 or hot-pressed and machined BN. The formulations have shelf lives of more than 1 year and form thick walls within 3 to 10 min.

Si_3N_4 composite materials with SiC whiskers are a most promising material for high-temperature application. It has been shown that the dispersion of high-strength SiC whiskers in a Si_3N_4 matrix material results in improved fracture toughness and other improved mechanical properties of the composite material.[53]

The SiC-whisker-reinforced Si_3N_4 composites have been manufactured by hot pressing only.[62] However, shaping methods (cold isostatic pressing and slip casting) are now receiving particular interest.

Uniform distribution of the whiskers in the Si_3N_4 matrix, high green densities, no whisker damage during shaping, and the possibility of obtaining preferred whisker orientations are attractive advantages of the slip-casting process.

The key to achieving success is simultaneous deflocculation of the aqueous suspension of both the nitride matrix powder and the carbide whiskers in the presence of additional metal oxide sintering aids. Finally the surface quality of the Si_3N_4 powders and the SiC whiskers combined with controlled ion concentration are of particular importance to maintain maximum colloidal stability of the aqueous suspension and produce slip-cast composites.

Aqueous Si_3N_4 slips containing up to 20 percent of β-SiC whiskers were investigated.[63] The green compacts produced were consolidated by plaster and pressure casting. The resulting microstructure showed a whisker alignment parallel to the mold surface. Other characteristics were measured (casting rates, shrinkage behavior, densities, and rheological effects) and indicated work for the future.[63]

Slip casting has also been applied to sialon material. Due to the stable nature of sialon powder, aqueous slips can be formed without risk of major compositional change. The slip is poured into an absorbent mold to provide thin-walled components.

Thin films of zirconia are being slip-cast in lieu of a gas-phase process. This new method makes it easier and cheaper to produce complex-shaped film of high quality. The researchers have used the slip-casting technique to form a gastight zirconia film 140 μm thick on an electrolyte fuel-cell element.

Aluminum titanate ceramic port liners, which are used for the thermal insulation of automotive cylinder heads, are produced in a precision slip-casting process. These ceramic port liners are cast into aluminum or iron cylinder heads, and the resultant geometrical shape, thermal shock resistance, thermal conductivity, and thermal expansion all meet the close tolerances demanded by the part.

Superplasticity—enormous ductility, typically in tension—has been put to good use over the past 10 years primarily in forming complex titanium- and aluminum-alloy components. Recently, however, this behavior also has been observed in fine-grain (less than 0.3-μm) YTZ, and researchers believe that this should facilitate slip casting the ceramic and further shaping consolidated preforms into complex near-net shapes.

It has been reported that a tensile elongation of 200 percent at 1450°C and a true strain rate of 2.8×10^{-4}/s was achieved. The 200 percent, however, is by no means the maximum that can be achieved, according to these researchers. Still greater ductility may be possible, and recently a maximum elongation of 800 percent was reported.

More recently, researchers found that superplasticity is retained when the material is reinforced with 20 percent alumina, which is significant because such modification is one of the ways to improve the toughness of typically brittle ceramics. The $20Al_2O_3$–YTZ exhibited the greatest ductility, 500 percent elongation at 1650°C and a strain rate of 8.3×10^{-4}/s.

Tape Casting

Strong efforts have been put forth in recent years to develop tape casting into a continuous-casting process. This entails the constant feed of the slip to a doctor blade that rests on a continuously moving carrier surface. The carrier moves progressively through a dryer for solvent removal.

Continuous tape casting involves near-total environmental control of the entire system downstream from the doctor blade. The continuous machine will, however, more closely approach production conditions than will a batch machine.[64]

There are basically three classes of carriers: rigid glass plate, continuous steel belt, and flexible plastic film. Ordinary plate glass made by the float-glass process is satisfactory for most tape casting. Continuous steel belts form a good surface for casting. Thin tape which dries quickly is easily cast on steel belts. Thicker tapes require very long belts for adequate production. Flexible films overcome most problems and are used for most thick-tape casting. Polyester is probably the most commonly used film. Cheaper films, such as polyethylene and acetates, may also be used at a sacrifice in strength.

Small batch-type tape casting is easily accomplished with a continuous flexible-carrier system. Most advantages of the full-production-scale methods and equipment are achievable with small laboratory experimentation. Scaleup to production does not require readjustment of the slurry composition or most other casting variables (Table 5.5).

Since temperature, inlet air flow, outlet exhaust, and casting speed are most easily controlled in a properly designed continuous tape-casting machine, this procedure is superior to any single batch-casting system.

In addition, there is no real limit to tape thickness "as cast." Drying time becomes the important variable. If the strength of the tape does not exceed the adhesion of tape to carrier, it will be impossible to remove the tape. Thick tapes will be stronger than thin tapes and should strip more easily.

A flexible-film, continuous tape-casting machine should therefore incorporate the following features: temperature-controlled slip supply tank with deairing vacuum system, slip agitation, pressure feed to doctor blade, micrometer, height-adjustable doctor blade, spool of flexible carrier film, speed-controlled carrier takeup, temperature-controlled dryer sections, controlled inlet air, controlled solvent exhaust, and green-tape takeup.

An example of the type of part that could be fabricated from this technique is an all-ceramic turbine-shroud seal for an advanced gas turbine machine. The hot face of the seal could be made of material with good thermal-cycling and high-temperature characteristics, whereas the cold side could be made of material with high structural strength and high fracture toughness. One combination of materials would be (1) yttria and partially stabilized zirconia (Y/PSZ) for the layers closest to the hot face of the turbine shroud seal, and (2) magnesium and partially stabilized zirconia (M/PSZ) for the "structural" layers.[65]

Slurry Pressure Casting

A slurry method was described earlier (Fig. 5.21).[43] With optimized process parameters, the following properties were obtained: a flexural strength of 1.25 GPa, a Weibull modulus of 30, an elastic modulus of 120 GPa, and a fracture toughness of 26 MPa · $m^{1/2}$.

Such slurry-type consolidation methods are preferred over dry-pressing techniques because the latter usually introduces agglomerates. Though injection molding is one alternative, the high binder content required can cause shrinkage problems. A molding technique being given considerable attention is pressure

TABLE 5.5 Effect of Casting Parameters on Casting Behavior*

Parameter	Solids content	Viscosity	Defect generation				Drying time
			Residual bubbles	Skinning and new bubbles	Cracking		
Increase amount of solvent	Decrease	Lower	Reduce	—	Possible increase		Increase
Increase slip temperature	—	Lower	Reduce	Possible increase	—		Decrease
Increase rate of solvent evaporation	—	—	—	Possible increase	Possible increase		Decrease
Increase casting rate	—	Increased shear	—	—	—		—
Increase air flow	—	—	—	—	Possible increase		Increase
Dispersed inorganics	—	Lower	—	Lower	—		—

*An optimized slip will, therefore, have (1) high solids content (lowest solvent content), (2) low viscosity to flow under blade but not so low as to flow off the carrier (aided primarily by good dispersion and elevated temperature), (3) solvent that will not encourage skinning and bubble entrapment, (4) a drying system that allows controlled gas removal, and (5) fast drying to increase drying rate.

Source: After Hyatt.[64]

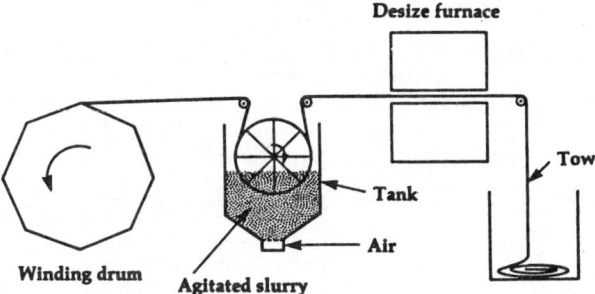
FIGURE 5.21 Slurry impregnation plus filament winding produces fiber-reinforced glass composite tape.[43]

filtration or slurry pressing, otherwise known as pressure casting. Pourable slurries with up to 64 vol % solids can be cast with the aid of polyelectrolytes, which increases the repulsive forces between particles. These polymers also help to prevent cracks that grow after unloading the part.

Since the removal of water from ultrafine ceramic slurry is a lengthy process in slip casting, the alternative being considered is pressure casting.

Working with SiC and combining pressure casting with sintering, researchers have produced material with an average strength of 430 MPa, a 30 percent improvement in strength. By switching to HIP instead of only sintering, the strength was further improved by 30 percent. This material also had an ultrafine microstructure (0.3 to 3 μm) and contained no internal defects.[66]

The main difference between traditional casting and pressure casting is that the cast formation in the pressure-casting process is based on the principle of pressure filtration. Therefore the basic concept of pressure casting is not entirely new. In fact, the ceramics industry has used this principle for many years in the form of the filter press. The pressure-casting process is directly comparable to filter pressing, with the basic difference being that the filter medium in the pressure-casting machine must both filter and shape the end product.[67]

With the high-pressure casting system, slip pressures up to 4 MPa are practical. It is evident that for such slip pressures, plaster molds would be unsuitable. Only a small fraction of this pressure would cause the molds to break, hence the use of a plastic material, polyelectrolytes, which combines high porosity and high mechanical strength; good elasticity has been developed. This material is able to operate successfully under high-pressure casting conditions.

Pressure casting also results in a more efficient dewatering of the part. The lower moisture content of the cast part minimizes the risk of deformation when the part is being removed from the mold. In addition, this means shorter drying times and less susceptibility to stresses.

The quality of the parts is considerably improved as well. Erosion of the surface of the molds is virtually eliminated. The pressure-cast parts have a much smoother surface and, above all, a consistently smooth surface. Embossments, edges, and borders remain constant, which means that the last cast from a mold will be as good as the first.

Variable body thicknesses also can be shaped far better than with traditional casting. The cast compression of the pressure-cast part is higher than with a normally cast one. This results in lower shrinkage during drying and firing, and thus better accuracy of the finished piece, as well as higher mechanical stability of the fired body.[67]

Busome and Pollinger[68] disclose the use of a pressure slip-casting technique used to eliminate the problem of sintering-aid segregation due to long casting times. The process steps used in the pressure slip casting of Si_3N_4 are described in Reed.[6] The advantages of pressure slip casting include reduction of the casting time required for a specific part and generation of more uniform green density in cast parts. By implementing pressure slip casting, the casting time required for a Si_3N_4 rotor has been reduced by 80 percent. As a result, production fabrication techniques are being developed for ceramic turbine rotors using pressure slip casting of Si_3N_4.[68]

Special Processes

Squeeze Casting. Squeeze casting is a well-developed process used to make low-cost, high-quality parts in almost any material. It has now been shown that it is feasible to insert ceramic reinforcement during casting that will substantially improve the properties of the finished product. In addition, metal inserts may also be placed in desired locations, and a designer using a computer-aided design approach has the flexibility to design castings with specific properties in different areas of the casting. This method of fabricating metal-ceramic composites should be low-cost and well suited to existing industrial methods using conventional manufacturing equipment. It is also amenable to automatic diagnostic quality-control procedures.[69]

In advanced diesel engines, the operating stresses in piston pin bosses and combustion bowls can now exceed the strength of conventional aluminum alloys. To address this problem, a proprietary squeeze-casting process has been developed to manufacture pistons. The mechanical properties of squeeze-cast pistons, such as tensile and fatigue strength, surpass those of conventional permanent-mold pistons and approach those of forgings. In addition, squeeze casting permits the reinforcement of selected areas of the piston with low-cost ceramic fibers.[70]

The heavy-duty diesel engine, used predominantly in transportation, off-highway, and construction applications, is currently experiencing a period of rapid technological change. Engine builders are considering the following design developments: (1) higher cylinder pressures, (2) increased fuel-injection rates and pressures, (3) combustion chamber geometry changes, (4) turbo compounding, (5) bottoming cycles, and (6) uncooled engines.

To satisfy these conditions, squeeze casting has provided a good manufacturing foundation for pistons. The application of pressure during the solidification of a casting (often referred to as liquid-metal forging) improves the material properties of the component by significantly reducing porosity and refining the microstructure. As a result, fatigue characteristics of squeeze-cast pistons are superior to conventional gravity-cast parts. However, operating environments for the new designs are creating additional challenges in terms of wear and thermal fatigue. These can be overcome with a new development in casting technology—selective fiber reinforcements.[71]

In the squeeze-casting process (Fig. 5.22) a measured quantity of modified and filtered metal is poured into a preheated die cavity located on the bed of a hydraulic press. The press is activated to close off the die cavity and pressurize the liquid metal. Pressure in excess of 70 MPa is maintained until solidification is complete. The near-net-shape casting is ejected from the die cavity.

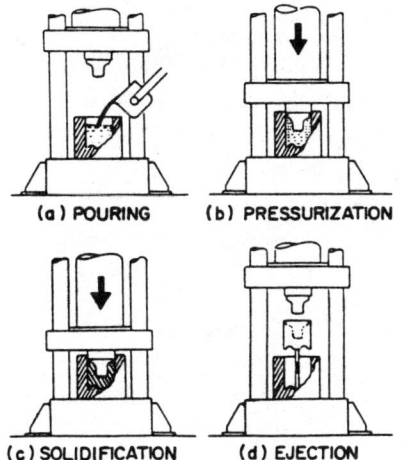

FIGURE 5.22 Basic steps in squeeze-casting process, illustrating production of a piston.[70]

The pressures produce a rapidly solidified, pore-free fine-grained part. The mechanical properties invariably exceed those of castings and generally fall midway between the longitudinal- and transverse-direction properties of wrought products. Costs are lower than forging because of cheaper starting materials, lower press tonnage, and less machining required. Compared with casting, any cost increase due to the need for a press is usually compensated for by improved material yield and, more importantly, by the higher rate of production made possible by rapid solidification in squeeze casting. Factors that influence casting quality, listed in decreasing order of importance, are melt quality and quantity, equipment and tooling, casting temperature, tooling temperature, time delay before pressurization, lubricant film thickness and its adherence, and pressure level and duration.

Manufacture of a selectively reinforced casting begins with the fabrication of a preform containing the required composition of fibers at a chosen volume fraction, molded into the desired shape. Preform manufacture is typically accomplished through slurry vacuum filtration technology. A preform may be a single component or an assembly in order to achieve the desired characteristics in the final casting.

During the pressurization cycle the liquid metal infiltrates the voids in the preform such that the reinforcing fibers become an integral part of the casting. It is important to note that the initial position of the preform does not change, nor is the structure of the preform altered by infiltration.

The two-dimensional orientation of the fibers is apparent and a scanning electron micrograph of the fractured surface shows no fiber pullout. This indicates excellent load transfer as a result of wetting of the fiber by the matrix alloy.

Fiber reinforcement provides considerable flexibility in controlling the final properties of a casting. The composition of the matrix alloy may be selected to provide the overall properties required for the casting.[70]

Based on some work that has been done,[69] a track shoe should be capable of being made as an aluminum squeeze casting. Two problems have to be overcome: (1) the strength of the casting needs to be increased by suitable reinforcement, and (2) the track pins will cause wear unless the holes are given adequate wear resistance. A schematic of the proposed method of manufacture is given in Fig. 5.23. Steel tubes are used to give the casting the required wear resistance on the track pin. The tubes are inserted in a fiber preform, which is molded to the shape of the track shoe. Four plugs are inserted in the steel tubes in the lower die cavity. During casting the upper die is closed into the lower die and clamped by the hydraulic press. The aluminum alloy is poured into the die cavity through the hold in the top die. The auxiliary hydraulic cylinder then applies pressure to the liquid metal and completes the casting. The finished casting is ejected by a lower hydraulic cylinder.[69]

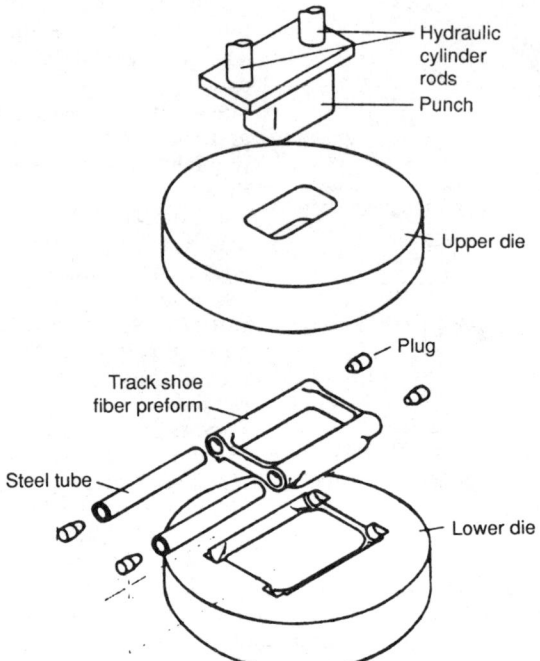

FIGURE 5.23 Proposed manufacturing method of steel tube and ceramic fiber-reinforced (SiC) aluminum squeeze casting.[69]

Another application where much of the automotive matrix composite effort has been directed is aluminum diesel pistons. These applications are aimed at combustion bowls, piston ring, and wrist pin areas.[72] The first commercial use of a metal matrix composite in the automotive field was with ceramic fiber in Toyota's 2L-T diesel piston in 1983. Toyota used ceramic fiber to improve aluminum wear resistance in the upper land piston ring groove.[73]

Retaining their strength and stiffness up to 350°C, SiC-fiber-reinforced aluminum composites are lightweight alternatives to titanium or steel parts. By combining filament winding and squeeze casting, components of fairly complex shapes can be produced. Squeeze infiltration of three-dimensional preforms or fabrics yields reinforced castings with perfect soundness as well.

Thus the filament-winding–squeeze-casting route has proved to be a highly efficient and versatile process. Fairly large components can be manufactured, and excellent properties can be achieved with the PCS (polycarbosilane)–SiC–Al system through proper microstructural design.

Hybridization is certainly one of the brightest concepts introduced in the field of continuous-fiber metal-matrix composites in recent years. It is an efficient but easy way to separate fibers from each other and avoid fiber-fiber contacts and defects. When a homogeneous distribution of fibers is effectively achieved, transverse strengths are raised significantly to as much as 300 MPa at room temperature, a value that could overcome the need for crossing fibers. This could allow simple designs of some parts submitted to highly directional loading.[74]

As a key material to be considered in the manufacture of squeeze-cast parts, PCS-SiC fiber aluminum composites appear worthy of consideration for demanding defense and aerospace applications, provided that their properties are effectively maximized via a tailored microstructure.

High specific strength, high modulus, and high wear resistance as well as thermal stability due to the excellent mechanical properties of whiskers are reflected in refractory-material whisker-reinforced aluminum alloys. In addition, three-dimensional strengthening and plastic deformation are possible for these composites, while they are difficult for continuous-fiber-reinforced metals.

Most studies on refractory-material whisker-reinforced aluminum alloys are related to SiC whiskers whose contact angle with molten aluminum seems to be smaller than that of other whiskers.[75] SiC-whisker-reinforced aluminum alloys have good mechanical properties. However, their poor machinability is a problem, which is due to high hardness of SiC. It is very difficult to machine the composite with ordinary tools, except diamond tools.

On the other hand, as the Vickers hardness of Si_3N_4 is about 1800, which is lower than that of SiC (2800), it might be easier to machine Si_3N_4-whisker-reinforced aluminum alloy composite than SiC-whisker-reinforced aluminum alloy composite.

The aspiration method[76] was used to make platelike preforms, and the composites were made by infiltrating molten metals into the preform by squeeze casting.

The tensile strength of the aluminum matrix composite at room temperature is about four times higher than that of aluminum. Although the strength of the composite decreased with increasing temperature, the strength was much higher than that of aluminum, even at 400°C. This result proves that the composite was significantly strengthened by whiskers.[75] In the case of 6061 matrix composite, the tensile strength of the composite is 490 MPa at room temperature, which is comparable to that of extruded SiC-whisker-reinforced 6061 composite. Although the strength decreased with increasing temperature, the effect of whiskers was significantly maintained up to 400°C.[75]

It is clear that Si_3N_4-whisker-reinforced aluminum alloys can be easily fabricated by squeeze casting, even though the contact angle of Si_3N_4 with molten aluminum is larger than 90°, and that the tensile strength of the composites is rather higher than that of SiC-whisker-reinforced aluminum alloys. In addition, because the hardness of Si_3N_4 is lower than that of SiC, it is possible to machine the Si_3N_4-whisker-reinforced aluminum alloy with ease using ordinary tools.[75]

Rheocasting. Aluminum alloy matrix composites reinforced by SiC short fibers (or whiskers) can be prepared by rheocasting, a process which consists of the incorporation and homogeneous distribution of the reinforcement by stirring within a semisolid alloy. Using this technique, composites containing fiber volume fractions in the range of 8 to 15 percent have been obtained for various fiber lengths (namely, 1, 3, and 6 mm for SiC fibers). This work is similar to that described earlier under compocasting. The aim of these developments is to overcome the difficulties related to compound preparation when the metal temperature is above the liquidus temperature.[77] Segregation and fiber rejection tend to reduce the reinforcement content, and despite the use of vigorous stirring, flocculation phenomena prevent any homogeneous distribution of the fibers in the composites. Nevertheless, decreasing the temperature gives rise to higher liquid-metal viscosity and enables the floccules to dissolve by shearing. As a result, the presence of another solid phase can increase the mechanical interaction and better disperse fibers. Hence, combining rheocasting techniques with fiber incorporation should allow the preparation of homogeneous fiber-matrix mixtures while giving the compound a sufficient fluidity for casting.[77,78]

Gel Casting. An entirely new process has now been disclosed which purportedly can make ceramics of almost any composition, in almost any shape and size, in a fraction of the time it takes for slip casting or injection molding.[79] Janney of Oak Ridge National Laboratory has discovered a way to do it with gel casting.

Using readily available starting materials and no special equipment, Janney claims he has achieved astounding results. Gel casting supposedly avoids all the pitfalls of processing time in slip casting and the thickness limitations of injection molding, and one has almost infinite control over final product geometry. Perhaps most importantly, the technique requires no new equipment for those currently using injection molding. In fact, the only equipment needed are the ceramic powder, monomer, and polymerization reagents (all commercially available), molds, and an oven.

A typical process proceeds as follows: The ceramic powder and a 10 vol % aqueous solution of vinyl monomer are mixed in a ratio of from 50:50 to 60:40 vol %. A dispersant such as Darvan-C is added and mixed well. The slurry is poured into a mold, and the free-radical initiator added. Polymerization and cross-linking are usually completed in less than 1 h. Sometimes it is desirable to carry out the polymerization below room temperature, in which case a catalyst is also added. In all, the initiator, dispersant, and catalyst take up less than 1 percent of the total volume. Since they are organic, like the matrix, they burn away in the next step.

At this point the gel has the consistency of hard gel candy and can be removed from the mold and manipulated easily. The gel is heated at the rate of 3°C/min in air in a humidity-controlled oven until the polymer burns out completely. This step can be as short as 2 to 3 h, even for very thick parts. Next the material is heated to 1300 to 1400°C at a rate of 5 to 10°C/min. After cooling, the part is ready for use or further processing.

The net result is that parts can be made in less than 1 day and the mold can be made of metal, glass, plastic, or wax, all with excellent results. The method seems to work for just about any ceramic powder, too. So far researchers have successfully tried Al_2O_3, ZrO_2, SiC, Si_3N_4, and powdered Si metal (in a nitriding process).

The method seems compatible in producing turbine rotors, gears, tensile bars, and any other detailed or oddly shaped part that can be molded.[79]

Electroacoustic Dewatering. Since the removal of water from ultrafine ceramic slurry is a lengthy process, the electroacoustic dewatering (EAD) technique may offer many advantages in the manufacture of monolithic ceramics and ceramic composite net shapes. Some of these benefits include better efficiency, higher green density and lower shrinkage, elimination of binder removal problems, uniform microstructure and green density, relatively inexpensive molds, fewer molds, longer mold life, and minimum handling. EAD combines electroseparation (an electric field is applied to the suspension, which moves the water away from the particles and causes agglomeration) with acoustics. The acoustic forces lower viscosity and surface tension, causing separation of the liquid from the particles and producing agglomeration.

This process can remove 10 wt % more moisture compared to conventional methods. It also has much faster dewatering rates when applied in vacuum than electroseparation or acoustics by themselves. The EAD process can be used to make ceramic shapes by selecting a filter material that can uniformly remove the water under an electroacoustic field. Furthermore, this filter can be used as a mold to provide the net shape during the dewatering process.[80]

Chemical Techniques. Chemical vapor deposition or infiltration techniques, or combinations thereof (CVD/CVI), also overcome problems of nonuniform fiber distribution and fiber thermal degradation.[81] Government-funded approaches involve fiber-reinforced ceramic composites for aerospace structural applications to 1400°C and above. Specifically, researchers are looking at reinforcing silicon-based ceramic matrices with continuous high-performance ceramic fibers. Their fabrication approach attempts to achieve high strength by preparing optimum matrix precursors, uniformly infiltrating the precursors into the fiber array, and consolidating and densifying the matrix without fiber-strength degradation, large shrinkage cracks, or porosity.

Other government laboratories are examining a chemical vapor infiltration technique for fabricating SiC-fiber-reinforced SiC. Low-density fibrous structures are infiltrated with vapors that deposit as solid phases on and between the fibers, forming the matrix of the composite. A combination of a thermal and pressure gradient (via a forced gas flow) is used to reduce infiltration times. Preforms can be made with one-dimensional reinforcement by aligning fiber tows or with two-dimensional reinforcement by stacking multiple layers of SiC cloth. Both types have moderate strengths and nonbrittle behavior.

Others have investigated the production of toughened ceramic composites by the simultaneous chemical vapor deposition of an SiC matrix and a dispersed phase of $TiSi_2$. Since the mechanical properties are controlled by the morphology of the dispersed phase, deposition of a coating in a fluidized bed is used to produce a finer, more uniformly dispersed second phase. Using this technique, fracture toughnesses approach 5.5 MPa · $m^{1/2}$.

A number of companies are now applying CVD/CVI technology to the production of a variety of shapes, for example, a method to make heat exchanger tubes. Initially, preforms are wound and then infiltrated, producing room-temperature strengths of up to 817 MPa which drop to 172 MPa at 1200°C. The fibers are precoated with pyrolytic carbon, and the material shows composite behavior with fiber pullout. On the other hand, a company makes ceramic composites out of alumina-boria-silica continuous fibers with a CVD SiC matrix.

Another combines chemical vapor infiltration and deposition on both woven cloths and graphite substrates to make SiC composites. The composite ceramics can be vacuumtight or very porous and require no machining, producing complex shapes with tight tolerances. These ceramics have been successfully tested in combustion environments up to 1400°C. Weaving of ceramic cloths has been combined with other shaping methods to produce a one-piece heat exchanger panel containing both headers and flow passages.

Spray Casting (Osprey* Process). Spray casting via the Osprey process is emerging as an attractive technology to produce near-net-shape components of a variety of alloys.[82] The process involves sequential stages of gas atomization and droplet consolidation on a substrate, as shown schematically in Fig. 5.24.

The alloy charge is melted in a crucible located on top of the spray chamber and exits through a nozzle in the bottom of the crucible. In the atomizing zone, the stream of molten metal is comminuted into a spray of droplets using nitrogen or argon gas. Droplets are cooled by the gas and accelerated toward the sub-

---

*Osprey is a registered trademark of Osprey Metals Ltd., South Wales, U.K.

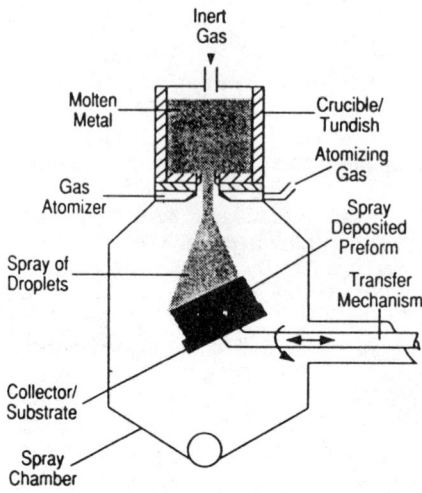

FIGURE 5.24 Schematic diagram of Osprey process.[83]

strate, which is positioned below the atomization zone. The droplets impinge and consolidate on the substrate to form a bulk-net or near-net shape.[83]

Composite materials can be manufactured via spray deposition either by injecting second-phase particulates into the spray of metal,[82,83] or by selectively reacting the droplets during flight using appropriate gaseous media. Currently being produced are aluminum-based composite materials reinforced with injected SiC particulates (10 to 50 μm).

Lanxide Processes (Dimox*). Ceramic composite components presently can be made up to 203 mm thick and heavier than 44 kg, but there are no real barriers to the production of larger sizes. Furthermore, proprietary techniques allow the forming of complex shapes with good-quality finish on a net- or near-net-shape basis.

To form a ceramic matrix product, molten metal in a refractory container is reacted with a gaseous oxidant, with the resultant solid reaction product growing outward from the metal surface (Fig. 5.25).

Examples of ceramic-metal matrix systems produced include Al_2O_3–Al, AlN–Al, TiN–Ti, and ZrN–Zr. The rapid oxidation reaction (Al_2O_3 grows up to 25 to 38 mm in 24 h in air) is promoted by the introduction of minor elemental or oxide powders as dopants, such as Mg in combination with Si, Ge, Sn, or Pb for the Al_2O_3–Al system.

By the addition of certain fillers or reinforcing materials adjacent to the surface of the molten metal (Fig. 5.26), various composites can be formed. Fillers may be in the form of particles, platelets, fibers, and other reinforcement shapes. Some of these composites are Al_2O_3–Al matrix infiltrated into such fillers as Al_2O_3, SiC, ZrO_2, and $BaTiO_3$; AlN–Al matrix infiltrated into Al_2O_3, AlN, AlN–Al_2O_3, TiB_2, and B_4C; TiN–Ti matrix infiltrated into TiC, TiN, TiB_2, and Al_2O_3; and ZrN–Zr with ZrN and ZrB_2 fillers.

The Dimox (directed metal oxide) process in its most basic form is used to grow ceramic composites via an oxidation reaction between a molten metal and an adjacent gaseous reactant (oxidant) through a preform of ceramic reinforcing materials. The metal is progressively drawn through its own oxidation product and the preform by capillary action to sustain the growth process into the preform, yielding a ceramic composite material that also typically contains some metal.

The reinforced ceramic technology is based on an approach to the directed oxidation mechanism. (As used here, oxidation describes the formation of a compound by the donation or sharing of electrons between a molten metal and one or more other elements.) This oxidation phenomenon is typically enabled by combining trace elements, or dopants, with the molten parent metal from which the composite is derived. The technique is generic and applies to numerous ceramic-

---

*Dimox is a registered trademark of Lanxide Corporation, Newark, Dela.

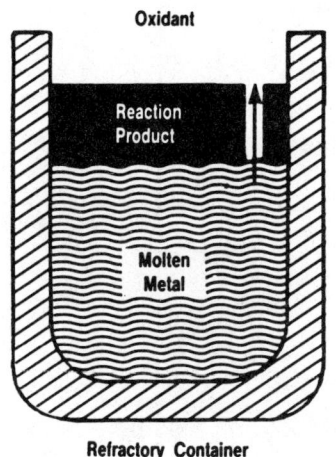

FIGURE 5.25 Formation of composite with oriented oxide growth from surface of molten metal, upon exposure to air oxidant.[84]

metal systems, including most of the oxides, nitrides, carbides, and borides of aluminum, silicon, titanium, zirconium, and hafnium.

The technique was originally used to make alumina-aluminum composites processed in air, with properties controlled by process time and temperature, alloy additives, and filler materials.[85] Table 5.6 reflects some properties for several current Lanxide products, which includes a new ceramic composite reinforced with boride platelets formed in situ, NX-3400. The composites are made by the directed reaction of a molten metal (such as zirconium, titanium, or hafnium) in an inert (argon) atmosphere through a shaped preform or powder bed of B_4C. The B_4C is completely consumed, forming a product that contains a metal diboride, a metal carbide, and a controllable amount of free metal, depending on the processing conditions. Some potential uses for ZrB_2 platelet-reinforced ZrC[86] are:

1. Rocket engine components: plates, nozzles, ducts
2. Wear parts: shaft seals, valve components, nozzles
3. Biomaterials: joint replacements, prosthetic devices

More uses for the other reinforced ceramic materials are discussed in Chap. 8.

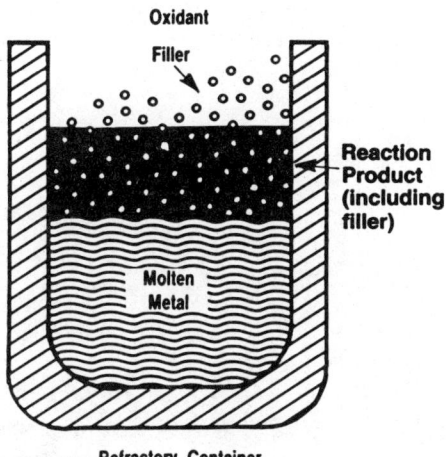

FIGURE 5.26 Formation of composite with oxide growth from surface of molten metal through a particulate filler.[84]

TABLE 5.6 Properties of Selected Lanxide Composite Products*

Reinforcement and matrix system	Tensile strength, MPa	Bend strength, MPa	Young's modulus, GPa	Fracture toughness, MPa·m$^{1/2}$	Density, g/cm^3	CTE, 10^{-6}/K	Compressive strength, MPa	Thermal conductivity, W/m·K	Rockwell hardness, R_A
NX-5201 (silicon carbide–reinforced aluminum; SiC–Al)	210	—	200	10	2.95	8.5	—	160	—
NX-5101 (alumina–reinforced aluminum; Al$_2$O$_3$–Al)	400	—	150	17	3.3	11.5	760	—	—
NX-1201 (silicon carbide–reinforced aluminum oxide; SiC–Al$_2$O$_3$)	150	—	313	6.3	3.32	5.4	1193	147	80
NX-3401 (zirconium diboride platelet-reinforced zirconium carbide; ZrB$_2$–ZrC)	—	800–900	300	16–18	6.2	7	—	50	—
NX-1010 (alumina–reinforced aluminum titanate; Al$_2$O$_3$–AlTiO$_3$)	—	20	14	—	2.6	1.9	330	1.0	—

*All materials are registered trademarks of Lanxide Corporation, Newark, Dela.
Source: After Ref. 85.

Primex*. The Primex process (pressureless infiltration of metal matrix composites) operates without pressure or vacuum. It involves the infiltration of molten aluminum into loose beds or compacts of reinforcing material. Two families of aluminum matrix composites have been developed. One is reinforced with SiC, the other by Al_2O_3. Currently prototype electronic components are being produced from aluminum-matrix composites using both SiC and Al_2O_3 as reinforcing materials.

There are two requirements for the Primex process: (1) the aluminum alloy used must contain magnesium, about 10 percent which will improve wettability, and (2) the atmosphere must contain N_2.

In this pressureless method of casting pore-free components, an ingot of the aluminum-magnesium alloy is placed on a preform (or loose particles) of ceramic reinforcing material such as aluminum oxide, silicon carbide, titanium diboride, magnesium oxide, or aluminum nitride. The alloy-ceramic assembly is heated at 1440 to 1800°C for a few minutes in a flowing nitrogen atmosphere. Infiltration takes place spontaneously, without the need for pressure or vacuum apparatus. In addition to the reinforcement in the fiber preform, aluminum nitride forms in the composite. The amount can be controlled by controlling the amount of nitrogen in the atmosphere. Increases in aluminum nitride content decrease the coefficient of thermal expansion and increase stiffness.

REFERENCES

1. J. B. Wachtman and M. G. McLaren, "Advanced Ceramics; Structural Materials with a Hot Future," *Manuf. Eng.*, pp. 56–60, Feb. 1985.
2. A. Roosen and H. K. Bowen, "Influence of Various Consolidation Techniques on the Green Microstructure and Sintering Behavior of Alumina Powders," *J. Am. Ceram. Soc.*, vol. 71, pp. 970–977, Sept. 1988.
3. "New Structural Materials Technologies; Opportunities for Use of Advanced Ceramics and Composites," Tech. Memo., Congress Office of Tech. Assessment, PB87-118253/KGC, OTA TM-E-32, Sept. 1986, 88 pp.
4. "Advanced Materials by Design," Congress Office of Tech. Assessment, OTA E351, June 1988, 335 pp.
5. M. M. Schwartz, *Composite Materials Handbook*, McGraw-Hill, New York, 1983, 672 pp.
6. J. S. Reed, *Introduction to the Principles of Ceramic Processing*, Wiley, New York, 1988, 486 pp.
7. R. Morrell, "An Introduction for the Engineer and Designer," in *Handbook of Properties of Technical and Engineering Ceramics*, Natl. Physical Lab., HMSO, London, May 1985, pt. 1.
8. E. P. Rothman, G. B. Kenney, and H. K. Bowen, "Potential of Ceramic Materials to Replace Cobalt, Chromium, Manganese, and Platinum in Critical Applications," M.I.T. Industrial Liaison Progr. Rep. 9-11-85, M.I.T., Cambridge, Mass., Final Rep., Contract 333-6530.2, Jan. 6, 1984, 284 pp.
9. A. Hendry, "Processing of Engineering Ceramics," *Powder Metall.*, vol. 31, pp. 20–22, 1988.

---

*Primex is a registered trademark of Lanxide Corporation, Newark, Dela.

10. *Net Shape Technology in Aerospace Structures,* vols. 1–3, Comm. on Net Shape Technol. in Aero. Struct., AF Studies Board, and Natl. Research Council, National Academy Press, Washington, D.C., 1986.
11. M. J. Edirisinghe and J. R. G. Evans, *Intl. J. High Technol. Ceram.,* vol. 2, no. 1, 1986.
12. J. A. Mangels and G. L. Meissing (Eds.), *Forming Ceramics,* vol. 9 of *Advances in Ceramics,* American Ceramic Soc., Westerville, Ohio, 1984.
13. E. Gugel, "Net-Shape Processing of Non-Oxide Ceramics," in *Proc. AGARD Conf.,* Florence, Italy, Sept. 1978, pp. 14-1–14-16.
14. B. C. Mutsuddy, "Injection Molding Research Paves Way to Ceramic Engine Parts," *Ind. Res. Dev.,* pp. 76–80, July 1983.
15. R. Ohnsorg, M. Teneyck, and T. Sweeting, "Development of Injection Molded Rotors for Gas Turbine Applications," ASME 86-GT-45, presented at the Intl. Gas Turbine Conf., Düsseldorf, Germany, June 8–12, 1986, 8 pp.
16. E. Lange and N. Müller, "P/M Injection Molding Techniques for Ceramic and Metal Parts," *PM Intl.,* vol. 18, pp. 416–421, 1986.
17. G. Bandyopadhyay and K. W. French, "Fabrication of Near-Net-Shape Silicon Parts for Engine Application," 86-GT-11, presented at the Intl. Gas Turbine Conf., Düsseldorf, Germany, June 8–12, 1986, 4 pp.
18. A. Krauth and K. Berroth, "Engineering Ceramics for Industrial Applications; Wear, Heat, and Automotive Technology," ASME 84-GT-260, presented at the ASME Mtg., Amsterdam, Netherlands, June 4, 1984, 7 pp.
19. D. W. Richerson, "Structure Property Processing Relationships to Net Shape Forming of Ceramics," in *Proc. Workshop II, Emerging Net Shape Technologies,* Santa Barbara, Calif., Mar. 27, 1985, pp. 63–92.
20. G. Bandyopadhyay, D. C. Smith, and A. E. Pasto, "Some Emerging Net Shape Fabrication Technologies at GTE," in *Proc. Workshop II, Emerging Net Shape Technologies,* Santa Barbara, Calif., Mar. 27, 1985, pp. 121–131.
21. J. A. Mangels, "Net Shape Forming of Ceramics for High Technology Applications," in *Proc. Workshop II, Emerging Net Shape Technologies,* Santa Barbara, Calif., Mar. 27, 1985, pp. 93–120.
22. F. H. Norton, *Elements of Ceramics,* Addison-Wesley, Reading, Mass., 1974.
23. T. Ueyama and N. Kaneko, in P. Vincenzini (Ed.), *High Tech Ceramics,* Elsevier, Amsterdam, 1987, p. 901.
24. T. M. Gainer and J. A. Carter, "Pressure Casting Sanitary Ware," *Am. Ceram. Soc. Bull.,* vol. 43, pp. 9–12, Jan. 1964.
25. R. W. Powers, "Ceramic Aspects of Forming Beta Alumina by Electrophoretic Deposition," *Am. Ceram. Soc. Bull.,* vol. 65, pp. 1270–1277, Oct. 1986.
26. P. G. Karandikar, "Manufacturing of Metal Matrix Composites by Compocasting and Gravity Infiltration and Their Characterization," M.S. Thesis, Univ. of Delaware, Newark, Rep. 89-07, June 1989, 144 pp.
27. R. A. Graham and A. B. Sawaoka (Eds.), *High Pressure Explosive Processing of Ceramics,* Trans. Tech. Publ., Switzerland-Germany-UK-USA, 1987, 400 pp.
28. W. J. Nellis, "Shock-Wave Science and Technology in Japan," *ONRFE Sci. Inform. Bull.,* vol. 14, pp. 99–104, 1989.
29. L. E. Murr, N. G. Eror, and A. W. Hare, "Fabrication of Metal/High-Temperature Superconductor Composites by Shock Compression," *SAMPE J.,* vol. 24, pp. 15–18, Nov.–Dec. 1988.
30. W. L. Frankhouser, *Advanced Processing of Ceramic Compounds, Dynamic Compaction, Self-Propagating High Temperature Synthesis, Plasmachemical Technology,* Noyes Publ., Park Ridge, N.J., 1987, 198 pp.

31. D. Raymond and T. Z. Blazynski, "Non-Metallic Materials under Shock Loading," in T. Z. Blazynski (Ed.), *Materials at High Strain Rates,* Elsevier Appl. Sci., London, 1987, pp. 71–131.
32. A. R. Rosenfield, "Ceramics," *Current Awareness Bull.,* DMIC, no. 190, pp. 5–6, June 1989.
33. *Metallurgical Applications of Shock Wave and High-Strain Rate Phenomena,* Marcel Dekker, New York, 1986.
34. *Adv. Mater.,* vol. 8, pp. 3–4, July 14, 1986.
35. N. N. Thadhani, "Shock Compression Processing of Powders," *Adv. Mater. Manuf. Processes,* vol. 3, pp. 493–549, 1988.
36. *Structural Ceramics,* NASA Conf. Publ. 2427, NASI. 5512427, 1986, 226 pp.
37. W. A. Sanders, J. D. Kiser, and M. R. Freedman, "Slurry-Pressing Consolidation of Silicon Nitride," *Ceram. Bull.,* vol. 68, pp. 1836–1841, Oct. 1989.
38. M. R. Freedman, W. A. Sanders, and J. D. Kiser, "Slurry Pressing of Silicon Nitride and the Effect of Powder Processing on the Properties of Slurry Pressed Sintered Silicon Nitride," NASA Tech. Briefs, LEW-14680, p. 54., Sept. 1988.
39. M. R. Freedman and M. L. Millard, "Improved Consolidation of Silicon Carbide," NASA TM-87243(N86-24836/NSP), NASA Tech. Briefs, LEW-14681, pp. 54–56, Dec. 1988.
40. L. M. Sheppard, "Ceramics at the 'Cutting Edge'," *AM&P,* vol. 3, pp. 73–79, Aug. 1987.
41. K. E. Burnfield and D. Schweizer, "Use of Cellulose Ethers for Rheological Control in Advanced Ceramics," EM-89-119, presented at the *Adv. Ceram. '89 Conf.,* Philadelphia, Penn., Feb. 20–22, 1989, 15 pp.
42. J. E. Schuetz, "Methylcellulose Polymers as Binders for Extrusion of Ceramics," *Ceram. Bull.,* vol. 65, Dec. 1986.
43. M. D. Sacks, *Dispersion and Rheology in Ceramic Processing,* vol. 21 of *Advances in Ceramics,* American Ceramic Soc., Westerville, Ohio, 1987.
44. M. D. Sacks, "Rheological Science in Ceramic Processing," in L. Hench and D. R. Ulrich (Eds.), *Science of Ceramic Chemical Processing,* Wiley, New York, 1986.
45. T. J. Whelan, "Improved Silicon Carbide for Advanced Heat Engines," NASA CR-182289, NAS3-24384, May 1989, 60 pp.
46. L. M. Sheppard, "Fabrication of Ceramics; The Challenge Continues," *Ceram. Bull.,* vol. 68, pp. 1815–1820, Oct. 1989.
47. R. W. Ohnsorg, M. O. Teneyck, and W. D. Friedman, "Injection Molding Development of Ceramic Turbine Components," ASME#89-GT-170, presented at the ASME Gas Turbine and Aeroengine Congress and Exposition, Toronto, Ont., June 1989, 7 pp.
48. A. F. Henriksen, "Intricate Ceramic Shapes by Low Pressure Injection Molding," presented at the ASM Ann. Mtg., Cincinnati, Ohio, Oct. 12–15, 1987.
49. B. C. Mutsuddy, "Emerging Methods of Forming Ceramics," presented at the ASM Ann. Mtg., Cincinnati, Ohio, Oct. 12–15, 1981.
50. W. J. Corbett and P. T. B. Shaffer, "Injection Molding of Advanced Ceramics," in *Proc. 19th Intl. SAMPE Tech. Conf.,* Oct. 12–15, 1987, pp. 545–553.
51. K. A. Anderson, Jr., J. P. Charkraverty, and J. S. Reid, "Injection Molding Advanced Ceramic Products," presented at the ASM Ann. Mtg., Cincinnati, Ohio, Oct. 12–15, 1987.
52. H. C. Yeh and H. T. Fang, "Improved Silicon Nitride for Advanced Heat Engines," NASA CR-179525, NAS3-24385, Feb. 1987, 113 pp.
53. S. T. Buljan, J. G. Baldoni, and J. Neil, "Dispersoid-Toughened Silicon Nitride Composites," ORNL Sub 85-22011-1, DE-AC05-84OR21400, Sept. 1988, 67 pp.

54. M. Scrinivasan and S. G. Seshadri, "Micro-Structural Design of Ceramic Matrix Composites," presented at the ASM Ann. Mtg., Cincinnati, Ohio, Oct. 12–15, 1987.
55. S. Bradley, "Characterization of SiC Whisker Toughened Silicon Nitride Composites," presented at the ASM Ann. Mtg., Cincinnati, Ohio, Oct. 12–15, 1987.
56. B. C. Mutsuddy, "Equipment Selection for Injection Molding," *Ceram. Bull.*, vol. 68, pp. 1796–1802, Oct. 1989.
57. K. Hunold, J. Greim, and A. Lipp, "Injection Moulded Ceramic Rotors—Comparison of SiC and Si_3N_4," *Powder Metall. Intl.*, pp. 17–21, Aug. 1989.
58. M. Guther and R. Liebich, "Injection Moulding of Ceramic Materials," *Silikattechnik*, vol. 40, pp. 62–65, 1989.
59. S. Oram, "Injection Molding of Ceramic Components Provides Flexibility in Metal Casting and Forming," *Ind. Heating*, pp. 42–43, Apr. 1987.
60. B. R. Patterson and C. S. Aria, "Debinding Injection Molded Materials by Melt Wicking," *J. Met.*, pp. 22–24, Aug. 1989.
61. E. P. Rothman and H. K. Bowen, "New and Old Ceramic Processes; Manufacturing Costs," Ceram. Proc. Res. Lab., M.I.T. Indust. Liaison Prog. Rep. 4-14-88, M.I.T., Cambridge, Mass., June 1986, 29 pp.
62. P. D. Shalek et al., "Hot-Pressed SiC Whisker/Si_3N_4 Matrix Composites," *Am. Ceram. Soc. Bull.*, vol. 65, pp. 351–356, 1986.
63. M. J. Hoffman et al., "Slip Casting of SiC-Whisker-Reinforced Si_3N_4," *J. Am. Ceram. Soc.*, vol. 72, pp. 765–769, May 1989.
64. E. P. Hyatt, "Continuous Tape Casting for Small Volumes," *Ceram. Bull.*, vol. 68, pp. 869–870, Apr. 1989.
65. J. D. Cawley, "Tape Casting as an Approach to an All-Ceramic Turbine Shroud Seal," NASA TM-87078 (N85-32333/NSP), 1987.
66. L. M. Sheppard, "Reliable Ceramics for Heat Engines," *AM&P*, vol. 2, pp. 54–66, Oct. 1986.
67. E. G. Blanchard, "Pressure Casting Improves Productivity," *Ceram. Bull.*, vol. 67, pp. 1680–1683, Oct. 1989.
68. B. J. Busome, Jr., and J. P. Pollinger, "Development of Silicon Nitride Engine Components for Advanced Gas Turbine Applications," 89-GT-259, ASME, DEN3-335, DEN3-336, June 1989, 6 pp.
69. M. A. H. Howes, "Ceramic-Reinforced MMC Fabricated by Squeeze Casting," *J. Met.*, pp. 28–29, Mar. 1986.
70. M. W. Toaz, R. R. Bowles, and D. L. Mancini, "Squeeze Casting Composite Components for Diesel Engines," *Ind. Heating*, pp. 17–19, Mar. 1987.
71. M. W. Toaz, "Near Net Shape Composite Castings with Tailored Engineering Properties," in *Proc. Adv. Comp. Conf.*, Dearborn, Mich., Dec. 24, 1985, pp. 8521–8608.
72. D. A Parker, "Ceramics Technology—Application to Engine Components," 21st John Player Lecture, *Proc. Inst. Mech. Eng.*, vol. 199, pp. 135–150, 1985.
73. S. B. Lasday, "Ceramic Fiber Utilization in Metal-Matrix Components," *Ind. Heating*, pp. 20–21, Mar. 1987.
74. X. Dumant, E. Beaugnon, and G. Regazzoni, "Designing the Microstructure of Squeeze-Cast Al Composites," *J. Met.*, pp. 46–51, Nov. 1989.
75. H. Matsubara et al., "Si_3N_4 Whisker-Reinforced Aluminum Alloy Composite," *J. Mater. Sci. Lett.*, vol. 6, pp. 1313–1315, Nov. 1987.
76. T. Imai et al., "Aspiration Method of Squeeze Casting," *J. Mater. Sci. Lett.*, vol. 6, p. 343, June 1987.
77. F. A. Girot, L. Albingre, and J. M. Quenisset, "Rheocasting Al Matrix Composites," *J. Met.*, pp. 18–21, Nov. 1987.

78. J. A. Cornie, *Proc. Microstructure and Processing of Metal and Ceramic Matrix Composites,* M.I.T. Fall Symp., M.I.T., Cambridge, Mass., Nov. 1986.
79. M. Janney, *World Report on Adv. Ceram.,* vol. 1, no. 13, p. 1, Nov. 1989.
80. B. C. Mutsuddy, "Ceramic and Ceramic Composite Net Shape by EAO," *J. Met.,* p. 53, May 1987.
81. L. S. Millberg, "The Search for 'Ductile' Ceramics," *J. Met.,* vol. 39, pp. 10–13, Nov. 1987.
82. D. Apelian, G. Gillen, and A. G. Leatham, in F. H. Froes and S. J. Savage (Eds.), *Processing of Structural Metals by Rapid Solidification,* ASM, Metals Park, Ohio, 1987, p. 107.
83. P. Mathur et al., "Process Control, Modeling and Applications of Spray Casting," *J. Met.,* pp. 23–28, Oct. 1989.
84. S. B. Lasday, "Unique Approach to Manufacture of Ceramic Composite Components," *Ind. Heating,* pp. 14–15, Apr. 1988.
85. *MMCIAC Current Highlights,* vol. 9, pp. 1–3, Sept. 1989.
86. S. B. Lasday, "Tailoring Properties of Platelet Reinforced Ceramics by Liquid Metal Oxidation Process," *Ind. Heating,* pp. 14–16, Aug. 1989.

BIBLIOGRAPHY

Alt, P., "Current State and Development of Cold Isostatic Pressing," *Keram. Z.,* vol. 42, pp. 151–154, 1990.

Andersson, C. A., "Ultimate Strengths of Fiber-Reinforced Ceramic Matrix Composites," presented at the 14th Ann. Conf. on Composites and Advances in Ceramics, Cocoa Beach, Fla., Jan. 14–17, 1990.

Bhaduri, S. B., A. Chakraborty, and J. J. Reddy, "Injection-Molding of Ceria-Stabilized Tetragonal Zirconia Polycrystals (Ce-TZP)," *J. Mater. Sci. Lett.,* vol. 9, pp. 209–210, Feb. 1990.

Boch, P., and T. Chartier, "Understanding and Improvement of Ceramic Processes; The Examples of Tape Casting," in *Materials Science Forum,* vols. 34–36, *Austceram 88, Conf. and Exhibition,* Sydney, Australia, Aug. 21–26, 1988, pt. 2, pp. 813–819.

Cannon, W. R., R. Becker, and K. R. Mikeska, "Interactions among Additives Used for Tapecasting," in *Advances in Ceramics,* vol. 26: *Proc. Intl. Symp. on Ceramic Substrates and Packages for Electronic Applications,* Denver, Colo., Oct. 18–21, 1987, pp. 525–541.

Carlstrom, E., et al., "Binder Removal from Injection Moulded Ceramic Turbocharger Rotors," *Sci. Ceram.,* vol. 14, pp. 199–204, 1988.

Doak, K. A., "Vacuum Hot Press Furnaces for Powder Compaction," *Met. Powder Rep.,* vol. 37, pp. 577–578, 581–582, Nov. 1982.

Du, H., B. Gallois, and K. E. Gonsalves, "Low-Temperature Metal-Organic Chemical Vapor Deposition of Silicon Nitride," *J. Am. Ceram. Soc.,* vol. 73, pp. 764–766, Mar. 1990.

Fareed, A. S., et al., "Mechanical Properties of 2-D Nicalon Fiber Reinforced Lanxide Aluminum Oxide and Aluminum Nitride Matrix Composites," presented at the 14th Ann. Conf. on Composites and Advances in Ceramics, Cocoa Beach, Fla., Jan. 14–17, 1990.

Haggerty, J. S., and Y. M. Chiang, "Reaction-Based Processing Methods for Ceramics and Composites," presented at the 14th Ann. Conf. on Composites and Advances in Ceramics, Cocoa Beach, Fla., Jan. 14–17, 1990.

Haunton, K. M., J. K. Wright, and J. R. G. Evans, "The Vacuum Forming of Ceramics," *Br. Ceram. Trans. J.,* vol. 89, pp. 53–56, Feb. 1990.

Hoffman, M. J., A. Nagel, and G. Petzow, "Processing of SiC-Whisker Reinforced Si_3N_4," in I. A. Aksay, G. L. McVay, and D. R. Ulrich (Eds.), *Proc. Sci. Adv. Ceram., Mater. Res. Soc. Symp.,* San Diego, Calif., Apr. 27–28, 1989, pp. 369–379.

Kennedy, S., "Binder Removal and Injection Molding," presented at the High Temperature Processing Equipment Mtg., Boston, Mass., Oct. 20, 1989.

Lackey, W. J., et al, "Ultrafine Microstructure Composites Prepared by Chemical Vapor Deposition," Contract N00014-87-K-0036, Ann. Rep. A-4699-2 for Jan.-Dec. 1988, 1989, 38 pp.

McCauley, J. W., "Historical and Technical Perspective on SHS," presented at the 14th Ann. Conf. on Composites and Advances in Ceramics, Cocoa Beach, Fla., Jan. 14–17, 1990.

Miyoshi, T., et al., "Characteristics of Hot-Pressed Fiber-Reinforced Ceramics with SiC Matrix," *Metall. Trans.*, vol. 20A, pp. 2419–2423, Nov. 1989.

Mortenson, A., et al., "Pressure Casting of Fiber-Reinforced Metals," in F. L. Matthews, N. C. R. Buskell, and J. M. Hodginson (Eds.), *Proc. 6th Intl. Conf. on Composite Materials*, Elsevier Appl Sci., London, 1987, pp. 2.320–2.329.

Mutsuddy, B. C., "Ceramic Injection Moulding," *Ind. Ceram.*, no. 839, pp. 436–441, 1989.

"Processing and Evaluation of Metal and Ceramic Matrix Composites; Opportunity Brief," presented at the M.I.T. Materials Processing Center Spring Symp., June 16, 1988, 28 pp.

Puczynski, J. A., S. Majorowski, and V. Hlavacek, "Engineering Scale-Up Principles of SHS Reactors," presented at the 14th Ann. Conf. on Composites and Advances in Ceramics, Cocoa Beach, Fla., Jan. 14–17, 1990.

Reh, H., "Economic Aspects of Dry Pressing," *Keram. Z.*, vol. 42, pp. 158–162, 1990.

Rice, R. W., "Microstructural Aspects of Fabricating Bodies by Self-Propagating Synthesis," presented at the 14th Ann. Conf. on Composites and Advances in Ceramics, Cocoa Beach, Fla., Jan. 14–17, 1990.

Rohatgi, P. K., and R. Asthana, "Solidification Processing of Metal-Matrix Composites," in *Adv. Struct. Mater. Proc.*, vol. 9, pp. 43–51, 1989.

Roosen, A., "Basic Requirements for Tape Casting of Ceramic Powders," *Ceram. Powder Sci. II, Trans.*, vol. 1, pp. 675–692, 1988.

Sheppard, L. M., "The Changing Demand for Ceramic Additives," *Ceram. Bull.*, vol. 69, pp. 802–806, 1990.

Stedman, S. J., J. R. G. Evans, and J. Woodthorpe, "Rheology of Composite Ceramic Injection-Molding Suspensions," *J. Mater. Sci.*, vol. 25, pp. 1833–1841, Mar. 1990.

Stinton, D. P., R. A. Lowden, and M. K. Ferber, "A Comparison of Fracture Toughness for CMCs Fabricated with Conventional and Forced Flow CVI," presented at the 14th Ann. Conf. on Composites and Advances in Ceramics, Cocoa Beach, Fla., Jan. 14–17, 1990.

Tierney, D., "The Use of Statistical Methods in the Manufacture of Structural Ceramic Components for Process Control, Monitoring and Improvement," presented at Ceramtec '90, Dearborn, Mich., June 12–13, 1990.

Tsao, I., and S. Danforth, "Flow Behavior of Injection Moldable $SiC/_w$/RBSN Composite; Viscosity, SiC_w Length and SiC_w Orientation," presented at the 14th Ann. Conf. on Composites and Advances in Ceramics, Cocoa Beach, Fla., Jan. 14–17, 1990.

Waku, Y., et al., "Mechanical Properties of Aluminum Alloy Composites Reinforced with New Continuous Si-Ti-C-O Fibers," *SAMPE Quart.*, vol. 20, no. 4, p. 47, 1989.

Wilkinson, D. S., "Advanced Structural Materials," in *Proc. Intl. Symp., 27th Ann. Conf. on Metallurgy*, Montreal, Que., Aug. 28–31, 1988, vol. 9, 318 pp.

Yu, D. Y. F., "Injection Molding of Polysilazane Preceramic Formulations," presented at the 14th Ann. Conf. on Composites and Advances in Ceramics, Cocoa Beach, Fla., Jan. 14–17, 1990.

CHAPTER 6
HIGH-TEMPERATURE PROCESSING AND CONSOLIDATION

6.1 INTRODUCTION

The firing of green-formed ceramic bodies at very high temperatures performs various functions. Depending on the material system, temperature, and atmospheric control, several material transport mechanisms come into play, resulting in particle-to-particle bonding, grain evolution and growth, pore shrinkage, and elimination. Thus the microstructure is evolved, fixing the resulting properties of the ceramic body. In addition, material shrinkage occurs as a result of densification, which may be isotropic or anisotropic. Controlled firing is a must to achieve defect-free high-density ceramics as well as ceramics with predictable microstructures and properties and, most importantly, ceramics within dimensional tolerances. Although several theories and mechanisms exist for specific ceramic systems, the industrial practice of controlled firing, also known as sintering, is mostly derived from practice and experience, and may even be specific to components manufactured within a given material system.

In addition to the thermodynamics and kinetics of densification of ceramic bodies, the mass being sintered, the variety of shapes, the loading arrangement, the distribution of the green body within the furnace volume, the type of heating source (such as electric resistance, induction heating, plasma, gas-fired, or microwave), the uniformity in temperature distribution, and the heating schedule, which includes the hold and cool down, all play major roles, interacting many times. The geometric arrangement of the sintering article, the temperature control system, and their variability can affect the sintering success. Often contradicting results may be obtained, precluding systematic conclusions, logical explanations, and recommended practice. Therefore, an extension of results obtained in small-scale laboratory sintering experiments to large-scale industrial manufacture is relatively difficult; extrapolation of the results of past experience and shop practice is crucial in achieving desired results.

Sintering is a process of microstructural change, which involves contributions from two ideal subprocesses: densification (replacement of free surface energy by grain-boundary energy) and coarsening (reduction of the extent of free surface or grain-boundary energy). The relative contributions of the two subprocesses depend on the processing variables, namely, temperature T, time t, composition, and particle size.

6.2 BASICS OF HIGH-TEMPERATURE CONSOLIDATION

It is generally considered that the final microstructure of a ceramic body is fixed largely by the practice followed prior to sintering. The final density is a function of the extent of successful green compaction which has been achieved.

The adhesion mechanisms between single particles and the agglomerates produced from them determine the subsequent extent of densification and the sintering process. Surface-active forming additives influence solid bridge formation, adhesive bonding, and glide-promoting effects.[1] In order to achieve the highest final density by a predetermined sintering practice, it is necessary to achieve as low a porosity in the green body as possible. The minimization of porosity in the green body is largely governed by the particle morphology and the particle-size distribution, which influence particle packing under compaction. The green compaction pressure needed to obtain the highest green density is usually obtained by experimentation. The requirement of high-density green compacts can also be understood from thermodynamical considerations of pore closure.[2] In addition, for nonuniformly dense green compacts, inhomogeneous dihedral angles for the pores can be expected. This leads to grain coarsening rather than densification.[3]

Ceramic Prefiring

In ceramic manufacture, green-forming aids are used to obtain good powder flow and compaction. These are plasticizers and lubricants, which control the viscosity of the ceramic mix, as well as reducing die-wall friction for ease of removal of the part. If these aids are not being removed in a slow and controlled fashion, their abrupt removal during initial sintering will cause catastrophic rupture of the ceramic, preventing densification.

In some instances prefiring is also needed to provide better strength for mechanical handling. The atmosphere control during prefiring is important to ensure the chemical surface characteristics suitable for required mass transport during subsequent sintering.

Sintering Mechanisms

The sintering of ceramics occurs by one or a combination of several mechanisms. Solid-state sintering involves solid-state reactions and is concerned with the mobility of diffusing additives in the lattice or at the grain boundaries. Liquid-phase sintering involves the formation of a compound which is liquid at sintering temperatures and facilitates the movement of sintering additive species. A combination of the two can also occur in certain systems.[4]

A third type of sintering process is vitrification. This is a heat treatment which produces enough viscous liquid at the firing temperature to fill completely the porous spaces in the original powder compact. The process is relatively inexpensive.

High-temperature consolidation and sintering begins with eliminating the pores which were introduced during particle packing. This process occurs by mass transport in the form of solid, liquid, or vapor, dictated by the system thermodynamics. Solid-state migration can occur through the grain, governed by the lattice diffusion D_l of the migrating atom, or by diffusion through the grain boundary at which the impurities usually segregate, D_b. Those impurities and compounds which form solid solutions with the host matrix would exhibit enhanced D_l. The selectively added sintering aids that form grain-boundary phases in the liquid phase during sintering by chemical reaction enhance the grain-boundary diffusion D_b. It is common knowledge that vapor-phase transport (such as surface diffusion) in general does not contribute significantly to sintering and densification.[3]

In solids with a high degree of covalent bonding, such as Si_3N_4 and SiC, the self-diffusivity is poor, and therefore volume and grain-boundary diffusion is slow. At low temperatures atomic mobility is too low for densification, and at temperatures high enough for appreciable atomic mobility these compounds tend to dissociate rather than consolidate.

The covalently bonded carbides are especially difficult to densify in the absence of sintering aids. The sintering aids function to form compounds that are compatible with the ceramic system, either forming a solid solution with the host matrix (such as AlN in SiC) or forming a liquid at sintering temperatures, acting as glue to join the grains together in addition to enhancing diffusion in grain boundaries.

The factors which determine pore elimination are related to the grain-boundary and grain-surface properties. The pores with grain boundaries as their surface (particle, before significant sintering occurs) will have a surface curvature that depends on the number of adjacent grains, termed the coordination number n_c, and the dihedral angle θ which is related to the ratio of grain-boundary energy to surface energy.[5]

When the equilibrium dihedral angle between two grains of the solid and their surfaces is less than 60°, the interface between solid and vapor is convex and will tend to move outward, thus inhibiting pore closure. If the angle is greater than 60°, then the concave interface will allow pore shrinkage. The critical energy ratio is given by

$$\gamma_{gb} = 2\gamma_s \frac{\cos \theta}{2} \tag{6.1}$$

such that the critical condition for pore closure and sintering (that is, the promotion of mass transport into the pore) is

$$\frac{\gamma_{gb}}{\gamma_s} = \sqrt{3} \tag{6.2}$$

where γ_{gb} is the critical grain-boundary energy and γ_s the critical surface energy.

In other words, the critical coordination number for pore closure and densification depends on the dihedral angle.[2] The greater γ_{gb}/γ_s, the greater the critical coordination number. Additives which segregate selectively to grain boundaries decrease γ_{gb} and lower the ratio γ_{gb}/γ_s. Thus the dihedral angle is increased, increasing the driving force for sintering. Contamination can alter these properties significantly, and hence raw materials with controlled purity as well as clean-room processing are a few of the requirements for obtaining sintered articles to theoretical density.

Given that the green-density requirements have been met and that prefiring has been accomplished with the porous body having a density greater than 50 to 60 percent, sintering and densification proceed in stages as the system temperature increases, as illustrated in Fig. 6.1.

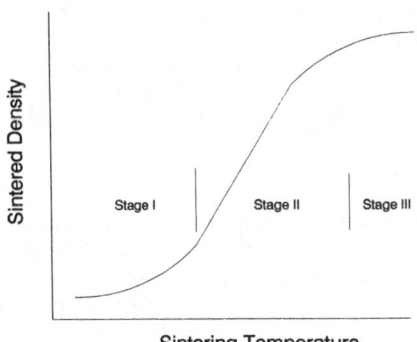

Figure 6.1 Schematic of densification of a ceramic with temperature rise in system.

During stage I, because of matter arriving through either the grain boundary or the lattice, neck growth occurs, leading to shrinkage of the pore and, hence, densification. However, at these relatively lower temperatures, the densification rate is not high enough. In stage II the same mechanism prevails, but rapid diffusion contributes to a higher densification rate. The grain boundaries developed during sintering act as vacancy sinks, permitting pore removal and densification. In stage III densification slows due to accelerated grain-growth kinetics predominating over the kinetics of pore elimination and shrinkage; but more exact reasons are not completely understood. This is because grain growth and densification are simultaneously occurring, interrelated phenomena, making distinction difficult. The grain-growth phenomenon, with grain-boundary motion being the driving force, depends on the local density. However, the kinetics of grain growth during later stages of sintering will be slower than that occurring during the subsequent heat treatment of a fully dense body.[6] Summing up, the decreasing sintering rate in stage III is due to grain-coarsening kinetics competing with densification kinetics.[7]

In industrial practice it is not uncommon to have ceramics sintered to lower than full density, but with excessive grain growth. In such cases, the neck growth during the initial stage, perhaps, occurred due to surface diffusion or by evaporation and condensation. Here particle coarsening will occur without densification. If the coarsening-to-densification ratio is high, then the surface free energy of the system can be minimized by the coarsening mechanism rather than by the densification mechanism, and hence, a high-density product is not reached.[8]

In extreme instances, for the less dense green compact, the temperatures reached might be high enough to activate the grain-growth phenomenon early so that densification is completely retarded, resulting in a nondensified body.

The sintering due to liquid-phase formation can be quite complex. For example, in the case of silicon nitride, densification additives react with surface silica and some form of the nitride to give an oxynitride which is a liquid at densification temperatures.[9] The formation of this liquid contributes to shrinkage and, at the same time, to phase transformation from α to β. The kinetics of densification was found to involve three successive stages of particle rearrangement, namely, solution, diffusion and precipitation, and coalescence. The crucial requirement in the understanding of the densification of kinetics and the development of microstructure in these systems is a thorough examination of the phase relationships and the reactions that can be expected.[10]

In the case of Si_3N_4, one way to promote sintering is to form a liquid phase at a temperature below that at which significant decomposition of Si_3N_4 occurs (that is, about 1820°C in air). The liquid phase can be achieved by mixing into the starting powder small amounts of additives, such as MgO, Al_2O_3, and Y_2O_3, which react with SiO_2 on the Si_3N_4 particle surface to form an oxynitride liquid at temperatures as low as 1600°C. Provided that the composition and viscosity are suitable, this liquid phase allows transport and rearrangement of the starting material at a greatly accelerated rate by a mechanism in which the starting Si_3N_4 particles dissolve into the liquid and the phase of Si_3N_4 is precipitated out as elongated interlocking grains. Upon cooling, however, the liquid phase solidifies into a grain-boundary phase that can become fluid, even unstable at temperatures greater than about 1000°C, thereby seriously degrading both the refractory and the mechanical properties of the Si_3N_4 material.[11]

In addition to these mechanisms, others may also contribute to densification. These may involve solid-state reactions, second phases, and phase transformations.[12] The driving forces for reactions and phase transformations can be greater than the surface and grain-boundary energies, thus dominating microstructural evolution. For example, the needle-shaped β-Si_3N_4 grains are known to result from the transformation of α-phase to β-phase, involving liquid-phase transport during densification and after most of the densification has been completed.[13]

The essential aspects of the surface and grain-boundary energy contributions can be cast in the form of compressive stress required for pore closure. The inherent grain-boundary contribution suggests that the sintering process can depend on the grain size.[14] In addition, the compressive stress for shrinkage depends on the pore size, the pore-size-to-grain-size ratio, and the pore coordination number.[15] The compressive stress increases during the sintering process until the grain size approximately equals the pore size. Thereafter the sintering pressure decreases, thereby contributing to the observed stage III behavior in Fig. 6.1.

Finally, high-temperature sintering and densification cannot be expected to "heal" all the flaws and pores contributed by inefficiencies in powder processing and powder compaction. For example, one of the most common problems in powder processing is the presence of powder agglomerates, classified as either soft or hard agglomerates. The soft agglomerates can usually be broken up during mixing and compaction, while the hard agglomerates remain during the entire process. These can essentially be considered as rigid inclusions, in the manner of Scherer,[16] which do not sinter at the same rate as the surrounding matrix, causing cracklike defects and partially sintered matrix in the vicinity of the agglomerate.

Selection of Sintering Aids

From the foregoing discussion, it is apparent that the primary objective in sintering is to raise the densification-to-coarsening ratio. Additives which form solid solutions are eminently suitable for this and essentially enhance the lattice diffusion. Examples are B_4C and AlN for SiC and MgO for Al_2O_3. Liquid-phase forming additives are very useful, especially those that form a transient liquid phase, with lifetimes long enough to sinter, and then dissolve in the final product or otherwise disappear via evaporation. For example, Stutz et al.[17] found that a simultaneous addition of boron and aluminum decreases the sintering temperature required for β-SiC. It was hypothesized that densification was enhanced by the formation of a liquid in the system Si–C–Al–B that may be only of a transient nature. Additives that enhance grain-boundary diffusion are also helpful. For example, Tajima and Kingery,[18] by using scanning transmission electron microscopy (STEM), found grain-boundary segregation of aluminum in both pressureless sintered and hot-pressed samples containing aluminum and carbon as the sintering additives.[19] Sintering of silicon carbides using liquid-phase additives has been receiving greater attention recently because of the possibility of using lower temperatures and shorter residence times, as well as for the nonoccurrence of exaggerated grain growth. Cutler and Jackson,[20] for example, densified silicon carbide using Al_2O_3 and Y_2O_3 additions at 1850°C with resulting grain sizes of less than 5 μm.

The sintering of SiC also requires the addition of sintering aids to the starting powder to promote a desired chemical reaction or the formation of a liquid phase. The usual aids are boron, carbon, or aluminum. The effects of these additions are not well understood (see Hannink et al.[21]), although there are indications that modifications to free surface–grain-boundary chemistry are important.

Fully dense SiC that has been sintered with aluminum additions exhibits an intergranular glassy phase according to Moussa et al.,[22] and this might suggest that aluminum aids the sintering process by forming a liquid phase.

To increase the densification rate, usually additives that enhance lattice diffusion D_l are chosen. Based on defect chemistry, these would be ions with a size close to that of the host to assist in the formation of solid solution. The amount added may be close to the solid-solution limit to maximize the effect.[8] In fact, maximum density of a sintered silicon carbide body was obtained at the concentration of maximum solid solubility of BN, BP, and B_4C additives.[23] Multiple additives which selectively enhance the mobility of impurities along the grain boundary and lattice mobility in lieu of promoting grain growth would be highly desirable.

One of the important aspects of sintering is to be able to control the growth of grains. Other than the grain coarsening required to form a fully dense body, sudden excessive grain growth may occur due to grain-boundary movement in some ceramics, when the hold time at sintering temperature exceeds the optimum. The control of grain growth is usually accomplished by adding single-phase inclusions[24] that have less than the critical size to avoid residual stresses. The inclusions essentially pin and increase the stress required for breakaway of the grain boundary. However, coalescence of the inclusions should be prevented; in other words, immobile inclusions are preferred. Other ways to control grain-boundary movement are the use of impurity clouds and phase partitioning.[24]

Due to the lack of reliable diffusion and creep data for structural ceramics at very high temperatures, the mechanisms of sintering are not completely understood for all cases. For example, controversy still persists with respect to the role of boron and carbon in sintering silicon carbide. It is known that both are required for densification and that boron alone does not densify silicon carbide.[25] Prochazka's suggestion that carbon deoxidizes the silica on the silicon carbide powder surface[5] and thereby increases the surface energy seems valid. However, the suggestion that boron segregates to the grain boundaries was found to be invalid. Maddrell[26] suggested that the role played by silica in inhibiting densification is very important in that the boron solubility is greater in silica that in SiC and that it segregates preferentially into the silica.

The rapid lattice diffusion of boron within the silicon carbide is generally accepted to be the major densification mechanism. The results of Murata and Smoak[23] confirmed the role of the solid-solution-forming additive in densification. In addition, the grain refinement role by carbon was also demonstrated. Boecker et al.[19] demonstrated that the addition of aluminum was beneficial in sintering α-SiC, although the exact mechanism was not determined. It was observed, however, that only 25 percent of the initial aluminum added remained in the sintered samples, perhaps indicating that transient liquid-phase formation contributes to sintering.

The role of silica is thus important in the pressureless sintering of silicon carbides. In this context, the selection of the correct densification aid will also depend on the type of silicon carbide used as the raw material. It is well known that powders with higher surface areas are more sinteractive than those with lower surface areas. However, oxygen pickup increases dramatically with the increase in surface area, thus inhibiting densification under normal conditions. Experiments by Hojo et al.[27] have indicated the amount of carbon required in silica reduction for densification to occur. It was shown that densification will occur for SiC powders having a C–O-to-Si ratio near 1.0.

Effect of Sintering Atmosphere

The sintering atmosphere, in addition to the selection of suitable additives, is especially critical for nonoxide ceramics, as can be deduced from the discussions on sintering mechanisms. The furnacing of oxide versus nonoxide ceramics to achieve densification can be quite different. The sintering atmosphere can consist either of solid bed, in which the articles to be sintered are imbedded, or of gaseous species surrounding the articles which are left free-standing, or both. The gaseous species can be introduced deliberately in the furnace system or can arise due to chemical reactions occurring in the system during densification. As mentioned previously, only lattice diffusion and grain-boundary diffusion promote sintering, while surface diffusion retards sintering. Evaporation and condensation of the diffusing species on the surface contribute to surface diffusion, which promotes coarsening in lieu of densification. However, suitable selective modification of the composition of the sintering atmosphere can promote the chemical reaction to occur in the desired direction by alteration of the chemical activity (partial pressure) of the reactants and products.

Prochazka et al.[28] have carried out a detailed study of the effect of the sintering atmosphere on the densification of silicon carbide or the lack thereof. The use of vacuum results in higher density as compared to the use of argon, which has limited diffusion in the SiC lattice at 2100°C. However, the dissociation of SiC under vacuum is enhanced, resulting in the loss of silicon from the surface. The silicon enrichment in the atmosphere inhibits densification by promoting coarsening in addition to nonuniform sintering. Separately, by influencing the atomic transport mechanisms, nitrogen in the sintering atmosphere also retards densification of SiC at temperatures at which argon sintering is possible. Subsequent work by Venkateswaran[29] has confirmed that nitrogen reacts with excess boron on the SiC surface, or the furnace atmosphere, forming BN, which dissolves near the surface region of SiC grains. As a result, the SiC surface is depleted of boron and densification ceases. As the sintering temperature is increased, the dissociation pressure of BN increases, thus allowing the presence of boron at the SiC surface in order for densification to occur. CO in the atmosphere also removes boron with the formation of oxides of boron, thus inhibiting densification.

In many instances, if the sintering temperature is high enough for the SiC dissociation to occur, then a black carbonaceous layer is formed on the surface, which may not be desirable. However, the desired articles can be enclosed in a closed crucible which maintains an equilibrium pressure of the atmospheric composition, resulting in a grey, shiny surface.

The controlled-atmosphere sintering for Si_3N_4 also requires a careful consideration of nitrogen pressures in order to prevent the surface decomposition of

Si_3N_4. It is not unusual to obtain sialon or silicon nitride bodies with extreme color variations within the specimen thickness due to chemical reactions that are specific to the silicon nitride or the sialon system.

6.3 HOT PRESSING

The application of pressure, simultaneously with temperature-assisted atomic movement for densification, assists to form new particle contacts conducive to sintering. The deformation-induced contacts decrease the need for coarsening. Depending on the system, the operative mechanisms are exactly the same as those described under pressureless sintering. Due to limitations in the process, this method is only suitable for simple shapes such as disks, plates, rods, and cylinders. Excellent surface finish is, however, achievable.[11]

Simultaneous application of pressure and heat can result in material with higher final densities and fine grain sizes. The improvement is due to an increased driving force for sintering through a combination of sintering mechanisms involving plastic flow (due to stresses set up at points of particle contact) and sintering mechanisms involving diffusion.

Hot pressing shares many similarities with cold pressing, for example, ceramic particulate is filled into a die and uniaxially pressed, except, of course, that in this case pressing is carried out at elevated temperature. Pressures can reach 400 MPa, and temperatures are typically about 1700°C for Si_3N_4 and 2000°C for SiC. At these high temperatures the die material (graphite) can also react with the ceramic itself, and a reaction inhibitor, such as BN, is used to coat the die.

A brief illustration of the ceramic cold-pressing process follows (Fig. 6.2). The material is crushed to a desired grain size and mixture after which the binding medium is added. The material is then cold-pressed and subjected to the sintering

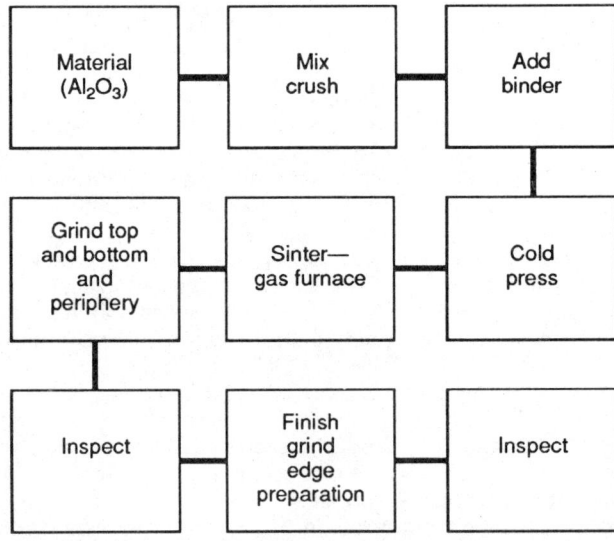

Figure 6.2 Ceramic cold-pressing process.

process. After sintering, the initial top, bottom, and peripheral grind operations are performed, followed by the first metallurgical inspection. This inspection is followed by the final grind operations and, lastly, the final geometric inspection.

In a typical hot-pressing process, the ceramic powder mixture is green-compacted to greater than 50 percent theoretical density, using metallic dies. The part is subsequently pressed by uniaxial application of pressure via graphite dies. Typically an inert atmosphere is maintained for hot pressing nonoxide ceramics, both for densification and because graphite dies are used. The graphite dies can be heated either inductively or resistively, and external graphite heating elements other than the dies themselves can also be used. Previous practice dictated by the system determines the temperature, pressure, hold times, and the temperature and pressure ramp schedule.

In one of the earliest works, Alliegro et al.[30] densified silicon carbide by hot pressing with separate additions of aluminum, iron, chromium, calcium, and lithium. Systematic work was later reported by Takeda and Nakamura,[31] who used less than one-half of the pressing pressures used by Alliegro et al. For the same hot-pressing pressure it was found that simple substances or compounds of aluminum, boron, or beryllium densified SiC to greater than 97 percent theoretical density, whereas the use of other additives, such as Si, P_2O_5, TiO_2, Cr, Fe, GaO, and Bi, did not result in full densification.

Hot pressing is also the preferred method of making cermets containing borides, the most notable of which is the TiB_2–BN composite used extensively as crucible material in the vacuum metallizing industry. Titanium diboride is able to withstand boiling aluminum at 1500°C, a quality possessed to only a slightly lesser extent by hexagonal boron nitride.

Apart from fusion casting of a few oxide ceramic products, hot pressing, or pressure sintering, is the only method available for shaping ceramics at high temperatures. Usually this does not lead to the final shape of a part, and therefore this should not be considered as a method of net-shape processing. However, the process involves plastic flow, and often reasonably close shapes can be achieved.

Further work might promote the use of pseudoisostatic methods,[32] where nonsinterable powder is used as the pressure-transmitting medium, the so-called powder vehicle process. Components produced by this technique include gas turbine rotor hubs and ball bearings.

Pseudoisostatic methods are particularly suitable for producing hot-pressed silicon nitride from reaction-bonded starting material. This method is used when the component has widely differing section thicknesses, as is the case with gas turbine rotor hubs.[32,33]

Hot pressing of reaction-bonded silicon nitride was developed as a process because it produced a fully dense material. The advantage of this process is the shorter densification time compared to direct hot pressing of powder. This method established the feasibility of producing particularly homogeneous components, a very severe problem with large parts. Another possibility with hot pressing is the production of duodensity rotors from Si_3N_4 using a reaction-bonded blade ring and hot-pressed hub. This concept is part of the development of the ceramic gas turbine.

According to Wachtman,[34] "grain boundary engineering" may be yet another promising toughening method, and Chinese researchers have obtained significant results for silicon nitride. By adding $Y_2O_3 + Al_2O_3$, or $Y_2O_3 + Al_2O_3 + La_2O_3$, before sintering, internal fibrous microstructures are produced between grains. When combined with postsintering or hot pressing, toughness can be improved by almost as much as 100 percent. When Al_2O_3 is added before hot pressing, SiC

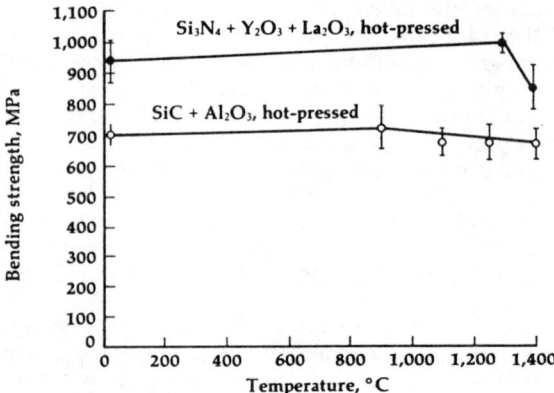

Figure 6.3 Various additives produce improved strength in SiC and Si_3N_4.[34]

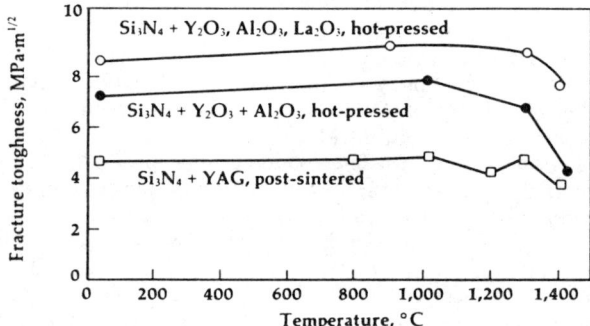

Figure 6.4 Addition of Y_2O_3 and Al_2O_3 additives is one technique to improve strength.[34]

can also maintain high-temperature strength, with almost no drop-off at 1400°C. Hot-pressed Si_3N_4 with additives shows some decrease at 1300°C. However, the question of how to maintain the strength over the long term remains (see Figs. 6.3 and 6.4).[34]

6.4 PRESSURELESS SINTERING

Pressureless sintering, as the name implies, involves densification without the application of other than atmospheric external pressure. If a suitable additive system can be identified for a ceramic system that can promote densification without pressure, then this is the preferred method since complex-shaped components can be densified rather easily in this manner. In this case, provided that appropriate powder selection and green forming have been accomplished, the uniformity in densification will be primarily governed by temperature gradients which are inherent in the furnace. Depending on the green density obtained by the efficiency of green forming and the amount of binders used, the amount of shrinkage during pressureless sintering will vary, but is usually in the range of 10 to 25 percent, typically around 16 to 18 percent. Obviously, the lower the amount of shrinkage, the better the ability to maintain dimensional tolerance and minimize the extent of warpage on thin parts and out-of-roundness on parts with circular cross-sectional geometry. In industrial practice, the majority of part rejection is attributed to the nonconformance to part dimensions.

Pressureless sintering is mostly done by batch furnacing with radiant heat from graphite heating elements (for sintering nonoxide ceramics), which are heated by resistance heating. The factors that govern the sintering success are type of furnace load, product mix, loading arrangement, amount of loading with respect to heating volume, and temperature gradient. The temperature gradients supplement the surface energy changes in driving and related processes influencing the surface, grain-boundary, or lattice diffusion of the controlling species.[35] Sintering schedules are usually established by prior density and dilatometry studies.

Continuous furnacing can also be done for pressureless sintering, especially for tubes, rods, and other continuously extrudable shapes. The engineering details of such a process can be quite complex, with separate but interactive controls for forming stage, drying stage, bake stage, presintering stage, and sintering stage. The optimization of temperature profile and residence time at each stage with the continuation of the process is an engineering challenge that is being met industrially.

Pressureless sintering has produced whisker-reinforced alumina, whereby strong complex shapes have been obtained and the process has been proven technically feasible. The difficult problems to be overcome include identifying niche applications where pressureless sintering might be competitive, such as for drill bits where only a small volume of costly whiskers is needed and performance requirements are high.

Recent work by Samanta et al.[36] utilized a combination of densification aids (Y_2O_3 and Al_2O_3) and sintering (pressureless and pressure) to enhance the properties of Si_3N_4. The pressureless sintering was performed in BN crucibles, and the two-step sintering process (pressureless sintering followed by pressure sintering) improved both density and hardness of sintered Si_3N_4. Fully dense Si_3N_4 near-net-shape products such as cutting tools and ball bearings can be produced by this method. For tough near-net-shape Si_3N_4 materials, Table 6.1 reflects the relative densities and hardnesses; the fracture toughness properties of the pressureless sintered Si_3N_4 were found to be comparable to those of hot-pressed Si_3N_4.

Other studies[37] have shown that with pressureless sintering of Si_3N_4 nearly full density can be achieved easily with an Li–Al silicate melt of the initial composition $LiAlSiO_4$.

Researchers are using pressureless sintering to achieve high-flexure strength in whisker-reinforced alumina. Sintered alumina matrix material containing 20 percent silicon carbide whiskers has been processed to more than 95 percent density, from where a hot-isostatic-pressing step boosts density to nearly 100 percent. Flexure strength is more than 500 MPa, while fracture strength is at least 7 MPa · $m^{1/2}$. Both strengths are anisotropic, the result of the fibers orienting themselves evenly but randomly rather than in the direction of pressing. To enhance whisker mobility, the researchers include proprietary additives in the start-

TABLE 6.1 Effect of Pressureless and Pressure Sintering on Relative Density and Hardness

Powder*	Particle size, μm	Y_2O_3:Al_2O_3, wt % ratio	Pressureless sintering conditions		Pressure sintering conditions (at 5-MPa pressure)		Pressureless sintering		Annealing at 1623 K for 33 h V hardness, kg/mm²	Pressure sintering	
			Temperature, K	Time, h	Temperature, K	Time, h	Theoretical density, %	V hardness, kg/mm²		Theoretical density, %	V hardness, kg/mm²
2	0.83	1:8	2058	5	2273	1	91.5	—	—	96.1	—
3	0.65	2:8	2058	5	2273	1	96.2	1260	—	98.9	1560
4	0.2	2:8	2058	5	2273	1	99.2	1350	—	99.9	1570
3	0.65	1:8	2058	5	2273	1	90.4	—	—	94.9	—
2	0.83	1:8	2098	1	2273	1	80.3	—	—	85.0	—
3	0.65	1:8	2098	1	2273	1	80.7	—	—	84.3	—
4	0.2	2:6	2098	1	2273	1	93.1	—	—	84.3	—
4	0.2	2:8	2098	1	2273	1	95.9	—	—	98.9	1430
4	0.2	2:8	2058	5	2273	1	99.2	1350	—	99.0	1570
3	0.65	2:8	2058	5	2273	1	96.2	1260	—	99.9	1570
2	0.83	2:8	2058	5	2273	1	98.6	1260	1460	98.9	1560
4	0.2	2:8	2058	5	2273	1	98.8	1395	1545	—	—
3	0.65	2:8	2058	5	2273	1	96.3	1345	1420	—	—
HPSN1	—	—	—	—	—	—	—	1350	—	—	—
HPSN2	—	—	—	—	—	—	—	1485	—	—	—

*Source of powders: Powders 2 and 3 from DENKA; powder 4 from UBE; HPSN1, HPSN2 are hot-pressed samples from Ford.

Source: From Samanta et al.[36]

ing composition. This new technology works on green compacts of any shape, but the process has a downside. The parts produced are thermally stable only to about 1000°C.

In other programs, researchers have used pressureless sintering to fabricate SiC-reinforced sol-gel glass ceramics. The gel, $LiO_2:Al_2O_3:2SiO_2$, is calcined, and particles are pressed into shapes and heat-treated to 1275°C, at which point densification is maximum and crystallization has not occurred. Silicon carbide loadings range from 20 to 40 percent, and the program goal is to develop a process that makes very large, fully dense shapes at low heating cost. The parts that the researchers now make at 1275°C are thermally stable to more than 1500°C.

Pressureless sintering is a suitable process for producing high-strength parts of SiC. Using an attritor mixing process, technicians have obtained SiC with strengths higher than 550 MPa from room temperature to 1400°C. Among the many variables experimented with to influence performance characteristics, a major contributor to strength turned out to be annealing in air. Other studies [38] have examined $SiC-Y_2O_3-Al_2O_3$. Yugoslavian researchers are investigating the use of pressureless sintering to lower temperatures of α-SiC and β-SiC. By adding 10 percent $Al_2O_3-Y_2O_3$ or $Al_2O_3-Dy_2O_3$, they achieved densities of 94 to 97 percent at 1950°C without pressure. A liquid phase in the material is the key contributor to good sintering.[39] Chinese experimenters have also examined pressureless sintering for Al_2O_3-TiC composites and found that these composites sintered with embedding powders can be hot isostatically pressed without encapsulation close to near theoretical density.[40]

6.5 REACTION SINTERING

This industrial process for manufacturing dense ceramic bodies has been well known for many decades, particularly for forming silicon-containing silicon carbide composite bodies. The term reaction sintering comes from the reaction between carbon and silicon to form SiC. Later on, the principles were applied to the manufacture of silicon nitrides by reacting silicon compacts with nitrogen-containing gas at higher temperatures in controlled fashion. The success of the process depends critically on establishing a sound understanding of the reaction kinetics.

By the reaction-sintering method of mixing and sintering silicon nitride, aluminum nitride, or aluminum oxide powders, a β-sialon composition has been produced. With this method, however, there is a limit to the degree of uniformity in mixing the raw-material powders, and since a liquid phase is involved in the sintering process, the disadvantage exists that the surplus glass phase remains in the sinter. As a result, the sinter may not have its full inherent properties, such as its excellent high-temperature strength and oxidation resistance.[41]

Most SiC powder is still made by the Acheson process, which essentially is the reaction of SiO_2 with carbon, with a posttreatment to make it suitable for sintering, although other syntheses are also used.

As is the case with Si_3N_4, SiC bodies can be prepared by sintering, hot pressing, and a form of reaction bonding. It is equally as difficult to sinter or hot press pure SiC as it is to sinter or hot press pure Si_3N_4. Again, sintering aids are used, but in the case of SiC, different aids and higher temperatures are required. For hot pressing, the usual additive is Al_2O_3, which may form a liquid with the passive oxide layer and provide a medium for liquid-phase sintering.[42]

Siliconizing

Siliconizing is a term used in the making of reaction-bonded (or sintered) silicon carbide.[43] Silicon metal has a melting point of 1410°C. In the process, varying sizes of silicon carbide grains are mixed with some form of carbon powders, such as graphite, and green-compacted to net shape by using binders and plasticizers. By carefully controlling the heating schedule, the article, which is embedded in silicon metal chips, is finally raised to temperatures above 1600°C. The molten silicon infiltrates through the capillary left by the forming aids and reacts with the carbon (graphite) to form "new" silicon carbide, which grows epitaxially on the original grit and bonds the silicon carbide together. The free silicon fills the excess porosity in the microstructure. The carbon in the initial body can be in several forms, such as particulate graphite or particulate carbon, or as carbon from the pyrolysis of resins or pitches. The siliconization can occur under both vacuum and inert atmospheres, with silicon infiltration performed either with molten silicon or with silicon vapor. Very little shrinkage occurs (less than 1 percent), primarily due to binder bake out, and therefore the method is very suitable for near-net-shape fabrication. However, excess silicon globules remain scattered throughout the surface, which need to be cleaned, typically by sand blasting.

The process has the advantage that by varying the initial grit size of the silicon carbide and the amount, one can tailormake varying Si/SiC compositions, depending on the final application. There are two primary disadvantages inherent to the process. One is thickness limitation governed by the inability of the silicon to penetrate to the interior of the part. The other is the occurrence of the so-called silicon rings around the periphery of the part because of the exothermic reaction and excess silicon left on the surface layer.

Nitriding

Nitriding of a silicon compact, in a controlled fashion, is an industrial process for the production of reaction-bonded silicon nitride. The green-formed shape, made by a variety of processes such as slip casting, extrusion, isopressing, and injection molding is converted to silicon nitride bodies with excellent thermal shock resistance and strength. The basic nitriding reaction can be written as

$$3Si(s) + 2N_2(g) \rightarrow Si_3N_4(s) \qquad (6.3)$$

Due to the nitriding reaction, a large amount of heat is generated, which requires that the external heating of the reacting mass be reduced in a controlled fashion in order not to exceed the melting point of silicon. A multistep process of interrupted heating cycles below the melting point of silicon at a maximum temperature of 1400°C was developed by Messier and Wong.[44] A solution proposed by Mangels[45] is now used commercially. It controls the nitriding reaction through a rate-controlled nitriding process by using nitrogen-demand information as the reaction proceeds. The essential features of the process include a constant nitriding rate over the entire range of the nitriding reaction and a hydrogen (helium)-containing nitrogen atmosphere to increase the thermal conductivity and control the reaction rate.

Nitriding has recently been applied to the prechamber of an automotive diesel engine, where injection molding and hot isostatic pressing have successfully produced reaction-bonded silicon nitride components (Figs. 6.5 and 6.6). Test results are listed in Table 6.2.[46]

Figure 6.5 Installation of metal prechamber into automotive diesel engine.[46]

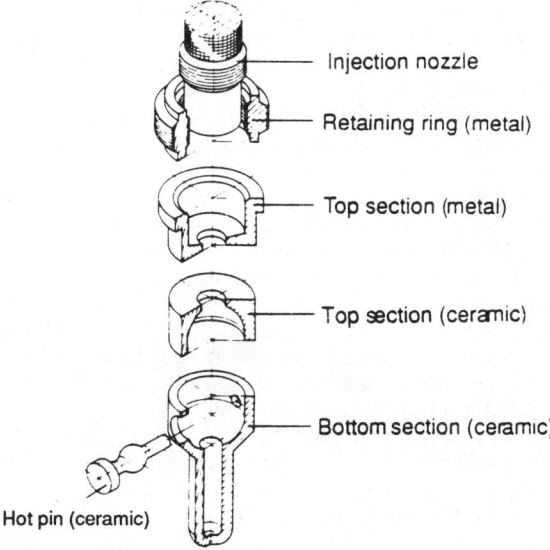

Figure 6.6 Exploded view of successfully tested prechamber arrangement.[46]

TABLE 6.2 Properties of Hot Isostatically Pressed Reaction-Bonded Silicon Nitride Determined by Different Test Procedures

Test procedure	Fracture strength, MPa	Weibull modulus	Young's modulus, GPa	Density, g/cm^3	Theoretical density, %	β-phase content, %
Four-point bending; 30-mm outer span, 10-mm inner span; samples diamond-ground	830	10	310	3.26	99.7	100
C-ring test; samples cut from precombustion chambers; surface as-fired	660	10	292	3.26	99.7	100
C-ring test; samples cut from precombustion chambers; oxidized at 1200°C for 60 h	650	9	290	3.26	99.7	100

Source: From Gasthuber et al.[46]

6.6 OVERPRESSURE SINTERING

Overpressure sintering, otherwise known as gas-pressure sintering (GPS), is now routinely practiced for sintering silicon nitrides to avoid the decomposition of silicon nitride. Instead of using one high pressure throughout the sintering cycle, a two-step approach seems to be favored. In the first step, a minimum amount of overpressure of nitrogen, sufficient to eliminate decomposition and minimize the nitrogen entrapment in the pores, is used. In the second step, sintering is completed with higher nitrogen pressures to completely eliminate the remaining closed porosity.

NGK of Japan manufactures the silicon nitride turbocharger rotor used in Nissan cars through a two-step process. In the first step, densification is performed for 2 h at 1950°C at a slight nitrogen overpressure of 2 MPa. Then in the second step, the parts are pressurized from 6 to 7.5 MPa at 2000°C for 1 to 2 h. Such a procedure was found to yield rotors of highest performance and reliability.[47] These turbochargers are supplied at a rate of 10,000 per month for use in production cars and have a failure rate of less than 1 ppm.

An advantage of the higher than normal sintering temperature is reduction of the amount of sintering additives, thereby enhancing high-temperature properties. It was realized, however, that the two-step sintering schedule depends very much on the type of silicon nitride raw powder used in the process.[48] The GPS process raises the thermal decomposition temperature of Si_3N_4. This reduces the porosity that results from decomposing Si_3N_4 and allows sintering temperatures up to 2000°C.

Densification of Si_3N_4-based materials at the usual sintering temperatures of 1700 to 1900°C has the unique problem of thermal decomposition of the materials. As a possible way to reduce the decomposition and enhance densification, high-atmosphere-pressure sintering was proposed in an investigation on sialon ceramic material by Kang et al.[49] During the infiltration of a glass into sintered β-sialon, large spherical pores (20 to 120 μm) were eliminated by being filled with the glass.

The benefit of high GPS was believed to be a suppression of the thermal decomposition of materials.

GPS was compared with conventional hot densification processes (sintering, sintering plus hot isostatic pressing, axial hot pressing) in producing TZP–ZrO$_2$ by Kessel et al.,[50] and it was shown that GPS test material under an Ar/O$_2$ atmosphere produced the highest strength values and very low standard deviations. It was also shown that GPS at 10 MPa under an O$_2$ atmosphere is at least equivalent to hot isostatic repressing of samples at 100 MPa.

Combustion sintering by gas pressure[51] has now been developed. The high-pressure combustion sintering (HPCS) can be classified as reaction sintering as well as GPS. It has the following advantages not exhibited in the conventional methods used for SiC or Si$_3$N$_4$:

1. Nearly full densification is possible.
2. Many material systems are applicable.
3. Reaction time is very short.
4. Long-time firing is unnecessary.

In this process, the reactant, encapsulated into a glass capsule, is pressurized after softening the capsule by heating. The combustion reaction is then initiated by exothermic heat of the surrounding ignition agent. By this process, the ceramic composites of titanium borides have been synthesized and simultaneously densified above 97 percent density from their starting compositions, according to Ohmura and Miyamoto.[51] This process is expected to be available for mass production of near-net-shape components for a variety of ceramic-based composites.

6.7 HOT ISOSTATIC PRESSING

Hot isostatic pressing (HIP) appears to be an attractive alternate method of producing high-quality structural ceramics. It can be used to produce components of complex shape, and it has been shown to be successful at closing up residual porosity in many materials. The application of pressure simultaneously with heat increases the driving force for sintering, and it may be possible to reduce, or even eliminate, the necessity for sintering additives. This improvement in microstructure due to reduced porosity and a less intergranular glassy phase, combined with the incorporation of SiC whisker reinforcement, may result in a ceramic of sufficiently high quality to be considered for use in gas turbine aero engines.

In this method, the pressure is applied to the ceramic powder compact or a predensified body uniformly in all directions by an inert gas medium at temperatures at which densification can occur with pore closure. The HIP process is ideally suited to ceramics such as nitrides, which decompose under normal pressures at lower than melting temperature. There are several ways by which one can utilize HIP in high-temperature processing. Other major advantages are the possibility of using higher pressures than are feasible with axial hot pressing in graphite dies, and the opportunity of using powder mixes with lower flux contents.

Researchers have developed several variations of HIP that can be applied to most materials. Sinter-plus-HIP requires that the part be presintered before HIP; therefore no canning is needed. Sinter/HIP, another variation, combines sintering and HIP into one cycle and also requires no canning.[52] The latter can be used for WC and ceramics, though it does not have furnace-design limitations. Pressure-assisted sintering (PAS) uses low pressures (7 to 14 MPa) to achieve full density. Though the process has equipment-cost advantages, it is only now being commercialized. Conversely, high-pressure HIP requires pressures of 310 MPa and achieves the same densification as other HIP processes, but with less structural damage. Though high-pressure HIP is currently limited by equipment size and cost, it has a wide variety of potential commercial applications. Increased interest, development work, and investigative programs[52,53] have been undertaken using the sinter/HIP, high-pressure sintering, isostatic green-defect healing, and rapid solidification-plus-HIP processes, which are covered in this chapter.

HIP advantages to material properties include the following:

It effectively densifies powders at lower temperatures, which results in a more uniform and finer-grained microstructure.

Combining this with the removal of residual flaws and porosity will result in improved mechanical properties.

It densifies difficult-to-sinter ceramic powders using lower levels of sintering aids, which in turn minimizes grain-boundary phases and their deleterious effects on high-temperature properties.

Since HIP is a near-net-shape process, it reduces or eliminates the need for machining, one of the primary causes of surface defects.

It provides a means to control or limit defect size, concentration, and distribution.

Table 6.3 compares the various advanced ceramic consolidation processes, and their suitability for the production of advanced ceramics. Improvements include elimination of the need for containerizing the part while also avoiding the need for separate processing steps, that is, sinter in one unit, HIP in another, and the requisite need for heating and cooling the same part twice. Densification and the properties of Al_2O_3-, ZrO_2-, AlN-, SiC-, and Si_3N_4-based ceramics are also improved. As a result, these developments will lead to the use of ceramics in advanced engines, nuclear devices, and structural components (such as nozzles). Symons and Danforth have reported on the use of HIP of laser-synthesized Si_3N_4 powder.[55] Other work covering HIP has been published elsewhere.[56-64] Some recently published reports attest to the activity in ceramic HIP,[65-71] and Fig. 6.7 reflects some of this work and improvements.

As was described earlier, the usual way to process ceramics by HIP is encapsulation, with gas pressure acting on the capsule (Fig. 6.8). Encapsulation methods can be grouped roughly into three categories. It is worth noting that encapsulation technology is an important commercial factor in a highly competitive market, often involving long periods of development by trial and error, and as a result, specific details are often confidential.

Components of simple shapes, such as cylinders, can be encapsulated in glass tubes or in welded metal sheet containers. Shapes of greater complexity can also be processed by HIP in this way if components are placed in an inert powder (such as BN), which acts as a medium to transmit pressure. For components of still greater intricacy, complex metal or glass capsules can be made, but the cost becomes very high.

A second encapsulation method involves coating the component with one or more layers of glass particles. The component is sealed off by heating under vac-

TABLE 6.3 Advantages, Disadvantages, and Applications of Hot Isostatic Pressing Methods

Method	Advantages	Disadvantages	Applications
Pressure impregnation	Near isotropic	Requires repeated cycles	Impregnate graphite preforms, porous ceramics Other composites
Diffusion bonding	No additives needed Complex shapes Improved bonding strengths Isotropic	Requires canning Dependent on bonding temperatures	Metal-ceramic and ceramic-ceramic bonding Metal-matrix composites
Powder consolidation	No presintering Low-temperature densification Less additives needed Minimal grain growth Loss of volatile species prevented Good dimensional control	Requires canning Lower packing density in autoclave Preforms may be difficult to handle Lower operating pressure if glass is used	Transparent ceramics $BaTiO_2$ dielectrics Silicon nitride structural ceramics Large alumina containers Superalloys Composites
Sinter plus HIP	No canning required Sintered parts are easier to handle Higher packing density in autoclave No die friction losses No geometric limitations Multiplicity of parts per cycle	Requires sintering to state of closed porosity Addition of sintering aids can degrade properties Excessive grain growth Surface defects are not removed Potential volatilization	Cutting tools—ceramic, carbides Multilayer capacitors Magnetic ferrites Cemented carbides
Sinter/HIP	No canning required Full density achieved in one cycle Lower cost, pressure Controlled grain size Improved surface finish	Furnace design limitations Higher temperatures	Cemented carbides Ceramic cutting tools Titanium alloys
Overpressure sintering	Low-temperature densification Less structural damage	Higher pressure Equipment size High cost	Refractory metals Ceramics

Source: From Lenoe.[54]

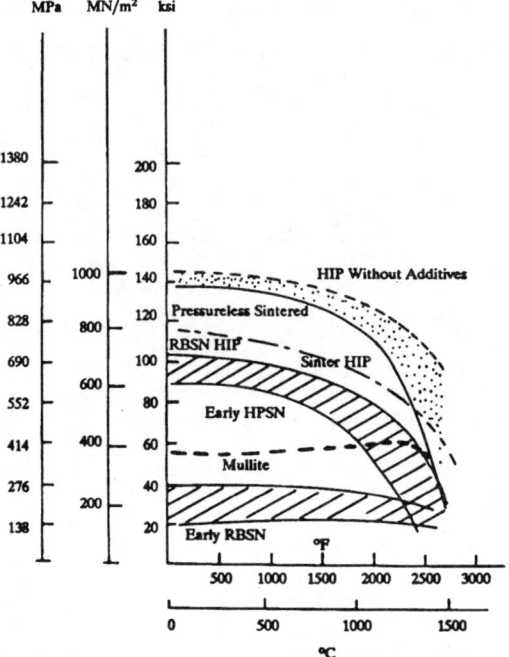

Figure 6.7 Trends in strength improvement for various ceramics.[54]

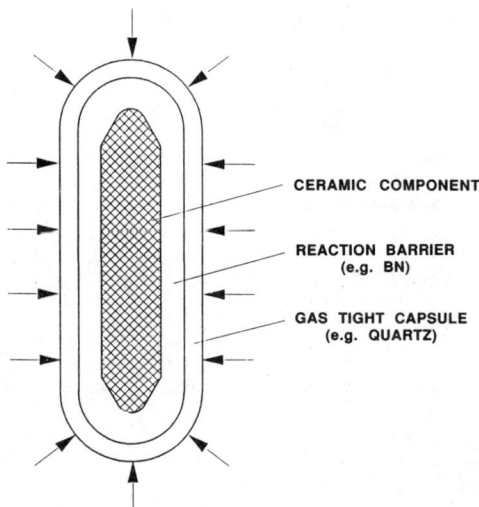

Figure 6.8 Schematic of encapsulated ceramic being processed by HIP.[11]

uum until the glass layer softens. This is known as the glass-particle method. A variant, known as the glass-bath method, involves packing the component into a glass powder bed and heating under vacuum until the glass liquifies and seals the component.

The third encapsulation method, the so-called sinter-canning technique, involves coating the component with a highly sinterable powder mixture and then sintering under vacuum to form a gastight skin. If for the coating a powder of similar composition to that of the component is used, a very good thermal expansion match results.

An alternative HIP route that has been applied to ceramics is simply to apply HIP components that have already been produced by some forming and sintering operation. (Note that past studies often refer to HIP after sintering by the somewhat misleading term of post-HIPing.) In this case, HIP is carried out to close up residual porosity and heal internal cracks. However, this method is effective only if the porosity is not surface-connected, and even so, pores containing gas will only be reduced in size and not eliminated completely.

Larker[56] was the first to report the successful application of HIP to consolidate Si_3N_4 powders, one a high-purity powder containing some amount of Y_2O_3 as a sintering aid, and the other a comparatively low-purity powder containing nothing that was added intentionally. In both cases, the powders were milled before cold isostatic pressing at 600 MPa into green compacts. The compacts were placed into a capsule of silica or borosilicate glass, and the capsules were then evacuated at elevated temperature and sealed.

HIP was carried out in two steps, and Larker claimed that an Si_3N_4 ceramic of "full" density was produced. Three-point bend strength tests were carried out, and the Si_3N_4 with Y_2O_3 as an additive exhibited a mean strength of 554 MPa when tested at 1370°C. The low-purity Si_3N_4 had a bend strength of 346 MPa at 1370°C, and of 655 MPa at room temperature. (Note that for ceramic materials the bend strength is often called the modulus of rupture.)

Clad HIP by Bumm et al.[57] with either a thin metal coating or a suitable metal-glass coating or covering is very common. This procedure requires the selection of appropriate canning material that does not react during high-temperature and high-pressure densification, and that also can be removed afterward without damaging the component being processed by HIP.

Uchida et al.[58] investigated HIP for eight high-purity Si_3N_4 powders over a range of temperatures, pressures, and times. No sintering aids were intentionally added, but the powder was milled with Al_2O_3 balls prior to cold pressing (to 150 MPa). The compacts were calcined for a few minutes at 1400°C under vacuum, placed in a BN capsule, and then placed in a high-silica glass capsule that was subsequently evacuated and sealed. This capsule was then placed into another capsule made of borosilicate glass, which was also evacuated and sealed. The capsules were treated with HIP, heated to the softening temperature of the glass under 2 MPa of pressure, and then the temperature and pressure were raised simultaneously to the desired conditions. The authors found that maximum density was attained after HIP in the range from 1650 to 1750°C, with higher temperatures resulting in decreased density due to the decomposition of Si_3N_4. The density was found to increase with increasing pressure, and the highest density attained at 200 MPa pressure was 94 percent theoretical.

An alternative method, proposed by Hardtl,[59] is the use of presintered bodies with closed porosity (approximately 93 to 95 percent theoretical density) which had been sintered previously by conventional methods. Currently, high-volume HIP processes use the presintered materials as well as "green HIP" after glass canning. Koizumi and Miyamoto[60] covered partially stabilized zirconia with powder of similar material during HIP in order to prevent contamination or chemical reaction of the processed material with the reduced atmosphere. However, Si_3N_4 was glass-encapsulated for HIP, yielding a Weibull modulus improvement of 65 percent.

Ishizaki[61] has recently considered the thermodynamics of the chemical reactions which occur during HIP and constructed phase diagrams correlating the Gibbs free energy to the partial pressures of the working gas. The effective application of pressure at a suitable temperature and then a further temperature increase for rapid densification, essentially the HIP cycle definition, will depend on the material and equipment system chosen. For Si_3N_4 it was found[62] that a higher pressurizing temperature can result in complete densification without gas penetration during HIP.

The atmosphere is also critical in HIP. Nitrogen is the preferable pressure transfer medium for nitrides compared to argon gas. Also, reaction between specimens and container crucible can be prevented by using suitable powder beds with silicon nitrides, such as BN.[63]

The practical considerations for HIP of ceramics, such as silicon carbides, at very high temperatures (>2000°C) involve reliable and long-life thermocouples. Recently HIP dilatometers were also available,[64] which should be helpful in studying densification kinetics during HIP and establishing reliable operating parameters.

Other tools being developed by researchers are HIP maps; these are used to determine the dominant mechanism of densification, which can be elastic, plastic yielding, power-law creep, or diffusion. When pressure is applied to the particles, deformations at the points of contact are at first elastic. The contact forces increase with a rise in pressure, causing plastic yielding, and contact areas develop. Once these contact areas support the forces without further yielding, time-dependent deformation processes determine the rate of further densification. These are power-law creep (which occurs in the contact zones) and diffusion (which occurs from a grain boundary to the void surface).

Finally, some of the most impressive aspects of the work going on are the strides that have been made in modeling efforts to describe the interrelationships of process parameters based on basic mechanisms active during HIP. The final products of modeling are so-called HIP maps, which are plots giving the relative density of the body treated by HIP as a function of time, temperature, and pressure.

Continuing refinements to HIP models include addressing bimodal starting powder particle-size distributions and applying finite-element analyses to both thermal and deformation distributions within the HIP body. The future can only bring the prospect of combining modeling with real-time HIP, that is, through the use of sensors and feedback, achieving the ability to control and adjust HIP parameters during an HIP run to maximize final part properties.

Sinter HIP

Although materials densified by HIP often meet required performance demands, economic considerations have prevented widespread adoption of technologies. Hot uniaxial pressing is normally limited to one or a few components per run, typically requiring many hours of operator and machine time per run. HIP can produce many components per run, but equipment cost, as well as gas, electric, and water consumption requirements all combine to produce a component that is prohibitively expensive for most large-scale applications.[72]

In the conventional HIP cycle shown in Fig. 6.9b there are two possible variations. The parts may be either containerized, making a separate sintering cycle unnecessary, or they may have been sintered previously to closed porosity in a separate unit, therefore making a container unnecessary. In either case, however, two major processing steps are required.

Using the sinter/HIP (S/H) process shown in Fig. 6.9a, it is possible to eliminate the need for containerizing the part and a separate sintering cycle, thus avoiding heating and cooling the same component twice. In an S/H cycle, the parts are sintered within the HIP unit to a closed-porosity state, whereby the part becomes its own container. Then high pressure is applied. Interestingly, for certain compositions, namely, Si_3N_4- and sialon-based ceramics, closed porosity is not necessary before applying high pressure. The high gas pressure during sintering actually promotes densification even in the open-porosity state. This phenomenon is typical of high-pressure sintering (HPS).

The potential for S/H and HPS is realized through the properties that have been achieved. Four-point flexural strengths as high as 2000 MPa have been obtained for yttria-stabilized tetragonal zirconia. This extremely high strength is significantly higher than even values obtained by sinter-plus-HIP processes. Excellent mechanical properties have been obtained for other ZrO_2-based ceramics as well.

S/H has been used to double the four-point bend strength of alumina (to 690 MPa) with a fast 100-min cycle and to make fully dense 15 vol % SiC-whisker reinforced alumina parts. Fully dense, highly translucent AlN has been obtained by S/H with the use of only 0.5 wt % Y_2O_3 or other metal oxides (for example, 0.5 wt % La_2O_3).

Flexural strengths of 730 MPa are obtained for monolithic alumina and 975 MPa for alumina toughened with small amounts of zirconia. Highly translucent alumina has been obtained as well, at temperatures much lower than typically used. Significantly, cycle times have been reduced to less than 100 min for Al_2O_3 and ZrO_2 ceramics.

Using the S/H process, Si_3N_4 has been fully densified with the addition of only 4 wt % Y_2O_3 sintering aids. This material should have excellent high-temperature strength and oxidation resistance, making it applicable for heat engine components and cutting tools. Flexural strengths of 1170 MPa and fracture toughnesses of 6.5 MPa · $m^{1/2}$ have been obtained for sialon reinforced with 15 vol % SiC whiskers. SiC has achieved >99% of theoretical density at 2000°C by means of

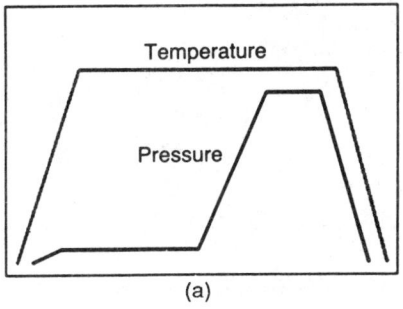 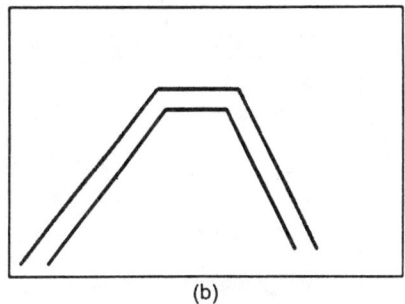

Figure 6.9 Temperature-pressure cycles. (a) Sinter/HIP or high-pressure sintering. (b) Conventional HIP.[73]

S/H. While optimizing material properties, overall cycle times for fabricating SiC have been reduced dramatically.

In applications, S/H offers an attractive alternative to HIP. Since S/H permits single-step loading, multistep processing of powder materials in a single furnace, the S/H system can be used to dewax (also called debinder or delube), presinter, sinter, overpressure consolidate, case harden (via plasma techniques), and cool rapidly (for short cycle).

For production loads and large batches, front-loading S/H systems are most effective, as they are designed for rapid and efficient processing of advanced materials in a production environment (Fig. 6.10). Automated control sequences, full safety interlocks, and user-friendly design are all standard. Temperature and pressure ratings are selected to match the material's process cycle. Computer interfaces for centralized control and data acquisition are optionally available.

It is estimated that there are approximately 100 overpressure consolidation furnaces (S/H type) in the world today. Applications are primarily for tungsten carbide mining tools. The S/H-type equipment shown in Fig. 6.10 is now being closely investigated for engineering ceramic and powder metal alloy production. It appears that S/H will replace the traditional HIP process in all applications where the product has closed surfaces and the sinter mechanism allows full density to be achieved at 150-bar overpressures or less.

In recent years the strength properties of Al_2O_3 ceramics have been markedly improved by the incorporation of ZrO_2 and the dispersion of metallic hard materials such as TiC or TiN. (See Chap. 4.) Kolaska et al.[74] conducted S/H tests on this material (30 percent TiC). Figure 6.11 shows the four-point transverse-rupture strength of samples containing 30 percent TiC, produced by the S/H process and by HIP in comparison with conventionally sintered samples. It was found that both HIP and the combined S/H process permit the production of dense samples of comparable microstructure. The transverse-rupture strength of the samples produced by the two processes is also comparable and more than 50 percent higher than with the sintered samples. This is mainly attributable to the removal of porosity.

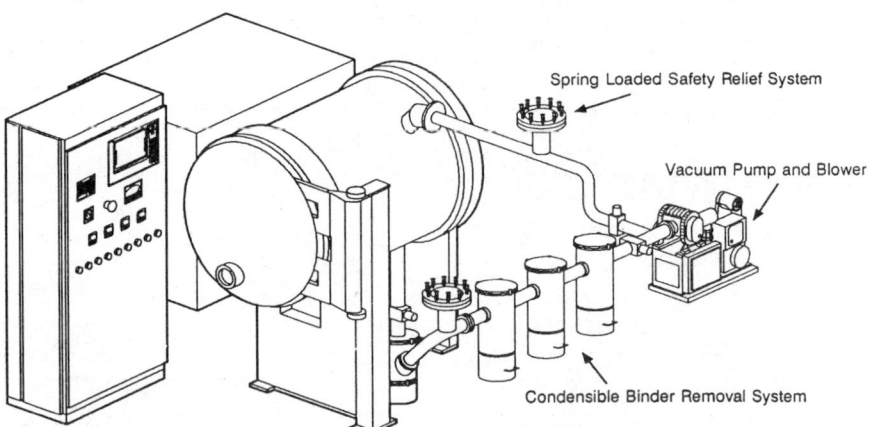

Figure 6.10 Front-loading production-size sinter/HIP system with safety closures for rapid loading and unloading.[72]

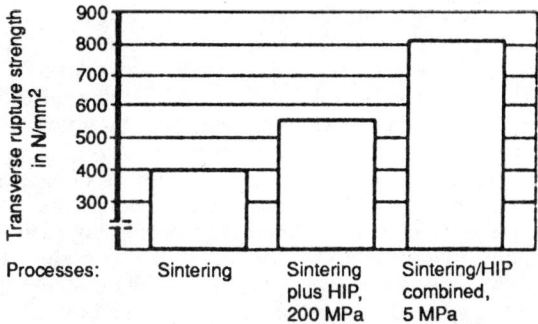

Figure 6.11 Transverse-rupture strength values of differently compacted Al_2O_3–ZrO_2 ceramics containing 30 percent TiC.[74]

By contrast, the combined S/H process resulted in a marked increase in the strength of ceramic material containing 5 percent TiC compared with HIP. This can be partly explained by the slightly more uniform microstructure. Other reasons for the strength increase may be found in the fact that the grain-boundary structure, which largely determines strength, is favorably influenced by an S/H process without intermediate cooling.

S/H was applied and produced a high-density fine-grain-size alumina–30%TiC cermet cutting tool material. The key to the success of S/H of this advanced ceramic material, which could previously only be produced by hot pressing, was in applying a heating rate exceeding 100°C for the temperature range above ~1550°C and minimizing the total sintering time above 1550°C. The rapid-rate process enabled the use of high sintering temperatures without inducing excessive grain growth and minimized the detrimental effect of high-temperature reactions between alumina and titanium carbide. Using a relatively slow heating rate up to an intermediate temperature to develop moderate strength in the material and then applying a rapid heating rate successfully avoided thermal fracture. Alumina–TiC cutting tools produced by the S/H process performed in machining tests at least as well as the best commercial grades of similar composition.[75]

Iturriza et al.[76] studied the densification of Si_3N_4 by a route that can be viewed as true S/H processing. Si_3N_4 powder mixes containing 3 wt % Y_2O_3 plus 0 to 3 wt % Al_2O_3, and mixes containing 6 wt % Y_2O_3 plus 0 to 6 wt % Al_2O_3 were milled together using Si_3N_4 balls. The powders were compacted using cold pressing to 100 MPa and then processed by S/H in three stages. The total time for the cycle was kept constant at 60 min, but the time spent at each stage was varied. The authors found that high density was achieved only if a sufficient time was spent at the first stage to allow sintering to closed porosity (that is, porosity- not surface-connected), and that the final density increased as the rate of pressurization decreased. However, in no case was the final density greater than 98 percent theoretical.

Greskovich[77] has successfully sintered Si_3N_4 green compacts to >99 percent theoretical density by a route that can be viewed as low-pressure S/H. High-purity Si_3N_4 powder was mixed with 7 wt % $BeSiN_2$ and milled together prior to cold isostatic pressing at 200 MPa. During processing, the powder was subjected to a controlled oxygenation in order to have a starting material that also contained about 7 wt % SiO_2.

Other workers have used N_2 overpressure of up to about 5 MPa (that is, 50 atm) to sinter Si_3N_4 (see Tani et al.[78]). However, at such comparatively low pressures the process is really gas-pressure sintering rather than HIP.

It has been shown at Gorham[79] that with Si_3N_4 material the process can increase the green density by as much as 10 to 15 percent above that obtainable via conventional slip casting and injection molding. The researchers believe, however, that their method is generalizable to any slip-cast, die-pressed, or injection-molded ceramic and metal powder parts, and can even be used to join dissimilar materials. Line, point, or planar defects introduced during casting or molding can be virtually eliminated during room temperature processing, even in parts as complex as a turborotor. This approach to controlling at the green-body stage could result in reduced sintering costs and the elimination of defects. Based on preliminary work with simple shapes, it should be possible to produce almost defect-free complex parts fabricated in several sections by simultaneously joining and densifying them.

Pulsed Sinter/HIP

A completely new HIP processing method of tremendous potential for several applications of advanced ceramics, called pulsed sinter/HIP, or PuSH,* is being developed and is based on the principles of rapid-rate sintering, rapid HIP, and sinter/HIP technology. The object of PuSH is to achieve closed porosity and then apply pressure, but to do it much more rapidly than by the sinter/HIP process.[73]

It is only necessary to obtain sufficient densities in order to achieve closed porosity on the surface of the part during sintering. If the sintering temperatures and times are sufficiently low, the resulting grain structure may be fine enough to give deformation or creep characteristics that allow complete densification by relatively low isostatic gas pressures. Another advantage of PuSH is to minimize the times available for unwanted reactions of materials. These might include reactions with the furnace atmosphere or contaminants, or reactions between toughening additives and the matrix. It may allow the processing of entirely new materials, such as rapidly solidified ceramics or composite combinations otherwise not possible. Lastly, but most important, is the effect on economics. One can obtain all the benefits of HIP, plus potentially others, at greatly reduced costs.

Pressure is commonly viewed as the factor that makes HIP expensive, but in reality it is the combined time and pressure used in the process that is costly. The costs of pressure, even high pressure, at short times are minimal. Expenditures can be further decreased by efficient utilization of the equipment. Smaller furnace loads processed more often reduce both equipment and per-cycle costs.

6.8 NOVEL METHODS

Many novel methods are being explored. These have special advantages in high-temperature processing and may be particularly applicable in the case of ceramics-matrix composites. However, wide industrial applicability is still emerging.

---

*PuSH is a registered trademark of Gorham International, Inc., Gorham, Maine.

Pressure-Assisted Sintering[73]

Investigation of the rapid-rate sintering portion of the PuSH process is being undertaken in a furnace designed with a very low thermal mass and with the heating elements and power source necessary to provide very quick thermal spikes in the pressure-assisted sintering (PAS*) process for powder metals. The furnace can range from high vacuum to 10.2 MPa, and has the capacity for extremely rapid heating rates. Furthermore, unlike other rapid-rate sintering methods, resistance heating, whether rapid or slow, can be easily used in HIP equipment.

Rapid pressurization has been used for some time in certain facilities for metals with large reservoirs of high-pressure gas, and innovative rapid-pressurization HIP equipment is becoming commercially available. Rapid heating rates, short times at temperature, rapid pressurization, and potentially high pressures have each independently proved to be beneficial. One can only anticipate the synergistic effect on properties of these combined parameters. The very fast processing times should inhibit grain coarsening, with the result being extremely fine and uniform microstructures. This condition in turn should improve mechanical properties significantly, and it may improve other properties as well. The extremely fine microstructure might also make it easier to deform or densify the part during isostatic pressing.

Rapid Sintering

It has been suggested that the control of the densification-to-grain-coarsening ratio can be better achieved by fast firing.[6] Zone sintering, in which the article can be pushed through a short hot zone (either a tub furnace externally heated by different means or an RF plasma), is beneficial in obtaining a dense body with a fine-grain microstructure.

Plasma Sintering

Three different plasma methods have been used for the sintering of various ceramics. In historical order, these are microwave-induced plasma (MIP),[80] hollow-cathode discharge (HCD),[81] and induction-coupled plasma (ICP)[82] sintering. The significance of any of these processes is the unusual rapid sintering rates achieved, uncommon to conventional densification methods. Rapid sintering of various ceramics has included Al_2O_3,[80–87] Y_2O_3–PSZ,[83] BeO_2,[80] HfO_2,[80] and ZrO_2.[88] Uniform fine microstructure and better mechanical properties are commonly observed in plasma-sintered ceramics with fractional densities ranging from 96 to 100 percent.[80–86]

Plasma-sintering experiments using HCD equipment (Fig. 6.12) of unshocked and shock-treated Al_2O_3-based composites have shown that Al_2O_3-$10ZrO_2$ composites can be sintered up to a fractional density of 94 percent in several minutes, where as limited sintering is achievable in 30 vol % SiC_w containing composites. Increasing the sintering time of whisker-containing composites did not increase their fractional densities. Mechanical properties of Al_2O_3-$10ZrO_2$ composites were not as good as conventionally sintered composites, which can be attributed to remnant porosity in unshocked and weaker interparticle bonding combined

---

*PAS is a registered trademark of Gorham International, Inc., Gorham, Maine.

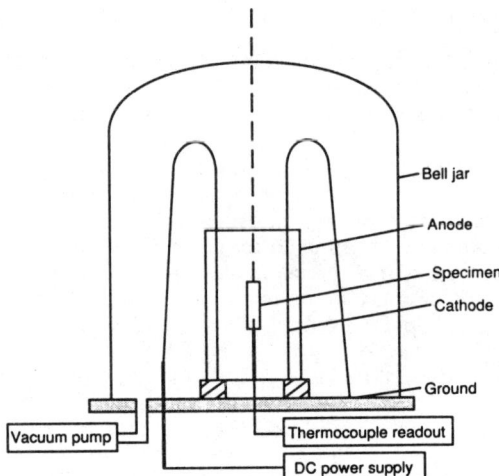

Figure 6.12 Experimental arrangement used for hollow-cathode-discharge plasma sintering.[89]

with higher porosity in shock-treated composites. Proper adjustment of plasma-sintering parameters has the potential of producing composites with mechanical properties surpassing those of composites fabricated by conventional routes.[89]

High-Frequency Sintering

Researchers at the University of Minnesota[90] have sintered small rods of stabilized and partially stabilized zirconia in less than 10 min at atmospheric pressure through radio-frequency (RF) energy. They used a converted industrial-heating RF oscillator-tube generator to create a plasma and adapted it for RF sintering; its effective power was 2 kW at 13.6 MHz. The researchers expect to increase both the power and the plasma temperature of their system in order to sinter larger and more complex shapes.

The present system heats samples to 2200°C by conduction so fast that the fine-grain structure is usually preserved throughout, especially in parts produced by HIP and in the radial direction. But temperature gradients in the axial direction are still a problem. Atmospheres, however, are not a problem: argon works well, as do diatomic oxygen and nitrogen, and mixtures of the two.

Finally in the sintering of composites, rapid sintering must be performed with special precautions to avoid thermal fracture, taking into consideration the differential properties such as Young's modulus, Poisson's ratio, thermal conductivity, coefficient of thermal expansion, and density.

Sinter Forging

Sinter forging is a new technique which differs from hot pressing in that the powder compact is not constrained by die walls.[91] The process requires an understanding of the shear and densification components of the shrinkage strain and their contribution to sintering. In sinter forging, the relative amounts of the radial and axial strains can be varied by changing the applied load. This allows control of the shear strain to either high or low values, which controls the processing path as shown schematically in Fig. 6.13.[91]

In this method the flaw (large pore) is purported to occur by the breakdown of large pores into many tiny pores, which then disappear due to densification. Ellipsoidal pores begin to close at lower shear strains than spherical pores, which require about 60 percent compressive strain. Sinter forging is especially suitable for composites with rigid inclusions. Near-net-shape forming has been demonstrated by Panda and Seydel[92] for magnesia-alumina spinel composite reinforced with silicon carbide fibers. The preforms were sinter-forged at 1580°C using a densification strain rate of 10^{-3}/s. Graphite dies were used with graphite foil lining of the die surfaces. The heating elements were of tungsten mesh, and argon gas was used as the atmosphere.

Slow Forging

This technique is proving to be the way to densify green pieces faster than by HIP. The idea is to use a molten glass material rather than a gas to transmit pressure and to do it at forging speeds while still maintaining isostatic conditions. In the laboratory it has proved fast and effective.[90] Researchers have consolidated greenware to full density in seconds at low energy cost in a forging press. They encapsulate a 50 percent dense compact in soft alumina or other ceramic together with chunks of borosilicate glass, a heating wire, and a thermocouple temperature sensor. The assembly is heated in a vacuum furnace to temperatures as high as 2500°C in order to melt the glass. This vacuum encapsulation covers everything inside the ceramic shell with glass, and although the vacuum is not strictly

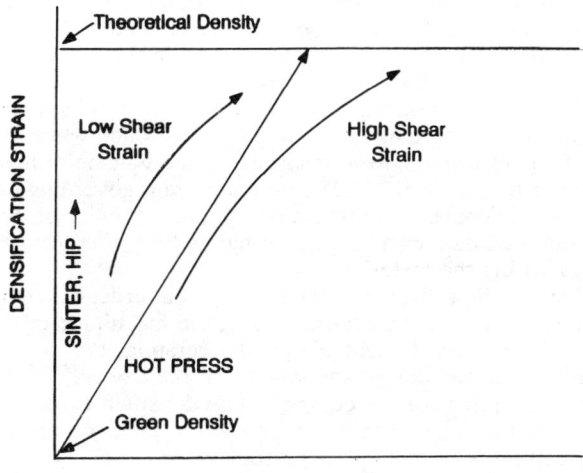

Figure 6.13 Schematic of processing path under various conditions.[91]

necessary, it is used to avoid gas problems and to obtain the best possible densification.

The entire assembly is placed in a cylindrical forging press and power is applied to the nickel or tungsten heating element. The tungsten wire heats the compact to 2000°C in situ under the glass in the ceramic shell. At this high temperature the glass conducts electrically and contributes to heating. As all heat is retained in the shell by the low-thermal-conductivity ceramic, the die is not thermally stressed and ordinary rather than tool steel may be used.

When the compact is at the right temperature, the shell is pounded by the forging pistons at a typical pressure of 2070 MPa. Densification is complete in seconds, and ceramics consolidated this way include PSZ, which densifies to 100 percent in seconds at only 1700°C, as well as SiC and Si_3N_4. A similar process, rapid omnidirectional consolidation (ROC), does not use a vacuum or a heating element, and it can get by with a minimal amount of heat (much less heat than other isostatic forging processes for ceramics).[90,93] Another variant on combining forging with HIP uses quartz sand as a pressure-transmitting medium, as well as graphite or ceramic.[90]

Microwave Sintering

The use of microwaves as an energy source for sintering has received considerable and supportive attention.[94–98] The energy conversion efficiency is particularly high, on the order of 95 percent, and this occurs with both reduced sintering times and sintering uniformity. As in many new processes, the earliest demonstration was using alumina, which has been confirmed recently at ORNL. At Los Alamos Laboratories boron carbide has been densified to >95 percent theoretical density by heating at 2000°C with 2.45-GHz microwave radiation.[98] Due to very low residence time, grain sizes are lower compared to those of the hot pressing process.

The microwave heating of a ceramic, which involves heating of the sample from within, is fundamentally different from other furnacing techniques in which heat is supplied external to the body. Due to the internal and volume heating, rapid and uniform heating of large shapes is possible. The reader is referred to an excellent review by Sutton[99] for more information on the characteristics and advantages of microwave processing of ceramics.

The critical aspect for utilizing microwave energy in high-temperature processing is the need for absorbing and "coupling" the microwave to interact with the ceramic. The microwave power absorbed by the volume of the ceramic provides the basis for the heating.[96,99] The two important governing parameters are the relative dielectric constant representing the polarizability of a material in an electric field, and the loss tangent representing the loss (or absorption) of the microwave energy within the material.

In practical application, these parameters are all interdependent and vary with heating and densification of the ceramic. Also, the magnitude of the microwave field depends on the geometry and size of the ceramic, the location within the microwave cavity, and the design and volume of the cavity.[96,99] In cases where the microwave absorption for the ceramic to be densified is low, additives with high conducting or magnetic phases can be used to enhance the microwave absorption, provided these do not act as poisons for sintering.

Processing. Microwave processing involves the use of microwave radiation (1 to 140 GHz) to couple to and heat a material. Both organic and inorganic coupling agents may be used. The coupling agent is needed because most oxides (many glasses and ceramics) are transparent to UHF microwaves at room temperature. The coupling agent absorbs the UHF microwave energy and generates heat, enabling the surrounding glass and ceramics to couple with the microwaves at higher temperatures. Materials with OH, CO, NO, or NH bonds usually make good coupling agents, including glycerol, nitrates, potassium silicate, sodium silicate, and several glasses. It is also important that the coupling agent bond well with the base material.

Sintering. The firing process that turns ultrafine powders into extremely hard materials is a critical factor in the production of advanced ceramics. An unsintered (green) ceramic material is softer than chalk; when the green body, which consists of millions of powder particles, is sintered at high temperatures, usually between 1000 and 2000°C, the particles fuse to become a solid material that has properties superior to many metals.

Sintering ceramics, ceramic matrix composites, metals, and metal matrix composites; joining ceramics to ceramics and ceramics to metals; and even synthesizing materials are some of the potential applications for microwave processing. It appears that microwave processing could prove to be more economical than conventional heating and sintering. As heating rates are as much as 50 times greater, the processing throughput rate could be much higher, and as the heating energy couples directly to the component being heated, considerably less power is needed.

Figure 6.14 shows the rapidity with which high-purity submicrometer Al_2O_3 powders (0.2-μm average particle size) can be sintered from 50 percent initial green density to 96 percent theoretical density. To reach the maximum target temperature of 1700°C from room temperature takes just 370 s. By comparison, it takes 20 h to conventionally sinter a similar Al_2O_3 green body to 96 percent theoretical density. When sintering an Al_2O_3 composite containing 10 vol % SiC whiskers from a 52 percent initial density to 77 percent theoretical density, it takes 320 s to reach the maximum target temperature of 1450°C from room temperature. The SiC whiskers appear unaffected by this treatment. With conventional processing, similar composites could only be sintered to 71 percent when heated to 1700°C in 2.5 h. Furthermore, there is reason to believe that because of the rapid heating rate, the microstructure of microwave-processed ceramics will be finer, with more uniform grain structures.

In microwave heating, microwave energy causes the molecules throughout the workpiece to vibrate. The result is that the workpiece heats uniformly without gradients of temperature even if the piece is irregularly shaped. In contrast, in radiant or convective heating, energy is absorbed only at the surface of the workpiece and must be transferred into the bulk of the material by conduction (Fig. 6.15).

No commercial furnace now available can offer a processing system for sintering and annealing at temperatures of up to 2000°C in vacuum, inert, or oxidizing environments, except the microwave. The basic component in microwave sintering of ceramics is a furnace cavity that is large in relation to the wavelength of the microwaves used. While ordinary household microwave ovens are powered by a low-power magnetron that typically operates at 2.45 GHz, the microwave source used by most researchers is a gyrotron oscillator capable of supplying 28 GHz at 200 kW continuous wave into a cylindrical cavity. Researchers have changed the ratio of the microwave wavelength to cavity size from 1:3 to 1:100 in order to achieve uniform heating without dead spots.

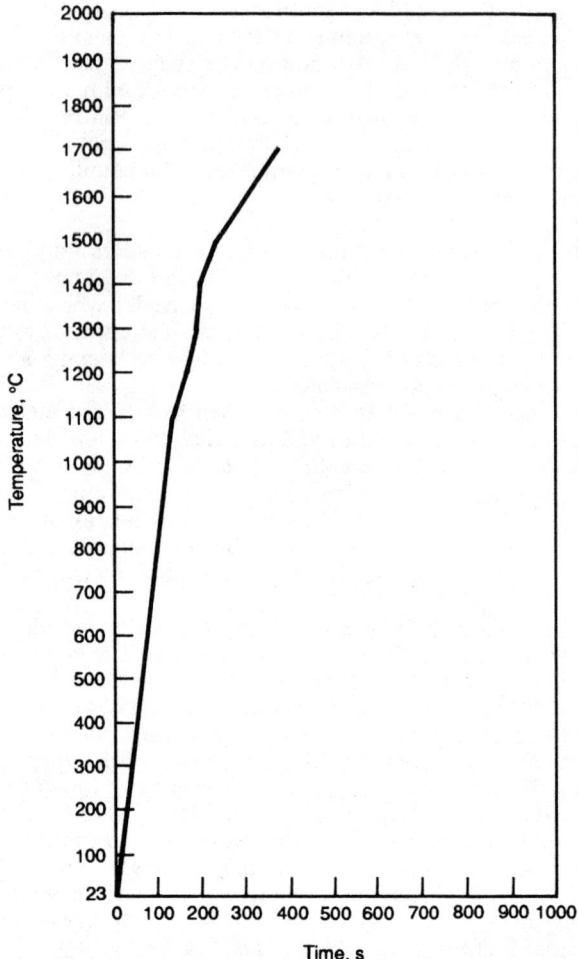

Figure 6.14 Temperature-time curve for Al_2O_3 specimen heated to 1700°C at 60 GHz.

This has allowed investigators to heat large irregular shapes to temperatures of 2000°C rapidly and evenly, and to produce volume-heating rates of up to 200°C/min for parts weighing 2 lb or more. They have found sintered workpieces having both fine grain size and extremely high densities that approach 100 percent of the theoretical maximum. They also compared microwave-sintered ceramics to conventionally sintered ceramics for density, fineness of grain, and fracture strength. The microwave-sintered samples were found to be equivalent even to ceramic materials that had been subjected to hot pressing, an expensive and time-consuming technique that combines high sintering temperatures with high pressures to exclude the formation of large grains and pores in the material during densification. Microwave sintering thus offers the possibility for combining the

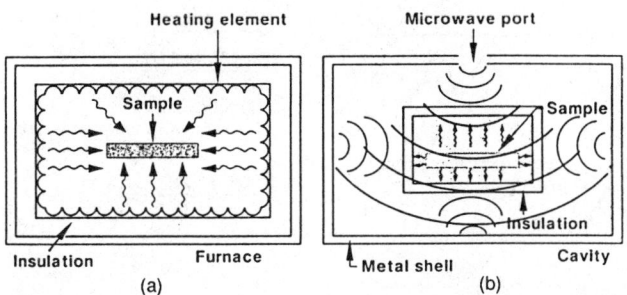

Figure 6.15 Heating patterns. (*a*) Conventional furnace. (*b*) Microwave furnace.[96]

microstructural advantages of hot pressing with the ability to make complex shapes, as in pressureless sintering.[95]

These results have led Kimrey and Janney[95,100–102] to believe that many ceramics that are difficult to heat at low frequencies will be easy to heat at higher frequencies. Alumina showed a factor of greater than 1000 increase in absorption between 1 and 100 GHz. This magnitude of increase may also be typical of many other ceramics, such as AlN and Si_3N_4. There may be a limit to the higher frequencies, however. The data suggest that at extremely high frequencies there is a decrease in a material's "skin depth" or degree of transparency to electromagnetic radiation, and scientists have found that microwave sintering works most effectively with ceramic materials that have a large skin depth. Al_2O_3 has a skin depth of several meters, and the radiation bounces through the Al_2O_3 workpiece over and over again, heating it uniformly throughout. With a shallow skin depth, microwaves at higher frequencies might heat only the surface, which will result in uneven heating analogous to conventional thermal processing. The uniformity of microwave heating is of special value when sintering ceramics with complex shapes, which are difficult to heat in conventional furnaces.

During sintering, the workpiece is surrounded by an insulating liner that prevents the migration of contaminating gases from the workpiece to the furnace wall. This liner also allows hot unloading of the workpiece, eliminating the extended cooling time required when using conventional furnaces.

Previous research of conventional heating has shown that significantly increasing the heating rate increases the ratio of the densification rate to the coarsening rate, thus improving the microstructure. However, with microwave processing a material is heated internally, and sintered large-volume ceramic components should have uniform microstructures and properties.[94,96] Further, since the heating takes place from the inside outward rather than in reverse, the internal temperature is higher than the surface temperature. The net effect is to reduce thermal stresses markedly, thus minimizing, and possibly completely eliminating, the cracking that normally occurs when large-volume ceramic components are conventionally processed.

Another characteristic of microwave processing is that different constituents and phases absorb microwaves at different rates. For this reason, depending on the composite system, such as zirconia particles in alumina or silicon carbide whiskers or fibers in alumina, the particles, whiskers, or fibers could heat up more rapidly than the surrounding matrix, or the reverse (Fig. 6.16). The possi-

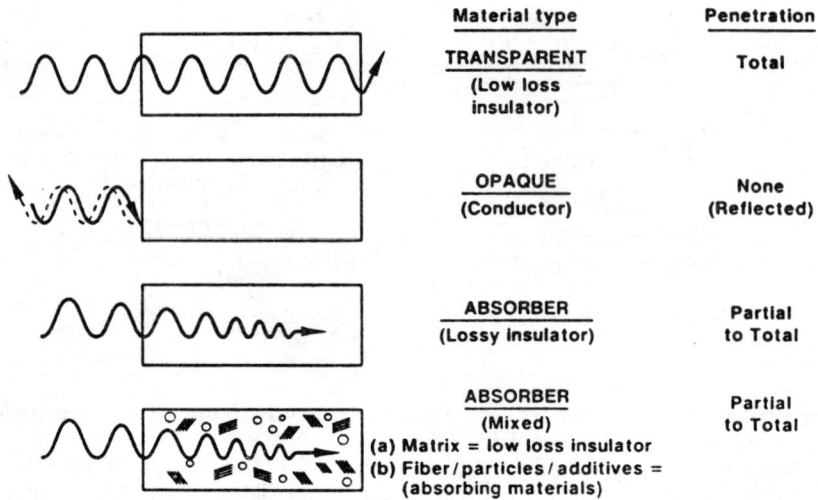

Figure 6.16 Interaction of microwaves with materials.[96]

bility of controlled, selective heating offered by microwave processing has promising implications for improving the properties of ceramic composites.

Application. Microwave sintering has been applied in the densification of small TiB_2 samples to nearly 90 percent density in minutes using only about 1 kW of 2.45-GHz microwave power. Investigators expect to approach 100 percent with improved microwave-cavity tuning. They have already achieved nearly complete density in less difficult to densify ceramics, including ZrO_2 and Al_2O_3. They use no sintering aids with their TiB_2, which they heat as high as 2090°C. Power is supplied by a commercial magnetron into a cubic cavity 2 ft on one side. Maximum magnetron power is 6 kW. One application of dense TiB_2 is armor.[103]

Not only have microwaves been used successfully for sintering, but microwave energy can be combined with sol-gel methods to produce fine powders.[94,104] Sheppard[94] cites the manufacturing of ceramic fixtures with microwaves. Productivity improved from 100 pieces per day requiring 16 workers, to 1600 pieces per day requiring 60 workers, unit cost dropped from $0.85 to $0.23, and the total number of molds was reduced by 75 percent.

However, there are numerous areas of unknowns yet to be explored. These include:

Converting the current batch process into a semicontinuous or continuous operation.

Understanding the science involved better, perfecting the processing techniques, and reducing the cost of HF microwave power from the current $30/W for a 15-kW unit to $10/W.

Developing larger ovens and improving techniques to produce the uniform fields required to process industrial-sized parts.

Being capable of applying uniform heating rates of up to 200°C/min, even when sintering 1000-g or larger ceramic pieces.

Developing ways of firing computer chip carriers and metal-ceramic composites with microwaves to avoid high sintering temperatures. Microwave sintering lowers temperature requirements, and thus would allow copper, the preferred conductor material, to be used in the future in lieu of molybdenum.

Self-Combustion Synthesis

Synthesis as well as consolidation of inorganic compounds can be performed taking into account the exothermic chemical reaction of certain systems. After initiation of the reaction, self-sustaining forward completion will occur through the reaction mixture in the form of combustion waves. Classic examples of combustion-synthesized ceramics are TiN, TiB_2, TiC, and SiC. The use of the exothermic reactions with advancing combustion wave front is commonly referred to as self-propagating combustion synthesis. Over the last 20 years, many borides, carbides, nitrides, intermetallics, and silicides have been made in the U.S.S.R. by this method.[105] This was discussed in Chaps. 4 and 5. In a typical example, the reactants are green-formed into the desired shape, with ignition initiated in one or several places. The resulting chemical reaction generates heat, and the propagation of the combustion front depends on the system characteristics. Controlled wave-front propagation, rather than runaway reaction, can be achieved by the addition of diluent and the use of large-size reactant powders.

Recent investigations by Adachi et al.[106] into high-pressure ceramic sintering (a combination of SHS and high-pressure sintering) has produced TiC ceramics from the reactants titanium and carbon fiber under 65 MPa and 3 GPa. Under 65 MPa porous TiC with a 50 percent density was produced, while dense TiC (>95 percent theoretical) was fabricated under 3 GPa. The mean grain size varied, depending on the mixing molar ratio of the reactants.

Other recent findings have concentrated on the simultaneous synthesis and densification of ceramic compounds. The synthesis and simultaneous sintering of the product, for example, the Ti–B–C system, was carried out by McCauley et al.,[107] Irving,[108] and Riley and Nüller.[109] Riley and Nüller have definitively shown that the mechanism of Ti + C powders reacting to form TiC is the movement of melted titanium to the solid carbon lattice.

Extensive porosity in the product is caused by this movement of the titanium, but additional voids are formed from the expulsion of contaminant gases during the combustion process. In order to eliminate these voids and the porosity in the material, the material must be compacted up to pressures of 0.41 GPa immediately after completion of the combustion. The results are densities greater than 95 percent and microhardness approaching the values obtained by conventional sintering techniques.

Riley and Nüller[109] combined two advanced techniques and consolidated simple shapes of TiC by first synthesizing them combustively and then compacting them explosively. Other researchers have also used the process with TiB_2 and with a composition of TiC and TiB_2.[110,111]

The point of the technique was to see whether or not SHS can be a low-cost route to inexpensive TiC and TiB_2. TiC shapes, especially, have some potential to become low-cost SHS products. The precursors of TiC are relatively cheap, and there is a theoretical possibility of it being combustively made without the need for hot-pressing. But two almost mutually exclusive conditions inhibit further progress. While the green shape must be contained during synthesis so that at least some consolidation pressure can be applied, it is also desirable that it be uncontained so that gaseous products of combustion can vent away safely. A fast combustion reaction does not always leave enough time for the impurities to escape from the heated region ahead of the shock-wave front. When this occurs, the material can explode. To negate this problem, some researchers have tried purging most impurities at 500°C, just before the combustion step, but this introduces other complexities.

These researchers consolidated the combustively synthesized shapes by compacting them explosively while they were still white hot from the combustion reaction and were at a temperature of about 2500°C. At that time they were about 50 percent dense. The explosive force densified them to more than 98 percent. The process is used only for simple shapes such as slabs and cylindrical sections.

Laser Sintering

The sintering and growth of ceramics in the systems of ZrO_2–HfO_2, PSZ–HfO_2, and Al_2O_3–WO_3 have been investigated using CO_2 laser beam as a heat source. According to Okutomi and Tsukamoto,[112] laser sintering has been a successful method and is very useful for the densification of oxides such as ZrO_2, HfO_2, and Y_2O_3, which have melting points above 2000°C. This method will be very useful and practical in the synthesis of oxide, nitride, carbide, and boride materials with their high hardness and melting points, and it will provide new information in the field of high-temperature chemistry and ceramic materials in a nonequilibrium state.[113–115]

6.9 OTHER PROCESSING PARAMETERS

High-Temperature Heat Treatments

In addition to normal sintering, selective high-temperature heat treatments are performed on manufactured ceramics in order to:

1. Relieve sintering stresses as well as postmachining stresses
2. Increase the homogeneity of surface stress distribution
3. Initiate appropriate phase transformation in the case of phase transformation-toughened ceramics
4. Crystallize the liquid phase responsible for full sintering so that a refractory grain-boundary phase is obtained with superior creep and oxidation resistance at elevated temperature. [For example, the Y–Si–Al–O–N liquid phase was crystallized in a sialon matrix as YAG (yttrium aluminum garnet; $3Y_2O_3 5Al_2O_3$).]

Transformation-Toughened Ceramics

The high-temperature processing of ceramics in which phase transformations occur during crack propagation leading to better toughness, for example, oxide ceramics such as zirconia and zirconia-toughened alumina, can be complicated, especially for complex components. In the zirconia system, the high-temperature cubic form is retained at certain volume fractions at room temperature by additions of MgO, CaO, or Y_2O_3, the remaining phases having monoclinic and tetragonal basic structure. The toughening process involves the transformation of tetragonal particles to monoclinic ones by the interaction with the stress field of the advancing crack. Factors such as size, shape, and the amount of tetragonal phase available for transformation are critical.

6.10 CONCLUSIONS

This chapter has presented an overview and given the recent status and understanding of high-temperature processing, including raw-material and microstructural factors governing densification by various industrially and laboratory practiced methods. The high reliability expectations of the engineered ceramics depend on the care and diligence with which these ceramics are densified to engineered components. Assuming that powder selection and green-forming methods have been exercised with care and control, a lack of densification and uniformity in properties in many instances arises from poor furnacing practice, such as furnace overload with respect to available heating volume, uncontrolled temperature gradients, furnace contamination from previous work load, and ill-defined packing of the greenware for sintering.

Intelligent processing, which would use sensors of various types, will be quite useful in monitoring densification and microstructure evolution as control ranges are established for the aforementioned variables. This is quite a challenge, since reliable sensors have to be developed which will operate under hostile environments of extremely high temperatures (>1500°C) and gaseous products of sintering reactions. In addition, calibration of these sensors and setting up useful ranges will require contributions by scientists with instrumentation and control skills. Finally, transfer of this intelligent processing technology to the manufacturing shop will require systematic education of shop personnel to take maximum advantage of new developments. This will be further discussed in Chap. 9.

REFERENCES

1. B. Frisch and W. R. Thiele, "Adhesion Effects of Intermediate Layers on the Densification of Ceramic Powders," *J. Adhesion,* vol. 22, pp. 81–95, 1987.
2. B. J. Kelle and F. F. Lange, "Thermodynamics of Densification; I: Sintering of Simple Particle Arrays, Equilibrium Configurations, Pore Stability, and Shrinkage," *J. Am. Ceram. Soc.,* vol. 72, pp. 725–734, 1989.
3. R. M. Cannon and W. C. Carter, "Interplay of Sintering Microstructures, Driving Forces, and Mass Transport Mechanisms," *J. Am. Ceram. Soc.,* vol. 72, pp. 1550–1555, 1989.
4. N. J. Shaw, "Densification and Coarsening during Solid State Sintering of Ceramics; A Review of the Models—1. Densification," *Powder Metall. Intl.,* vol. 21, pp. 16–21, June 1989.
5. S. Prochazka, "The Role of Boron and Carbon in Sintering Silicon Carbide," in P. Popper (Ed.), *Special Ceramics,* vol. 6, Proc. BCRA, 1975.
6. T. K. Gupta, "Possible Correlation between Density and Grain Size during Sintering," *J. Am. Ceram. Soc.,* vol. 55, pp. 276–277, 1972.
7. F. F. Lange, "Thermodynamics of Densification; II: Grain Growth in Porous Compacts and Relation to Densification," *J. Am. Ceram. Soc.,* vol. 72, pp. 735–741, 1989.

8. R. J. Brook, "Fabrication Principles for the Production of Ceramics with Superior Mechanical Properties," *Proc. Brit. Ceram. Soc.,* pp. 7–24, 1982.
9. S. Hampshire and K. H. Jack, "The Kinetics of Densification and Phase Transformation of Nitrogen Ceramics," in *Special Ceramics,* vol. 7, *Proc. Brit. Ceram. Soc.,* no. 31, p. 27, 1981.
10. F. F. Lange, "Silicon Nitride Polyphase Systems; Fabrication, Microstructure, and Properties," *Intl. Metal. Rev.,* vol. 247, pp. 1–20, 1980.
11. T. M. Maccagno, "Processing of Advanced Ceramics Which Have Potential for Use in Gas Turbine Aero Engines," Natl. Aeronautical Estab., Aero. Note NAE-AN-58, NRC 30057, Ottawa, Ont., Feb. 1989, 41 pp.
12. H. K. Bowen, "Basic Research Needs on High Temperature Ceramics for Energy Applications," *Mater. Sci. Eng.,* vol. 44, pp. 1–56, 1980.
13. J. B. Iskee and F. F. Lange, in R. A. Fulrath and J. M. Pask (Eds.), *Ceramic Microstructures '76,* Westview Press, Boulder, Colo., 1977, 669 pp.
14. R. Raj, "Analysis of the Sintering Pressure," *J. Am Ceram. Soc.,* vol. 70, pp. C-210–C-211, 1987.
15. S. T. Lin and R. M. German, "Compressive Stress for Large-Pore Removal in Sintering," *J. Am. Ceram. Soc.,* vol. 71, pp. C-432–C-433, 1988.
16. G. W. Scherer, "Sintering with Rigid Inclusions," *J. Am. Ceram. Soc.,* vol. 70, pp. 719–725, 1987.
17. D. H. Stutz, S. Prochazka, and J. Lorenz, "Sintering and Microstructure Formation of β-Silicon Carbide," *J. Am. Ceram. Soc.,* vol. 68, pp. 479–482, 1985.
18. Y. Tajima and W. D. Kingery, "Grain-Boundary Segregation in Aluminum-Doped Silicon Carbide," *J. Mater. Sci.,* vol. 17, pp. 2289–2297, 1982.
19. W. Boecker, H. Landferman, and H. Hausner, "Sintering of Alpha Silicon Carbide with Additions of Aluminum," *Powder Met. Intl.,* vol. 11, pp. 83–85, 1979.
20. R. A. Cutler and T. B. Jackson, "Liquid Phase Sintered Silicon Carbide," in *Proc. Intl. Gas Turbine Ceramics Conf.,* Las Vegas, Nev., 1989.
21. R. H. J. Hannink et al., *J. Mater. Sci.,* vol. 23, pp. 2093–2101, 1988.
22. R. Moussa, J. L. Chermant, and F. Osterstock, "Creep and Creep Rupture of HP-SiC Containing an Amorphous Intergranular Phase," in R. E. Tressler and R. C. Bradt (Eds.), *Deformation of Ceramic Materials II,* Mater. Sci. Res. Series, vol. 18, Plenum Press, New York, 1984, pp. 617–629.
23. Y. Murata and R. H. Smoak, "Densification of Silicon Carbide by the Addition of BN, BP, and B_4C, and their Correlation to Their Solid Solubilities," in *Proc. Intl. Symp. of Factors in Densification and Sintering of Oxide and Non-Oxide Ceramics,* Hakone, Japan, 1978, pp. 382–399.
24. F. F. Lange, "Powder Processing Science and Technology for Increased Reliability," *J. Am. Ceram. Soc.,* vol. 72, pp. 3–15, 1989.
25. H. Endo, M. Ueki, and H. Kubo, "Hot Pressing of SiC-TiC Composites," *J. Mater. Sci.,* vol. 25, pp. 2503–2506, June 1990.
26. E. R. Maddrell, "Pressureless Sintering of Silicon Carbide," *J. Mater. Sci. Lett.,* vol. 6, pp. 486–488, 1987.
27. J. Hojo et al., "Effect of Chemical Composition on the Sinterability of Ultrafine SiC Powders," *Comm. Am. Ceram. Soc.,* p. C-114, 1983.
28. S. Prochazka, C. A. Johnson, and R. A. Giddings, "Atmosphere Effects in Sintering of Silicon Carbide," in *Proc. Intl. Symp. of Factors in Densification and Sintering of Oxide and Non-Oxide Ceramics,* Hakone, Japan, 1978, pp. 366–381.
29. V. Venkateswaran, The Carborundum Co., personal communication, 1985.
30. R. A. Alliegro, L. B. Coffin, and J. R. Tinkepaugh, *J. Am. Ceram. Soc.,* vol. 39, p. 385, 1956.

31. Y. Takeda and K. Nakamura, "Effects of Additives on Microstructure and Strength of Dense Silicon Carbide," in *Proc. 23d Japan Cong. on Materials Research—Non-Metallic Materials,* 1980, pp. 215–219.
32. E. Gugel, "Net-Shape Processing of Non-Oxide Ceramics," in *Proc. AGARD Conf.,* Florence, Italy, Sept. 1978, pp. 14-1–14-16.
33. A. Krauth and K. Berroth, "Engineering Ceramics for Industrial Applications; Wear, Heat, and Automotive Technology," ASME 84-GT-260, ASME Meeting, Amsterdam, June 4, 1984, 7 pp.
34. J. Wachtman, "Moving Toward Tougher Ceramics," *AM&P,* vol. 133, pp. 45–46, Jan. 1988.
35. D. Beruto, R. Botter, and A. W. Searcy, "Influence of Temperature Gradients on Sintering; Experimental Tests of a Theory," *J. Am. Ceram. Soc.,* vol. 72, pp. 232–235, 1989.
36. S. K. Samanta, H. Wada, and L. Wang, "Pressure Sintering of Si_3N_4 System for Near-Net Shape Forming," *J. Eng. Mater. Technol., Trans. ASME,* vol. 111, pp. 176–182, Apr. 1989.
37. D. J. Kim, P. Greil, and G. Petzow, "Formation and Characteristics of Silicon Nitride-Lithium Aluminum Silicate Ceramics; Part I, Sintering and Microstructure," *Adv. Ceram. Mater.,* vol. 2, pp. 817–821, Oct. 1987.
38. M. Omori and H. Takei, "Preparation of Pressureless-Sintered $SiC-Y_2O_3-Al_2O_3$," *J. Mater. Sci.,* vol. 23, pp. 3744–3749, Oct. 1988.
39. E. Kostic, "Sintering of Silicon Carbide in the Presence of Oxide Additives," *Powder Metall. Intl.,* vol. 20, no. 6, pp. 28–29, 1988.
40. Y. W. Kim and J. G. Lee, "Pressureless Sintering of Alumina-Titanium Carbide Composites," *J. Am. Ceram. Soc.,* vol. 72, pp. 1333–1337, Aug. 1989.
41. M. E. Millberg, "Ceramics for Cars," *Chem Tech.,* no. 9, pp. 552–558, Sept. 1987.
42. R. A. Cutler and T. B. Jackson, "Liquid Phase Sintered Silicon Carbide," in *Proc. 3d Intl. Symp. ACS,* Las Vegas, Nev., Nov. 27–30, 1988, pp. 309–318.
43. P. Kennedy and B. North, "Microstructure and Properties of an Ultra-Fine Grained Reaction-Bonded β-Silicon Carbide," in *Special Ceramics,* vol. 7, *Proc. Brit. Ceram. Soc.,* no. 31, pp. 9–17, 1981.
44. D. R. Messier and P. Wong, "Kinetics of Nitridation of Silicon Compacts," *J. Am. Ceram. Soc.,* vol. 56, pp. 480–485, 1973.
45. J. A. Mangels, "Effect of Rate-Controlled Nitriding and Nitriding Atmospheres on the Formation of Reaction-Bonded Si_3N_4," *Ceram. Bull.,* vol. 60, pp. 613–617, 1981.
46. H. H. Gasthuber et al., "Hot Isostatically Pressed Reaction-Bonded Silicon Nitride Prechambers for the Diesel Engine," *Ceram. Bull.,* vol. 68, pp. 2104–2108, Dec. 1989.
47. K. Katayama et al., "Development of Nissan High Response Ceramic Turbocharger Rotor," SAE Paper 861128, 1986.
48. K. Yabuta et al., "Development of Gas Pressure Sintered Silicon Nitride and Its Automotive and Industrial Application," SAE Paper 890423, presented at the Ind. Cong. and Exposition, Detroit, Mich., 1989.
49. S. J. L. Kang et al., "Elimination of Large Pores during Gas-Pressure Sintering of β-Sialon," *J. Am. Ceram. Soc.,* vol. 72, pp. 1166–1169, July 1989.
50. H. U. Kessel, H. Kolaska, and K. Dreyer, "Manufacture and Properties of Gas-Pressure Sintered Zirconia," *Powder Metall. Intl.,* vol. 20, pp. 35–39, Oct. 1988.
51. S. Ohmura and Y. Miyamoto, "Gas-Pressure Combustion Sintering of TiB_2 Based Composites," in *Proc. 3d Intl. Symp. ACS,* Las Vegas, Nev., Nov. 27–30, 1988, pp. 247–259.
52. L. M. Sheppard, "Expanding HIP's Horizons," *AM&P,* vol. 1, pp. 37–41, Oct. 1985.
53. L. M. Sheppard, "Predicting HIP's Future," *AM&P,* vol. 2, pp. 24–30, Mar. 1986.

54. E. M. Lenoe, "Survey of Hot Isostatically Pressed Ceramics," *ONFRE Sci. Inform. Bull.*, vol. 13, no. 2, pp. 33–63, Apr.–June 1988.
55. W. Symons and S. C. Danforth, "Hot Isostatic Pressing of Laser Synthesized Silicon Nitride Powder," in *Proc. 3d Intl. Symp. ACS*, Las Vegas, Nev., Nov. 27–30, 1988, pp. 67–75.
56. H. T. Larker, "HIP of Ceramics; Technology and Applications Today," in T. Garvare (Ed.), *Proc. of Hot Isostatic Pressing—Theories and Applications*, Centek Publ., 1987, pp. 19–26.
57. H. Bumm, F. Thummler, and P. Weimar, "Isostatic Hot-Pressing; A New Technique in Powder Metallurgy," *Ber. Deut. Keram. Ges.*, vol. 45, pp. 406–412, 1969.
58. N. Uchida, M. Koizumi, and M. Shimada, "Fabrication of Si_3N_4 Ceramics with Metal Nitride Additives by Hot Isostatic Pressing," *J. Am. Ceram. Soc.*, vol. 68, pp. C-38–C-40, 1985.
59. K. H. Hardtl, "Gas Isostatic Hot Pressing without Molds," *Ceram. Bull.*, vol. 54, pp. 201–207, 1975.
60. M. Koizumi and Y. Miyamoto, "Sintering and HIP," in *Ceramics Databook*, Gordon & Breach, 1987, pp. 190–200.
61. K. Ishizaki, "Thermodynamics for HIP'ing of Ceramics—Proposed HIP Phase Diagrams," in *Proc. 2d Intl. Conf. on Hot Isostatic Pressing—Theory and Applications*, Gaithersburg, Md., 1989.
62. T. Kito et al., "Influence of the Pressurizing Temperature on the Densification during Post-HIP'ing of Silicon Nitride," in *Proc. 2d Intl. Conf. on Hot Isostatic Pressing—Theory and Applications*, Gaithersburg, Md., 1989.
63. K. Homma et al., "HIP Treatment on Non-Oxide Ceramics," in *Proc. 20th Japan Cong. on Materials Research*, Soc. of Materials Science, Japan, 1982, pp. 213–217.
64. S. Das Gupta, Electroceramics, Toronto, Ont., personal communication, 1990.
65. P. K. Mehrotra, "Hot Isostatic Pressing of Ceramic Metalcutting Tools," *Met. Prog.*, pp. 506–509, July/Aug. 1987.
66. W. A. Sanders, "Strength of Hot Isostatically Pressed and Sintered Reaction Bonded Silicon Nitrides Containing Y_2O_3," NASA TM 101433, Jan. 1989, 10 pp.
67. M. O. Teneyck, R. W. Ohnsorg, and L. E. Groseclose, "Hot Isostatic Pressing of Sintered Alpha Silicon Carbide Turbine Components," *Trans. ASME*, vol. 109, pp. 290–297, July 1987.
68. O. Yeheskel, Y. Gefen, and M. Talianker, *Mater. Sci. Eng.*, vol. 78, pp. 209–216, 1986.
69. H. Okada, "Fabrication of Dense Si_3N_4 by Hot Isostatic Pressing," in P. Vincenzini (Ed.), *High Tech Ceramics*, Materials Science Monographs 38A, Elsevier, Amsterdam, 1987, pp. 1023–1032.
70. L. Pejryd, *Adv. Ceram. Mater.*, vol. 3, pp. 403–405, 1988.
71. J. Heinrich, E. Backer, and M. Bohmer, *J. Am. Ceram. Soc.*, vol. 71, pp. C-28–C-31, 1988.
72. W. L. Kovacs, "Advances in Sinterhipping," in *Proc. Advances in Ceramics '89*, Philadelphia, Penn., Feb. 20–22, 1989, 10 pp.
73. M. C. Mecray, "Advanced Ceramic Consolidation by HIP Processing," *Ind. Heating*, pp. 15–17, June 1988.
74. H. Kolaska, K. Dreyer, and G. Schaal, "Use of the Combined Sintering HIP Process in the Production of Hardmetals and Ceramics," *Powder Metall. Intl.*, vol. 21, pp. 22–28, Apr. 1989.
75. M. Lee and M. P. Borom, "Rapid Rate Sintering of Al_2O_3–TiC Composites for Cutting-Tool Applications," *Adv. Ceram. Mater.*, vol. 3, pp. 38–44, Jan. 1988.
76. I. Iturriza, J. Echeberria, and F. Castro, "Sinter-HIP of Silicon Nitride Ceramics with

Oxide Additions," in *Proc. Intl., Conf. on Hot Isostatic Pressing of Materials*, Royal Flemish Soc. of Eng., Antwerp, 1988, pp. 5.27–5.33.
77. G. Greskovich, *J. Am. Ceram. Soc.*, vol. 64, pp. 725–730, 1981.
78. E. Tani et al., *Am. Ceram. Soc. Bull.*, vol. 65, pp. 1311–1315, 1986.
79. "Critical Advances in Materials and Processing," *M.I.T. Press J.*, M.I.T., Cambridge, Mass., pp. 3–4, 1986–1987.
80. C. E. G. Bennett, N. A. McKinnon, and L. S. Williams, *Nature, Appl. Sci.*, vol. 217, pp. 1287–1288, 1968.
81. L. G. Cordone and W. E. Martinsen, *J. Am. Ceram. Soc.*, vol. 65, p. C-380, 1972.
82. D. L. Johnson and R. R. Rizzo, "Plasma Sintering of β"-Alumina," *Am. Ceram. Soc. Bull.*, vol. 59, pp. 467–472, 1980.
83. K. Upadhya, "An Innovative Technique for Plasma Processing of Ceramics and Composite Materials," *Ceram. Bull.*, vol. 67, pp. 1691–1694, 1988.
84. J. Kim and D. L. Johnson, "Plasma Sintering of Alumina," *Bull. Am. Ceram. Soc.*, vol. 62, pp. 620–622, 1983.
85. D. L. Johnson et al., "Sintering of α-Al_2O_3 in Gas Plasmas," in W. D. Kingery (Ed.), *Advances in Ceramics*, vol. 10, American Ceramic Soc., Westerville, Ohio, 1984, pp. 656–665.
86. M. E. Brodwin and D. L. Johnson, "Microwave Sintering of Ceramics," in *Dig. IEEE MTT Intl. Microwave Symp.*, vol. 1, New York, May 1988, pp. 287–288.
87. E. L. Kemer and D. L. Johnson, *Am. Ceram. Soc. Bull.*, vol. 64, pp. 1132–1136, 1985.
88. P. C. Kong, Y. C. Lau, and E. Pfender, in G. L. Messing and H. Hausner (Eds.), *Ceramic Transactions, Ceramic Powder Science II*, American Ceramic Soc., Westerville, Ohio, 1988, pp. 939–946.
89. M. Bengisu and O. Inal, "Plasma Sintering of Explosive Shock Activated AlO-Based Ceramic Composites," *Ind. Heating*, pp. 33–36, Dec. 1989.
90. *World Report on Advanced Ceramics*, pp. 1, 2, 5, Sept. 1989.
91. K. R. Venkatachari and R. Raj, "Enhancement of Strength through Sinter Forging," *J. Am. Ceram. Soc.*, vol. 70, pp. 314–320, 1987.
92. P. C. Panda and E. R. Seydel, "Near-Net-Shape Forming of Magnesia-Alumina Spinel/Silicon Carbide Fiber Composites," *Ceram. Bull.*, vol. 65, pp. 338–341, 1986.
93. R. R. Irving, "Near Net Shapes; Closing in on Materials Waste," *Iron Age*, pp. 31–46, Nov. 5, 1984.
94. L. M. Sheppard, "Manufacturing Ceramics with Microwaves; The Potential for Economical Production," *Ceram. Bull.*, vol. 67, pp. 1656–1661, 1988.
95. B. Swain, "Microwave Sintering of Ceramics," *AM&P*, vol. 3, pp. 76–82, Sept. 1988.
96. M. E. Brodwin and D. L. Johnson, "Applicators for Microwave Sintering of Ceramics," *Sprechsaal*, vol. 122, pp. 1152–1153, 1989.
97. A. J. Klein, "Processing with Microwaves," *AM&P*, vol. 1, pp. 36–39, Dec. 1985.
98. J. D. Katz et al., "Microwave Sintering of Boron Carbide," in W. H. Sutton, M. H. Brooks, and I. J. Chabinski (Eds.), *Microwave Processing of Materials*, vol. 124, Materials Research Soc., Pittsburgh, Penn., 1988, pp. 219–226.
99. W. Sutton, "Microwave Processing of Ceramic Materials," *Ceram. Bull.*, vol. 68, pp. 376–386, Feb. 1989.
100. M. A. Janney and H. D. Kimrey, "Microwave Sintering of Alumina at 28 GHz," in *Advances in Ceramics*, vol. 23, American Ceramic Soc., Westerville, Ohio, 1988.
101. H. D. Kimrey, M. A. Janney, and P. F. Becher, "Techniques for Ceramic Sintering Using Microwave Energy," in *Proc. 12th Intl. Conf. on Infrared and Millimeter Waves*, 1987.
102. M. A. Janney and H. D. Kimrey, "Microstructure Evolution in Microwave Sintered

Alumina," in *Sintering of Advanced Ceramics,* American Ceramic Soc., Westerville, Ohio, 1989.
103. *World Reports on Advanced Ceramics,* vol. 1, p. 8, Mar. 1989.
104. B. P. Barnsley, "Microwave Processing of Materials," *Met. Mater.,* vol. 5, pp. 633–636, 1989.
105. Z. A. Munir, "Synthesis of High Temperature Materials by Self-Propagating Combustion Methods," *Ceram. Bull.,* vol. 67, pp. 342–349, 1988.
106. S. Adachi et al., "Fabrication of Titanium Carbide Ceramics by High-Pressure Self-Combustion Sintering of Titanium Powder and Carbon Fiber," *J. Am. Ceram. Soc.,* vol. 72, pp. 805–809, May 1989.
107. J. W. McCauley et al., "Simultaneous Preparation and Self-Sintering of Materials in the System Ti-B-C," *Ceram. Sci. Proc.,* vol. 3, pp. 538–554, 1982.
108. R. R. Irving, "SHS Ceramic Process R&D Heightening," *Metalwork. News,* pp. 18–20, Mar. 6, 1989.
109. M. A. Riley and A. Nüller, "Low Pressure Compaction of SHS (Self-Propagating High-Temperature Synthesis) Prepared Ceramics," Ballistic Res. Lab., Cumberland, Md., BRL MR3574, Mar. 1987, 19 pp.
110. O. Yamada, Y. Miyamoto, and M. Koizumi, "High-Pressure Self-Combustion Sintering of TiC," *J. Am. Ceram. Soc.,* vol. 70, pp. C-206–C-208, 1987.
111. M. Koizumi, "R&D Today, Applications Tomorrow," in *Proc. Intl. Conf. on Hot Isostatic Pressing,* Lulea, Sweden, 1987, pp. 287–296.
112. M. Okutomi and K. Tsukamoto, "Sintering and Growth of New Ceramics Using a High Power CO_2 Laser," *Appl. Phys. Lett.,* vol. 44, pp. 1132–1134, June 15, 1984.
113. W. R. Cannon et al., "Sinterable Ceramic Powders from Laser-Driven Reactions," *J. Am. Ceram. Soc.,* vol. 65, pp. 324–336, 1982.
114. M. Okutomi, "Sintering of New Oxide Ceramics Using a High Power CW CO_2 Laser, Part 1: ZrO_2-Y_2O_3-HfO_2 Ceramics Synthesis with Laser Sintering Method" (in Japanese), *J. High-Temp. Soc. (Japan),* vol. 12, no. 3, pp. 109–120, 1986.
115. M. Okutomi, "Synthesis of New Ceramic Using CO_2 Laser," *Techno Japan,* vol. 20, no. 1, 1987.

BIBLIOGRAPHY

Adlerborn, J., et al., "Development of High Temperature, High Strength Silicon Nitride by Glass Encapsulated Hot Isostatic Pressing," *Mater. Des.,* vol. 8, p. 229, 1987.

Arias, A., "Effect of Oxide Additions and Temperature on Sinterability of Milled Silicon Nitride," NASA TP-1644(N80-21532), Apr. 1980.

Bauer, R. E., "Sinter-HIP Furnaces—Sintering and Compacting in a Combined Cycle," in *Proc. 1988 Intl. P/M Conf.,* Orlando, Fla., June 5–10, 1988, pp. 91–99.

Baumgartner, H. R., "Improved Reaction Sintered Silicon Nitride," NASA CR-135291, Contract NAS3-19723, Mar. 1978.

Bennison, S. J., "Microstructure Control During Sintering of Monolithic Structural Ceramics," presented at the ASM Ann. Mtg., Indianapolis, Ind., Oct. 3, 1989.

Borbidge, W. E., and P. T. Whelan, "Reaction Bonding for Structural and Wear Applications, in C. C. Sorrell and B. Ben-Nissan (Eds.), *CSIRO, AUSTCERAM 88,* Sydney, Australia, Aug. 21–26, 1988, *Materials Science Forum,* vols. 34–36, 1988, pt. 1, pp. 439–443.

Bordia, R. K., and R. Raj, "Hot Isostatic Pressing of Ceramic/Ceramic Composites at Pressures <10 MPa," *Adv. Ceram. Mater.,* vol. 3, pp. 122–126, Mar. 1988.

Bordui, D., "Third Generation Silicon Nitride," TE89-150, presented at the SME Cutting Tools and Applications Conf., Altamonte Springs, Fla., Feb. 15–16, 1989, 10 pp.

Bugajski, W., and M. Nocun, "Perspectives for Obtaining Ceramic Materials by the Sol-Gel Method," *Szklo Ceram.*, vol. 39, no. 3/4, pp. 65–67, 1988.

Cameron, C. P., et al., "A Comparison of Reaction Versus Conventionally Hot Pressed Ceramic Composites," presented at the 14th Ann. Conf. on Composites and Advances in Ceramics, Cocoa Beach, Fla., Jan. 14–17, 1990.

Castro, F., and I. Iturriza, "HIP of Si_3N_4 and Si_3N_4 + 1 wt % Y_2O_3 to Full Density," *J. Mater. Sci. Lett.*, vol. 9, pp. 600–602, May 1990.

Chang-Chun, G., X. Yuan-Luo, and C. Li-Min, "Enhancing Thermo-Mechanical Properties of Si_3N_4 with Non-Toxic Non-Oxide Additive," in P. U. Gummerson and D. A. Gustafson (Eds.), *Modern Developments in Powder Metals,* vol. 21, *Proc. Intl. Powder Metallurgy Conf.*, Orlando, Fla., June 5–10, 1988, pt. 4, pp. 619–625.

Dalton, R. C., I. Ahmad, and D. E. Clark, "Combustion Synthesis of Materials Using Microwave Energy," presented at the 14th Ann. Conf. on Composites and Advances in Ceramics, Cocoa Beach, Fla., Jan. 14–17, 1990.

Das, P., "Effect of Temperatures on the Sintering of Zirconia Based Mixed Oxide Powder Compacts," presented at the Intl. Conf. on Sintering of Multiphase Metal and Ceramic Systems, Intl. Inst. for Sci. of Sinter & P/M Assoc. of India, New Delhi, Vaduz, Liechtenstein, Jan. 31–Feb. 3, 1989.

Dauskardt, R. H., "Transformation Toughening and Its Implication for Crack Growth," presented at the ASM Ann. Mtg., Indianapolis, Ind., Oct. 3, 1988.

De, A. S., I. Ahmad, D. E. Clark, and E. D. Whitney, "Effect of Green Microstructure on Microwave Processing of Alumina," presented at the 14th Ann. Conf. on Composites and Advances in Ceramics, Cocoa Beach, Fla., Jan. 14–17, 1990.

Druschitz, A. P., and J. G. Schroth, "Hot Isostatic Pressing of a Presintered Yttria-Stabilized Zirconia Ceramic," *J. Am. Ceram. Soc.*, vol. 72, pp. 1591–1597, Sept. 1989.

Ekstrom, T., "Sialon Ceramics Sintered with Metal Oxide Additives," presented at the Intl. Conf. on Sintering of Multiphase Metal and Ceramic Systems, Intl. Inst. for Sci. of Sinter & P/M Assoc. of India, New Delhi, Vaduz, Liechtenstein, Jan. 31–Feb. 3, 1989.

Ferrando, W. A., "Processing and Use of Zirconium Based Materials," *Adv. Mater. Manuf. Processes,* vol. 3, pp. 195–231, 1988.

Frisch, A., "Hot Isostatic Pressing—Theories and Applications," *Powder Metall. Intl.,* Practice and Mgmt. Repts., pp. 37–43, Oct. 1989.

Gugel, E., "On the Sintering of Silicon Carbide," NASA-TM-88402, Mar. 1986, 13 pp.

Haggerty, J. S., "Sinterable Ceramic Powders from Laser-Heated Gases," MIT EL 88-001, N00014-82K0350, FR 4/82-5/86, Feb. 1988.

Hodge, E., "Processing on Near-Net Shapes of Ceramics by CIP, Sinter/HIP and HIP," presented at the ASM Ann. Mtg., Indianapolis, Ind., Oct. 3, 1988.

Holcombe, C. E., and N. L. Dykes, "Importance of 'Casketing' for Microwave Sintering of Materials," *J. Mater. Sci. Lett.*, vol. 9, pp. 425–428, Apr. 1990.

Honma, K., "Hot Isostatic Pressing of Ceramics," NASA TM77931, Oct. 1985, 10 pp.

Iturriza, I., et al., "Densification of Silicon Nitride Ceramics under Sinter/HIP Conditions," *J. Mater. Sci.*, vol. 25, pp. 2539–2548, May 1990.

Jahn, P., "Processing of Reaction Bonded Silicon Nitride," presented at the High-Temperature Process Equipment for Advanced Ceramics Mtg., Boston, Mass., Oct. 20, 1989.

Jain, H., O. Parkash, and M. Y. Xu, "Sintering of a Ceramic at Very Low Temperatures," *J. Am. Ceram. Soc.*, vol. 72, pp. 2176–2177, Nov. 1989.

Kim, J. Y., et al., "Analysis of Hot Isostatic Pressing of Presintered Zirconia," *J. Am. Ceram. Soc.*, vol. 73, pp. 1069–1073, 1990.

Larker, H. T., "HIP of Ceramics: Technology and Applications Today," in T. Garvare (Ed.), *Proc. Intl. Conf. Swedish Inst. of Produc. Eng. Res.*, Lulea, Sweden, June 15–16, 1987, pp. 19–26.

Lee, R., et al., "Highly Oriented Fiber Reinforced Ceramic Composites," CPS 89-004, AFOSR TR 89-053, F49620-88C0104, FR 8/88-1/89, Mar. 1989.

Lehman, R., "Hot Pressing of Advanced Ceramics and Composites," presented at the High-Temperature Process Equipment for Advanced Ceramics Mtg., Boston, Mass., Oct. 20, 1989.

Marion, J. E., C. H. Hsueh, and A. G. Evans, "Liquid-Phase Sintering of Ceramics," *J. Am. Ceram. Soc.*, vol. 70, pp. 708–713, Oct. 1987.

Meek, T. T., et al., "Temperature Measurement for Microwave Processing of Advanced Ceramics," in I. A. Aksay, G. L. McVay, and D. R. Ulrich (Eds.), *Proc. Science of Advances in Ceramic Materials Research Soc. Symp.*, San Diego, Calif., Apr. 27–28, 1989, vol. 155, pp. 267–273.

Melzer, D., "Isostatic Pressing—Historical Summary and Present State of Development," *Bergakademie Freiberg, Silikattechnik,* vol. 40, no. 2, pp. 65–68, 1989.

Miyamoto, Y., "New Ceramic Processing Approaches Using Combustion Synthesis under Gas Pressure," *Ceram. Bull.,* vol. 69, pp. 686–690, Apr. 1990.

Mraz, T., et al., "Continuous Production of Fine Pure Refractory Powders through a High Temperature Rotary Kiln," presented at the ASM Ann. Mtg., Indianapolis, Ind., Oct. 3, 1988.

Mukerji, J., "Pressure and Pressureless Sintering of Silicon Nitride with Selected Liquids," presented at the Intl. Conf. on Sintering of Multiphase Metal and Ceramic Systems, Intl. Inst. for Sci. of Sinter & P/M Assoc. of India, New Delhi, Vaduz, Liechtenstein, Jan. 31–Feb. 3, 1989.

"Nonsintering Process for Producing High-Strength Ceramic Materials," JETRO, no. 8, Nov. 1989.

Ovri, J. E. O., and T. J. Davies, "Pressure Sintering and Mechanical Strength of Silicon Nitride," presented at the Intl. Conf. on Sintering of Multiphase Metal and Ceramic Systems, Intl. Inst. for Sci. of Sinter & P/M Assoc. of India, New Delhi, Vaduz, Liechtenstein, Jan. 31–Feb. 3, 1989.

Pechenik, A., A. J. Pyzik, and D. R. Beaman, "Rapid Omnidirectional Compaction of YO Stabilized Tetragonal Zirconia," in I. A. Aksay, G. L. McVay, and D. R. Ulrich (Eds.), *Proc. Science of Advances in Ceramics Materials Research Soc. Symp.,* San Diego, Calif., Apr. 27–28, 1989, vol. 155, pp. 267–273.

Prohaska, S., "Sintering of Silicon Carbide," G.E. Tech. Inform. Series, Rep. 73, CRD325, 1973.

Ritter, J. E., et al., "High-Strength, Reaction-Bonded Silicon Nitride," *J. Am. Ceram. Soc.,* 1988.

Sarkozy, R., "Chemical Vapor Deposition and Chemical Vapor Infiltration Processes," presented at the High-Temperature Process Equipment for Advanced Ceramics Mtg., Boston, Mass., Oct. 20, 1989.

Selkregg, K. R., et al., "Microstructural Characterization of Silicon Nitride Ceramics Processed by Pressureless Sintering, Overpressure Sintering, and Sinter/HIP Cycles," presented at the 14th Ann. Conf. on Composites and Advances in Ceramics, Cocoa Beach, Fla., Jan. 14–17, 1990.

Shaw, N. J., "Densification and Coarsening during Solid State Sintering of Ceramics; A Review of the Models—II: Grain Growth," *Powder Metall. Intl.,* vol. 21, pp. 31–33, Oct. 1989.

Shaw, N. J., "Densification and Coarsening during Solid State Sintering of Ceramics; A Review of the Models—III: Coarsening," *Powder Metall. Intl.,* vol. 21, pp. 25–29, Dec. 1989.

Sheldon, B. W., and J. S. Haggerty, "The Nitridation of High Purity Laser-Synthesized Silicon Powder to Form Reaction-Bonded Silicon Nitride," presented at the 12th Conf. on Composites and Advances in Ceramics (ACS), Cocoa Beach, Fla., Jan. 1988.

Shen, W. M., and R. V. Sara, "Characteristics and Processability of Gas Phase Synthesized and Direct Nitridation Si_3N_4 Powders," *Ceram. Powder Sci. II, Trans.,* vol. 1, pp. 1042–1050, 1988.

Sheppard, L. M., "Spotlight on SiC," *AM&P,* vol. 132, pp. 33–35, Oct. 1987.

Uematsu, K., K. Itakura, and M. Sekiguchi, "Grain Growth during Hot Isostatic Pressing of Presintered Alumina," *J. Am. Ceram. Soc.,* vol. 72, pp. 1239–1240, July 1989.

Weiss, D., "Reaction Hot Pressing," presented at the High-Temperature Process Equipment for Advanced Ceramics Mtg., Boston, Mass., Oct. 20, 1989.

Whalen, T. J., W. Trela, and J. R. Baer, "Injection-Molded Pressureless-Sintered Silicon Carbide; Improvements with Statistically Designed Experiments," presented at the ASM Ann. Mtg., Indianapolis, Ind., Oct. 3, 1988.

Zeng, J., et al., "Hot Isostatic Pressing and High-Temperature Strength of Silicon Nitride-Silica Ceramics," *J. Am. Ceram. Soc.,* vol. 73, pp. 1095–1097, 1990.

CHAPTER 7
FABRICATION AND MANUFACTURING METHODS

7.1 INTRODUCTION

The previous chapters covered the conventional and unconventional techniques for net-shape forming of ceramic parts. Dry pressing, cold isostatic pressing, and slip casting are some of the techniques used in green-state processing. Free sintering, host isostatic pressing, and hot pressing are processes used for high-temperature consolidation. While these and other processes have been used successfully with structural ceramics, they have not been sufficiently reliable in making new materials such as composites. The development of these and other new techniques is in its infancy. Eventually a combination of them may be practiced to make parts of high quality and unusual microstructures.

7.2 MANUFACTURING METHODS

In the past, because of their hardness, ceramics were produced to final size and shape without machining. However, new applications require precision, which cannot be obtained without machining. The hardness of ceramics makes traditional abrasive machining slow and expensive; complex shapes may be impossible to machine. In order to achieve the finishes for today's and tomorrow's products, various processes are used:

1. Mechanical techniques
    a. Diamond grinding
    b. Cutting
    c. Piercing
2. Chemical techniques
    a. Etching
3. Electrical techniques
    a. Electric discharge
4. Radiation techniques
    a. Laser beams
    b. Electron beams
    c. Ion beams
    d. Sparks

To achieve the enormous economic potential predicted for advanced ceramic products, it is generally agreed that there will have to be breakthroughs in lowering production costs and in improving both reproducibility in manufacturing and reliability in use. Machining is an essential processing step that influences each of these breakthrough requirements. It is needed to produce the required geometry, tolerances, and surface finish not achievable by most forming methods. In addition, machining techniques can be used to eliminate surface flaws, microcracks, and subsurface damage, which can limit the performance of an advanced ceramic part.[1]

To date, relatively little information exists about the effect of machining on the microstructures of advanced ceramic materials. As a result, numerous symposia and conferences have been held on machining advanced ceramics, with the objective of facilitating the development of approaches that integrate materials, design, and process engineering considerations in order to accelerate the commercialization of advanced ceramics.

The average surface roughnesses R_a as achieved by standard machining techniques and by new methods developed specifically for advanced ceramics are listed in Table 7.1. Using new machines, procedures, and tooling, grinding can be adapted to achieve finishes in the superpolishing region.

The relation between machining techniques, surface condition, and component properties has important implications for the performance of a ceramic in use, and it has been demonstrated that improving the surface finish does not necessarily result in greater strength. For example, using a standard metal grinding "spark" technique with ceramics can cause severe residual stress and subsurface damage. These problems can, however, be avoided while very low R_a values are being achieved by lapping or by new micromachining techniques. Finally it should be pointed out that optimum results require a systems approach, involving the careful matching of machine, workpiece material, and abrasive tool.[2-5]

Machining Methods

Until recently the only advances that had been made on the ancient methods were improved abrasives and substitution of power machinery for hand labor, whereas the basic processes of machining by mechanical abrasion remained the same.

The semiconductor industry, which uses glass and ceramics as substrates for devices, required methods that would drill, cut, and finish these thin small ceramic plates with great precision and without damage, and, according to

TABLE 7.1 Average Surface Roughnesses Obtained with Standard and Newly Developed Machining Techniques

Surface roughness R_a, μm	Machining operation
7–90	Milling
0.09–7	Grinding
0.009–0.09	Lapping
<0.009	Superpolishing

Source: From Ref. 2.

Firestone,[6] beckoned to industry for new and improved abrasionless machining methods.

Machining is the application of energy to the workpiece to remove stock by creating new surfaces. The main sources of machining energy are mechanical, chemical, or thermoelectric. Mechanical energy is the oldest source. It is used for traditional abrasive machining (AM), abrasive-jet machining (AJM), and ultrasonic abrasive machining (UAM), as well as for hydrodynamic machining (HDM), which is abrasionless. Chemical energy is used for chemical machining (CM), the oldest abrasionless machining method, and for the hybrid between chemical and thermoelectric energy, electrochemical machining (ECM). All thermoelectric methods are abrasionless and include electric-discharge machining (EDM), electron-beam machining (EBM), laser-beam machining (LBM), and ion-beam machining (IBM)[6] (Fig. 7.1).

Machining methods produce new surfaces by two processes: (1) generation, in which energy from a point source is focused on the workpiece and, by relative movement, generates a surface of the desired configuration, and (2) transfer, in which a pattern is used to configure the energy from a diffuse source. The generation methods work like an end mill on metal, while the transfer methods are like printing. The fundamental machining process, materials limitations, and the environmental requirements of each method are listed in Table 7.2.

Generally the process of abrasive machining is restricted to simple parts that are flat or have axial or radial symmetry. Machining complex or asymmetrical parts may be nearly impossible and prohibitively expensive, and therefore nontraditional abrasive and abrasionless methods free ceramics from this limitation. Hydrodynamic and abrasive-jet machining work like a jigsaw and are best for cutting slots and grooves or trepanning large holes. Ultrasonic abrasive machining can make holes with precise contours by transferring the pattern of the

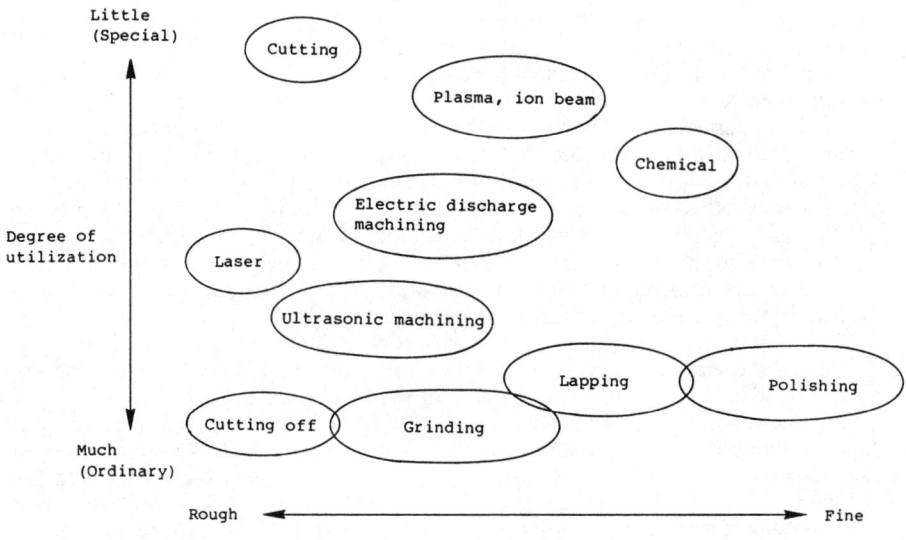

FIGURE 7.1 Degree of utilization of various material-removal processes and machining accuracy.

TABLE 7.2 Machining Methods

Machining method	Process	Material	Environment
Abrasive (AM)	Generation	Any	Ambient
Hydrodynamic (HDM)	Generation	Any	Ambient
Abrasive jet (AJM)	Generation	Any	Ambient
Ultrasonics (UAM)	Transfer	Any	Ambient
Chemical (CM)	Transfer	Glass	Reagent
Electrochemical (ECM)	Transfer	Conductive	Reagent
Electric discharge (EDM)	Generation	Conductive	Dielectric
Electron beam (EBM)	Generation	Any	Vacuum
Laser beam (LBM)	Generation	Any	Ambient
Ion beam (IBM)	Generation or transfer	Any	Vacuum

Source: From Firestone.[6]

tool to the workpiece, whereas chemical machining transfers patterns of small holes and slots well.

Electrochemical machining is faster than chemical machining. However, its use is limited to SiC and a few other conductive ceramics since it requires materials with electrical resistivities of less than 300 $\Omega \cdot$ cm. Electric-discharge machining drills small holes, cuts slots, contours surfaces, and makes pockets, that is, cavities with straight sides and flat bottoms, but it is also limited to the few conductive ceramics. Laser-beam machining is excellent for drilling small holes, while ion-beam machining is very slow, but can produce surfaces of the highest quality. The latter increases the strength of the machined parts over the as-fired strength while other machining processes usually decrease the strength.

The applications and results possible with abrasionless machining methods are listed in Table 7.3. Abrasive-jet machining is also included, although technically an abrasive method, since it is a variation of an abrasionless process.[6] Improvements in abrasionless machining methods, which expand their uses and overcome present limitations, are continually being developed and are discussed later in this chapter.

Hydrodynamic machining uses a high-velocity fluid jet, either alone or augmented with abrasive particles (AJM), to erode material. The fluid is usually plain water or water with a small amount of polymer added to reduce the spread of the jet. The method is often referred to as water-jet machining. There are three types of fluid jets: pulsed, cavitating, and continuous. Since ceramics fracture easily, pulsed jets are convenient for removing ceramics which are bonded to metal. Cavitating jets are used primarily under water, while continuous jets can be used with or without entrained abrasive particles.

The continuous-fluid jet is a nonwearing tool with no frictional drag in the cut, and there are virtually no forces at right angles to the jet. It is a line-cutting device. The kerf is insensitive to the dwell of the fluid jet, although there is a slight enlargement when abrasive particles are in the fluid. There are no heat effects nor large mechanical forces, so that even the most thermal-shock-sensitive, weakest ceramics can be cut without damage. The power requirements are related to the pressure (24 to 680 MPa) and nozzle diameter (0.003 to 0.50 mm). The cutting rate is a complex function of the process and material parameters. In general, the rate increases with increasing pressure and decreasing nozzle diameter, stand-off distance, and depth of cut. The hardness function of the ceramic is not a major fac-

TABLE 7.3 Comparison of Machining Methods*

	HDM	AJM	CM	ECM	EDM	EBM	LBM	IBM
				Application†				
Holes:								
Standard (>0.005 in)	Good	Good	Poor	Poor	Fair	Poor	Fair	Poor
Small (0.005–0.001 in)	—	—	Good	Good	Fair	Good	Good	Fair
Micro (<0.001 in)	—	—	—	—	—	Fair	Good	Good
Shallow (L/D < 20)	Good	Good	Good	Good	Fair	Fair	Fair	Fair
Deep (L/D > 20)	Poor	Fair	—	—	Fair	Poor	Poor	—
Through cutting:								
Shallow (L/D < 20)	Good	Good	Poor	Poor	Good	Good	Good	Poor
Deep (L/D > 20)	Fair	Good	—	—	Good	Poor	Poor	—
Contouring	Poor	Poor	—	—	Fair	—	—	Good
Pocketing	Fair	Fair	Poor	Poor	Good	—	—	Fair
				Results‡				
Speed	Fast	Fast +	Slow	Slow	Fast	Fast +	Fast +	Slow–
Maximum precision	0.010 in	0.020 in	0.001 in	0.001 in	0.0005 in	0.001 in	0.001 in	>0.0001 in
Finish	Poor	Good	Good	Good	Good	Good	Good	Very good
Effect on strength	—	—	+ or –	—	—	—	+ or –	+++

*Acronyms are defined in Table 7.2.
†L/D = length-to-diameter or length-to-depth ratio.
‡– decreases strength; + increases strength, compared to as-fired part.
Source: After Firestone.[6]

tor, but porosity is, and hard, porous ceramics cut faster than soft, dense ceramics. Polymers may be added to reduce the dispersion of the jet and abrasives may be entrained in the jet to increase the cutting rate and depth of cut in dense, hard ceramics. The most common abrasives are quartz and garnet for softer ceramics and SiC and Al_2O_3 for harder ceramics. For very hard ceramics, BC abrasive is used, but the cutting rates are very slow.

The kerf width is a function of the nozzle diameter and the porosity and grain size of the workpiece. For small nozzle diameters and fine-grain, dense ceramics, the kerf may be only 0.50 mm wide. For porous materials, the kerf may be as large as the nozzle diameter plus two grain sizes. The jet spreads as it leaves the nozzle, so the kerf is larger at the bottom than at the top. The spread may be reduced by adding long-chain polymers, such as polyethylene oxide, to the water.

Chemical machining is the controlled solution of the workpiece by a chemical reagent. The areas to be machined are defined by a chemically resistant, removable mask. Since the solution proceeds radially from the openings in the mask, the resulting undercut limits this process to shallow holes and pockets. The process parameters are ceramic material properties, reagent type and concentration, agitation, temperature, and time.

Although there are reagents which will dissolve almost all ceramics, chemical machining, as opposed to chemical polishing, is restricted to glass by the lack of reagents that will dissolve polycrystalline ceramics uniformly at the low temperatures where masking can be used.

Electrochemical machining uses an electric field to supply additional energy to the machining process. It can only be used with carbides or other ceramics with a resistivity of 300 $\Omega \cdot$ cm or less. However, the additional energy allows polycrystalline conductive ceramics to be machined uniformly at room temperature. The process can severely degrade the strength of ceramics and is not used very often.

Electric-discharge machining is the controlled melting and vaporization of the workpiece by multiple electric arcs between workpiece and tool, which causes a melting of the material. There is a thermally affected zone beneath the machined surface, which produces residual stress and cracks. This method is often used to introduce atomically sharp cracks in metal-fatigue specimens. It can be used for ceramics if they are conductive (resistivity of 300 $\Omega \cdot$ cm or less), but is not often used due to the severe degradation in strength which can result.

Electron-beam machining vaporizes material by focusing a beam of electrons onto the workpiece in a vacuum. It is primarily a through-cutting, hole-drilling process. The beam may be positioned electrically at traverse rates of 762 mm/s. Holes may be drilled in thin ceramic substrates at the rate of 12,000 per second. The holes can be cylindrical, tapered, or with irregular cross sections.

Since electron-beam machining is a thermal process, there is a melted zone around the cut, which produces residual stresses and may induce cracking. The strength of thin ceramics (<1 mm) is only slightly decreased, but thicker ceramics may have to be heated to avoid shattering during machining, and if they survive machining, they are severely weakened. Thick ceramics may be machined in the unfired condition; the binder vaporizes during machining, which prevents excessive heating, and firing heals any cracks formed.

Laser-beam machining focuses an intense beam of light from a laser onto the workpiece to vaporize material. It competes directly with electron-beam machining and has supplanted it in many applications. The laser beam is positioned mechanically, and it is therefore not as fast as electron-beam machining. But no vacuum chamber is required, so that workpiece loading is faster and there is no limit on size. In addition, the equipment is less expensive.

Lasers operate in two modes, pulsed and continuous. The pulsed mode is used for hole drilling, while the continuous mode is used for cutting, although it may be used for drilling also. Laser-beam machining is indifferent to the hardness of materials. For example, graphite and diamond are cut with almost equal speed.

The process parameters are beam intensity, traverse speed, and ceramic properties, while the rate of machining is controlled by the rate at which the material, which is melted and vaporized, is removed by the beam. Removal occurs by thermal convection and beam pressure, and the rate may be greatly increased if a gas jet is used to blow away the material.

The thermal gradients around the laser cut may cause cracking and fracture, so that thick ceramics may have to be heated or machined in the unfired state. However, thin laser-machined ceramics may be stronger than diamond-machined ceramics since heating may produce beneficial residual stresses.

Ion-beam machining is the gentlest of all machining processes for ceramics. It uses a stream of ions to sputter material atom by atom from the surface of a ceramic in a high vacuum. Argon is the normal gas used for the beam, but reactive gas beams can be used to produce surface coatings while machining. There is no damage to the ceramic, and the surface is left atomically clean. Two types of ion beams have been used: (1) focused, which operate like electron beams, and (2) diffuse, which use masks and operate like chemical machining. The ion-beam machining rate is determined by a number of complex parameters, but it is always very low. The highest rate to date has been 0.13 mm/h.

The strength of ion-beam-machined ceramics is greater than that of as-fired ceramics due to the complete removal of strength-reducing surface flaws.

In conclusion, abrasionless machining methods have become a useful addition to traditional abrasive machining techniques for ceramics. For the machining of many, thin ceramics, they have completely supplanted abrasive methods.

Abrasive Machining

Most ceramic materials are machined using an abrasive machining principle due primarily to the hardness and brittleness of the ceramic and its abrasive action when machined, causing dulling of conventional metalworking tools.

Flat surfaces of ceramic workpieces are either ground with a revolving grinding wheel, lapped with a free-rolling abrasive, or polished with a fine abrasive. The normal limitations of grinding with a revolving grinding wheel are accuracy, chipping, grinding lines in the surface finish, breakage of brittle components, and work-holding limitations.[7]

Ceramics can be finished using diamond abrasives in a variety of processes. When the ceramic is ground or machined in its unfired state, it is called green machining. Finishing of partially sintered ceramic is called white machining; finishing of fully densified ceramic material is often referred to as diamond grinding or hard machining. Flat surfaces and contours are finished using diamond powders, compounds, or slurries through a process called lapping. Superfinishing of internal surfaces is called honing. Polishing often involves loose or coated abrasives applied to the work surface against a flexible material or backing.

The range of particle sizes available for use in all these operations is fairly extensive. The control of shape, size, quantity, and method of use of the diamond abrasives has a pronounced effect on the chip-generation process and hence the performance results of the finished ceramic component.

According to Indge,[7] the abrasive machining process utilizes a loose abrasive as a cutting tool. It is known by more than one name and, depending on what industry is involved, it has been called lapping, precision polishing, grinding, free abrasive machining, and superfinishing.[3]

Lapping. There are two basic abrasive machining processes, lapping and polishing. Although there may be a number of different explanations of each process, depending on the application, when lapping or polishing ceramic-type materials, certain conditions usually exist. Lapping takes place when abrasive grains in a liquid vehicle, often known as a slurry, are guided across the surface to be lapped and backed up by a lapping plate.

When a relative motion is induced and pressure applied, the sharp edges of the grains are forced into the ceramic material to be lapped and either make an indentation or cause the material to chip away microscopic particles. Even though the abrasive grains are irregular in size and shape, they are used in large quantities, and thus a cutting action takes place continuously over the entire workpiece surface that comes in contact with the abrasive slurry backed by a lapping plate.

The depth of the marks will determine the roughness or smoothness of the surface, which is denoted by R_a or R_t, and is usually measured with a surface analyzer and read in micrometers. The larger and harder the abrasive grains, the rougher the finish. Conversely, the finer the abrasive grains, the smoother the finish.

When lapping sintered-type ceramic materials, it has been found that a rolling abrasive cutting action causes the sintered particles to break away from the bond. There often is a limit to the smoothness that can be obtained by lapping, even when very fine abrasive grains are used.

Polishing. The process to precision-polish ceramic-type materials is based on the basic principle of applying a fine abrasive powder to a polishing tool and, with relative motion between workpiece and tool, improve the surface finish on the workpiece. Softer ceramic materials such as glass are usually polished with a soft abrasive such as cerium oxide, and the polishing tool is a fabric or other soft material such as pitch. In principle, the abrasive powder used for polishing is mixed with a fluid and applied evenly to the polishing tool.

A harder abrasive is embedded into a firmer material, causing a portion of the abrasive particles to protrude from the tool. The abrasive acts as a scraping cutting tool, shearing off microscopic peaks of the surface being polished. This cutting action of multiple abrasive cutting edges produces a randomly polished, reflective surface. The smoothness of the surface will depend on the size and quality of the abrasive and the technique used.

Process Variables. Designers have strived to overcome some of the variables in this type of processing, and perhaps the three most difficult procedures to control are: (1) keeping the wheel flat, (2) applying a uniform and predictable pressure, and (3) applying and maintaining a uniform and consistent flow of abrasive. Due to these variables, lapping and polishing have often been considered an art, and such processes rely a great deal on operator judgment. Considerable strides have been made in today's modern single-wheel lapping and polishing machines to reduce these variables and provide more control.

Grinding. It is generally stated that grinding produces damage in the ceramic parts, and according to Subramanian,[8] it would be more appropriate to state: "Grinding is a chip-generation process. The shape, size, and efficiency of this chip-generation process determine the geometry and quality of the surface produced. The rate and efficiency of production of chips determine the economics of the grinding process."

The most efficient diamond-abrasive finishing method requires a systems approach that addresses four key input aspects of the grinding process: machine tool, wheel selection, work-material properties, and operational factors. Figure 7.2 shows this systems approach to developing an efficient precision-grinding process. Inadequate attention to details in any one of these system input variables can influence the grinding process results.[8]

Taking into account the interaction of these variables, Thomas et al.[9] conducted an investigation to evaluate the effect of grinding, downfeed rates, grit size, and so forth, on the surface finish and strength of hot-pressed Si_3N_4, since this material was commercially available and considered for future applications. Results indicated a decrease in strength resulting from an increased downfeed rate. The strengths were highest for materials machined at 25-μm downfeed rate.

Higher strengths and better surface finish can be obtained by machining with a finer-grit wheel (320 grit instead of 150 grit). While good surface finish is a necessary characteristic of a ceramic, it is not in itself an adequate guarantee of improved mechanical properties of the component.

Nakano et al.[10] report that sialon ceramics are one group of materials which

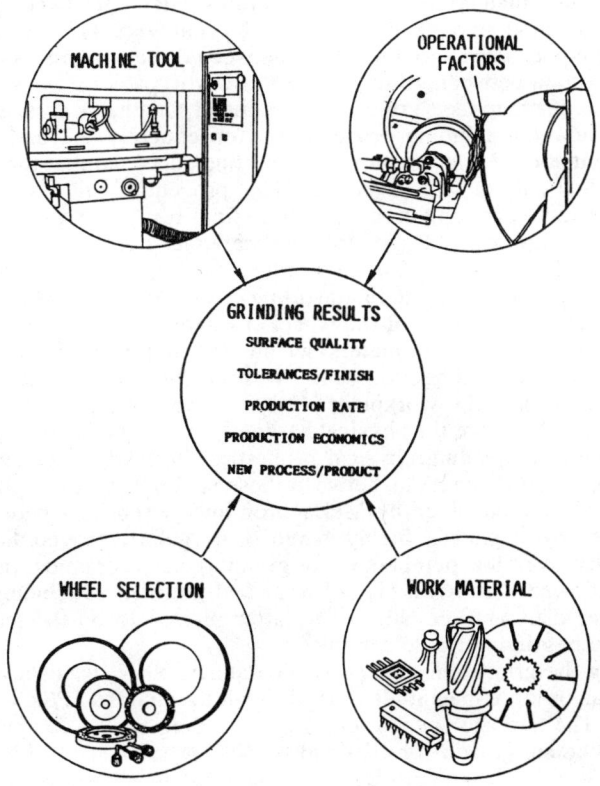

FIGURE 7.2 Grinding-system variables.[8]

are difficult to be cut off by conventional diamond machining because the diamond wheel loses its cutting ability with time due to a pileup of ceramic chips around the diamond grains. A newly developed dressing technology consists of the continuous removal of these chips during the cutting-off process, in which a glass-fiber-reinforced plastic plate is used as a dresser. Using this technique, sialon ceramics can be cut off under stable conditions at higher cutting speeds, for longer cutting times, and with a smaller cutting force than previous no-dresser cutting methods. Also there is less wear of the diamond wheel compared to conventional dressing methods using green carborundum (GC) or white alundum (WA) dressers.[10] Since this continuous-dressing method with the fiber dresser is a new technique, it still involves many unknowns. The optimum combination or matching of workpiece machining conditions and fiber-dresser conditions is considered especially important.

Creep-Feed Grinding. The two most important parameters in the process of surface grinding are infeed and feed rates. The product of these parameters is the specific material-removal rate, which is a measure of the workpiece volume removed per wheel width per second. In simplified terms, the material-removal rate can be described as the workpiece area that is in contact with the grinding wheel within a specific time unit.

In conventional machining, the specific material-removal rate is almost always achieved by using small infeed and high feed rates. High infeed and low feed rates also enable the same amount of material to be removed. This process is called creep-feed grinding. The major difference between these two processes is the size of the contact area between grinding wheel and workpiece.

In the case of creep-feed grinding, the contact length, or area, is essentially larger than for other grinding procedures. Consequently the material-removal rate is distributed over a larger number of cutting edges, so that the depth of cut of the individual cutting edge is smaller. This procedure will have considerable effect on both the machining process and the final results.[11,12]

In grinding complex profiles in structural ceramics it is common to use creep-feed grinding. This process offers significant advantages over conventional surface grinding in productivity and final workpiece quality.[13–15] Creep-feed grinding usually involves a single pass of the wheel at relatively low table speeds (centimeters per minute rather than meters per minute) and high infeeds. Ultimately, the efficiency of any grinding process lies with the dynamic interactions between the abrasive grain and the workpiece. Only two abrasives, diamond and cubic boron nitride (CBN), have the physical hardness necessary to grind ceramics. As a result of their outstanding physical properties, both are often referred to as superabrasives. The interactions between these superabrasives in grinding Si_3N_4 and PSZ have been examined by McEachron and Lorence,[13] who showed that diamond A, a low-toughness, highly included, irregularly shaped diamond abrasive, is the best overall superabrasive for grinding these ceramics. In both cases, diamond A showed the lowest crystal wear and the highest grinding ratio compared to other diamonds or CBN. The latter proved to be the least effective superabrasive in grinding these ceramics.

Comparing the grindability of the two ceramics, Si_3N_4 was shown to be approximately an order of magnitude more difficult to grind than PSZ with regard to wheel wear. The average specific energy in grinding Si_3N_4, compared to PSZ, was only 80 percent greater for diamond A.[13]

Conclusions. The technology to produce flat surfaces by abrasive machinery with a loose rolling abrasive has evolved from basic, simple hand lapping to simple machine lapping and polishing. This labor-intensive process has been challenged over the years with the advance of precision surface-grinding equipment. Grinders have been able to remove stock at a faster rate, utilize higher horsepower, process more workpieces per load, and achieve acceptable accuracies. As demands for closer tolerances become greater and materials become more exotic, grinding processes are now being challenged by sophisticated precision lapping and polishing machines. According to Indge,[7] when products require sizes measured in increments of 10 millionths of an inch (0.25 μm), controlled lapping and polishing processes are the recognized methods of production machining.[16]

Abrasive machining, which consists chiefly of cutting, grinding, lapping, and polishing, accounts for more than 70 percent of the total removal work of ceramics (Table 7.4).[17,18] It is, however, accompanied by over 86 percent of all problems resulting from such ceramic processing. This shows that while in the machining of ceramics, abrasive machining is the most basic and popular process, it still has many problems to overcome. The main problems are machining efficiency, machining expenses, tool cost, and machining accuracy.

A possible future trend for ceramics processing is toward a higher degree of ultraprecision machining, ultrafine processing, an increase in work size, further complexity in shape, and integration with different kinds of materials. For these purposes, efforts need to be focused in the directions shown in Fig. 7.3.[19]

Drilling. In spite of the popular conception that ceramic and glassy materials are difficult to machine, Wilson et al.[20] have shown that Bioglass* and Cervital* materials can be drilled and shaped to smooth final finishes and function well as implants.

Although the experiments of Wilson et al. have concentrated on the shaping of a middle-ear prosthetic device, many other applications of the Bioglass material are envisaged, particularly in orthopedic, dental, and maxillofacial applications, and their initial findings with the ear will have relevance in these other areas too.

Ricci[21] evaluated the drilling of metal-matrix composites of fiber-reinforced SiC_f–6061-T4 Al, particulate-reinforced SiC–7079-T6 Al, and B_4C_p–AZ61A–Mg with polycrystalline-diamond-tipped twist and spade drills, diamond-plated twist and core drills, and the abrasive water-jet hole-cutting process. The diamond-tipped twist drills outperformed all the other drills. Core drills were found to be viable alternatives for the production of larger holes in high-volume fraction composites. Plated twist drills were viable alternatives for low-volume fraction particulate composites. Spade drills failed due to low edge strength. Abrasive water-jet hole cutting was successful for rough, large-diameter holes.

The failure of diamond-coated tools used on metal-matrix composites was found to be due to diamond glazing by the hard and abrasive reinforcement material and loading by the soft metallic matrix. It was determined that the machinability and the cutting rate of metal-matrix composites can be predicted by using the rule of mixtures and the machinability data for the individual components of a composite. High-volume-fraction reinforcement composites were found to necessitate techniques and tooling similar to those for diamond grinding. In contrast, low-volume-fraction reinforcement composites required tool geometries similar to those used for workpiece materials which deform plastically and readily form chips.

The following drilling parameters were recommended at the conclusion of the study. Drilling speeds and penetration rates for 30 vol % SiC–Al and 45 vol % SiC_f–Al composites were 64 and 38 mm/min, respectively, for 12.7-mm-diameter

---

*Bioglass and Cervital are registered trademarks of the University of Florida, Gainesville.

TABLE 7.4 Machinability of Sintered Ceramics with Diamond Tools

Material	Sawing	Turning	Milling	Grinding	Drilling
Hot-pressed Si_3N_4	Yes	No	No	Yes	Yes
HIP Si_3N_4	Yes	No	No	Yes	Yes
Liquid-sintered Si_3N_4	Yes	No	No	Yes	Yes
Hot-pressed SiC	Yes	No	No	Yes	Yes
HIP SiC	Yes	No	No	Yes	Yes
Reaction-sintered SiC	Yes	No	No	Yes	Yes
Liquid-sintered SiC	Yes	Possibly	Possibly	Yes	Yes
Hot-pressed B_4C	Yes	No	No	Yes	Yes
Hot-pressed TiB_2	Yes	No	No	Yes	Yes
Hot-pressed BN	Yes	Yes	Yes	Yes	Yes
TiB_2–BN composite	Yes	Yes	Yes	Yes	Yes
BeO	Yes	Yes	Yes	Yes	Yes
Al_2O_3	Yes	Possibly	Possibly	Yes	Yes
MgO	Yes	Possibly	Possibly	Yes	Yes
ZrO_2	Yes	Yes	Yes	Yes	Yes
Mica-based glass ceramics	Easy	Easy	Easy	Easy	Easy
Al_2O_3–ZrO_2–SiC whiskers	Relatively easy	Relatively easy	Relatively easy	Relatively easy	Relatively easy

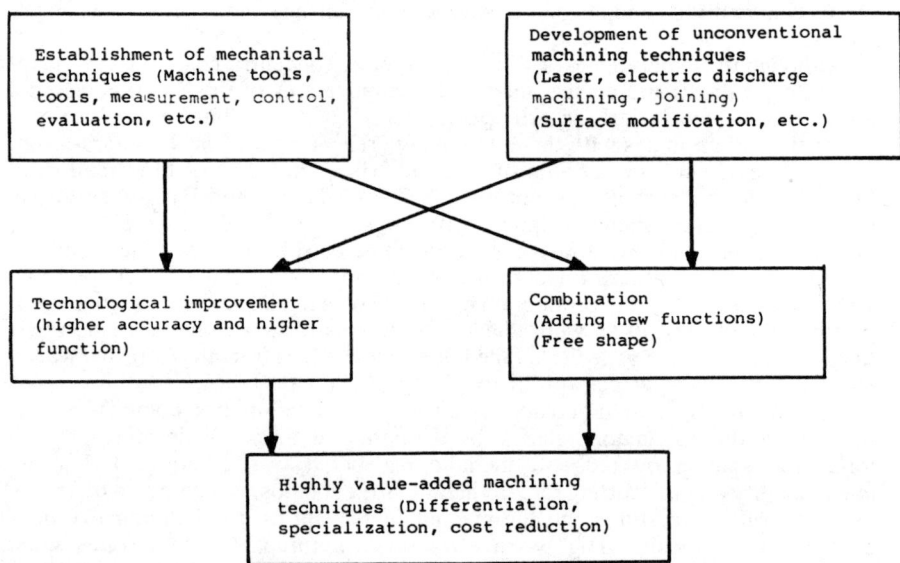

FIGURE 7.3 Trends of new-materials machining techniques.[19]

polycrystalline-diamond-tipped, negative-rake twist drills. The initial cost of the tipped drills is, however, extremely high.

Diamond-core drills were noted by Ricci as requiring frequent dressing, while low speeds and high feeds were recommended for these drills. Intermittent flushing of the cut area was required for high-volume-fraction reinforcement composites.

Diamond-plated twist drills were found to be unsuccessful when applied to high-volume-fraction fiber-reinforced composites. Unlike the other drills tested, plated twist drills may not be sharpened. However, their cost is only 10 percent that of a diamond-tipped drill.

Duran[22] has developed an off-the-shelf ultrasonic drill which has been fitted with a new device that feeds slurry to the cutting tip, allowing the drill to cut advanced ceramics and other hard materials faster and deeper. Currently ultrasonic drills can cut 1.58 mm deep into a hard ceramic before the cutting rate drops off asymptotically, and they quit altogether after about 8.38 mm. This glandlike device continuously feeds a thick slurry to the drill's diamond-tipped head without disrupting the ultrasonic action, which is the key to the operation.

Ultrasonic drills, which rotate at between 2000 and 3000 rev/min, vibrate at about 20,000 Hz with an amplitude on the order of 0.03 mm. The drill attachment feeds slurry under pressure to the bottom of the cut. The liquid slurry, which contain B_4C abrasive particles, enhances the cut while cooling the cutting area and carrying away debris.

Another rotary ultrasonic machine was devised for drilling holes in ceramics and is being produced by two Japanese companies, Kuroda Precision Industries Ltd. and Ultrasonic Engineering Company Ltd. The machine has a main spindle speed of 1100 rev/min when drilling holes up to 10 mm in diameter and 21 mm deep, and its mean feed speed is 7 mm/min. Its machining surface accuracy is within 1 μm, its roundness within 0.01 mm, and its cylindricity within 0.05 mm.

The unit, according to these companies, can be fitted to existing machine tools, reducing costs to one-fifth that for buying a new purpose-built device. The unit works by directing a 21,000-Hz radio-frequency signal at the tip of a diamond grinding wheel to generate vibrations. These vibrations reduce the grinding resistance by one-third to one-half, thus increasing the machining speed.

Cracks and a recast layer can form when pulsed lasers drill holes in ceramics. Those can be serious defects in some ceramic components, and the need to avoid them limits the range of laser machining applications. One way around the problem is to submerge the part being machined in water.

Morita et al.[23] report that their technique, utilizing a 1.06-μm ND-YAG laser, can be used to successfully scribe, cut, and drill without damaging the inherent characteristics of the ceramic. They have drilled defect-free holes in submerged Si_3N_4 with 100-ns Q-switched pulses at a 1-kHz repetition rate. Power densities in the experiments were below 71 MW/cm^2. Both a limited repetition rate and a short pulse duration were essential to avoiding recast layers. Flaws appeared when the repetition rate was raised to 10 kHz, or when 1-ms-long pulses were used.

They attribute their success to sublimation of Si_3N_4 into silicon and nitrogen. If all the material sublimes, no melt can form to deposit a recast layer. From the results they infer that Q-switched pulses at modest repetition rates keep silicon pressure from reaching the saturation level. At pressures above the saturation level, the molten silicon would appear on the processing surface.

Supporting the conclusion, they noted a similar effect for AlN ceramics, which sublimate much like Si_3N_4. On the other hand, Al_2O_3 ceramics cannot be processed without creating a recast layer because they melt.

Milling. A protective silica material that can withstand 1371°C has been machined into tiles of varying shapes, sizes, and thicknesses. These tiles have been used to protect more than 70 percent of the Space Shuttle's exterior surface. According to Vaughn,[24] there are 30,000 tiles used in the Space Shuttle, and each differs dimensionally. Therefore, an automated computer-aided manufacturing process was devised, developed, designed, and implemented.

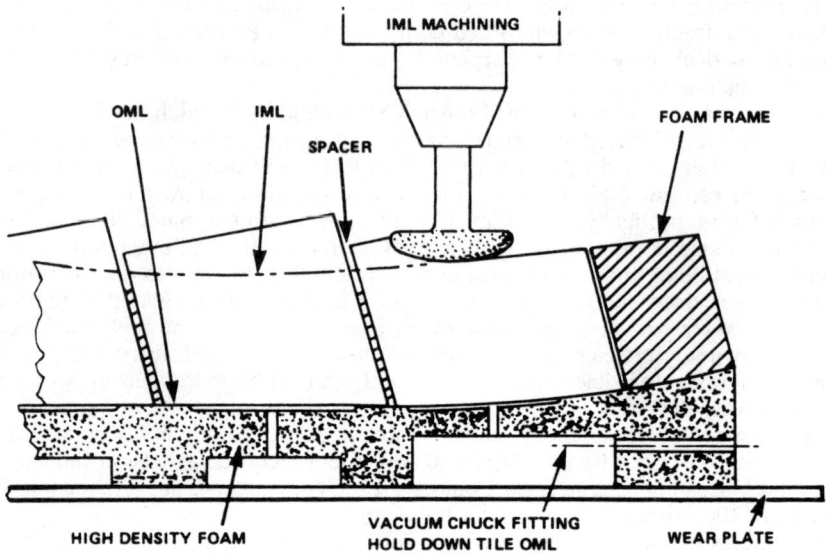

FIGURE 7.4 Tiles held by vacuum and milled with specially made diamond tools.[24]

The machining operation selected was similar to a grinding operation in that it used an abrasive-coated cutter. According to Vaughn, these cutters (Fig. 7.4) were developed by Lockheed out of necessity since commercially available diamond-coated wheels were found to be unsatisfactory for close tolerance work. Testing to determine optimum spindle speeds and feeds with these cutters showed that both could be as high as the machine tool was capable of producing. A 25.4-mm cutter was run at 14,000 rev/min, full width, at 127-mm depth at a rate of advance of 3048 mm/min. An 8000-rev/min spindle speed was selected to optimize cutter wear and tile finish. Currently about 43 different cutter sizes and shapes are being used to generate the tile configurations.

Hydrodynamic or Abrasive-Jet Machining

The process uses a high-velocity fluid jet, either alone or with abrasive particles, to erode material. The jet can be used in the pulsed, cavitating, or continuous modes, the latter for most ceramic applications. This type of machining is best for cutting slots and grooves or trepanning large holes.

Figure 7.5 is a schematic of an abrasive water-jet nozzle, illustrating its basic components according to Hashish.[25,26] The independent parameters involved in an abrasive water-jet machining operation can be divided into two general groups. The first group is related to the abrasive water jet itself and includes the hydraulic, abrasive, and mixing parameters, while the second group is related to the specific machining applications, such as slotting, turning, milling,[27] and drilling.

Sheppard[28] reports that the University of Rhode Island uses a high-pressure abrasive water jet for machining 85, 94.5, and 99.9 percent Al_2O_3. The system consists of a robotically controlled water jet that can handle pressures up to 207 MPa with feed rates of 0.11 to 1.35 kg/min. Piercing experiments studied the effect of water pressure and abrasive flow rate to eliminate the effect of nozzle traverse speed.

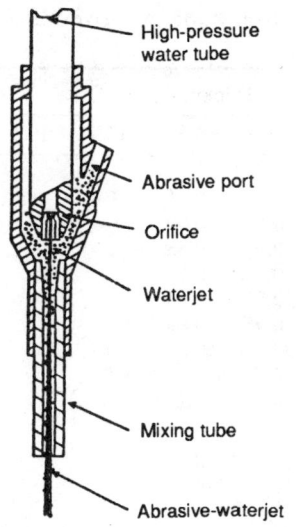

FIGURE 7.5 Abrasive water-jet nozzle components.[26]

The piercing time decreases substantially as the water pressure increases. Garnet abrasive is as effective as Al_2O_3 or SiC abrasive for lower-strength ceramics. However, for high-strength ceramics, the flow rate of SiC abrasive is eight times faster than that of garnet. For hard ceramics, zero abrasive flow rate produces no penetration. The piercing or cutting efficiency generally increases as the abrasive flow rate increases. However, there is a critical flow rate beyond which the penetration efficiency decreases as the flow rate increases. The piercing or cutting efficiency depends on both the abrasive and the workpiece material.

Surface roughness is a measure of cutting efficiency. The traverse speed plays a major role in controlling the smoothness of the finished surface. Decreasing the speed reduces the surface roughness. Al_2O_3 abrasive at higher traverse speeds achieves the same surface finish on 85 percent Al_2O_3 as garnet abrasive at much lower speeds. Therefore, the material-removal process can be optimized by control of the traverse speed and position of the jet as a function of the nozzle wear and tolerance.

A water-jet unit built by Kawasaki Heavy Industries Company Ltd. is in use in Japan and is able to slice ceramics. The cutter ejects, at almost the speed of sound, a slurry containing 70-μm particles of Al_2O_3 at a pressure of up to 2500 bar.

Existing types of jet cutters normally use sand grains about 800 μm in diameter. The nozzles are easily damaged and lose their efficiency after an hour or so. The new cutter is made of carbon tungstate reinforced with cobalt, and will last longer since the abrasive material is much smaller in size. The machine will cut ceramics and can be attached to a precision processing unit along with computer-controlled machines or robots. Kawasaki thus sees the machine as being of key importance in automated processes in the future.

Cutting. Linear cutting tests were conducted to determine cutting rates and surface finishes using different abrasive water-jet parameters and to cut tensile and fatigue test specimens. Also, the linear cutting tests were used to evaluate candidate nozzle materials, according to Hashish.[29] Table 7.5 summarizes the results, and it should be observed that the maximum cutting traverse rate is primarily controlled by the matrix material. For example, the cutting rates for the Al–SiC composites are slightly lower than those for aluminum. One should also observe the slight change in maximum cutting speed as the percentage of the harder reinforcing fiber increases. The cutting rates for ceramic composites are relatively slow due to the use of garnet as an abrasive material. When SiC or Al_2O_3 abrasives are used, the cutting rate will increase by between 200 and 500 percent over the rates obtained with garnet. The tougher the ceramic, the larger the percentage of improvement. However, the use of these effective abrasives will wear out the tungsten carbide nozzle

TABLE 7.5 Typical Through-Cutting Traverse Rates (mm/s) with Abrasive Water Jets for Different Composites*

Material	\multicolumn{7}{c}{Thickness, mm}
	0.79	1.60	3.18	6.36	12.7	19.1	50.8
Ceramic matrix composites:							
Toughened zirconia		0.9		0.65	0.4	0.25	
Dense zirconia			0.80	0.70			
SiC fiber in SiC		1.1	0.6	0.45			
ZrO$_2$–MgO			0.8	0.7			
Al$_2$O$_3$–CoCrAly (80%/20%)			0.95	0.65			
Al$_2$O$_3$–CoCrAly (60%/40%)			0.95	0.65			
C-glass	100	90	80	60.5	40	20	6
Al$_2$O$_3$–SiC (7.5%)†			2.7	1.4			
SiC–TiB$_2$ (15%)			0.29	0.15			
Metal matrix composites:							
Mg–B$_4$C (15%)	70	30	15	10		4	
Al–SiC (15%)	70		17	10	5		
Al–SiC (25%)				9.5	5		
Al–Mullite (5%)	75	35	20	12	7.5	5	2.5
Al–Al$_2$O$_3$ (15%)	65	28	15	8	4		

*Cutting conditions: water-jet pressure 345 MPa; water-jet orifice diameter 0.299 mm; mixing-tube diameter 0.762 mm; garnet mesh 80.
†SiC abrasives.
Source: From Hashish.[29]

in less than 5 min, rendering the process impractical. This area of ceramic composite machining requires further investigation.[29,30]

Turning. In turning with abrasive water jets, a workpiece is rotated while the jet is continually fed in an *X–Y–Z* pattern.[26] The material that the jet sweeps is converted to very fine debris, contrary to chip formation in conventional machining.

Turning tests were conducted on two types of metal-matrix composites, Mg–B$_4$C (15 percent B$_4$C) and Al–SiC (20 percent SiC). For example, the machining rates for Mg–B$_4$C are lower by about 38 percent than those for aluminum to obtain the same waviness. With solid tools, the machining rates of the Mg–B$_4$C and Al–SiC composites will be about 15 to 20 times slower than those for aluminum.

The repeatability and accuracy of the abrasive water-jet turning process depend on the control and steadiness of many parameters.[29] As a result, strategies have been developed that may be used for improving the surface finish of a roughly turned part:

Multipass finishing by traversing the jet without lateral feed. The surface waviness of turned Mg–B$_4$C can be reduced from 122 to 64 µm with two additional passes.

Use of finer abrasives with additional passes. It has been found that finer abrasives may increase the surface waviness and reduce the surface roughness.

Use of softer abrasives.

Finishing with slurried abrasives. Again, the result is an improvement in surface roughness.

Milling. In abrasive water-jet milling the main objective is to be able to produce a cavity with a controlled depth. In conventional milling, this depth is determined geometrically by the feed of the tool. With an abrasive water jet (or any streamlike tool) the jet-material interaction process is the depth-determining factor. Many parameters control this interaction process and, accordingly, the milling results.

To improve the surface finish of a milled pocket, the milling strategy may include the use of different abrasive materials or sizes in the different phases of machining. For example, the use of hard abrasives would be suitable for fast material-removal rates, while soft abrasives may be suitable for finishing. Observations on the milling of metal, according to Hashish,[27] also apply to metal-matrix composites and indicate that the depth uniformity control should be planned at the beginning of the milling process, contrary to turning, where the surface finish can easily be changed with subsequent passes.

Drilling. Hole drilling can be accomplished with any of the following methods:

1. Piercing: suitable for small-diameter holes
2. Kerf cutting (in a circular path): suitable for large-diameter holes
3. Milling (using a mask): suitable for blind holes

Techniques of hole piercing with abrasive water jets depend largely on the target material. Piercing techniques for glass, for example, are different than those for ductile metals. Thus it is expected that piercing conditions will be different for each class of composites.

Hole shaping is greatly dependent on jet structure, target material, and stand-off distance. The sensitivity to jet structure increases as the material resistance to drilling increases. For example, kidney-shaped holes have been observed in tough and hard materials such as toughened ZrO_2 or WC. This suggests that the distribution of abrasives in the jet is not uniform. The difficulty of controlling uniformity of hole diameter increases as the hole length increases. However, for relatively short holes, the dwelling time after drilling through is an important parameter for diameter control. Better holes with minimal damage can be produced in TiB_2–SiC composite by abrasive water jet at low stand-off distances.[31]

Comparison of Methods. A comparative study[30,31] between diamond-saw and abrasive water-jet cutting of Al–SiC and Al_2O_3–SiC composites indicated that abrasive water jets are significantly faster (over 20 times for Al_2O_3–SiC). However, surfaces produced with diamond saws have a 0.3-μm surface finish compared to 3 to 4 μm for the abrasive water-jet technique.

Electric-discharge machining can be used with conductive silicides, borides, nitrides, carbides, metal-matrix composites, and ceramic-matrix composites. Cutting rates with electric-discharge machining are about one to two orders of magnitude slower than with abrasive water jets, but they produce more precise cuts. Surfaces produced with abrasive water jets are of higher quality than surfaces produced with electric-discharge machining.[29] Holes 0.25 mm in diameter can be drilled with the latter process in SiC–TiB_2 composites, and cutting rates in that same material range between 0.05 and 0.2 mm/s.

Ultrasonic Abrasive Machining

Technological breakthroughs come from the ability to try things that seem almost impossible. Ultrasonic abrasive machining, or ultrasonic machining, has been providing the necessary assistance to permit breakthroughs in advanced ceramics since their emergence. By combining the capability to machine limitless numbers of shapes in hard, brittle materials with the virtual absence of stress, ultrasonic machining provides the technology to experiment and innovate with what is often the fastest and possibly the only way to get new designs off the ground.

Ultrasonic machining is an abrasive process that has certain advantages over conventional grinding. Sometimes called impact grinding, it is a mechanical process that uses a high-frequency transducer. As the electric energy is converted into mechanical motion, a low-amplitude vibration is produced. The vibration is transmitted to the toolholder, which results in a linear physical movement or "stroke" of typically 0.02 mm at a rate of 20,000 cycles/s.

This action produces a microscopic chipping at a steady penetration rate. A constant flow of abrasive slurry is used under controlled pressure. A 20 to 50 percent concentration of B_4C or SiC are the most commonly used abrasives. The slurry is recirculated to provide fresh abrasive and to remove abraded particles. The particle size of the grit determines the finish, the size of the cavity in relation to the tool, and the cutting rate (Fig. 7.6). This was determined by Bullen Ultrasonics, Inc.,[28] in a series of tests. Vacuum can also be used to improve abrasive flow.

When the ultrasonically induced vibrations are combined with an abrasive slurry, ultrasonic machining creates accurate cavities of virtually any shape. This machining process is nonthermal, nonchemical, and nonelectrical and creates no

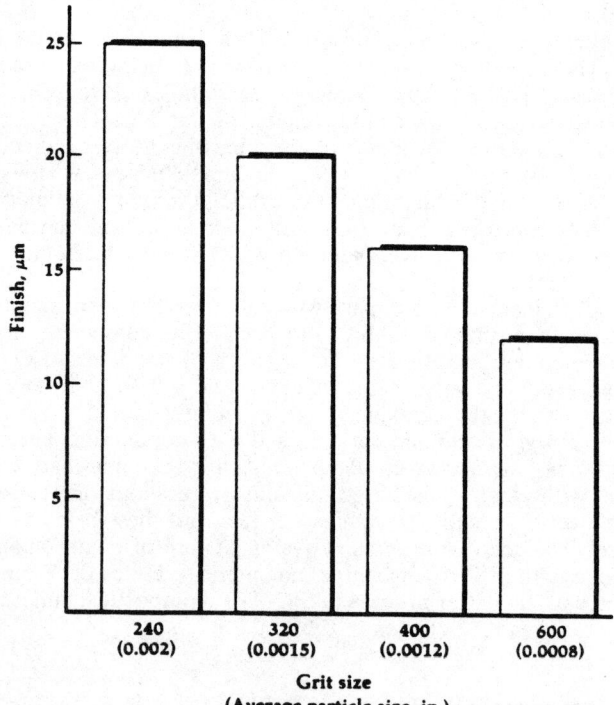

FIGURE 7.6 Particle size (grit size) plays major role in final surface finish.[28]

change in the metallurgical, chemical, or physical properties of the workpiece. Typical materials machined are Al_2O_3, SiC, Si_3N_4, piezoceramics, fused quartz, glass, borosilicate glass (Pyrex), sapphire, and carbides. Ultrasonic vibrations are produced by coupling an electronic generator, a transducer package (either magnetostrictive or piezoelectric), a transmitting connecting body, and a toolholder and tool combination. The generator converts typical line voltage into the voltage and frequency required to energize the transducer coupled to it.

Most transducers are connected to a transmitting connecting body. This connecting body is designed to resonate at the same nominal frequency as the transducer and to provide a means to connect the toolholder. In most cases, a threaded stud is used on the end of the connecting body to connect the toolholder. Figure 7.7 illustrates an ultrasonic transducer and toolholder assembly.[32]

The abrasive is a key element of ultrasonic machining. Commonly used abrasives are B_4C, SiC, and Al_2O_3. B_4C is the hardest abrasive and will last the longest; it is recommended for machining engineering ceramics (such as SiC, B_4C, Si_3N_4, Al_2O_3, ZrO_2). SiC is used frequently for materials such as glass, quartz, and single-crystal silicon.

Two theories have been proposed to explain the machining mechanism of the tool tip: (1) abrasive grains accelerated by a vibrating tool strike a workpiece and crush the surface; and (2) when the tool vibration hits abrasive grains between the tool and the workpiece, high stress occurs at the contact point, crushing the workpiece surface (Fig. 7.8).

In ultrasonic machining, the speed of the tool is so fast that it is difficult to observe and to make a judgment about which theory is true. However, today theory (2) seems to hold true, for when the tool is placed a short distance from the workpiece, the machining speed drops sharply.

Today ultrasonic machining is used throughout industry. It enables tool shapes to be transferred to the work surface, which is a reason why the method is an efficient one for processing ceramics. Automotive applications range from ceramic engine components to Al_2O_3 pallets used in an automated production line for making diodes. Electronic applications vary from machining ceramic substrates to drilling holes in borosilicate glass for the sensor industry. Microwave substrates with holes ultrasonically machined are readily metalized, thus permitting superior yields and high dependability.

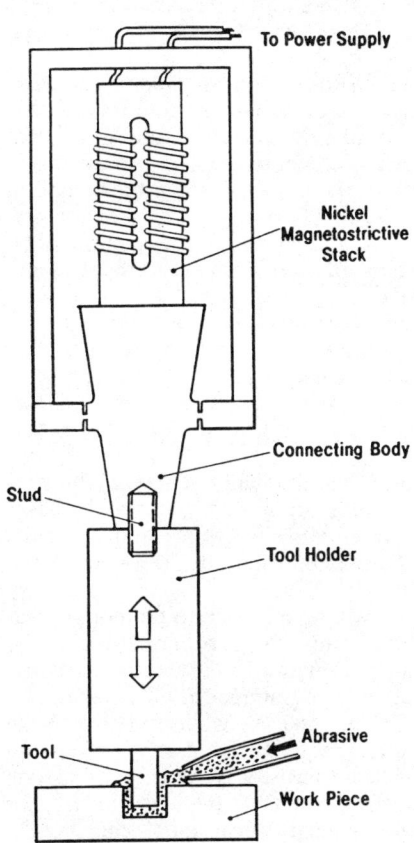

FIGURE 7.7 Ultrasonic transducer and toolholder assembly.[32]

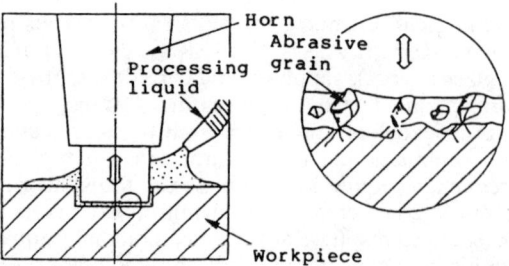

FIGURE 7.8 Machining mechanism.[19]

Advantages and Future. Because of its nonthermal characteristics, ultrasonic machining produces virtually stress-free machined surfaces. Thus the quality of an ultrasonic cut is such that there is less stress and damage to the workpiece than with other processes.

Ceramic manufacturing often entails high tooling prices and long delivery times. In contrast, ultrasonic machining provides an alternative method to supply parts sooner and to reduce tooling charges. Changes are easily incorporated; thus, designs can be changed quickly.

Tooling can be designed to accommodate a virtually limitless diversity of cuts. Through-holes, blind holes, slotting, shaping, and odd-shaped contours are possible. In addition, ultrasonic machining can frequently hold finer tolerances than typical pressing methods. Reflective and metallized parts can be ultrasonically machined without damage to surrounding areas. Hole patterns can be located in the existing metallization pattern.

Ultrasonic impact grinders are built to serve a great variety of requirements, from laboratory use to high-volume production. In their simplest forms, the machines have manually operated axis movement.

According to Moreland and Moore,[32] in many applications the cutting tool can be allowed to cut at its own rate. This is true when (1) the amplitude of the cutting tool is greater than the abrasive particle size being used, (2) the physical strength of the cutting tool is sufficient to handle the ultrasonic vibration as well as the tool pressure used to feed it into the workpiece, and (3) the depth of cut is less than 7 mm.

The cutting-tool feed rate can also be controlled manually. This may be necessary when using very sensitive miniature cutting tools. Feed rates as low as 4 μm/s may be necessary. Periodic manual reversing of direction can also let the abrasive flush under the cutting tool, a procedure that may be necessary when the cut is deep enough to create abrasive starvation.

Numerical control of axes by computer allows the machine to be more versatile, especially in production applications. Feed rates can be programmed to any combination required for the material being machined and the depth of cut desired. Automatic tool-wear compensation can be programmed if necessary.

Ultrasonic transducers now range from 300 to 2400 W, while the size of the toolholder used governs the size of the transducer required. Solid-state power supplies with variable frequency control use the latest transistor technology to operate efficiently. Automatic tracking of the toolholder's resonant frequency adjusts for changes in toolholder temperature and cutting-tool length because of wear.

Ultrasonic Vibration Grinding. Ultrasonic vibration grinding is a composite machining method that combines diamond-wheel grinding and ultrasonic machining.

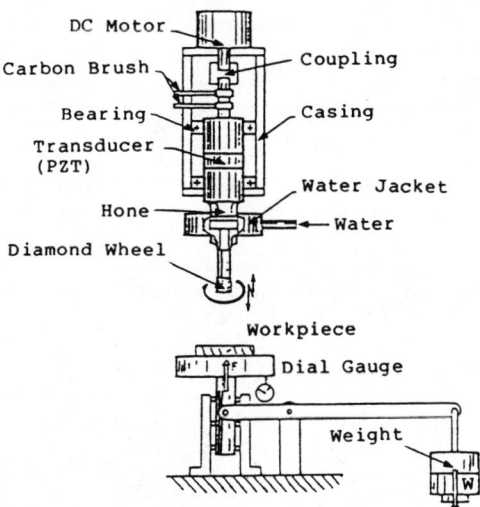

FIGURE 7.9 Construction of ultrasonic grinding unit.[19]

In ultrasonic vibration grinding, the ultrasonic vibration most often used is about 20 kHz and the tool tip amplitude ranges from 15 to 50 μm. Figure 7.9 shows the construction of an ultrasonic vibration grinder.[19] The main unit is composed of a vibration system, a dc motor to turn the vibration system, bearings, and a water jacket to supply grinding liquid. The vibration system consists of an oscillator to generate ultrasonic vibration, a connected horn, and a diamond wheel fixed at the tip. Gravity and a hydraulic system are used to feed the workpiece similarly to ultrasonic machining.

One of the differences between ultrasonic machining and ultrasonic vibration grinding is in the machines. In ultrasonic machining there are floating abrasive grains, no rotating parts, and machining of unsymmetrical holes is possible. In ultrasonic vibration grinding there are fixed abrasive grains, existence of rotating parts, and only concentric grinding is possible. The two processes are similar in that only a very small amount of work surface is crushed by the abrasive grains, and neither method affects conductivity or causes processing deformation and decomposition of the work surface.

When ultrasonic vibrations are applied to a grinding wheel with 20-kHz frequency and 10-μm amplitude, the acceleration reaches about 16,000 times that of gravity. Thus even a fine chip is exposed to a large force and frees itself from the grinder surface. Therefore, using ultrasonic vibration is making rotation driving of a grinding wheel compact.

Electric-Discharge Machining

The efficiency of traditional cutting processes is limited by the mechanical properties of the material and the complexity of the workpiece geometry. Electric-discharge machining, being a thermal erosion process, is not subject to these constraints. According to König et al.,[33] it is a reproducible shaping process in which the form of the tool electrode is mirrored in the workpiece. Electrically conductive materials are eroded by means of electric discharges. Thermal erosion is the separation of solid, liquid, or gaseous particles under the influence of heat. The required separation energy is generated by spatially discrete, high-frequency electric discharges (sparks) between tool and workpiece. The erosive process takes place in a dielectric liquid medium. A gap is maintained between tool and workpiece, preventing contact of the two electrodes. There is no direct physical contact between the electrodes, and therefore no mechanical stress is placed on the workpiece, so this is an ideal method to shape hard, brittle, and refractory materials.

With the manufacturing importance of the process, discovered 40 years ago by Lazarenko and Lazarenko,[34] and with the addition of computer numerical control in the late 1970s, many dramatic advances occurred that enabled electric-discharge machining to go from exotic and rare to a routine technology for hard metals.

In recent years, concerted development of engineering ceramics has led to the manufacture of electrically conductive electrodischarge-machinable ceramics. This process results in almost no alteration of the most important material properties such as hardness and temperature resistance.

Some ceramics are electrically conductive without further treatment. The advantages of the electric-discharge machining process applied to electrically conductive materials opens a wide range of opportunities to the design engineer. In addition to the generation of extremely complex geometries through sinking and cutting, electric-discharge processing can be performed regardless of a material's hardness and strength, and the forces exerted on the workpieces are negligible.

There are two types of conductive ceramics machined with electric-discharge machining: one consists of conductive materials, and the other is conductive because it is a mixture of dielectric (or semiconductive) materials and electric conductive materials.

Typical ceramics that consist of only conductive materials are carbides, such as TaC, TiC, ZrC, and SiC (reaction sintering); borides, such as TiB_2 and ZrB_2; and nitrides, such as TiN and ZrN.

Ceramics which consist of conductive and nonconductive materials are Si_3N_4–TiN, sialon–TiN, ZrO_2–CbC, Si_3N_4–SiC, whiskers, and so on. Electric-discharge machining can be applied to ceramics, including single phases, and to composites of ceramics and ceramic and metal if the electrical resistivity is below about 100 $\Omega \cdot$ cm.

Finishing advanced ceramics into intricate shapes with the required tolerance is an extremely difficult task, but electric-discharge machining can produce a mirror finish, even on ceramics, giving a surface roughness $R_a < 0.3$ μm, with tolerances in the micrometer range.[35–37]

There are two types of electric-discharge machines, each with different applications. The die-sinking machine (Fig. 7.10), also known as the ram-type or vertical-erosion machine, can be used for tapping, cutting holes, and helical machining and is capable of cutting or sinking very complicated shapes into a workpiece. In die sinkers, the workpiece is the cathode and the shaping tool is the anode. Die materials range from brass, steel, and other improved alloys to specially made graphite, which should be fine-grained and isotropic to last longer. Heavy hydrocarbons and kerosene are the most common dielectrics used in this conventional form of electric-discharge machining.

Wire machining uses a thin metallic wire under tension to cut like a jigsaw (Fig. 7.11). Some erosion of the wire is allowed since the wire is continuously being wound between spools as it travels through the material, but excessive wire erosion is the major cause of wire breakage. This breakage is the dominant prac-

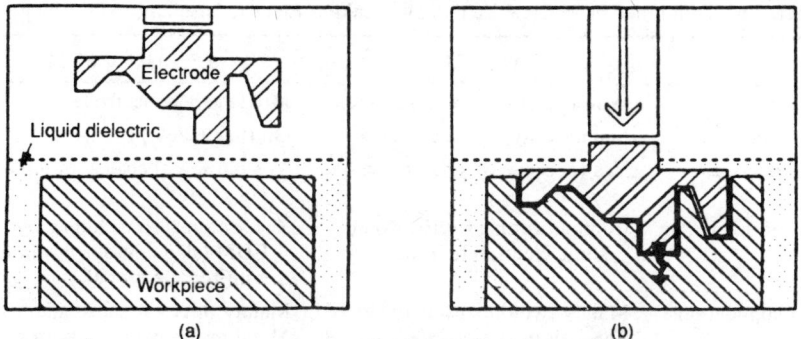

FIGURE 7.10 Die-sinking electric-discharge machine. (*a*) Before EDM. (*b*) After EDM.[35]

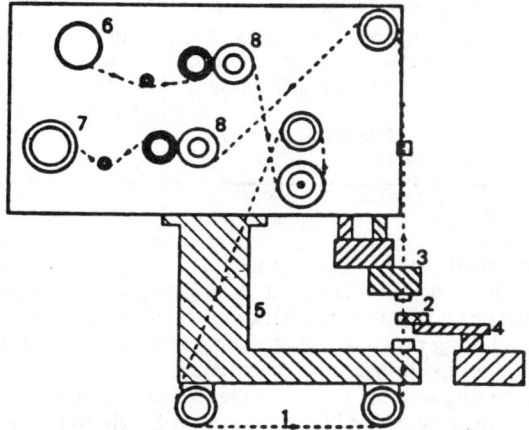

FIGURE 7.11 Setup for wire-cutting electric-discharge machine. Components: 1—wire cathode; 2—workpiece (anode); 3—wire-guide table; 4—workpiece table; 5—machine body; 6—wire spool; 7—used-wire spool; 8—wire-tension adjusters.

tical and economic problem of wire-cut electric-discharge machining. For this type of machining, water is usually used as the dielectric, and brass-, steel-, or molybdenum-based wire is used for the electrode.

Table 7.6 shows the difference between die-sinking and wire electric-discharge machining. König et al.[33] point out that when die sinking uses a plunger electrode, it is identical to the wire process. Any material that can be wire-cut can also be machined by die sinking. With die sinking, however, complete control is more difficult to achieve, especially when complex workpieces are involved. A straight cylindrical wire electrode is generally easier to control. Tables 7.7 and 7.8 show the main technological parameters for some types of engineering ceramics.

In addition it should be noted that in wire electric-discharge machining, the relation of the surface roughness to the cutting rate differs by material, and tests have shown that the wire process causes degenerated layers and microcracks on

TABLE 7.6 Difference between Wire and Die-Sinking Electric-Discharge Machining

	Wire EDM	Die-sinking EDM
Electrode	Wire of 0.03–0.3-mm diameter	Mold-shaped electrode
Insulating liquid	Mainly water (nonflammable)	Oil (flammable)
Electrode wear	Yes (because of small pulse width)	Yes but low (because of large pulse width)
Type of machining	Contour cutting by EDM and some electrolytic erosion	Shape-transfer machining by EDM (also contour machining with rod electrode)
Main application	Mainly two-dimensional: Through-type metal molds Electrodes for EDM Parts manufactured in small amounts Minute machining	Mainly three-dimensional: Metal molds having a bottom Hard-to-grind materials Minute machining
Characteristics	Unmanned machining possible Special electrodes unnecessary, hence shorter manufacturing period Required time for machining easily predicted	Both two- and three-dimensional machining possible Machining of complicated three-dimensional shapes possible

the surface of the workpiece, resulting in a decrease in bonding strength. The bending strength, however, is recovered by grinding off degenerated layers.

Tests have further shown that the bending strength of the workpiece also decreases in the die-sinking process. The bending strength can, however, be recovered through grinding, as in discharge wire cutting.

Strength tests on SiC specimens cut by electric-discharge machining show that material strength following a rapid cutting is comparable to that after grinding under conventional industrial conditions. Thus König et al.[33] expect that a loss of component strength following electric-discharge machining can be avoided by optimizing the parameters and using multicut technology.

Ricci et al.[44] cut TiB_2 and SiC successfully by the wire electric-discharge process. B_4C workpieces could not be cut in the monolithic state; however, when a 3-mm-thick brass strip was placed on the top surface, cutting was possible. Si_3N_4 could not be cut in either the monolithic state or when a 6-mm copper sandwich structure was placed about the workpiece in an attempt to induce spark generation and initiate cutting.

Success in cutting the B_4C is believed to be due to the effects of the local increase in electrical conductivity because of increased electron mobility and possibly the formation of new surface phases, at the higher temperatures present during machining.

Flexure tests comparing wire-cut TiB_2 with diamond-ground samples showed that the electric-discharge machined samples exhibit a 23 percent reduction in strength versus the diamond-ground samples. The reasons for this decrease are not conclusive, but are thought to be due to a combination of surface roughness and tensile stresses as well as the lowering of the surface fracture toughness due to porosity and chemical alteration.

TABLE 7.7 Die-Sinking Electric-Discharge Machining

Range of resistivity, $\Omega \cdot$ cm	Material	Main characteristics				Conditions for machining			Results of machining		
		Density, g/cm^3	Bending strength, kg/mm^2	Breaking strength, MN/m$^{3/2}$	Resistivity, $\Omega \cdot$ cm	Electrode	Machining liquid	Others	Cutting rate, mm^3/min	Surface roughness, R_{max}, μm	Electrode wear ratio,* %
10^{-1}–10^{-2}	Si$_3$N$_4$–SiC whiskers[38]	2.92	65		30–50	Pure copper	Kerosene	Applied voltage 110–120 V Duty factor 50%	~3	10–50	(−) 8.5 (+) 2.3
10^{-0}–10^{-1}	SiC, sintered at atmospheric pressure[39]	2.92	30.7		0.8	Pure copper	Kerosene	Current peak value 1–8 A Duty factor 50%	0.91–1		(+) 0–40
10^{-2}–10^{-1}	SiC, reaction-sintered[39]	3.08	40.6		0.08	Pure copper	Kerosene	Current peak value 1–20 A Duty factor 50%	0.01–1		(+) 0–18
10^{-3}–10^{-2}	Conductive ZrO$_2$[40] Conductive Si$_3$N$_4$[41]	Specific gravity 6.52 3.60–3.89	120 85–100	7.5 4.3–4.8	6×10^{-3} 5×10^{-3}	Silver tungsten Pure copper	Kerosene	Current peak value 25 A Pulse width 5–200 μs Duty factor 50%	0.45–0.55 15–35	~10 5–22	(−) 8–15 Electrode (+)
10^{-4}–10^{-3}	Conductive sialon[42]	4.00	85	5.0	7×10^{-4}	Pure copper	Diamond EDF	Applied current 1–4 A Pulse width 5–15 μs Pulse interval 15–60 μm	11–14	4–7	(−) 4–11
10^{-5}–10^{-4}	ZrB$_2$[43]	5.40	55	4.2	1.5×10^{-5}	Pure copper Brass Copper tungsten Silver tungsten	Kerosene	Applied voltage 210 V Applied current 14.5 A Pulse width 2.5 μs Pulse interval 35 μm Duty factor 6.7%	0.4–0.9	4–6	(−) 20–100 (+) 110–1100

*(+), (−) Electrode polarity.
Source: From Refs. 38–43.

TABLE 7.8 Wire Electric-Discharge Machining

Range of resistivity, $\Omega \cdot$ cm	Main characteristics				Conditions for machining				Results of machining		
	Material	Density g/cm^3	Bending strength, kg/mm^2	Breaking strength, MN/m$^{3/2}$	Resistivity, $\Omega \cdot$ cm	Wire	Machining liquid	Applied voltage	Cutting rate	Surface roughness R_{max}, μm	Others
10^{-3}–10^{-2}	Conductive ZrO$_2$[40]	Specific gravity 6.52	120	7.5	6×10^{-3}	0.2-mm-diameter brass	—	55 V	0.8–1.5 mm^3/min	21–27	
	Conductive Si$_3$N$_4$[41]	3.60–3.89	85–100	4.3–4.8	5×10^{-3}	0.2-mm-diameter brass	Deionized water	—	5 mm thickness, 3 mm/min	10–20	
10^{-4}–10^{-3}	Conductive sialon[42]	4.00	85	5.0	7×10^{-4}	0.2-mm-diameter brass Wire polarity (−)	Conductivity 2 mA	32 V Pulse width 0.3–1 μs Pulse interval 10–30 μs	10 mm thickness, 90–165 mm^2/min	10 mm thickness, 10–12	Clearance 30–40 μm
10^{-5}–10^{-4}	ZrB$_2$[43]	5.40	55	4.2	1.5×10^{-5}	0.2-mm-diameter brass Wire tension 850 g	Ion-exchanged water	100–140 V Pulse width 1.0–2.0 μs Pulse interval 3.5–4.0 μs Duty factor 20–35 %	10–20 mm^2/min	8–15	

Source: From Refs. 40–43.

Even though material-removal rates were low for the ceramics tested, wire electric-discharge machining should be considered a viable alternative for the shaping of complex ceramic contours, since this is extremely difficult if not impossible by standard abrasive grinding methods. However, it is believed that the electric-discharge machining of small-surface critical components will require secondary finish machining or annealing operations to assure removal of any deleterious effects on fracture strength and toughness properties.[44,45]

Future of the Process. There are problems which require resolution and investigation for conductive ceramics:

Thickness of the degenerate layer on the surface and influence of the layer on material characteristics need to be determined for different materials and under different conditions to understand the minimum amount of layer to be removed.

Electrode wear should be reduced to improve cutting rate, reduce surface roughness, and understand the amount of allowance for machining.

Machining characteristics and their relations with material characteristics and machining need to be considered and differences among machines should also be taken into account.

Methods are needed to make ceramics conductive while not weakening the characteristics of a material even under high-temperature oxidative conditions.

Applications must be developed where conductive ceramics can outperform metals and cemented carbide in temperature ranges where the weakening of material characteristics does not occur.

Electric-Discharge Grinding

When electric-discharge machining is used with a grinding wheel, it is known as electric-discharge grinding. Here current is passed through the insulated, but conductive grinding wheel toward the metal conductive workpiece. Between the conductive wheel bond and the workpiece are the nonconductive abrasive grains and a dielectric fluid. Since neither conducts electricity, an electric potential is built up in the wheel bond, and it eventually bursts through the dielectric in the form of a spark. When this spark impinges on the workpiece, it literally blasts away a minuscule portion of the surface. As this is done anywhere from 150 to 500,000 times per second, the workpiece surface is eroded away. The detritus is carried off in the dielectric fluid, which is continuously circulated in the gap between the wheel and the workpiece.

Since electric energy does about 90 percent of the work and the abrasive grains the other 10 percent, there is less wheel wear and the job is done much faster. Gettelman[46] points out that when the workpiece is ceramic, the task of grinding becomes quite different in the combined-energy process.

Primarily the electrolytic fluid (10 percent solution NaOH) is directed toward the wheel-workpiece interface, similar to the way grinding fluid is used in any normal grinding operation. Also, as in any purely mechanical grinding operation, it is unlikely that any fluid actually gets to the wheel surface because of the air disturbance generated by the rapidly spinning wheel surface that blocks the fluid.

Just as with electric-discharge machining, the spark concentrates into a spot of heat at ultrahigh temperatures. But in this case the spot of heat is adjacent to the workpiece rather than being concentrated in the workpiece itself. The result is similar though, in that a small crater is eroded in the workpiece, which contributes to the metal removed by the grinding wheel. Hence the term COMMEC, which means a combination of mechanical, electrical, and chemical energies.

Gettelman further states that the greatest practical benefit of electric-discharge grinding is the wheel-cleaning effect that unloads the wheel as it cuts. Experienced grinders of ceramic materials report that wheel loading seriously impedes the normal abrasive process, especially in production applications. The improved grinding accuracy that results, as well as the less frequent dressing cycles, are some of the primary economic justifications for the process.[46]

The process is used most frequently where fine shapes such as narrow slots or steps are ground into electronic and other specialized components that are made of glass, ceramic, or composites. Some of the slots may be as small as 0.7 mm.

Future of the Process. Combined-energy machining has spawned a minor Japanese industry in manufacturing research. Clearly, the Japanese feel that they are dealing with a process that has some real significance for the future of manufacturing. Ceramics promise great gains in the performance of bearings, electronics, and high-temperature engines.

Manufacturing these hard and brittle materials, especially cutting and finishing them, is part of the holdup to their more extensive use. Thus alternatives to conventional grinding and lapping receive more than usual attention by researchers working in the ceramics field.

The answers may lie in such hybrid processes as this combination of electric-discharge machining, electrochemical machining, and conventional grinding. The COMMEC process, which is the practical application of combined-energy machining, has been shown to solve major problems in ceramic grinding. The principle behind it may be obscure, but the benefits are ample proof that machining ceramics is worth an extra measure of exploration and research.[46]

Laser-Beam Machining

Ceramics are expected to expand their applications as functional and structural materials. However, they have drawbacks such as machining difficulties because of their high hardness and the microcracks caused by machining. Laser-beam machining is a noncontact thermal process, exhibiting excellent beam concentration and position control, which means that machining is much more flexible than with conventional methods. It has the following advantages:

> One laser-beam system can be used for various applications, such as drilling, cutting, welding, etching, surface processing, and material composition.
>
> Because laser-beam machining is a heat process, it can be applied to nonmetallic materials. This type of machining is effective for processing ceramic materials.
>
> Laser beams can be remotely controlled very easily because they have good directivity and suffer little attenuation through air. Therefore simultaneous machining can be done at more than one machining station, or through time-sharing machining by mechanical optical control. Note that a laser beam can pass through various atmospheres, including air or a vacuum.
>
> Laser beams can be easily operated and applied to automation. Since the

laser-beam process is nonreactive, the setting mechanism of the workpiece can be simplified to machine a complex shape.

However, because of high heat during the machining process and the use of high-intensity energy, the process has the following disadvantages in machining ceramics:

A decomposed layer is formed at the processed surface, which has a different structure from the matrix.

Ceramics have poor ductility, and cracks can occur in the processed part because of heat distortion.

Figure 7.12 shows the basic design of a laser processing machine, consisting of a laser oscillator, a mechanical optical system, and a workpiece system. Each system has many controllable factors which influence processing characteristics. The mechanical optical system includes the beam shape and diameter, divergence angle, focal length and depth, defocusing, spot diameter, shield and assist gas (composition, flow amount, direction), and the nozzle shape and position. The workpiece system consists of material, shape, dimension, surface condition, atmosphere, and travel speed.

Figure 7.13 shows the popular kinds of lasers used for processing. Their features are listed in Table 7.9. YAG and CO_2 lasers are the most popular lasers because a continuous, high-speed repetitive oscillation is possible and oscillation efficiency is high. Ruby lasers and glass lasers are used less because they have poor oscillation efficiency and low-speed repetitive oscillation. However, the Ar^+ and excimer lasers are currently expanding into new applications, such as microetching (discussed later in this chapter) and laser chemical-vapor deposition.

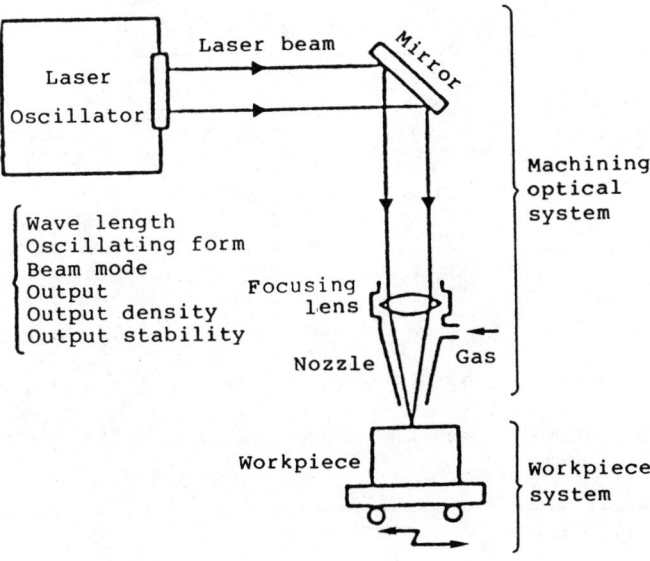

FIGURE 7.12 Basic mechanism of laser-beam machining apparatus and conditioning factors.[19]

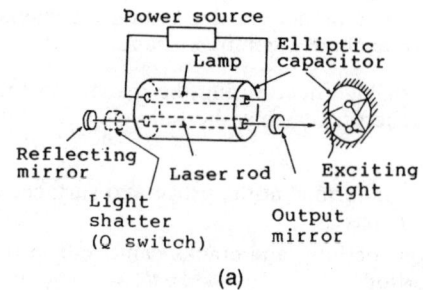

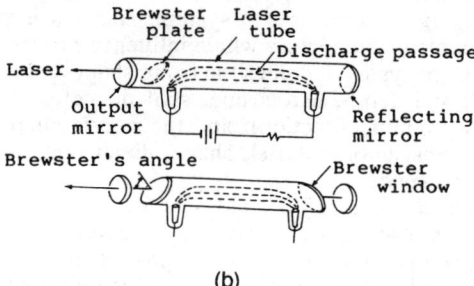

FIGURE 7.13 Various types of lasers.[19] (a) Solid laser. (b) Gas laser. Top—inner mirror type; bottom—outer mirror type.

TABLE 7.9 Features of Lasers Illustrated in Fig. 7.13

Kind of laser		Wave length, μm	Output range		Applications
Solid laser	Ruby	0.694	Pulse	0.5 mJ–400 J	Drilling, balancing
	Glass	1.06	Pulse	1 mJ–100 J	Drilling
	YAG	1.06	Pulse	0.1 mJ–100 J	Drilling, welding
			CW	0.1 W–1 kW	Scribing, marking, trimming, annealing
	Alexandrite	0.73–0.78	Pulse	to 10 J	Drilling, annealing
Gas laser	CO_2	10.6	Pulse	0.2 mJ–10 kJ	Drilling, cutting, welding, surface processing
			CW	1 W–20 kW	
	Ar ion	0.35–0.51	CW	5 mW–40 W	Annealing, etching, metal plating
	Excimer	0.19–0.78	Pulse	5 mJ–1 kJ	Micromachining, surface processing, etching, graphic art

Source: From Nagawa et al.[19]

Laser processing is classified according to laser power density and operational time. Power density correlates with the temperature at the machining part surface. When the power density is low, processing (surface thermal processing) progresses in the solid phase for the part irradiated by the laser beam. When power density is high, melt processing takes place at the liquid phase to effect laser cutting, laser welding, and laser cladding.

Machining and Cutting. Because of the extremely high wear resistance of ceramics, the removal capacity in cutting, even using diamond tools, is very low. For this reason the main geometry of ceramic workpieces must be given in the blank. In general, shaping as a means of cutting is only grinding. It is expensive and a time-consuming process, according to Tönshoff and Emmelmann.[47]

Since the laser has gained wide acceptance among ceramic machining processes in the microelectronics industry, we must examine the temperature effects and thermal stresses induced by this process.[47-49]

The critical temperature cycle during laser processing of ceramics is defined by the following factors, according to Tönshoff and Emmelmann[47,49]:

1. External stress
    a. Amplitude of temperature cycle
    b. Speed of temperature cycle
    c. Surface-heat transfer coefficient
    d. Induced oxidation processes
2. Geometric factors
    a. Size
    b. Form
3. Characteristics of material
    a. Elastic parameters (modulus of elasticity, modulus of shear, Poisson's ratio)
    b. Breaking strength
    c. Coefficient of thermal expansion
    d. Heat and thermal conductivity

Beside the geometry and the properties of the laser-machined ceramic workpiece, the external stress also influences the critical value of the temperature gradient in the material. In laser processing, the amplitude and speed of the temperature cycle can be influenced by the type and amount of laser energy and the environmental conditions. Among other factors, the thermal stress resistance depends on the critical speed of thermoshock. For a mild heat transfer the damage resistance is higher than for a hard thermoshock. Therefore the initiation and propagation of cracks induced by residual stress can be influenced by the heat transmission at the cut face. For this, supplementary additives can be involved in the cut process with lasers. In general a ceramic suitable for laser machining must have low parameters of elasticity and thermal expansion as well as a high breaking strength and a high thermal conductivity.

There are three commonly used lasers in laser-beam machining: CO_2, Nd:YAG, and excimer. Each type causes different thermal effects during laser-beam machining of ceramic materials.

Due to the absorption of the material, which depends on the wavelength, the type of emitted laser radiation determines the quantity of captured energy. In ceramic materials, radiation of CO_2 lasers (10.6 μm) is absorbed more than 90 percent, whereas due to the lower wavelengths of Nd:YAG (1.06 μm) and excimer lasers (0.2 to 0.4 μm), less than 10 percent captured energy of the emitted laser power can be detected.[50] This little absorption, as well as the excellent focusing properties of low-wavelength radiation, create a small heat-affected zone and thus minimize the thermal damage of laser-processed ceramic cut faces. With adapted CO_2 and Nd:YAG lasers, processing the depth of crack-damaged zones at the cut face can be minimized to 60 to 100 μm.[50] According to Srinivasan and Leigh,[51] in excimer-laser machining thermal effects are nearly negligible, as the removal is by photochemical ablation. The high energy of the emitted photons in the ultraviolet range creates electrically excited or ionized states and causes atoms and molecules to move.

Excimer Laser. Excimer lasers are high-pressure gas-pulsed lasers using an active medium composed of various rare gases (such as argon, krypton, and xenon) and halides (such as chlorine and fluorine). In contrast to other high-power lasers for applications in the field of material processing, excimer lasers emit radiation in the near ultraviolet region, and the laser wavelength can be varied by the composition of the gas mixture.

Excimer lasers operate exclusively in the pulse mode, by pulse rates of a few up to 500 Hz and pulse durations ranging from 10 to 200 ns.

In contrast to CO_2 and Nd:YAG lasers, ceramic processing based on excimer lasers only allows small removal rates. In spite of the high quality of excimer-laser machining, since the obtainable removal rate is so small, this is not an economically worthwhile cutting method (0.01 μm/J · cm^3), but it is well suited to microprocessing.[52,53]

CO_2 Laser. Due to their high available beam power (<25 kW) and long pulse durations, continuous-wave CO_2 lasers offer good qualifications for cutting ceramic materials of high thickness with high feed rates. Nevertheless, the low beam intensity of 10^7 W/cm^2, the high absorption in ceramic materials (>90 percent), and large diameters in the working focal spot produce large heat-affected zones, which are critical for very brittle materials. As Si_3N_4 can be cut in the continuous-wave mode with a high feed rate of 250 mm/min, this material is very suitable for CO_2-laser cutting because of its low temperature for sublimation, high heat conductivity, and low content-melting phase.[49]

Nd:YAG Laser. Compared to CO_2 lasers, Nd:YAG lasers offer a lower beam power (<400 W) and a smaller pulse duration. Nevertheless Nd:YAG lasers can also cut very brittle materials, with the predominant removal method being sublimation and small kerf widths, due to their high beam intensity of 10^9 W/cm^2 and good focussing behavior. Therefore, the smaller thermal load of the Nd:YAG laser allows for the cutting of SiC, which cannot be machined by CO_2 lasers without crack damage.[54]

In the future, machining with Nd:YAG lasers will gain further importance and increased use as it permits unsophisticated three-dimensional processing with fiber optics.

Comparison of Alternative Processes. In Table 7.10 Tönshoff and Emmelmann[49] compare various competitive machining processes. Mechanical machining with diamond tools has produced cuts on ceramic materials of high thickness (<200 mm) at high feed rates with even profiles of the cut face, but the high mechanical loads and the resulting thermal loads have to be compensated by additional processing with cooling liquids. Only the diamond-wire processes achieve cuts with lower thermal loads but smaller feed rates. The abrasive water jet is able to drill, but it cannot be applied for very hard materials. Ultrasonic machining allows cutoffs for ceramics of high thickness, but should use symmetrical tools and thus is very inflexible for form-cutting applications

TABLE 7.10 Machining Processes for Advanced Ceramics*

Process	Tool	Application	Example
Cut-off grinding	Diamond cutting wheel	Cutting of plate and bar materials	Al_2O_3 (10 mm), 500 mm/min
Diamond wire	Welded diamond or CBN wire	Form cutting	Si_3N_4 (5 mm), 15 mm/min
Water jet	Water with materials of abrasion	Form cutting and drilling (limited hardness)	Al_2O_3 (8 mm), 150 mm/min
Ultrasonics	Sound transmitter with lapping abrasive	Closed symmetrical finished borders	Al_2O_3 removal rate, 410 mm^3/min
Wire erosion	Electrode of Cu, graphite, W, and dielectric	Form cutting (only conductive material)	Si SiC (10 mm), 60 mm/min
Laser	CO_2, Nd:YAG, or excimer laser	Form cutting and drilling	Al_2O_3 (15 mm), 10–20 mm/min

*Tolerances and surface finish depend on parameters of process and material.
Source: From Tönshoff and Emmelmann.[49]

when the lot size is small. Wire erosion, however, allows for flexible form cutting of materials of high thickness with small mechanical and thermal loads, but can only be used for conductive materials.

Therefore CO_2, Nd:YAG, and excimer lasers have to be chosen for their suitability concerning the thickness and the properties of the workpiece as well as the required quality of cut. Very flexible form cutting and even drilling are possible for thermally insensitive ceramic workpieces of high thickness or very brittle, thin materials that require a high-quality cut.

A series of comparative tests were conducted with fully dense reaction-sintered Si_3N_4 and PSZ, and early results indicate that hot machining with a CO_2 laser beam may be an alternative to machining the ceramics with diamond abrasives. Comparing typical production material-removal rates of 1.76×10^3 mm^3/min using diamond abrasives, Firestone[6] and Vesely et al.[55] found that reaction-sintered Si_3N_4 can be machined as much as 10 times faster at 996°C, PSZ as much as 30 times faster at 663 and 996°C without fracture. Besides being slower, machining with diamond abrasives weakens the ceramics, which are candidates for load-bearing engine components.

The study by Vesely et al.[55] also included SiC, but the limited number of samples available precluded quantitative evaluation. However, SiC, according to the researchers, is similar to Si_3N_4 in laser-beam machining.

Vesely et al.[55] and Firestone[6] heated the samples because they could not be machined at room temperature without cracking. The hot tooling consisted of a furnace for heating them in air and a four-axis vertical lathe. The furnace also contained areas to preheat the samples and to store them after machining so that heating and cooling rates could be controlled.

Si_3N_4 samples were machined at beam powers of 1.2 and 4 kW, the PSZ at 2, 4, and 6 kW. The former were also used to assess the effect of machining distance, or the distance from the sample to the beam focal point. Changing this distance from zero to 111.5 mm reduced the material-removal rate appreciably, especially if the gas shield was not used. Scanning or machining speeds were 25.5, 49.2, and 70.8 mm/s for the Si_3N_4; 153, 185.2, and 254 mm/s for PSZ.

Increasing the beam power, reducing the scan speed, and, in the case of PSZ, increasing the sample temperature increased the material-removal rates. The fastest rates for Si_3N_4 were 1.35×10^3 mm^3/min at 2 kW and 49.2 mm/s with the sample at the focal-point distance; with the machining distance 111 mm away from the focal point, the rates were 1.56×10^3 and 1.84×10^3 mm^3/min at 4 kW and 49.2 and 70.8 mm/s, respectively. For PSZ, the fastest rate at 663°C was 4.87×10^3 mm^3/min (4 kW and 254 mm/s) and at 996°C, 2.24×10^4 mm^3/min at the same beam power and scan speed.

Finally, Yamamoto and Yamamoto[56] reported on their studies of laser cutting conditions of Si_3N_4 ceramics. A thin Si_3N_4 plate can be cut with high speed by increased laser output power. The kerf width of the thin plate is also narrow. On the contrary, to cut a thick Si_3N_4 plate, the heat input has to be increased by slowing down the cutting speed because the plate fractures due to thermal shock.

A heat-affected layer is formed on the kerf surface. This layer contains metallic silicon to which Si_3N_4 decomposes. When O_2 is used as the assist gas, SiO_2 is formed on the top surface of the heat-affected layer.

The flexural strength as cut decreases by about one-third of the strength of the sintered Si_3N_4. This is thought to be caused mainly by the microcrack induced on the sintered Si_3N_4 surface due to thermal shock (Fig. 7.14). When the laser-cut plate is annealed to 1400°C for 1 h in N_2 atmosphere, the flexural strength regains up to about 80 percent of the sintered Si_3N_4 strength, even with the kerf roughness as cut, because the microcrack tips are blunted by the liquid phase.

Drilling. Generally ceramics are hard and brittle. Therefore to make a hole in a ceramic, a very hard drill such as a diamond drill or a cemented carbide tool must be used. If a hole must be punched first, it should be performed while the material is soft before sintering. The former method is time-consuming and tools wear down, while the latter has poor precision and accuracy because the work shrinks by several tenths of a percent during sintering. Laser processing appears to be a

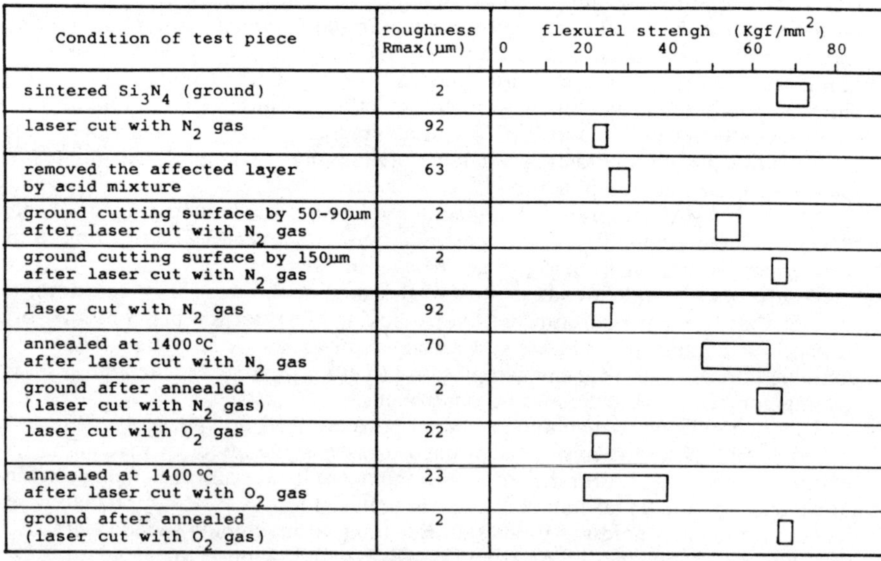

Condition of test piece	roughness Rmax (μm)	flexural strength (Kgf/mm^2)
sintered Si_3N_4 (ground)	2	~65
laser cut with N_2 gas	92	~20
removed the affected layer by acid mixture	63	~22
ground cutting surface by 50-90μm after laser cut with N_2 gas	2	~55
ground cutting surface by 150μm after laser cut with N_2 gas	2	~70
laser cut with N_2 gas	92	~22
annealed at 1400°C after laser cut with N_2 gas	70	~55
ground after annealed (laser cut with N_2 gas)	2	~60
laser cut with O_2 gas	22	~22
annealed at 1400°C after laser cut with O_2 gas	23	~40
ground after annealed (laser cut with O_2 gas)	2	~65

FIGURE 7.14 Flexural strength of laser-cut Si_3N_4.[56]

successful approach to solve this drilling application. In the past, a ruby laser was used to punch Al_2O_3. Today a CO_2 or a Nd:YAG laser is used together with pulse control to produce a hole. By moving the focal point according to the depth of the drilling hole, several laser-beam pulses are irradiated and the molten product is blown off by a gas jet which is coaxial to the laser beam.

When making a very small hole, the position of the laser beam is fixed, transferring the dimension and shape of the beam to the workpiece. To make a round or oval hole, the beam penetrates the material and moves to another spot.

When laser-beam machining in the liquid or gas phase is being applied to drilling, the most important problem is the removal of the molten product.[56-58] YAG lasers have been used to drill thick plates of more than 6 mm, because pulses with short pulse width and high energy can be obtained by Q switching. The ceramic chips were eliminated by raising the temperature of the material to a gas-phase state. However, CO_2 lasers have a long pulse width and small pulse energy. Thus as the ceramic chips are generated, they are generally removed in a liquid phase and clogging occurs easily. Therefore this laser is applied only to thin plates of less than 2-mm thickness.

It is recommended that to utilize lasers for drilling, a set opening ratio should be established. Round or oval holes with large diameters are advisable, because for small holes the distance between holes is small and a workpiece may be cracked easily. YAG lasers remove a very small amount at a time, and as the hole becomes larger, the processing time also increases. CO_2 lasers remove a large amount and the pulse frequency can be raised. Therefore, the processing time remains almost the same for any size hole. Further, YAG lasers precisely control the sectional form of a round hole while drilling a straight hole, but the processing time is long.

According to Affolter and Schmid,[54] in laser processing of ceramics with CO_2 and Nd:YAG lasers, the thermal damage of the cut face can be minimized to cracks of less than 100-μm depth by suitable process control.

Shaping. The development of reliable lasers that operate at wavelengths strongly absorbed by many materials of technological interest has led to their application in three-dimensional shaping operations. Complex shapes of hard ceramic materials such as Si_3N_4, SiC, sialon, and Al_2O_3 have been produced in the laboratory by laser shaping.[59-63]

Wallace and Copley[64] report that two approaches for laser shaping have been proposed: (1) overlapping shallow multiple grooves produced by a single beam to systematically remove layers, and (2) deep grooving with two orthogonal beams to remove sections with square or rectangular cross sections. In the former approach to laser shaping, used by Wallace and his associates, grooves with uniform cross section were formed on the workpiece source by scanning it with a laser beam at constant beam speed and power. By applying a continuous feed motion perpendicular to the scan direction, the grooves could be overlapped to remove planar layers with orthogonal, cylindrical boundaries. By applying an intermittent feed motion perpendicular to the freshly machined surface to either the workpiece or the focusing lens, successive layers were removed to produce stepwise approximations of three-dimensional shapes.

In the case of Si_3N_4, material removal results from decomposition to form liquid silicon and gaseous products. The total partial pressure of the gaseous products exceeds 1 atm, causing the liquid silicon to be ejected as small droplets.[64]

This study also determined the effects of beam speed and feed on the material-removal rate and the arithmetic average surface roughness in laser shaping by the overlapping multiple-groove (OMG) approach. Having verified the feasibility concept of the process, the researchers determined that it might be feasible to shape Si_3N_4 with a CO_2 laser and that it might also be more economical to remove material by laser versus conventional grinding methods. An effort was initiated to compare laser-beam machining with diamond grinding of an airfoil shape.[64] In reviewing and analyzing the data one can conclude that substituting laser shaping for diamond-point grinding could reduce the cost by 64 to 89 percent.[64]

Etching. The application of laser energy to etching technology is attracting attention as a promising future field of application.[65] It aims at finer micromachining through etching by making use of laser-induced chemical reactions, which can be divided into (1) photochemical reactions by electronic excitation and (2) thermochemical reactions by heat absorption and oscillation excitation.

The former mainly uses short-wavelength lasers (such as the excimer laser) having high photon energy for easy dissociation of molecular bonds. This technology is being promoted as a promising method for semiconductor lithography. The latter method heats the surface of the material by laser-beam irradiation. In processing by thermal reactions with processing liquids and reactive gases, Ar, YAG, and CO_2 lasers in the visible to infrared zones are mostly used.

Initially the possibility of etching by irradiating a laser beam onto Al_2O_3–TiC ceramics immersed in an aqueous solution of KOH was shown by the IBM Corporation, and similar work was disclosed in Japan.[19] Nogawa et al. conducted laser-etching experiments of HIP-sintered Al_2O_3–TiC.

Observations by Nogawa et al., who were trying to understand the laser-scanning conditions for low- and high-power beams, are presented in Fig. 7.15, which shows a proposed processing model.[19]

Recent work by Tsuchiya et al.[66,67] has shown that in a reactive gas atmosphere, high temperatures generated by laser irradiation cause a thermochemical reaction, forming a reaction layer and deforming the surface layer. If the material is highly volatile, efficient removal is possible. This was tried with a CO_2 laser-beam etching of ZrO_2 and Si_3N_4 ceramic materials in CF_4 gas.

Composite Machining (Laser plus Ultrasonics)

In composite machining, a material is machined not with a single machining method, such as in laser machining, electric-discharge machining, or grinding, but with a composite method in which more than two machining methods are combined. This combination utilizes the characteristics of each method for improving the performance and effectiveness of machining as a whole.

Fine ceramics are used in many areas as structural materials because of their superior durability and wear resistance. As a result, the requirements for machining advanced ceramics are becoming higher and the demand for machining two- or three-dimensional shapes is increasing. This is especially important for better finishing to prevent cracks on machined surfaces as well as to improve the effectiveness of machining.[19]

In the past, grinding, electric-discharge machining, and laser-beam machining have been used for these advanced ceramics and the following problems were encountered:

1. *Grinding:* Although simple-shaped materials such as plates and cylinders are easily machinable, there are many limitations on machining other shapes.

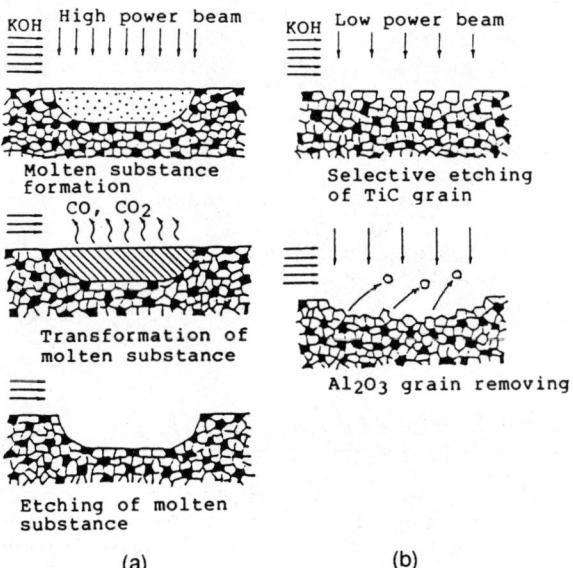

FIGURE 7.15 Processing mechanisms of laser-assist etching for Al_2O_3–TiC.[19] (a) Using high-power beam. (b) Using low-power beam.

2. *Electric-discharge machining:* While there are few limitations on the shapes of materials, nonconductive materials are not machinable.
3. *Laser-beam machining:* This process does not have the shortcomings of methods 1 and 2, and cutting out two-dimensional shapes is also possible. However, cracks occur easily when this method is used, thereby reducing the strength of ceramic materials. The machining of three-dimensional shapes is also very difficult.

In composite machining, to prevent the reduction of material strength, a workpiece is premachined with a laser, then finished with numerically controlled ultrasonic machining, using abrasive grains to produce a finishing equivalent of grinding.

This machining method is especially effective for ceramics which cannot be machined with the electric-discharge method. Using laser beams, two-dimensional shapes have been cut out, and contour sweeping corresponding to the machining depth required has been done for three-dimensional shapes. Subsequently, with ultrasonic machining, molten lumps produced by laser machining are crushed into small pieces, and sweeping traces of the laser are removed, producing the desired three-dimensional shapes. The machine is numerically controlled, giving a two- or three-dimensional pattern by moving the table on which the workpiece is placed. Ultrasonic oscillatory machining is performed with a formed tool, and machining of an arbitrary locus is done with a simple shaped tool.

Effects of Composite Machining. A sintered Al_2O_3 ceramic, which is nonconductive and cannot be machined with electric-discharge machining, was sweep-machined by laser beam and assisted by ultrasonic machining. The ultrasonic process was done on both a laser-machined surface and a ground surface. The machining speed was four to five times faster for the former than for the latter.

In this ultrasonic process the water in which the abrasive grains were suspended was poured over a vibratory tool during machining, as is done conventionally to grind machined surfaces. In addition, the vibratory tool, which is smaller than the conventional tool, provides the oscillatory motion.

In ultrasonic machining with an oscillatory motion, the machining fluid and abrasive grains are supplied smoothly between the tool and the workpiece, which results in a better removal of chips, leading to faster machining speeds and higher machining precision.

7.3 MACHINING WITH CERAMIC CUTTERS

In his introduction to the ASTME publication *Cutting Tool Material Selection* nearly a decade and a half ago, Shaw[68] wrote: "As the third 'law' of history indicates, new tool materials should be expected to appear in response to new requirements." This "law" has not been violated. Today's production engineer has available a wide variety of tool materials from which to choose. The production and economic advantages of utilizing the proper tool material in any machining operation are considerable.

With the exception of metal-cutting tools, all other cutting tools can be considered ceramic in nature and are products of the ceramic industry. Cermets, one type of ceramic, are composites of ceramics and metals; sintered carbides are included in this category.[69]

Because hard materials are used to cut softer materials and because ceramics are inherently hard, an interest in the use of ceramics for cutting tools developed in Germany around the beginning of the century (1905).[70] Cemented carbides were introduced in Europe in 1926. During World War II, carbides were used to increase the cutting capability of the earlier high-speed steel tools. This was followed by the introduction of Al_2O_3-coated WC.

Ceramic cutting tools did not become commercially available in the United States until 1954. It is apparent that since the end of the 1960s the strengths of ceramic tools have been improved through a better control of the microstructure and by the addition of additives. A better understanding was initiated, and as a result mechanical, thermal, and chemical properties of ceramics were obtained and adapted to ceramic cutting tools. In addition, composition variables gave rise to the cermets, and additions of titanium, TiC, and WC were made to improve the strength of Al_2O_3 cutting tools.

Harder materials such as polycrystalline diamonds (PCD) and cubic boron nitride (CBN) were developed for use during the 1972 to 1974 time period. CBN, Al_2O_3–TiC ceramics, and the newer ceramics based on Si_3N_4 are very promising because of their wear resistance and effective application to the machining of nickel- and cobalt-based superalloys in the jet engine industry. Polycrystalline diamonds find excellent application in machining abrasive materials such as silicon-aluminum automotive alloys and glass- and carbon-fiber-reinforced composite materials. The history of ceramic cutting tools has evolved from the use of pure Al_2O_3 to the tailoring of new compositions.

Major Barriers

There are several barriers to the development and diffusion of advanced ceramic cutting tools. Technological barriers include the reliability and cost of these versus other types of cutting tools. Other barriers are institutional in nature and involve the questionable reputation of ceramics resulting from improper early use.

Technical Barriers. Reliability problems stem from the brittle behavior of ceramic cutting tools. This results in inadequate predictability with regard to the timing of failure or breakage in a tool. Premature, or unexpected, breakage of a cutting tool increases the costs of machining. Moreover, the reliability problem with advanced ceramic cutting tools may be more serious than with carbide cutting tools, both because failures in an advanced ceramic tool are less predictable and because the damage caused by unexpected breakage is greater for an advanced ceramic tool. A number of critical areas for further R&D exist:

The relationship between tool failure and cutting speed for ceramic tools is complex and poorly understood. The toughness of ceramic materials tend to increase with increasing temperature. Therefore, reducing the cutting speed gives no assurance that tool life will be longer or more predictable. In fact, it has often been observed that catastrophic failure occurs less frequently at higher cutting speeds.

A better understanding of fracture mechanisms to improve the fracture toughness of advanced ceramic cutting tools is needed. These mechanisms have been and are studied in the structural ceramics field, and the lessons learned can be transferred to the cutting tool field. Research is needed in compositional variations to increase hardness and to gain a better understanding of the chemical behavior of ceramic cutting tools in use with various metals. A reliable, standardized measurement test, however, is needed. Once the fracture problem is addressed and understood, there remains the problem of high-temperature plastic flow of the ceramic, which causes wear by flow-induced nucleation. The development of material structures with increased flow stress is also desirable.

If the ceramic cutting tool is to be used to increase productivity, it has to be capable of withstanding high temperatures and thermal shock, and it must not be subject to corrosive reaction with the workpiece.

The development of more wear-resistant advanced ceramic coatings for cemented carbides and high-speed steel tooling is needed.

Measurement-Related Barriers. Standardized test methods and standard reference data are lacking throughout the ceramic industry. Besides many critical input parameters being unavailable and the questionable reliability of existing data, the establishment of a computer database for the design and selection of tooling systems, and particularly for the design of coating systems as well as the characterization of the bulk properties of ceramics, is required. Required areas include:

Adhesive bond strength
Thermodynamic properties
Modulus of elasticity
Coefficient of thermal expansion
Fracture toughness

There have been no clear-cut scientific selection criteria developed to match the specific workpiece with the optimum and most cost-effective type of cutting tool. Most often users follow the recommendations of a vendor or the guidelines in handbooks, and they use their own trial-and-error experience to arrive at the proper choice of cutting tool.

The scientific approach would involve all of the measured data and the criteria developed as the fallout of all test results for the user to be able to systematically arrive at the best choice for a specific machining application.[70]

Institutional Barriers. Machine tool users initially suffered many negative experiences with advanced ceramic cutting tools. Poor performance resulted and the principal problem was reliability. Premature failure occurred frequently, and as a result many users became particularly resistant to the idea of trying ceramic cutting tools again. However, ceramic tools can behave well when used properly and are most productive (Fig. 7.16). The demand for these tools has increased, particularly in computer-aided, high-speed machining applications.

Cutting-Tool Materials and Properties

A new variety of ceramic cutting-tool materials was introduced in recent years and is now used commercially in a wide range of metal-cutting applications, including turning and milling. These materials comprise three main categories: metal oxides, metal nitrides, and metal carbides. They can be variously combined, as shown in Fig. 7.17. Table 7.11 lists typical compositions and other details.

According to Gruss,[71] whether the cutting tool can handle higher cutting temperatures or cutting speeds depends on the material's properties. The bending strength of the cutting material at high temperatures is obviously important. In general, the carbides have the highest bending strength at room temperature (Fig. 7.18).

Gruss[71] and others[72] state that hardness and fracture toughness are two other critical properties. Hardness indicates wear resistance. The higher hardness of the carboxides, as shown in Fig. 7.19, permits the turning and milling of steel hardened to 64 on the Rockwell C scale HR_C. On the other hand, the oxides, Si_3N_4, and cermets are limited to the machining of materials hardened to about 38 HR_C. High fracture toughness is needed for interrupted cutting. The fracture toughness of cermets is similar to that of cemented carbides (Fig. 7.20), permitting the use of cermets for interrupted cuts in turning and milling.

Si_3N_4-based ceramics, generally used for machining gray cast irons, also have sufficient toughness for milling applications. However, the low toughness of the oxides and carboxides allows for interrupted cuts only at very low feed rates. Consequently ZrO_2 is added to Al_2O_3 to improve toughness.

Because oxides and carboxides have very low thermal-shock factors according to Gruss,[71] Whitney and Vaidyanathan,[73] and Drozda,[74] coolant can be used in finishing cuts only under the following maximum conditions: cutting speed 7.6 m/s, feed rate 0.025 cm/rev, and depth of cut 0.051 cm. Cermets have similar thermal-shock resistance; therefore cooling is applicable for finishing cuts, threading, and grooving, but is not recommended for roughing operations. Because Si_3N_4-based ceramics have good thermal-shock resistance, coolant can be used in roughing operations.

Commercially available ceramic cutting tools comprise four material catego-

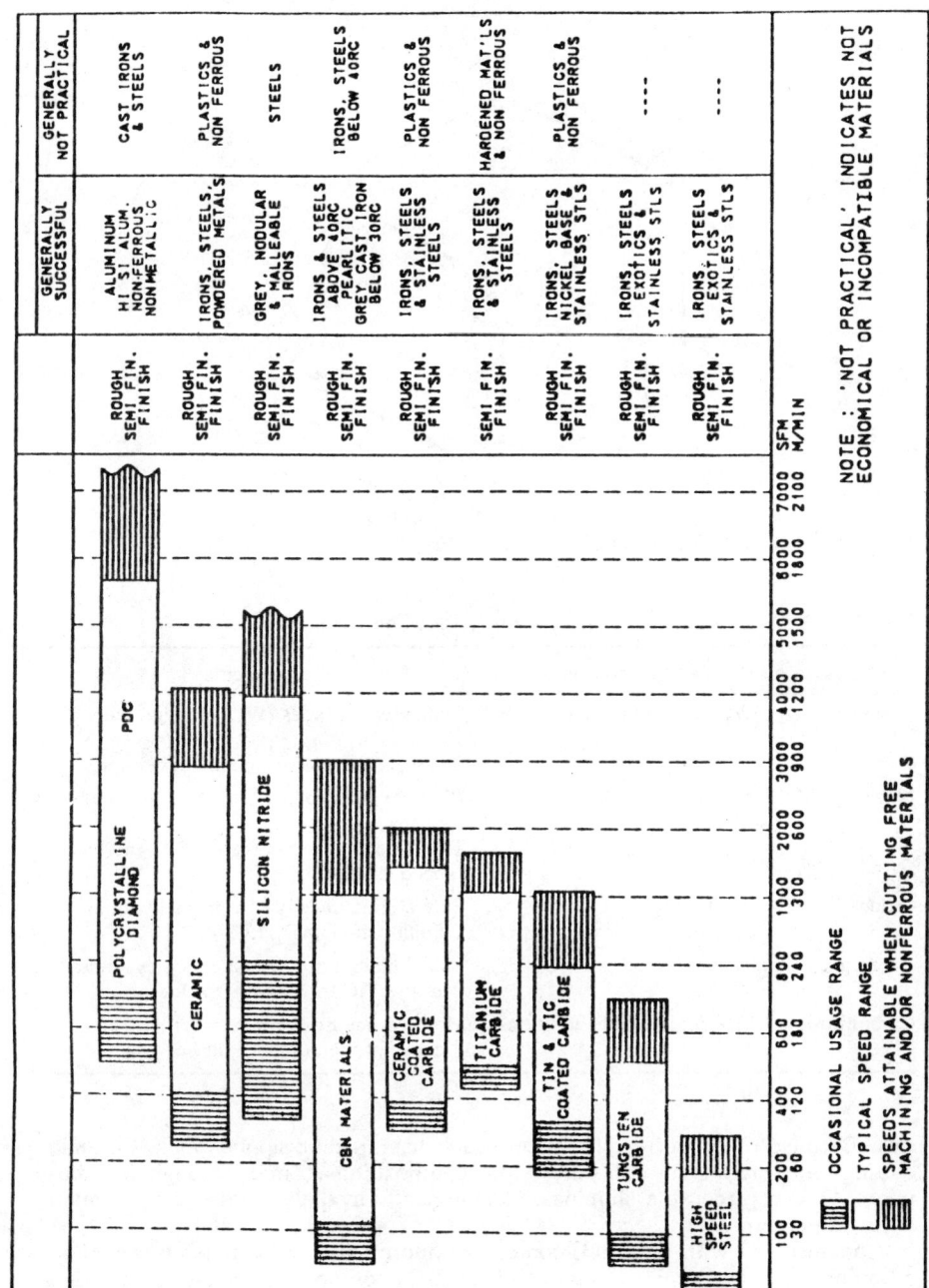

FIGURE 7.16 Spectrum of cutting-tool materials and guidelines for use.

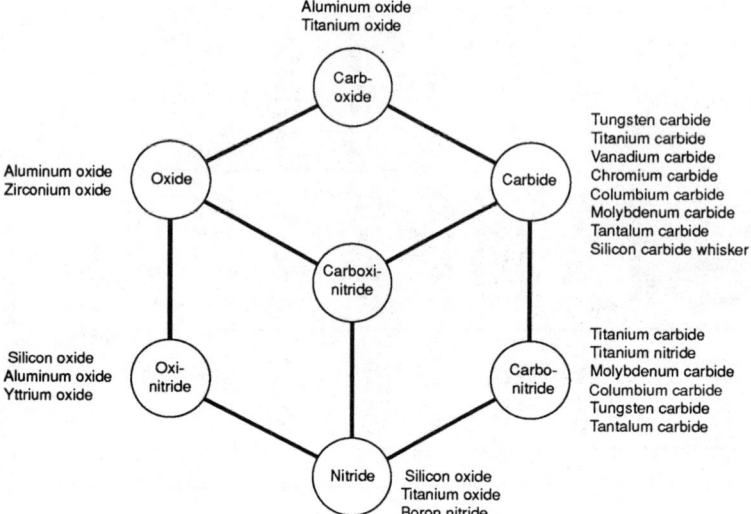

FIGURE 7.17 Major categories of cutting-tool materials.

TABLE 7.11 Wide Range of Ceramic Cutting-Tool Compositions

Category	Compositions	Comments
Carbides	WC, TiC, MoC, NbC	Cemented carbides (WC) are sintered with metals
		SiC-whisker reinforcements in ceramics
Nitrides	BN, sialon, TiN	BN is expensive because of high manufacturing costs
		TiN is used as a coating for WC and high-speed steel
Oxides	Al_2O_3, ZrO_2	Pure alumina has been replaced with alumina containing $\leq 15\%$ ZrO_2
Carboxides	TiC dispersed in Al_2O_3	Called black hot-pressed ceramics, contain $\leq 40\%$ TiC to improve hardness
Carbonitrides	TiN with carbide additions	More commonly known as cermets because of metal binder phase

ries. Because of property differences, each has specific applications. Ceramic cutting tools originated with pure Al_2O_3 compositions. Zirconia-toughened alumina (ZTA) replaced the aluminas. Commercially available grades now contain up to 12 percent ZrO_2.

Compositions with low ZrO_2 content are more suitable for finish turning and grooving, whereas compositions with high ZrO_2 content are preferred for rough turning. Applications for the latter include turning of gray cast irons, nodular cast iron, and malleable cast iron hardened to 300 on the Brinnell scale HB at cutting

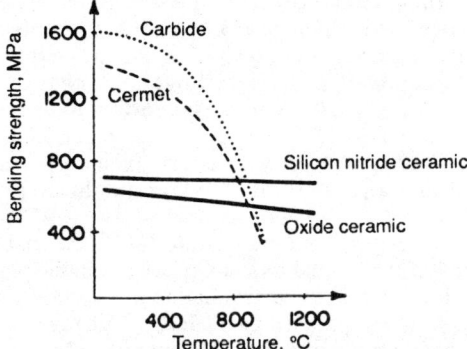

FIGURE 7.18 Hot strength of cutting-tool materials.

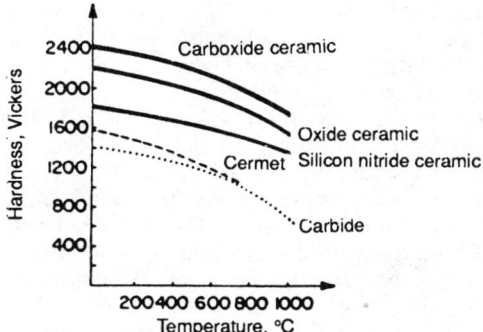

FIGURE 7.19 Hot hardness of cutting-tool materials.

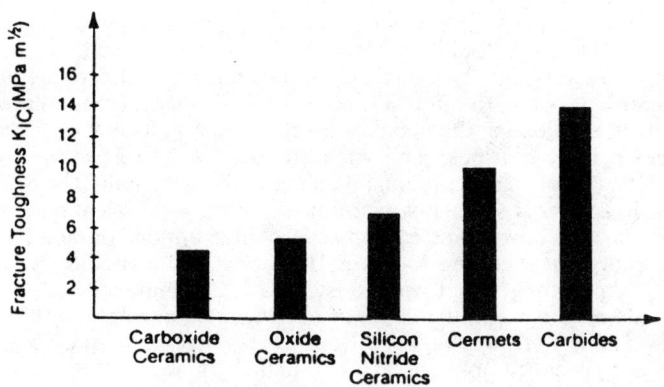

FIGURE 7.20 Fracture toughness of cutting-tool materials.

speeds to 15 m/s. Turning of carbon steel, alloy steel, and tool steel hardened to 38 HR_C at cutting speeds to 18 m/s also is possible with this grade.

Toughening of Al_2O_3 cutting tools via phase changes exploits martensitic-type transformations, a specific type of crystallographic change that is an energy-absorbing mechanism. Al_2O_3–ZrO_2 composite cutting tools have been developed that are much more resistant to fracture and thermal shock than conventional Al_2O_3 cutting tools.[73] Metastable tetragonal ZrO_2 is incorporated in an Al_2O_3 matrix and increased strength results from the stress-induced phase transformation of the ZrO_2 particles.

A hot-pressed ZrO_2–Al_2O_3 ceramic cutting tool[74] insert is being used by an aircraft engine manufacturer to machine a variety of engine parts made of Inconel 718. Often many of the company's parts must be production-machined in the fully heat-treated condition, some as hard as 48 HR_C. Cutting speeds were increased from 34 m/min for cemented carbide to 244 m/min, which is nearly double the speed of conventional hot-pressed ceramics and a sevenfold improvement over cemented carbide.

Higher cutting speeds now mean that operations involving long cuts do not have to be stopped for tool changes. Surface finishes are improved due to higher cutting speeds. Feed rates are comparable to rates formerly used with carbide cutting tools. The depth of cut is unchanged except for severe roughing cuts. Even chip removal is easier because at higher cutting speeds chips are broken into flakes, reducing chip control problems.

The second category includes Al_2O_3–TiC materials. In order to increase the strength of Al_2O_3-based cutting tools, in the late 1960s Japanese researchers added TiC in varying amounts, typically 30 wt %. This resulted in what is often called a composite or black ceramic, and increased the transverse rupture strength (breaking strength) of the basic Al_2O_3 ceramic. These ceramic materials have been very successful in turning hard cast irons and heat-treated steels hardened to 64 HR_C to very fine surface finishes and tolerances normally obtained only by grinding. Parts machined with this category of ceramic cutting tools range from small gear parts for the automotive industry to large form rolls for steel mills. The cutting speed is 1.3 m/s, the feed rate is 0.25 mm/rev. Productivity can double with this type of material compared to cemented WC.

Ceramics which combine Al_2O_3 and TiC are known by many names in addition to hot-pressed and black ceramics. Other terms used to describe such ceramics include cermets, composite ceramics, and modified ceramics. Hot-pressed ceramics have found application in the replacement of conventional carbide in many jobs. For example, Al_2O_3–TiC hot-pressed ceramic inserts have replaced carbide tooling to effect a productivity increase and an improvement in surface finish, eliminating the need for a subsequent polishing operation.

Carbon steel key bearings, requiring a surface finish of 2 μm rms, historically were machined using carbide tooling on a vertical boring mill. The 813-mm bearing plates, in spherical segments, were mounted on a spherical part for machining. Bolt holes and dowel pins caused severe interruptions of cut. The feed rate was 0.25 mm/rev at a cutting speed of 166 m/min. By switching to hot-pressed Al_2O_3–TiC, the cutting speed was boosted to 355 m/min, the feed rate was reduced to 0.20 mm/rev, and the depth of cut was held constant at 0.38 mm. As a result, the cutting time was reduced from 13 to 7.7 min. With carbide, each triangular insert produced only one and one-half workpieces. With ceramics, each 80° diamond-shaped insert produced four workpieces. Consequently actual tooling costs were reduced. Finally, because the surface finish was held consistently to specification, no polishing was required.

The higher toughness and thermal-shock resistance of Si_3N_4-based ceramics, compared to Al_2O_3, permits the rough turning and milling of cast irons under severe conditions. These conditions include heavy interruptions, variations in depth of cut, rough scale, or the use of a coolant. High-temperature nickel-based alloys can also be turned quite efficiently with this type of cutting tool. However, steels are unsuitable for machining with Si_3N_4 ceramics. An example of an application with this material involves the milling of a housing made of gray cast iron. The cutting speed was 13 m/s and the feed rate 0.15 mm per tooth. This is the third group of ceramic cutting tools.

These Si_3N_4 cutting tools can increase horsepower utilization without increasing tool force. According to Drozda,[74] five times the speed at a given horsepower results in one-fifth the cutting force. So five times the speed allows five times the horsepower utilization with about the same tool force. Still further gains are possible due to improved plastic flow caused by increased temperature in the cutting zone.

Field performance reports show a variety of benefits. In machining cast iron brake drums, for example, the Si_3N_4 tooling provided 4.7 times the tool life, while slicing the cutting time to 25 percent of that possible with traditional hot-pressed ceramic. When tested against carbide at low speed (229 m/min), Meehanite cast iron bushings were machined in half the cutting time with a sixfold increase in tool life. When milling a cast iron machine base, the metal-removal rate was doubled while the tool force was reduced by 70 percent of that exerted by hot-pressed ceramic.[75]

Other applications are equally dramatic. Inconel 718 lug castings are machined using a Si_3N_4-based ceramic with a 90 percent reduction in cutting time and an 84 percent reduction in total machining cost.

The workpieces, measuring up to about 305 mm in diameter, are machined on a 50-hp (37-kW) vertical turret lathe. In the past, this lug machining operation was performed using two different uncoated carbide inserts making three 127-mm passes. The operation formerly took 22.5 min, cutting at 30 m/min to about 0.89 mm depth at 0.30 mm/rev. Today, with the use of Si_3N_4 inserts, the operation cuts at speeds of up to 267 m/min. The job now takes about 2.2 min, running at 1.78-mm depth of cut.[76–78]

Si_3N_4 tools are also effective when compared with Al_2O_3. In one application, a gray cast iron automotive brake disk increased feed rate over Al_2O_3 from 0.38 to 0.51 mm/rev and provided an average tool life of 400 pieces per cutting edge versus 150 for Al_2O_3.

Another class of Si_3N_4-based cutting tools are called sialons. These are solid solutions of metal oxides such as Al_2O_3, Y_2O_3, MgO, BeO, and other materials in a β-Si_3N_4 crystal matrix.

Field tests have shown sialon cutting tools to be successful in the machining of cast iron, nickel-based superalloys, and aluminum-silicon alloys. In these tests the sialon tools exhibited longer tool life (at both low and high cutting speeds) than the cutting tools traditionally used for these materials.

Cutting tools are the leading edge of commercial ceramics, and breakthroughs there have a way of working back to other applications where wear and heat resistance are important. So it is heartening to see how researchers have solved sintering problems that occur when Si_3N_4 cutting tools are strengthened with TiC.

While TiC–Si_3N_4 tools do have excellent properties, the two materials react to form nitrogen gas when sintered. The released gas forms pores, which concentrate stresses and weaken the tool. The solution found was to coat TiC particles with thin TiN coatings. The TiN film keeps the TiC from reacting with Si_3N_4 during sintering, so no nitrogen gas or by-products are formed. The resulting tool is easier to sinter and performs better than TiC–Si_3N_4 tools.

The fourth and final grouping of tools are SiC whiskers, which are now commercially available from various manufacturers as reinforcements for cutting tools. The whiskers improve toughness, strength, thermal-shock resistance, and reliability. These reinforcements can be added to oxide, nitride, and carboxide ceramics. Commercially available grades contain fewer than 30 percent of whiskers, resulting in higher toughness and better thermal conductivity.

The high cost of SiC whiskers is still an obstacle, but is expected to decrease with improved manufacturing processes and increased market growth. Such whisker-reinforced tools are applied mostly in the machining of high-temperature alloys, resulting in significant productivity increases. The rough turning of a turbine rotor made of Incoloy 901 (40 HR_C) was machined using round inserts 12.7 mm in diameter at a cutting speed of 4.5 m/s and a feed rate of 0.15 mm/rev. Stock removal was about five times faster than with conventional cemented carbides.

Considered a reinforced Si_3N_4 ceramic, a new material recently introduced holds the promise of dramatic performance. This material is a ceramic-ceramic composite. The unusual ceramic material, WG-300, is said to have almost twice the fracture toughness of traditional hot-pressed ceramics and a significant increase in resistance to thermal shock.

Improved performance of the material is attributed to a very uniform dispersion of single crystal "whiskers" of SiC. The so-called whiskers measure about 0.02 μm in diameter and about 1 μm in length. Hexagonal in shape and randomly distributed, these fine strands reportedly serve as reinforcing rods, distributing stress within the matrix. Also the high thermal conductivity of the SiC whiskers conducts heat away from the cutting edge, thereby reducing thermal gradients and the resulting stress.

Al_2O_3–SiC composite inserts generally contain 30 to 45 percent SiC whiskers. Extensive physical, mechanical, and performance data on Al_2O_3–SiC whisker material are now available.[71,73] It has been reported[79] that the "characteristic strength" of hot-pressed Al_2O_3 bodies increased from about 504 to 690 MPa as the content of SiC whiskers increased from 0 to 30 wt %. Adding SiC whiskers, followed by hot pressing, can more than double Al_2O_3's fracture toughness to a value of 8.7 MPa · $m^{1/2}$.[80] In actual metal cutting of alloys such as Inconel 718, composite tools featured a performance up to three times better than conventional ceramic tools and eight times that of carbide.[81,82]

Other cutting ceramic materials should be noted and include cubic BN. Cubic BN can be used effectively in machining powder-metal parts and yields acceptable tool life of 15 times that of coated carbides.

A BNX4 grade of cubic BN has excellent wear resistance and good retention of edge sharpness in machining iron-based powder metal and cast iron. According to Nakai et al.,[83] these properties have been achieved by improving the ceramic binder materials and choosing the right size and type of cubic BN particles.

Turning powder-metal diesel-engine valve-seat inserts containing dispersed carbide particles for high wear resistance was not practical with carbide inserts because of very rapid flank wear. Cubic BN inserts, having a sharp edge and run dry, cut 60 of these parts in tests at 100 m/min, a feed of 0.1 mm/rev, and a 0.2-mm depth of cut. The new grade BNX4 showed about half the flank wear of a previously marketed cubic BN grade. In a tool-life test on gasoline-engine valve-seat inserts, BNX4 cubic BN inserts machined 3000 pieces, 15 times more than were completed by carbide inserts.

Other cubic BN abrasives have been used for advanced ball-bearing grinding and superfinishing bearing components[84] as well as for cylinder liners, engine bores, and brake disk rotors.[85]

Future Use of Ceramic Tool Materials

The advances in understanding the relationships between ceramic microstructure and properties, together with improved processing techniques, have led to the existing range of ceramic tool materials with significantly superior combinations of wear resistance and toughness in the materials of even 5 years ago. Ceramic materials are still intrinsically brittle, however, and in the context of modern automated manufacture it is essential that the performance of any tool be reliable in terms of tool life, security against breakage of the edge, and chip control, particularly in finishing operations where the swarf may be a few tenths of a millimeter wide and as little as a tenth of a millimeter thick and coming off the workpiece at a speed of several hundred meters per minute. In order to break the swarf into manageable lengths, cemented carbide inserts have complex chip-breaker forms which are produced when the powder is pressed to shape. Ceramic pressing technology does not, as yet, allow such complex forms to be created.

Some of the future processing needs, according to Kennedy and Skaar,[86] include the following:

> Grinding inherently creates surface and subsurface damage, and the extent of this damage must be quantified before it can be minimized. General areas needing attention include improved dimensional tolerance, finer finish, complex geometries, and part reproducibility in production quantities. It is also important to define the relationships among the grinding parameters and between these parameters and the desired grinding properties so that machining algorithms can be developed for different workpiece materials and grinding conditions.

> One major research need for polishing systems is to determine optimum combinations of machines and polishing materials, and the conditions of their use. No single source is presently available which can be consulted to determine polishing grits, machines, and speeds and feeds for an arbitrary ceramic material and finish. This is due in part to an absence of fundamental research, and in part to the lack of codification of existing research. Present and new combinations of abrasives, machines, and machining conditions need to be studied together with applicable ceramic materials, and the results need to be organized and published.

> Another research need is to study the fundamental mechanisms of polishing, including electrochemical interactions between slurry, grit, and the workpiece. While ceramic materials are usually inert, chemical reactions are known to occur in some combinations of slurry, grit, and workpiece. Such reactions can be either detrimental or beneficial to the surface finish of the workpiece, but in either case they are not well known.

> A third significant research need is for surface and subsurface characterization to be applied to polishing. The ability to measure surface properties has improved with the advent of acoustic-wave measurements and selected-area electron-channeling techniques. Specific needs exist for rapid reliable surface finish measurement, particularly for applications in statistical process control.

Postmachining treatments need to be developed which will improve the strength of the ceramic significantly in service environments. Also, the mechanisms of strength enhancement need to be identified. A combination of these approaches would have a high payoff, because some strength improvements may be recognized in the short term and the implementation is expected to be relatively low in cost.[86-90]

7.4 JOINING OF STRUCTURAL CERAMICS

By applying principles and basic requirements known for joining various materials regardless of joint complexity, one can successfully join glass-metal, ceramic-metal, and ceramic-ceramic components.[91-96] The fundamental requirements for strong glass-metal and ceramic-metal assemblies can be summarized as follows:[91]

1. Maximum strength dependent on the development of chemical bonding at all interfaces
2. Consideration of physical factors (stress and stress gradients) as affected by properties and chemical reactions, such as matching coefficients of thermal expansions and determining stress gradients by composition gradients and interfacial zone microstructures

Chemical bonding at interfaces is achieved by one of two methods:[91]

1. Formation of an intimate atomic contact interface
    a. Solid-solid assembly: Pressure, resulting in physical adjustment at interface by localized deformation
    b. Solid-liquid assembly: Wetting, resulting in penetration of irregularities; spreading by reaction or pressure
2. Reaction to reach chemical equilibrium at interface, saturation with compatible phase
    a. Metal-metal and ceramic-ceramic assemblies: Solution reactions and saturation at interface; diffusion bonding
    b. Glass-metal and ceramic-metal assemblies: Solution of oxide layer on preoxidized metal, redox reactions; saturation at interface with substrate oxide, product redox reactions; formation of compound at interface compatible with both phases

Joining facilitates the use of structural ceramics by providing means for manufacturing structures which cannot be made in one piece, or which can be made less expensively, or otherwise more satisfactorily, by joining. Joined ceramics can be stronger, either by combining stronger components or by introducing compressive stresses (from strong fibers, as in composites, or from metal parts in tension). Joined parts can be made with shapes and tolerances not readily achieved otherwise.

Perhaps the most salient application for ceramic joining is in the attachment of ceramic components functioning at high temperatures to structures or moving parts which must withstand stresses or temperature gradients too great for ceramics and must therefore be metals.

The design of systems with ceramic joining requires detailed information on the joints, including the materials in the layers of the joints. Analyses of joint and system performance can be made if appropriate material properties are known.

These include thermal expansion, with hysteresis and annealing effects (such as devitrification), viscosity, elastic moduli, strength, fracture toughness, creep, and fatigue—all as functions of temperature. The availability of joining techniques and of good information on the joint properties will influence designs in the development of systems utilizing high-temperature structural ceramics.

Ceramic-metal and ceramic-ceramic components have been joined by at least one of three major techniques:

1. Mechanical joining, used in both traditional and new applications such as tying in of furnace roof refractories with metal hooks or dog bones and the clamping of Space Shuttle leading edges. Press and shrink fitting is another type of mechanical joining which is widely used in mass-production processes.
2. Direct joining, achieved by pressing together very flat mating surfaces to achieve diffusion bonding, as in the joining of sapphire to columbium during the fabrication of high-pressure sodium lamps. Experimental work on fusion welding using electron-beam, laser, and imaging arc techniques has also met with some success with high-melting-point systems.
3. Indirect joining, the most common method of achieving high-integrity joints using a wide range of intermediate bonding materials such as organic adhesives, glasses or glass-ceramics, oxide mixtures including cements and mortars, or metal. Metal intermediates are used as solid-state diffusion bonding agents or as brazes with or without pretreatment of the ceramic surfaces to render them wettable.

Whatever process is used, the formation of successful joints depends on the achievement of intimate contact between the workpieces, the conversion of these contacting surfaces into an atomically bonded surface, and the ability of this interface to accommodate thermal expansion mismatch stresses generated during cooling after fabrication or temperature changes in operational conditions. The successful fabrication of metal-ceramic interfaces is not easy, and the difficulties are often impractical rather than matters of engineering design or economics. Thus producing metal-ceramic contact by fusion welding is usually, but not always, impractical because of marked differences in the refractoriness of the workpiece materials, and the nonwettability of many ceramics by common metals and alloys presents problems that must be overcome when brazing is used for indirect joining.

The technology of joining ceramics to metals has progressed steadily since its beginnings in the early 1930s. Like other joining processes, the joining of ceramics to themselves and to other materials was long considered to be more an art than a science. Much has been accomplished in the intervening years to establish a joining technology based on sound fundamental procedures and an understanding of the reactions that occur during joining.

One must realize that in the joining of ceramics for most applications, ceramics are not used alone. Rather, ceramic components are part of larger assemblies. Therefore the ceramic must be joined to more conventional materials in the assembly to function properly. Broad research in joining ceramics to metals, glasses, and other ceramics could have a decisive impact on a future use of monolithic ceramics, coatings, and composites. The key to joining is an understanding of the surface properties of the two materials and of the interface between them. In general, the interface is a critical point of weakness in discrete ceramic components such as those in heat engines, in ceramic coatings on metal substrates, and in ceramic fibers in composites. The principal needs in this area are in the strengthening and toughening of joints, an understanding of their high-temperature chemistry, and improved resistance to corrosion in the environments of interest.

Before going further, the adhesive bonding of ceramics should be mentioned, especially from the aspect of adhesion.[97] This is covered later in this section.

Key Parameters of Joining

The basic requirements for strong assemblies are chemical bonding and minimal stress differentials at the interfaces, with favorable stress gradients in the interfacial zones, as stated by Pask.[91] In fact, Pask[91] and others[93,96] feel that one can generalize that any two phases can form an acceptable assembly with a chemical bond if they are at stable chemical thermodynamic equilibrium at their interface, whether or not the bulk phases are at equilibrium, provided they are also compatible physically.

In joining, the formation of an intimate or true interface is the first requirement. Such an interface is one in which atomic contact exists either with van der Waals attractive forces or with an electronic structure across the interface, that is, a chemical bond. An intimate solid-liquid interface can be formed and recognized easily if the liquid wets or spreads, thereby penetrating irregularities at the solid surface. The second, more critical, requirement in all cases is the presence of a stable chemical thermodynamic equilibrium at the interface. The simplest reaction is the solution of one phase by the other to form an immediate equilibrium saturation at the interface. A continuation of the reaction is associated with diffusion into the bulk. The overall reaction is thus referred to as diffusion bonding.[91] The various theories, experiments, and technology developments were compiled recently in Schwartz.[95]

Wetting and Spreading. The shape of a drop of molten braze filler metal on a solid ceramic surface is determined by gravity and interacting forces of solid-liquid interfacial energy γ_{SL}, solid-vapor interfacial energy γ_{SV}, and liquid surface tension γ_{LV} (Fig. 7.21).[98] This balance of interfacial tensions is characterized at equilibrium by the Young equation

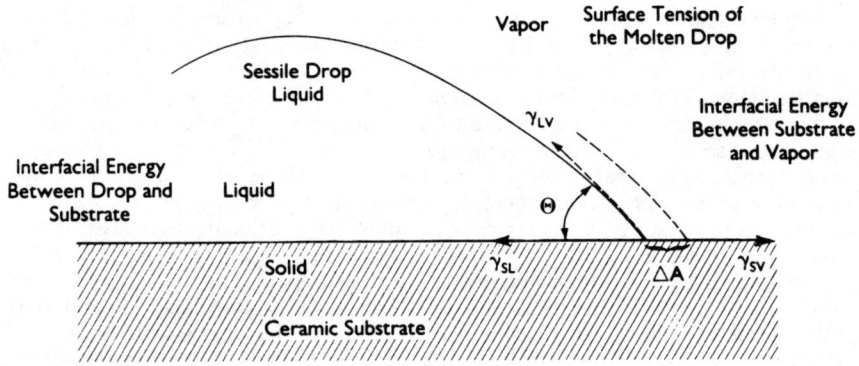

FIGURE 7.21 Surface energy forces acting on sessile drop.[91,95,98]

FABRICATION AND MANUFACTURING METHODS

$$\gamma_{SL} - \gamma_{SV} + \gamma_{LV} \cos \theta = 0 \qquad (7.1a)$$

or

$$\gamma_{SL} = \gamma_{SV} - \gamma_{LV} \cos \theta \qquad (7.1b)$$

The angle between the solid surface and the tangent to the liquid surface at the contact point, that is, the contact angle θ, may vary from 0 to 180°. If γ_{SL} is high, the liquid tends to form a ball having a small interface area. If γ_{SV} is relatively high, the drop tends to spread. The effect of the liquid-vapor interfacial energy (surface tension) is not as straightforward. If only the liquid surface energy is decreased, the contact angle decreases (that is, wetting is increased) for initially wetting drops ($\theta < 90°$), but it increases for initially nonwetting drops ($\theta > 90°$), as illustrated in Fig. 7.22.

In Eqs. (7.1) it can be seen that θ is greater than 90° when γ_{SL} is larger than γ_{SV}, as shown in Fig. 7.23a, and the liquid drop tends to spheroidize. If the contact angle θ is less than 90°, the reverse is true, as shown in Fig. 7.23b, and the liquid drop flattens out and wets the solid. If the balance is such that $\theta = 0$ and greater wetting is desired, θ should be as small as possible so that $\cos \theta$ approaches unity and the liquid spreads over the solid surface.

These considerations show the importance of surface energies in brazing. For a more detailed analysis of adherence, brazing wettability, and the associated energy or work of adhesion, consult the literature.[91,98-102]

The normal technological procedure of achieving chemical bonding and favorable stress gradients in glass-metal assemblies is to preoxidize the metal and apply the glass. According to Tomsia and Pask,[103] the molten glass wets and dissolves the oxide at that temperature. The glass at the oxide interface immediately becomes saturated because the solution rate of the oxide is faster than the diffusion rate of the dissolved oxide into the bulk glass. Chemical bonding is thus realized.

The concept of similarity is also used to promote molten metal-solid oxide wetting and bonding by metallizing the ceramic surface. The most widely used process[104-106] uses molybdenum and "debased" Al_2O_3, where the grains are held together by a glassy binder phase. Electroless plating and various vapor-deposition processes have also been used to metallize oxide and nonoxide

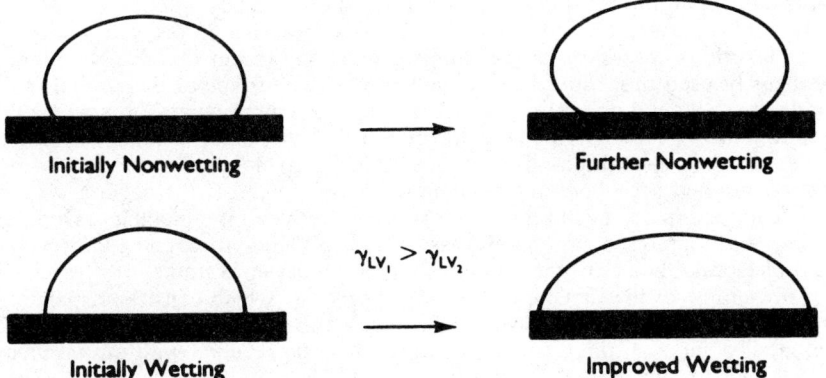

FIGURE 7.22 Effect of contact angle of decreasing liquid surface tension γ_{LV} for initially nonwetting drop (top) and initially wetting drop (bottom).[95]

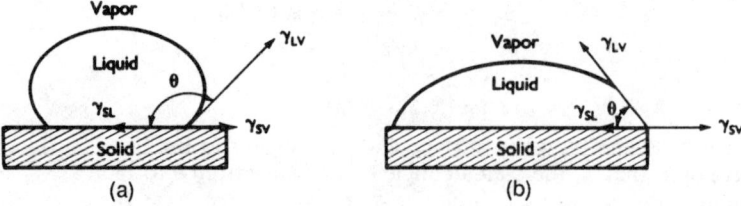

FIGURE 7.23 Sessile drops and interfacial energies. (*a*) Contact angle >90°; no wetting occurs. (*b*) Contact angle <90°; wetting occurs.[95]

ceramics. The second most common process is probably TiH_2 activation, in which the hydride is applied as a powder before vacuum brazing, dissociates at 350 to 550°C, and can form a wettable titanium coating. In some practice the hydride is used to fill the ceramic-metal capillary gap rather than merely coating the ceramic. Hence dissociation produces a wettable gap structure of loosely touching titanium powder that acts as a wick and dissolves in the liquid to form an active metal braze in situ. This dissolution effect is also exploited in the use of intermediate-layer foils or coatings, which alloy with the metal workpiece to form a eutectic melt capable of wetting and bonding to the ceramic as, for example, in the joining of copper to Al_2O_3 using a titanium foil that forms a Cu–27 Ti eutectic melting at 880°C.

Ceramic and Metals versus Glass and Metals. More complicated technological problems generally arise in joining ceramic and metal solids than in glass-metal sealing where the glass itself is in a liquid state during sealing. But, according to Klomp,[107] the same basic requirements exist for an acceptable assembly: chemical bonding across the interfaces and microstructures in the interfacial zones with favorable stress patterns. The need for a favorable stress pattern and for the presence of a liquid phase at the interface primarily for fabrication reasons has been the dominating factor in the design of various assemblies that use some intermediate material for joining.

The simplest configuration and approach is the direct solid joining of flat specimens. Materials tested include α-Al_2O_3–Cb bonded at 1700°C and Cu–Al_2O_3 bonded between the melting point of copper and the CuO_2 eutectic.

Instead of having an active metal as part of the braze in order to realize the redox reactions necessary for spreading and an equilibrium chemical reaction, an extensively used procedure is to metallize first the surface of the ceramic to be bonded to the metal or another metallized ceramic component. The braze is then used to join two metal surfaces with which it is compatible, wets those surfaces, and undergoes a solution-diffusion reaction. Two metallizing processes most commonly used are illustrated schematically in Fig. 7.24. In the first type the chemical bond is achieved by a redox reaction between the metallic coating and the ceramic. In the second type the assembly is obtained through an intermediate glass phase that bonds to the metallizing and the ceramic grains.

An example of the first type is the TiH_2 process, which consists of painting a slurry of TiH_2 onto the ceramic, drying, and firing at a reduced pressure. The hydride decomposes, the titanium undergoes a redox reaction with the ceramic at the interface to form a bond, and the bulk titanium powder sinters to form a coating.

The second type is referred to as the MoMn process and results in the formation of a molybdenum coating. A paint of molybdenum and manganese powder,

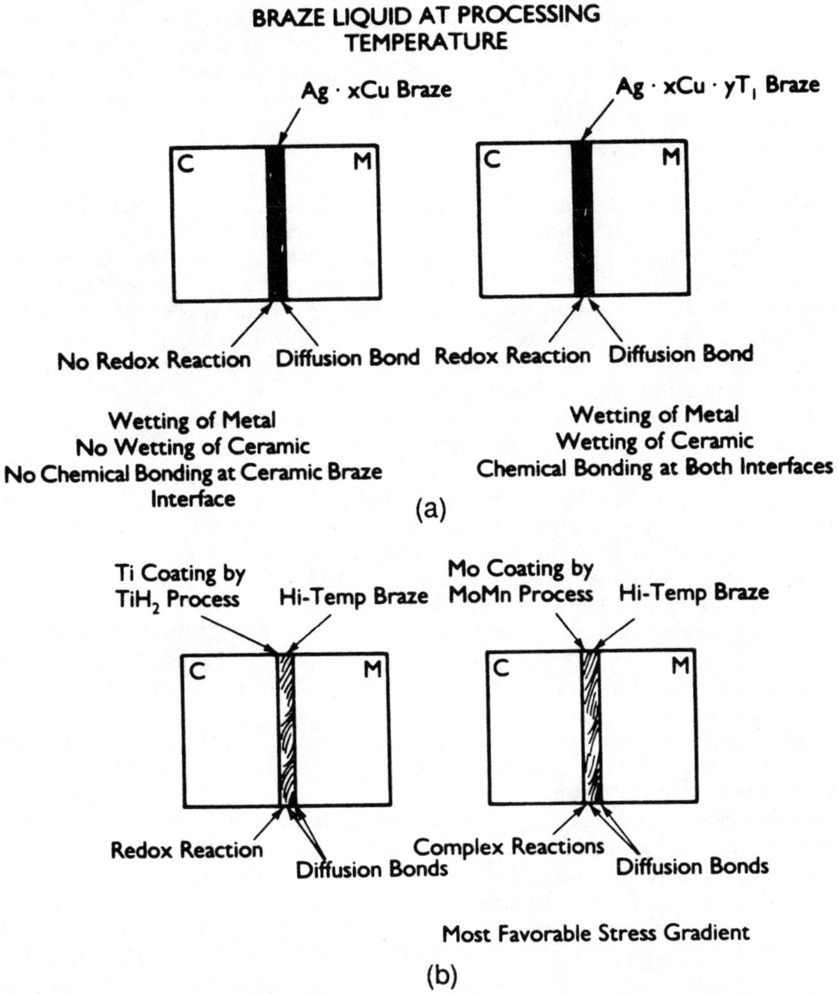

FABRICATION AND MANUFACTURING METHODS 7.53

BRAZE LIQUID AT PROCESSING TEMPERATURE

Ag · xCu Braze

C | M

No Redox Reaction — Diffusion Bond

Wetting of Metal
No Wetting of Ceramic
No Chemical Bonding at Ceramic Braze Interface

Ag · xCu · yT$_1$ Braze

C | M

Redox Reaction — Diffusion Bond

Wetting of Metal
Wetting of Ceramic
Chemical Bonding at Both Interfaces

(a)

Ti Coating by TiH$_2$ Process — Hi-Temp Braze

C | M

Redox Reaction
Diffusion Bonds

Mo Coating by MoMn Process — Hi-Temp Braze

C | M

Complex Reactions
Diffusion Bonds

Most Favorable Stress Gradient

(b)

FIGURE 7.24 Schematic illustrating several conditions under which brazes are utilized. (a) Braze in contact with unmetallized ceramic. (b) Braze in contact with metallized ceramic.[95]

or their oxides, is applied to the ceramic, generally Al$_2$O$_3$ or BeO, and fired in H$_2$ with a controlled dew point so that manganese is present as MnO and molybdenum as a metal. The MnO reacts with the ceramic grains and the liquid glassy phase to form a controlled amount of glassy phase with the proper viscosity.[108] The molybdenum sinters to form a porous coating into which the glassy phase permeates and interlocks mechanically. In addition, the glass at the interface reacts with the molybdenum and becomes saturated with MoO, thus forming a chemical bond with molybdenum. The glass also wets and bonds to the ceramic grains since they are compatible. The molybdenum coating is generally electroplated with nickel to provide a clean and continuous surface as well as one on which the braze spreads easily. A similar process is also followed by tungsten metallizing, with an appropriate bonding glass composition.

The second type of metallization process is generally more reliable and used more extensively. Although it is not normally recognized, the increased reliability is suggested to be most likely due to a more favorable stress pattern across the interface because of the formation of a thicker interfacial zone with a more extended and graded microstructure.

Wetting and Contact Angles. Experimental and evaluation studies have consumed and occupied the time of scientists and researchers. They have reported on analyses and the determination of what makes wetting effective, what role surface cleanliness has, about surface roughness versus smoothness, and on how to improve wetting, especially as new ceramic materials and grades are developed. New braze filler metals must be analyzed and produced under rigid conditions of cleanliness to remove traces of impurities, and with implementation of in-process controls in cleaning, processing, and brazing.[109–129]

A good example is Mizuhara and Huebel.[109] A 90° contact angle of the braze filler metal (62Ag–22.2Cu–14.5In–1.3Ti) to the ceramic is desired so that the braze filler metal wets the ceramic surface but does not flow. This is a desirable trait, similar to the Mo–Mn surfacing process, where the braze filler metal will wet only where the surfacing material is applied on the ceramic. However, if a low contact angle or flow on the ceramic is necessary, this can be achieved by using a highly active metal-content alloy. Most high-titanium-content alloys are excessively hard to brittle and also have higher yield strengths, thereby limiting their use in joining ceramic-metal systems with mismatched thermal expansion. The molten filler metal contact angle to the base metal will range from 0°, where the filler metal blushes or flows over the metal substrate surface, to 90°, where the molten filler metal stays where it melts.

The blushing effect is not a simple phenomenon. For example, the braze filler metal can alloy with the substrate to form a lower-melting-temperature alloy, which results in deep erosion followed by severe blushing (flowing) of the new lower-melting-temperature alloy composition. The opposite effect occurs when the braze filler metal readily alloys with the base metal to form a higher-melting-temperature alloy and results in sluggish braze formation and no flow.

Another variable is the surface finish of the base metal. For example, all variables being constant, the filler metal flow can be at a minimum on smooth finishes around 0.127 μm, but can be extensive on sandpaper-scratched surfaces of 0.76 to 1.27 μm.

The joining of two different materials with the same braze filler metal contact angle is simple. Examples are Al_2O_3 to 410 stainless steel, Al_2O_3 to Al_2O_3, and copper to Kovar. The joining of two materials with different brazing filler metal contact angles, as for Al_2O_3 to Kovar, is more difficult.

The added difficulty of joining two materials having different contact angles can be attributed in many cases to the mass of the ceramic being much greater than that of the metal member, which results in the metal becoming much hotter than the ceramic at any given time. This effect causes the braze filler metal to blush over the metal, thus leaving insufficient braze filler metal to wet the ceramic. To overcome this difficulty, temperature control of the parts to be brazed becomes very important.

Johnson[119] describes a method of joining Si_3N_4 using active braze filler metals, which is relatively simple and produces joints with the highest strengths (773

MPa) reported for Si_3N_4. The failures in the high-strength joints initiate in the Si_3N_4, not in the joint or interfacial area.

Until recently the use of metals as joining materials for Si_3N_4 has been restricted to such metals as silicon and aluminum. However, active-metal brazes that have been used extensively for joining oxide ceramics, particularly Al_2O_3,[120,121] have recently been applied to joining Si_3N_4, according to Loehman et al.[122] Wetting studies by Mizuhara[121] indicated that active metals were required to wet Si_3N_4, that metals that wetted Si_3N_4 adhered when cooled, and that complex titanium phases segregated at the Si_3N_4 interface. The changes in wetting angle with time for Si_3N_4 and various alloys with and without titanium point out that the wetting angle decreases to zero (or complete wetting) rapidly for those alloys containing substantial amounts of titanium.[122] The strongly adherent interface between Si_3N_4 and titanium containing alloys and the complete wetting in short times indicate the suitability of these alloys as joining materials for Si_3N_4.[119]

Bonding Ceramics

Most designers' reasoning for using ceramics is based on the following current concept: Use ceramics only on parts where their characteristics are needed on a metal-based structure. As it is difficult to produce complex, large products with ceramics, it is necessary to start with simple, small parts. Therefore it is important to be able to bond ceramics to metal, or to bond ceramics with each other. Figure 7.25 classifies ceramic bonding methods. The reason so many methods have been proposed and tried is that ceramic bonding technology has been applied widely from electronics to nuclear, aerospace, automotive, and mechanical structures.

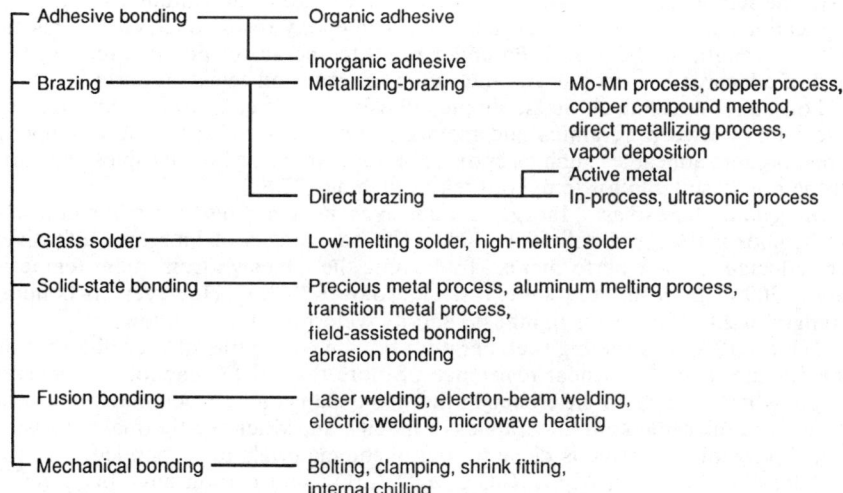

FIGURE 7.25 Classification of bonding methods for ceramics.

Adhesives and Adhesive Bonding. Bonding or cementing offers an attractive means of making an essentially stress-free joint for use when high temperatures and vacuum tightness are not involved. Cold-cure resins and adhesives adhere well to ceramics because of the relative surface roughness, which can be improved by grit blasting if necessary. Adhesion to glasses is also good. Hot-cure resins, particularly the more brittle varieties, should be used with caution because of the effects of thermal expansion mismatch between the resin, the ceramic, and other materials on cooling to ambient. For example, if a cure at 150°C is required, the joint is stress-free at that temperature, but the high coefficient of thermal expansion of the resin causes it to go into tension on cooling. The joint may crack or cause unacceptable distortion in thin-section components, such as thin ceramic tiles bonded to metal sheet.[3]

Conventional ceramics are not particularly difficult to bond, as is evidenced by the fact that bathroom tiles have been bonded to walls and broken panes and flower pots have been repaired for decades with easy-to-use household adhesives. However, in the demanding applications that are being considered for high-performance ceramics, where flaws so small as to defy detection can be the initiation sites for failures, it is important to have joints of quality comparable to that of the ceramics.

For optimum performance the glue-line thickness may need control according to manufacturers' recommendations, but in any event, moderate out-of-flatness of the ceramic-component face may be compensated for by variations in adhesive thickness.[3] Properties inherent in the ceramic material rather than the design or the cost of the operation have to be overcome.

Care must be taken that the adhesive chosen is suitable under the conditions of use. For example, some organic adhesives may swell in oil or water, and the designer must avoid a hydraulic bursting effect in confined spaces.

Organic Adhesives. An organic adhesive can bond almost all kinds of materials. It is therefore often used for bonding not only ceramics of electronic devices, but also structural ceramics to metal. An organic adhesive has many bonding factors, which causes the large dispersion in its bonding strength. The heat-resistant temperature is assumed to be 200 to 250°C maximum for practical applications as a structural material.[130,131] The epoxy resin adhesive ensures excellent strength of about 80 MPa under room temperature. For this reason, this kind of adhesive has been used quite often for abrasion-resistant parts.[131]

For many mechanical joints, simple adhesive bonds can produce strong, reliable joints between ceramics and metals or other ceramics. For low temperatures, organic adhesives such as epoxies or anaerobic acrylic adhesives can form strong bonds suitable for temperatures as high as 177°C.

Inorganic Adhesives. Inorganic adhesives in a narrow sense include silicates, phosphates, and colloidal silicas. The advantages of inorganic adhesives are reflected in their performance; by heating the adhesive from room temperature to 300°C the resultant joint is resistant to over 1000°C. However, its bonding strength of 20 MPa, its air tightness, and its waterproofness are low.

Table 7.12 shows the high-temperature bonding strengths of four different inorganic adhesives.[132] A heat resistance of more than 500°C cannot be achieved except when there is no difference in thermal expansion between materials, such as between the same kind of ceramics or metals, and when the thermal expansion of the inorganic adhesive is close to that of the materials to be bonded.

Adhesives based on Al_2O_3, SiO_2, MgO, ZrO_2, and carbon have been developed for 2205°C service temperatures and above. These ceramic-based adhesives are available in one- and two-part systems, their physical form is similar to that of

TABLE 7.12 High-Temperature Bonding Strength (kg/cm²)

Adhesive	Material combination	Temperature, °C				Thermal expansion, $10^{-7}/°C$	
		100	500	700	1100	Adhesive	Material
A	Tungsten-tungsten	103	100	100	—	40	46
B	Titanium-titanium	81	38	—	—	80	84
C	Alumina-alumina	230	210	220	190	80	78
D	Alumina-stainless steel	14	—	—	—	105	Alumina 78 Stainless steel 164

Source: From Yamada et al.[132]

organic adhesives, and their viscosity can range from thin paintlike consistencies to thick thixotropic solutions. Application can be performed by painting, spraying, screen printing, dipping, or injection, using conventional adhesive dispensing equipment.

Most of the currently formulated production adhesives contain an organic binder that melts at relatively low temperature and causes a reaction with the ceramic base. Cure temperatures, depending on the adhesive and the level of performance desired, may range from 121 to 649°C. The coefficients of thermal expansion of the adhesives can be matched to those of the substrates for which they are intended to minimize stresses during heating and cooling.

Ceramic adhesives are suitable for bonding metals or ceramics that do not require a hermetic seal or a high degree of reliability. For example, a graphite-based adhesive is used in bonding graphite molds for use in glass-to-metal sealing. A fixture holds nickel-iron parts and glass preforms together in a 900°C wet reducing atmosphere. The graphite adhesive has a coefficient of thermal expansion matching that of the fixture to prevent thermal shock and is used because it bonds the graphite while not being wetted by molten glass from the parts being bonded.

In another application, an aircraft sensor operating in a 593°C engine environment is potted with an MgO-based adhesive. The ceramic adhesive is used because it resists vibration and maintains high dielectric strength at elevated temperatures.

In general, adhesives are not considered suitable for ceramic applications as automobile engine and spacecraft components for the following reasons:

1. Engineering advantages of ceramics are primarily their ability to withstand high temperatures and corrosive chemicals. Organic adhesives clearly cannot match this endurance, although the ceramic-based adhesives can survive conditions similar to those that the substrates can withstand.

2. While adhesives can produce strong bonds, the joints are rarely hermetic. Bonds are not continuous, and the result is porosity that allows gases to penetrate, which can contaminate many electronic assemblies.

3. Organic adhesives can outgas in vacuum conditions, causing contamination of vacuum-packaged electronic components and devices used in space. As a result, higher-temperature bonding techniques are required for these applications and the bonding techniques are often difficult to apply without special methods and designs.

High-performance ceramics must often be bonded to other ceramics, both because certain shapes cannot be molded and because smaller, simpler shapes can be molded with better quality control. The objective has always been to obtain a joint resembling the homogeneous material as closely as possible and to minimize discontinuities that can weaken the parts.

For best results, adhesive joints should be designed so that tensile and shear strengths are minimized. Butt joints should be avoided in favor of concentric or lap joints. When materials with mismatched coefficients of thermal expansion are joined, the ceramic, or the material with low coefficient, should be on the outside of the metal, or material with the higher coefficient. This ensures that the joint will be in compression rather than in tension when the assembly is heated. Clearances between mating parts should be 0.05 to 0.020 mm. Smaller clearances do not allow uniform distribution of the adhesive in the joint, whereas larger clearances can cause the joint to fail in shear within the adhesive.

Ceramic-to-Ceramic Joining. The four basic joining technologies for ceramic-to-ceramic joining are:

1. Mechanical interlocking
2. Brazing
3. Welding
4. Ceramic adhesive and sealant bonding

Mechanical Interlocking. Mechanical joining of ceramics to ceramics is a very simple and efficient technique, particularly where tensile stresses can be kept low. A common example of mechanical joining is that of furnace refractories, which use mechanical interlocking or metal brackets. For applications where tensile loads are excessive, soft materials such as lead or copper can be used to minimize stress concentrations. Kutzer[133] stresses that he has basic information on stress concentrations and temperature limitations pertaining to mechanical joining. Threaded joints are sometimes used for joining ceramics in low-stress situations.

Mechanical joints have been used in the development of ceramic heat exchangers. Pietsch[134] refers to an EPRI-funded study in which mechanical joints were developed for attaching ceramic tubes (SiC) to a ceramic manifold. A ceramic heat exchanger has now been designed, tested, and is marketed using a mechanical ceramic-to-ceramic joint. The basic design for higher-pressure applications uses a high-density SiC insert with a ball-and-socket joint and a variable-load bellows.

Among the mechanical interlocking types that may be used with as-fired dimensions are bolting, threaded parts, crimping, clamping, gluing and cementing, and glaze and glass joints. Methods that need more accurate dimensional control, that is, ground surfaces, include shrink fitting, metallizing, soldering, brazing, vacuumtight clamp sealing, and diffusion bonding.[3]

1. *Bolting and rivets:* Direct bolting to rigid substrates is to be avoided if possible because of the risk of local fracture around the bolt head. If bolting is required, then soft washers (such as lead or aluminum) are recommended to absorb local loads. Wherever possible, bolts with large heads or with hard washers under the heads should be used to spread the load over a second, soft washer in contact with the ceramic (Fig. 7.26). Direct contact between the various types of spring and shakeproof washers and the ceramic surface is to be avoided.

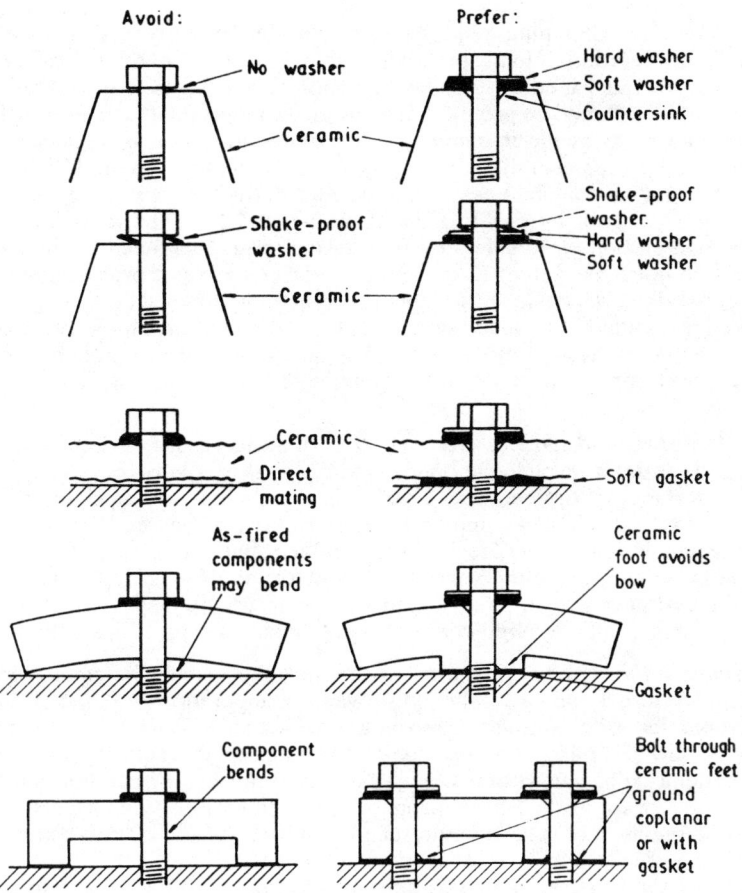

FIGURE 7.26 Techniques of bolting ceramics to other components.[3]

2. *Threads:* Threads in or on ceramic materials should not be used as a means of rigidly fastening a ceramic component into an assembly. An alternative approach is to fix a metal stud or nut into a ceramic component with a suitable adhesive or cement. For example, a hexagonal-head bolt may be retained by gap-filling solder or an organic adhesive, which acts as a soft load spreader, cushioning the bolt-head rotation. The internal shape of the hole should provide a suitable key for the adhesive. If a threaded metal or plastic insert is used, the insert should project above the ceramic surface so that when the bolt is tightened, the insert takes the load and is not pulled out of the ceramic by the tension in the thread.

3. *Crimping:* Thin metal edges can be crimped or rolled onto ceramic components, but only with caution. Normally this is done only on high-strength ceramics such as Al_2O_3 and requires careful process control. An example is the rolled lip which fixes the insulator into the metal housing of an automotive spark plug.

4. *Clamping:* Clamping requires many of the safeguards described under bolting, consideration of load spreading, and avoidance of bending loads. Clamping is the preferred method for glass components, using soft gasket materials. Mating surfaces should be ground flat or to the same profile (such as cone joints). A useful design principle in clamping a ceramic with a bolt is to take the compressive load on the ceramic as far away from the edges of the bolt hole as possible. This may be done by chamfering the edge of the bore, by using a large load-spreading washer, or by profiling the clamping surface of the metal or ceramic. Always tighten bolts to a safe specified torque to give sufficient clamping pressure but do not overtighten. If loosening of bolts becomes a problem, one of the temporary adhesives designed for this function should be used.

Riveting should be avoided because of the unknown forces involved. If riveting is essential, radial serrations in the ceramic around the rivet hole help prevent the rivet from turning relative to the ceramic and reduce the risk of breakage in assembly.

5. *Adhesives and glazing:* Keying on both surfaces of a cement joint is important to ensure a mechanical bond. A very rough so-called sand glaze is used on joint surfaces of the ceramic. Glazing is a technique appropriate for joining most types of ceramic and some of the more refractory glasses. For small pieces with reasonably flat mating faces, it is necessary simply to coat each face with a suitable glaze slip, place the surfaces in contact, and fire to fuse the glaze and bond the two pieces together. It is most often used with dissimilar ceramic materials, or in assembling large complex pieces such as stacked insulators.[3]

Ceramic Adhesives and Sealants. Ceramic adhesives are being used in many high-temperature bonding applications where temperatures exceed 376°C and where organic-based adhesive systems are no longer effective. The ceramic industry is now producing unique high-strength materials requiring high-temperature ceramic adhesive systems. For the successful use of ceramic adhesives, one must be aware of their basic properties and limitations, the designs that produce good bonds, any special environment criteria, and the manufacturing methodology in order to permit their economic use.

Many of the criteria applied to adhesive bonding can be applied to ceramic adhesive designs. The most important factors to consider are:

1. Stress analysis
2. Coefficient of thermal expansion
3. Clearance between mating parts
4. Ceramic property limitations and environmental factors

1. *Stress analysis:* Figure 7.27 illustrates typical stress loading for good ceramic design. Since ceramics exhibit poor tensile and shear strength, it is desirable to change the configuration of the glue line to relieve some of these stresses. Parenthetically, this change produces a longer glue line with reduced unit pressures on the bonded joint.

2. *Coefficient of thermal expansion:* This is an extremely important factor to consider since elevated temperature change (thermal shock) is implicit in most ceramic adhesive applications. Consideration must be given as to the coefficients of thermal expansion of the adhesive and the parts to be joined. Where possible, they should match closely.

Figure 7.28 illustrates a typical design problem: joining ceramic to metal for

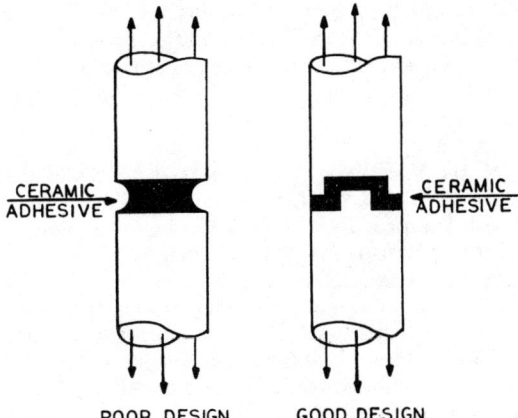

FIGURE 7.27 Typical stress loading for ceramic adhesive design.[135]

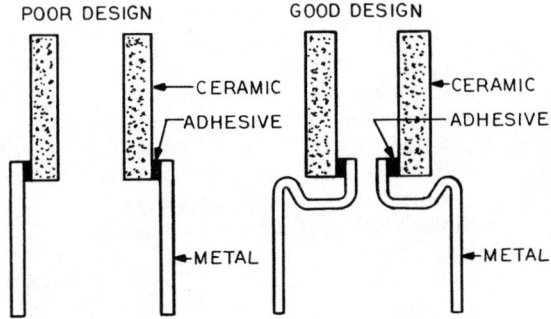

FIGURE 7.28 Designs for bonding ceramic to metal.

use at some elevated temperature. The "good" design takes into account the mismatch in the coefficients of thermal expansion and loads the (ceramic) adhesive glue line in compression. Another trick, which can be used in bonding mismatched materials together, is to purposely thin out the metal and even take steps to anneal it at the bond area. This will reduce some of the peel forces which might cause the bonded joint to fail.

3. *Clearance between mating parts:* Physical laws that apply to adhesive bonding with organics apply in a similar fashion to ceramic adhesives. Clearances between mating parts (calculated at operating temperature) should be 0.05 to 0.20 mm. Should the glue line get below 0.05 mm, the ceramic particles will not offer uniform adhesion to the surfaces being bonded. If they get too thick (above 0.20 mm), the joint will fail in shear within the adhesive itself.

4. *Ceramic property limitations and environmental factors:* Some of the ceramic property limitations and environmental factors affecting ceramics and their subsequent joining with adhesives are discussed in Schwartz[135] and in Chaps. 2 and 3.

Adhesives, Cements, and Mixtures. Various novel and innovative approaches have been developed and used by manufacturers to form close bonds. One manufacturer has developed a process which uses a metallic powder mixture of 95 wt % Sn, a CO-containing atmosphere, and a transition-metal additive that allows virtually any ceramic to be bonded to metal.[136]

Thin layers of oxynitride glass have been used successfully to form strong bonds between Si_3N_4 parts.[137,138] Joint strengths as good as the base ceramic can endure temperatures up to 1100°C. The method developed uses a Si–Ca–Y–Al–O–N oxynitride glass. The process produces strong joints without applying pressure. Maximum joint strengths (four-point bending) reached 62.3 kg/mm². These impressive strengths were achieved with joint layers averaging 12 μm. Researchers tried varying the nitrogen contents and the processing temperature to achieve optimum thermal expansion matches. Some complex reactions and diffusion of the bonding layer components take place at the 1400 to 1700°C processing temperatures. Yttria, for example, develops in layers 50 μm beneath the interface and CaO has been found diffused up to 200 μm into the Si_3N_4 substrate.[137]

Another study found that glasses in the system Y–Al–Si–O–N were effective in joining Si_3N_4. Three types of joints were observed: closed joints, glass-filled joints, and emptied joints. Voids were observed in the joint region when 15 vol % amorphous grain-boundary Si_3N_4 was joined at 1725°C with a glass (high yttrium, low aluminum). Reductions of yttrium and aluminum in the glass during joining were observed in glass-filled joints. Emptied joints were observed for the 5 vol % grain-boundary Si_3N_4 joined at 1725°C.[138]

Joining Si_3N_4 having high additive content and large amounts of glassy grain-boundary phase was dominated by interaction of the bonding material with the grain-boundary phase. Joining Si_3N_4 having low additive content and a low amount of glassy grain-boundary phase was dominated by flow of glass across the surface. The ability to join Si_3N_4 with oxynitride glass was dependent on the chemical nature of both the Si_3N_4 and the bonding glass.[138]

An extremely popular ceramic adhesive for Al_2O_3 is a thixotropic single-component adhesive paste that can be applied in situ since it requires no heat cure. The adhesive is dimensionally stable and chemically inert to 1649°C and will not outgas under vacuum.[135,139] The adhesive paste can be applied by brush or syringe and sets up in 2 h at room temperature or in 1 h at 121°C. This latter curing process increases the moisture resistance, too. Among the adhesive uses are chemical mixing tanks and plastic processing equipment where the temperature exceeds 649°C. A Ceramabond* 569 adhesive handles the task. This ability has made the adhesive effective in bonding ceramic bases to quartz high-intensity lamps and in other situations where ceramics, glass, quartz, graphite, refractories, and metals are bonded. When there is an extreme difference in the coefficients of thermal expansion, it may be desirable to "grade" the bonded joint by using two cements. In the case of bonding quartz to steel, for example, coat the steel with one adhesive in the Ceramabond series and then use a second member of that same cement family to make the actual bond.[135]

A novel ceramic-metal bonding process resistant to high heat and capable of making high-strength, hermetic bonds between barium silicate glass and Inconel X-750 superalloy has been developed. A thin skin of chromium oxide (a ceramic oxide) and nickel titanate forms. The surface of the ceramic melts, forming an oxide with the same crystalline structure as the chromium oxide; the oxides bond. Because the skin

---

*Ceramabond is a registered trademark of Aremco Products, Ossining, N.Y.

on the metal part has a strong bond to the metal, the metal-ceramic bond does not break on cooling or under thermal and pressure cycling.

Level detectors for nuclear-reactor cooling water made with the process performed successfully when cycled from ambient to 649°C. The process can produce composites that combine the resistance to high temperatures and corrosive environments of ceramics with the fatigue resistance of metal to make parts for gasoline engines, heat exchangers, and structures.

A ceramic adhesive, consisting of fused silica and a bonding agent of magnesium phosphate, was developed to bond gap fillers to low-density ceramic insulating tiles used in the thermal-protection system of the Space Shuttle. The adhesive, which has a coefficient of thermal expansion compatible with that of the tiles, is unaffected by extreme temperatures and vibrations. Assuring direct bonding of the gap fillers to the tile sidewalls, the new adhesive obviates the expensive and time-consuming task of the removal, treatment, and replacement of tiles.

Special Methods and Joining Media. Joining Si_3N_4 ceramics has been attempted with the use of some joining agents, such as ZrO_2,[140] a mixture of SiO_2–CaO–MgO–Si_3N_4,[141] a mixture of Y_2O_3–La_2O_3–MgO–Si_3N_4,[142] an MgO–Al_2O_3–SiO_2 glass composition,[143] and kaolin with some fluoride.[144] However, the presence of these foreign materials resulted in the formation of glassy phases at the joining interface and consequently caused a degradation of the high-temperature strength of the joined parts.

Recently, joining processes of Si_3N_4 ceramics by hot isostatic pressing, uniaxial hot pressing, and reaction bonding without any extra materials on the joining interface have been reported.[145] In these cases, the joined components were expected to maintain the same strength as the original body at high temperatures even above 1200°C. However, the effect of the joining conditions on the joining strength has not been published.

Nakamura et al.[145] attempted to join Si_3N_4 ceramics by means of polyethylene film as a joining agent. They used a hot-pressing furnace in a nitrogen atmosphere and varied the joining conditions such as temperature, pressure, and holding time. The joining strength of the specimens was measured by four-point bending tests on test pieces prepared from the joined specimens.

The effects of the joining conditions of hot-pressed Si_3N_4 ceramics containing Al_2O_3 and Y_2O_3 as sintering aids on the joining strength showed that the joining strength was increased with increases in joining temperature, joining pressure, and holding time. The highest joining strength attained was 567 MPa at room temperature. Furthermore, the joining strength maintained almost the same value at temperatures up to 1200°C in air. Si_3N_4 ceramic components joined with this process could be used at elevated temperatures as high as those used for bulk Si_3N_4 ceramics.[145]

Bates et al.[146] have attempted to develop joining techniques for Si_3N_4 and SiC to achieve optimum high-temperature strength and to avoid the problems associated with earlier methods. Their techniques consisted of joining green parts using an interlayer of the same nominal composition as the green adherend pieces, followed by codensification of the aggregate body.

Three joint interlayers were evaluated by Bates et al.[146]: self-bonded, slip, and injection-molding compound. The interlayer of the self-bonded joints was formed by the bare, flat surfaces of the parent material placed in contact. The slip and injection-molding compounds consisted of the 4 wt % Y_2O_3–Si_3N_4 composition of the parent material as a powder suspended in their respective media. The advantage of the slip layer is that the slip, being plastic, conforms itself to the asperities of the green surface, so that minor defects introduced by machining, such as scratches, can be filled. For all joining techniques, densification was accomplished by glass encapsulation HIP.

Two joining techniques were used with the SiC material. The first one, similar to that used for Si_3N_4, used a slip between two flat surfaces of green billets. After application of the slip, the two parts were pressed together by hand and cold isostatically pressed. The joined green SiC parts were densified by pressureless sintering. This is referred to as the slip/CIP method.[146]

The second technique consisted of grinding and polishing two dense SiC parts with diamond pastes of increasing fineness to a mirrorlike finish and was unsuccessful.

Current capabilities in the manufacture of structural ceramics are bound by the limitations of the forming process. The potential for structural ceramics has not been entirely realized because only small sizes and simple geometries can be manufactured reliably and cost-effectively. The work by Bates et al.[146] and others[147-151] are attempts toward overcoming such manufacturing limitations.

Zdaniewski and Kirchner[152] successfully joined Al_2O_3 by hot pressing with polystyrene or mica sheets at temperatures of 1250 and 1350°C. The polystyrene decomposed during hot pressing and created reducing conditions in the joints which enhanced interfacial diffusion and bonding. The measured fracture toughness and flexural strength of such joints approached values for monolithic Al_2O_3. Joints prepared under similar conditions, but without polystyrene, showed much lower K_{Ic} and strength values.[152]

A process capable of producing high-strength glass-ceramic seals compatible with Inconel 718 has proved successful. The process was used to make hermetic seals of the insulative glass-ceramic to the Inconel alloy used in two explosive actuators, and the corrosion-resistant seals were able to withstand high temperatures up to 700°C. Future uses appear to be reactor instrumentation, well-logging instruments, experimental high-temperature testing with liquid sodium, and reactor safety.[152-154]

Metal Solders and Ultrasonic Soldering. The metal indium is a metallic solder used for bonding because it has a low melting point of 157°C and it successfully wets the surface of a ceramic when molten. In addition, it can bond ceramics by pressing under room temperature.[155] This method is used for sealing window glass.

Ultrasonic soldering is a process used to solder ceramics directly with solders such as Pb–Sn by applying ultrasonic vibration. In this method, the addition of elements likely to react with ceramics, such as zinc and tin, as well as abrading and agitating the interface by ultrasonic vibrations are also utilized to enhance bonding.

Glass Solders. Normally an oxide mixture is inserted between ceramic materials or between ceramic and metal, which are bonded together, and the oxide mixture is fused to form glass. Depending on the applicable temperature range, low- and high-melting glass solders are currently available.[155]

1. *Low-melting glass solder:* The low-melting solders have good electrical characteristics, corrosion resistance, and air tightness. Due to the low melting temperature of the solders, ceramics can be bonded relatively easily, whereby the material to be bonded is slightly influenced thermally. For this reason, these solders are used widely in the electronics industry. When bonding with glass solder, the thermal expansion coefficient of the glass solder must correspond to that of the materials to be bonded. However, because it is better that a slight compression stress be applied after bonding, the thermal expansion coefficient of the glass solder is selected so as to be slightly smaller than that of the materials to be bonded (Fig. 7.29).

The glass solders are classified into crystalline and noncrystalline types, depending on their state after bonding. Most noncrystalline solders are based mainly on B_2O_3–PbO–ZnO or B_2O_3–PbO–SiO_2.[156] Their bonding temperatures are 300 to 700°C. Although the thermal expansion coefficient and the bonding temperature can be selected separately to some extent, generally the former increases as the melting point decreases. Crystalline solder has better heat resis-

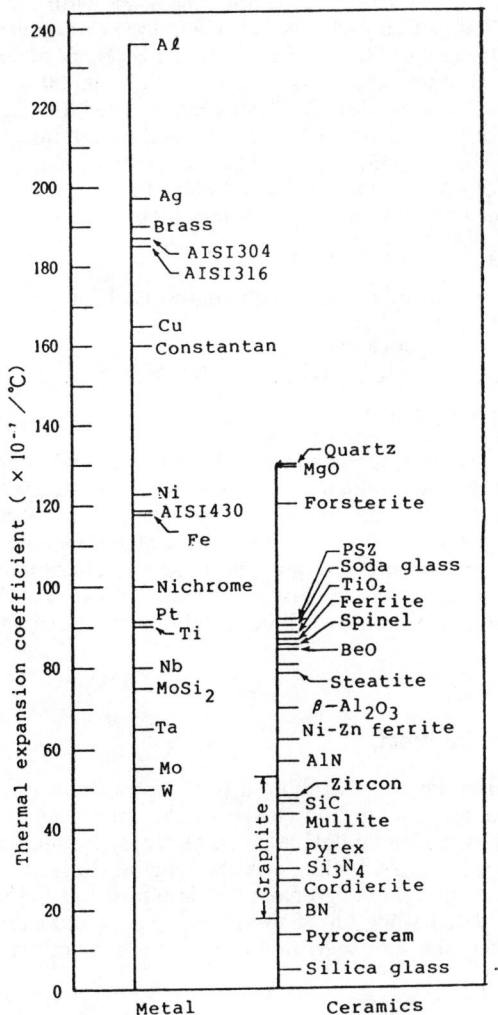

FIGURE 7.29 Thermal expansion coefficients of some ceramics and metals.[156]

tance, strength, and electric characteristics than noncrystalline solder, and its thermal expansion coefficient is smaller than that of a glass of a corresponding composition.[156] Typical crystalline solders having low melting points are $ZnO-B_2O_3-PbO$ and $ZnO-B_2O_3-SiO_2$, with the coefficient of thermal expansion of the former being larger than that of the latter.

2. *High-melting glass solder:* The heat resistance of a ceramic-to-ceramic joint is required to match that of the ceramics body in order to make large or complex products from some small or simple parts. A variety of high-melting solders have been used, including Al_2O_3-based $Al_2O_3-CaO-MgO-SiO_2$ and $Al_2O_3-MnO-SiO_2$.[155]

3. *Other glass solders:* The glass solders $CaO-SiO_2-TiO_2$ and $La_2O_3-Y_2O_3-Al_2O_3$ have been used successfully in bonding Si_3N_4 with bending strengths of 250 and 290 MPa, respectively. A mixture of a halogen compound and kaolin provides a bending strength of 300 MPa for bonding Si_3N_4 or sialon.[157]

Procedures for joining other glass and ceramic-to-metal seals have been developed by Honnell and Stoddard.[158] An electrical feedthrough was developed and a procedure for fabricating vacuumtight metal-to-ceramic ring seals between Inconel 625 and $MgO-3Y_2O_3$ tubes metallized with a glass ($CaO-29Al_2O_3-35 SiO_2$) containing 50 vol % molybdenum filler. Palniro 1 (Au-25Pd-25Ni) was found to be the most reliable braze for joining Inconel to metallized $MgO-3Y_2O_3$ bodies.

Conclusion. There are three basic technologies used for ceramic bonding:

1. Cement and adhesive sealants
2. Vitreous ceramic sealants
3. Crystalline ceramic sealants

Adhesive ceramic sealants such as mortar have been used with a high degree of success and reliability for industrial building and furnace applications. Cement and adhesive sealants include such bonding materials as phosphoric acid and calcium-aluminate cements. Vitreous ceramic sealants include glazes and other ceramic materials that form stable glasses on firing. These materials have been used to seal sapphire windows to glasses,[159] while Wood et al.[160] discuss the development of a compliant interlayer joint design for potential use in ceramic heat exchangers.

Diffusion Bonding (Indirect)

A fundamental understanding is required of the properties of metal-ceramic interfaces and, as a premise, a knowledge of the structure of the heterophase boundary and of characteristic defects such as steps, facets, and dislocations at or close to the interface. In addition, possible chemical reactions, reaction products, chemical gradients, and segregation of impurities at the interface must be analyzed and evaluated since all of these microstructural parameters, namely, equilibrium or nonequilibrium segregation, may have an influence on the properties of the interface.

Solid-State Bonding with Interlayers. This process is intended to bond together ceramics or ceramics and metal in the solid state by reacting them under high temperatures and pressures with various types of interlayers.[161–172] The bonding methods using the interlayer are classified largely into:

1. Precious-metal process
2. Aluminum process
3. Transition-metal process

Precious Metal Process. The technique usually uses a foil of precious metals, such as platinum, gold, palladium, or silver, which is inserted between the materials to be bonded and heated at a temperature below its melting point under a relatively low pressure in an air atmosphere. Platinum, for example, enables one to achieve a bond of excellent airtightness and high heat resistance.

Aluminum Process. Aluminum is likely to react with ceramics because it is relatively active, and it is easily placed between bonded surfaces because of its softness. In addition, because it has the advantage of lessening residual stress between bonded materials, it is used widely as an interlayer for bonding ceramics to metal. However, the disadvantage of aluminum is its low heat resistance.

For example, using a 0.6-mm-thick interlayer of an aluminum alloy with both surfaces cladded by an Al–10%Si alloy, Si_3N_4 and steel were bonded together at 610°C, a pressure of 9.8 MPa, in a time of 1.8×10^3 s, and under a vacuum of 2 to 4×10^{-2} Pa. When 5-mm-thick Kovar was selected as the interlayer for thermal stress relaxation, a bending strength of 200 MPa was found at room temperature. However, the bending strength dropped rapidly for temperatures over 327°C.

Solid-state diffusion bonding was found to be a promising route for the achievement of good ceramic-to-metal junctions (PSZ on cast iron and steel). The use of aluminum as interlayer metal resulted in a maximum tensile strength of 60 MPa. It was found that 0.5-mm foils were more effective than thin foils due to a higher deformation ratio, and the maximum strength was reached after relatively short times (15 min) at 600°C under 25 MPa.[167]

In another study Wicker et al.[168] found that a solid-state bonding technique (thermocompression bonding) could join ZrO_2, sialon, cordierite, Si_3N_4, and α-SiC on Nimonic 80 alloy by means of an intermediate gasket of aluminum. Conclusions drawn from this study pointed out that[168]:

Thermocompression bonding with aluminum gasket requires fully dense ceramics.

Polished surfaces lead to better results than surfaces in the as-fired state; this is less severe for oxide ceramics.

For nonoxide ceramics, vacuum leads to better results than a reducing atmosphere, while oxide ceramics need the reducing atmosphere.

Reproducible surface states are necessary to obtain reliable seals.

Transition-Metal Process. This process is a solid-state bonding method using iron, nickel, cobalt, or columbium, whereby a reaction layer is found on the bonding interface. For example, bonding of columbium and Al_2O_3 is often used for analyzing the microstructures of the bonding interface because their thermal expansion coefficients are similar. A low-columbium oxide CbO_x is formed on the interface, maintaining lattice matching. A variety of the bonding processes which have characteristic procedures have been publicized.[162,167,169,170,172,173]

One of these examples is the pressureless-sintered eutectic bonding of

Si_3N_4.[169] Using a nickel interlayer, bonding occurred in the temperature range of 1000 to 1300°C, such that a eutectic melt of silicon and nickel was formed at the interface. The silicon reacted with and diffused deeply into the nickel. Nickel reacted with oxide additives in the Si_3N_4, namely, Y_2O_3 and Al_2O_3, to form compounds at the interface. When bonding occurred between 1100 and 1200°C, the interfacial strength was more than 200 MPa in bending. Finally, the heat and oxidation resistance of the bond above 600°C were better than could be achieved through bonding by active-metal brazing with an Ag–Cu–Ti laminate foil.[169]

In an effort to evaluate the use of ceramics for machinery components to be used at elevated temperatures, Yamada et al.[170] developed a method of diffusion bonding SiC or Si_3N_4 to Nimonic 80A. The composites were produced by the insert metal-bonding method, using varying thickness of nickel, tungsten, Kovar, and copper. Tests showed that a sound composite with low thermal stress and free from cracks was obtained if bonding was performed using nickel as the insert metal for bonding and a low-thermal-expansion metal (tungsten or Kovar) and soft metal (copper) as the insert metal to reduce the thermal stress. The tensile strength of this composite at room temperature was 98 MPa.

The bonding method by Yamada et al. was applied to marine diesel-engine components.[170] From the results of an exhaust-valve damage simulator test, the exhaust valve in which Si_3N_4 was used as the face was expected to have a burnout resistance 10 to 50 times higher than the conventional valve (SUH3/Stellite).[170] This high burnout resistance means that the exhaust valve may be used without maintenance during the dock-to-dock interval (about 2 years). This resulted from the Ni–Kovar–Cu insert joint.

In Fig. 7.30 a SiC ceramic bonded fuel nozzle is solid-state diffusion-bonded to the atomizer seat, and its wear resistance is several tens of times higher than that of the conventional fuel nozzle. This wear resistance means that the atomizer seat may be used without maintenance during the dock-to-dock interval.

In order to exploit the thermomechanical properties of engineering ceramics in advanced heat-engine applications, it is necessary that they be joined to the ex-

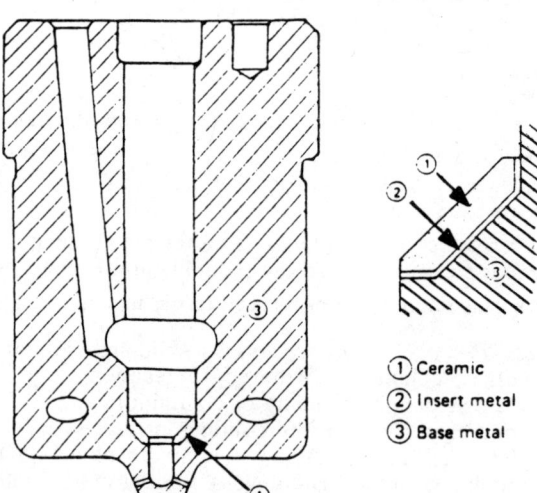

FIGURE 7.30 Ceramic bonded fuel nozzle.[170]

isting metallic support structure. As a result refractory metal interlayers (columbium, molybdenum, and tungsten) of various thicknesses were placed on Inconel 600 (IN600) and reaction-bonded SiC joints and bonded. It was found that the most effective combination was an IN600–Cb–Mo–reaction-bonded SiC joint with the columbium and molybdenum thicknesses being 3 mm, resulting in a reduction in the residual stress levels by a factor of 8. However, it was found that using refractory metal–reaction-bonded SiC joints reduced the stresses much more effectively while providing a simpler joint geometry according to McDermid et al.[171] Larker et al.[172] found that HIP was a useful method to obtain good conditions for solid-state diffusion bonding between Si_3N_4-based ceramics and superalloys. They also found that silicon oxynitride (Si_2N_2O) is more stable than Si_3N_4 against diffusion of reactive metals such as titanium and columbium from the superalloy. Finally they concluded that the properties of Si_2N_2O (higher coefficient of thermal expansion, lower Young's modulus, and better chemical stability) might be useful as a cost-effective intermediate joining layer for HIP joining of Si_3N_4 to superalloys for high-temperature applications such as engine components and refractory metal interlayers.

Electrostatic Bonding. Electrostatic bonding is similar to diffusion bonding in that the bonding parameters include pressure, temperature, and time. However, a voltage is also applied across the components, causing ionic conduction and thereby inducing a very strong electrostatic attraction between them. The technique has been used to join metals and semiconductors to borosilicate or, similarly, glass. According to Dunn,[174] this technique has been used successfully to produce hermetic seals between β-Al_2O_3 and aluminum without any intermediate filler materials.

Brazing and Metallizing. Brazing and metallizing of ceramics to form joints with metals finds many uses, especially in small-scale and electronic applications such as Al_2O_3 to Fe–Ni–Co alloy for vacuum-tube production. The molybdenum or Mo–Mn metallization layer is painted onto a ceramic, allowing a subsequent braze layer to wet. Brazing temperatures of 1580°C are typical. At this temperature the glassy phase in an Al_2O_3 begins to mix with the metallizing mixture of semisintered molybdenum or Mo–Mn. The literature discloses several theories relating to this mechanism, including capillary action.[173,175,176]

Other studies of brazing[177] include an investigation into the use of an amorphous filler material between Al_2O_3 and Fe–Ni–Co alloy and the effect of brazing conditions on the resulting shear strength. These results showed that when using a Cu–Ti braze filler, the joint strength was dependent on the amount of joining between Al_2O_3 and the intermediary TiO_2.

Some excellent work was performed in the active alloy sealing of an Al_2O_3 ceramic to a copper stud via a Ti–Ni eutectic metallization for use as a high-reliability transistor package. The Ti–Ni metallization was chosen in lieu of the more widely used Mo–Mn metallization because of batch-to-batch variations with the latter technique. The process used 0.5 μm of titanium evaporated onto the ceramic followed by 7 μm of nickel. This was followed by heating to 1000°C for 2 min in dry H_2. This heating bonded the titanium to the ceramic by forming a liquid phase of the Ti–Ni eutectic between the bulk of the nickel layer and the ceramic. The bulk of the nickel layer remained solid as the proportion of nickel was far higher than required for the 71.5Ti–28.5Ni eutectic composition with its melting point of 955°C. Close control of this postevaporation heating was found necessary. When too short a time or temperatures below 955°C were used and did not allow the reaction to form the eutectic, a weak joint between the metallization and the ceramic resulted. After this postevaporation heating, the joining was completed by brazing with a 72Ag–28Cu braze filler metal at >780°C.

Another unique metallizing process has been developed and applied at temperatures as low as 850°C. This makes it especially suitable for materials such as Mg–PSZ that cannot withstand temperatures above 1000°C. The braze filler metal consists of 95 percent Sn, with the balance containing carbide or carbonyl formers and other alloying elements. The process consists of the formation of a chemical bond with the surface of the ceramic, which will continue to wet the surface even when it becomes molten. While the basic braze filler metal melts at relatively low temperatures, it is capable of alloying with brazing metals during joining to form ternary alloys with high melting points.

Metallizing Processes. Metallizing, in which a thin metallic layer is bonded to a ceramic substrate, is often used as an intermediate step when more massive metal members are to be joined to the ceramic. Metallizing processes improve the wettability of ceramic surfaces for conventional braze filler metals, and the metal member can then be joined to the metallized ceramic substrate. Metallizing processes are still the most widely used techniques for producing ceramic-to-metal seals.[178]

Metallizing procedures were originally developed to improve the wettability of ceramic surfaces by conventional low-temperature filler metals. Later investigators found that some active metals and their alloys or compounds (such as titanium and zirconium) would wet unmetallized ceramic surfaces under certain conditions. Although variations of the so-called active-metal process have been used commercially to produce ceramic-to-metal seals, they have not been accepted to the extent characterized by the metallizing-brazing concept of joining these materials. In reviewing developments in this area, it should be emphasized that metallizing is a surface preparation for ceramics, not a joining process.[179]

To ensure the production of reliable ceramic-to-metal seals, most metallized surfaces are coated with nickel, copper, or other metals. The metals are usually deposited by electroplating. In some cases, however, the coatings are produced by reducing oxides of the desired metal. These coatings perform several functions, depending on the method used to produce the ceramic-to-metal seal (Table 7.13).

1. *Sintered metal-powder techniques:* In this process, finely divided metal powders are combined with a suitable binder to form a suspension or paste that can be painted on the ceramic surface. The coating is sintered to the ceramic (and the binder removed) by heating the ceramic at a high temperature in a controlled atmosphere. Usually molybdenum or tungsten is used in high-temperature applications; rhodium, iron, nickel, and chromium have also been used. Certain metals and metal oxides may be added to the metal powders in order to improve adhesion between the sintered metal coating and the ceramic.[3,95,101,179–181]

The most widely known process of this type is the "moly-manganese" process in which MoO_2 powder with manganese or MnO powder is milled to produce particulate typically 1 to 2 μm in size. (Sometimes metal powders rather than oxides are used.) A suspension is prepared using a binder such as nitrocellulose, so that a paint layer 10 to 25 μm thick can be applied to the base ceramic. The fine powders are usually 80 percent Mo and 20 percent Mn. The ceramic substrate is then fired in a wet H_2–N_2 atmosphere at a temperature of 1000 to 1800°C. Some of the oxides are reduced to metal and others combine with themselves and the ceramic substrate (and glass phases of the ceramic) to form a viscous melt. This

TABLE 7.13 Summary of Metallizing and Brazing Methods

Method	Metallized layer type	Suitable ceramics	Metallizing materials in finely divided form	Metallizing temperature, atmosphere	Additional layers required	Suitable solders or brazes
1	Ag, Ag/Pt	Hard glasses, most ceramics	Mixtures of $PtCl_4$, Ag_2O, Ag, Pt	500–900°C in air	None	Sn, Pb, or Pb solders*
2	Ag, Ag/Pt + fluxing glass	Hard glasses, most ceramics	Mixtures of $PtCl_4$, Ag_2O, Ag, Pt + <20% soft glass or flux, e.g., lead borate	500–900°C in air	Ag, Ag/Pt	Sn, Pb, or Pb solders*
3	Ag + CuO	Most ceramics	Ag_2O + <10% CuO, Cu_2O	>940°C in air	Ag, Ag/Pt	Sn, Pb, or Pb solders*
4	Mo + Ag	Most ceramics	Mo + ~10% Ag_2O	1300°C in dry H_2	Ni†	Cu, Cu/Ag, Ag
5	Mo + glass	Most ceramics	Mo + 10–20% glass	1200–1300°C in dry H_2	Ni†	Cu, Cu/Ag, Ag
6	Ni + glass	Most ceramics	NiO + ~10% glass	1300°C in dry H_2	Metallized layer needs buffing	Solders, Cu/Ag
7	Cu, Cu alloys	Most ceramics	CuO (+ alloying oxides)	1100°C in air +900°C in H_2	None	Solders
8	W or Mo (+ Mn or Fe)	Oxide ceramics BeO, Al_2O_3 (debased type)	W or Mo (+ Mn or Fe compounds, e.g., 80% Mo, 20% Mn)	1450–1650°C in wet H_2	Ni†	Cu/Ag, Au/Cu
9	Ti or Zr‡ (active metal joints)	Oxide and some nonoxide ceramics	Ti and Zr, or more commonly TiH_4, ZrH_4, or other compounds	>1000°C in inert atmosphere	None	Zr, Ti eutectic brazes
10	Mo + Ti	Pure Al_2O_3	Mo + TiN or TiC	1450–1900°C in wet H_2	Ni†	Cu/Ag, etc.

*Excessive reaction between solder and metallized layer can be prevented by an additional electroplated layer of copper on the metallizing, or by using a solder with a high silver content, e.g., Cu/Ag.
†Nickel coating can be achieved by electroplating, or by a second coating as in method 6, but without glass.
‡This process can be done by direct brazing in vacuum with an "active metal" braze, using optionally TiH_4 or ZrH_4 as fluxes to wet the ceramic, i.e., a one-stage process.
Source: After Morrell.³

melt coats the partially sintered metal phase, thoroughly wets the ceramic substrate, and solidifies to form a glassy phase upon cooling. Actually during this sintering process, manganese is oxidized by H_2O to MnO, so that it reacts with SiO_2 and with the foreign matter in Al_2O_3 and generates a fused-glass fluid. This fluid quickly fills in the pores in molybdenum particles and accelerates the sintering of the molybdenum particles to form a metallic molybdenum layer. The glassy phase contracts less than the ceramic substrate upon cooling and hence is placed with the metal phase in compression. The glassy phase thus remains strongly adherent to both the partially sintered metal and the ceramic substrate. The resistant metallized layer, typically about 10 μm thick, may be plated with a 2- to 4-μm-thick layer of nickel or copper to improve wetting during subsequent brazing. Ceramic-to-metal tensile bond strengths of 70 MPa or higher are typical.

The process developed by Nolte and Spurck[179,181] has been widely accepted by industry as a standard method to metallize ceramic surfaces. Numerous variations have been developed to extend the usefulness of the process.[104,109,182–186] These include W–Mn and Mo–SiO_2 processes. The process works very well with PSZ using a Mo–MnO–SiO_2–Al_2O_3 mixture (Table 7.14).

Another significant coating technique is the copper–copper compound process. In this process a metallic layer is formed by heating a copper plate placed on an oxide ceramic at about 1100°C in air. The molten copper reacts with and adheres to the ceramic, and after cooling, the oxide layer on the surface of the ceramic is removed to produce a metallic surface. At the interface, CuO–Al_2O_3 is formed, so that Al_2O_3 and copper are bonded together through this reaction layer. Spots seen in the copper layer indicate a part whose copper density is lower because the percentage of oxygen absorbed during melting in the air is high.

In fact according to the copper compound process, copper sulfide paste with 10 percent kaolin has been coated on ceramics, and subsequently they have been sintered at a temperature of 1000 to 1300°C. Because the surface of the ceramic is copper oxide in this state, the following metallization methods are used. One is to form a metallic silver layer by spraying silver carbonate at about 900°C; another is to form a metallic copper layer by reduction by hydrogen gas or a wet reducing agent. It has been reported that the copper compound process has been applicable with Si_3N_4 as well as oxide ceramics. A third method is direct metallization. After a braze filler metal-powder paste containing an active metal is coated onto a ceramic, they are both sintered at a temperature higher than the melting temperature of the filler metal in a vacuum or inert atmosphere to realize metallization. Ag–Cu–Ti is often used as the braze filler metal.

In addition, the direct metallizing processes with iron, nickel, cobalt, and chromium or their alloys have been used on Si_3N_4 successfully.[122,187–192] With these processes, after the powder of these metals or alloys was coated onto Si_3N_4, they were sintered at about 1300°C in a vacuum and metallized. When the metallized Si_3N_4 and molybdenum were brazed with a silver-based brazed filler metal, the shear strength was about 160 MPa. The mechanism of metallizing was considered the result of the reaction between the silicon generated by slight decomposition of Si_3N_4 in a high-temperature vacuum and metal or alloy powder.

2. *Active-metal brazing techniques:* The most active metal in this technique is titanium. It chemically diffuses into the ceramic, bonding to atoms in the material. For example, titanium fired on Al_2O_3 joins to the oxygen atoms, "stealing" them from the Al_2O_3, because titanium demands the oxygen more than the aluminum. This technique of gettering has been used for 40 years in the brazing of honeycomb structures by retort brazing.[95,109,178,189,191,193–199]

TABLE 7.14 Comparison of Metallization Techniques

Technique	General chemistry	Typical atmosphere	Firing temperature, °C	Some uses	Comments
Refractory metallization	Mo-Mn, W often followed by Ni plating	H_2, $H_2 + N_2$ (forming gas)	1300–1600	Package circuitry, seals for oxides	Adherence is aided by presence of glassy phase; some success with W directly on AlN; high-temperature oxidation concern
Active-metal brazing	Ti-dispersed in Ag-based brazing filler metal	Inert (vacuum, Ar)	850–1000	Sealing vacuum components and feedthroughs; general brazing of oxides and nonoxides	Single-step brazing; concern regarding oxidation resistance at high temperature
Direct-bond copper	Cu with controlled surface oxide	N_2 or N_2 + trace O_2	1066–1084.5	High-conductivity circuitry on Al_2O_3, BeO, mullite	Some success on AlN; concern regarding high expansion of Cu
Intragene	Sn + transition elements	CO or N_2	580–1000	Soldering to ceramics and high-temperature metals; brazing to Al_2O_3, ZrO_2, and graphite	High degree of wetting of substrates; not for high-temperature brazing to nitrides

Source: From Intrater.[178]

Cusil or Incusil active-metal brazing are two varieties derived from the basic method (Tables 7.15 and 7.16). Indium, when used, promotes wetting or spreading over the substrate. It also lowers the joint reflow temperature and, thus, expansion stresses. Both Cusil and Incusil contain small amounts (usually up to 2 wt %) of titanium metal. During heating, the titanium migrates toward the interface. It promotes adherence of the braze to the substrate and acts as an intermediary between the pieces being joined and the Cusil or Incusil matrix. Specially devised techniques have been developed by various manufacturers and suppliers of braze filler metal preforms (powder, foil, paste).[178] Tables 7.15 and 7.16 reflect the braze filler metals which are manufactured in a variety of shapes and preforms.

Active-metal brazing has been around since the 1940s, when titanium hydride in an organic solvent was coated onto ceramic seals, which were then sandwiched around a silver (or gold or copper) filler metal and brazed together in vacuum. The trouble was that under high vacuum the excess titanium in the filler metal formed a brittle joint, while under low vacuum most of the titanium formed brittle oxides and nitrides, and little was left for the actual active brazing.

One way around the problem was to take titanium wire or foil and clad it with a filler metal. While that protects the titanium from the furnace atmosphere until the filler metal starts to melt, so more titanium is available for brazing, the filler metal itself was not ductile enough to yield to stresses caused by the different rates of thermal expansion for the ceramic and the metal.

The problem was solved by the development of a series of Cu–Ag–Ti and Cu–Ag–In–Ti brazing filler metals. Like cladding, alloying protects the active element until the filler metal starts to melt. Because the titanium was protected, less was needed (1.25 to 2.0 wt %), and so the resulting braze was less hard and ductile enough to form strong, reliable ceramic-to-ceramic and ceramic-to-metal hermetic bonds without prior metallization (Fig. 7.31).

Active-metal brazing is used for microwave feedthroughs, electric bushings, and attaching Si_3N_4 turbochargers to steel shafts, among other applications. Unlike refractory metallizations, active-metal brazing works readily on nitrides and carbides. Joints can also be made in one step on a variety of oxides, tellurides, and borides.

Figure 7.32 shows a Si_3N_4 rotor and steel turbocharger. The rotor is mass-produced by injection molding a powder mixture of the ceramic, Y_2O_3–$MgAl_2O_4$ sintering aids, and binder, then sintering.

To join the rotor to the steel turbocharger shaft, a sleeve of iron-based superalloy Incoloy 903 with layers of a ductile, low-temperature-melting brazing filler metal (BAg-8) on the inside diameter, is first fused to the rotor, and then the sleeve is electron-beam-welded to the shaft.

As a face material for aluminum tappets, the ceramic has demonstrated 30 times the wear resistance (against chilled cast-iron cams) of current chilled cast-iron tappets. The composite tappet, a hypereutectic aluminum-silicon alloy, which is interchangeable with those currently used, is also two-thirds lighter in weight.

3. *Intragene:* Another process, Intragene, shown in Table 7.14, is a proprietary process.[178] It is an example of a direct chemical-bonding technique. Adding a few weight percent of transition-metal powders to tin and heating them in a nonoxidizing atmosphere produces a direct, strong chemical bond to many ceramics and metals.

Tin-based alloys, or other metals with low melting points, may be used in place of tin. Because of tin's ductility, metal-to-ceramic and ceramic-to-ceramic joints can be made reliably, even between parts that have greatly different thermal expansions.

TABLE 7.15 Selection Guide to Metal-Ceramic Braze Filler Metals

	Carbon and low-alloy steel	Stainless steel	Tool steel	Nickel, cobalt alloys*	Copper†	Nickel†	Titanium, zirconium alloys	Refractory metals	Ceramic, metallized	Carbon‡	Tungsten carbide
Ceramic, metallized	Copper Copper-gold Cusil Nicusil Palcusil Silver	Copper Copper-gold Palcusil Silcoro	—	—	Copper-gold Cusil Georo Gold-tin Incusil Nicoro Nicusil Palcusil Silver	Copper Copper-gold Cusil Nicoro Nicusil Palcusil Silcoro	—	—	Copper-gold Cusil Incusil Silver	—	—
Carbon†	Ticusil	Ticuni Ticusil	Ticuni Ticusil	Ticuni Ticusil Nicusiltin Palcusil	Ticuni Ticusil	Ticuni Ticusil	Ticuni Ticusil	Ticusil	—	Ticuni Ticusil	Ticusil
Tungsten carbide	Cocuman Copper Nicuman Nicusiltin	Cocuman Copper Nicoro Nicuman Nicusiltin	Cocuman Nicoro Nicuman Nicusiltin Palcusil	Cocuman Nicoro Nicuman Nicusiltin	Copper-gold Nicuman Nicusiltin	Cocuman Copper Copper-gold Nicoro Nicuman Nicusiltin	Cocuman Gapasil Ticuni Ticusil	Cocuman Gapasil Nicuman Ticusil	—	Ticusil	Cocuman Copper Nicuman Silver

*Corrosion- and heat-resistant alloys.
†Includes alloys.
‡Graphite and diamond.
Source: From Weymueller.[196]

TABLE 7.16 Brazing Materials for Ceramics

Name	Composition, %	Liquidus, °F	Solidus, °F
Copper (BCu-1)*	100 Cu	1981	1981
Nicoro (BAu-3)*	62 Cu, 35 Au, 3 Ni	1886	1832
Cu-Au(1)	65 Cu, 35 Au	1850	1814
Cu-Au(2) (BAu-1)*	62.5 Cu, 37.5 Au	1841	1805
Cu-Au(3)	60 Cu, 40 Au	1832	1796
Cocuman	58.5 Cu, 31.5 Mn, 10 Co	1830	1645
Cu-Au(4)	50 Cu, 50 Au	1778	1751
Silver	100 Ag	1760	1760
Ticuni	70 Ti, 15 Cu, 15 Ni	1760	1670
Nicuman 23	67.5 Cu, 23.5 Mn, 9 Ni	1751	1697
Nicoro 80	81.5 Au, 16.5 Cu, 2 Ni	1697	1670
Nicuman 37	52.5 Cu, 38 Mn, 9.5 Ni	1697	1616
Palcusil 15	65 Ag, 20.3 Cu, 14.7 Pd	1652	1562
Silcoro 75	75 Au, 20 Cu, 5 Ag	1643	1625
Gapasil 9	82 Ag, 9 Ga, 9 Pd	1616	1553
Palcusil 10	58.5 Ag, 31.8 Cu, 9.7 Pd	1566	1515
Ticusil	68.8 Ag, 26.7 Cu, 4.5 Ti	1562	1526
Silcoro 60	60 Au, 20 Cu, 20 Ag	1553	1535
Palcusil 5	68.5 Ag, 26.8 Cu, 4.7 Pd	1490	1485
Nicusiltin 6†	62.5 Ag, 29 Cu, 2.5 Ni, 6 Sn	1476	1275
Nicusil 3	71.15 Ag, 28.1 Cu, 0.75 Ni	1463	1436
Cusil (BAg-8)*	72 Ag, 28 Cu	1436	1436
Incusil 10	63 Ag, 27 Cu, 10 In	1346	1265
Incusil 15	61.5 Ag, 24 Cu, 14.5 In	1301	1166
Georo	88 Au, 12 Ge	673	673
Au-Sn	80 Au, 20 Sn	536	536

*AWS specification.
†AMS 4774A.
Source: From Weymueller.[196]

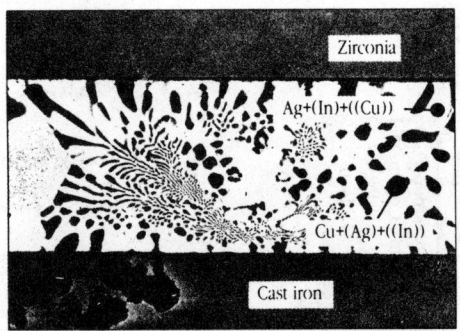

FIGURE 7.31 Zirconia-to-cast-iron joint brazed with copper-silver eutectic filler metal.

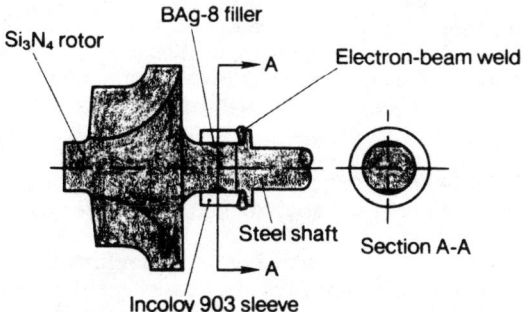

FIGURE 7.32 Rotor joined to turbocharger.

In contrast to Mo–Mn metallization, Intragene metallization requires no plating, abrasion, or reprocessing after initial processing to solder or braze to it. The initial metallization layer may be postbrazed onto Al_2O_3 or ZrO_2 in air, with an oxyhydrogen flame. This makes it easier to fixture parts and gives the process an advantage over active-metal joints, which may deteriorate at elevated temperatures due to reaction with oxygen and nitrogen in the air.

This method achieves direct chemical bonding with most common oxide ceramics and glasses, as well as to carbides, nitrides, graphites, diamonds, and the like. However, the bond formed on certain nitrides, carbides, and extremely low-expansion materials, such as most glasses, may not be strong enough to be postbrazed to high-expansion metal parts.

In applications that do not require the strength of brazing, the tin-based version can be used as an effective direct-solder method. This works for ceramics as well as for high-temperature metals and alloys such as chromium, molybdenum, tungsten, silicon, and stainless steels. Such applications include sealing laser and spectroscopy windows, making electrical contact to graphite, and heat-sink attachment to ceramic electronic substrates such as Al_2O_3 and AlN, according to Intrater.[178]

Being tin-based, the method is relatively inexpensive. It does not require prealloying the tin with the transition elements. The technique is fairly recent, having been developed in the 1970s. However, it is gaining acceptance as a direct method to solder to ceramics and graphites without damaging substrates.

4. *Vapor-deposition techniques:* Vapor-deposition processes used to metallize ceramic substrates can be divided into two classes: physical-vapor deposition (PVD) and chemical-vapor deposition (CVD). In physical-vapor deposition the coating is identical in composition to the source material, since intentional chemical reaction is not involved. Such techniques include sublimation and vaporization (often referred to as vacuum metallizing), sputtering, ion plating, exploding wire or foil techniques, and plasma-spray coating.[104,187–189]

Vapor coating Si_3N_4 surfaces with a 1-μm-thick layer of titanium, Santella[188] has shown an effective way of promoting wetting and bonding during the brazing of this important structural ceramic at temperatures up to 1130°C. With brazing filler metals of Ag–Cu and Au–Ni (brazing temperatures of 790 and 970°C, respectively), the titanium coating appeared to act as a reaction barrier on the Si_3N_4 surface, shielding the ceramic from direct contact with the liquid braze filler metal. At 1130°C the titanium appeared to dissolve in a Au–Ni–Pd braze filler metal and react with the Si_3N_4, but remained localized at the Si_3N_4 surface.

Flexure testing of Si_3N_4–Si_3N_4 joints established that the preferred fracture path was through the ceramic and indicated that joint strength was not limited by adhesion of the titanium coating to the Si_3N_4 surfaces or by the strength of the filler metal. Joint strengths as high as 474 MPa were obtained with a Au–Ni–Pd braze filler metal. For a Ag–Cu braze filler metal, joint strength as high as 334 MPa was obtained at 400°C. Porosity in joints made with the Au–Ni–Pd braze filler metal may have limited their strength and contributed to the relatively large scatter in the strength data.

Brazing of Si_3N_4 to A286 steel and to titanium confirmed that adhesion of the titanium coating to the Si_3N_4 surfaces was excellent. The shear strengths of Si_3N_4–A286 and Si_3N_4–Ti joints were low, and undoubtedly the result of high residual stress levels in the joints due to the large thermal expansion coefficient mismatches between Si_3N_4 and the metals.

Other variations include sputtering the coating of an active-metal film directly onto the ceramic, that is, ZrO_2, prior to conventional brazing. Here, active-metal braze filler metals based on the Cu–Ag eutectic with additions of titanium, with or without indium, have been evaluated (Fig. 7.31).

Some newly designed automotive engines require joining of PSZ to nodular cast iron (NCI) in which both materials are heat-treated and the joints are formed at relatively low brazing temperatures, around 732°C, to avoid loss of properties of the zirconia and the iron.

In this active-substrate brazing technique, Hammond et al.[194] use a material that promotes wetting of the brazing filler metal, which is vapor-coated onto the surface of the ceramic in a preliminary treatment designed to optimize surface reactivity and adhesion. The PSZ is vapor-deposited with a 0.3- to 1-μm-thick layer of titanium, brazed at 723 to 735°C; total brazing cycle 22 min. A BVAg-18 braze filler metal (Ag–30Cu–10Sn) with good strength and ductility was used. A shear strength of 138 MPa was achieved along with tolerance to thermal cycling and shock.

By adding a transition piece, strain on the ceramic was minimized. Titanium was selected as the transition piece because its coefficient of thermal expansion is nearly identical to that of PSZ, and it has excellent ductility, which minimizes strain between PSZ and the titanium due to differences in the coefficients of thermal expansion. Mismatch between the NCI and the titanium was accommodated in the titanium.[195] To enhance wetting of the NCI, an electroplated coating of copper was applied.

From the foregoing it is evident that significant interest has been shown in the use of physical-vapor-deposition techniques for bonding metallic coatings to ceramics. Very thin, tenacious coatings are possible in a short time with relatively state-of-the-art equipment using these techniques without excessive heating of the ceramic substrate. Joint tensile strengths of 35 to 140 MPa have been obtained using physical-vapor-deposition metallizing techniques.

Chemical-vapor deposition is the deposition of elements or compounds in massive form or as a coating by chemical reaction of the vapors of suitable compounds, usually at a heated surface.

General Points on Brazing Ceramics

1. Success in producing a good brazed joint depends on the following:
    a. The braze and its compatibility with the metallized surface and the metal component

b. Stresses produced in a joint by its geometry and the thermal expansion mismatch between component parts
   c. Adherence of the metal coating to the ceramic
   d. Quality of the joint
2. The choice of metallizing and brazing materials is determined essentially by the requirements of the end use, in particular the temperature and the corrosive nature of the environment. Combinations of materials may be limited because of wetting or compatibility problems.
3. The stresses produced in a joint depend on the following:
   a. Thermal expansion mismatch between ceramic, braze, and metal (Fig. 7.32)
   b. Relative thickness of ceramic and metal, and geometry of joint (Fig. 7.33)
   c. Mechanical properties of both metal and braze; in particular, their ability to relax stresses by deformation in thin sections of soft metal
   d. Brazing temperature
4. The strength of the joint depends on the following:
   a. Preexisting joint stresses
   b. Properties of ceramic, braze, and metal
   c. Integrity of interfaces
5. The adherence of the metallized layer on the ceramic depends on a large number of practical variables:
   a. Type of ceramic and its surface finish
   b. Type of metallizing
   c. Type of chemical bond
   d. Particle size of the metal powder and its composition (it may contain a number of components, including metals, reducible metals, oxides, glasses, carbides, and hydrides)
   e. Thickness of coating
   f. Heating cycle and atmosphere used to make the bond

From the foregoing points it should be clear that considerable work has been done and is required to make a successful ceramic-to-metal joint.

Braze Filler Metals. Most braze filler metals do not wet ceramics easily unless their surfaces are treated in a manner to promote wetting. Such difficulties are to be anticipated when one recalls that oxide ceramics comprise the largest group of these structural materials. Even a small percentage of aluminum or titanium in a superalloy presents wetting problems unless brazing is conducted in a vacuum or another expedient is used to prevent the formation of oxides during the heating cycle.

Although the metallizing of ceramic surfaces is costly and time-consuming, the brazing of metals to such surfaces is a relatively straightforward process because the metallized layer ensures wettability of the ceramic by the filler metal. However, certain metals and hydrides possess the ability to wet ceramic surfaces that have not been metallized, and active-metal and active-hydride processes based on this characteristic have been developed for producing ceramic-to-metal joints and seals.

The joining of ceramics to metals with the active-metal or active-hydride processes dates back to the middle 1940s, when titanium hydride was used for this purpose.[95,104,183,198,199] Since these early investigations, joining ceramics to metals by the active-metal or active-hydride process has advanced significantly. The

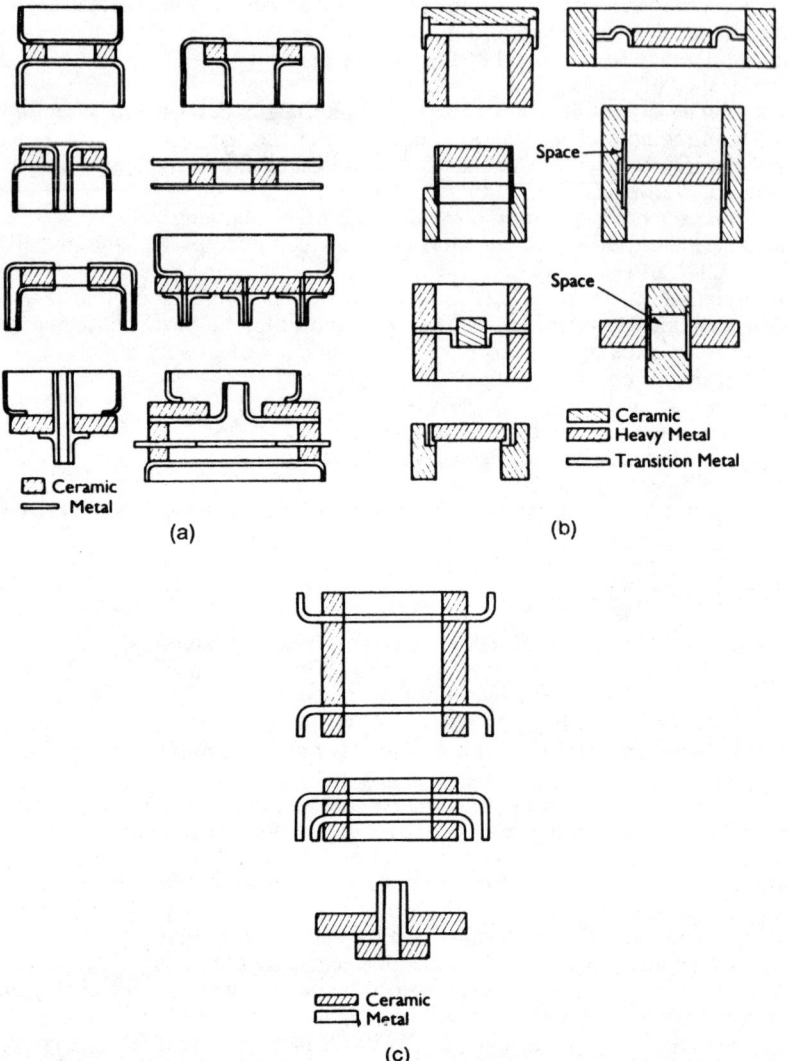

FIGURE 7.33 Ceramic-to-metal joint configuration.[95] (a) Butt and lap seal joint designs. (b) Joint designs for transitions to thick all-metal members. (c) Backup of ductile metal seal with blank ceramic.

strengths of joints made by this process are as great now as those obtained with joints made by the Mo–Mn process.

The concept of fabricating ceramic-to-metal joints and seals by the active-metal or active-hydride process was first applied in the electronics industry. In recent years, however, these joining processes have found other uses to meet the need of high-temperature vacuumtight seals in the nuclear and aerospace industries.

Fox and Slaughter[200] reveal the use of experimentally developed active-metal alloys for producing ceramic-to-ceramic and ceramic-to-metal joints, some of which are potentially useful in nuclear reactor technology. The filler metals 68Ti–28Ag–4Be and 49Ti–49Cu–2Be were originally developed for joining graphite to metal. However, studies indicated that good wetting and flow occurred between 49Ti–49Cu–2Be and Al_2O_3, BeO, and UO_2.[201]

New active braze filler metals are constantly undergoing changes and modifications in composition to meet the ever-demanding requirements to permit metals to be joined to ceramics without the ceramic materials being metallized. Some of these silver-based filler metals (Cusil and Incusil) are ductile and adaptable to brazing metals to such materials as Si_3N_4, PSZ, transformation-toughened Al_2O_3, and SiC, as well as many other refractory materials (Tables 7.15 and 7.16).

Four different braze filler metals for direct brazing of ceramics to metals and ceramics have recently been developed and made available to the industry. The developers believe that these fillers, which are ready for commercialization now, could revolutionize the design of complex assemblies used in high-temperature applications, such as advanced heat engines and heat exchangers. Such parts are currently too difficult to mold by conventional ceramic processes.

One braze filler metal, Cu–26 Ag–29 Ti, has been designed to bond Al_2O_3 to Al_2O_3, or PSZ to either NCI, which is commonly used in heat engines, 40-46 stainless steel, or commercially pure titanium. Bonding is done at 1000°C.

The three other braze filler metals are used to join Si_3N_4 to Si_3N_4, A286 superalloy, pure titanium, or TZM, a molybdenum alloy, at a variety of temperatures. Ag–0.3Li–28Cu is used at 790°C; Si–18Ni–4Au operates at 970°C; and Au–25Pd–25Ni melts at 1130°C. Before its use, Si_3N_4 is usually coated with a micrometer-thick layer of titanium.

Microwave Heating. Microwaves are electromagnetic waves considered as a brazing heat source. Many different types of devices have been developed to produce microwaves, but for industrial applications in the United States, the magnetron operating at 2.45 GHz is the most popular.[202] Regardless of the production device used, the microwaves, once generated, are conducted by an electromagnetic transmission line to an applicator, in which the materials to be joined are placed.

Because microwave heating causes a material to heat itself, it will heat rapidly, ensuring high thermal efficiency. For example, ceramics were bonded by microwave heating using a microwave oscillator with a frequency of 6 GHz. Al_2O_3 round rods of 3-mm diameter whose ends were butted together were heated and bonded under a pressure of 0.6 MPa and at a temperature of 1850°C for 3 min, and a bending strength of 450 MPa was recorded. Other recent accomplishments include joints made in 3 to 10 min using about 100 W of power and pressures of 0.6 to 9 MPa in mullite and Si_3N_4. This brazing was performed at 1720°C, both with and without an intermediary material (flux), with room-temperature strength equal to that of the unjoined material. Strength retention for the joined material, in cases where it has been measured, matches that of the unjoined material up to 800°C.[202–204] In addition, a method for performing in situ acoustic nondestructive evaluation of the joint as it is being formed has also been developed and demonstrated.[204]

There are several potential benefits of using microwave energy to join ceramics, all of which derive from the fact that the electromagnetic field penetrates the material, heating it internally. This leads to a rapid temperature rise, since one does not have to wait for the heat to be conducted from the surface to the interior.[203]

The joining of ceramics using microwave energy was first reported by Meek and Blake.[205] They formed a glass seal between two thin Al_2O_3 plates by surrounding the combined materials with a fibrous Al_2O_3 insulation and placing the package in a conventional home-type (700-W) microwave oven. At full power, temperatures of 700 to 800°C were reached and bonding was achieved in 99 min. The method was later extended to form a ceramic-glass-metal seal.

The following conclusions can be drawn for this type of heating and joining media:

Sintered 92 to 96 percent purity Al_2O_3 ceramics have been directly joined together by microwave heating. The joint strengths were equal to their original strengths after heating more than 3 min. Moreover, the high-temperature strength of directly joined Al_2O_3 rods did not decrease up to 800°C, as was the case with the original strength.

Indirect joining of 99 percent purity Al_2O_3 or Si_3N_4 containing yttrium was possible by using an intermediate of a ceramic sheet having a large dielectric loss factor. Maximum joint strengths with suitable intermediates reached 70 to 90 percent of those of base materials. High joint strength was obtained for Al_2O_3, where the composition of the sintering aids in the intermediate was similar to that in the base material.

For Al_2O_3 joining, regardless of direct or indirect joining, the joined boundary line was not observed, and there was little difference in microstructure between pre- and postjoining. This result suggests that sintering aids in the grain-boundary phase were preferentially heated and melted without melting Al_2O_3 particles.

In the case of Si_3N_4 joining, the joined boundary line was also not observed, but an anomalous layer, in which yttrium was lacking, occurred in the vicinity of the joined boundary. It is believed that the existence of this layer caused the joint strength of Si_3N_4 to decrease.[202–204,206]

It is quite clear that engineering-quality joints of both oxide and nonoxide ceramics can be made in very short times using modest microwave power and compression, with very little surface preparation. The increased densification, implied by the increased hardness and fracture toughness, and the scanning electron microscopy (SEM) studies in the joint region of the materials joined together with the microscopic homogeneity of the joint interface suggest a joining mechanism: melting and diffusion of grain-boundary phases.[203]

One important class of materials which are currently under investigation[203] are ceramic-matrix composites. The following characteristics of the microwave-joining process bode well for its eventual adoption by industry: (1) modest power requirements, (2) short joining time, (3) densification of the joint region, and (4) potential for on-line process control. Item (4) is a direct result of the in situ nondestructive evaluation technique incorporated into the joining equipment and apparatus.

As a final point Palaith and Silberglitt[203] note that the use of microwave energy offers the potential for entirely new controls over the joining process. It is not only impurities in the material that determine its heating rate, but also, more precisely, the density and distribution of the microwave absorbers.[203]

Direct Bonding by Fusion Welding

Fusion welding of ceramics is accomplished by filling the joint with molten material obtained by melting the edges of the components, or with melt from a filler of similar materials, just as in the welding of metals. These welding techniques are summarized in the literature.[93,95,96,101,111,207–210]

Fusion is widely used to join glasses and metals, and its application in bonding ceramics to metals would be a major technological advance. Complete contact would be achieved, and joints might be designed for use at temperatures up to the workpiece melting points. However, the range of materials for which fusion may be useful would seem to be very limited. There should be ideally a close match of melting points and thermal contraction characteristics of not only the metal and the ceramic, but also the complex material formed in the weld pool. However, this similarity is rarely achievable in practice and, further, some ceramics (such as BN, SiC, and Si_3N_4) sublime or decompose before melting,[93] while others, such as MgO, vaporize rapidly when molten. Again, on cooling, disruptive phase transformations may occur. Nevertheless, some technologists have been successful in welding certain combinations. For example, Wallace and Copley[64] detail the use of an electron beam, because of its small spot size and good atmospheric control, to produce a prototype electrical feedthrough by fusing wires of molybdenum, columbium, tungsten, or Kovar (the Fe–Ni–Co low-expansion alloy) to 96 percent Al_2O_3 insulators. The successful fusion of Al_2O_3 to molybdenum has also been attributed to Dring (in work reported in Pattee et al.[211]). Welding was successfully achieved at fairly high accelerating voltages (~90 kV) and low currents (~2 mA).[210] Rice[207] has described an extensive study of ceramic welding that included the production of an Al_2O_3–Ta couple using an electron beam and Gr–W, and ZrO_2–Mo, ZrO_2–Cb, or ZrO_2–Ta couples using arcs.

Fusion welding can be divided into electromagnetic laser-beam welding, electron-beam welding, arc welding, ultrasonic welding, and friction welding.[96] Some investigators have considered these techniques to be only marginally successful in the bonding of ceramics, mainly due to resultant stress fracture.[111,210] The fractures are probably due to thermal stresses developed from temperature gradients during heating. A solution may be to heat the weld zone area or even heat the entire workpiece to minimize thermal variations. Uniform cooling will also reduce the risk of thermal shock. Electromagnetic laser-beam welding is a versatile technique, which is primarily laser welding. It requires a vacuum chamber. Rice[207] has shown that welds on Al_2O_3 bodies were usually cracked because of a larger resultant grain size in the weld area. However, other reports exist of the successful electron-beam welding of ceramics.[64,210,211]

Ultrasonic welding is a solid-state bonding process for joining metals by introducing high-frequency vibratory energy into overlapping workpieces. An intermediate material may be used to promote bonding.[212,213] Engineers used ultrasonic joining techniques to produce a hermetically sealed transistor package.[214] The base of the package was a 94 to 96 percent Al_2O_3 wafer, and the required seal and thermal patterns on the ceramic wafer were provided by Mo–Mn metallizing. After metallizing, the patterns were nickel- and copper-plated and then solder-dipped. The transistor was in the ceramic wafer. After assembly, a brass cap was soldered to the wafer to produce a sealed unit. Since flux could not be used due to the danger of contaminating the transistor surface, ultrasonic methods were used for joining and were successful.

Laser welding of ceramics is restricted to fusible oxide ceramics. However, bonding of highly pure, dense Al_2O_3 plates using the CO_2 laser with the long oscillation wavelength[215] have been reported. A major problem with the laser welding of ceramics is that thermal cracks have occurred when ceramics, whose resistance to thermal shock is poor, were heated or cooled rapidly. Other problems of laser welding have included the generation of bubbles and the enlargement of crystalline grains.

Another unusual joining technique utilizes laser-activated brazing.[216] This new technique of joining Si_3N_4 ceramics uses a CO_2 laser beam for local heating of the ceramics and the mixture of refractory ceramic powders as a brazing filler material. The narrow gap of the contact-fitting joint (3 to 5 μm) is filled with the laser-activated molten braze material by capillary attraction.

Initially the localized laser-beam heating will elevate the temperature of Si_3N_4 ceramics and heat up to brazing temperatures, 1680 to 1850°C, without appreciable deterioration of the mechanical strength of the ceramics at the furnace temperature of 1100°C. At the furnace temperature of 1100°C, the joining of Si_3N_4 is attained by using a refractory filler material of Y_2O_3–La_2O_3–MgO–Si_3N_4 in a very short time period (100 to 150 s) due to the high fluidity and reaction rate of the molten filler material. A joint strength of 550 MPa has been attained with the brazed joint characterized by the narrow interlayer containing polygonal particles of Si_3N_4.

Another unusual direct ceramic-to-metal joining technique is friction welding. The Welding Institute has shown the ability to friction-weld a 3-mm-diameter aluminum stub to an Al_2O_3 chip carrier. For this application (heat sinking of an electronic circuit) the technique could prove extremely useful.

The Frauenhofer Institute and Kuka Welding Technology[217] have been experimenting with friction welding as a step toward the large-scale production of ceramic-metal compounds. The technique involves rotating one element of a compound against the other while the second is held firmly. The friction, in combination with compression, welds the elements together in only a few seconds. It has successfully welded aluminum alloy parts to parts of Al_2O_3 and ZrO_2. The compound parts are intended for motors and machines.

Arc welding, restricted to electrically conducting materials, has been tried on some carbides, borides, and refractory metals to graphite.[208]

7.5 COATINGS

In the broad category of advanced ceramics, high-performance ceramic coatings constitute materials which combine such properties as wear, erosion, corrosion, and high-temperature resistance. With these coatings, many metals may be used in applications for which they would otherwise be unsuitable. The promise of performance improvements is the main force driving the continued development and commercialization of high-performance ceramic coatings, which are generally applied to cast iron, steels, superalloys, titanium alloys, tungsten carbides, carbon-carbon composites, and even ceramics.[218–230]

Although nitrides are becoming more commonly used, oxides and carbides are the most frequently utilized materials for high-performance ceramic coatings, and the most popular application techniques include thermal spraying (Fig. 7.34), wet process, and physical- and chemical-vapor deposition. In general, wet processes and thermal spraying are preferred for depositing oxide ceramics, while physical- and chemical-vapor deposition are the processes of choice for nonoxide ceramics. Emerging coating technologies include sol-gel processing, ion-assisted processes, and combination processes.[218–230]

One technique not mentioned was used by Gotman and Gutmanas,[224] whereby satisfactory coatings were produced on Si_3N_4 surface by treating it in metal powders. The coatings usually consisted of two layers, the top layer being a nitride and the inside layer a silicide of the metal in which Si_3N_4 was treated. Both ni-

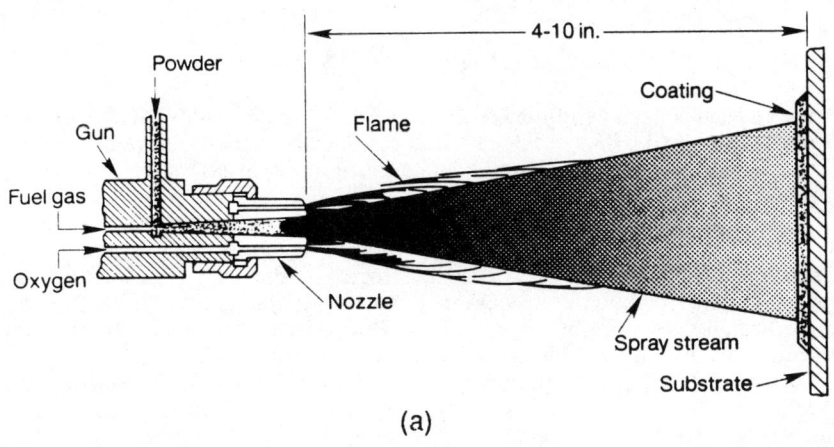

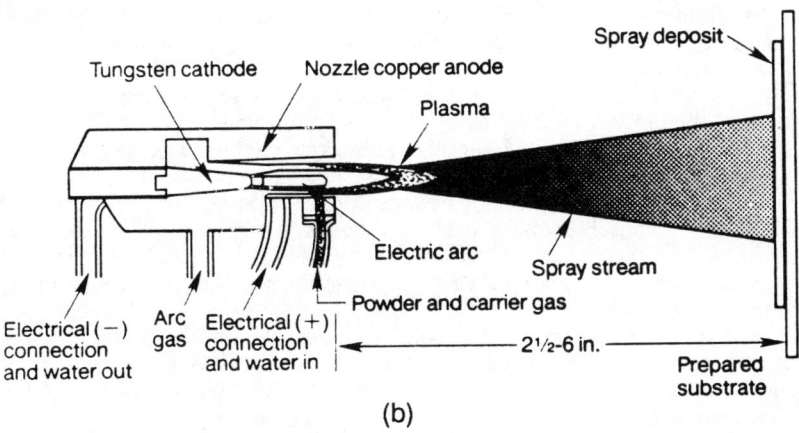

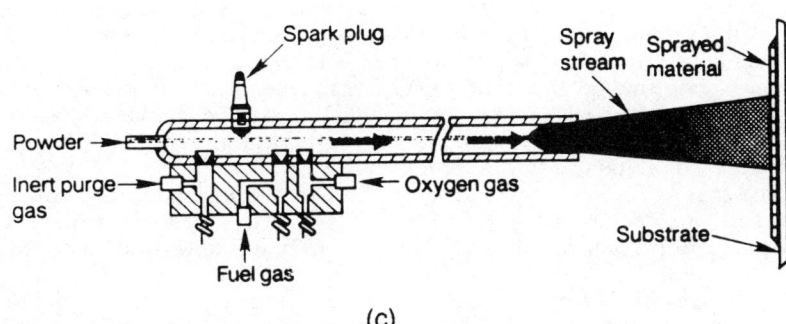

FIGURE 7.34 Application of thermal spray coatings.[219] (*a*) Powder flame spray gun. (*b*) Plasma torch. (*c*) Detonation gun.

trides and silicides thus obtained were metal-rich compounds. Multilayer coatings were obtained by successive treatments of Si_3N_4 in two different metal powders. The method has been shown to be feasible for Si_3N_4 as well as for other nitride ceramics.

Rapidly solidified (RS) ceramic coatings are formed by very fast cooling processes that permit the materials to retain their fine microstructures, making them tougher and more durable than ceramics formed from highly purified powders. By applying RS technology to ceramics, parts manufacturers will be able in the future to fabricate more homogeneous ceramics. In coating applications, impure starting materials are used which dramatically cut product costs.

Naval Research Laboratory investigators[231] worked with ZrO_2-based alloys in developing RS ceramic coatings. They rapidly solidified ZrO_2 alloy compositions on fixed and moving substrates using laser melt quenching and plasma-spraying processes. ZrO_2 was alloyed with 8 wt % Y_2O_3 and was rapidly solidified as a thin coating on room-temperature substrates. The result was a uniform distribution of grains on the order of 5 to 10 μm.

Future work at NRL has involved melting impure ZrO_2 by a skull melt process, which is a cold crucible technique whereby the impurities of the ZrO_2 vaporize during the 3000°C melt. Subsequently the mixture is either vaporized by plasma spraying or just rapidly cooled by a proprietary method, and then ground into fine powder. The powder is then incorporated into the plasma spray. It is hoped that this process will be able to go into pilot-plant operation.[231]

Widespread use of ceramic coatings in aircraft engines was the major breakthrough which spurred acceptance of the coatings for other demanding applications.[226] Current high-performance ceramic coatings are used for carbide tool inserts;[232] diesel-engine parts; land-based turbines;[228,230] carbon-carbon composites; heat exchangers; wear-resistant parts for the textile, paper, and petroleum industries;[222] repair of worn-out machinery; and numerous aerospace and military applications.[220,221,223,226,227,229]

In terms of the value of only the coating, the world market for ceramic-coated structural applications was estimated at $1.7 billion in 1993. Coatings applied by physical- and chemical-vapor deposition account for nearly two-thirds of the total. Construction (coated glass), metal fabricating (cutting and forming tools), and military applications represent about 70 percent of the market.

Currently in the United States about 2 million aircraft-engine components are coated with thermal spray techniques, and approximately 600,000 components are coated using chemical vapor deposition. The ceramic-coating service market for aircraft-engine components during manufacture is estimated to be $190 million for 1990. Ceramic physical-vapor-deposited coatings have been found to be the most durable for fan blades and have begun to appear in engines. The ceramic-coating service market for aircraft-engine maintenance and overhaul is presently estimated to be about $8 million. Potential aerospace markets include high-emissivity coatings for thermal protection systems in the Space Shuttle and the aero-assisted orbital transfer vehicle (AOTV).

The coating service market for metal-cutting tool inserts utilizes coatings by chemical-vapor deposition of TiC, TiN, and Al_2O_3 90 percent of the time, and the remaining 10 percent includes physical-vapor deposition and ion-assisted techniques.

Ceramic coatings have provided improved performance in racing cars for several years. Recently some of the major automobile manufacturers have been using ceramic coatings for exhaust manifolds. A few heavy-duty diesel engines use ceramic coatings on the piston dome and exhaust manifolds. This automotive market segment is not expected to grow very significantly until the late 1990s. Other promising engine markets include land-based turbines and marine engines.

There are, however, certain technical and economic issues that remain to be resolved, including high costs and consistency in producing high-quality coatings. Also, the technique used to deposit a ceramic coating greatly influences microstructure and performance.

The four most prominent coatings in use today[232] include TiC, TiN, titanium carbonitride, and Al_2O_3 (Table 7.17). Selection of the best coating for the application depends on the workpiece material and its condition, the machine tool (it must be very rigid and have ample power), and the requirements regarding tool life and surface finish.

Although a specific type of coating is more advantageous than the others in certain situations, TiN long has been the number one coating applied to high-speed steel and carbide cutting tools requiring sharp edges. However, the TiN coating material does not work well on cutting tools used to machine any of the titanium alloys or high-nickel alloys, or many of the nonferrous materials. The same holds true for the other titanium-based coatings.

There is a new coating now available, B_4C. It is black instead of gold, applied in a layer 0.007 ±0.001 mm thick. Tools coated with the B_4C show performance improvements on the soft, long-chipping workpiece materials, including aluminum alloys, brass, and stainless steel. It also works very well on titanium materials, cast iron, very hard steels, and hardfacing alloys that generate short chips.[233]

Chemical-Vapor Deposition

Chemical-vapor deposition (CVD) has become an established technique and an essential technology in the manufacture of materials and coatings. CVD coatings are used to stop wear, corrosion, and erosion of such products as metal-cutting tools, turbine-engine components, and bearings. CVD coatings also protect against oxidation at extreme temperatures and facilitate joining of carbon-carbon composites. In still other cases, the process is used to manufacture or densify composites and other materials that are difficult to make by other methods.

CVD coatings are deposited by causing chemical reactions to occur at the

TABLE 7.17 Typical Substrate-Coating Material Combinations

Substrate	Al_2O_3	ZrO_2	TiC	B_4C	SiC	TiN
Steels	x	x	x	—	—	x
Superalloys	—	x	—	—	—	x
Titanium alloys	—	—	x	—	—	x
Graphite	x	—	x	x	x	x
SiC	x	—	—	—	—	x
Si_3N_4	—	—	—	x	x	—
Carbon/carbon composites	—	—	—	x	x	x
Nylon	x	x	—	—	—	—
Polyester	x	x	—	—	—	—
Polyimides	x	x	—	—	—	—

Source: From Chan and Wachtman.[218]

surface of the material being coated, governed by the thermodynamics of the substances. In its simplest form, the process begins with a gaseous or vaporizable precursor or combination of substances, containing the elements to be deposited. These vapors are introduced to the reactor containing the object to be coated. The surface of the object, or the entire reactor, is heated to a temperature at which the precursors react or decompose. The solid reaction products are deposited on the surface of the object as well as on any other surface at the same temperature, while the gaseous products are exhausted to be replaced by fresh reactants.

The reactions typically take place at high temperatures, which limits the substrate materials to those that can withstand the temperatures. The rate of deposition is determined primarily by the gas-flow properties on the substrate surface rather than by the thermodynamics of the reaction.

CVD is not a line-of-sight coating process, which makes it ideal for coating complex shapes such as cutting tools, tubes, and other parts with holes or recesses. A large number of small components can be coated in a single batch, helping to keep costs at a minimum.

Disadvantages. The primary drawbacks to CVD are the high temperatures involved (typically at least 700°C) and the need to develop a different chemical process for each type of coating. However, a wide variety of coatings is commercially available, and developments in plasma-activated CVD are helping to lower the deposition temperatures.[232]

Applications. The most widespread application of CVD coatings outside of the electronics field is on metal-cutting tools. CVD coatings can increase the carbide tool life by 100 to 1000 percent or more, while allowing removal rates up to six times as high. As a result, approximately 85 percent of carbide tools are coated by CVD.

The coatings are best used on substrates such as tool steels, martensitic stainless steels, and cemented carbides. Lower-temperature metals, such as high-speed steels, can be coated, but require a subsequent vacuum heat treatment to restore the hardness that is lost during coating.

CVD is used frequently to apply oxidation-resistant coatings to carbon and carbon-carbon composites because of its ability to deposit a uniform, impervious coating on complex shapes. SiC coatings, deposited by the decomposition of methyltrichlorosilane (CH_3Cl_3Si), provide oxidation protection up to 1650°C, but crack as the composites cool, allowing oxygen to penetrate if glass-forming additives are not added to the carbon.

The CVD of SiC from CH_3Cl_3Si was studied by Besmann[234] in the preparation of a two-phase coating of B_4C–BN as a potential wear coating, because of its likelihood of having a high fracture toughness resulting from its composite nature and inherent lubrication resulting from the presence of BN.

Carbon nozzles used for propulsion are coated with hafnium oxide (HfO_2) for more durable oxidation protection, up to 1925°C, while a process has also been developed for depositing iridium, which can protect substrates up to its 2400°C melting point. This coating on a thrust chamber allows the optimum mix ratio of nitrogen tetroxide–monomethyl hydrazine propellant to be used without damaging the chamber.[232]

Plasma-Activated CVD or Plasma-Assisted CVD. Reducing the temperature of the CVD process is an important goal of producers of CVD coatings. One method for doing this is plasma-activated CVD, in which the chemicals are excited by an electric field to produce a plasma that increases the rate of reaction at lower temperatures. For example, a company uses plasma-activated CVD to manufacture Si_3N_4 microlenses for fiber-optic couplers. The Si_3N_4 is coated on a glass substrate etched with circular recesses in the shape of the lenses, after which it is polished flush with the glass surface.

The plasma-activated CVD process is carried out at about 300°C, compared with 850°C for thermal CVD. The low temperature helps avoid cracking that could result from mismatched thermal expansion coefficients between the lenses and the Pyrex substrate.

A WC coating is applied at temperatures between 325 and 525°C. Compared with other low-temperature CVD WC coatings, which have a columnar microstructure, this coating offers far better resistance to erosion and adhesive, sliding wear.[232]

Akin to plasma-activated CVD is sputtering. The state of the art of sputtering and plasma-assisted CVD technology has advanced to the point where production coating machines are being operated reliably for long periods of time.

Sputtering. Sputtering represents an increasingly attractive physical-vapor deposition (PVD) process for coating large irregularly shaped parts. Conventional diode sputtering techniques have been in use for over 25 years. However, it is the development of magnetron sputtering technology over the past 10 years that has taken sputtering out of the laboratory and made it a viable large-scale industrial coating process. Triode sputtering machines can operate under conditions similar to magnetrons and provide high-quality coatings at high deposition rates. Triode sources, however, are limited for many applications by the lifetime of their electron emitters and difficulties of scaling to large sizes.

Sputtering permits the deposition of virtually any material for which a target is available. Unfortunately target fabrication is not always a simple matter. Therefore sputtering technology could benefit from improvements in target fabrication methods, particularly where large target sizes are involved.

Unlike electroplating, PVD technologies are significantly limited by the fact that the coating species are not significantly influenced by electric fields. Sputtered coating atoms move in a line-of-sight trajectory following this last gas-phase collision, if any, before encountering the substrate surface. This can cause severe shadowing effects for deposition into deep cavities on a substrate. Shadowing problems represent one of the most significant limitations to the use of PVD technologies for the coating of substrates of complex shapes.[235-237]

The most significant recent advance in sputtering technology has been the development of magnetron sputtering.[235] The magnetron provides an extremely high-density plasma, which is confined to the area near the sputtering target (cathode) and can be scaled up to very large dimensions. Currently production magnetron sputtering systems use cathodes up to 4 m wide. Magnetrons can also provide extremely rapid deposition. Rates as high as 4 μm/min have been achieved in small test magnetrons.

Laser-Assisted CVD. Laser-assisted CVD is an entirely new technique, which is currently under development at a number of laboratories. It offers the potential for direct writing of coatings on limited areas of substrates. In addition, the photolyzing laser energy can be tuned to select a given species and discriminate against competing gas-phase reactions which would dominate in conventional or plasma-assisted CVD.[238]

In conclusion, plasma-based techniques have become an essential element in the fabrication of modern high-technology equipment. Net-shape parts, particularly for gas-turbine engines, have been fabricated by a large number of techniques. These techniques permit improved mechanical properties and enhanced cooling capabilities. Ultimately, however, the applications for many net-shape parts are limited by the thermal or chemical stability of the materials of which the parts are constructed. Under such conditions the lifetime of the part can, in many cases, be extended by the application of a protective coating. The range of coating compositions, their properties, and the low deposition temperatures that are possible with plasma-based techniques make sputtering and plasma-assisted CVD attractive for many such applications.

Physical-Vapor Deposition

Of all the established coating processes, most attention is now on the previously mentioned CVD and PVD. PVD is a generic term that describes a number of processes that operate in the low-temperature range of 204 to 510°C.

Coatings applied to carbide and high-speed-steel tools are usually TiN or TiC, both of which improve tool life. Al_2O_3 coatings are also applied to carbide tools and inserts. Other coatings provide chemical and thermal barriers and are applied by PVD processing. PVD coatings exert compressive residual stresses within the coating and do not degrade the transverse rupture strength of the carbide tools as do CVD process coatings.

Another advantage of PVD is the ability to coat uniformly over sharp cutting edges and to do so without degradation of edge strength. This is particularly beneficial for milling applications, which place a great deal of stress on the cutting edge.

As favorable as the PVD coatings are, Santhanam[239] stated that "they would not replace CVD coatings to any great extent in the near future," at least for cemented carbide tooling. For one thing, honing the edge of inserts overcomes many of the objections to the CVD process, which apparently is less costly than other processes.

Based on further technical information from developers, however, it appears that the pendulum may swing in favor of PVD coatings. Researchers have reported that "PVD coated inserts experience greater reliability and predictability of tool life for the end user."

Sputtering, evaporative electron-beam melting, and cathodic arc deposition are the main PVD coating processes.[240–244] Sputter ion plating (SIP) is based on coating methods developed for the semiconductor industry and has a cycle time of 6 to 8 h. One advantage is that a variety of parts can be coated during the same cycle. Regarding cutting edges, Podob[241] recommends the use of rounded edges to protect the coating, whether the coating was on carbide or on high-speed-steel tools.

Another PVD coating is B_4C, which is used on cutting tools in a vacuum at 121°C less than 0.01 mm thick. The B_4C coating (almost twice the hardness on the Knoop scale over TiN coating) has no advantage over TiN coatings in machining many ferrous materials. But in machining aluminum and titanium alloys, as well as cast iron and stainless steel, it has been advantageous. Coatings have been applied to inserts, shank tools, carbide-tipped tools, saws, and other cutting tools and wear parts.

Tool-life improvement for the B_4C coated tools over uncoated high-speed-steel tools in machining aluminum has been from 3 to 21 times, in machining titanium from 2 to 72 times, and in drilling aircraft titanium-aluminum laminate 9 times.

Sputter Ion Plating. This technique uses a glow discharge to sputter titanium from an extended source, whereupon it reacts with nitrogen to form TiN. Sputter ion plating has the following advantages related to both the execution and the application of the process:

1. Good throwing power
2. Stationary fixturing
3. Mechanically simple construction
4. Ability to coat uniformly components of different sizes and shapes as one workload
5. Ability to coat large components
6. Excellent replication of the substrate surface finish
7. Excellent gold color
8. High equipment uptime
9. Minimum maintenance

Figure 7.35 is a schematic of the sputter ion plating process. In principle, the chamber can be of any size to accommodate components of any diameter or

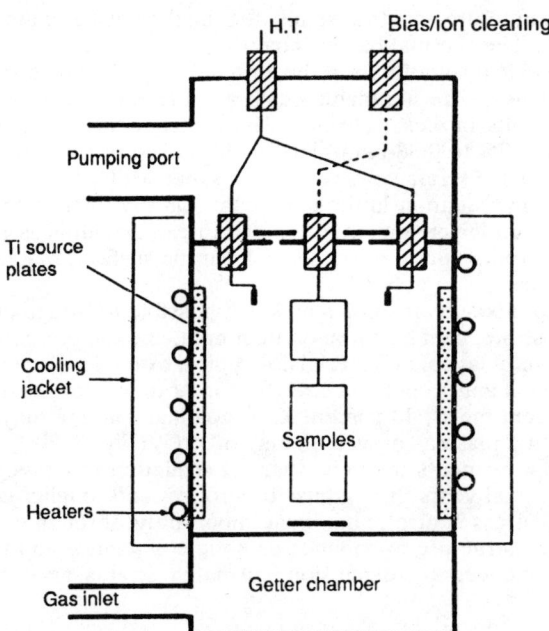

FIGURE 7.35 Schematic view of sputter ion plating system.[241]

length, provided adequate titanium source electrodes are distributed within the work chamber.[241]

TiN coated drills by sputter ion plating have demonstrated up to tenfold improvement in life compared with approximately 80 percent of this improvement in life retained after resharpening. Alternatively, a coated drill may be used at a 50 percent higher cutting speed than a steam-tempered drill with about the same life. The real benefit on inexpensive tools such as drills lies not in the improved economics from longer tool life, but rather in higher productivity and less downtime. Furthermore, the quality of holes drilled (in terms of surface finish) is also improved significantly, leading to the additional engineering benefit that a subsequent reaming operation may in some instances be omitted.

Electron-Beam PVD. The electron-beam PVD process provides a completely different structured ceramic. In plasma-spray processes, each particle is melted as it passes through the plasma flame. When the molten particle comes in contact with the workpiece, it "splats." A buildup of splats results in what is known as the characteristic splat structure of the plasma-spray process.

In the electron-beam PVD process the particles are first vaporized. They then condense on the workpiece, producing a columnar structure. According to Stearns,[226] Miller,[230] and Irving,[245] this high-technology vacuum process has a long way to go to be fully competitive with the plasma flame. The plasma-spray process can accomplish in minutes what might take hours with the electron-beam process. In addition, large part sizes can be accommodated by a plasma-spray system.

Therefore, why consider this coating system? What place is there for thermal-barrier coatings in future engine systems, that is, the adiabatic diesel engine? It is very significant. The thermal-barrier coatings for diesel-engine components are actually technological spinoffs from the abradable tip clearance seals used in aircraft turbine blades. The abradable seals are the building blocks for plasma-sprayed components in diesel engines. The diesel-engine coatings are 10 times thicker than the 0.25- to 0.38-mm-thick coatings used in gas-turbine engines.

"On the thinner aircraft coatings," points out Miller,[230] "the prime failure mechanisms are in oxidation in the bond coat. On the thicker coatings in diesel engines, due to the lower operating temperatures, oxidation is not of as much concern. That permits you to do some tricks at the surface, mainly in mixing ceramics with metals."

The ability to mix such materials makes it possible to bridge the gap between the metallic substrate, be it cast iron or aluminum, and the protective ceramic top coat through several layers of materials. A typical example is a coating consisting of 85 percent metal and 15 percent ZrO_2 on top of the metal substrate. On top of that is a 60 percent metal, 40 percent ZrO_2 coating. The multilayered coating is then topped with a plasma-sprayed coating of ZrO_2 (Fig. 7.36).

Dramatic improvements in cyclic thermal spallation life have been achieved over the past several years through use of stronger and tougher ceramic compositions, residual stress control, and, most importantly, through careful control of the ceramic microstructure to enhance ceramic compliance and thus reduce the level of stress induced by expansion mismatch strains according to several researchers.[228,245-249]

A structure produced by electron-beam PVD consists of individual free-standing ceramic columns, each of which is very tightly bonded to the metal substrate but is essentially free to separate from adjacent columns as the underlying substrate thermally expands relative to the ceramic. While the cyclic strain tol-

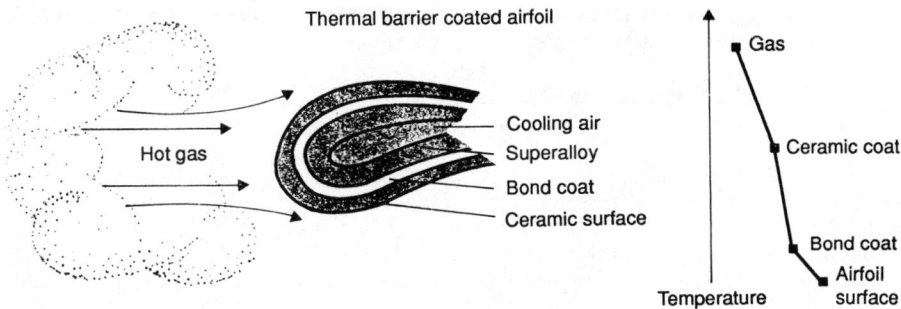

FIGURE 7.36 Performance of thermal-barrier coatings.[245]

erance benefits of this structure have been known for some time,[250] it is only recently that electron-beam PVD process technology has advanced to the point of producing this coating with reproducibly high levels of relative cyclic thermal durability. As a result of improved basic understanding of electron-beam PVD ceramic failure modes, the cause of infant mortality failure has been clearly identified, and improved reactive evaporation processing[251] has been developed to eliminate the problem. Laboratory burner rig screening tests indicate that the relative durability increase for the improved electron-beam PVD can be about 8 times the plasma-sprayed coating durability.[246,247]

For the thermal-barrier coating systems it is evident that 17 percent Y_2O_3 is required to fully stabilize the cubic structure of ZrO_2 throughout the 816 to 1204°C operating temperature range used for gas turbines, while around 12 percent is sufficient to maintain the cubic phase metastably at room temperature. The partially stabilized ZrO_2 containing 6 to 8 percent Y_2O_3 (PYSZ) had been reported to outperform all other systems.[228,247–249,252] Some recent testing[253] shows that ZrO_2 stabilized with ytterbia (Yb_2O_3) is better than the aforementioned ZrO_2–Y_2O_3. The best performance was achieved by ZrO_2–$12Yb_2O_3$ on a Ni–17.5Cr–6.6Al–0.22Yb base coat. That combination endured nearly 600 cycles to failure in a furnace atmosphere as the temperature was repeatedly altered between 301 and 1107°C. That is a 60 to 90 percent gain over the $6.1Y_2O_3$- and $8Y_2O_3$-stabilized ZrO_2 applied to the yttrium-bearing base coat (Ni–16.5Cr–5.92-Al–0.17Y). Less but nevertheless substantial improvement was achieved by the Yb_2O_3-modified ZrO_2 on the yttrium-bearing base, and this is only the beginning to demonstrate the necessary combination of durability, reliability, and predictability for prime reliable application of electron-beam PVD on ceramics.[245–249,253]

Laser. A new ceramic coating technique using a CO_2 laser was developed to make a layer of high-temperature solid electrolyte for a fuel cell. The laser beam was focused parallel to the surface of the substrate to which melted ceramics were deposited by a powder gun. This method has the following advantages:

The substrate is not damaged because the laser beam does not hit it.

The coating of various materials having a high melting point is possible because of high power intensity.

These coated layers have high purity because they are made in a vacuum or inert-gas atmosphere.

Coating conditions (such as laser power, substrate temperature, and feeding velocity of coating materials) are set independently.

By using laser irradiation, Tsukamoto et al.[254] successfully made YSZ layers both on an Al_2O_3 surface and on a copper plate. The structures of these materials were not found to be changed before and after the processing. More efforts are required in such areas as the supplying of raw materials, irradiation methods, heating substrates, and coating environmental conditions.

A CO_2 laser was also used by Yasunaga et al.[255] as a surface-treatment technique, which enabled them to form hard ceramic layers on metal surfaces using chemical reactions between painted powders and reactant gases blown at the laser-irradiated area. A 2-kW continuous-wave CO_2 laser irradiated on a pure aluminum specimen on which various kinds of metal and ceramic powders were painted. Reactive gases were blown simultaneously to the laser-irradiated area, where ceramic layers such as Al_2O_3, Al_4C_3, AlN, TiN, and Cr_2O_3 were formed, dispersing or adhering to the molten aluminum subsurfaces. The wear resistance of the ceramic-layer formed specimens increased dramatically to over 10^3 to 10^4 times that of the uncoated aluminum specimens.

Ion-Beam Sputtering and Ion-Assisted Deposition. These two activated processes for thin-film ceramic coatings are relatively new techniques along with PVD and CVD and are capable of producing graded interface coatings. Their applications are usually in the visible and ultraviolet regions.[256]

Nondestructive Testing Processes for Coatings

The measurement and evaluation of ceramic coatings have become critical and most significant areas of development. Researchers and scientists are evaluating a variety of noncontacting, nondestructive sensor systems to measure the on-line production of thin-film ceramic coatings on metals and ceramics.

The technique uses photothermal radiometry to monitor the surface uniformity of materials, detect flaws, and measure the thermal resistance of ceramics. It uses a modulated laser beam to periodically heat a thermal "hot spot" on the surface of the coating. As the thermal wave is partly reflected by the metal substrate, an infrared sensor aimed at the same spot picks up temperature fluctuations and measures the coating's thermal resistance.

Ceramic coatings of oxide, carbide, boride, and nitride measuring a few micrometers to a millimeter in thickness can support temperature differentials of several tens to several hundreds of degrees Celsius, and therefore the thermal conductivity of these coatings depends on both the materials and the method used to deposit them on metal surfaces.

REFERENCES

1. J. B. Wachtman et al., "Japanese Structural Ceramics Research and Development," FASAC Tech. Assessment Rep. (TAR), SAIC, McLean, Va., and San Diego, Calif., July 1989.
2. Material Process Rep., M.I.T., vol. 2, pp. 6–7, Aug. 1987.
3. R. Morrell, "An Introduction for the Engineer and Designer—Part I," in *Handbook of*

Properties of Technical and Engineering Ceramics, Her Majesty's Stationery Office, London, May 1985.
4. K. Oczos, "Advances in the Processing of Ceramic Engineering Parts," *Keram. Z.,* vol. 41, pp. 411–415, 1989.
5. E. C. Skaar, "Research Needs in Ceramic Machining," Clemson Univ., NSF Eng. 88015, DMC 8612943, 1988, 26 pp.
6. R. F. Firestone, "Lasers and Other Nonabrasive Machining Methods for Ceramics," in *Proc. Advanced Ceramics Conf.,* Feb. 1987, pp. 133–138.
7. J. J. Indge, "Flat Precision Machining of Ceramic Materials," *Production,* pp. 59–61, Dec. 1986.
8. K. Subramanian, "Precision Finishing of Ceramic Components with Diamond Abrasives," *Ceram. Bull.,* vol. 67, pp. 1026–1029, June 1988.
9. B. Thomas et al., "Effect of Machining Parameters on the Surface Finish and Strength of Hot Pressed Silicon Nitride," *Powder Metall. Intl.,* vol. 20, pp. 39–44, May 1988.
10. T. Nakano, K. Abe, and H. Kubo, "Newly Developed Dressing Technology for Cutting-Off Sialon (Si-Al-O-N) Ceramics," *Ann. CIRP,* vol. 38, pp. 327–330, June 1989.
11. W. König and J. Wemhöner, "Optimizing Grinding of SiSiC," *Ceram. Bull.,* vol. 68, pp. 545–548, Mar. 1989.
12. W. König and M. Popp, "Precision Machining of Advanced Ceramics," *Ceram. Bull.,* vol. 68, pp. 550–553, Mar. 1989.
13. R. W. McEachron and S. C. Lorence, "Superabrasives and Structural Ceramics Creep-Feed Grinding," *Ceram. Bull.,* vol. 67, pp. 1031–1036, June 1988.
14. H. O. Juchem and B. A. Cooley, "Creep Feed Grinding—A Review," *Ind. Diamond Rev.,* vol. 6, pp. 313–319, 1985.
15. K. Subramanian and P. P. Keats, "Parametric Study on Grindability of Structural and Electronic Ceramics: Part I," presented at the ASME Winter Mtg., Machining of Ceramic Materials and Components, Miami Beach, Fla., Nov. 17–22, 1985.
16. S. F. Wayne and S. T. Buljan, "Role of Thermal Shock on Tool Life of Selected Ceramic Cutting Tool Materials," *J. Am. Ceram. Soc.,* vol. 72, pp. 754–760, 1989.
17. R. Kopp, "A Guide to Grinding Advanced Ceramics," *MAN,* pp. 22–26, Feb. 1990.
18. F. Mason, "Job Shop Machines Ceramics, Glass," *Am. Mach.,* pp. 51–53, Mar. 1990.
19. H. Nogawa et al., *Ceramics Processing—State of the Art of R&D in Japan,* Am. Soc. of Metals, Metals Park, Ohio, 1988, 400 pp.
20. J. Wilson, G. E. Merwin, and L. L. Hench, "Machining Bioglass® Implants," *SAMPE J.,* pp. 6–8, May/June 1985.
21. W. S. Ricci, "Machining Metal Matrix Composites," AMTL, SME-MR 87-827, presented at Composites in Manufacturing Conf., Long Beach, Calif., Dec. 7–10, 1987, 28 pp.
22. E. Duran, *Mach. Des.,* p. 16, Nov. 9, 1989.
23. N. Morita et al., *Lasers and Optronics,* p. 31, Aug. 1988; *Appl. Phys. Lett.,* June 6, 1988.
24. R. L. Vaughn, "Skin Milling for the Space Shuttle," *Manuf. Eng.,* pp. 80–82, Mar. 1980.
25. M. Hashish, "Optimization Factors in Abrasive-Waterjet Machining," in K. P. Rajurkar (Ed.), *Research and Technical Development in Nontraditional Machining,* ASME PED-V34, ASME Ann. Winter Mtg., Chicago, Ill., 1988, pp. 163–180.
26. M. Hashish, "Turning with Abrasive-Waterjets—A First Investigation," *ASME J. Eng. Ind.* vol. 109, pp. 281–290, 1987.
27. M. Hashish, "Milling with Abrasive-Waterjets—A Preliminary Investigation," in *Proc. 4th U.S. Waterjet Conf.,* Berkeley, Calif., 1987, pp. 1–10.

28. L. M. Sheppard, "Machining of Advanced Ceramics," *AM&P,* vol. 132, pp. 40–48, Dec. 1987.

29. M. Hashish, "Machining of Advanced Composites," in M. Taya and M. Ramulu (Eds.), ASME PED-V35, ASME MD V12, ASME G 00496, ASME Ann. Winter Mtg., Chicago, Ill., 1988, 62 pp.

30. W. S. Ricci et al., "Abrasive Waterjet Cutting of Ceramics and Organic Based Materials," Army Lab. Command, MTL-TR-87-31, FR 7/87, July 1987.

31. G. Hamatani and M. Ramulu, "Machinability of High Temperature Composites by Abrasive Waterjet," in M. Taya and M. Ramulu (Eds.), ASME PED-V35, ASME MD V12, ASME G 00496, ASME Ann. Winter Mtg., Chicago, Ill., 1988, pp. 49–62.

32. M. A. Moreland and D. O. Moore, "Versatile Performance of Ultrasonic Machining," *Ceram. Bull.,* vol. 67, pp. 1045–1047, June 1988.

33. W. König et al., "EDM Short-Cuts Ceramic Machining," *MAN,* pp. 19–24, Dec. 1989.

34. B. R. Lazarenko and N. I. Lazarenko, "Machining by Erosion," *Am. Mach.,* vol. 91, pp. 120–121, 1947.

35. N. F. Petrofes and A. M. Gadalla, "Electrical Discharge Machining of Advanced Ceramics," *Ceram. Bull.,* vol. 67, pp. 1048–1052, June 1988.

36. J. C. Quinlan, "What's New in EDM," *Tool Prod.,* vol. 52, pp. 40–49, Dec. 1987.

37. R. F. Firestone, "Abrasionless Machining Methods for Ceramics," in B. J. Hockey and R. W. Rice (Eds.), *The Science of Ceramic Machining and Surface Finishing II,* Spec. Publ. 562, Natl. Bureau of Standards, 1979.

38. N. Tamari et al., "Electric Discharge Machining of Composite Ceramics of Si_3N_4–SiC Whiskers," *J. Ceram. Ind. Assoc.,* vol. 94, pp. 1231–1235, Dec. 1986.

39. M. Nakamura, "Electric Discharge Machining of Ceramics Sintered SiC," *J. Elec. Mach.,* vol. 19, pp. 1–11, 1987.

40. T. Sasaki et al., "Electric Discharge Machining of Ceramic (Conductive Zirconia)," *Mach. Tech.,* vol. 35, pp. 67–71, Feb. 1987.

41. E. Kamijo et al., "Silicon Nitride Ceramics Which Are Electric Discharge Machinable," *Sumitomo Electric,* vol. 125, pp. 178–185, Sept. 1984.

42. S. Kobayashi et al., "Practical Application of Conductive Ceramic Starts: Conductive SiAlON Has Machinability as Good as Steel," Suppl., *Eng. Ceram.,* pp. 104–109, 1986.

43. K. Manba et al., "Electric Discharge Machining of Composite Sintered ZrB_2," *J. Ceram. Ind. Assoc.,* vol. 94, pp. 214–223, Jan. 1986.

44. W. S. Ricci, W. R. Blumenthal, and H. A. Skeele, "Electrical Discharge Machinability of Ceramics," in K. P. Rajurkar (Ed.), *Research and Technical Development in Nontraditional Machining,* ASME PED-V34, ASME MD V12, ASME G 00496, ASME Ann. Winter Mtg., Chicago, Ill., 1988, pp. 281–295.

45. W. S. Ricci et al., "Wire-Cut Electrical Discharge Machinability of Ceramics," Army Lab. Command, MTL-TR-87-52, FR 9/87, Sept. 1987.

46. K. Gettelman, "Electrical Assist for Grinding Ceramics," *Mod. Mach. Shop,* pp. 70–75, Sept. 1987.

47. H. K. Tönshoff and C. Emmelmann, "Laser Processing of Ceramics," *Ann. CIRP,* vol. 38, pp. 199–205, June 1989.

48. H. K. Tönshoff and H. Semrau, "Laser Beam Machining in New Fields of Application," in K. P. Rajurkar (Ed.), *Research and Technical Development in Nontraditional Machining,* ASME PED-V34, ASME MD V12, ASME G 00496, ASME Ann. Winter Mtg., Chicago, Ill., 1988, pp. 249–259.

49. H. K. Tönshoff and C. Emmelmann, "Laser Cutting of Advanced Ceramics," *Ann. CIRP,* vol. 38, pp. 219–222, June 1989.

50. A Kleppe, "Snijden Lassen met Plasmastrahlen en Lasers," *Metaal Kunstof,* vol. 22, p. 42, 1984.
51. R. Srinivasan and W. Leigh, "Ablative Photodecomposition Action of Far-Ultraviolet-Laser Radiation," *J. A. Chem. Soc.,* no. 104, pp. 6784–6785, 1982.
52. K. J. Schmatjko and G. Endres, "Feinbearbeitung von Keramik mit dem Excimer-Laser," *Fachber. Metallbearb.,* p. 294, Apr. 1987.
53. C. Emmelmann et al., "Excimer-Hochleistungslaser in der Materialbearbeitung," *Lasermagazin,* vol. 3, Apr. 1987.
54. P. Affolter and H. G Schmid, "Processing of New Ceramic Material with Solid State Radiation," *Proc. SPIE,* vol. 801/87, Paper 32, 1987.
55. E. J. Vesely et al., "Laser Beam Machining of Ceramic Diesel Engine Components," IIT Res. Inst., IITR-M 01633-13, TACOM TR-12960, DAAE 07-84C-R107, FR 9/84-3/86, Sept. 1986.
56. J. Yamamoto and Y. Yamamoto, "Laser Machining of Silicon Nitride," in *Lamp '87,* pp. 297–302.
57. C. Emmelmann and H. Semrau, "Keramikbearbeitung mit dem Laser," *Lasermagazin,* pp. 8–14, Mar. 1987.
58. V. Bödecker, H. Semrau, and C. Meyer, "Neuartige Anwendungen für den YAG-Laser," presented at Anwendungsforum Lasermaterialbearbeitung in der Automobilindustrie, Bremen, BIAS, Sept. 1987.
59. S. M. Copley, "Laser Machining," in M. B. Bever (Ed.), *Encyclopedia of Materials Science,* vol. 4, Pergamon/M.I.T. Press, Cambridge, Mass., 1986, pp. 2511–2512.
60. S. M. Copley, "Shaping Ceramics with Lasers," in S. K. Samanta et al. (Eds.), *Interdisciplinary Issues in Materials Processing and Manufacturing,* vol. 2, ASME, 1987, pp. 631–636.
61. S. M. Copley, "Laser Applications," in R. I. King (Ed.), *Handbook of High Speed Machining Technology,* Chapman and Hall, London, 1985, pp. 387–416.
62. S. M. Copley et al., "Shaping Materials with Lasers," in M. Bass (Ed.), *Laser Materials Processing,* North-Holland, Amsterdam, 1983, pp. 297–336.
63. M. Schwartz (Ed.), *Engineering Applications of Ceramic Materials,* ASM Intl. Source Book, Am. Soc. of Metals, Metals Park, Ohio, 1985.
64. R. J. Wallace and S. M. Copley, "Shaping of Silicon Nitride with a Carbon Dioxide Laser by Overlapping Multiple Grooves," *J. Eng. Ind.,* vol. 3, pp. 315–321, Nov. 1989.
65. *Welding, Brazing, and Soldering, Metals Handbook,* 9th ed., vol. 6, Am. Soc. of Metals, Metals Park, Ohio, 1983, pp. 10, 59, 647–671, 1064–1066.
66. H. Tsuchiya et al., *J. Jpn. Soc. Prec. Eng.,* vol. 53, p. 1765, 1987.
67. H. Tsuchiya and H. Göto, *Kikai to Kogu,* vol. 32, p. 66, 1988.
68. M. C. Shaw, in H. J. Swinehart (Ed.), *Cutting Tool Material Selection* ASTME Dearborn, Mich., 1968, p. 8.
69. E. D. Whitney, "New Advances in Ceramic Tooling," SME Tech Paper, MRR 76-15, 1976; and in D. W. Richerson (Ed.), *Ceramics Applications in Manufacturing,* SME, 1988, pp. 87–104.
70. "Technological and Economic Assessment of Advanced Ceramic Materials—A Case Study," vol. 6, Charles River Assoc., PB85-113132, prepared for NIST, NBS GCR84-470-6, Aug. 1984, 65 pp.
71. W. W. Gruss, "Ceramic Tools Improve Cutting Performance," *Ceram. Bull.,* vol. 67, pp. 993–996, June 1988.
72. S. J. Burden et al., "Comparison of Hot-Isostatically-Pressed and Uniaxially Hot-Pressed Alumina-Titanium-Carbide Cutting Tools," *Ceram. Bull.,* vol. 67, pp. 1003–1005, June 1988.

73. E. D. Whitney and P. N. Vaidyanathan, "Engineered Ceramics for High Speed Machining," in J. A. Swartley-Loush (Ed.), *Proc. ASM and SCTE Conf. on Advanced Tool Materials for Use in High Speed Machining,* Scottsdale, Ariz., Feb. 25–27, 1987, pp. 77–82.
74. T. J. Drozda, "Ceramic Tools Find New Applications," *Manuf. Eng.,* pp. 110–115, May 1985.
75. D. Bordui, "Third Generation Silicon Nitride," SME-TE 89-150, presented at Altamonte Springs, Fla., Feb. 15–16, 1989, 10 pp.
76. E. Raia, "New Ceramic Makes Better Cutting Tool," *High Technol.,* pp. 66–67, Dec. 1985.
77. S. T. Buljan and S. F. Wayne, "Silicon-Nitride-Based Composite Cutting Tools: Material Design Approach," *Adv. Ceram. Mater.,* vol. 2, pp. 813–816, Apr. 1987.
78. C. W. Beeghly and A. F. Shuster, "Application-Specialized Ceramics: A Silicon Nitride for Machining Gray Cast Iron," in J. A. Swartley-Loush (Ed.), *Proc. ASME and SCTE Conf. on Advanced Tool Materials for Use in High Speed Machining,* Scottsdale, Ariz., Feb. 25–27, 1987, pp. 91–99.
79. D. Agranov, presented at the CIRP Conf., Israel, Aug. 1986.
80. A. J. Klein, *AM&P,* vol. 9, pp. 26–34, 1986.
81. J. D. Christopher and N. Zlatin, "New Cutting Tool Materials," SME Tech Paper MR 74-101, 1974; and in D. W. Richerson (Ed.), *Ceramics Applications in Manufacturing,* SME, 1988, pp. 105–109.
82. E. R. Billman et al., "Machining with Al_2O_3-SiC-Whisker Cutting Tools," *Ceram. Bull.,* vol. 67, pp. 1016–1019, June 1988.
83. T. Nakai et al., *Am. Mach.,* pp. 13, 15, Feb. 1989.
84. N. Matsumori, "CBN Abrasives Advance Ball Bearing Manufacture," *Manuf. Eng.,* pp. 70–72, Nov. 1988.
85. T. J. Broskea, "High Speed Machining of Gray Cast Iron with Polycrystalline Cubic Boron Nitride," in J. A. Swartley-Loush (Ed.), *Proc. ASM and SCTE Conf. on Advanced Tool Materials for Use in High Speed Machining,* Scottsdale, Ariz., Feb. 25–27, 1987, pp. 39–47.
86. W. J. Kennedy, Jr., and E. C. Skaar, "Improving the Machining of Ceramics," SME-MS 89-813, presented at Nontraditional Machining, Orlando, Fla., Oct. 30–Nov. 2, 1989, 16 pp.
87. D. G. Flom, "Manufacturing Technology for Advanced Metal Removal Initiatives (AMRI)," vol. I: "Executive Summary," F33615-80-C-5057, AFWAL-TR-85-4044, 84-SRD-039, May 1985, 22 pp.
88. P. M. Noaker, "At the Cutting Edge: Rethinking Strategies," *Production,* pp. 53–56, July 1988.
89. C. Wick, "Ceramic Cutting Tools Update," *Manuf. Eng.,* pp. 81–87, Apr. 1988.
90. L. A. Brakhman et al., "Super-Hard Material and Ceramic Cutting Tools in the Auto Industry," *Soviet Eng. Res.,* vol. 4, no. 7, pp. 57–58, 1985.
91. J. A. Pask, "From Technology to the Science of Glass/Metal and Ceramic/Metal Sealing," *Ceram. Bull.,* vol. 66, pp. 1587–1592, Nov. 1987.
92. A. P. Tomsia and J. A. Pask, "Chemical Reactions and Adherence at Glass/Metal Interfaces: An Analysis," *Dent. Mater.,* vol. 1, no. 2, pp. 10–16, 1986.
93. M. G. Nicholas and D. A. Mortimer, "Ceramic/Metal Joining for Structural Applications," *Mater. Sci. Technol.,* vol. 1, pp. 657–665, 1985.
94. B. J. Dalgleish, M. C. Lu, and A. G. Evans, "The Strength of Ceramics Bonded with Metals," *Acta Metall.,* vol. 36, pp. 2029–2035, 1988.
95. M. M. Schwartz, *Ceramic Joining,* Am. Soc. of Metals, Metals Park, Ohio, Apr. 1990, 183 pp.

96. R. L. Tallman et al., "Joining Silicon Nitride Based Ceramics: A Technical Assessment," EG&G Idaho, EGG-SCM-6572, DE84011356, DOE Cont. DE-AC07-761DO1570, Mar. 1984.
97. C. E. Lewis, "Putting Ceramics Together," *Mater. Eng.*, pp. 31–34, Feb. 1988.
98. *Brazing Manual,* 3d ed., Am. Welding Soc., Miami, Fla., 1976, pp. 262–263.
99. *Welding, Brazing and Soldering Metals Handbook,* 9th ed., vol. 4, Am. Soc. of Metals, Metals Park, Ohio, 1983, p. 956.
100. *Welding Handbook,* 7th ed., vol. 2, Am. Welding Soc., Miami, Fla., 1978, pp. 377, 404, 414, 416.
101. M. Schwartz, *Brazing,* Am. Soc. of Metals, Metals Park, Ohio, 1989, 195 pp.
102. K. Miyoshi, "Adhesion in Ceramics and Magnetic Media," NASA TM 101476, Mar. 1989, 14 pp.
103. A. P. Tomsia and J. A. Pask, "Kinetics of Iron-Sodium Disilicate Reactions and Wetting," *J. Am. Ceram. Soc.,* vol. 64, pp. 523–528, Sept. 1981.
104. H. E. Pattee, "Joining Ceramics to Metals and Other Materials," WRC Bull. 178, Welding Research Council, New York, 1972.
105. J. T. Klomp, "Interfacial Reactions between Metals and Oxides during Sealing," *Bull. Am. Ceram. Soc.,* vol. 59, pp. 794–799, 1980.
106. M. Ueki, M. Naka, and I. Okamoto, "Joining and Wetting of CaO-Stabilized ZrO_2 with AlCu Alloys," *J. Mater. Sci.,* vol. 23, pp. 2983–2988, 1988.
107. J. T. Klomp, "Ceramic-Metal Reactions and Their Effect on the Interface Microstructure," in J. A. Pask and A. G. Evans (Eds.), *Ceramic Microstructures '86, Role of Interfaces,* Plenum, New York, 1987, pp. 307–317.
108. D. M. Mattox and H. D. Smith, "Role of Manganese in the Metallization of High-Alumina Ceramics," *Am. Ceram. Soc. Bull.,* vol. 64, pp. 1363–1367, 1985.
109. H. Mizuhara and E. Huebel, "Joining Ceramic to Metal with Ductile Active Filler Metal," *Welding J.,* vol. 65, no. 10, pp. 43–51, 1986.
110. R. E. Loehman et al., "Why Metals Adhere to Si_3N_4," *J. Am. Ceram. Soc.,* 1987, submitted.
111. M. Vilpas, "Joining of Ceramics for High Temperature Applications," NASA TT-20030, N87-29678, NTIS HC A03/MF AO1, Oct. 1987; transl. of "Korkeissa Lampotiloissa Kaytettavien Keraamisten Materiaalien Liittaminen," Tech. Res. Center of Finland, Espoo, Rep. UTT-TIED-481, pp. 1–37, Aug. 1985.
112. S. Noda et al., "Improvement for Adhesion of Thin Metal Films on Ceramics by Ion Bombardment and the Application to Metal-Ceramic Joining," *J. Mater. Sci. Lett.,* vol. 5, p. 381, 1986.
113. J. Lottgers, "Investigations into the Direct Brazing of Ceramic to Metal Joints in Oxidizing Atmosphere or under Inert Shielding Gas," *Schweissen Schneiden,* pp. 19–23, Mar. 1982.
114. H. Mizuhara and K. Mally, "Ceramic-to-Metal Joining with Active Brazing Filler Metal," *Welding J.,* vol. 64, no. 10, pp. 27–32, 1985.
115. A. J. Moorhead, H. M. Henson, and T. J. Henson, "The Role of Interfacial Reactions on the Mechanical Properties of Ceramic Brazements," ECUT Program, DOE Contract DE-AC05-84 OR21400 with Martin-Marietta Energy Systems, Inc., 1984.
116. D. P. Kramer and W. E. Moddeman, "Chemistry of Glass-Ceramic to Metal Bonding for Header Applications," MLM-3556, UC-25, EG&G Mound Applied Technologies, U.S. Dept. of Energy Contract DE-AC04-88DP43495, Nov. 30, 1988, 22 pp.
117. B. J. Dalgleish, M. C. Lu, and A. G. Evans, "The Strength of Ceramics Bonded with Metals," *Acta Metall.,* vol. 36, pp. 2029–2035, Aug. 1988.
118. J. P. Hammond, S. A. David, and M. L. Santella, "Brazing Ceramic Oxides to Metals at Low Temperatures," *Welding J.,* vol. 67, no. 10, pp. 227s–232s, Oct. 1988.

119. S. M. Johnson, "The Formation of High Strength Silicon Nitride Joints by Brazing," D88-1208, SRI Intl., Sept. 1987, 16 pp.
120. M. G. Nicholas and R. M. Crispin, "The Role of Titanium in Active Metal and Activated Brazing of Alumina," in *Proc. 2d Intl. Colloq. on the Joining of Ceramics, Glass and Metals,* Bad Nauheim, Germany, Mar. 27–29, 1985.
121. H. Mizuhara, "Vacuum Brazing Ceramics to Metals," *AM&P,* vol. 131, Feb. 1987.
122. R. E. Loehman et al., "Bonding Mechanisms in Si_3N_4 Brazing," *J. Am. Ceram. Soc.,* vol. 73, pp. 552–558, Mar. 1990.
123. I. Miyamoto et al., "Joining of Si_3N_4 Ceramics by Laser-Activated Brazing," in *LAMP '87,* pp. 237–242.
124. K. Suganuma et al., "Ceramic Surfaces and Surface Treatments," in *British Ceram. Proc.,* no. 34, p. 273, 1984.
125. M. V. Goodyear and A. Ezis, "Joining of Turbine Engine Ceramics," in *Proc. 4th Army Materials Technology Conf., Advances in Joining Technology,* Brookhill Publ., Chestnut Hill, Mass., 1976, pp. 113–154.
126. R. W. Rice, "Joining of Ceramics," in *Proc. 4th Army Materials Technology Conf., Advances in Joining Technology,* Brookhill Publ., Chestnut Hill, Mass., 1976, pp. 69–111.
127. J. Selveran, E. M. Dunn, and S. Kang, "Microstructural Examination of Ceramic-Metal Joints Brazed with Alloys Containing Palladium," presented at the 21st AWS Intl. Brazing and Soldering Conf., Anaheim, Calif., Apr. 24, 1989.
128. E. Lugscheider and W. Tillman, "Development of New Active Filler Metals in the System Ag-Cu-Hf," presented at the 21st AWS Intl. Brazing and Soldering Conf., Anaheim, Calif., Apr. 24, 1989.
129. B. Wielage, D. Ashoff, and M. Turpe, "Application of Advanced Ceramics," presented at the 21st AWS Intl. Brazing and Soldering Conf., Anaheim, Calif., Apr. 24, 1989.
130. H. Oogi, *J. Jpn. Welding Soc.,* vol. 56, no. 2, p. 91, 1987.
131. *NIKKEI New Material,* p. 41, May 11, 1987.
132. K. Yamada, T. Hashimoto, and Y. Koumi, *Bull. Ceram. Soc. Jpn.,* vol. 11, p. 785, 1976.
133. L. G. Kutzer, "Joining Ceramics and Glass to Metals," *Mater. Des. Eng.,* pp. 106–110, Jan. 1965.
134. A. Pietsch, "Coal Fired Prototype High Temperature Continuous Flow Heat Exchanger," EPRI/AF-684, Feb. 1978.
135. H. Schwartz, "Ceramic Adhesive for High-Temperature Bonding," SME-AD-86-865, presented at Adhesives '86, Baltimore, Md., Sept. 8–10, 1986, 22 pp.
136. A. Intrater, Adv. Tech., Inc., private communication, June 1990.
137. *Adv. Mater.,* vol. 12, p. 6, Feb. 12, 1990.
138. D. N. Coon, R. L. Tallman, and R. M. Neilson, Jr., "Hot Isostatically Pressed Si_3N_4–Si_3N_4 Joints Bonded with Oxynitride Glass," *Adv. Ceram. Mater.,* vol. 3, pp. 154–158, 1988.
139. B. T. Lyons, R. Outcalt, and K. Hubbard, "Ceramic Adhesive Encapsulator Improves Heat Reliability," *Adhesives Age,* pp. 18–20, June 1989.
140. P. F. Becher and S. A. Halen, *J. Am. Ceram. Soc. Bull.,* vol. 58, p. 582, 1979.
141. Y. Owada and K. Kobayashi, *J. Ceram. Soc. Jpn.,* vol. 92, p. 693, 1984.
142. Y. Tamari et al., *J. Ceram. Soc. Jpn.,* vol. 93, p. 154, 1985.
143. Y. Ebata and M. Kinoshita, *J. Ceram. Soc. Jpn.,* vol. 90, p. 714, 1982.
144. M. L. Mecartney, R. Sinclair, and R. E. Loehman, *J. Am. Ceram. Soc.,* vol. 68, p. 472, 1985.

145. N. Nakamura et al., "Joining of Silicon Nitride by Hot Pressing," *J. Mater. Sci.*, vol. 22, pp. 1259–1264, 1987.
146. C. H. Bates et al., "Joining of Non-Oxide Ceramics for High-Temperature Applications," *Ceram. Bull.*, vol. 69, pp. 350–356, 1990.
147. S. M. Johnson and D. J. Rowcliffe, "Silicon Nitride Joining," SRI Intl., F 49620-81-K-0001, AFOSR TR-83-114, Mar. 1983, 36 pp.
148. J. R. Hellman, "Projects within Center for Advanced Materials," Pennsylvania State Univ., GRI 88-0181, AR 6/87-5/88, June 1988, 366 pp.
149. S. M. Johnson and R. H. Lamoreaux, "Innovators in Ceramic Research," *Ceram. Bull.*, vol. 68, pp. 1431–1434, Aug. 1989.
150. A. F. Erickson, J. C. Nablo, and C. Panzera, "Bonding Ceramic Materials to Metallic Substrates for High-Temperature Low-Weight Applications," ASME 78-WA/GT-16, Dec. 10–15, 1978, 8 pp.
151. H. Matsuoka and H. Kawamura, "Development Status of Isuzu Ceramic Engine," Isuzu Ceramics Res. Inst. Ltd., and private communication with S. Gota, Isuzu Motors America, Nov. 30, 1988, 31 pp.
152. W. A. Zdaniewski and H. P. Kirchner, "Joining of Alumina Ceramic by Inducing Localized Reducing Conditions," *J. Am. Ceram. Soc.*, vol. 70, pp. C-4–C-6, Jan. 1987.
153. D. L. Sheppard, "Improved Glass-Ceramic to Metal Bonds for Pyrotechnic Header Applications," MLM-3386, UC-25, Oct. 6, 1986, 17 pp.
154. D. P. Kramer and W. E. Moddeman, "Chemistry of Glass-Ceramic to Metal Bonding for Header Applications," MLM-3556, UC-25, Nov. 30, 1988, 22 pp.; also W. E. Moddeman et al., "Metal Modification for Improved Glass-Ceramic to Metal Seals; 1. Auger Results on Ion Implantation and Vapor Deposition Films," MLM-3479, UC-25, Jan. 29, 1988, 26 pp.
155. H. Takashio, *Ceramics Bonding Technology*, Intl. Publ. Corp., Tokyo, Japan, 1985, pp. 33, 125, 141.
156. *Glass Handbook*, Asakura Shoten, Tokyo, Japan, 1987, p. 143.
157. Y. Ebata, R. Hayami, and M. Kinoshita, *Eng. Mater. Jpn.*, vol. 30, no. 7, p. 85, 1982.
158. R. E. Honnell and S. D. Stoddard, "Development of a High Temperature Ceramic-to-Metal Seal for an Air Force Weapons Laboratory Laser," LA-10884-MS, UC-25, Mar. 1987, 16 pp.
159. A. Anderson and E. E. Stepp, "Glass-to-Sapphire End Window Seals," *Rev. Sci. Instrum.*, vol. 33, pp. 119–120, 1962.
160. K. E. Wood et al., "Development of a Ceramic Tube Heat Exchanger with Relaxing Joint," DOE Rep. FE-2556-30, June 1980.
161. J. Godziemba-Maliszewski and R. Lison, "Metal-Silicon Carbide Bonds for Elevated Temperature Application," in W. Kraft (Ed.), *Proc. Intl. Conf. on Joining Ceramics, Glass and Metal,* Bad Nauheim, Germany, pp. 433–440, 1989, IIW Doc. IA-372-89.
162. G. L. White and P. J. Oakley, "Industrial Ceramics—A Survey of Materials, Applications, and Joining Processes," Welding Inst., Abington, Cambridge, U.K., 302/1986, June 1986, 23 pp.
163. J. Cawley, "Bonding of Ceramics," Edison Welding Inst., Mar. 1988.
164. D. Hauser, "Ceramic-Metal and Ceramic-Ceramic Joining;" in *Proc. North American Welding Research Seminar,* Edison Welding Inst., Oct. 1987.
165. P. Batfalsky, J. Godziemba-Maliszewski, and R. Lison, "Difference between Diffusion-Welded and Brazed Metal-to-Ceramic (SiC) Joints," in W. Kraft (Ed.), *Proc. Intl. Conf. on Joining Ceramics, Glass and Metal,* Bad Nauheim, Germany, 1989, pp. 81–89, IIW Doc.-IA-375-89.
166. J. Godziemba-Maliszewski, R. Lison, and P. Batfalsky, "Thermal Stresses at Diffusion Bonding Metal-Ceramic (SiC) Joints," in W. Kraft (Ed.), *Proc. Intl. Conf. on*

Joining Ceramics, Glass and Metal, Bad Nauheim, Germany, 1989, pp. 13/1–13/7, Doc.-IA-373-89.
167. C. Colin et al., "Solid-State Bonding of Partially Stabilized Zirconia to Metals," in *Proc. 3d Intl. Symp. on Ceramic Materials and Components for Engines,* Las Vegas, Nev., Nov. 27–30, 1988, pp. 492–502.
168. A. Wicker, Ph. Darbon, and F. Grivon, "Solid State Bonding of Metal to Ceramic," in *Proc. Intl. Symp. on Ceramic Components for Engines,* Japan, 1983, pp. 716–720.
169. K. Suganuma, K. Nühara, and T. Fujita, "Solid State Bonding of Silicon Nitride with a Nickel Interlayer," *J. Less-Common Met.,* vol. 158, no. 1, pp. 59–69, 1990.
170. T. Yamada et al., "Diffusion Bonding SiC for Si_3N_4 to Nimonic 80A," *High Temp. Technol.,* vol. 5, pp. 193–200, Nov. 1987.
171. J. R. McDermid, M. D. Pugh, and R. A. L. Drew, "The Joining of Reaction-Bonded Silicon Carbide to Inconel 600 Using Refractory Metal Interlayers," in D. S. Wilkinson (Ed.), *Proc. Intl. Symp. on Advanced Structural Materials,* Montreal, Que., Aug. 28–31, 1988, Pergamon, New York, pp. 169–175.
172. R. Larker, B. Loberg, and T. Johansson, "Diffusion Bonding Reactions between Silicon Nitride, Silicon Oxynitride and Incoloy 909 by Hot Isostatic Pressing," in *Proc. 3d Intl. Symp. on Ceramic Materials and Components for Engines,* Las Vegas, Nev., Nov. 27–30, 1988, pp. 503–512.
173. M. E. Twentyman, "Mechanism of Glass Migration in the Production of Metal-Ceramic Seals," *J. Mater. Sci.,* vol. 10, pp. 765–776, 1975.
174. B. Dunn, "Field-Assisted Bonding of Beta-Alumina to Metals," *J. Am. Ceram. Soc.,* vol. 62, pp. 545–547, Nov./Dec. 1972.
175. M. E. Twentyman, "The Effect of Experimental Variables on the Structure of Seals to Degassed Alumina," *J. Mater. Sci.,* vol. 10, pp. 777–790, 1975.
176. M. E. Twentyman, "The Use of Metallizing Paints Containing Glass or Other Inorganic Bonding Agents," *J. Mater. Sci.,* vol. 10, pp. 791–798, 1975.
177. M. Naka et al., "Influence of Brazing Conditions on Shear Strength of Alumina-Kovar Joint Made with Amorphous $Cu_{50}Ti_{50}$ Filler Metal," *Trans. Jpn. Welding Res. Inst.,* Dec. 1983.
178. J. Intrater, "The Challenge of Bonding Metals to Ceramics," *Mach. Des.,* pp. 95–100, Nov. 23, 1989.
179. H. J. Nolte and R. Spurck, "Metal-Ceramic Sealing with Manganese," *Telev. Eng.,* vol. 1, no. 11, pp. 14–16, 18, 39, 1950.
180. H. J. Nolte, "Metallized Ceramic," U.S. Patent 2,667,432, Jan. 26, 1954.
181. R. F. Spurck et al., "Use Metallizing Tape for High Quality Ceramic-to-Metal Seals," *Ceram. Ind.,* vol. 79, no. 3, pp. 88–91, 94, 1962.
182. J. T. Klomp and P. J. Bolden, "Sealing Pure Alumina Ceramics to Metals," *J. Am. Ceram. Soc.,* vol. 49, pp. 204–211, 1970.
183. K. H. Kohl and P. Rice, "Electronic Tubes for Critical Environments," Tech. Rep. TR-57-434, Contract 33(616)-3460, Stanford Res. Ctr., Menlo Park, Calif., Mar. 1985.
184. L. A. Tentarelli, J. M. White, and R. W. Buck, "Low-Temperature Refractory Metal-to-Ceramic Seals," Final Rep. ECOM-03734-F, Contract DA-36-039-AMC-03734(E), Sperry Rand Corp., Gainesville, Fla., Apr. 1966.
185. P. F. Varadi and R. Dominiguez, "Tungsten Metallizing of Ceramics," *Am. Ceram. Soc. Bull.,* vol. 45, pp. 789–791, 1966.
186. M. E. Staumanis and A. W. Schlechten, "Titanium Coatings on Metals and Ceramics," *Metallurgia,* vol. 10, pp. 901–909, 1956.
187. M. Santella, "Brazing of Titanium-Vapor-Coated Silicon Nitride," *Adv. Ceram. Mater.,* vol. 3, pp. 457–462, 1988.
188. M. Santella and L. C Manley, "Strength and Microstructure of Titanium-Vapor-

Coated Silicon Nitride Braze Joints," in *Proc. 3d Intl. Symp. on Ceramic Materials and Components for Engines,* Las Vegas, Nev., Nov. 27–30, 1988, pp. 513–525.

189. J. D. Cauley, "Joining of Ceramic-Matrix Composites," *Ceram. Bull.,* vol. 68, pp. 1619–1623, Sept. 1989.

190. R. R. Kapoor and T. W. Eagar, "Oxidation Behavior of Silver- and Copper-Based Brazing Filler Metals for Silicon Nitride/Metal Joints," *J. Am. Ceram. Soc.,* vol. 72, pp. 848–854, Mar. 1989.

191. H. Mizuhara, E. Huebel, and T. Oyama, "High-Reliability Joining of Ceramic to Metal," *Ceram. Bull.,* vol. 68, pp. 1591–1599, Sept. 1989.

192. S. Kang et al., "Issues in Ceramic-to-Metal Joining; An Investigation of Brazing a Silicon Nitride-Based Ceramic to a Low-Expansion Superalloy," *Ceram. Bull.,* vol. 68, pp. 1608–1617, Sept. 1989.

193. E. Lugscheider, M. Boretius, and R. Lison, "Active Brazing of Silicon-Carbide to Steel Using a Thermal-Stress Reducing Metallic Interlayer," IIW-Doc.-IA-356-88/OE & Doc.-I-860-88/OE, presented at the 19th AWS Intl. Brazing and Soldering Conf., New Orleans, La., Apr. 19–21, 1988, 14 pp.

194. J. P. Hammond, S. A. David, and M. L. Santella, "Brazing Ceramic Oxides to Metals at Low Temperatures," *Welding,* vol. 67, pp. 227s–232s, Oct. 1988.

195. D. Hauser, "Brazing Ceramic Oxides to Metals," *Current Awareness Bull. (Metals and Ceramics Inform. Center),* no. 189, pp. 1–2, May 1989.

196. C. R. Weymueller, "Braze Ceramics to Themselves and to Metals," *Welding Des. Fab.,* pp. 45–48, Aug. 1987.

197. C. W. Johnson, *Manual of Metal-to-Ceramic Sealing Techniques,* Publ. NA-27-0001, Contract AF33(602)-2371, Sperry Gyroscope Co., Great Neck, N.Y., May 1963.

198. R. Bondley, "Metal-Ceramic Brazed Seals," *Electronics,* vol. 20, no. 7, pp. 97–99, 1947.

199. C. S. Pearsall and P. K. Zingeser, "Metal to Nonmetallic Brazing," Tech. Rep. 104, M.I.T. Res. Lab. of Electronics, Cambridge, Mass., Apr. 5, 1949.

200. C. W. Fox and G. M. Slaughter, "Brazing of Ceramics," *Welding J.,* vol. 43, pp. 591–597, 1964.

201. D. A. Canonico, N. C. Cole, and G. H. Slaughter, "Direct Brazing of Ceramics, Graphite and Refractory Metals," *Welding J.,* vol. 56, pp. 31–38, Aug. 1977.

202. H. Fukushima, T. Yamanaka, and M. Matsui, "Microwave Heating of Ceramics and Its Application to Joining," *J. Mater. Res.,* vol. 5, pp. 397–405, Feb. 1990.

203. D. Palaith and R. Silberglitt, "Microwave Joining of Ceramics," *Ceram. Bull.,* vol. 68, pp. 1601–1606, Sept. 1989.

204. D. Palaith et al., "Microwave Joining of Ceramics," in W. H. Sutton, M. H. Brooks, and I. J. Chabinsky (Eds.), *Microwave Proc. of Materials,* vol. 124, Materials Res. Soc., Pittsburgh, Penn., 1988, pp. 255–266.

205. T. T. Meek and R. D. Blake, "Ceramic-Glass-Metal Seal by Microwave Heating," U.S. Patent 4 529 857, July 16, 1985.

206. A. J. Klein, "Processing with Microwaves," *AM&P,* vol. 1, pp. 36–39, Dec. 1985.

207. R. W. Rice, "Welding of Ceramics," NRL Rep. 7085, July 6, 1970.

208. R. W. Rice, "Joining of Ceramics," in *Advances in Joining Technology,* Brookhill Publ., Chestnut Hill, Mass., 1976, pp. 69–111.

209. W. M. Phillips, "Metal-to-Ceramic Seals for Thermionic Converters, A Literature Survey," JPL/Rep. 32-1420, Nov. 1969.

210. H. A. Hokanson, S. L. Rogers, and W. I. Kern, "Electron Beam Welding of Alumina," *Ceram. Ind.,* vol. 81, pp. 44–47, Aug. 1963.

211. H. E. Pattee, R. M. Evans, and R. E. Monroe, "Joining of Ceramics and Graphite to Other Materials," NASA SP-5052, Battelle Memorial Inst., Columbus, Ohio, 1968.

212. M. L. Dring, "Ceramic-to-Metal Seals for High-Temperature Thermionic Convert-

ers," Red Bank Div., Tech. Doc. Rep. TDR-63-4109, Contract AF33(657)-10038, Bendix Corp., Eatontown, N.J., Oct. 1963.
213. J. J. Metelkin, A. Y. Makarkin, and M. A. Pavlova, "Welding Ceramic Materials to Metals," *Welding Prod.,* vol. 14, no. 6, pp. 10–12, 1967.
214. H. Scheffer et al., "How to Ultrasonically Seal Hermetic Transistor Packages," *Ceram. Ind.,* vol. 79, no. 6, pp. 50–52, 64, 1962.
215. K. Tsukamoto et al., in *Proc. Spring Mtg. Japan Soc. of Prec. Engineering,* 1979, p. 19.
216. I. Miyamoto et al., "Joining of Si_3N_4 Ceramics by Laser-Activated Brazing," in *LAMP '87,* pp. 237–242.
217. Schweissanlagen & Rotor GmbH, *World Rept. on Adv. Ceramics,* vol. 1, p. 9, June 1989.
218. K. K. Chan and J. B. Wachtman, Jr., "Expanding World of Ceramic Coatings," *Ceram. Ind.,* pp. 24–26, Sept. 1987.
219. E. S. Hamel, "Ceramic Coatings; More Than Just Wear Resistant," *Mater. Eng.,* pp. 30–34, Aug. 1986.
220. R. C. Novak, "Processing Aspects of Plasma Sprayed Ceramic Coatings," ASME 88 GT 289, June 1988, 5 pp.
221. W. J. Lackey et al., "Ceramic Coatings for Advanced Heat Engines—A Review and Projection," *Adv. Ceram. Mater.,* vol. 21, pp. 24–30, Jan. 1987.
222. S. R. Levine and R. A. Miller, "Ceramic-Coated Metals Can Survive Contact with Hot Working Fluid," *R&D,* pp. 122–125, Mar. 1984.
223. "Fast Glazing of Alumina/Silica Tiles," NASA Tech Brief, NTIS Tech Notes, p. 1240, Nov. 1986.
224. I. Gotman and E. Y. Gutmanas, "A New P/M Method for Coating Si_3N_4 Ceramics," *Powder Metall. Intl.,* vol. 21, pp. 30–33, Dec. 1989.
225. "Ceramic Coatings for High-Performance Applications," GB-111, BCC Inc., Norwalk, Conn., Mar. 1988, 177 pp.
226. C. A. Stearns, "Thermal-Barrier Coatings," *Aerospace Am.,* pp. 27–31, May 1987.
227. J. Desmaison, N. Roels, and P. Belair, "High-Temperature Oxidation-Protection CVD Coatings for Structural Ceramics; Oxidation Behaviour of CVD-Coated Reaction-Bonded Silicon Nitride," *Mater. Sci. Eng.,* vol. A121, pp. 441–447, 1989.
228. T. N. Rhys-Jones and F. C. Toriz, "Thermal Barrier Coatings for Turbine Applications in Aero Engines," *High Temp. Technol.,* vol. 7, pp. 73–81, May 1989.
229. M. R. Gruninger, "Sol-Gel Ceramics as Coatings," *Ceram. Ind. Intl.,* vol. 98, no. 1074, pp. 22–24, 1989.
230. R. A. Miller, "Ceramic Thermal Barrier Coatings for Electric Utility Gas Turbine Engines," NASA TM-87288, N86-22687, presented at the 3d Berkeley Conf. of the National Association of Corrosion Engineers, Jan. 29–31, 1986.
231. "Rapidly Solidified Ceramic Coatings," in *Tech. Insights, Futuretech,* no. 68, Oct. 24, 1988, 12 pp.
232. C. F. Lewis, "Complex Coatings with CVD," *Mater. Eng.,* pp. 35–38, June 1989.
233. B. West, "Boron Carbide Tool Coatings Make Their Entry," *Modern Machine Shop,* pp. 76–79, Jan. 1989.
234. T. M. Besmann, "Chemical Vapor Deposition in the Silicon-Carbon and Boron-Carbon-Nitrogen Systems," ORNL/TM-10884, UC-25, ECUT Prog. & Dept. of Energy under Contract DE-AC05-840R21400, Nov. 1988, 21 pp.
235. J. A. Thornton, *Thin Solids Films,* vol. 80, no. 1, 1981.
236. E. McClanahan, presented at the AVS Sputtering Workshop, San Diego, Calif., Apr. 7–9, 1984.
237. J. E. Greene and S. A. Barnett, *J. Vac. Sci. Technol.,* vol. 21, p. 285, 1982.
238. J. G. Eden et al., in *Proc. Mater. Research Soc. Symp.,* vol. 17, p. 185, 1983.

239. A. T. Santhanam and G. P. Grab, "Innovations in Coated Carbide Cutting Tools," in *Proc. ASM Conf. on Advanced Tool Materials for Use in High Speed Machining,* Scottsdale, Ariz., Feb. 25–27, 1987, pp. 67–76; *Met. Powder Rep.,* vol. 42, pp. 840–845, Dec. 1987.
240. J. B. Pond, "Cutting Tool Coatings; As User Experience Grows, They're Proving Their Worth," *Metalwork. News,* pp. 12–13, Aug. 31, 1987.
241. M. Podob, "Process and Engineering Benefits of Sputter Ion Plated Titanium," in *Proc. ASM Conf. on Advanced Tool Materials for Use in High Speed Machining,* Scottsdale, Ariz., Feb. 25–27, 1987, pp. 1–9.
242. H. Herman, "Plasma-Sprayed Coatings," *Sci. Am.,* vol. 259, pp. 112–117, Sept. 1988.
243. T. Okada, H. Hamatani, and T. Yoshida, "Radio-Frequency Plasma Spraying of Ceramics," *J. Am. Ceram. Soc.,* vol. 72, pp. 2111–2116, 1989.
244. C. d'Angelo and H. E. Joundi, "Reliable Coatings Via Plasma Arc Spraying," *AM&P,* vol. 135, pp. 41–44, Dec. 1988.
245. R. R. Irving, "NASA Lewis Research Center Studying Plasma-Sprayed Ceramic Coatings," *Metalwork. News,* pp. 18, 20, Sept. 21, 1987.
246. K. D. Sheffler and D. K. Gupta, "Current Status and Future Trends in Turbine Application of Thermal Barrier Coatings," ASME 88-GT-286, NAS3-23944, presented at the Gas Turbine and Aeroengine Cong. and Exposition, Amsterdam, June 5–9, 1988, 9 pp.
247. W. J. Brindley and R. A. Miller, "TBCs for Better Engine Efficiency," *AM&P,* vol. 136, pp. 29–33, Aug. 1989.
248. F. C. Toriz, A. B. Thakker, and S. K. Gupta, "Thermal Barrier Coatings for Jet Engines," ASME 88-GT-279, presented at the Gas Turbine and Aeroengine Cong. and Exposition, Amsterdam, June 5–9, 1988, 9 pp.
249. S. Alperine, "Molten Salt Induced High Temperature Degradation of Thermal Barrier Coatings," O.N.E.R.A., AGARD Structural and Materials Panel, 68th Mtg., Ottawa, Ont., CP-461, Apr. 23–28, 1989, pp. 6-1–6-17.
250. D. S. Duvall, "Processing Technology for Advanced Metallic and Ceramic Turbine Airfoil Coatings," in *Proc. 2d Conf. on Advanced Materials for Fuel Capable Heat Engines,* Palo Alto, Calif., EPRI-RD-2369-SR, 1987, pp. 6–102, 198.
251. R. F. Bunshah, "Reactive Evaporation," in B. N. Chapman and J. C. Anderson (Eds.), *Science and Technology of Surface Coatings,* Academic Press, New York, 1974.
252. J. Schienle and J. Smyth, "High Temperature Coating Study to Reduce Contact Stress Damage of Ceramics," Final Rep. Garrett Turbine Eng. Co., ORNL/Sub/84-47992/1, Contract DE-AC05-840R21400, Mar. 1987, 77 pp.
253. *Am. Mach.,* p. 29, Oct. 1988.
254. K. Tsukamoto et al., "A New Ceramics Coating Technique Using CO_2 Laser," in *LAMP '87,* pp. 491–495.
255. N. Yasunaga et al., "Ceramic Layer Formation on Metal Surfaces by Gas Reaction Assisted CO_2 Laser Treatment," in *LAMP '87,* pp. 485–490.
256. A. E. Ledger, "Activated Processes at OCLI," vol. 3: "Appendix: Emerging Net Shape Technologies," Workshop, Santa Barbara, Calif., Mar. 27–29, 1985, Nat. Academy Press, Washington, D.C., 1986, pp. 499–515.

BIBLIOGRAPHY

Agranov, D., "Silicon-Nitride Based Materials for Modern Machining," presented at the ASM Intl. Annual Mtg., Indianapolis, Ind., Oct. 3, 1989.

Amos, A., and B. Barnea, "Superabrasive Cuts Honing Costs," *Am. Mach.,* pp. 107–109, Jan. 1983.

Aparicio, R., T. J. Anderson, and M. D. Sacks, "Chemical Vapor Infiltration and Atomic Layer Deposition of TiC$_x$ on Ceramic Substrates," presented at the 14th Ann. Conf. on Composites and Advanced Ceramics, Cocoa Beach, Fla., Jan. 14–17, 1990.

Asai, H., U. Fumio, and N. Iwase, "Titanium Nitride-Molybdenum Metallizing Method for Aluminum Nitride," *IEEE Trans. Components, Hybrids, Manuf. Tech.,* vol. 13, pp. 457–461, June 1990.

Bachin, V. A., E. A. Goritskaya, and I. E. Tikhonova, "Diffusion Bonding Aluminum-Magnesium Alloys to Ceramic Materials," *Svarochnoe Proizvodstvo,* vol. 33, no. 11, pp. 17–19, 1986; *Welding Intl.,* no. 8, pp. 762–764, 1987.

Balaguer, J., and D. W. Walsh, "A Survey of Ceramic Joining Techniques," presented at the ASM Intl. Annual Mtg., Indianapolis, Ind., Oct. 3, 1989.

Barbier, F., C. Peytour, and A. Revcolevschi, "Microstructural Study of the Brazed Joint between Alumina and Ti-6Al-4V Alloy," *J. Am. Ceram. Soc.,* vol. 73, pp. 1582–1586, June 1990.

Bardui, D., "Hard-Part Machining with Ceramic Inserts," *Ceram. Bull.,* vol. 67, pp. 998–1001, June 1988.

Bates, C. H., et al., "Joining of Non-Oxide Ceramics for High-Temperature Applications," *Am. Ceram. Soc. Bull.,* vol. 69, pp. 350–356, Mar. 1990.

Batfalsky, P., J. Godziemba-Maliszewski, and R. Lison, "Strength Investigations on Ceramic (SiC)-to-Metal Joints," *CF Intl.,* vol. 65, pp. 464–468, 1988.

Bennett, M. J., "New Coatings for High Temperature Materials Protection," UKAEA, Harwell, Gt. Br., AERE-R-11387, AGR-CWG/P(84)72, Aug. 1984, 25 pp.

Bergstrom, R. P., "Getting a Grip on Cutting Tools," *Manuf. Eng.,* pp. 116–120, Jan. 1986.

Bose, P. P., "Honing with CBN," *AM Auto. Manuf.,* pp. 66–69, Jan. 1988.

Brookes, K. J. A., "Guide to the World's Advanced Cutting-Tool Matls.," *Am. Mach.,* pp. 63–68, Mar. 1990.

Brown, S. D., and G. P. Wirtz, "Anodic Spark Deposition: A Novel Approach to Ceramic Coatings," presented at the ASM Intl. Ann. Mtg., Indianapolis, Ind., Oct. 4, 1988.

Brown, S. D., B. Blackwell, and G. P. Wirtz, "Critical Examination of the Deceleration Method for Adherence Determination of Ceramic Coatings," presented at the ASM Intl. Ann. Mtg., Indianapolis, Ind., Oct. 4, 1988.

Buergel, R., and R. Buergel, "Evaluation of Ceramic Thermal Barrier Coatings for Gas Turbine Engine Components," Brown, Boveri U.C.I.E., Mannheim, Germany, FR 8/86, Aug. 1986, 53 pp.

Butt, D. P., et al., "Effects of Plasma-Sprayed Ceramic Coatings on the Strength Distribution of Silicon Carbide Materials," *J. Am. Ceram. Soc.,* vol. 73, pp. 2690–2696, Sept. 1990.

Cales, B., C. Martin, and P. Vivier, "High Strength Ceramic Components for Electrical Discharge Machining," in V. J. Tennery (Ed.), *Proc. 3d Intl. Symp. on Ceramic Materials and Components for Engines,* Las Vegas, Nev., Nov. 27–30, 1988, pp. 1189–1201.

Carius, A. C., "What You Should Know about CBN Grinding," *Iron Age Metalwork Intl.,* pp. 23–26, Dec. 1982.

"CNC Versatility for Machining Ceramics—with Ceratech," *Aircraft Eng. & Aero. Tech.,* p. 21, Mar. 1988.

Dibble, M. A., "Coatings Cover New Ground," *Mach. Des.,* pp. 40–48, June 22, 1989.

Dini, J. W., "Developments and Trends in Electrodeposition," UCRL-97328, DE88 009822, presented at the SAMPE 2d Intl. Conf., Dayton, Ohio, Aug. 2–4, 1988, 9 pp.

Duh, J. G., W. S. Chien, and B. S. Chiou, "Wettability in ZrO_2-Ni and ZrO_2-Cu Bonding," *J. Mater. Sci. Lett.,* vol. 8, pp. 405–408, 1989.

Federer, J. I., "Alumina Base Coatings for Protection of SiC Ceramics," *J. Mater. Eng.,* vol. 12, no. 2, pp. 141–149, 1990.

Forecast Issue, 1988, *AM&P,* vol. 133, pp. 30–38, Jan. 1988.

Gheyee, R., and C. L. Beatty, "Inorganic Coatings via R. F. Plasma Processing," presented at the 14th Ann. Conf. on Composites and Advanced Ceramics, Cocoa Beach, Fla., Jan. 14–17, 1990.

Goncharov, V. I., et al., "Intensification of the Milling Process of Refractory Materials Using a New Surface Active Substance," *Refractories* (Ukraine), vol. 29, no. 3/4, pp. 140–142, 1988.

Gottschall, R. J., "Recent Activities in Ceramic and Semiconductor Sciences in Japan," Div. of Materials Science, Dept. of Energy, Washington, D.C., vol. 49, p. 17, Sept. 1987.

Graham, R. A., and A. B. Sawaoka, *High Pressure Explosive Processing of Ceramics*, PRO Books, Rockport, Mass., 1987, 412 pp.

Gruninger, M. F., "Characterization of Thermal Barrier Ceramic Powders," presented at the ASM Intl. Ann. Mtg., Indianapolis, Ind., Oct. 4, 1988.

Gurumoorthy, B., et al., "Lifetime Predictions for a Ceramic Cutting Tool Material at High Temperatures," *J. Mater. Sci.*, vol. 22, pp. 2051–2057, June 1987.

Gutman, I., and E. Y. Gutmanas, "Joining of Al_2O_3 to PM Iron and Nickel Alloys by Cold Sintering," in *Proc. 1988 Intl. Powder Metallurgy Conf.*, Orlando, Fla., June 5–10, 1988, vol. 18, pp. 675–688.

Gyarmati, E., W. Kesternich, and R. Forthmann, "Joining Silicon Carbide with the Aid of Thin Ti-Layers," *Ceram. Forum Intl./Ber. Deut. Keram. Ges.*, vol. 66, no. 7/8, pp. 292–297, 1989.

Hatschek, R. L., "New Ceramics Rev up Cutting Speed," *Am. Mach.*, pp. 110–112, Jan. 1983.

Huber, R. F., "How Cutting Tools Can Help You Make a Quick Buck," *Production*, pp. 80–87, May 1986.

Iino, Y., et al., "Joining of Ceramic to Metal by Use of an Electric Discharge Machined Surface," *J. Mater. Sci Lett.*, vol. 8, pp. 493–495, 1989.

Imanaka, O., "Recent Progress in Ceramic Machining in Japan," in S. Somiya and R. C. Bradt (Eds.), *Fundamental Structural Ceramics, 4th Japan-U.S. Seminar on Basic Ceramics Conf.*, Terra Scientific Publ., Tokyo, 1987, pp. 297–325.

Irving, R. R., "Sophisticated Coating Can Lengthen Service Life of Gas Turbine Blades," *Metalwork. News*, p. 26, Nov. 2, 1987.

Isecki, T., and T. Yano, "Brazing of SiC Ceramics with Active Metals," in C. C. Sorrell and B. Ben-Nissan (Eds.), *AUSTCERAM 88,* Sydney, Australia, Aug. 21–26, 1988, *Mater. Sci. Forum*, vols. 34–36, pt. I, pp. 421–425.

Jack, D. H., "Hard Materials for Metal Cutting," *Met. Mater.*, vol. 9, pp. 516–520, Sept. 1987.

Janeway, P. A., "Britain Gears Up for Push in Advanced Ceramics," *Ceram. Ind.*, Feb. 22–29, 1985.

Jiang, C. C., T. Goto, and T. Hirai, "Preparation of Titanium Carbide Plates by Chemical Vapour Deposition," *J. Mater. Sci.*, vol. 25, pp. 1086–1093, 1990.

Johnson, G. A., "Beneficial Compressive Residual Stress Resulting from CBN Grinding," presented at the SME 2d Intl. Grinding Conf., Philadelphia, Penn., June 10–12, 1986.

Johnson, G. A., "Cool-Grinding Abrasives Preserve Fatigue Life," *Mater. Eng.*, p. 52, Oct. 1986.

Jordan, D. W., and K. T. Faber, "Interfacial Properties of a Y_2O_3-ZrO_2 Thermal Coating," presented at the 14th Ann. Conf. on Composites and Advanced Ceramics, Cocoa Beach, Fla., Jan. 14–17, 1990.

Kalish, H., and L. Peters, "Aluminum Oxide Coating on Cutting Tools and the Resulting Properties," *Met. Powder Rep.*, vol. 44, pp. 846–850, L.31350, 1989.

Kamota, S., M. Sakai, and K. Tagashira, "Bonding of Sintered Alumina and Mild Steel Using Thermal Spray Coatings," *J. Mater. Sci. Lett.*, vol. 8, pp. 553–554, 1989.

Karunanithy, S., "Melt Infiltration and Reaction at the Fiber/Matrix Interface During the Brazing of a Fiber-Reinforced Ceramic to Metal," *J. Am. Ceram. Soc.*, vol. 73, pp. 178–181, Jan. 1990.

Kempfer, L., "Ceramics; Toughening up for the Future," *Mater. Eng.*, pp. 45–48, Jan. 1990.

Kertesz, J., et al., "Machining Titanium Alloys with Ceramic Tools," *J. Met.*, pp. 50–51, May 1988.

Klomp, J. T., "Thermodynamics of Ceramic-Metal Interfaces," in S. D. Peteves (Ed.), *Designing Interfaces for Technological Applications: Ceramic-Ceramic, Ceramic-Metal Joining (Comm. of Europ. Communities)*, Petten, Netherlands, Apr. 20–21, 1988, pp. 127–144.

König, W., and M. Popp, "High Precision Machining of Advanced Ceramics!," in V. J. Tennery (Ed.), *Proc. 3d Intl.. Symp. on Ceramic Materials and Components for Engines*, Las Vegas, Nev., Nov. 27–30, 1988, pp. 1159–1169.

Krafft, F. G., and P. Grieb, "Grinding with Plated-CBN Wheels," *Am. Mach.*, pp. 122–125, Apr. 1982.

Kraft, W., "Joining Ceramics, Glass and Metal," DGM Metall. Inform., 1989, 411 pp.

Kulischenko, W., "Abrasive Jet Machining," SME Tech. Paper, MR 76-694, in D. Richerson (Ed.), *Advanced Ceramics Conf., Ceramic Applications in Manufacturing*, SME, Dearborn, Mich., 1986, pp. 176–184.

Layton, R. E., and Y. Higano, "Low Pressure Chemical Vapor Deposition System for Surface Treatments Improves Product Properties," *Ind. Heating*, pp. 16–20, July 1990.

Lewis, C. L., "Researchers Tackle Reliability of Ceramics," *Mater. Eng.*, pp. 23–28, July 1987.

Lewis, M. H., "Ceramics to Be Joined 10 Years from Now," in S. D. Peteves (Ed.), *Designing Interfaces for Technological Applications: Ceramic-Ceramic, Ceramic Metal Joining (Comm. of Europ. Communities)*, Petten, Netherlands, Apr. 20–21, 1988, pp. 271–290.

Loehman, R. E., "Ceramic Joining in the U.S.," in S. D. Peteves (Ed.), *Designing Interfaces for Technological Applications: Ceramic-Ceramic, Ceramic-Metal Joining (Comm. of Europ. Communities)*, Petten, Netherlands, Apr. 20–21, 1988, pp. 235–245.

Lugscheider, E., and W. Tillman, "Development of New Active Filler Metals in a Ag-Cu-Hf System," *Welding J.*, pp. 416s–420s, Nov. 1990.

Lumby, R. J., "The Preparation, Structure, and Properties of Commercial Sialon Ceramic Materials," in J. A. Swartley-Loush (Ed.), *Proc. ASM and SCTE Conf. on Advanced Tool Materials for Use in High-Speed Machining*, Scottsdale, Ariz., Feb. 25–27, 1987, pp. 185–200.

Mader, W., and M. Ruhle, "Electron Microscopy Studies of Defects at Diffusion-Bonded Nb/Al_2O_3 Interfaces," *Acta Metall.*, vol. 37, pp. 853–866, Mar. 1989.

Mason, F., "Milling Cast Iron with Silicon-Nitride Inserts," *AM Auto. Manuf.*, pp. 67–68, Dec. 1986.

Mevrel, R., "High Temperature Protective Coatings, Recent Trends," presented at the AGARD Structure and Materials Panel, 68th Mtg., Ottawa, Ont., CP-461, Apr. 23–28, 1989, pp. 12-1–12-7.

Miller, R., "Plasma Sprayed Thermal Barrier Coatings," presented at CERAMTEC 90', Dearborn, Mich., June 12–13, 1990.

Moreland, M. A., "Ultrasonic Advantages Revealed in 'The Hole Story,'" in D. Richerson (Ed.), *Advanced Ceramics, Conf., Ceramic Applications in Manufacturing*, SME, Dearborn, Mich., 1986, pp. 156–162.

Morita, M., K. Suganuma, and T. Okamoto, "Fracture of Silicon Nitride Joined with an Aluminum Braze," *J. Mater. Sci.*, vol. 22, pp. 2778–2782, Aug. 1987.

Naka, M., T. Saito, and I. Okamoto, "Solid State Bonding of SiC to Nb," in S. Somiya et al. (Eds.), *Sintering '87, Proc. 4th Intl. Symp. on Science and Technology of Sintering*, Tokyo, Japan, Nov. 4–6, 1987, Elsevier Appl. Sci., London, pp. 1373–1378.

Nakao, Y., K. Nishimoto, and K. Saida, "Improvement in Bonding Strength of Si_3N_4-Mo Joint by Controlling Reaction Layer Thickness," Osaka Univ., IIW Doc. IX-1594-90, July 1990, 18 pp.

Nastasi, M., et al., "Friction and Wear Studies in N-Implanted Al_2O_3, SiC, TiB_2, and B_4C Ceramics," *J. Mater. Res.*, vol. 3, pp. 1127–1133, Nov./Dec. 1988.

Nicholas, M. G., "Joining Structural Ceramics," in S. D. Peteves (Ed.), *Designing Interfaces for Technological Applications: Ceramic-Ceramic, Ceramic-Metal Joining (Comm. of Europ. Communities)*, Petten, Netherlands, Apr. 20–21, 1988, pp. 49–76.

Nicholas, M. G., et al., "Some Observations on the Wetting and Bonding of Nitride Ceramics," *J. Mater. Sci.*, vol. 25, pp. 2679–2689, June 1990.

Nicoll, A. R., and S. Rangaswamy, "Plasma Spray Deposition of Ceramic Coatings," presented at the ASM Intl. Ann. Mtg., Indianapolis, Ind., Oct. 4, 1988.

Novak, R., "Thermally Sprayed Ceramic Coatings," presented at CERAMTEC 90', Dearborn, Mich., June 12–13, 1990.

Orsini, P. G., et al., "Effect of Different Oxidation Degrees on the Structure and Properties of Stabilised Zirconia Plasma Spray Coatings," in S. Meriani and C. Palmonari (Eds.), *Zirconia '88, Advances in Zirconia Science and Technology*, Bologna, Italy, Dec. 16–17, 1988, Elsevier Appl. Sci., London, pp. 229–237.

Ostyn, K., "Current Research on Ceramic Joining in Europe," in S. D. Peteves (Ed.), *Designing Interfaces for Technological Applications: Ceramic-Ceramic, Ceramic-Metal Joining (Comm. of Europ. Communities)*, Petten, Netherlands, Apr. 20–21, 1988, pp. 265–269.

Pak, J. J., M. L. Santella, and R. J. Fruehan, "Thermodynamics of Ti in Ag-Cu Alloys," *Metall. Trans.*, vol. 21B, pp. 349–355, Apr. 1990.

Petrofes, N. F., and A. M. Gadalla, "Processing Aspects of Shaping Advanced Materials by Electrical Discharge Machining," *AM&P*, vol. 3, pp. 127–153, 1988.

"Plasma Flame Achieves High-Temp Ceramic Coatings," *Des. News*, p. 34, Apr. 10, 1989.

Pond, J. B., "New Tools Change Cutting Economics," *Iron Age*, pp. 33–38, Feb. 7, 1986.

Pond, J. B., "Creep-Feed Grinding Gets Second Chance," *Iron Age*, pp. 50A1–50A9, Nov. 7, 1986.

Raj, R., "Bonding at Metal-Ceramic Interfaces in Hybrid Materials," Cornell Univ., AFOSR TR-88-0768, AFOSR86-0321, FR 9/86-11/87, June 1988.

Ramulu, M., and M. Taya, "EDM Machinability of SiC_w/Al Composites," *J. Mater. Sci.*, vol. 24, pp. 1103–1108, 1989.

Rathner, R. C., and D. J. Green, "Joining of Yttria-Tetragonal Zirconia Polycrystal with an Aluminum-Zirconium Alloy," *J. Am. Ceram. Soc.*, vol. 73, pp. 1103–1105, Apr. 1990.

Ridealgh, J. A., R. D. Rawlings, and D. R. F. West, "Laser Cutting of Glass Ceramic Matrix Composite," *Mater. Sci. Tech.*, vol. 6, pp. 395–398, Apr. 1990.

Ruhle, M., and W. Mader, "Structure and Chemistry of Metal/Ceramic Interfaces," in S. D. Peteves (Ed.), *Designing Interfaces for Technological Applications: Ceramic-Ceramic, Ceramic-Metal Joining (Comm. of Europ. Communities)*, Petten, Netherlands, Apr. 20–21, 1988, pp. 145–195.

Santella, M. L., J. A. Horton, and J. J. Pak, "Microstructure of Alumina Brazed with a Silver-Copper Titanium Alloy," *J. Am. Ceram. Soc.*, vol. 73, pp. 1785–1787, June 1990.

Schiepers, R. C. J., et al., "Interaction between SiC and Metals," in G. DeWith, R. A. Terpstra, and R. Metselaar (Eds.), *Euro-Ceramics*, vol. 3: *Engineering Ceramics*, European Ceramic Soc., Maastricht, June 18–23, 1989, pp. 3.424–3.428.

Schmid, T. E., and R. J. Hecht, "High Technology Ceramic Coatings—Current Limitations/Future Needs," *Ceram. Eng. Sci. Proc.*, vol. 9, pp. 1089–1094, 1988.

Shane, M., and M. L. Mecartney, "Sol-Gel Synthesis of Zirconia Barrier Coatings," *J. Mater. Sci.*, vol. 25, pp. 1537–1544, Mar. 1990.

Sharp, P., "Machining of Ceramics," *Ceram. Ind. Intl.*, vol. 98, no. 1073, pp. 26–29, 32, 1989.

Sheppard, L. M., "Hot New Applications for Ceramics," *AM&P*, vol. 1, pp. 39–43, Nov. 1985.

Shichi, Y., et al., "Study of Joining Interface between Si_3N_4 and Metal by Active Metal," *J. Ceram. Soc. Jpn.*, vol. 97, pp. 1354–1357, Nov. 1989.

Smeltzer, C. E., and A. G. Metcalfe, "Joining of Ceramics for High Performance Energy

Systems," DOE/ET/15359-Tr(DE 82005243), Contract AC22-79ET15359, FR 8/1/79-5/31/81, June 30, 1981, 131 pp.

Somiya, S., E. Kanai, and K. I. Ando, "Ceramic Components for Engines," in *Proc. 1st Intl. Symp.,* Hakone, Japan, Oct. 17–19, 1983, 611 pp.

Stewart, D. A., D. B. Leiser, and S. M. White, "Response of Ceramic Coatings to Dissociated Nitrogen and Air Flows," presented at the 14th Ann. Conf. on Composites and Advanced Ceramics, Cocoa Beach, Fla., Jan. 14–17, 1990.

Streckert, H. H., and T. D. Gulden, "Ceramic Fiber Coatings by Gas Phase and Liquid Phase Processes," presented at the 14th Ann. Conf. on Composites and Advanced Ceramics, Cocoa Beach, Fla., Jan. 14–17, 1990.

Strife, J. R., and R. D. Veltri, "Si_3N_4 Coated C-C for Advanced Turbopropulsion Applications," presented at the 14th Ann. Conf. on Composites and Advanced Ceramics, Cocoa Beach, Fla., Jan. 14–17, 1990.

Suga, T., "Current Research and Future Outlook in Japan," in S. D. Peteves (Ed.), *Designing Interfaces for Technological Applications: Ceramic-Ceramic, Ceramic-Metal Joining (Comm. of Europ. Communities),* Petten, Netherlands, Apr. 20–21, 1988, pp. 247–263.

Suganuma, K., T. Okamoto, and K. Kamachi, "Influence of Shape and Size on Residual Stress in Ceramic/Metal Joining," *J. Mater. Sci.,* vol. 22, pp. 2702–2706, Aug. 1987.

Szekely, J., "Mathematical Models in New Process Development," *J. Met.,* pp. 16–21, Feb. 1990.

Taylor, T. A., "Plasma Sprayed YSZ Ceramic Coatings," presented at the ASM Intl. Ann. Mtg., Indianapolis, Ind., Oct. 4, 1988.

Teeter, F., "Benefits of Titanium Nitride Coated Tube Forming Tooling," FC 89-174, SME, presented at Tube Making and Fabricating Techniques, Dearborn, Mich., Mar. 28–31, 1989, 16 pp.

Tennery, V. J. (Ed.), *Ceramic Materials and Components for Engines, 3d Intl. Symp.,* Las Vegas, Nev., Nov. 27–30, 1988, Am. Ceram. Soc.

"Thermal-Spray Technology in Japan," *Adv. Mater. Processes,* vol. 136, no. 6, L.31343, pp. 15–19, 1989.

Thompson, E., "Joining Ceramics," presented at Vacuum Industrial High-Temperature Processing Equipment for Advanced Ceramics, Boston, Mass., Oct. 20, 1989.

Timmel, P., "CVD as Ceramic Processing Technique," *Keram. Z.,* vol. 41, pp. 259–262, 1989.

Tönshoff, H. K., and S. Bartsch, "Wear Mechanisms of Ceramic Cutting Tools," *Ceram. Bull.,* vol. 67, pp. 1020–1025, June 1988.

Troczynski, T. B., et al., "Advanced Ceramic Materials for Metal Cutting," in D. S. Wilkinson (Ed.), *Advanced Structural Materials, Proc. Intl. Symp.,* vol. 9, Montreal, Que., Aug. 28–31, 1988, Pergamon, New York, 1989, pp. 157–168.

Wachtman, J. B., and L. M. Sheppard,"Inorganic Thin Films: Expanding Applications," *Ceram. Bull.,* vol. 68, pp. 91–95, Jan. 1989.

Wang, P., et al., "Microstructure and Phase Investigations of Ceramic Thermal Insulation Coatings," *Keram. Z.,* vol. 41, pp. 657–662, 1989.

Whitney, E. D., "Modern Ceramic Cutting Tool Materials," *Powder Metall. Intl.,* vol. 15, pp. 201–205, 1983.

Wilkinson, D. S. (Ed.), *Advanced Structural Materials, Proc. Intl. Symp.,* Montreal, Que., Aug. 28–31, 1988, Pergamon, New York, 1989, 318 pp.

Wu, Y. C., and J. G. Duh, "Eutectic Bonding of Nickel to Yttria-Stabilized Zirconia," *J. Mater. Sci. Lett.,* vol. 9, pp. 583–586, Apr. 1990.

Yi, H. C., and J. J. Moore, "Self-Propagating High-Temperature (Combustion) Synthesis (SHS) of Powder-Compacted Materials," *J. Mater. Sci.,* vol. 25, pp. 1159–1168, 1990.

Yoshino, Y., "Role of Oxygen in Bonding Copper to Alumina," *J. Am. Ceram. Soc.,* vol. 72, pp. 1322–1327, Aug. 1989.

CHAPTER 8
APPLICATIONS FOR CERAMIC MATERIALS AND PROCESSES

8.1 INTRODUCTION

Looking ahead, the decade of the 1990s will be a period of uncertainty during which the people will have multiple options and will exercise them with intelligence and creativity. There will be greater challenges for industry and organized labor to be more forward-looking in analyzing the future, anticipating change, and mitigating the traumas of change.

Today, as well as in the future, we can expect that advanced materials will create new sources of influence and strength.[1,2] Federal investment in materials technologies will have high leverage because it creates demands for new tools and equipment (Fig. 8.1).

One of these material technologies that has created considerable excitement is the wide variety of high-performance ceramic materials, which are made in a way very similar to that used for conventional ceramics. However, they all differ by having substantially higher performance than traditional ceramics. This performance sets stringent requirements on phase composition, purity, and microstructural control, requirements that, in turn, demand sophisticated and highly reproducible processing. The result is a family of high-value-added materials, generally at a much higher cost per pound than traditional ceramics.

Ceramics can be organized into categories by their materials and by the functions they perform. These functions, which are listed in Table 8.1, depend on the special properties that can be designed into the ceramics. The properties, in turn, depend on the fundamental characteristics of the chemical bonding.

8.2 CERAMICS BY FUNCTION AND MATERIAL

Chemical Properties

Microporous materials such as Al_2O_3 and ZrO_2 are involved in coal gas filters for sulfur removal, high-temperature or corrosion-resistant membranes for chemical processing, chemical sensors, and liquid metals.

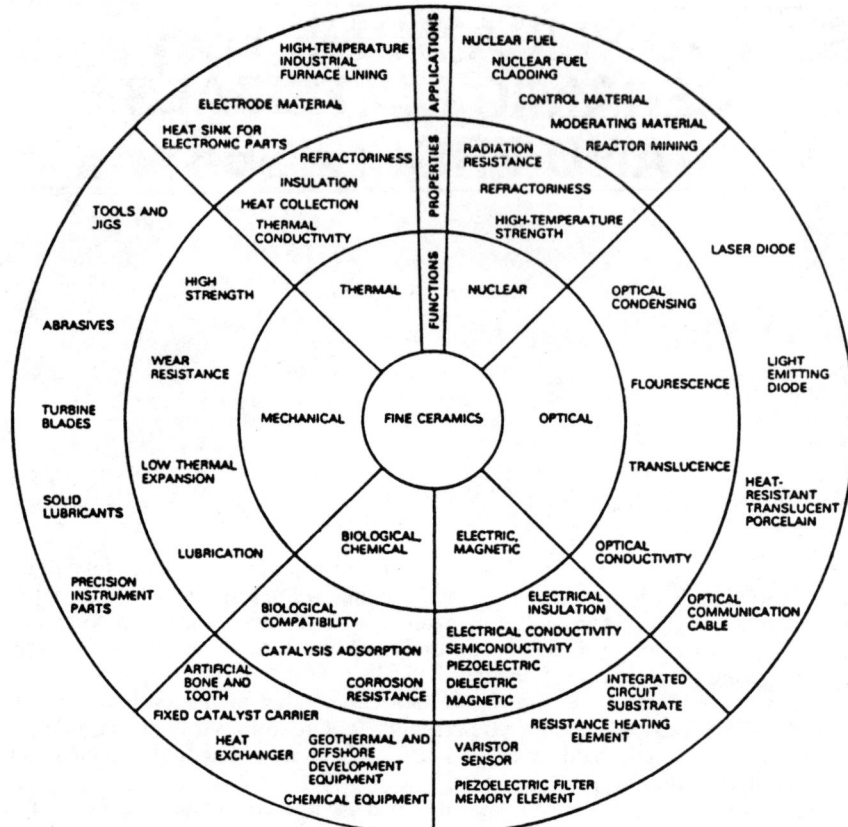

FIGURE 8.1 Functions, properties, and applications of advanced ceramic technology.[3,4]

Electronic, Electromechanical, Magnetic, and Optical Properties

The most rapidly growing category of highly specialized task-specific ceramic materials derives directly from the mixed ionic and covalent chemical bonding, which makes most ceramics good electrical insulators, but allows for modifications in semiconductors, conductors, and superconductors. Each ceramic material is tailored compositionally and structurally to have unique performance-related properties.

Electrical insulators make up a large category of uses, including the complex three-dimensional electronic substrate wiring arrangements widely used with electronic chips. Ceramic dielectrics are used extensively in capacitors; production is measured in terms of a billion units per year. The semiconducting capability of ceramics is the basis for various specialized sensors and devices such as high-frequency filters for signal selection and absorbers of radio-frequency energy for stealth applications. High-intensity electrical sources for electron microscopes use ceramic electronic emitters. Overvoltage protection devices are composed of strongly nonlinear ceramic materials. Solid ceramic electrolytes form the basis of fuel cells and high-performance batteries, while electrolytes for severe chemical and thermal conditions utilize special conducting ceramics.

TABLE 8.1 Advanced Ceramics by Function and Materials

Function	Material
Chemical/processing functions	
Coal gas filters for removal of sulfur, etc.	Microporous Al_2O_3
Membranes for chemical processing	ZrO_2
Chemical sensors	ZrO_2–mullite liquid-metal filters
Thermal functions	
Fibrous superrefractories	SiO_2–Al_2O_3
Thermal-barrier coatings	ZrO_2
High-heat-conduction paths	Diamond, BeO
Electronic functions	
Insulators	
Electronic substrates	Al_2O_3
Dielectrics for capacitors	$BaTiO_3$
Semiconductors	SiC
Filters	Pb–Zr–Ti–O
Absorbers of radio frequency	Mg–Al–Si–O
Electron emitters	Na–Al–O
Overvoltage protection	LaB_6
Solid electrolytes	ZnO
Electrodes, translucent	SnO_2
Superconductors	123, 2112, etc.
Electromechanical functions	
Transducers	Pb–Zr–Ti–O
Micropositioners	Pb–Zr–Ti–O
Pickups	Pb–Zr–Ti–O
Magnetic functions	
Soft magnets (recording heads, magnetic tape)	Zn–Mn–Fe–O
Hard magnets	Sr–Fe–O
Motor parts	
Optical functions	
Light sources	Al_2O_3–Cr lasers
Light guides	Silica
Light detectors	Ba–Na–Nb–O infrared
Reflectors	Sn–O for plate glass
Memory systems	Pb–La–Zr–Ti–O
Shutters	Pb–La–Zr–Ti–O
Frequency doublers	
Mechanical functions	
Cutting tools	Monolithic ceramics
Engine parts	Si_3N_4
Pump parts	SiC
Dies	Transformation-toughened Al_2O_3
Valves	Transformation-toughened ZrO_2
Ceramic composites	SiC-whisker-reinforced Al_2O_3
	SiC-fiber-reinforced glass-ceramic
	SiC-whisker-reinforced iron, aluminum
Tribological functions	
Bearings	Monolithic Al_2O_3, SiC
Seals	Carbide, nitride, oxide coatings
Guides	Diamond coatings

Source: From Wachtman.[1]

Ceramic superconductors represent an entirely new technological advance. Ceramic transducers, micropositioners, and pickups represent electromechanical functions ranging from the serious use in submarine warfare to the scientific cutting edge in scanning tunneling microscopes to such light-hearted consumer products as singing birthday cards.

Magnetic and optical functions also derive from the fundamental chemical bond characteristics, which provide the ability to tailor ceramic structures and compositions to specific magnetic or optical properties. Small magnets, many of them made of ceramics, are exemplified by recording heads, magnetic tapes, and motor parts. Optical functions are an exploding technology with revolutionizing consequences for the fields of information handling and communication. Most ceramics have the kind of distribution of electron energies that make them transparent and give them vital roles as laser light sources, optical waveguides, and light detectors. The coming generation of optical memory systems with enormous storage capacity—many using ceramics—will make a qualitative difference in information handling.

Mechanical and Tribological Properties

The impact of ceramics is threefold. First, ceramic coatings on cobalt-bonded WC have greatly increased cutting speeds and tool life. Typical coatings are TiC and TiN as well as multilayer coatings such as $TiC-Al_2O_3$ and $TiC-TiN-Al_2O_3$. Typical cutting speeds have increased from 2.5 m/s for cemented carbide to 5.1 m/s for coated cemented carbide. Ceramic coatings are used for the following finishing products:

> Abrasives and grinding materials: bonded disks and wheels [using, for example, SiC, TiC, Ti(C,N), Ti_3AlC]
>
> Machining aids: (1) ceramic and cemented carbide tool bits [using, for example, TiC, Ti(C,N), $(Ti,Cr)B_2$, (Ti,Ta)C, Si_3N_4, TiC with Mo/Re binder, WC with Co binder, W_2C, diamond, Al_2O_3, CBN]; and (2) surface coating of tools [using, for example, Si_3N_4, TiC with Mo/Re binder, WC with Co binder, W_2C]
>
> Polishing compounds: grits and pastes [using, for example, TiC, Ti(C,N), Ti_3AlC]

Second, monolithic ceramic cutting tools are also coming into increasing use. The older families of ceramic cutting tools are based on Al_2O_3 and are used mostly for cutting cast iron. The newer families involve either whisker reinforcement (typically, with SiC whiskers) or alternate matrix materials (typically Si_3N_4 or sialon). Cutting speeds as high as 51 m/s are foreseen for Si_3N_4 composites.

Third, great advances in the use of diamond in cutting and grinding are occurring. Several technologies are coming together to offer real breakthroughs. The new methods for synthesizing diamond by chemical-vapor deposition now offer synthetic diamond coatings at reasonable cost. Fine diamond crystals can now be made on a simple torch using a cooled substrate from which the diamonds are then removed. Another exciting advance is the development of cast iron bonded diamond grinding tools. Japanese investigators have demonstrated material-removal rates of up to 100 times previous practice in the grinding of hard materials such as Si_3N_4. This is important because previous grinding costs have been a serious obstacle to wider use of hard ceramics as parts in machinery and other devices. The new grinding tools are made by a process for sintering cast iron containing diamonds which does not cause the diamonds to convert into graphite and which yields dense, strong compacts.

TABLE 8.2 Performance Objectives for Advanced Ceramics

Classification	Objective
High-strength materials	≥1200°C in air after 1000 h holding: 　Weibull modulus ≥20 　Average tensile strength ≥30 kg/mm^2 1200°C in air after 1000 h continuous loading: 　Creep rupture strength ≥10 kg/mm^2
Corrosion-resistant materials	≥1300°C in air after 1000 h holding: 　Weibull modulus ≥20 　Corrosion resistance (weight gain) ≤1 mg/cm^2 　Average tensile strength ≥20 kg/mm^2
Wear-resistant materials	Room temperature: 　Wear resistance 10^{-4} mm^3/kg · mm 　Surface flatness ≤2 μm 800°C in air after 1000 h holding: 　Weibull modulus ≥22 　Average tensile strength ≥50 kg/mm^2

Source: After Niesz et al.[5]

Ceramics for heat engines offer a potential market comparable in size to the presently large and growing electronic ceramics market. Possible involvement includes friction- and wear-reducing applications, uses where reduced mass is beneficial, applications for high-temperature strength, and uses involving heat management. Some uses on an exploratory low-volume basis in production automobiles include cam-follower pads made of Si_3N_4, sialon, and ZrO_2-toughened Al_2O_3 and turbocharger rotors made of Si_3N_4 and SiC.

Applications for heat engines now in the stage of laboratory development and expected to come into practical use later in this decade include valve guides, valve seats, and valves made of Si_3N_4, sialon, or ZrO_2-toughened Al_2O_3; piston pins made of Si_3N_4 or sialon; piston rings made of Si_3N_4, sialon, or SiC; exhaust system coating made of Al_2O_3; and an exhaust manifold liner made of mullite or SiO_2.

Although property requirements for advanced ceramics will vary considerably with specific products, some generic requirements can be given as examples for many high-temperature structurals such as engine components. Those listed in Table 8.2 have been established as specific 10-year program objectives by the Engineering Research Association for High-Performance Ceramics in Japan. Emphasis is placed on strength and rigidity up to 1200°C, corrosion resistance in air up to 1300°C, and wear resistance up to 800°C.

It should be further noted that in other specifications, minimum toughness properties at room temperature often are included among mechanical requirements. Values of K_{Ic} on the order of 5.5 MPa · m$^{1/2}$ or greater are typical of such requirements.

8.3 PROGNOSTICATION OF CERAMIC APPLICATIONS

Figure 8.2 gives an estimated timetable for the introduction of ceramic products in various categories.

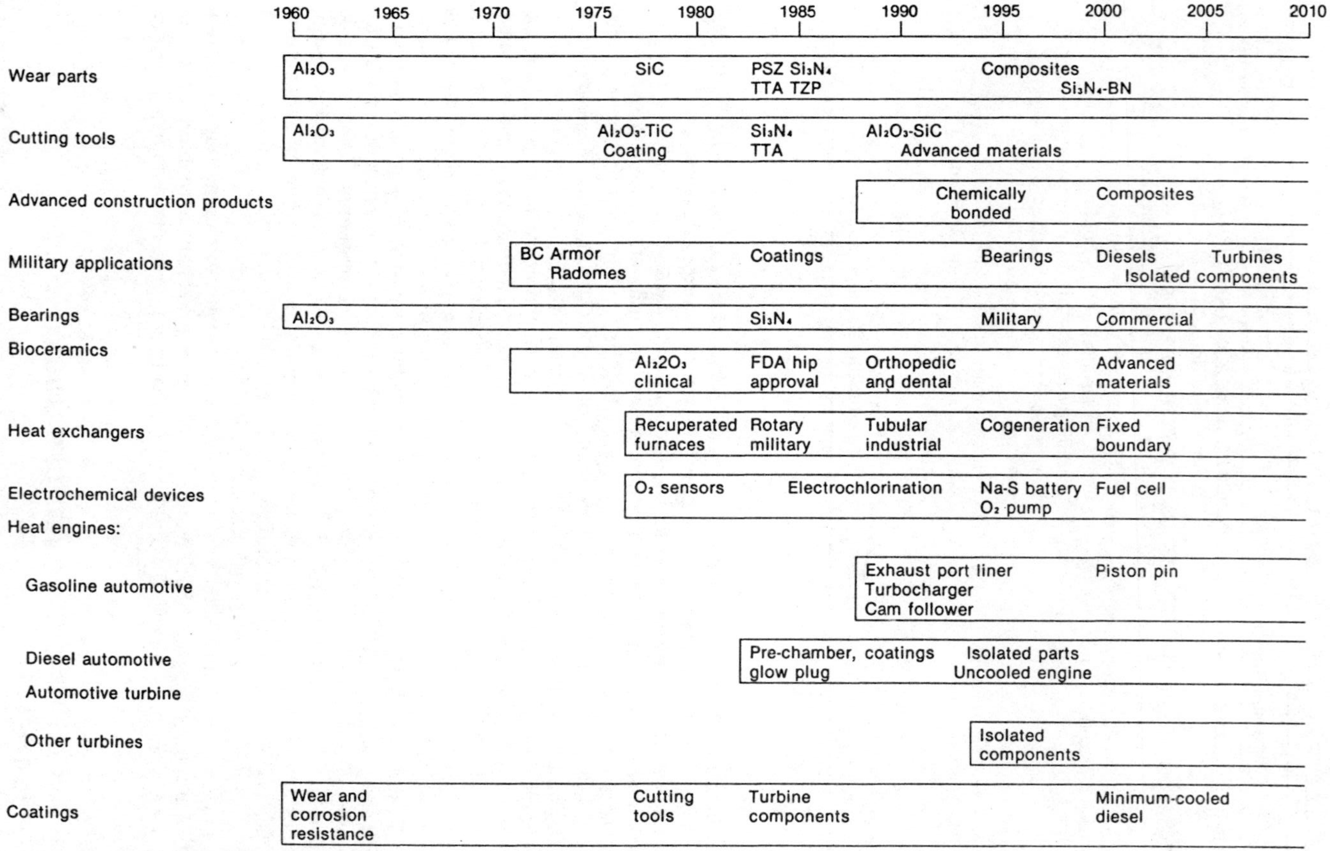

	1960	1965	1970	1975	1980	1985	1990	1995	2000	2005	2010
Wear parts	Al_2O_3				SiC	PSZ Si_3N_4 TTA TZP		Composites		Si_3N_4-BN	
Cutting tools	Al_2O_3			Al_2O_3-TiC Coating		Si_3N_4 TTA	Al_2O_3-SiC Advanced materials				
Advanced construction products								Chemically bonded	Composites		
Military applications				BC Armor Radomes		Coatings		Bearings	Diesels	Turbines Isolated components	
Bearings											
Bioceramics	Al_2O_3					Si_3N_4		Military	Commercial		
					Al_2O_3 clinical	FDA hip approval	Orthopedic and dental		Advanced materials		
Heat exchangers					Recuperated furnaces	Rotary military	Tubular industrial	Cogeneration Fixed boundary			
Electrochemical devices					O_2 sensors	Electrochlorination		Na-S battery O_2 pump	Fuel cell		
Heat engines:											
Gasoline automotive							Exhaust port liner Turbocharger Cam follower			Piston pin	
Diesel automotive						Pre-chamber, coatings glow plug		Isolated parts Uncooled engine			
Automotive turbine											
Other turbines								Isolated components			
Coatings			Wear and corrosion resistance		Cutting tools	Turbine components				Minimum-cooled diesel	

FIGURE 8.2 Estimated timetable for implementation of ceramic components in structural application categories.[3]

Current Production

The established ceramics of Al_2O_3, Si_3N_4, and SiC are in production in the application categories of wear, cutting tools, bearings, and coatings. Substantial growth in ceramics production is expected to occur over the next 25 years by growth of the overall markets, by achieving an increase in the market share, and by spin-off applications.

Ceramics are in limited production in Japan in discrete engine components such as turbochargers, glow plugs, and precombustion chambers, while current U.S. military applications include radomes, armor, infrared windows, and heat sources.

Much of the near-term funding is directed toward the development of new or improved ceramic or ceramic-matrix composite materials. Key objectives are to achieve improved toughness, higher reliability, and decreased cost.

Near-Term Production

Near-term production is projected in advanced construction products, bearings, bioceramics, heat exchangers, electrochemical devices, isolated components for internal combustion engines, and military applications. The technology feasibility has generally been demonstrated, but scaleup, cost reduction, and design optimization are required. Although much of the feasibility demonstration has occurred in the United States, foreign industry and government-industry teams have aggressive programs to commercialize the near-term applications. Large markets are at stake; for example, the world market for all biocompatible materials has been projected to be as high as $6 billion by the year 1995[7]; ceramics could capture 25 to 30 percent of this.[8]

Long-Term Applications

Some potential applications of ceramics require the solution of major technical and economic problems. These high-risk applications are categorized as long-term (more than 15 years away). The ultimate payoff may be large, but it is impossible to predict confidently that the problems will be overcome to achieve the benefits.

Long-term applications include the automotive gas-turbine engine, the advanced diesel, some electrochemical devices such as fuel cells, some heat exchangers, and some bearings. A variety of other turbines, especially those for aircraft propulsion and utility-scale power generation, should also be categorized as long-term.

Substantial design, material property, and manufacturing advances are necessary to achieve production of applications in the long-term category. Advancement currently is being driven by government funding. In many of these categories, military usage will most likely predate commercial use.

8.4 MARKETS FOR STRUCTURAL CERAMICS

Estimated U.S. markets by the year 2000 for several ceramic materials are illustrated in Fig. 8.3.

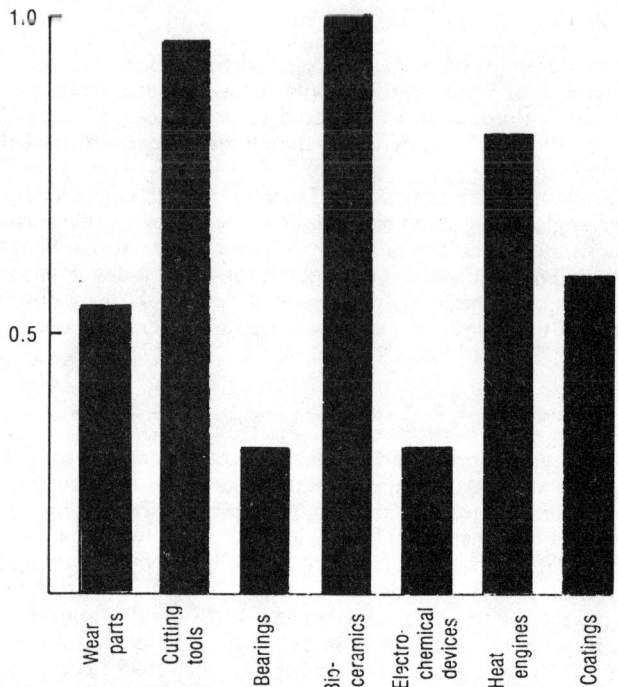

FIGURE 8.3 Projected U.S. markets for structural ceramics in the year 2000 (billions of dollars).[3]

Wear and Abrasive Parts

Wear Parts. Wear parts include such applications as seals, valves, wear pads, grinding wheels, dies, grinding media for ball mills, mechanical seals, slurry pump rotors, guide rollers, textile yarn guides, high-temperature and sand-blast nozzles, ceramic liners, and scrapers. The Department of Commerce has estimated that by the year 2000 ceramics could capture roughly 6 percent of the wear-parts market, currently dominated by WC cermets and specialty steels. With the total market estimated at $9 billion, the ceramic portion would be $540 million.[9]

Why ceramics for wear? Wear is a major cause of equipment failure and downtime in the military, in industry,, and in commercial products. Associated friction is a major consumer of energy. Therefore, ceramics have been targeted as a means of reducing wear and friction and avoiding the chemical and galvanic attack to which bonded WC and steels are vulnerable. The potential advantages of ceramics compared to metals and polymers include high hardness, chemical stability, ability to be ground with a very smooth surface to high tolerances, strength retention over a broad temperature range, and low cost.

Wear parts are an elusive category of small pieces, and there is a difficulty in defining a wear part, which leads to wide variations in market-size estimates. Ceramics probably accounted for $30 million (United States) and $60 million (worldwide) of wear parts in 1985.[10]

APPLICATIONS FOR CERAMIC MATERIALS AND PROCESSES   8.9

The mineral, chemical, and oil and gas recovery industries are key users of high-end pump seals. These are generally made in custom batches by small job shops. Al_2O_3 now accounts for 30 to 40 percent of that market, SiC for 15 to 20 percent, and carbides for most of the rest.

Nozzles for sandblasters and processing equipment require good wear and abrasion resistance. This is a much smaller market than bearings or seals. Designs now encase ceramic wear surfaces inside a tougher shell.

The ceramic materials currently used for wear resistance and other industrial applications are Al_2O_3, Si_3N_4, PSZ, TTZ, SiC, Al_2O_3 stabilized with ZrO_2, B_4C, and BN. For various industrial applications, the materials need to possess a combination of properties such as wear resistance, hardness, stiffness, and corrosion resistance.[11,12]

Grinding Media for Ball Mills. Grinding media are the balls or cylinders that are tumbled in a ball mill to achieve particle-size reduction of the powder being milled. Size reduction is achieved as the particles are pinched between adjacent balls and against the mill wall.

Since particle-size reduction takes place when the material being ground is sandwiched between the media, it is best to choose the smallest feasible grinding media combination, as smaller media provide many more contacts than larger media. The shapes of the media depend primarily on the material being ground, and on the work to be done. Generally, the ball shape is preferred for converting the large-size feed materials into finely ground products. Natural shapes with sliding mortar and pestle action are best when there is little requirement for primary size reduction but more need for fine product. Rod media are used when a uniformly fine product of limited size range is required.

Sandblast nozzles have been made out of PSZ and a variety of ceramic-based materials. Sintered Al_2O_3 nozzles cost less, but have shorter life and thus result in more labor because they must be replaced more frequently. Hot-pressed B_4C has a higher cost of fabrication, but a much longer life than either Al_2O_3 or cobalt-bonded WC cermet because of its extreme hardness.

Dies. During hot extrusion of metals, the die and tooling are subjected to severe conditions of pressure and temperature. PSZ is a ceramic material that possesses the necessary mechanical and thermal properties to withstand extrusion conditions and provide good dimensional control and long die life. PSZ, due to its unique microstructure, exhibits the thermal stability, thermal-shock resistance, and fracture toughness necessary to survive extrusion conditions without brittle fracture typical of other ceramics. It has also a nonwetting, nonabrasive surface, which gives improved surface finish to the extruded product. PSZ ceramic dies are currently used for hot extrusion of copper, aluminum, and brass tubes and rods and for steel rods.[13,14]

A number of mills are currently using TTZ dies on a production basis for Be–Cu, Cu–Ni, brass, and bronze tubing and rod products. Since switching to ZrO_2 dies in its copper heat-exchanger tubing plant, for instance, one foundry claims an increase in die life of 3.5 times. Improvements in surface finish, dimensional consistency, and overall production costs are also reported. Based on available data, TTZ inserts average between 500 and 600 tubes per die.[14]

ZrO_2 dies permitted a smooth press operation with excellent surface finish quality and elimination of surface defects at a reduction ratio of 113:1 in the extrusion of Be–Cu 25 alloy with steel die materials.

TTZ is now being adapted to the forming dies for two-piece aluminum and steel beverage cans. This ceramic composite is used for wear surface inserts in can-making dies.

Ceramic-inserted tools are also lighter. A typical carbide-steel punch weighs 3.5 to 4.8 kg; the same ceramic-steel punch weighs 1.4 kg. In addition, the ceramic composite has been used in whisker-reinforced cutting tools to shape superalloy nozzle inserts and wear-resistant pump components.

Ceramic Liners and Scrappers. Ceramic liners are used as the lining material for the sliding surface of conveying or storage equipment of mineral products. In these applications both corrosion- and wear-resistance properties are important. An example is the handling of drilling muds, where ceramic liners can extend service life and reduce operating cost. The use of ceramics as liner materials provides the following advantages:

It substantially eliminates theft of components for reclaim value.

It reduces the risk of shortage of components due to a shortage of critical materials.

It lowers operating costs due to longer service life and reduced downtime in critical environments.

Some of the applications are mud pumps, cyclones, ball mills, and chutes.

The development of ceramic liners and gas seals for rotary engines by a Japanese car and engine manufacturer may have a long-term benefit for rotary aviation engine programs in the United States. The use of ceramic material in the rotary engine offers greater benefits than in other engines by lowering friction and conserving thermal energy. One possible use would be in the 350-hp-class stratified charge rotary engine under development in the United States.

Ceramic scrappers are used, for example, for the scrapping of foreign objects sticking to a conveyor.

Slurry Pumps. High-pressure slurry pumps used for transporting liquid suspension of SiC whiskers failed due to erosion of the stainless-steel valves. Realizing a need for higher wear resistance in the valves, developers turned to SiC-whisker-reinforced Al_2O_3 to make the valve seats, thereby extending the pump life by 45 times. The wear resistance and fracture toughness of the composite virtually eliminated erosion.

Check ball valves in oilfield subsurface pumps are subjected to harsh environments, which take a toll on the valve components. Valve life can be extended by replacing balls and seats made of traditional materials with ones made of ZrO_2 and Si_3N_4, which provide improved strength and durability. HIP further enhances the strength and toughness of the ceramic components.

Heat- and corrosion-resistant TTZ balls extend the service life of ball and check valves, pumps, and ball slide bearings in a variety of corrosive and high-temperature applications. The balls undergo a microstructural phase change when subjected to impact or abrasion, which toughens the stressed area by creating a crack-arresting compressive zone. The transverse rupture strength and fracture toughness of the balls is increased by up to 25 percent under these conditions, so that they will not crack, break, or deform.

Ceramic Seals. Ceramic seals are finding more industrial applications due to a combination of properties such as hardness, low friction when machined to a fine surface finish, high resistance to corrosion, and higher-temperature capability than materials such as rubber, nylon, and Teflon. Carbon graphite is one of the best seal materials. Other ceramic seal materials finding applications are Al_2O_3, SiC, and Si_3N_4.

Face seals essentially provide a seal at a rotating interface, which prevents the passage of liquids or gases on one side of the seal to the other. For instance, the compressor seal in an automotive air conditioner seals halogenated hydrocarbons and oil at pressures up to 1.7 MPa and surface speeds of 549 m/min. Graphite against Al_2O_3 provides a low cost, reliable seal for this application.

A more severe application is the main rotor bearing seal in jet engines. It must seal the oil-lubricating system from 0.8-MPa hot air at temperatures up to 593°C and surface speeds up to 6096 m/min. Graphite impregnated with other materials to increase strength and oxidation resistance is required for this application.

Another severe application is in the recovery of crude oil by the saltwater pressure system. Saltwater is pumped into the ground at about 17 MPa to force crude oil out of the rock formations so that it can be recovered in adjacent wells. The face seal in the pump must survive the 17-MPa pressure plus temperatures up to 315°C plus surface rub speeds of 1524 m/min.

Face-type seals are used in many applications, including sand slurry pumps (which pump approximately 35 percent solids), chemical processing and handling, fuel pumps, torque converters, washing machines, dishwashers, and garbage disposals.

The primary coolant pump in nuclear reactors for power generation requires a highly reliable first-stage mechanical seal. Al_2O_3 has been used for this application; however, hot-pressed Si_3N_4 is replacing this material. SiC has also been used as seal material for nuclear reactor primary coolant pumps.

Hot-pressed Si_3N_4 has successfully been used as seal material. Single-piece Si_3N_4 seal rings, 673 mm in diameter, last three to five times as long as segmented rings, which tend to work loose in operation. These segmented rings were formerly made of Al_2O_3. Hot-pressed Si_3N_4 has a higher density than pressed and sintered Si_3N_4 and is stronger than hot-pressed SiC at temperatures as high as 1370°C. It is lighter than ZrO_2, WC, or Al_2O_3 and has outstanding thermal-shock resistance (quenching from 800°C into cold water with no damage).

A nuclear submarine on a world-ranging 6-month mission cannot stop to replace stern propeller tube seals. To increase the life of such seals by a factor of 5 (to 10 years), Si_3N_4 was hot-pressed, developed, tested, and machined to extreme tolerances; it is now in use.

Wire and Cable Uses. Wire-drawing products using advanced ceramics include wire-drawing capstans, pulleys (sheaves), rolls and sleeves, guides, eyelets, and dies. The capstan is the heart of a wire-drawing machine, which is used for extending thin wires of iron, nickel, copper, and noble metals. It is generally made in the form of a multistage pulley, where a solid ceramic ring bolted between two separate metal flanges makes capstans especially cost-effective. Normally, Al_2O_3 is used and the average life of ceramic wire-drawing capstans is seven to ten times longer than that of tempered-steel capstans and three to five times longer than that of carbide-coated capstans.

Bolt-together, molded polyurethane-ceramic, and solid-ceramic pulleys are essentially dynamic wire guides. The wire drives these lightweight pulleys, designed to provide the lowest possible amount of inertia. The bolt-together type is useful for high-speed applications. Molded polyurethane-ceramic pulleys are less expensive than the bolt-together pulleys. Solid-ceramic pulleys are recommended for low-speed applications in which the presence of corrosives limits the life of the wear parts.

Guide rollers are used for determining or changing the passing direction of wire before and after a capstan. Standard ceramic guides currently available include eyelets, pigtails, rods, tubes, and tension devices. Wire drawing dies are usually made of TTZ due to its excellent fracture toughness, resistance to abrasion and corrosive wear, and good thermal characteristics.

Can-seaming rolls made of PSZ, which have roll lives five to ten times longer than alternative tool materials, are now available. The inherent hardness and corrosion resistance of PSZ make it ideal for applications where tool steels, WC, and TiN fail. It also has high strength and impact resistance. Effective in a variety of can-tooling applications, PSZ tooling is especially useful for the manufacture of two-piece, tin-free steel cans, virtually eliminating galling when using can stock containing pin holes or bare spots in the enamel coating.

Valves. Designing eight rocket exhaust valves to fit into a space the size of a hockey puck is difficult, and most of that difficulty centers on the materials needed to withstand the high temperatures and pressures. Engineers are examining ceramics for the valve structure for these hypervelocity projectiles, which would be launched from space-based platforms and travel at approximately 30,000 km/h to intercept ballistic missiles.

A set of ceramic valves has successfully logged more than 8000 km of city and highway driving in a 2.5-L four-cylinder automobile engine. The Si_3N_4 valves have been shown to allow higher engine speed and horsepower by improving valve-train stability, reduce engine noise during idling, lengthen maintenance intervals by improving wear rates, and save fuel by reducing valve-train friction.

Si_3N_4 automotive valves have been used in a stock-car racing automobile and the valves are said to boost engine power 6 to 7 percent. Ceramic valves such as spool valves are not exclusively for demanding applications. Due to their long life and moderate cost, they are suited to more mundane uses as well. In the future, this concept will likely expand to a wider variety of sizes and to other types of valves.[15]

Si_3N_4 was chosen as the candidate material to form part of the valve for a medium-speed diesel engine. The fabrication procedure was by isopressing, green machining, low-pressure sintering, and final machining. Pugh et al.[16] showed the feasibility of integrating a Si_3N_4 disk into a standard metal engine valve assembly. This enabled the engine to be operated effectively for up to 74 h.[16,17]

Current trends in spark-ignition engines are toward multivalve engines capable of operating at relatively high speeds to maximize performance and fuel economy at acceptable emissions levels. Structural ceramics appear to offer potential as an alternative valve material to the steels, superalloys, and titanium alloys currently in use. Progress to date on materials development, component design, component performance, and engine testing indicates that Si_3N_4 valves will operate satisfactorily.

Figure 8.4 reflects the components of a push-rod engine. Valve motion is usually controlled by a cam-operated valve-train system consisting of a cam, a tappet, a rocker arm, a push rod, and a spring. A finite-element analysis of SiC, PSZ, and Si_3N_4 valves under steady-state and transient state conditions showed that Si_3N_4 offered the best material properties. High-quality sintered Si_3N_4 valves passed extensive dynamometer testing and field trials in a 1987 passenger car. Potential advantages offered by these valves include reduced seat insert and

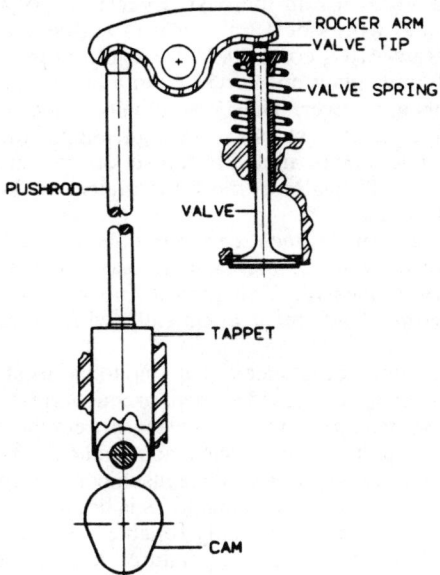

FIGURE 8.4 Elements of valve train in pushrod engine.[17]

valve guide wear, improved valve-train dynamics, increased engine output, and reduced friction loss using lower spring loads.[17,18]

Cutting Tools

Ceramics have demonstrated a capability as a cutting tool, especially in competition with WC–Co cermets (ceramic-metal composites) as inserts for metal turning and milling operations. The advantage of ceramics compared with carbides is retention of high hardness, strength, and chemical inertness to temperatures in excess of 1000°C. This allows use of the ceramics at much higher machining speeds than can be tolerated by carbides. However, the ceramics have lower toughness than the carbide materials, and have only been used successfully in the limited operations of turning and milling.

Throwaway cutting tools are inserts which shape rotating metal workpieces by chipping or shaving them. Cast iron and superalloys account for about half of all materials machined, and for materials most commonly machined with ceramics. Throwaway tools account for about one-third of the total insert market.

Compared with carbides, ceramics retain hardness and strength better at temperatures over 600°C, run longer without wear or abrasion, and creep (deform) less. They excel under high-heat conditions created by more productive high-speed machining. Several studies reflect estimates that ceramic tools could reduce machining costs by $530 million by 2000.[3,9,19]

There are several classes of cutting tools. Al_2O_3-based tools trace their pedigree back many years. The addition of TiC or PSZ, or both, and reinforcement with SiC whiskers overcame the poor fracture toughness and thermal-shock resistance that limited Al_2O_3 use in the past. Even now, Al_2O_3 cannot be cooled without fracture. Al_2O_3 works best on cast iron and superalloys. According to Rothman et al.,[20] if raw Al_2O_3 costs \$0.73/lb, Al_2O_3 tools could be produced for \$3 in a small (50,000-unit) plant and for \$1.60 in a larger (350,000-unit) facility.

Si_3N_4 cutting tools are newer. They are actually made of sialon and are used where mechanical and thermal stresses are high. Applications include finishing cast iron for auto-engine blocks, nickel-based superalloys for aerospace and corrosion use, and hard steels. Sialon is more thermally shock-resistant than Al_2O_3 and can be used with coolant.

Al_2O_3 and sialon account for most ceramic tools. Cubic BN, an extremely inert, high-temperature material, is increasingly used for hardened tool steel and cast iron, but is very expensive. Cemented diamond tools are used with Al–Si alloys, brass, and bronze, but react chemically with nickel and ferrous-based alloys.[21]

Several factors will limit acceptance of ceramic tools. First, they cost more than carbide tools. Cost estimates are based on high production rates and lower-cost raw materials than are now available. In the early stages, ceramic tools will be more expensive. Second, they require more powerful and rigid cutting machines that do not vibrate and fracture the ceramic inserts. Since machines are replaced only gradually, the size of the potential market for ceramic tools is limited.

Third, ceramic tools are still not highly reliable. Computer numerical control (CNC) sensors still cannot predict tool failure, and broken tools can ruin a workpiece. Fourth, the practice of coating carbide and high-speed steel inserts with hard ceramics offers many of the advantages of monolithic and composite ceramic inserts. TiN and TiC coatings both improve hardness, wear, temperature resistance, and chemical inertness and lower friction. Diamond and cubic BN coatings may soon be available too.

It is estimated that the throwaway insert market will reach \$950 million by 2000, assuming that the number of turning and milling machines powerful and rigid enough to use ceramics will grow 40 percent by 1995 and 60 percent by 2000. If ceramic cutting tools (whose performance will merit a price premium) penetrate 60 percent of that potential market, then the business could be worth from \$340 to \$450 million by 2000.

Refinement of the microstructure versus property relationship for ceramic materials has resulted in significant improvement of the performance of ceramic cutting tools over the last 20 years (Fig. 8.5).[22] With the introduction of advanced ceramic cutting tools, at least an order of magnitude in speed increase over early carbon-tool steels has been achieved.

The biggest family of presently utilized ceramic cutting tools is still based on Al_2O_3. Room-temperature rupture strength of modern pure Al_2O_3 tools reaches 700 MPa. However, low fracture toughness and thermal conductivity of pure Al_2O_3 excludes use of this material in high-thermal-shock environments. Advances in ceramic-ceramic composites research has allowed for a consecutive introduction of improved Al_2O_3-based materials. Their bend strength (0.7 to 1 GPa) substantially exceeds 0.2 GPa, the typical strength of traditional ceramics. The combination of strength and toughness suggests that the average processing flaw size in modern ceramic tool materials does not exceed 10 μm. Thus the primary processing defects are no longer a source of catastrophic failure. In advanced ceramic tools, wear-generated stress concentrations set the critical conditions for fracture.

An overview of typical room-temperature properties of ceramic and carbide tools appears in Table 8.3. The figures of merit represent the thermal-shock re-

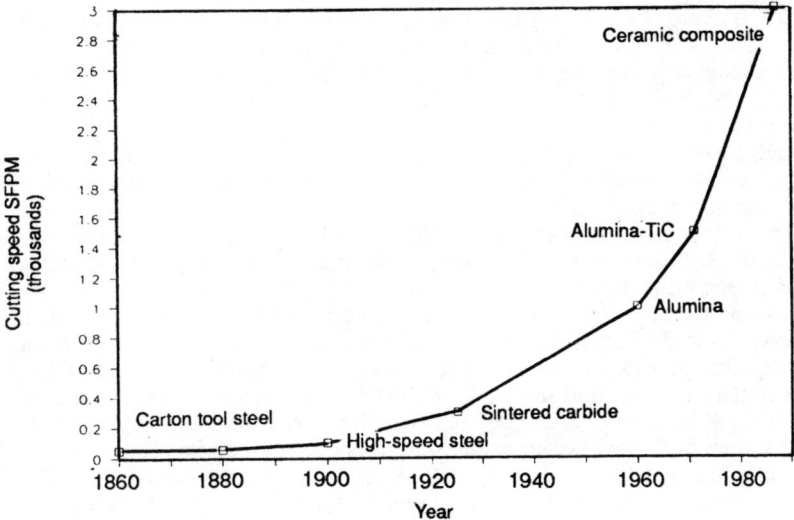

FIGURE 8.5 Historical evolution of cutting tool.[22]

TABLE 8.3 Typical Room-Temperature Properties of Ceramic and Cemented Carbide Cutting-Tool Materials

Material	Strength S, GPa	Young's modulus E, GPa	Hardness H, GPa	Fracture toughness K_c, MPa·m$^{1/2}$	Thermal conductivity L, W/m·K	Thermal expansion coefficient A, 10^{-6}/°F	Figures of merit R, W/m$^{1/2}$	M, GPa$^{5/4}$·m$^{3/8}$
Al$_2$O$_3$	0.7	450	17	3	17	7	16	9
TiC–Al$_2$O$_3$	1	420	20	4	20	7	27	13
ZrO$_2$–Al$_2$O$_3$	1	400	17	6	14	7	30	16
SiC$_w$–Al$_2$O$_3$	0.8	400	19	5–9	20	6	50	17
Si$_3$N$_4$	0.8	300	15	5–7	25	3	160	15
Carbide	1.5	550	17	10–13	100	7	220	26

Source: From Troczynski et al.[22]

sistance R and the abrasive-wear resistance M. The values of R and M are obtained from combinations of the basic parameters according to the following formulas:

$$R = \frac{\sqrt{K_c L}}{EA}$$

$$M = H^{1/2} K_c^{3/4}$$

Cutting-Tool Materials

Al_2O_3 Base. Al_2O_3 cutting tools have been produced by high-temperature sintering of very fine (1- to 3-μm range) Al_2O_3 grains in the presence of additives such as MgO.[23] The main limitation to Al_2O_3 cutting tools is their lack of toughness. These tools are used for uninterrupted cuts at low feed rates and depths in cast iron machining.

HIP of Al_2O_3-based and Si_3N_4-based cutting-tool materials has resulted in reduced porosity, a higher density, sometimes higher hardness, and a more consistent fracture resistance.[24]

Other recent tool material developments include a ZrO_2-toughened Al_2O_3, which has a broad application range, including the middle range of steels, hardened steels, chilled irons, and cast irons.

This new grade of tool material exhibits phase transformation strengthening, a characteristic that before had been unheard of in ceramics. Phase transformation strengthening plays an interesting role in ceramics. If a crack starts to form, the energy that is running at the front of the crack causes the material just ahead of the crack to change phase. The energy is then dissipated and the crack stops.

Another Al_2O_3 developed material is a fiber-reinforced fine ceramic formed by blending SiC whiskers and ZrO_2, with Al_2O_3 acting as the base material. The resulting composite is said to have very good physical properties, with a fracture toughness of 8 MPa and a three-point bending strength of 1600 MPa.

Ceramic-Coated Carbide. Since 1979, substantial effort has been directed toward the development of improved wear-resistant coatings (such as TiC, Al_2O_3, TiC–Al_2O_3, TiC–TiCN, TiC–Al_2O_3–TiN) and special substrates for increased toughness (cobalt-enriched layer and improved bonding coating). Criteria for selecting a coating for a cutting tool are shown in the literature.[23,25,26]

Ceramic-coated carbide inserts consist of a hard refractory composite ceramic (4 to 7 μm) deposited on a sintered-carbide substrate via chemical-vapor deposition. In general, ceramic-coated carbide inserts can be used at speeds double or even triple those normally achieved with uncoated carbide.

Continued growth of coated materials will dominate the future of cutting tools. Most major manufacturers of cutting tools either have their own coating technology or contract with a supplier. New development efforts are focused on advanced coating technology.

Al_2O_3–TiC. Addition of TiC in varying percentages, typically 30 wt %, in the Al_2O_3 increases the strength of the cutting tool. The addition of TiC provides improved toughness and thermal-shock resistance. This allows the tool to be operated at much higher cutting speeds than conventional cemented or coated carbides.

Al_2O_3–SiC-Whisker-Reinforced. Rice husks are revolutionizing the cutting-tool industry. The husks are made into SiC whiskers that are embedded in a ceramic composite that is stronger and more wear-resistant than metal-cutting tools. The rodlike whiskers deflect the growth of cracks, making the composite much less brittle even at temperatures as high as 1000°C. These cutting tools can be operated 10 times faster than metal tools, making it possible to machine extremely tough superalloys. Even at the higher speeds, the tools last up to seven times longer than those from conventional materials.

Si_3N_4. High-boring of automotive cylinder bores in six different cast iron engines is being accomplished using Si_3N_4 tools. The bores are rough-machined with high-speed "stitchers," or single-spindle tools that move from one bore to another like a stitching needle across a garment.

A relatively tough ceramic, Si_3N_4–sialon, consisting of elongated "whisker-like" grains of β-Si_3N_4–β'-sialon in a glassy or crystalline binder phase, usually containing yttrium, has a set of material properties that permits combinations of high speeds and feeds, depths of cut, and interruptions. The β-Si_3N_4–β'-sialon materials appear best for machining grey cast irons.

In summary, Si_3N_4 tools are tougher than Al_2O_3 tools. The principal application of Si_3N_4 tools is in the machining of high-temperature alloys and in the turning of cast iron, where the tool is subjected to high mechanical and thermal stresses. The machining of cast iron is the second largest application area after steel.

Other Ceramic Cutting Materials. SiC, not considered for use in cutting tools because it may react with steel and cast iron, is now being evaluated for its effectiveness as a cutting-tool material.

Another example of a new ceramic material is transformation-toughened Al_2O_3. This material consists of a very fine dispersion of tetragonal-phase ZrO_2 in Al_2O_3. When a stress is applied to the material and a crack tries to form, the tetragonal ZrO_2 particles adjacent to the stress transform to monoclinic ZrO_2. This transformation is accompanied by a volume expansion of the local ZrO_2 particles, which places the surrounding Al_2O_3 in compression. This localized compressive zone inhibits a crack from propagating and gives the material high toughness and strength. This transformation toughening is similar to martensitic toughened steels, and the ceramic has been referred to in the literature as "ceramic steel."

The transformation-toughened Al_2O_3 appears to be less susceptible to chipping and fracture than other ceramic cutting inserts. One end user reports successfully using this material for turning Inconel 718 at 215 m/min at a depth of cut of 2.54 mm, compared with only 25 m/min for formerly used uncoated carbide.

Bearings

Ceramics have been tested and proven in ball, roller, and plain bearings (bushings). The potential to resist fatigue, corrosion, high-temperature, and loss of lubrication better than metals makes ceramic materials attractive for ball and roller bearing use. The high hardness and the excellent high-temperature performance of ceramics are attractive for critical roller-bearing applications, such as in high-thermal-efficiency engines. In addition, the lower density of ceramics permits lower centrifugal loads, less skidding at high speeds, and, consequently, results in longer bearing component life than conventional bearing steels or alternative superalloys or cermets. Si_3N_4 shows outstanding performance in the roller-bearing application because of the material's low friction coefficient, high wear resistance, high compressive strength, and extremely long fatigue life. Contrary to what many engineers expect, the failure mode of ceramic rolling elements is not catastrophic. Other ceramic materials used for rolling bearings are Al_2O_3 and SiC.

The current state of processing and fabrication techniques for ceramic wear parts (rather than raw-materials costs) leads to the higher initial costs than those for metal parts. Ceramics can, however, provide longer service life than metals. Ceramic bearings tend to be consumed primarily by petroleum and chemical industries, which have demanding performance standards (for example, in corrosive environments) and in specialized aerospace applications, all of which can absorb the high initial cost.

Worldwide sales of ceramic bearings are currently in the range of $3.5 to $6 million. By the mid-1990s, this market could reach $85 to $175 million in the United States and perhaps $375 to $650 million worldwide.

Both all-ceramic and hybrid bearings (ceramic rolling elements in steel race) have proved to be effective in five major application areas to date:

1. High temperature, high speeds: all-ceramic bearings
2. Low temperature, high speeds: hybrid bearings
3. High acceleration conditions: all-ceramic or hybrid bearings
4. Corrosive and hostile environments: all-ceramic or hybrid bearings
5. Ultraprecision conditions: ceramic or hybrid bearings

The selection of Si_3N_4 for rolling elements and races is based on its ease of manufacture and the following findings:

Si_3N_4 in rolling contact forms elastohydrodynamic films similar to steel.

Dry traction forces of Si_3N_4 against itself or steel are lower than steel against steel.

Rolling traction forces of Si_3N_4 in lubricated contact are identical to those of steel.

Wettability of Si_3N_4 with oils is comparable to that of steel.

Si_3N_4 components have an endurance life greater than steel components.

Bearings containing Si_3N_4 balls run significantly cooler than all-steel bearings when lubrication is limited.

Smaller centrifugal forces of the low-density Si_3N_4 balls increase high-speed bearing life.

Conventional grinding machinery can be used to finish ceramic bearings.

Si_3N_4 bearings appear to be as reliable as high-quality steel bearings.

Studies of the type listed here[27-33] indicate the many advantages of Si_3N_4. This material is therefore an excellent choice for high-speed bearings where the centrifugal force generated by rolling elements at the outer race becomes greater than the application loads.

In the past, ceramics had poor rolling-contact fatigue life. Recent improvements in Si_3N_4 purity and grain structure have improved high-stress fatigue life. Under high test loads, the rolling-contact fatigue life of Si_3N_4 bearings is now equal to or better than that of M-50 steel. Since the usable strength is also highly dependent on quality, the additives as well as the process control of powder purity, density uniformity, and surface flaws are extremely important.

Some ceramic bearings use MgO as a sintering additive. During sintering, MgO builds up a glassy phase on the Si_3N_4 grain boundary. This glassy phase, or enstatite, loses its strength at 800°C and melts at around 1500°C.

However, one U.S. manufacturer uses Y_2O_3 and other rare-earth compounds as sintering additives with its Si_3N_4 roller bearing. Y_2O_3 forms a crystal phase during sintering and increases the melting point to 1850°C. Consequently, the bending strength at 800°C is about 300 MPa greater than with Si_3N_4 bearings containing MgO (Fig. 8.6).

Bearing Applications

1. When a pump company eliminated seals in a magnetic drive pump, it found that the bearings took a beating. However, after switching to bearings of SiC,

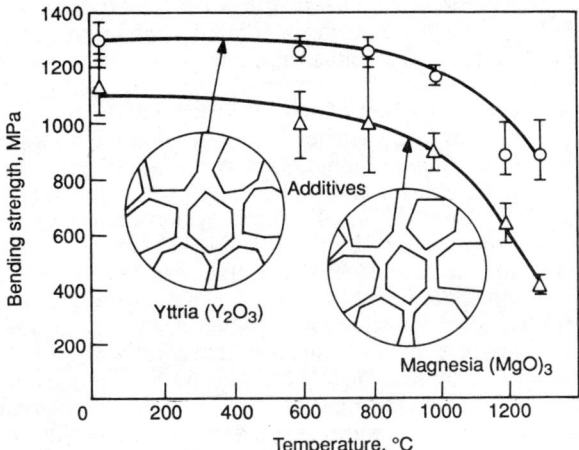

FIGURE 8.6 Additives strengthen bearings.

service life on the pumps increased by 150 to 200 percent. Mechanically sealed pumps proved to be the weak link in systems that transfer hazardous chemicals at temperatures of up to 204°C and viscosities of up to 100 cP. A new generation of sealless pump was designed where the bearings play a critical role because they support the pump shaft, and therefore the bearing system had to be strengthened. For this new breed of pumps a sintered single-phase SiC was selected for the journal and thrust bearings.

2. Si_3N_4 balls used as tool-spindle bearings are reported to have outstanding wear resistance and perform well without lubrication or where lubrication is limited. The balls can withstand high-temperature exposure to 1400°C, have an expansion coefficient only 25 percent that of steel, and are inert to virtually all environments, including liquid metals. They feature good fracture toughness, and have a roundness diameter tolerance of 0.13 μm and surface roughness of 0.02 μm. Applications besides tool spindles are in the chemical processing, nuclear, automotive, and aerospace and defense industries.

3. Current mast bearings for a helicopter are made of VIM VAR M-50 tool steel. The lives of these bearings are relatively short because of the micropitting mode of failure. The exact cause of this micropitting is unknown. However, it is believed that a significant difference in hardness between the two contacting surfaces (ball and inner ring and ball and outer ring) will eliminate the micropitting mode of failure. A test program was initiated to replace the existing VIM VAR M-50 mast bearing in the helicopter transmission with a hybrid bearing where the balls were made of Si_3N_4 and the inner and outer rings were made of VIM VAR 52100 bearing steel. Si_3N_4 weighs only 40 percent as much as the steel, has a modulus of elasticity 1.5 times higher than steel, does not corrode, is stable up to 1371°C, and has a hardness of approximately $R_c 80$ as compared to $R_c 60$ for the bearing steels. In rolling contact fatigue tests it has been shown that Si_3N_4 rods in contact with 52000 steel balls exhibit a life at least five times longer than that of VIM VAR M-50. As a result, a 35 percent life-cycle cost reduction in reduced downtime and cost of replacement items is expected to result from the installation of this type of bearing.

4. Another application of ceramic bearings is for instruments. Since most instrument bearings fail by a wear mechanism rather than by rolling contact fatigue, the high hardness and low friction coefficient of ceramics promise substantial performance improvement in many applications. Most instrument bearings are not subject to excessive loads and are not designed with unusual ring geometries, thereby reducing the risk of catastrophic ring failure, a concern in some larger bearings. For hybrid and miniature bearings, two types of bearings are available: all-ceramic and ceramic hybrid. Hybrid bearings use ceramic balls with races of conventional materials such as stainless steel or beryllium copper. All-ceramic bearings are made using races and balls of Si_3N_4. The range of design configurations available include radial and angular contact designs, as well as full complement bearings with fill slots or relieved races. The currently available Si_3N_4 all-ceramic bearings are costlier. They are applied only where current bearings are unsuccessful or where performance improvements outweigh initial cost considerations. In many applications, hybrid bearings may be a preferred alternative to the more costly all-ceramic bearings. They clearly offer performance improvements in some applications, although the positive results have not been universal.[23]

5. One of the recent potential applications for Si_3N_4 is as a replacement for M-50 steel bearings in aircraft turbine engine systems. The fatigue life of the M-50 bearing steel currently used in the main-shaft bearing support system is one of the key considerations that limits the maximum time between overhauls. The bearing fatigue life is strongly affected by the rotational speed and resulting centrifugal loads that are developed within the bearing. Substitution of lighter-weight Si_3N_4 rolling elements would result in lower centrifugal loads and longer fatigue life, and, equally important, may eventually allow higher operational speeds. The ability to operate to higher rotational speeds at higher temperatures than possible with M-50 steel makes the substitution of Si_3N_4 in this application extremely attractive. Other examples of high-speed or high-acceleration applications where the lighter-weight Si_3N_4 can be easily justified are high-speed machine-tool grinding spindles, turbocharger-turboalternator systems, and high-temperature air motor actuator assemblies such as those used on aircraft thrust reversers. Other applications for Si_3N_4 bearings might include photographic coating machines and other chemical processing equipment exposed to highly corrosive environments; satellite gyro-system and X-ray tube bearings which operate in a vacuum, often at extreme temperatures, and where lubrication is minimal; and applications in severe abrasive environments where Si_3N_4 bearings provide substantially higher wear resistance and significantly longer life.

6. Ruby sapphire and ZrO_2 are other ceramics with good bearing properties. For example, ZrO_2 balls get stronger when subjected to impact or abrasion. In the region of stress, transverse rupture strength increases by as much as 25 percent.

Coatings

Ceramic coatings provide a variety of benefits, including abrasion resistance, thermal protection, corrosion resistance, and high-temperature lubrication. Applications include ultrahard coatings for cutting tools, thermal insulation and lubricating coatings for adiabatic diesel engines and cooled gas turbines, and bioactive glass coatings for metal orthopedic implants. The list could be expanded to include other sectors such as mining (drills), utilities (turbine-generator sets, heat exchangers), agriculture (plows and tillers), and aerospace (bearings, power transfer assemblies, and actuator drive systems).

The availability of advanced ceramic coatings is expected to be a significant benefit to the U.S. economy. The value of the market for ceramic coatings is not easily assessed, because the range of applications is so wide. One estimate is for a $1 billion market worldwide for all coating materials, about 60 to 70 percent of which is domestic. This estimated market includes jet-engine, printing, chemical, textile, and tool and die applications. This list could be further expanded to include wear parts, bearings, biomaterials, heat exchangers, and automotive components in the future. Ceramic coatings should be considered an extremely important technology for extending the performance of metal components, and, in some cases, coated-metal structures may be an excellent alternative to monolithic ceramics. (See Chap. 7.)

Bioceramics

Bioceramics, or ceramics for medical applications such as dental or orthopedic implants, represent a major market opportunity for ceramics in the future. The overall worldwide market for biocompatible materials is currently about $4.5 billion, and this is expected to double or triple in the next decade.[7] Ceramics could account for 25 to 30 percent of this market.[8]

Bioceramics may be grouped into three categories: nearly inert, surface active, and resorbable.[34] Nearly inert ceramics can be implanted in the body without toxic reactions. These materials include Si_3N_4-based ceramics, transformation-toughened ZrO_2, and transformation-toughened Al_2O_3. Many of the dental materials, including glass-ceramic or ceramic crown materials, can also be grouped under this category.

Surface-active ceramics form a chemical bond with surrounding tissue and encourage ingrowth. They allow the implant to be held firmly in place and help prevent rejection due to dislocation or to influx of bacteria. Surface-active ceramics which will bond to bone include dense hydroxyapatite, surface-active glass, glass-ceramic, and surface-active composites.

The function of resorbable bioceramics is to provide a temporary space filler or scaffold, which serves until the body can gradually replace it. Resorbable ceramics have been used to treat maxillofacial defects, for filling periodontal pockets, as artificial tendons, as composite bone plates, and for filling spaces between vertebrae, in bone, above alveolar ridges, or between missing teeth. An early resorbable ceramic was plaster of paris (calcium sulfate), but it has been replaced by trisodium phosphate, calcium phosphate salts, and polylactic acid–carbon composites.[35]

Any new material intended for use in the body must undergo extensive testing before it is approved. Preclinical testing, clinical studies, and followup may take as long as 5 years to complete. However, ceramics have been in clinical use for some 15 years and are gaining acceptance. Industry interest in bioceramics has been increasing since 1980. As an example, the work of direct chemical bonding through ionic and covalent forces can occur with a new generation of bioactive glass and glass-ceramic materials. The range of bioactivity varies as a function of the relative weight percentage of the primary constituents.[23]

Bioglass in Dental Applications. Bioglass-coated Al_2O_3, stainless steel, and Vitallium find application as dental implants which have load-bearing characteristics. Coated Al_2O_3 devices have excellent mechanical behavior. Considerable attention has been given to the use of bioglass-coated implants of Al_2O_3 or metal in orthodontics. Researchers in studies have inserted blade-shaped implants of Al_2O_3, coated with surface-active glass, into rhesus monkeys to produce maxillary expansion with good results. It is expected that research and new developments in the use of bioglass-coated dental implants are likely to continue.

Al_2O_3 for Artificial Teeth. Al_2O_3 has been used in several dental components (Fig. 8.7).[36] A ceramic pin is screwed into the jawbone and a crown is then cemented to it, giving a construction similar to that of a natural tooth. For these applications it is essential that the ceramics be strong and wear-resistant. It is particularly important that they be highly resistant to corrosion by fluids in the body, primarily that there are no toxic or irritant effects, and also that the materials do not deteriorate in use.

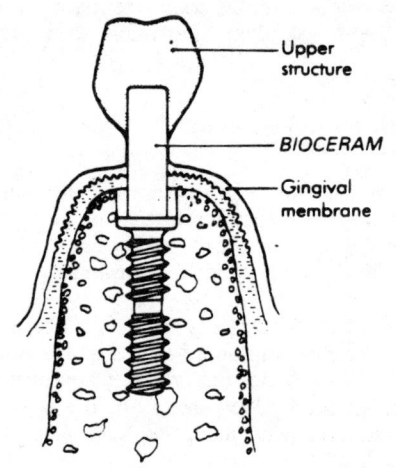

FIGURE 8.7 Structure of artificial tooth made from Al_2O_3 ceramic.[36]

In another approach gold and ceramic-fired, high-density Al_2O_3 are used to make artificial tooth roots that are implanted into human patients. To create the implant, Al_2O_3 artificial roots are implanted in bone in either fresh or healed extraction sites. Once implanted, the roots stabilize for approximately 2 to 3 months. During this time, dense bone grows around the artificial roots. A gold post and core specially fabricated for each root are then cemented into the implant to hold a gold crown. Artificial roots may be more effective than fixed bridges as a procedure for replacing lost teeth.[37]

A ZrO_2 orthodontic bracket has been developed at a Japanese dental university.[38] The high-strength ZrO_2 Torayceram has nearly the same gloss as natural teeth and so does not ruin the appearance of the mouth. Furthermore, it does not cause allergies resulting from metal ions, and its high strength and rigidity ensure good workability of the product. The bracket, which was produced using injection-molding technology, is said to have the high strength, surface smoothness, biocompatibility, and the insulation properties characteristic of ZrO_2. It is three times as strong as an orthodontic bracket made from high-purity Al_2O_3 and presents virtually no problem with regard to causing tooth caries.[39,40]

Artificial Hip Joints and Other Applications. Artificial hip joints are leading examples of how improved ceramics are replacing other materials. Single-crystal Al_2O_3, a form with 816-MPa flex strength and 3.808×10^5-MPa modulus, can be polished to give a low-friction abrasion-resistant extremely durable implant. Thousands of these devices have been implanted in people in Europe. The stem portion of the ceramic joint is usually metal—titanium or a superalloy.[41–43]

Implantable electronic stimulators and pacemakers, too, are being fabricated from Al_2O_3. The ceramic is formed into hermetically sealed lightweight cases for the device's electronic components.

Ceramic-metal leadthroughs for electrodes for nerve and heart pacemakers need to be bonded hermetically to the electronic module to prevent any penetration of body fluids. Gold is used as the filler metal for the bond between Al_2O_3 and titanium. Prior art bonds of this nature were hard and brittle due to titanium diffusion into the gold during the bonding process. Therefore, during the cooling cycle, differences in the thermal contraction of titanium and the ceramic could lead to cracks in the latter. Now a wider solder gap between the titanium and the ceramic has been developed and the soldering process is kept sufficiently short to prevent diffusion of titanium through the entire width of the solder joint. Part of the joint, therefore, remains ductile gold, which can absorb the cooling stresses.

Al_2O_3 has been used successfully for ear-bone replacements and jawbone reconstruction. Figure 8.8 shows a porous-Al_2O_3 valve body machined from a cylindrical blank consisting of a network of interconnecting cylindrical pores that allow tissue ingrowth. The new glass-ceramic Ceravital[23,44] shows promise as a coating for metal-base implants and as a middle-ear prosthetic device.

Heat Exchangers

Increasing the efficiency of thermal conversion systems requires the use of higher operating temperatures, and ceramics offer great potential for doing this. Significant obstacles must be overcome, however, before the full potential of ceramics can be realized. These problem areas include the economic production of consistently reliable turbine components, fuel combustors, regenerators, and other heat exchangers. In each of these applications, the fundamental understanding of thermal and mechanical response to operating conditions needs to be greatly expanded. This is especially true now that resource conservation requires engineering design to be more efficient and less reliant on overdesign. At the same time, more stringent safety, emission, and antipollution standards call for a high level of confidence that predicted performance in all engineering systems will, in fact, be the result.[23]

Ceramic heat exchangers are of great interest because they can utilize waste

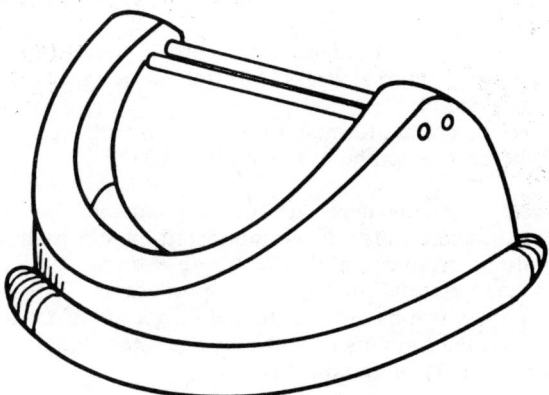

FIGURE 8.8 Sketch of Al_2O_3 valve body.[44]

heat to reduce fuel consumption. Heat recovered from the exhaust of a furnace is used to preheat the inlet combustion air, so that additional fuel is not required for this purpose. The higher the operating temperature, the greater the benefit. Ceramic systems have potential for greater than 60 percent fuel savings.[45]

Ceramic heat exchangers may be used in a variety of settings, including industrial furnaces, industrial cogeneration, gas-turbine engines, and fluidized-bed combustion. Ceramic heat exchanger systems can be classified in several ways, namely, according to configuration (tubular, finned-plate, cellular), size, mode of operation (stationary or rotary), pressure, temperature, and type of application (industrial, residential, military, cogeneration, vehicular, solar).

Ceramic heat exchangers are in production in Japan to reduce fuel consumption of glass-melting furnaces and for roofing-tile furnaces and in the United States for some foundry furnaces.

The heat exchangers in current production are low-pressure. They operate at close to atmospheric pressure, can be of relatively simple design, and can tolerate lower-strength ceramics.

The low-pressure heat exchangers typically are used for industrial furnace heat recovery where heat from the exhaust gases is used to preheat combustion inlet air. A typical system can preheat the inlet air to around 1050°C using approximately 1300 to 1400°C exhaust gases. In a clean environment (natural gas or petroleum-fired furnace), a lifetime of at least 5 years is expected.

High-pressure heat exchanger systems typically have internal pressures in the range of five to ten times higher than atmospheric. The ceramics must be better (lower porosity, higher strength) to withstand higher stresses, and attachment designs must be more sophisticated to minimize leakage.

Heat-Exchanger Materials. The major material being used and evaluated for industrial heat exchangers is SiC, primarily in tubular form. SiC has high thermal conductivity, excellent high-temperature and corrosion resistance, and moderately good thermal-shock resistance. Although the thermal-shock resistance is substantially better than that of most oxide ceramics, it is marginal for some heat-exchanger applications and requires very careful design to minimize thermally induced strains.

A second important limitation of SiC (and other candidate ceramic materials) is fabricability in the required size. Scaleup is required to produce the size components needed for industrial heat exchangers. SiC is presently limited to temperatures below 1200°C.

Si_3N_4 has been considered for higher temperatures, but the higher thermal expansion results in higher thermal stress. Substantial design modification is necessary in order to use Si_3N_4.

Other mixed-oxide ceramic formulations for heat exchangers are in development, which can be used at temperatures up to 1400°C.

Fabrication Methods. Ceramic heat-exchanger components have been fabricated by a variety of techniques. Early tubes, fabricated by slip casting, were expensive and could only be produced in limited length. Most tubes are now made by extrusion, which has potential for large-scale production. Extrusion techniques have been developed, which have the potential for reasonable cost as well as improved performance (due to thinner cell walls, a greater number of cells per unit area, and better uniformity of cell dimensions).

Economics. Government support has been necessary to accelerate the development of ceramic materials and system technology for heat exchangers, in spite of the design projections of significant fuel savings and short payback time. The material manufacturers, system designers, and end users have all considered the risks too high to invest their own funds in the development and implementation of a system. Specific concerns include the high installed cost, which represents a significant financial risk to the user for a technology that is not well proven; the fact that many potential end users are in segments of industry that presently are depressed; and the fact that designs vary according to each installation, leading users to want a demonstration relevant to their particular situation.

If the current high fuel prices persist in the next decade, this could increase the activity in a widespread implementation of ceramic heat exchangers for waste-heat recovery.

Recuperators

The use of ceramics in recuperators, a type of heat exchanger that allows a furnace to operate more efficiently by recycling its waste heat to preheat incoming combustion air, has received a considerable amount of attention. To recover the waste heat coming off furnaces, the recuperator has to survive in the temperature range of 1093 to 1649°C.

Ceramic recuperator materials overcome the difficulties faced by metallic alloy recuperators in corrosive, hot exhaust gas streams. Metallic recuperators can be used only in processes with clean stack gases or in systems that can bypass the recuperator during intermittent fluxing operations. Processes with continuous fluxing or corrosive stack gases have excellent potential for substantial savings by recycling thermal energy back to the process with a ceramic recuperator that can withstand the rigorous operating conditions.

Figure 8.9 shows a schematic of a SiC ceramic-tube recuperator module. This design has been developed and the technology proven, and the manufacturing expertise and skills required to fabricate a reliable ceramic heat-recovery system have been put in place. The ceramic-tube recuperator uses multiple self-contained tube modules to constitute a unit sized to meet the customer's site requirements. A typical module is fabricated of 12 to 16 SiC ceramic tubes (each with one end closed), which form a bayonet-style heat-recovery unit (Fig. 8.10).[46] Other firms have developed ceramic heat recuperators of different designs using mixed-oxide ceramic formulations developed specifically for compact heat-exchanger applications. The low expansion and high resistance to thermal stress of this unique material allow reliable operation at routine temperatures of 871 to 1399°C with a melting point higher than 1705°C.[23]

The main constraints as technology barriers to the expanded use of ceramics in heat-exchanger applications include the need for improved material properties such as thermal-shock resistance and resistance to certain corrosive environments, the development of joining and sealing technologies, and low confidence in the reliability of the ceramic components.[36,47-53] The use of ceramics is still not cost-effective vis-à-vis metal alloys in some heat-exchanger designs, especially for large industrial furnace and power systems, which are constructed from an array of thin-walled ceramic tubes. Processing technologies are still being refined to provide high reproducibility for long (greater than about 244 cm) tubes.

Electrochemical Devices

Though not strictly structural applications of ceramics, devices in this category utilize ceramics for both their electrical and structural properties. Typically, the

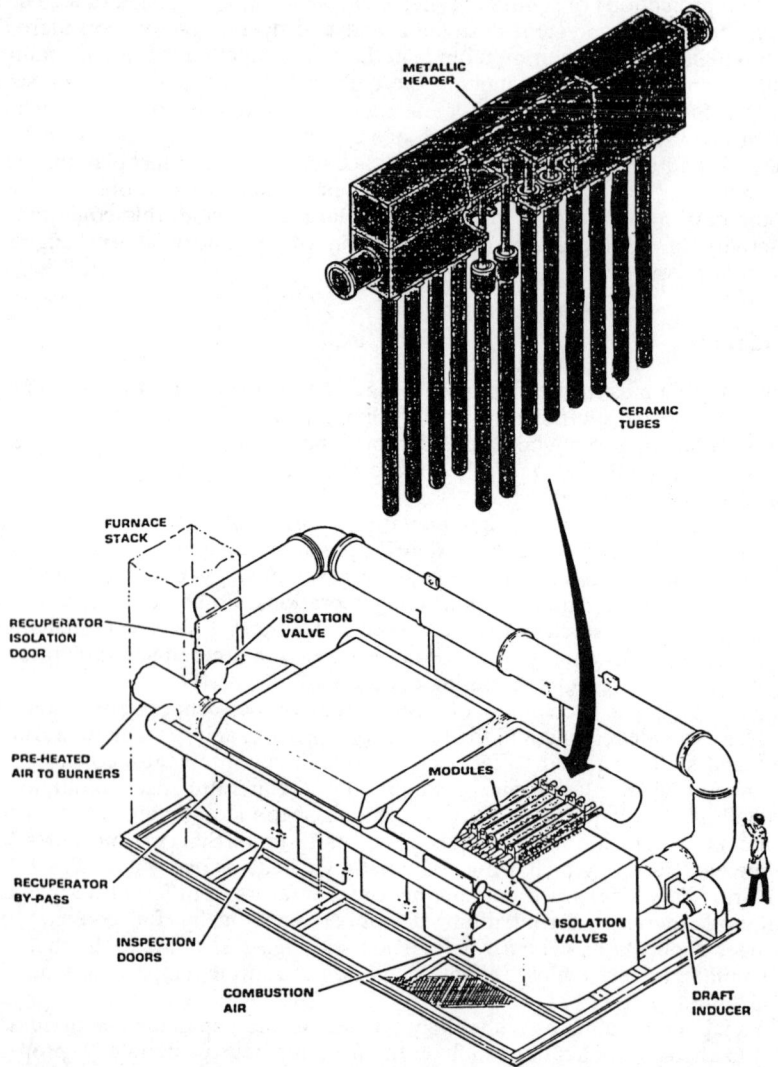

FIGURE 8.9 Ceramic-tube heat recuperator module. (*Courtesy of Solar Turbines, Inc.*)

ceramic, such as ZrO_2 or $\beta\text{-}Al_2O_3$, serves as a solid-phase conductor for ions such as oxygen or sodium. Examples include oxygen sensors,[54] oxygen concentration cells, solid oxide fuel cells, the sodium sulfur battery, sodium heat engine, and electrodes for metal and electrochlorination cells. As a group, these applications could comprise a market of over $250 million for ceramics by the year 2000.[2,23,55]

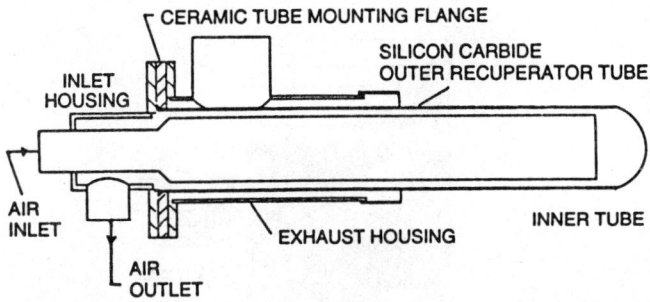

FIGURE 8.10 Schematic of SiC recuperator.[46]

Oxygen Concentration Cells. A new cell configuration promises to raise the efficiency of oxygen production in solid-electrolyte ZrO_2 cells. The new cell structure would ensure a more uniform feed pressure over the electrolyte surface.

ZrO_2-membrane electrolytic cells may eventually replace cryogenic equipment in producing oxygen for such industrial applications as steel manufacturing. The ZrO_2 process has no moving parts, requires less electric energy, and can use waste heat or coal to maintain the operating temperature of 375 to 1200°C.

In Fig. 8.11 impervious electrodes form the top and bottom layers of a sandwich composed of a feedstock distributor, porous ceramic spacers, and a ZrO_2 membrane. The spacers are made of mixed-metal oxides that are electrically conductive so that they can apply the requisite voltage to the membrane. The distributor is made of a similar material.

Channels communicate with the adjacent ceramic spacer through many narrow holes. The porous ceramic spacer contains a multitude of passageways that help to distribute the feedstock more evenly to the ZrO_2 membrane below it.

Oxygen ions migrate through the membrane and, reconstituted as oxygen atoms and molecules, enter the lower ceramic spacer and exit through a port for storage or use.[55,56]

Fuel Cells. A U.S. laboratory[57] has developed a 100-kW cell at half the weight of engines suitable for powering standard-size automobiles, which could provide the same power. A lab model, relying heavily on advanced ceramics, if scaled up could make possible not just practical electric cars but also electrically powered helicopters and even airplanes. The proposed 100-kW unit with the equivalent power of a 132-hp internal combustion engine would be small in size, measuring only 2.33 cm in a more or less cubic shape. Even with cooling systems, the fuel cell would weigh only 135 kg. By relying solely on electrochemical reactions, fuel cells are highly efficient. They can attain a 50 percent efficiency; internal combustion engines deliver only 30 percent efficiency. Several ceramics were applied in the design. In a layered structure resembling a corrugated box, ceramics perform as cathodes and electrolytes. The cathode is a layer of lanthanum manganite doped with strontium. The anode is cobalt or nickel. Yttria-stabilized ZrO_2 performs as an electrolyte and is sandwiched between the anode and cathode materials. Connecting a number of the sandwiched corrugated units together is accomplished with a fourth layer made of lanthanum chromite doped with magnesium. In that configuration the cells are in an electrical series. The unit operates at 1000°C, turning the fuel into gases that oxidize to create the electron flow (Fig. 8.12).

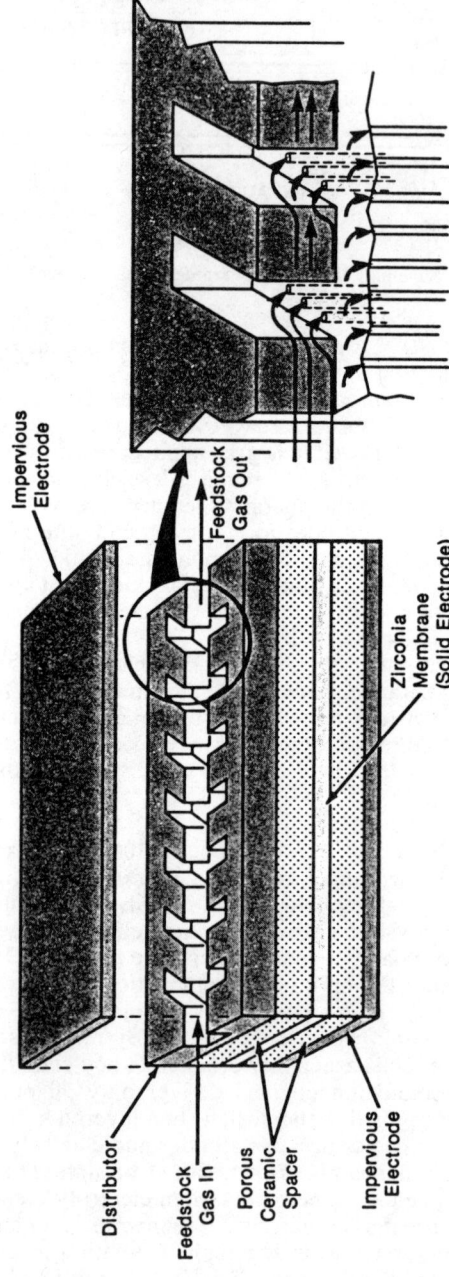

FIGURE 8.11 Multilayer cell structure over ZrO_2 electrolyte membrane.[56]

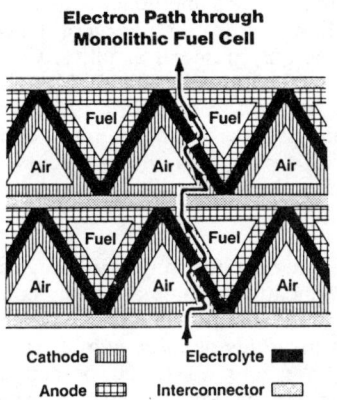

FIGURE 8.12 Four discrete ceramic materials used for fuel cell.

High-Temperature Batteries. The potential role of ceramics in high-temperature, high-performance batteries can be illustrated by examining the possible uses of ceramics in the Na–S battery. While ceramics play a role in other advanced batteries [such as Li–Al–FeS(FeS$_2$)], their most extensive use to date is in the Na–S system. Three components of the battery either require or might use ceramic materials: the separator, the feedthroughs, and the positive-electrode current collectors.

The Na–S cell is an electrochemical device of unique design in that the electrodes are liquid sodium and sulfur (or sodium polysulfide), while the electrolyte or separator is a solid ceramic which is permeable to sodium ions and essentially impermeable to electronic charge carriers.[23]

Two major applications are forecast for the Na–S battery: electric vehicles and electric-utility load-leveling–energy-storage facilities.

Na–S cells have been built and operated in various laboratories in the United States, Great Britain, Europe, and Japan. While technical feasibility has clearly been demonstrated, several problems remain, such as cell lifetime, reproducibility, fabrication by economical techniques using inexpensive raw materials, and scaleup to commercial-size batteries.

In the Na–S system, ceramic materials have been used in three major areas of cell assembly: the electrolyte, seals, and the cathodic current collector or container. The most favored candidates for the electrolyte are β-Al$_2$O$_3$ and β″-Al$_2$O$_3$. The β″-phase, in polycrystalline form, is favored because of its significantly higher conductivity.

Seal materials for the β″-Al$_2$O$_3$ system consist of α-Al$_2$O$_3$ disks or tubes, which are glass-sealed (borosilicate glass) to the electrolyte. Finally, in an effort to overcome corrosion problems encountered by metals in the sulfur electrode, some work has begun on investigating the application of a corrosion-resistant semiconducting oxide ceramic (such as rutile or TiO$_2$ doped with either Cb$_2$O$_5$ or Ta$_2$O$_5$) for use as the current collector in sodium core cells.[23]

Military and Space Applications

Production of ceramics for military applications is projected to expand substantially during the next 25 years.[3,9,19,23,44,55,58,59] Near-term growth is expected for armor, radomes, infrared windows, bearings for missiles, and rocket nozzles (carbon-carbon composites and ceramic-coated carbon-carbon composites). New applications are likely to be laser mirrors, gun-barrel liners, rail gun components, and turbine and diesel engine components. Ceramics and ceramic composites in many cases offer an "enabling" capacity, which will allow applications or performance that could not be achieved otherwise.

Diesels. In military diesels, ceramics provide much the same benefits as in commercial diesels. Of particular interest to the military is the elimination of the cooling system to achieve smaller packaging volume and greater reliability. Considerable progress has been made through the use of ceramic coatings. Monolithic ceramics have also been tried, but have only been successful in a few components and require further development. A military diesel with minimum cooling achieved primarily with ceramic coatings could be produced within 5 years. Engines containing more extensive ceramic components are not likely to appear before 1995 to 2000.[3,9,19,23,55]

Turbines. Turbine engines are in widespread use in the military for aircraft propulsion, auxiliary power units, and other applications. They are being considered for propulsion of tanks, transports, and other military vehicles. Ceramics have the potential to enable advanced turbines to achieve a large increase in performance—as much as 40 percent more power and 30 to 60 percent fuel savings.[3,9,19,23,55] In addition, they offer lower weight, longer range, decreased critical cross section, and decreased detectability.

Design and material technologies are available in the United States to produce high-performance ceramic-based turbine engines for short-life applications such as missiles and drones. Furthermore, it appears that these engines have potential for lower cost than current superalloy-based short-life engines.[23,55] Longer-life engines will require considerable development to demonstrate adequate reliability. This development must address both design and materials in an iterative fashion. While the use of ceramic thermal-barrier coatings in metal turbines is well under way, new turbines designed specifically for ceramics are not likely to be available before the year 2000.

Armor. Ceramic armor was developed after it was observed that a plate of brittle material backed by a resilient one had equivalent stopping power (energy-absorbing capacity) to metal armor. The use of ceramic armor was accelerated by the need for lightweight high-performance armor to protect U.S. helicopter crews during the Vietnam War. It was used as a protection against small-arms armor-piercing projectiles. The ceramic used in this case was B_4C.

Hot-pressed B_4C is the best ceramic armor material produced in commercial quantities, but its high price restricts its use to premium applications such as helicopters (armor for seats, control systems, and crew members) where its low density is performance-effective. Al_2O_3 has the disadvantage of higher density than B_4C, but is relatively inexpensive and is widely used as ceramic armor for land-based personnel and vehicles. Hot-pressed TiB_2 has been developed to provide protection for vehicles against very heavy threats such as armor-piercing antitank ammunition. Table 8.4 summarizes the applications for the various monolithic ceramics.

Ceramic armor is a composite material. The ceramic material on the front face destroys the projectile which disintegrates, its kinetic energy being dissipated as heat and in the generation of fracture surfaces in the ceramic. The backing material is normally glass-reinforced plastic or laminated Kevlar, which traps the fragments of both the projectile and the ceramic. The greater the rigidity of the backing, the higher is the stopping power of the ceramic tile. When the projectile struck B_4C, it was shattered by its high hardness. The energy of this impact was absorbed by localized fracture of the B_4C. The momentum of the debris was then absorbed by the Kevlar in much the way that a baseball glove deforms and

TABLE 8.4 Applications of Monolithic Ceramics in Armor

Material	Advantages	Disadvantages	Applications	
			Current	Potential
Al_2O_3	Least expensive armor ceramic	Slightly inferior ballistic performance compared to other monoliths	Personnel armor; aircraft undersides; aircraft/ helicopter crew and component protection	Hovercraft; armored vehicles
B_4C	Hardest and least dense ceramic	Most expensive armor ceramic	Helicopter crew protection	—
SiC	—	—	Aircraft undersides	Armored vehicles
TiB_2	Densest armor ceramic; space-efficient	—	—	Heavy armor

adsorbs the momentum of a baseball. The fabric spall shield prevents chips and particles from rebounding and causing secondary damage.

A plate of B_4C about 0.64 cm thick with a similar backing of fiberglass can stop a 0.30-caliber armor-piercing projectile. A much heavier layer of steel would be required to defeat the same armor-piercing projectiles.[60]

Ceramic ballistic performance depends strongly on its method of manufacture. For Al_2O_3, hot pressing and HIP are both superior to sintering. Hot pressing is the only currently feasible alternative for both B_4C and TiB_2. For SiC, sintering is superior to hot pressing, which is, in turn, superior to reaction sintering. The manufacturing processes used for each of the ceramics and composites mentioned previously are described in Ref. 61.

B_4C continues to be the major armor material. TiB_2 and Al_2O_3 have proved adequate for some applications and their use has grown. Research is being conducted to locate alternate materials and to reduce raw-material and fabrication cost. SiC has been developed by some companies as a potential material. Fiber-reinforced composites such as carbon or SiC fibers in a matrix of glass or $LiAlSiO_3$ ceramic appear to have considerable potential, especially to yield a very lightweight armor system (Table 8.5).

Besides the helicopter applications with B_4C one-piece armored seat bottoms, heavier armor systems have been developed for military land vehicles using SiC.[62,63] A prototype sandwich composite armored hull has been designed and built as a single shell for a Marine Corps vehicle. The fiberglass hull was postcured in an autoclave. Subsequently Al_2O_3 ceramic tiles were bonded to the fiberglass structure with a conventional liquid adhesive, followed by the bonding of an aluminum cover. The three layers were allowed to cure simultaneously. This composite armored hull has two distinct advantages over conventional metal hulls: (1) it will protect troops better against small arms fire and (2) its lightweight design will enable the vehicle to travel at higher speeds.

A SiC ceramic composite has a new mission, namely, to protect aircraft hydraulics and other critical systems from bullets and flak. The SiC material covers

TABLE 8.5 Applications of Ceramic-Matrix Composites in Armor

			Applications	
Material	Advantages	Disadvantages	Current	Potential
Al_2O_3–SiC	—	Inferior to monolithic Al_2O_3	—	—
TiC–Ni	—	Plate-size limitations	—	—
Compglass*	Multihit capability; potential low cost	Harder matrix needed	—	Structural armor sections
B_4C–Al	Less expensive B_4C replacement	—	—	Armor
Lanxide†	Multihit capability; projected low cost	Plate-size and quality limitations	—	Lightweight armor

*Borosilicate glass reinforced with graphite or SiC (United Technologies Research Center, East Hartford, Conn.).
†Alumina containing residual aluminum and reinforced with fibers or particulates (Lanxide Corporation, Newark, Dela.).

each component with a snug fit. The SiC armor ceramic hosts an advanced composite backing. The system provides ballistic protection from small arms and automatic weapons at less than half the weight of conventional steel armor.

Radomes and Infrared Windows. Radomes are essentially a protective covering and window for electronic guidance and detection equipment on missiles, aircraft, and spacecraft.

Metals are opaque to the electromagnetic waves in the radar, infrared, and optical ranges. "Window" materials for missiles and advanced aircraft should allow transmittance of radar and infrared wavelengths, yet be also able to withstand thermal shock and high-velocity impact with rain drops.

MgO, Al_2O_3, and fused SiO_2 are transparent to ultraviolet wavelengths and a portion of the infrared and radar wavelengths. SiO_2 and Al_2O_3 have accounted for most of the prior production. Si_3N_4 appears to have improved characteristics and has gained a portion of the market. Typically high-purity SiO_2 radomes are slipcast, sintered, and NC ground. These unique fabrication processes produce extremely uniform mechanical and electrical properties, resulting in superior design performance. The Si_3N_4 ceramic radomes are used for millimeter-wave seeker and hypervelocity applications.

New materials that have dual-mode (radar and infrared) capability, with the lowest possible permittivity, for missile sensor windows include MgO–Al_2O_3 Spinel, β'-sialon, toughened ZrO_2, SiO and germanium mullites, and aluminum oxynitride. These materials illustrate the interesting combinations of optical and mechanical properties which are required for advanced applications.[64]

The radomes, also known as electromagmetic windows, shield missile sensors from high temperatures and harsh weather conditions. The material typically used to make radomes, sintered fused SiO_2, offers excellent electromagnetic properties, but like most ceramics, it is brittle and highly susceptible to rain erosion.

In the past, attempts to toughen brittle materials such as fused SiO_2 by adding ceramic fiber usually have been limited to hot-pressing methods, since the material tends to lose density when subjected to slip casting. Now engineers have been able to modify the material microstructure so that a crack developing through the part would run into a fiber and follow that fiber along a tortuous path. This means that more energy is needed to break the part.

Researchers doubled the ceramic's work of fracture—a measure of toughness—by using shorter fibers. Typically, they selected fibers with lengths measuring 100 times the fiber diameter, that is, a 100:1 length-to-width ratio. But the team at the Georgia Institute of Technology prepared chopped fibers with lengths measuring only 20 to 30 times their diameter.

A sample of the fused SiO_2 material, reinforced with 12 vol % aluminum borosilicate fiber, demonstrated twice the toughness of a normal sample, with only a minimal loss of density and improved impact resistance.

Bearings for Missiles. The current candidate ceramic for missile bearings is hot-pressed Si_3N_4. Material requirements are high strength and toughness and as near to zero porosity as possible. The bearings are cut and ground from solid blocks and are therefore quite expensive. HIP is being developed to improve properties and to decrease cost by allowing near-net-shape fabrication.

Ceramic bearings have potential for use in missiles that require a long shelf life. They are inert to the low-temperature corrosion which could incapacitate metals. Ceramics also have other capabilities of benefit to advanced missile bearings: high-temperature stability, low specific gravity, and the ability to operate for moderate lengths of time under lubrication-starvation conditions or with no lubrication.

Laser Mirrors. Space-borne lidar systems, with applications to aeronomy and tropospheric and stratospheric research, require large collecting optics to obtain reasonable signal levels for accurate measurements. Large-aperture mirrors require tight fabrication tolerances, high thermal conductivity, low thermal expansion coefficients, high stiffness, and low areal mass. The technology for fabricating lightweight Si–SiC lidar mirrors by chemical-vapor deposition has been developed for small-diameter (7.5-mm) mirrors. The process has been scaled to much larger (1- to 1.5-m-diameter) mirrors.

Mirrors are produced in a high-temperature furnace by depositing, to a predetermined thickness, SiC onto a master mandrel of graphite having a surface figure that is the negative of the actual mirror figure. A lightweight honeycomb graphite structure is bonded to the substrate, and the combination is reloaded into the CVD furnace in which SiC is deposited onto the structure. The SiC bonds to the SiC substrate and the exposed surfaces of the honeycomb to form a rigid and lightweight backup structure. It is then turned around and remounted in the CVD furnace for chemical-vapor deposition of a thin faceplate of silicon on the substrate. A high-quality, very lightweight mirror is produced.

Beryllium is rare, costly, and toxic, but the metal's strength, stiffness, and light weight have made it a preferred material for high-performance mirrors in space and avionic systems. Advances in fabrication technology are creating a new alternative, SiC ceramic which, according to Bolch and Drake,[65] may be up to 50 percent cheaper for some quantity applications.

Siliconized SiC, sometimes referred to as reaction-sintered or reaction-bonded SiC, is currently the form of SiC most often used for optical applications. Using a process known as Ceraform, siliconized SiC mirrors can be produced with a front wall, honeycomb center, and optional back wall for added stiffness, all in a single monolithic structure. The process freezes a SiC slurry in the mold, removes the mold, then freeze-dries the material before firing.

Solid SiC is denser than beryllium, with mass of about 2.92 g/cm^3 versus 1.85 g/cm^3 for beryllium. The ceramic's advantage is in its stiffness. Metal mirrors used in aircraft or spacecraft are not solid blocks of metal, but "light-weighted" structures with ribs supporting a surface skin on the front and often on the back, too. Because SiC is stiffer, ribs made of it can be thinner than those in a beryllium mirror, but retain the same strength. The smaller overall volume of SiC in the mirror leads to equivalent or lower mass than a similar-sized beryllium component.

Gun-Barrel Liners. Ceramic gun-barrel liners have been investigated as more economic or more wear-resistant alternatives to the existing hard metal liners. The properties of ceramics are well matched to those required for a machine-gun barrel liner, such as high melting point (>1400°C), high modulus of rupture (>350 MPa) and Young's modulus (>210 GPa), high hardness, and low density, combined with low thermal conductivity. Thermal-shock resistance is not very important, but it is difficult for ceramic liners to match the relatively high coefficient of thermal expansion of the metal gun barrel. Other environmental exposures include hot gases, high wear, and high stresses.

Table 8.6 shows various representative materials. Test results indicated that α-SiC successfully survived a 1000-round, single-shot smoothbore demonstration, but the glass composite eroded excessively.[64] The remaining ceramic materials are undergoing testing and results are not available. The initial limited success with α-SiC indicates the potential for ceramic gun barrel liners.

Antenna Windows. The results of trade studies are given in G. Campbell et al.[66] The Ford Aerospace Corporation evaluated approaches for using advanced ceramic composite materials (Si_3N_4, BN, and SiO_2) for large, terminal fix sensor (TFS) radar antenna windows for maneuvering reentry vehicles (MaRV).

Turbine and Diesel Engines. Studies[67] have recently been concluded in which components were identified for advanced Space Shuttle main engines (SSME)

TABLE 8.6 Properties of Potential Gun-Barrel Liner Materials

Material	Melting point, °C	Modulus of rupture at 20°C, MPa	Coefficient of thermal expansion, 10^{-6}/°C	Young's modulus at 20°C, GPa	Hardness, kg/mm^2	Thermal conductivity, W/m · K	Density, g/cm^3
α-SiC	2800	397	4.0	379	2800	0.83	3.2
Sintered Si_3N_4	2800	698	3.5	296	2700	0.29	3.2
PSZ	2700	620	10.5	207	1100	0.017	5.6
Glass-graphite	600	827	4.7	13.8	500	—	2.0
Gun steel	1450	1172	15.0	207	400	7.5	8.0

Source: From Newland.[64]

that could benefit from the use of structural ceramics. Sintered Si_3N_4 was selected as the prime candidate material based on its superior resistance to thermal shock. Thermal-shock conditions were found to be considerably more severe in SSME-type systems than in gas turbines designed for terrestrial uses. The conclusion was that properly designed ceramic components will be viable in high-temperature advanced SSME-type engines (Fig. 8.13).

Turbine blades, stator vanes, and other elements in the hot-gas flow path of rocket engines for space missions beyond this decade are candidates for structural ceramic materials. These must be capable of operating for longer periods of time, withstand more duty cycles, and be more efficient than present engines, and they must meet the severe requirements of advanced rocket engines.

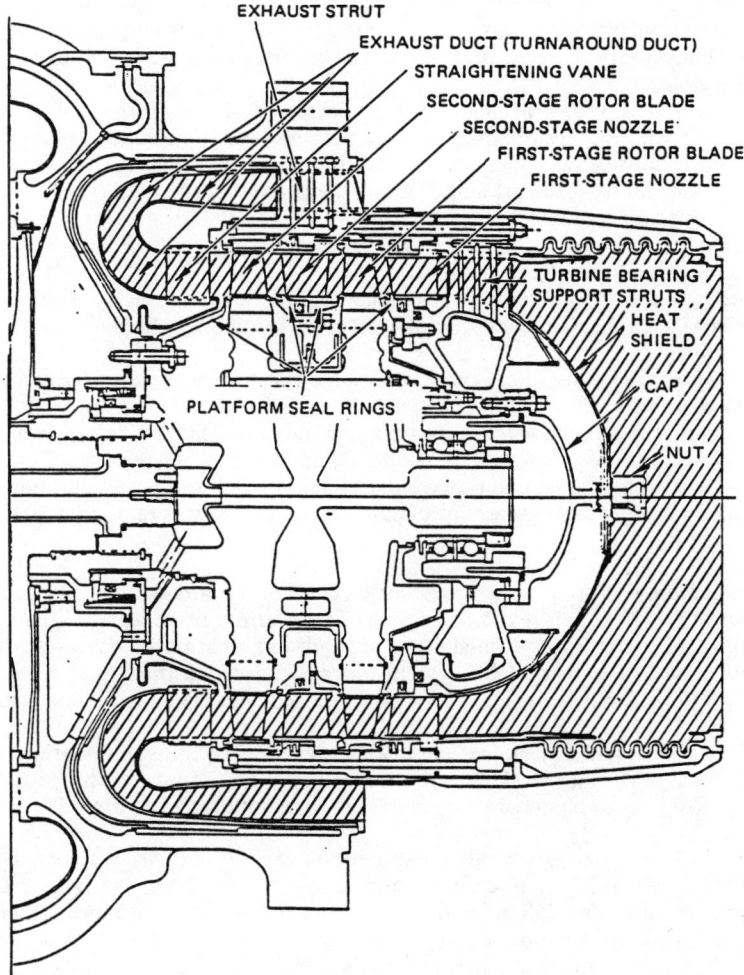

FIGURE 8.13 Potential components for ceramics in hot-gas path of turbine section of Space Shuttle main-engine high-pressure fuel turbopump.[67]

First-Stage Nozzle Vane. A segmented nozzle design is favored for a ceramic stator, and the primary candidate material is sintered Si_3N_4. The nozzle vanes would be made by injection molding plus sintering. HIP could be used as a postsintering process to ensure maximum density and strength for the product.

Rotor Blades. Demonstration SSME Si_3N_4 turbine blades have been injection molded and sintered. Fatigue behavior of the Si_3N_4 must be considered since turbine blades undergo cyclic loads. However, the most critical aspect will be thermal shock and attachment design, where stress concentrations must be kept low.

Heat Shield. The heat-shield candidate materials include sintered Si_3N_4, sintered SiC, and siliconized SiC. The heat shield shape can be formed by a modified injection-molding technique or by slip casting. Fabrication is feasible but represents a high risk due to the thin wall.

Bearing Cap. The cap candidate materials include sintered Si_3N_4, sintered SiC, and siliconized SiC. Fabrication of the general shape is feasible, but redesign would be necessary to avoid highly stressed features such as bolt holes and sharp thread roots.

Turbine-Blade Platform Seals. The turbine-blade platform seal candidate materials include sintered Si_3N_4, sintered SiC, and reaction-bonded Si_3N_4. These rings can be fabricated by injection molding, isostatic pressing, or slip casting. Fabrication is state of the art since similar rings have been produced for automotive gas-turbine programs.

Exhaust-Straightening Vanes. These exhaust vanes could be made of sintered Si_3N_4 and injection-molded.

Exhaust Duct. The ceramic materials include sintered Si_3N_4, sintered SiC, and siliconized SiC. The duct would have to be formed by slip casting. Fabricating ceramic shapes of this size and geometry would present a high technical risk, and therefore this would be a most difficult piece of hardware to fabricate from ceramic materials.

In summary, several of the turbine components can be made from high-performance ceramic materials at moderate to low risk. Experience has shown that ceramic components can replace metallic hardware only after careful and often major redesign. Assessment of candidate hot-gas path ceramic components is summarized in Table 8.7.

Space Vehicles. When the Space Shuttle reenters the atmosphere, surface temperatures of 1472°C can result in friction. The thermal-protection systems (TPS) used for the Space Shuttle must have a good combination of strength, thermal stability, and low density. The system currently used on the Space Shuttle Orbiter, a rigidized fibrous silica insulation, is a reusable TPS with density of 0.14 g/cm^3.

A new generation of rigidized, lightweight fibrous composite materials, called high thermal performance (HTP), provides significant improvements in the thermal stability and mechanical properties, while maintaining required TPS characteristics.[23]

HTP technology uses various combinations of SiO_2 and Al_2O_3 fibers, BN and SiC. These materials are sintered together at high temperature to produce a rigid, fibrous, reusable insulation with good thermal-shock resistance, high-temperature capability, low thermal conductivity, and higher strength than the original rigid surface insulation (RSI) materials. HTP can be tailored to achieve specific mechanical characteristics (tensile or compressive strength), thermophysical

TABLE 8.7 Summary of Ceramic Components in Hot Gas Path of Advanced Space Shuttle Main Engine Type Reentry Vehicle

Component	Structural loads — Mechanical	Structural loads — Thermal	Current attachment method	Fabrication risk	Development priority	Primary concerns	Comments
First-stage nozzles	Moderate	High	Lugs	Low	1	Contact stresses; transient thermal stresses	Primary candidate for ceramics; structural analysis shows that ceramics are viable
Rotor blades	High	High	Root	Low	2	Contact stresses; transient thermal stresses; lack of fatigue data	Centrifugal loads will be lower than for metal blades
Heat shield	Moderate	High	Nut	Moderate	4	Contact stresses; pressure loading	Promising candidate for ceramics but would require detailed structural analysis and some redesign
Cap	High	High	Flange	Moderate	—	High loads, flange attachment	High risk for a ceramic component with moderate reward; a metal cap will be satisfactory because it is cooled by LH_2 and it is not exposed directly to hot gases
Turbine-blade platform seals	Moderate	High	Floating screw	Low to moderate	5	Contact stresses; rubbing on metal parts	Feasible but requires redesign of configurations and attachment considerations
Exhaust straightening vanes	Moderate	Moderate to high	Welded	Low	3	Method of attachment; accommodating thermal expansion	Low risk, provided a suitable attachment method can be designed
Exhaust duct	Moderate	Moderate to high	Welded	Very high	—	Fabrication; attachment	Present configuration would be very difficult to fabricate
Inlet and exhaust struts	High	Moderate to high	Integral with surrounding structure	Low	—	Tenacity of ceramic coatings on metal struts	Existing cooled metal design could be retained by protecting struts with thin ceramic thermal barrier
Inlet and exhaust strut cans	Low	High	Welded	Not feasible	—	Thin, complexly shaped	Cans would be replaced with ceramic thermal barrier

Source: From Ref. 67.

characteristics (thermal expansion or thermal conductivity), and RF characteristics (dielectric constant).[9,23,68–72]

The TPS of the Space Shuttle features ceramic materials in several different forms. RSI consisting of several different fibrous fused-silica-composite materials covers 90 to 95 percent of the exterior surface area of the vehicle. Reinforced carbon-carbon shapes shield the remaining surface area of the vehicle, where maximum strength and thermal (1590°C) protection are needed.

Black tile, known as high-temperature reusable surface insulation (HRSI), protects over 50 percent of the exterior of the vehicle against the 1260 to 1370°C temperatures that must be quickly radiated during reentry. HRSI tile, according to NASA, must have a 100-mission life at these temperatures. On the other hand, white-colored heat-reflecting tile, known as low-temperature reusable surface insulation (LRSI), covers only 10 percent of the vehicle. The latter tile must be capable of a 100-mission life at 650°C. A high-purity, fused-silica blanket that is also white in color covers the remaining surface area of the vehicle where temperatures are the least. Being blanket-type material, it must be strengthened by being sewn with quartz thread and then rigidized with a colloidal silica.[68]

Radiators. The heat generated by future space power systems, such as those being considered for manned Mars missions, may be removed by a new lightweight radiator. The rotating bubble-membrane radiator is made from plastics, metals, and ceramic fabrics. It is more than 80 percent lighter than comparable all-metal radiators (Fig. 8.14).

The bubble is made up of two layers. A thin inner liner of either metal or plastic, depending on the fluid, contains the working fluid. The outer structural layer is made of advanced ceramic fabrics such as fused SiO_2, graphite, aluminum borosilicate, or SiC, and these ceramics are used in cloth form. They have a strength-to-weight ratio factor of 10 greater than metals. So if one can find a ceramic cloth that survives the outer-space environment, one can reduce the weight of any component made from it by 10.

Heat Engines

The advantages of using ceramics in advanced heat engines have been widely publicized. These include increased fuel efficiency due to higher engine operating temperatures, more compact designs, and reduction or elimination of the cooling system.[73] Generally there are two schools of thought regarding the use of ceramics in engines. Some are of the opinion that ceramics should only be used for those components where heat resistance is critical, while others contend that the whole engine should consist of ceramic components. In any case, application of ceramic materials does not always involve a simple one-for-one exchange with metal components, and new engine designs are often required.[74]

Ceramics are being considered in three general categories: discrete components such as turbochargers in metal reciprocating engines; coatings and monolithic hot-section components in advanced diesel designs; and all-ceramic gas-turbine engines.

Currently, "a conventional engine converts up to 40 percent of its heat into energy," says H. Kawamura of Isuzu Motor Company, "a ceramic engine would convert 50 percent." Even more fuel would be saved because the engine would not need a cooling system. "The engine would do away with radiators and fans," he claims. "Fewer moving parts means less energy expended." And with the

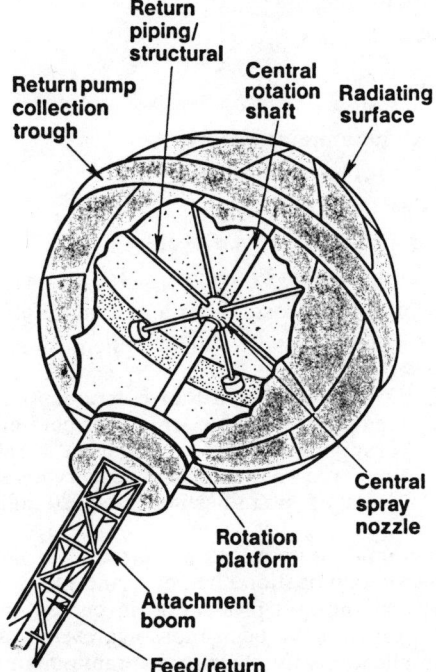

FIGURE 8.14 Rotating bubble-membrane radiator.

world engine market now worth $80 billion, "even 10 percent of that would be huge."

All of the engines under development (diesel, gas turbine, Stirling, Rankine) would benefit from the use of ceramic materials to either replace or coat metallic parts to help achieve their projected goals of fuel economy and reduce emission levels.

The potential benefits of the use of advanced ceramic materials in heat-engine technology include:

Greater fuel economy

Utilization of low-quality fuels

Compatibility with synthetic fuels

Reduction in use of strategic materials

Potential cost reduction

Reduced maintenance

Applicability to both commercial and military requirements

Elimination of cooling systems

Resistance to corrosion and erosion

Improved emission control

Wide range of properties
Would lead to new industries

In addition, other benefits include:

Higher operating temperature capability
Lower weight inertia (in heat engines)
Lower friction losses
Smaller engines for a selected power level
Lower heat rejection to the environment
Significant impact on system costs (both initial and life cycle)
Good wear and corrosion resistance (unique chemical stability)

A number of sources have predicted that components for heat engines will constitute the largest area of growth for structural ceramics over the next 25 years.[3,9,23,48,49,59,73-80] Compared to the world market at $80 billion in 1990, the U.S. Department of Commerce has estimated conservatively a U.S. market of $875 million by 2000,[3,9] while other studies indicate $920 million to $1.3 billion by 2000.[76]

Some structural ceramic components are already in limited production for heat engines. Ceramic precombustion chambers and glow plugs for diesels, and ceramic turbochargers are now in production in current-model Japanese cars. Japanese ceramic companies have been more aggressive than U.S. industry in finding commercial applications for structural ceramic materials in order to gain production experience and develop a market for powders.

To reach the projected $1 billion sales for heat-engine components in the year 2000, the growth to this level would require material and design technology breakthroughs as well as manufacturing scaleup and cost reduction. It has been suggested by some advocates of advanced ceramics that a major benefit attributable to the use of ceramics in heat engines is a reduction in the consumption of critical materials. While ceramics are likely to substitute for "critical" materials such as chromium, cobalt, and nickel to some extent, the resultant reduction in U.S. consumption of these materials is expected to be relatively minor. The use of these critical materials in superalloys for heat engines typically accounts for a small fraction of total U.S. consumption. Thus even if ceramics were to replace these materials entirely in heat-engine applications (which is unlikely), the critical material savings would be relatively small. Moreover, superalloys are used mostly in aerospace engine applications, whereas ceramics are unlikely to penetrate this application area to a significant degree until well into the next century.[76]

Technological Barriers. There are a number of major technological barriers that must be overcome before structural ceramics technology can achieve full-scale commercial application. These barriers affect the cost, reliability, and reproducibility of structural ceramic heat-engine parts. Technological advances may be required at nearly every major processing stage in order to surmount these barriers. For example, improved characterization of ceramic powders would provide important leverage in overcoming the reliability and reproducibility problems, as would integrating powder characterization with studies of the effect of process parameters on final ceramic properties. Better ceramic powders that are free of agglomerates or clusters of particles and that exhibit better particle-size distribution would also facilitate overcoming the reliability and reproducibility problems.

To improve reliability, there are three general approaches:

1. Improvement in design methodology. Through continued design improvements, the brittleness of ceramics must be accommodated and, therefore, the operating stress and other requirements put on the material lowered.

2. Improvement of predictive capabilities. Through much improved methods of nondestructive evaluation, proof testing, and prediction of time-dependent properties, effective means must be provided to eliminate catastrophic failures both at zero time and during a predictable service life.

3. Improvements in materials and processes. Through greatly improved understanding of the interrelationships between processing science and material behavior, materials for ultimate use must be provided which offer greater safety margins in properties compared to design requirements in addition to much fewer variations in properties.

Technological advances in the component-processing stages are also important. At the consolidation stage, the key issue involves developing methods to form complex shapes economically without causing agglomeration in order to improve component reliability. Binder burnout can take up to 2 weeks with some current techniques. Greater control of sintering is also required, particularly for the large commercial-scale furnaces that will be required when the technology moves from the laboratory to the production line. Improved control of sintering would help improve product yields and hence lower costs.

Several key measurement and standards-related barriers are:

1. Improved nondestructive evaluation techniques would be very useful to the industry in developing and controlling production processes. Advances that would permit improved detection of the size and shape of defects, allow complex parts to be evaluated, and accelerate the speed of the test would all serve to make nondestructive testing a more commercially useful tool. Processes must be optimized experimentally, using nondestructive testing or failure analysis to determine where the flaws occur, and modifying the tooling to prevent them. For the most critical applications, the processes cannot at present be controlled adequately to prevent strength-limiting flaws, which can be as small as 50 μm.

2. New techniques for modeling fracture mechanisms would help in developing a better understanding and prediction of failure.

3. Chemical analysis methods to better detect concentration and spatial distribution of elements such as oxygen, nitrogen, carbon, and boron would facilitate optimization of the production process and product performance.

4. New and improved process-control instrumentation such as sensors would assist in identifying, measuring, and controlling key process parameters and thereby help control the performance properties of the ultimate product.

5. Standard reference data for time-dependent properties such as crack propagation, creep, and fatigue would aid in the understanding of the behavior and performance of ceramic engine parts.

6. Standardized testing methods and standard reference materials for measuring properties of the final ceramic part, such as strength and crack propagation, would facilitate both the comparability of research results and the sale of materials and components by reducing uncertainties with regard to material and component properties.[76–81]

Global Perspective. The "ceramic fever" of the early 1980s has abated somewhat and the enthusiasm held by many has been replaced by a more cautious approach. In an earlier survey[82] conducted by Argonne National Laboratories, 1990 was projected as the year for the introduction of rotors, rocker arms, cam followers, and exhaust port liners for light-duty vehicles. They also indicated that valve guides and valve seats would not appear until 1992 and 1993, respectively. Even further down the road (not until 1995) would be valves, piston caps, rings, pins, and cylinder liners. Similar predictions were made for heavy-duty vehicles, except for rotors (1992) and valve guides (1993). The piston for both types of vehicles would not be introduced until the year 2000. Finally, ceramic gas-turbine engine applications would be introduced by 2015 and would reach a $3 to $4 billion market in 20 years.

Gasoline Engines. The automotive internal combustion engine offers a vast market for materials. Current engine designs are considered by automotive companies to be mature, reliable, and cost-effective. Very few incentives for change exist. Cost reduction remains a significant incentive, but this is extremely difficult to satisfy for a new material, such as ceramics, whose introduction may require redesign for adjacent parts, retooling, and modification of the production line.

Ceramics considered for gas engines which have potential for production include exhaust port liners, cam followers, and turbocharger components.

U.S. automotive companies do not appear to have the same level of confidence as the Japanese that a ceramic turbocharger market will develop. This is an institutional barrier based on perceived risk.[3,9,23,76] One study has estimated that if a ceramic rotor price of $15 can be reached, and if performance and reliability are acceptable, a worldwide market of $60 million could be generated for ceramic turbocharger rotors by the year 2000.[83,84]

The ceramic turbocharger has a significance far beyond its contribution to the performance of the engine. It is regarded as a forerunner technology to far more ambitious ceramic engines, such as the advanced gas turbine. Design, fabrication, and testing methods developed for the turbocharger are expected to serve as a pattern for subsequent ceramic engine technology efforts.[84,85,86]

The fundamental change in ignition systems that began when microprocessor control arrived in 1977 is leading to distributorless setups with more precise spark control. Fuel systems, which were likewise revolutionized by electronically controlled fuel injection earlier in this decade, are entering a new era of refinement as airflow becomes a major concern of engine designers.[87]

Although current multivalve engines use four valves per cylinder, the benefits multiply with even more valves. Multivalve technology will maintain its dominance well into the next century. Despite their complexity, turbochargers are well suited to small engines because they are "on-demand" systems. Ceramic engines are expected to overcome the problem of high fuel consumption in currently available gas turbines, permit the use of a variety of fuels, emit a cleaner exhaust, and perform without oscillation and noise. Thus the driving force for increased use of structural ceramics in automobiles is principally the possibility of increasing engine efficiency or performance. A list of ceramic components and materials is given in Table 8.8.

TABLE 8.8 Engine Components and Related Material

System	Accessories or components	Remarks
Compound gasoline engine*	Adiabatic	SN, AlO
	Compound gasoline engine†	ZrO_2 coating
Gasoline engine	Piston†	SN or SC
	Piston pin†	
	Cylinder	
Gasoline-engine assisting components	Swirl chamber‡	Uncharged gasoline engine, SN‡
		Full ceramics, SN‡
		Charged gasoline engine, SN‡
		SN† or SC†
	Turbocharger‡	SN‡
		SN† or SC†
		SN†
	Glow plug‡	SN‡
	Valve†	SN†
	Cam†	SN† or SC†
	Tappet†	SN†
	Rocker arm‡	SN‡
		SN† or SC†

*Gasoline engine: SN–Si_3N_4, SC–SiC, AlO–Al_2O_3.
†Not yet commercialized.
‡Commercialized, installed in passenger car.
Source: After Wachtman et al.[88]

A Japanese and a U.S. firm developed the following ceramic parts for use in gasoline (g) and diesel (d) engines:

Valve spring retainer (g, d)
Glow plug (d)
Fuel-injection nozzle (d)
Swirl chambers, upper and lower (d)
Cylinder liner top (d)
Piston cap (d)
Rocker arm pad (g, d)
Cam (g, d)
Exhaust port liner (g, d)
Valve and valve seat (g, d)
Piston ring (g, d)
Piston pin (g, d)

Although specific items are in use, including swirl chambers, glow plugs, and turbocharger rotors in some models, there is a general concern that many items are at present too expensive for use in automobiles.[88]

The European automobile industry is now using SiC ceramic water-pump seal faces.[89] Two cars are using the SiC seals in the V-8 engines of their 1991 lines; plans call for subsequent application in their diesel engines as well.

The water-pump seal faces are made from α-SiC (Hexoloy* SA). This material has exceptional properties, beneficial for heat-, wear-, and corrosion-resistant applications, and is produced by pressureless sintering of submicrometer-size powder. The powder is mixed with nonoxide sintering aids. In general, complex shapes can be formed from this material by dry or isostatic pressing, injection molding, extruding, or slip casting. The water-pump seal faces are formed in automated dry presses. The sintering process results in a single-phase fine-grain SiC product with virtually no porosity and a density greater than 99 percent theoretical.

With the current trends in spark ignition engines toward multivalve engines capable of operating at relatively high speeds to maximize performance and fuel economy at acceptable emissions levels, engineers are examining structural ceramics.[90]

Spark ignition engines use poppet valves to control both the flow of air into each cylinder and the removal of exhaust gases after each combustion cycle. The materials used for each valve are different, principally reflecting the higher temperature properties required for the exhaust valve.

The replacement of a metal valve by a lighter-weight valve reduces the mass of the valve train by the difference in weight of the two valves. A typical automotive stainless-steel exhaust valve, such as 21-2N stainless steel, has a weight of 100 g, whereas the corresponding Si_3N_4 or SiC valve is approximately 60 percent lower in weight.

The trend in today's automotive industry is toward improving performance and durability in both spark ignition and diesel engines (Table 8.9). For example, in severe-duty engines, especially diesel engines, nickel- or cobalt-based alloys have been used in valve seat inserts to extend the engine operating temperatures and service life above that of ferrous materials. Recently ceramic monoliths and composites have shown considerable promise in reducing valve-train wear in internal combustion engines, thus extending the service operating range.

Ceramic-matrix-composite exhaust valve seat inserts have been produced to near-net shape using two Lanxide Al_2O_3 matrix composite systems. One contained Al_2O_3 filler particles, while the other contained SiC particles. The composite systems were found to have hot hardness values significantly greater than severe-duty metallic valve seat inserts. The composite valve seat inserts were tested successfully for 100 h of full-load operation in a single-cylinder diesel engine.[91]

Turbochargers. The turbocharger is a compressor which boosts the intake air pressure to the engine cylinders and allows the engine to produce a significantly higher power output than a normally aspirated engine.

At the end of a shaft is an impeller which boosts the intake pressure, at the other end is the rotor spun by the engine exhaust flow. The main concern in the turbocharger using conventional metal alloys is turbolag. During low-end acceleration, when extra engine power is most urgently needed, these turbochargers fail to deliver boost promptly because of delays in the exhaust pressure overcoming the metal rotor's high mass media. Figure 8.15 schematically shows a turbocharger fitted with a ceramic rotor and ceramic bearings.

---

*Hexoloy is a registered trademark of Carborundum Corporation.

TABLE 8.9 Key Milestones in Development of Ceramic Heat-Engine Technology

Milestone	Estimated date of commercial introduction	Estimated percent of fuel savings*
Uncooled diesel	Early 1990s	10–15
Adiabatic (minimum-heat-loss) diesel	Early to middle 1990s	20–25
Gas turbine	Middle to late 1990s	25–30 (automobiles) 15–20 (trucks)
Minimum-friction engine	21st century	30–35

*Percentage reduction in fuel requirement per kilometer.
Source: After Ref. 76.

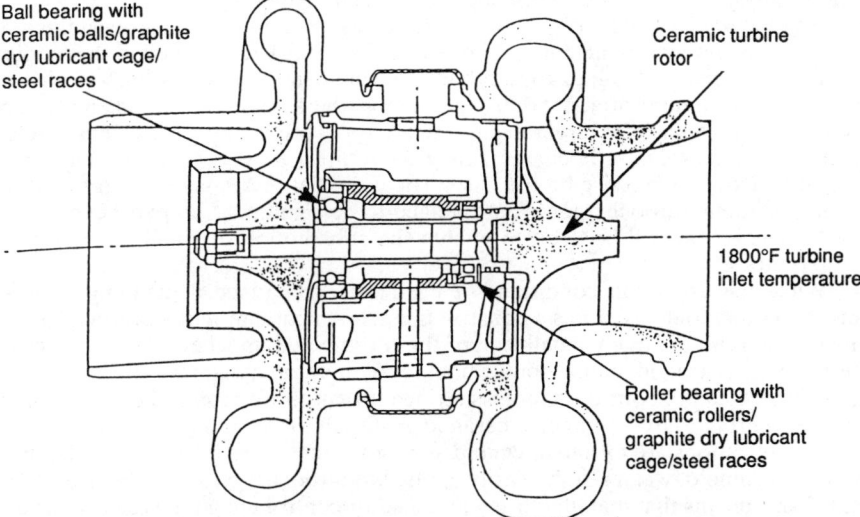

FIGURE 8.15 Schematic of modified T46 turbocharger with Si_3N_4 turbine rotor and bearings.[23,55]

Ceramics offer significant advantages over metal counterparts when the ceramic properties are used properly.[92–99] Thus far the turbine rotors appear to be a metal copy. Again, the very properties that make ceramics valuable in turbocharger applications are:

Low coefficient of expansion
High strength and temperature capability
Light weight for response characteristics
Lack of strategic materials
Burst containment

Reducing the rotor mass would improve turbocharger performance. Si_3N_4 has been considered by many companies as an excellent substitute for the metal alloys used at present for the weight-paring job. An experimental Si_3N_4 rotor weighs 60 percent less than its typical metal counterparts.

Several companies have been developing turborotors. However, companies are undecided whether the Si_3N_4 or the SiC version is the material for production. For example, a turborotor made of Si_3N_4 was burst-tested to 197,000 rev/min, 31 percent over design speed.[23]

In addition to rotor weight reduction by over 50 percent, which results in more rapid acceleration of the rotor to operation speed during vehicle acceleration, a second benefit is associated with the amount of burst protection required in case of a rotor failure. Metal rotors burst into large pieces that require a heavy metal housing to contain them. Ceramics fracture into very small fragments that can be contained within a thinner metal housing or a ceramic housing, either of which would be lighter and less expensive than current metal housings.

Considerable progress has occurred in ceramic turbocharger rotor development during the last 5 years, especially the ceramic fabrication technology to produce high-quality rotors by both slip casting and injection molding.[55,88,94,97–100]

Why Turbocharger Emergence? Turbochargers are used to increase the performance and power of an engine. For example, it has been estimated that adding a turbocharger to a 1.8-liter four-cylinder engine can increase its horsepower by 79 percent to a total of about 150. This allows the substitution of smaller, more fuel-efficient engines for larger engines without reducing power. The commercialization of ceramics in turbochargers may allow turbochargers to diffuse more rapidly than would otherwise be the case. The potential fuel savings from the diffusion of ceramic turbochargers is not judged to be significant, in part because not all turbochargers will be used to enable the substitution of smaller engines for larger ones.[76]

While the operation conditions of a turbocharger place demanding requirements on materials used in its construction, the maximum temperature of about 1000°C and the relatively small size of the turbine and impeller wheels make the turbocharger a good initial application for structural ceramics. Because these characteristics are considerably less stringent than those required of an automotive gas-turbine (AGT) engine, ceramic materials undergoing development for AGT applications are excellent candidates for turbocharger rotors. The AGT processing technology can also be easily applied to turbocharger rotors because their small size means that manufacturing and quality-control techniques are not as difficult as those for AGT components.[101]

Major changes in the automobile industry in the last decade have increased the use of technological innovations to improve vehicle efficiency. Smaller and more space- and energy-efficient vehicles have created a new emphasis on smaller, high-output engines. The turbocharger is a well-understood technology used to recover waste heat from engines; its use improves power output and engine efficiency. In addition, the need for increased economy in manufacturing has led to fewer engine "families" and more commonality of vehicle platforms. Thus, economic and environmental pressures have created conditions that are right for the increased use of turbochargers.

An important parallel development has been the widespread application of advanced electronics and computers for controlling ignition and fuel distribution in modern engines. These sophisticated, programmable engine-control systems are compatible with turbocharger systems and are ideally suited for managing the complex map of engine operating conditions to obtain optimum fuel efficiency and performance. In fact, such a sophisticated approach to engine control is necessary for the successful, widespread application of turbochargers, given today's constraints on fuel economy and emissions.

For in-house turbocharger production to be economically viable in any company, the production volume should be at least 250,000 units per year. When a production process for ceramic rotors can be developed that is reproducible, highly controllable, and capable of high yields, costs can be expected to be competitive with those of conventional metal rotors. The cost to produce a metal alloy rotor casting for a small automotive turbocharger today is approximately $8. A complete rotor assembly (with shaft) after balancing is estimated to cost about $15. Although ceramic rotor assemblies will initially cost several times that, the price is likely to fall after the first few years to approximately $25 per rotor assembly and to approach that of a conventional metal assembly within 5 years after the start of large-scale production (injection molding and slip casting).

For an automobile manufacturer with enough demand for turbochargers and sufficient resources, producing turbochargers in-house makes good business sense. So far, no single automobile company (United States or Japan) can be said to have a large lead in the race to develop ceramic turbochargers. Each major company has developed about the same level of technology and knowledge. Thus overall management philosophy, marketing strategy, and willingness to accept risk become the dominant factors in determining how fast the production of ceramic turbochargers will be initiated.

The use of ceramics in turbochargers is likely to be evolutionary, beginning with one or two components in small units and eventually making up the entire assembly, that will be produced in various sizes and for numerous applications. As the number and type of ceramic applications in turbochargers grow, the demand for turbocharged vehicles is projected to be more than 6,000,000 worldwide by 1992 and in excess of 1,500,000 in the United States.[101] This accounts for about 15 percent of gasoline engines and 20 percent of diesel engines. Part of the large increase in turbocharged gasoline engines is attributable to the development of several new engines that were designed from the beginning with turbocharging in mind. Ceramic turbochargers could capture more than 75 percent of this market, depending on price, performance, and reliability.[101]

Finally a major technical barrier that must be overcome is the joint attachment between the ceramic wheel and the steel shaft. The joint represents a significant portion of the assembly cost and can have reliability problems. Another problem stressed is foreign object damage. Manufacturers are looking at the effect of the particle size on turbine wheel damage and are evaluating centrifugal particle traps for turbine housings.

According to Sheppard,[82] a Japanese manufacturer has successfully addressed the joint-attachment problem in the development of its second-generation turbocharger rotor. The ceramic rotor is joined to the metal shaft by either active brazing or shrink fitting, which produces either a chemical or a mechanical bond. With the brazing method, a titanium alloy is used in combination with a laminated multibuffer layer of nickel and tungsten alloy. This layer, developed through finite-element analysis, reduces the residual and thermal stresses between the ceramic rotor and the metal shaft.

Microstructural analysis reveals that the titanium diffuses into the Si_3N_4 boundary, forming a titanium-rich phase. Si_3N_4 decomposes into silicon and nitrogen forming Ti–N and Ti–Si compounds in the bonded boundary, with a total thickness less than 1 μm. The chemically bonded parts have survived a variety of tests, including bending, torsion, low temperature, vibration, and engine tests, the tests totaling 10,000 h.

With shrink fitting, the bond must possess enough strength to withstand the elongation of the metal shaft during operation. This result is accomplished by clamping the ceramic rotor shaft in two places, each having a different diameter. Clamping the portion with the larger diameter gives the rotor sufficient bending strength, while clamping the portion with the smaller diameter gives the rotor sufficient torsional strength. Both bonding methods have sufficient torsional strength for acceptable reliability and durability.[82,96,102]

Sheppard[103] also reported that what is claimed to be the world's first production car equipped with twin ceramic turbochargers is the 1990 Nissan Skyline, which is being produced at rates in excess of 2000 turbochargers per month. Toyota Celica and MR-2 models have also had turbochargers introduced into their sport-type engines.

A flexural strength of over 700 MPa, a temperature of 1900°C, and a Weibull modulus of 20 were achieved by carefully controlling the sintering conditions. The optimum combination of sintering aids was determined to be 3 wt % each of Y_2O_3 and $MgAl_2O_4$. Injection molding was found to be the best fabrication method for mass production as long as suitable binder content (minimum wax and resin) and molding parameters (nozzle size, injection rate, and temperature) were chosen. Reliable joining of the ceramic part to the metal shaft was achieved by using double buffer layers of ductile metal formed by "full-filling" the ductile metal into the metal sleeve.[103]

Gas-Turbine Engines. In addition to work aimed at developing parts for conventional gasoline engines, much research is under way around the world to design entirely new engine systems that take greater advantage of the properties of ceramics. Promising candidates include the uncooled diesel, the "adiabatic" or minimum-heat-loss engine, and the automotive gas-turbine engine. Turbine engines have heretofore been uncompetitive with diesel or gasoline engines in automotive uses because of the high cost of the superalloys required to withstand very high turbine inlet temperatures, and the insufficient material strength and oxidation-limited engine operating temperatures. The greater temperature resistance and lower cost of ceramic materials have transformed the gas-turbine engine into a potentially viable engine for automobiles.

The major incentive for the use of ceramics in turbines is the possibility of operating the engine at turbine inlet temperatures up to about 1371°C, compared with superalloy designs, which are limited to about 1038°C with cooling. This temperature difference translates into an increased thermal efficiency of around 40 to nearly 50 percent.[3,9,55,103–106] Power increases of 40 percent and fuel savings of around 10 percent have been demonstrated in research engines containing ceramic components. Other potential advantages include reduced engine size and weight, reduced exhaust emissions, and the capability to burn alternative fuels, such as powdered coal.

Structural ceramics are an "enabling technology" for the automotive gas turbine, that is, without the use of ceramics, an automotive turbine cannot be designed and manufactured that can compete in cost or performance with current gasoline and diesel engines. Extensive design, materials, and engine efforts have been made over the past 15 years, and these efforts have resulted in significant progress in design methods for brittle materials, the properties of Si_3N_4 and SiC materials, fabrication technology for larger and more complex ceramic components, nondestructive evaluation and proof testing, and engine assembly and testing.

New ceramic materials may finally make small low-cost gas turbines a reality for applications such as automobiles, trucks, and buses within 10 years. High-temperature ceramic heat exchangers and hot wall combustors may allow the use of "dirty" fuels such as powdered coal and heavy oils. Rotating ceramic components may be lower in cost and capable of higher-temperature operation than metallic equivalents. Ceramic combustion linings could be used currently in piston engines to reduce jacket losses and increase thermal efficiency.

Government-sponsored development programs were initiated in the 1970s with several engine and automotive companies and continued through the 1980s. In addition to testing ceramic components for automotive applications, new fabrication techniques of these components increased their flexural strength. The use of HIP found its way into the program. New ceramic materials such as Y_2O_3 with Al_2O_3 or alone and BeO or $BeSiN_2$ with Si_3N_4 were also developed during this period.

AGT 100. The literature[3,9,23,55,77,94,97,102–111] reflects the 6-year project and the results of designing a 300-hp automotive gas turbine with an all-ceramic hot section. Two engines were built, and both failed suddenly after running for a few hours on a test stand, owing to extensively fractured rotors and stators. Foreign-object impact was deemed the most probable cause of the fracturing in one engine, and either that or fracture from strength degradation appeared to cause failure of the other engine. The rotor in the AGT 100 engine was made from sintered SiC (Table 8.10).

AGT 101. The literature[3,9,23,55,59,77,79,82,95,96,100,103–105,112,113] shows the results of this 6-year ceramic-engine project. In the project, a 100-hp automotive gas-turbine engine featuring an all-ceramic hot section was designed for operation with a turbine-inlet temperature of 1371°C and a rotor speed of 100,000 rev/min.

TABLE 8.10 Ceramics Used in AGT 100 Program

Component	Material system	Trade name
Regenerator disk	Aluminum silicate (AS)	
Regenerator bulkhead	Lithium aluminum silicate (LAS)	
Combustor assembly	Sintered alpha silicon carbide (SA SiC)	Hexoloy SA
Vane	Sintered silicon nitride (SSN)	AY6, PY6
Scroll	Siliconized silicon carbide (Si SiC)	NC-430
Rotor	SSN	SN22M, SN250
Outer backplate	SA SiC	Hexoloy SA
Inner backplate	Glass-ceramic matrix/SiC fiber	BMAS-III
Rotor	SA SiC	Hexoloy SA
Combustor	SiC/SiC	Cerasep
Thermal barrier/shim	Plasma-sprayed zirconia (PSZ)	ZN-40
Inner backplate	SSN	SNW-1000, AY6
Vane	SA SiC	Hexoloy SA
Coupling and piston rings	Reaction-bond silicon carbide (RB SiC)	Refel, PS-9242
Combustor body	SiC	C600
Coupling	RB SiC	SC2
Backplates	SSN	SN250

Source: From Hamano et al.[110]

Following extensive development and testing of ceramic components, an engine was built and run on a test stand. After operating successfully for a total of 85 h with turbine-inlet temperatures up to 1204°C, rotor speeds up to 70,000 rev/min, and accumulating five starts, the engine failed suddenly from extensive fracturing of the stator and rotor. The failure was attributed to foreign-object impact, believed to be carbon deposits that came from the combustor. Subsequently a rotor of the same ceramic fractured after 12 h of engine operation at 1204°C. This fracture appeared to be the result of service-incurred strength degradation. These results are remarkably similar to those obtained by the AGT 100 engine program, except that the AGT 101 was made from sintered Si_3N_4.

ATTAP. As a result of the two previously sponsored and completed programs, the Department of Energy evolved a 5-year program to continue the ceramic technology effort initiated in the AGT and Ceramic Technology Project programs, such that the United States will not lag behind world competition. The Advanced Turbine Technology Applications Project (ATTAP) program is to continue to advance the technological readiness of the automotive ceramic gas-turbine engine through the development of processes to fabricate and test components from the improved (next-generation) ceramic materials, expand the experimental database of ceramic properties, expand analytical tools needed to support the ceramic industry, evaluate the reliability and durability of the next generation of high-temperature ceramics, and assess low-cost ceramic processing and manufacturing technology.[23,55,79,93,95,98,113] The ATTAP engine program, also known as the AGT-5 ceramic automotive engine, is expected to have an operable life of 3500 h at temperatures to 1371°C.

The AGT-5 is a two-shaft all-ceramic regenerative unit that uses axial turbine and gasifier components. The engine would provide better acceleration and at least 30 percent greater fuel efficiency than current General Motors four-cylinder engines.

A high-performance Si_3N_4 material has been used for AGT 101 and AGT-5, and improved rotors have been made with a pressure-assisted slip-casting process that has cut casting time for the AGT 101 rotor by 80 percent. Before, many primary components were made of SiC; however, now Si_3N_4 has become the basic ceramic for those components. Unfortunately there is still a considerable falloff in high-temperature strength with Si_3N_4, but with new types of Si_3N_4 better strengths are obtained than with SiC.[95,113,114] Figure 8.16 shows this improvement in Si_3N_4 material.

Materials and processes have been selected and include the following components:

1. *Turbine rotor:* This component is considered the engine's critical component because of its high-temperature requirements, geometric complexity, and susceptibility to impact damage. Initially experiments have been conducted with pressure slip casting of fracture-resistant Si_3N_4 materials for radial turbine rotors. In addition, rotor development activity has two alternative objectives: to improve injection-molding techniques for axial turbine rotors and to toughen Si_3N_4 material systems by applying SiC-whisker reinforcement. Tests have shown that this composite material can be successfully densified by HIP with a resultant increase in strength. Another material considered for axial rotors and formed by injection molding is a sintered α-SiC material that is widely used for seal rings in car water pumps.

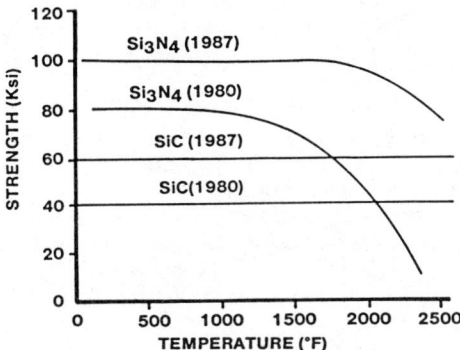

FIGURE 8.16 Improvements in ceramic component material strengths and capabilities.[110]

2. *Turbine stator:* Another complex shape, the stator has 19 segments, each with contours that must be precisely controlled for aerodynamic performance. The material selected was Si_3N_4 processed by HIP. It was found that dimensional control could be attained through a thorough understanding of shrinkage rates, residual stress, forming stresses, and control of densification schedules and hot-zone temperature variation.

3. *Ceramic wave spring:* A process being developed uses extruded thin sheets of α-SiC which are then stamped into shape. Maintaining the spring's sinusoidal waveform during densification will require creative furnace supports and precisely controlled hot-zone temperatures.

4. *Transition duct:* Injection molding of Si_3N_4 has been used to produce the transition duct, which derives its name from the transition of hot gas flow from the combustor discharge to the larger-diameter turbine stator inlet.

5. *Turbine backshroud:* This highly stressed part will require a new type of reinforced ceramic to work reliably at the elevated temperatures for ATTAP testing. One process to be tried involves ram pressing of SiC toughened by dispersed particles of TiB_2. Since it is 50 percent stronger than direct sintered SiC, the material offers the bonus of being electrodischarge machinable. This stock-removal technique could reduce the cost of a final component significantly when coupled with net-shape-forming development.

6. *Ceramic scroll assembly:* Efforts here are directed toward slip casting of the previously mentioned sintered α-SiC.

7. *Regenerator disk:* Research on this material is examining ways to extrude alumino-silicate regenerator disks for the AGT-5. The extrusion process should vastly improve uniformity, strength, and cost relative to the spiral wrapping methods developed for metal turbines.

Nondestructive evaluation methods such as X-ray computed tomography, optical holography, and nuclear magnetic resonance imaging will play a central role in both quality assurance and process refinement during ATTAP.

Nondestructive evaluation techniques will be used not only to detect flaws in the finished parts during the ATTAP program, but rather to improve each step in the fabrication process. For example, tomography can be used to look at an injection-molded rotor in the green state and find out where the density variations are, then go back and change the molding parameters or make other adjustments to improve consistency.

Voids, iron wires, and other seeded defects can be placed in select materials to determine resolution and detection limits for current nondestructive testing techniques, as well as to examine the potential of new testing technologies. Nondestructive testing is discussed in Sec. 8.5.

Other Turbine Applications. Isolated ceramic components are likely to occur in other gas-turbine engines well before they appear in an automotive gas turbine. The most likely components for early implementation are the combustor liner, transition liners, stator vanes, and support structures. The most likely applications for early implementation are small generator units ranging from 10 to about 500 kW. Application to unmanned military aircraft (missiles, drones) could also occur in the early 1990s with a focused development effort.

By 1995 isolated ceramic components could be applied to a truck-size gas turbine for off-road vehicles and military transport vehicles. By 2000 prototype ceramics could be demonstrated in a tank engine and in auxiliary power units for aircraft. Use of ceramics in aircraft propulsion turbine engines is not expected until well after 2000.

Many of the barriers and concerns stated for an automotive gas turbine apply also to other gas turbines. Reliability improvement and cost reduction will be required before widespread use is possible.

1. *Combustors:* In principle, gas turbines can burn any fuel that will sustain a flame. In current practice, however, the overall economics of turbine cycle efficiency and component life dictate that only pure hydrocarbon fuels, such as natural gas distillates, be used. The ability of combustors to burn "dirty" fuels will probably always exceed the tolerance of gas turbines to accept them. Thus a strategy of advanced combustor development aimed at low emissions and multifuel capability must be guided by the tolerance of hot parts (vanes and blades) toward damaging contaminants. Since the availability of coal, gas, and synthetic fuels in large quantities is at least a decade away, natural gas and petroleum distillates will continue to be the dominant gas turbine fuels for the next 10 to 15 years. Even with these "pure" fuels, there are serious difficulties in meeting EPA emissions standards with current combustor design. The pollutants of concern include NO_x, CO, HC_n, smoke, and particulates.

Dirty fuels complicate combustion designs further because they introduce particulates as well as nitrogen compounds. Coal, gas, synthetic fuels, and crudes usually contain inorganic substances such as salts, sulfides, and oxides, which can either fuse to form a slag or vaporize and subsequently condense in the turbine as the combustion gases cool. By various routes, the particulates contribute significantly to the corrosion and erosion of the gas-turbine hot parts as well as to the undesirable high levels of pollutants.

Combustors currently used in large stationary gas turbines are an outgrowth of aviation gas-turbine practice which emphasizes compactness and minimum weight.[23]

The maximum turbine-inlet temperature, which is set by the temperature limits of the vanes, is the centerline temperature rather than the lower average combustor outlet temperature. Thus the flatter the temperature profile or pattern factor, the higher the thermodynamic efficiency of the turbine.

The potential of hot-wall combustors for achieving lower emissions, lower temperature profiles, and higher heat rates has been recognized for a long time, but the formidable materials problems have been a severe deterrent. Ideally one would like to have a hot-wall structure that could sustain inner-wall temperatures of between 1400 and 1670°C, be thermal-shock-resistant, have good corrosion resistance at gas velocities of 100 to 230 m/s, have outer-wall temperatures of less than 560°C, and be able to withstand mechanical shock and vibration. Metals are ruled out because their melting points are too low and they have limited resistance to oxidation. Structural ceramics have the required refractoriness and oxidation resistance, but they are sensitive to thermal shock. However, the advances being made in optimal designs for brittle materials and in the fabrication of structural ceramics indicate that a satisfactory hot-wall combustor can be developed eventually. Hot-wall combustors not only offer the potential for very low emissions and a means of achieving very high turbine-inlet temperatures with a variety of fuel types, but also can increase the heat rate of the gas turbine by decreasing the pressure drop across the combustor.

Recently combustion tests of a gas-turbine ceramic combustor were completed in Japan.[115] It was found that a pseudocoal gas provided a highly satisfactory combustion efficiency over a broad range (100 percent combustion efficiency at a load condition of over 50 percent). There was no damage to the combustor's ceramic, which is much more resistant to heat than metals, but is highly brittle (Fig. 8.17).

2. *Auxiliary power unit (APU):* The next generation of small gas turbines used as compact APUs in aircraft and mobile ground power applications will achieve higher power density than current installations by operating at substantially higher turbine-inlet temperatures. Future aircraft and armored vehicles will

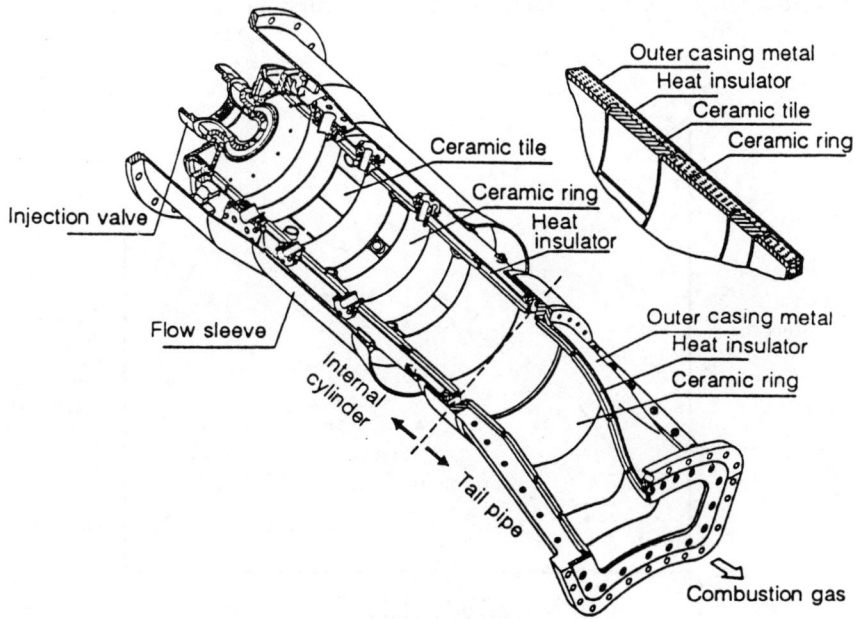

FIGURE 8.17 Structural view of gas-turbine combustor.[115]

require more secondary power with little or no allowance for a corresponding increase in size and weight of the APU.

An advanced small gas-turbine APU is currently under development[116] and will be used to investigate the feasibility of increasing the engine shaft power output by operating at higher turbine-inlet temperature and shaft speed. Two alternate design philosophies have been selected in this program, high-performance structural ceramic materials on the one hand and an air-cooled turbine nozzle in combination with a multialloy turbine wheel on the other.

During the past 35 years, the intensive materials research—primarily driven by the needs of the gas-turbine industry—has resulted in the development of numerous superalloys and a steady growth in their high-temperature capabilities. These improvements are reflected by the increase of turbine-inlet temperatures in uncooled small gas turbines over the years, as shown in Fig. 8.18. Most recently, the use of advanced superalloys resulted in peak operating temperatures at 1000 to 1030°C. Although further improvements are possible, it appears that the future growth in turbine-inlet temperature for uncooled-metal-alloy turbines will level off around 1090°C. Further increases will come slowly and may result only from the application of cooling techniques.

The alternate paths of high-temperature structural ceramics and high-temperature metal alloys with cooling or multialloy construction for increased turbine-inlet temperature are diverse with respect to design, design methods, manufacturing methods, capabilities, and limitations.

The ceramic path represents a high-risk alternative with much greater potential benefits to engine performance and the potential for cost saving. However, the development of the design, manufacturing, and inspection methods will require substantial investment. Successful completion of ceramic component demonstrations may result in the use of ceramics in nonmission critical and unmanned small gas-turbine applications during the first half of the 1990s.[116]

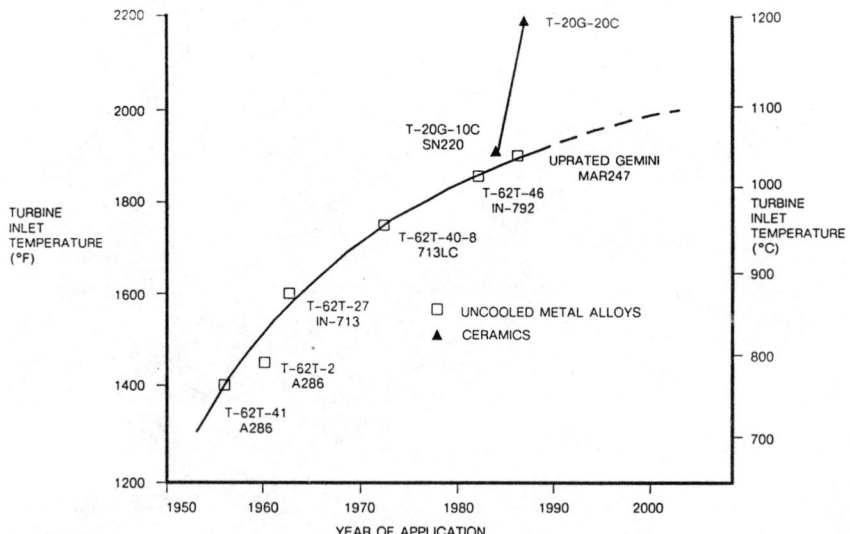

FIGURE 8.18 History of turbine-inlet temperature in gas-turbine engines.[116]

Foreign competition in gas-turbine engines and engine components has primarily come from Japan. Research and development related to the manufacture of ceramic gas turbines for use in automobiles is a national project under MITI.[117] This project involves automobile manufacturers, petroleum companies, ceramic companies, and auto parts manufacturers. The aim of R&D is to produce a 100-kW ceramic gas turbine with certain target performances, one of which is a thermal efficiency of 40 to 42 percent requiring a turbine-inlet temperature of 1300 to 1400°C. Consequently one of the research thrusts is to develop a variety of heat-resistant ceramic materials suitable for use in ceramic gas-turbine high-temperature components. In parallel with the effort to develop this automotive ceramic gas turbine is a program to develop a high-efficiency 300-kW gas turbine for use in electricity generation. This gas turbine has a proposed turbine-inlet temperature of 1350°C. One research goal is to develop heat-resistant ceramic components with minimum strengths of 400 MPa at 1500°C.

Automobile manufacturers in Japan have been performing research on heat-resistant ceramics for over 10 years. Some of the results from this research have been applied in their reciprocating engines to Si_3N_4 components, such as glow plugs and rocker arms. Furthermore, Si_3N_4 ceramic turbochargers have been installed in numerous vehicles.

According to Lenoe,[118] Nissan Motor Company Ltd. has been working since 1963 on designing, developing, establishing manufacturing techniques, and implementing stress analysis methods for gas-turbine ceramic components. Components built and tested include Si_3N_4 seal rings, combustion chambers, turbine shrouds, axial turbine rotors, radial turbine nozzles, and radial turbine rotors. Using these capabilities, Nissan developed a Si_3N_4 ceramic turbocharger, which was released in a passenger car in 1985. As of November 1987 more than 60,000 ceramic turbochargers were in service in three Nissan models.[118] Other Japanese engine groups are pursuing similar engine components.[4,78,88,96,102,111,118]

Summary. The automotive gas turbine would be a more revolutionary application of ceramics than the diesel. The diesel engine is a familiar technology and incorporation of ceramics can occur in stages, consistent with an evolutionary design. The gas turbine, on the other hand, represents a completely new design, requiring completely new tooling and equipment for manufacture. Because of the remaining technical barriers to ceramic gas turbines and the fact that they represent a complete departure from current designs, it is unlikely that a ceramic gas-turbine passenger car could be produced commercially before the year 2010.[1,3,9,19,20,76,77,82] In view of this long development time, it appears possible that this propulsion system could be overtaken by other technologies, including the ceramic diesel. One factor which would favor the turbine engine would be dramatically increased fuel costs. In the case that traditional fuels became scarce or expensive, the turbine's capability to burn alternative fuels could make it the power plant of choice in the future.

The outlook for ceramic heat engines for automobiles appears to be highly uncertain. The performance advantages of ceramic engines are more apparent in the larger, more heavily loaded engines in trucks or tanks than they are in automobiles. Ceramic gas turbines and adiabatic diesel designs do not scale down in size as efficiently as reciprocating gasoline engines.[3,9] Thus, if the trend toward smaller automobiles continues, reciprocating gasoline engines are likely to be favored over advanced ceramic designs.

However, motoring's loss will probably be aviation's gain. As far as the aviation industry is concerned, the ceramic glass turbine will have immediate and significant impact. The ability to generate high-temperature gases not only guarantees efficient combustion and fuel economy, but also provides significant increases in exhaust thrust—a factor meaningless for automobiles but useful for airplanes.

Thus even if the ceramic automobile engine fails to materialize in the short run, researchers insist that the R&D effort will inevitably produce a far more efficient aircraft propulsion system.

Certainly, the economics are far more favorable. Aircraft engines are typically produced in low volume at high prices, allowing the elevated cost of ceramic components to be easily absorbed. According to Patiky,[114] an automobile engine costs just $3 per horsepower, while a jet engine costs $250 per horsepower because of the far finer engineering tolerances required. Hence, material costs represent a much lower fraction of aeronautical costs. In addition, the thrust increases possible in a ceramic jet engine mean a lot more power for the money. Engineers thus forecast that ceramic engines will be lofting planes within 15 years.

Diesel Engines

There are several levels at which ceramics could be incorporated in diesel engines, as shown in Table 8.11. The first level involves a baseline diesel containing a ceramic turbocharger and discrete ceramic components. The second level adds a ceramic cylinder and piston and eliminates the cooling system. The ceramic used at this level would provide high-temperature strength rather than thermal insulation. The uncooled diesel engine is expected to be the first new engine system designed to take advantage of the properties of ceramics to be commercialized. The uncooled diesel may become commercially available in the early 1990s. It is anticipated that the uncooled diesel will result in fuel consumption savings of 10 to 15 percent compared to current diesel-engine technology.

The third level would utilize ceramics for thermal insulation in the hot section as well as in the exhaust train. Turbocompounding would be used to recycle energy from the hot exhaust gases to the drive train. The adiabatic diesel engine could be introduced commercially during the mid-1990s or beyond. These engines are expected to provide fuel consumption savings of approximately 20 to 30 percent compared with current diesel-engine technologies.

The fourth level would use advanced minimum-friction technology to improve the properties of the engine.

The four levels listed place different demands on the ceramic materials. Levels 1 and 2 require a low-cost, high-strength material, but without insulating properties; sintered SiC or Si_3N_4 would be possibilities here. It has been suggested that level 2 represents the best compromise for light-duty ceramic diesels such as those in automobiles.[9,23,55,59,74,78,95,96,103,105,106,119-122] The ceramic components selected and used first for discrete hot components include[16,91,123,124] glow plugs, headliners, exhaust port and manifold, precombustion and vortex chambers,[125] turbochargers, intake heaters, and valve-train wear parts (Fig. 8.19).

One of the initial introductions was by Mazda with its 1200-cm^3 diesel and Si_3N_4 vortex chamber, which is capable of meeting U.S. particulate emission standards without electronic fuel injection and other complex systems. Other uncooled ceramic engines have used Si_3N_4 or SiC pistons and cylinders. Isuzu, the first automaker to use ceramics (1981), plans to introduce a ceramic engine in the early 1990s.[78,120] A prototype has been test driven over 160,000 km at speeds

TABLE 8.11 Future Diesel-Engine Technology Development

Technology level	Engine configuration	Potential ceramic components	Potential payoffs
1	State-of-the-art engine, turbocharged	Turbocharger Valve-train components Prechamber, glow plugs	Improved performance Reduced cost? Early manufacturing experience
2	Uncooled, nonadiabatic (no water or air cooling, no turbocompounding)	Turbocharger Valve-train components Piston, cap Cylinders, liners	Reduced weight—efficiency gain Gives option to improve aerodynamics—efficiency gain Reduced maintenance Reduced engine system cost? Flexibility of engine replacement
3	Adiabatic turbocompound	Turbocharger Turbocompound wheel Valve-train components Piston, cap Cylinders, liners Exhaust-train insulation	Very significant reduction in specific fuel consumption Improved aerodynamics Reduced maintenance
4	Minimum-friction technology (could be combined with 1, 2, or 3)	Air bearings High-temperature rings High-temperature bearings Nongalling wear surfaces Low-friction liquid, lubricant-free bearings	Lower specific fuel consumption

Source: From Katz.[73]

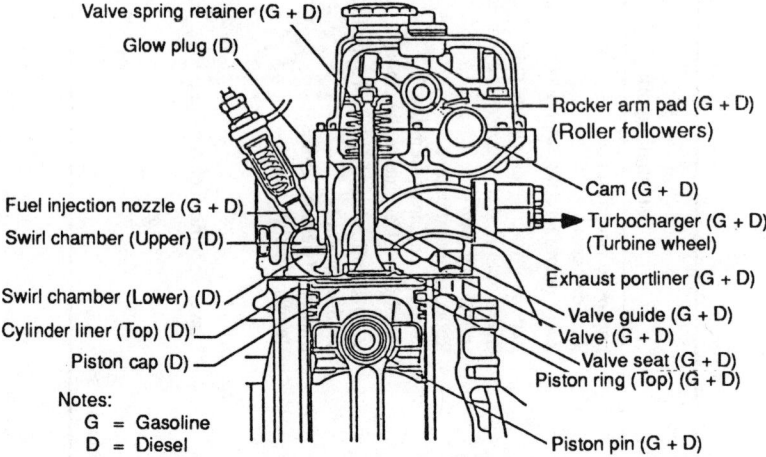

FIGURE 8.19 Ceramic-engine components for diesel-engine application.[102]

up to 115 km/h. Power output and fuel economy are up to 30 percent greater compared to conventional diesels, and engines are 27 percent lighter than metal engines.

Level 2 would require an insulating ceramic, probably ZrO_2 or a ZrO_2-based composite. This is the level where the most significant improvements in fuel efficiency would be realized. The current ZrO_2 and Al_2O_3–ZrO_2 transformation-toughened ceramics are not reliable at the high stress of the piston cap and do not have a low enough coefficient of friction to withstand the sliding contact stress of the rings against the cylinder liner. These materials do seem to have adequate properties, however, for other components, such as the head plate, valve seats, and valve guides.[3,9,23,55,78,95,105,106,111]

Emission requirements will likely affect the size of the diesel market for passenger cars and trucks. Diesel engines generally produce a high level of particulate emissions. The higher operating temperatures of the adiabatic diesel could reduce emissions or allow emission control devices to operate more efficiently. The market for ceramics could also be affected in another way: a major candidate for diesel emission control is a ceramic particle cap. However, such a trap is likely to be expensive and could raise the price of the automobile to an unacceptable level.

The third-level engine is uncooled and retains maximum possible heat content in the exhaust gas. This heat content is recovered from the exhaust gas through the use of a turbocompound concept. A diesel-engine prototype developed by Cummins Engine Company and U.S. Army Tank and Automotive Command (TACOM),[126] a nonturbocompounded version of the ADE, has been installed in a 5-ton military truck, and the engine has 361 parts less than the standard diesel engine. These parts are mainly from the cooling system and include the radiator, fan, and water pump. The removal of these parts has resulted in a reduction in the net weight of the engine by 153 kg and the engine has accumulated over 15,000 km of successful road testing.[3,76,123]

Industry sources indicate that commercial introduction of this engine will not occur until the mid-1990s at the earliest. It is expected that the adiabatic engine will be about 20 to 25 percent more fuel-efficient than a conventional diesel engine.

The entire combustion area of the engine is insulated by a thin coating of ZrO_2 and the vehicle has accumulated over 12,900 km of road testing.

Ceramic coatings may be an alternative to monolithic ceramics in diesel applications. ZrO_2 coatings can be plasma-sprayed onto metal cylinders to provide thermal insulation. The current state-of-the-art thickness of ZrO_2 coatings is 0.6 to 0.76 mm. It is estimated that thicknesses of 3.2 mm will be required to provide thermal insulation comparable to monolithic ZrO_2. The coating is not as impermeable nor as resistant to thermal shock as the monolithic ZrO_2. However, the coated metal part has higher strength and toughness than the all-ceramic part.

Ceramic Materials. In evaluating the many new materials now available commercially or in development, none meet the desired specifications of the adiabatic engine. The material possessing the nearest compromise to the requirements is PSZ.

It should be noted that PSZ possesses very good insulation properties and a compatible coefficient of expansion to match the cast iron engine components. Most current component designs for the adiabatic engine are composed of shrink interference fitted PSZ materials with cast iron. Major components include the piston crown, cylinder-head hot plate, valve seat liners, and cylinder liner.

The Mg–PSZ has exceptionally good fracture toughness and thermal-shock resistance. However, aging properties at elevated temperatures leave some things to be desired.[127–129] The MgO-stabilized PSZ tends to revert back to the original monoclinic phase from the tetragonal. Other disadvantages of PSZ would be its high density and cost, especially when Y_2O_3-stabilized PSZ is considered. Although the Y_2O_3-stabilized PSZ has performed satisfactorily, no long-term durability aging properties have been measured. Therefore, in the future, as ceramic technology advances, the newer materials which could spearhead the material list for the adiabatic diesel must be prioritized. The candidate heat-engine ceramics and coatings include the following[127]:

1. Bulk ceramics
    a. Partially stabilized ZrO_2 (PSZ)
    b. PSZ + metal or oxide dispersoid
    c. Al_2O_3 + HfO_2 dispersoid
    d. Pressureless sintered Si_3N_4
    e. Al_2O_3 + ZrO_2 dispersoid
    f. Al_2O_3 + metal dispersoid
    g. Mullite
    h. Mullite + metal or oxide dispersoid
    i. Low-thermal-expansion ceramics
    j. Si_3N_4–Si_3N_4 composites (fiber–matrix)
    k. SiC–Si_3N_4 composites
    l. SiC–SiC composites
2. Ceramic coatings
    a. ZrO_2 base and HfO_2 base
    b. Non-ZrO_2 base
    c. Boride
    d. Carbide
    e. Nitride

Ceramic Coatings. An alternative to the monolithic ceramic used in composite adiabatic engine components is the ceramic coating. There are many new ceramic coating techniques which are quite attractive alternatives to the monolithic ceramic approach.

The most popular of the ceramic coatings is the plasma spray of ZrO_2 onto a metal substrate with suitable bond coating. ZrO_2 with Y_2O_3 additive, that is, ZrO_2–$0.08Y_2O_3$, as a thermal-barrier coating and Ni–16.8Cr–5.8Al–11.8Y_2O_3 bond coatings offer an attractive thermal-barrier coating system for an adiabatic engine.[127–130]

One of the difficulties with plasma spray is the inability to coat a thick layer in order to establish a large thermal barrier. However, improvements in plasma-spray coating have been obtained by increasing the power level and arc gas composition used during plasma spraying.

Finally it has been shown that the insulating properties of coatings are quite good when compared to their monolithic counterparts because of the high degree of porosity in most coatings as well as the thermal conductivity of common monolithic insulating ceramics in the plasma-sprayed form.[127–133]

The property of the low coefficient of friction of a ceramic in its unlubricated state is used in level 4, which is known as the minimum-friction engine (MFE). The goals of the MFE project are to incorporate all the advantages of the ADE into the engine, to further improve the fuel economy, and to reduce mechanical friction of the engine by 50 percent in order to eliminate the engine oil system.[127,132] Figure 8.20 shows a diagram of the adiabatic engine with the minimum-friction concept. To achieve the 50 percent target reduction, an engine friction-reduction technique was used which consisted of:

Oilless engine operation

Gas-lubricated piston and cylinder liner

Ceramic main, crank, and wrist pin bearings

Solid lubricant gears, rocker arm bearings, and so on

The properties of ceramics which are essential for the MFE application are:

Low coefficient of expansion

High stress capability

High temperature and strength

Low friction property

Low wear rate

A recent study by Charles River Associates[76] predicts that the uncooled ceramic diesel-engine system will be the first to be commercialized. It projects that the initial introduction will be in the early 1990s, and could account for 5 percent of the new engines manufactured in 1995. The ZrO_2 materials in development mentioned can be used for level 3 technology, where the greatest fuel efficiencies are expected. It remains to be seen whether the elimination of the cooling system will provide sufficient incentives to U.S. automakers to commercialize level 2 ceramic technology. However, Japan, Great Britain, France, and Germany in particular have very active programs, both in material and in diesel-engine development[3,9,23,76–78,95,96,103,105,106,111,118–121,123,124,126–128,132] (Table 8.12).

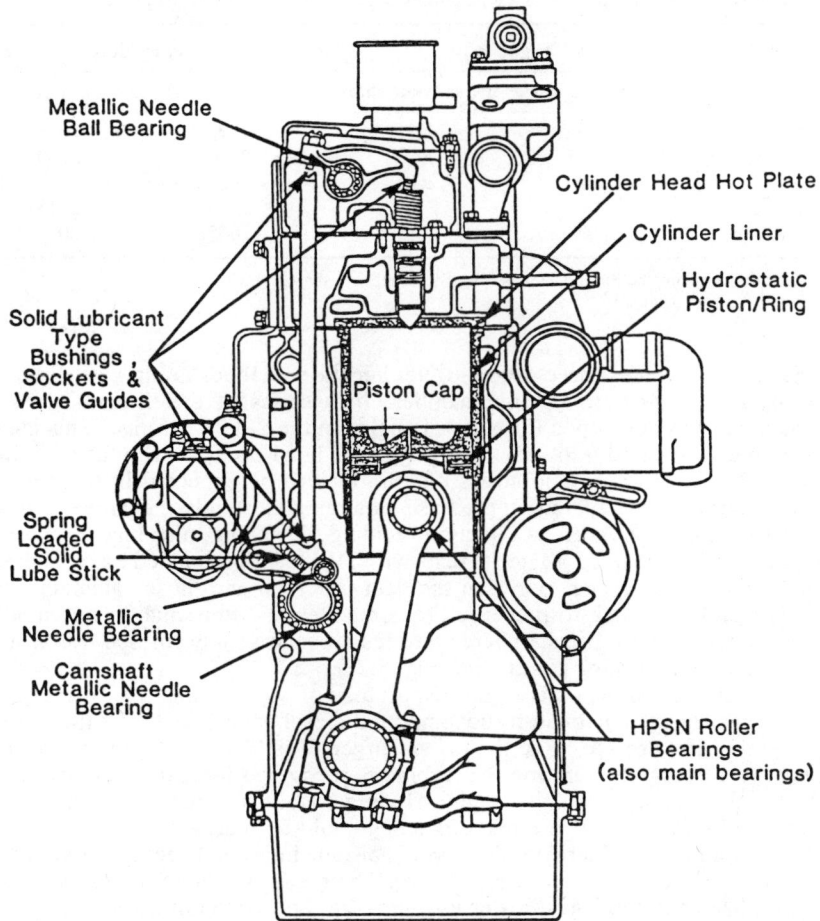

FIGURE 8.20 Schematic of minimum-friction engine.[127,132]

Ceramic Components. A variety of new technologies is expected to be developed in the future to further improve the thermal efficiency and output of automotive engines. Many technological issues will have to be resolved to achieve higher reliability and durability as well as the ability to accommodate alternative fuels, which might be used in the future (Table 8.13).[134]

High-technology ceramics may be able to contribute in some form to new and improved automotive technologies. However, in applying ceramics it is important not to regard them simply as replacements for certain metal components. Rather, their effect on the total system, including potential gains, losses, and significance, must be identified clearly. Gradually it has become clear, through trial application of ceramics to some engine components and studies of the engine performance, that we should not expect an immediate and rapid expansion of ceramic applications (Table 8.13).

TABLE 8.12 Projected Market Development for Ceramic-Intensive Engines

Engine type	Light duty		Heavy duty	
	Introduction*	5% market share	Introduction*	5% market share
Diesel	1995	2005	1995	2005
Spark ignition	1996	2005	2000	2010
Rotary	2000	2005	2005	2010
Adiabatic	2000	2010	2005	2015
Gas turbine	2003	2025	2005	2015

*Introduction into the market means 1% of total market share.
Source: From Taguchi.[96]

For diesel engines, three different fuel injection methods are in use: direct injection, swirl chamber, and prechamber. In a series of experiments a Si_3N_4 prechamber was developed for a diesel engine in the 2- to 3-L range. This chamber can be assembled into an existing cylinder head of a car without any alterations. The temperature and stress distributions to be expected in the ceramic components were calculated on the basis of a metallic standard chamber. Different processing steps, such as injection molding, dewaxing, nitriding, encapsulation, and HIP, were used. Material tests with HIP reaction-bonded Si_3N_4 samples proved that the difference between the data observed and those calculated previously was big enough to preserve the components. After small design modifications some ceramic prechambers were tested successfully for 300 h in a fired engine under all working conditions (Fig. 8.21).

This type of results points to the importance of utilizing the latent heat of the exhaust gas. In the adiabatic turbocompound diesel engine,[127,131,132] the hot exhaust gas is first used to drive the turbocharger. It is then sent to the power turbine, where energy for driving the turbine is recovered from the exhaust gas to increase the efficiency of the engine. It is generally estimated that if friction losses can be minimized, thermal efficiency could be increased to as high as 50 percent. Combustion characteristics would also be improved because combustion would take place at a higher temperature. The result would be increased power output, cleaner exhaust gas, lower engine noise, and no cooling system.

Several Japanese companies are also working on the development of a turbocompound engine (Table 8.14). These include automakers such as Nissan, Toyota, Isuzu, Hino, and Komatsu, Ltd.[96] Hino Motors has reported that the use of ceramics in the cylinder, piston, and other combustion chamber components could possibly result in the optimum adiabaticity for improving engine performance.[136]

Toyota has reported on the development of a two-piece articulated Si_3N_4 ceramic piston crown and aluminum-alloy skirt. From finite-element analyses and test measurements the stress levels of the main loads were determined. The component strength of the ceramic piston crowns was ascertained from fracture tests, and the capacity to withstand fast fracture was assessed. Test results indicate that the materials in the components were statistically equivalent to those in the test specimens, and that the analytical approach was correct. They also indicated that the piston crowns were of high quality, and a 300-h engine test was conducted without fracturing of the ceramic crowns.[137]

The U.S. Ford Motor Company has also evaluated ceramic pistons in heat engines, and one method being investigated to insulate the combustion chamber is

TABLE 8.13 Summary of High-Technology Ceramics for Adiabatic Engines

Component	Compressive strength	Strength	Fracture strength	Low friction	Light weight	Insulation	Resistance Wear	Resistance Heat	Resistance Erosion	Expansion coefficient	Potential ceramic
Piston	x	x		x	x		x	x	x		Si$_3$N$_4$, PSZ, TTA
Piston crown	x	x		x	x		x	x	x		Si$_3$N$_4$, PSZ, TTA, coating
Fire deck	x	x		x			x	x			Si$_3$N$_4$, PSZ, TTA
Cylinder chamber	x	x	x		x		x	x			Si$_3$N$_4$, PSZ, coating
Prechamber				x		x	x				PSZ, Si$_3$N$_4$
Valve	x	x		x	x	x	x	x			SSN, PSZ, composite
Valve seat		x		x	x	x	x				PSZ, SSN
Valve guide			x	x		x	x				PSZ, SSN, SiC
Exhaust ports				x		x	x				ZrO$_2$, Si$_3$N$_4$, TiO$_2$, Al$_2$O$_3$
Manifolds				x		x	x				ZrO$_2$, Si$_3$N$_4$, TiO$_2$, Al$_2$O$_3$
Tappets	x	x		x		x		x			PSZ, SiC, Si$_3$N$_4$
Wear pad	x	x		x		x		x			PSZ, SiC, Si$_3$N$_4$
Piston pin	x	x		x		x		x			PSZ, SiC, Si$_3$N$_4$
Mechanical seals	x	x	x		x		x				SiC, Si$_3$N$_4$, PSZ
Rings	x	x			x	x					SSN, PSZ, coating
Ceramic bearings	x	x	x	x		x	x			x	SSN

Source: From Sheppard.[130]

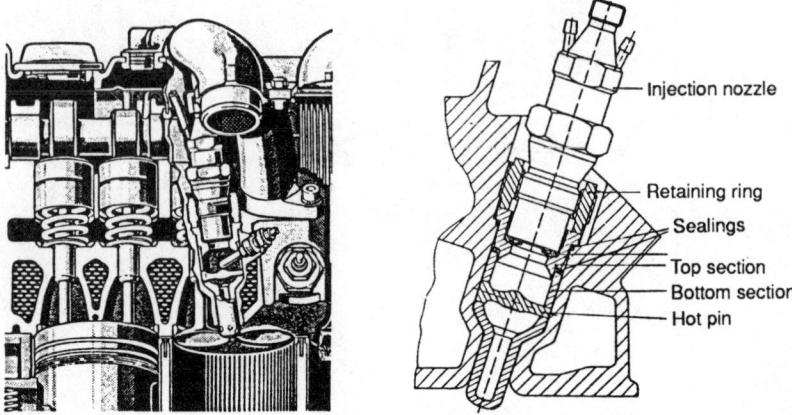

FIGURE 8.21 Installation of metal prechamber into diesel car engine and arrangement of ceramic prechamber in cylinder head.[135]

to use a ceramic insulating material inserted into a metallic shell. The prime ceramic material choice for this inserted-type engine is PSZ because of its low thermal conductivity, high strength, and a thermal expansion close to that of metals for ease of attachment.[138]

In order to design ceramic components successfully, such as piston bowl inserts, thermal, stress, and reliability analyses of the components were investigated. The use of Weibull theory to predict the reliability of a component is well established. However, a modified Weibull method was developed to aid in reliability calculations.[136] This method was used to analyze a PSZ piston bowl design and to determine what the material properties would have to be to make this design work.[139]

Isuzu has a 1.8-L passenger car mounted with a ceramic turbocompound engine, operational in 1992–1994.[140] This prototype vehicle has been run continuously for 6 h and attained a maximum speed of 140 km/h[82] (Fig. 8.22).

Recommendations. The use of ceramics in engines with all its associated benefits will not come easy. There are numerous problems that need to be resolved:

1. High-temperature lubrication
2. High-temperature, high-strength, and insulative materials
3. Ceramic bearings
4. Gas-lubricated bearings
5. Low-cost fabrication
6. Low-cost finishing and machining
7. Solid lubricants
8. Better ceramic coatings

Of these problem areas, items 1, 5, and 6 will probably present the greatest challenge.[142]

Higher-temperature, higher-pressure thermodynamic cycles will be the way for heat engines of the future. Since these temperatures and pressures are beyond

TABLE 8.14 Japanese Applications for Ceramic-Engine Components

Manufacturer	Type of engine	Type of material	Components
Isuzu	1.6-L 4-cylinder spark ignition	Si_3N_4, SiC	Piston rings, cylinder liners, valves, camshaft, cam lobes, turbocharger rotor
	1.8-L 4-cylinder diesel*	$Al_2(TiO_3)_3$	Exhaust port, exhaust manifold, turbocharger scroll
		Si_3N_4	Cylinder liners, piston rings, valves, turbocharger rotor
Mazda	3.0-L V6 DOHC 24-valve diesel	Si_3N_4	Piston pin, others same as above
	2.0-L 4-cylinder DOHC 16-valve spark ignition	Si_3N_4	Intake valves
Nissan	"Feather Concept" 3.0-L V6 DOHC 24-valve spark ignition	Si_3N_4	Turbine rotor, intake and exhaust valves
	Naturally aspirated CAX 2.0-L 4-cylinder DOHC, 16 valves	Si_3N_4	Intake valves
Toyota	GTV II regenerative gas turbine	Ceramic, specific composition unknown	Gasifier and turbine rotors, gasifier and power turbine scroll assemblies, combustor, turbine vanes, scroll backplates, regenerator, regenerator seal platform duct, thermal barriers

*3000 h of testing have been accumulated.
Source: From Sheppard.[103]

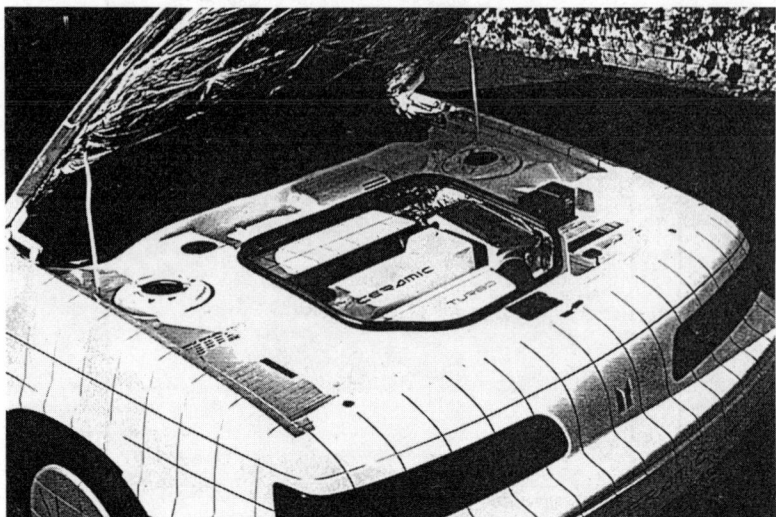

FIGURE 8.22 Car equipped with ceramic engine. Coolantless engine does not need radiator in vehicle and engine weight has been reduced by 15 percent.[141]

the current limit of metal, high-strength, high-temperature ceramics will be the material of the future. The manufacturing technology, the evaluation technology, and the applications technology of ceramic materials are many; but they appear to be tenable.

In many ceramic materials, expansion coefficients and conductivity can be varied to meet the needs of the applications. Some of the current favorite monolithic ceramics and coatings have satisfactorily met the needs of the adiabatic diesels. There are bound to be more new materials in the future. However, for the present, the materials that have met the requirements of an adiabatic engine for various adiabatic diesel components were summarized in Table 8.13.

Electronic Ceramics

Conventional ceramics have enhanced many important aspects of our modern technological society. The next generation of ceramics—advanced ceramics and ceramic composites—promises to play a further enabling role for still higher performance devices. These ceramics and ceramic composites can be tailored to have unique combinations of electrical, optical, mechanical, and chemical properties that make them essential and irreplaceable in many engineering applications.

At present, electronic applications dominate the world market for high-technology ceramics. Electronic ceramics are essential to the electronics industry. Ceramics for electronic applications are high-band-gap, largely insulating ceramics and glasses in which a broad range of dielectric, elastic, optical, thermal, electrical-conductivity, piezoelectric, and pyroelectric-tensor properties can be manipulated and controlled to close tolerances. In most applications the ceramic does not stand alone, but is bonded to a metal (for electrodes and conductors) or a polymer, which requires precise control and understanding of the interface. Often the metal and ceramic parts are in closely spaced lamellar structures, as for multilayer interconnection packages or multilayer capacitors, and must be coprocessed through the whole ceramic forming and firing process to yield defect-free monolithic hermetic packages.

For integrated-circuit packaging, glass-bonded Al_2O_3 ceramics are widely used. The green ceramic is usually tape-cast with a suitable organic vehicle, and molybdenum metallization inks are screen-printed for the interconnection wiring. Automation is used extensively in inspection, punching, printing, and lamination. Precise control of shrinkage on firing is essential for the more complex packages. A high degree of perfection has been achieved in this technology. Future-generation integrated-circuit packages will require ceramics with higher thermal conductivity as the size of the package shrinks to increase computation speed.

Although Al_2O_3 is the dominant, traditional ceramic substrate material, its low thermal conductivity makes it vulnerable to being replaced by other materials that can dissipate heat from an electronic package more rapidly. Despite their much higher cost, SiC, AlN, and BN are challenging Al_2O_3, particularly for the more critical applications.[143]

BeO,[144] the second most common packaging material, is also threatened by some of the newer materials, but for different reasons. The thermal conductivity of BeO is highest of all the packaging ceramics (250 W/m · °C), but its toxicity to personnel has made it unwelcome in the workplace. Taking the place of BeO for some substrates and multilayer semiconductor packages is the nontoxic AlN, with a thermal conductivity approaching 200 W/m · °C, which satisfies the requirement for all but the most stringent applications. Both of these ceramics are at the high end of the cost range.

Other application areas where BeO is being replaced by other materials such as AlN and BN are power-transistor mounting wafers and conductive epoxy adhesives. The adhesives, with conductivities of about 18 W/m · °C, are used for bonding integrated-circuit chips to ceramic substrates.

Thermal conductivity varies considerably with temperature among the electrical ceramics. For example, values for BeO are very high at room temperature, but they drop drastically at higher temperatures. In contrast, conductivities of AlN and BN fall only slightly with increasing temperature, and that of SiC is nearly linear, rising slightly with temperature.

AlN in solid shapes, and AlN and SiC in substrates, are nontoxic, pressureless sintered ceramics with excellent heat dissipation and electrical resistivity characteristics. They have a mechanical strength 1.5 times greater than traditional Al_2O_3 or BeO.

Applications. Applications include:

Resistor cores

Support switches

Power transistors

Traveling-wave-tube components

Laser tubes

Crucibles

Multichip arrays (metallized AlN and SiC substrates)

Thin-film applications (AlN substrates polished to a 3-μm surface)

BN ceramic powders and coatings have been introduced for thermal-mechanical and metal processing applications. These include:

Mold washes and ladle coatings

Mold release agents

Thermal filler material

High-temperature lubricants

Heat sinks

Dielectric components

Microwave windows

Horizontal continuous-casting break rings for metal processing

Surface-Mount Technology and Printed-Circuit Boards. Surface-mount technology's increasing popularity has forced users, suppliers, and developers to consider numerous issues that previously existed only on the periphery of the electronic interconnection marketplace. Although surface-mount technology helped to reduce electronic equipment size significantly, it also presented the industry with connectivity, manufacturing-process, and material challenges. One area strongly affected by surface-mount applications involves printed-circuit-board computer-aided design tools. Original systems developed on the assumption of dual-in-line packaging technology must continually adapt their databases and design tools to new features, such as complex surface pads, surface pin-hole absence, blind and buried vias, fine pin pitch, and increased pin counts.[145–147]

The design of ceramic packages for microelectronics constitutes a very special subset of electronic product design. Packages lie directly between boards and chips, not only physically, but also in their design.

Cofired multilayer ceramic packages are used to house integrated circuits, crystals, hybrid circuits, and other similar devices when the highest levels of package integrity and system performance are required. These packages provide hermeticity, shock resistance, thermal conductivity, and mechanical strength that cannot be achieved in other package types such as plastic and cerdip.

Typical cofired multilayer printed-circuit-board ceramic packages are shown in exploded views in Figs. 8.23 and 8.24. Each layer of the "board" is formed from a plane of ceramic material, typically Al_2O_3, and is usually between 254 and 762 μm thick. The through vias are a composite of a refractory metal, such as tungsten or molybdenum, and a ceramic glass in holes through the ceramic plane. The conductive traces are formed from a matrix of tungsten or molybdenum grains in a ceramic-glass binder. The individual planes of the package are punched, vias are filled, and traces are screen-printed. The planes are then laminated to one another under heat and pressure. The parts may be cut out or scribed for subsequent breaking after firing, and are then fired at temperatures exceeding 1650°C. The exposed metal portion of the package is plated with nickel. Metal parts such as pins, leads, and seal rings are then brazed on, and, finally, nickel and gold plating are applied. Firing the entire composite package simultaneously (cofiring) is the key to achieving hermeticity and integrity.

Although ceramic substrates have been used in wide-ranging hybrid applications for years, their use as an alternative to epoxy glass multilayer boards is fairly recent. Ceramic-based designs became popular with the introduction of leadless ceramic chip carriers (LCCC). These high-pin-out and high-pin-pitch surface-mount components are primarily selected for military applications and are often the choice package in very-high-speed integrated-circuit (VHSIC)-related programs. A ceramic substrate's fundamental advantage is its close thermal coefficient of expansion with leadless chip carriers, its thermal dissipation characteristics, and its ability to accommodate fine-line, high-interconnection-density imaging techniques.

Thick-film ceramic substrates are fabricated on a 96 percent Al_2O_3 base. Their thickness typically varies between 25 and 60 mm, and they are available in sizes as large as 323×10^2 mm^2. Depending on the application, the interconnect network (traces, vias, and surface pads) is established using a thick-film paste of silver, gold, copper, or their alloys.[148–150]

Thin- and Thick-Film Ceramic Metallization. Ceramic metallizations and ceramic-to-metal bonding technologies are used for such applications as the laying down of circuitry on insulating substrates, creating hermetic enclosures, establishing electrical contact, and fastening structural ceramics to metal components for high-strength assemblies.

Thin-film metallizations are generally less than 12 μm thick and, more typically, less than 1000 Å. These are coatings generated through vacuum-deposition techniques such as sputtering, evaporation and condensation, or chemical-vapor deposition. These techniques are used, for example, to manufacture magnetic coatings for hard computer disks, reflective surfaces for CDs, and for creating integrated-circuit devices. These films rely on mechanical and van der Waals bonding to provide adhesion to the substrate in question.

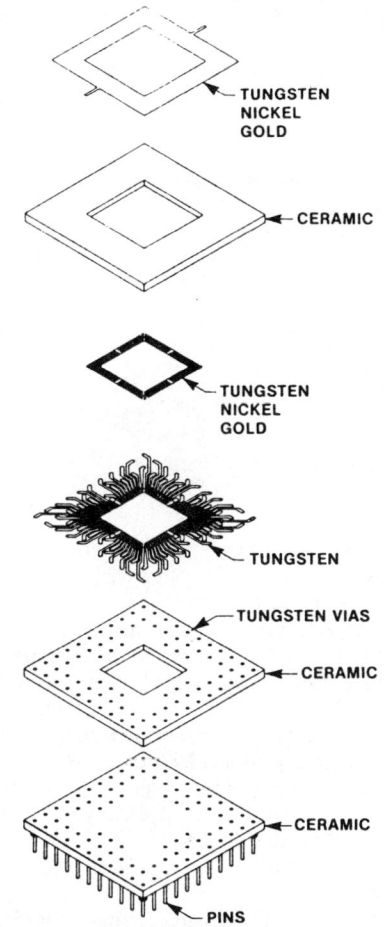

FIGURE 8.23 Exploded view of pin-grid-array (PGA) package.[147]

Upon deciding that a thick-film ceramic-metal joint is desired, the engineer must consider the following questions in order to select a proper joining method, according to Intrater[151]:

1. What mechanical loading, as well as chemical and thermal environmental conditions, will the final assembly encounter during use?
2. Will the joint simply need to provide for chemical attachment, or must it also provide an active role in the final assembly (such as a conductive path)?

Four premier techniques for metallizing (that is, joining to) ceramics are: (1) refractory metallization, (2) active metal brazing, (3) Intragene metallization, and (4) gas-metal eutectic bonding. These were discussed in Chap. 7.[151,152]

Ceramic researchers are targeting several areas of electronic ceramics for im-

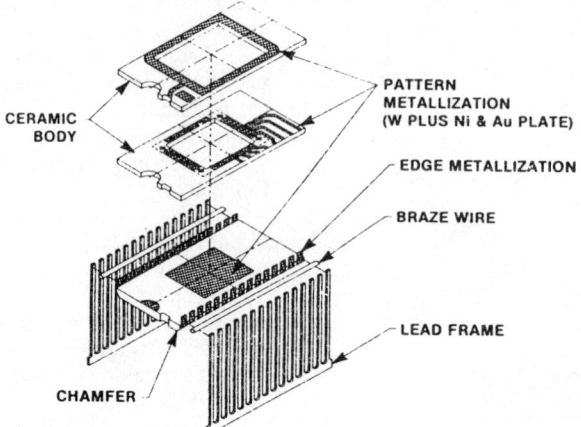

FIGURE 8.24 Exploded view of dual-in-line package (DIP).[147]

provement. One notable direction is that of developing greater toughness. Microstructure control of grain boundaries is one approach; other techniques involve phase transformation and surface-treatment toughening.

Considerable research is also under way in composites. Ceramic composite integrated-circuit packages are designed to take advantage of specific properties of various materials. For example, a substrate application can combine a high-thermal conductivity AlN with a ceramic of low dielectric constant, such as cordierite or mullite. The resulting material provides the high-speed signal path and power capability required for contemporary integrated circuits.

With integrated circuits getting smaller and denser, the relatively low thermal conductivity of Al_2O_3 is no longer acceptable for many new applications. AlN, because of its much higher conductivity as well as its closer match to silicon's thermal expansion characteristics, is expected to take over many of the new integrated-circuit substrate needs.

BN, with a moderate (but nearly constant with temperature change) thermal conductivity and a low dielectric constant is almost transparent to microwaves and has a low dissipation factor. As a dielectric material, it serves as microwave windows for high-frequency satellite applications. In addition, BN's nonabrading nature makes it useful as a thermally conductive potting-compound additive because it does not damage delicate parts.

Superconductivity. Unusual ceramic materials that lose all electrical resistance at temperatures above the boiling point of liquid nitrogen and whose magnetic field and current density can be lowered to certain levels have captured the interest of engineers and scientists worldwide.[153–156]

Superconductivity is now in the world spotlight. Research had previously centered on metallic substances, but these do not exhibit superconductivity unless cooled to near absolute zero (−273°C) and are hardly practical because of the cooling problems involved. When the focus of this research widened to include ceramics, however, the critical temperature was raised by more than 100°C in a matter of a year, and hopes of practical implementation rose quickly.[157]

The new superconductor substances recently discovered are having an enormous impact in many different areas and are bringing about a new industrial revolution. How large of a market, then, are these superconductor applications likely to create, and which companies will be dominant in it? Also, will these applications be significant enough to ignite a true industrial revolution?[158-160]

The new superconductors are relatively easy to make; conventional ceramic processing equipment is sufficient. An understanding of the chemistry of ceramics is required to obtain the proper phases, since isolation of the proper phase was a key part of the discovery. Many of the formulas are published, although many are covered by patents.[161]

Future Applications. Superconductivity, wherein the electrical resistance of a material drops to zero, has many envisioned applications if materials can be developed that possess the property at reasonable temperatures. For example, large coils of superconducting material can be used by power companies to store excess electricity generated during off hours, which is to be used to supplement the supply during peak hours. Because the coils have no resistance, currents continue to circulate indefinitely within the coils. Transformers also can be made more efficient and able to convert more power in a smaller size, since maintenance-free coils do not heat up.

Extremely powerful magnets can be made with superconductive materials, with no wasted energy in the coils. Such magnets made from conventional superconductors are already in use in particle accelerators, nuclear-magnetic-resonance (NMR) instruments for medical examinations, and experimental trains that are levitated by magnetic fields.

The materials are limited in current-carrying abilities. To be useful, they must be able to carry several thousand amperes per square centimeter. Like all materials, they lose their superconductivity in the presence of a strong magnetic field. However, the new materials will remain superconducting in stronger magnetic fields than can metals.

Long transmission lines have the obstacle of requiring costly cooling equipment and thermal insulating materials, and are heavier and more bulky than conventional power lines.

Greater promise in the short run is in the area of microelectronics. The Josephson junction, which occurs in superconducting materials, can perform as a fast switch at speeds of less than 10 ps with a power dissipation of less than 10^{-6} W. The promise of high-temperature superconductivity is expected to bring about faster computers with more compact circuitry and more memory.[162-167]

Miscellaneous Ceramic Applications. With the development and use of advanced ceramics many more applications are appearing, and a revolution in the fields of material science and technology is under way.

Ceramic Wafer Seal. This is a low-leakage high-temperature flexible sliding seal, which is used to control high-temperature gases at high pressures on hypersonic engines and two-dimensional turbojet exhaust-gas nozzles. One application is the control, without coolant, of the 1371°C gases, which will flow at pressures of 690 kPa from the Mach-25 National Aerospace Plane (NASP). Preventing leakage of these explosive hydrogen-oxygen mixtures is paramount in preventing the possible loss of the entire aircraft.

The unusually compliant serpentine seal of engineered SiC ceramic materials can accommodate sidewall distortions as large as 4.7 mm in only 457 mm of wall.

Ceramic Honeycomb Panels. Ceramic honeycomb panels serve as lightweight, heat-resistant structural members. Depending on the choice of ceramic materials, the panels are expected to withstand temperatures as high as 1800°C.

The honeycomb structure is made by vapor-depositing a ceramic on a fabric substrate woven in a honeycomb pattern and then eliminating the substrate by oxidizing it. The fabric can be made of a loosely woven polymer such as polyacrylonitrile. It is impregnated with an organic binder, such as a phenolic resin, for stiffness.

In one version, the fabric honeycomb (Fig. 8.25) is placed in a reactor and pyrolyzed at a temperature of 700 to 1100°C. Then at a temperature of 900 to 1100°C, trichloromethylsilane vapor is introduced and decomposes, depositing a layer of SiC evenly on the fabric. The reaction is allowed to continue until the weight of the honeycomb has increased by 100 to 300 percent. A greater amount of deposited ceramic would make removal of the substrate difficult. The ceramic material is not limited to SiC; other reactants may be used to yield SiB, Si_3N_4, or BN, for example.

As the fabric and binder oxidize, they pass through the pores of the ceramic as gas, leaving a microstructure of voids. The voids can be filled with ceramic by further chemical-vapor deposition in the reactor. The ceramic filler can be the same as or different from the base ceramic. In addition, the hexagonal holes of the honeycomb can be filled with the same or a different ceramic. Filling the holes reduces convective and radiative transfer of heat and increases the interlaminar shear strength of the honeycomb panel.

Ceraform Mirrors. These ultrastiff, lightweight, thermally stable mirrors are used in airborne and space applications and provide advantages over processes for making beryllium, aluminum, or custom metal matrix mirrors. Ceraform mirrors couple the superior properties of SiC, notably its high modulus, with a precision forming process that incorporates a honeycombed closed-back monolithic construction. This produces a higher section stiffness than is provided by the open-backed designs of mirrors made from materials other than SiC.

The result is a lighter, stiffer mirror. In one case, the Ceraform process produced a mirror with an 89 percent section stiffness increase and a 37 percent weight reduction for the same section as a beryllium mirror.

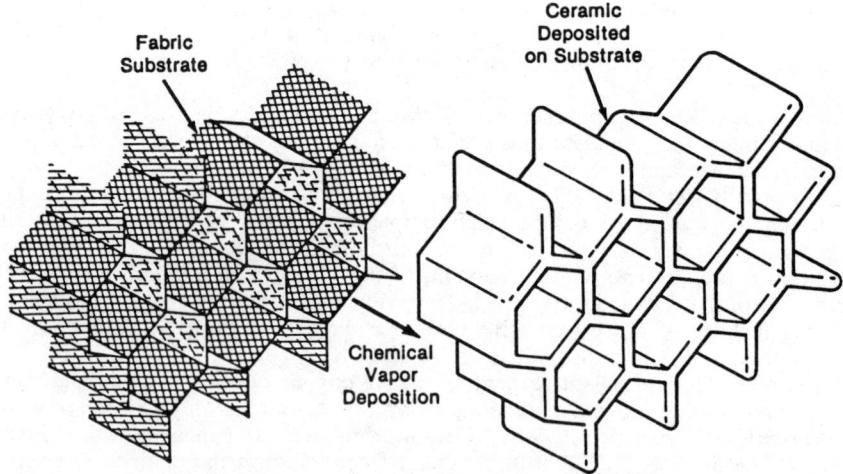

FIGURE 8.25 Polymer fabric woven into honeycomb substrate for CVD ceramic.

Ceramic Mitts. Mitts made in part from Nextel fabric allow items at temperatures up to 1204°C to be handled for short periods. The outer layer of the mitts is made of Nextel 312 fabric, a weave of continuous Al_2O_3–BO_2–SiO_2 fibers. The mitts are for handling diffusion-furnace tubes during semiconductor production. They replace conventional aluminum-coated, asbestos, or Kevlar mitts that wear out more quickly at these elevated temperatures.

Castable Ceramic Tooling. Thermo-Sil Castable 120 is a ceramic used to form tools in superplastic forming and diffusion bonding of titanium, aluminum, and stainless steel. The material is a fused SiO_2 ceramic that maintains dimensional stability throughout all superplastic temperature ranges. Most recently a tool of this material was used by British Aerospace PLC to superplastically form an aluminum cockpit floor for a European fighter aircraft.

Another high-temperature ceramic is Cera-Fab* 635,[168] a two-component alumino-silicate-based castable and machinable ceramic with temperature resistance to 1149°C, which is now used as a replacement for the asbestos fixtures used to hold stainless steel pins during annealing in a tunnel kiln, as stand-offs, welding jigs, furnace parts,[169] induction-heating tools, sealing molds, and high-temperature instrument components.

Ceramic Gage Blocks. Hard, stiff, light, and stable high-performance ceramics are attractive materials for use as gaging standards in industry. The material ZrO_2 suggests that ceramic standards can retain their accuracy for far longer than metal gages. They are hard, scratch-resistant, and offer greater toughness than gage blocks normally have. Their corrosion-proof, glasslike finish helps them wring tight for consistent accuracy.[170]

Injection Barrel Liners. Al_2O_3 or ZrO_2 are now being evaluated as injection-molded barrel liners for the Xaloy C series of thermoplastics. These ceramic materials give superior wear resistance and are able to withstand pressure at temperatures up to 499°C. Temperatures over 427°C are now required to mold some of the newer "exotic" thermoplastics, and these ceramic liners, first of a kind, are being tried and evaluated.

Ceramic Sports Equipment. A golf putter made of ceramics has made an initial successful entry into the golf market. The putter is made of TTZ ceramic material. The TTZ powder is dry-pressed into the putter shape and machined for greater flatness than standard alloy putter heads. Experts have noted that the solid ceramic putter head, attached to a graphite composite shaft, does not crack with age or use, as do alloy putters. The ceramic putter is 20 percent lighter than conventional putters, giving golfers greater putting power.

Spark Igniter. The spark igniter of the engine used by several wide-bodied jets is now equipped with an insulator made of a BeO ceramic material. The switch to the insulator produced from this 99 percent BeO material was made to improve reliability and service life under the extremely severe conditions seen by this part. It provides for higher resistance to thermal shock and spark erosion than any of the other high-performance ceramic materials previously used for the part. These key properties are combined with high mechanical strength and stiffness and excellent dielectric properties. The result is an insulator that has a service life more than 5 times that obtained from any previous materials, including Al_2O_3.[171]

"DaVinci Noire". ZrO_2 was selected as a material for a special wrist watch which would look luxurious while providing protection for the intricate works. The ceramic is harder than steel, unbreakable, and extremely corrosion-resistant,

---

*Cera-Fab is a registered trademark of Aremco Products, Ossining, N.Y.

yet its molecular structure results in the same clarity and angle of refraction as a diamond, providing a unique appearance.

Flame Lance. A Si_3N_4 flame lance has replaced a stainless steel cleaning system in an industrial fluidized-bed waste-heat-recovery unit. The lance cleans the lower side of a ceramic distributor plate in a heat-recovery unit on an aluminum-remelt furnace.

Hydrocyclones and Pumps. Hydrocyclones and pumps for mining slurries can now be lined with a ceramic-matrix composite that lasts up to 20 times as long as conventional ceramics, metals, or polymers. Approximately 90 percent SiC in an Al_2O_3 matrix provides hardness and wear resistance that last months longer than other materials.

To make Alanx CG896, porous preforms of SiC in the required shape are placed in contact with the molten aluminum alloy, then heated in a furnace. The metal oxidizes through the preform, forming a net-shape fully dense composite.

Spray Equipment. Electrostatic spray equipment uses controlled resistance cable rated at 200 MΩ to connect the power unit to the spray device. Previous cable designs used ten 20-MΩ resistors in the line, which were subject to failure due to high voltage drops.

This new cable incorporates a continuous resistance core of ceramic fibers, eliminating the need for resistors altogether. Using Nicalon SiC fibers, the resistance is maintained in each cable regardless of length by varying the number of strands in the fiber bundle. The continuous resistance core also eliminates voltage spikes, which are a major cause of damage to electronic components in the equipment.

Ceramic Spheres. Residential and commercial builders have long searched for an insulator that is virtually fireproof while producing no health or environmental threats. Researchers believe they have found the answer in tiny ceramic bubbles. The bubbles, which are formed in a liquid slurry, are heated to form hollow ceramic spheres that are lightweight, durable, and capable of withstanding temperatures hotter than 1760°C. By comparison, fiberglass insulation withstands temperatures only up to 427°C.

The ceramic spheres, which are as small as 1.5 mm in diameter, can also be bonded together to form boards to insulate superhot equipment in steel or petrochemical plants. It is believed that hollow bubbles of different sizes have many uses; for example, as a support for catalysts in chemical manufacturing.

Ceramic Helical Expander. A chemical helical expander consists of meshing male (convex land) and female (concave groove) helical rotors in a tightly fitted housing. Expansion (or compression) ratios up to 1:6 are practical, although 1:3 is more common. A promising application is recovering mechanical energy from the exhaust of the so-called adiabatic diesel engines now under development. The expander's low shaft speed and flat efficiency curve make it more suitable than turbines for compounding a piston engine.[172]

Cutting Blades. ZrO_2 cutting blades have extremely sharp edges, resist corrosion and oxidation, and last 90 times longer than steel blades. Suggested applications are in the areas of textile, paper, plastic film, and specialty materials.

Ceramic-Metal Chute. A ceramic-metal composite chute for use in hard-rock processing that contains nine slabs, each measuring 863.6 by 508 by 25.4 mm and weighing nearly 44 kg, has recently been poured. These slabs represent the largest single pieces of ceramic-metal composite material ever produced, and tests have shown that the composite, Alanx CG896, wears over 10 times more slowly than comparable ceramic and metal materials.

The chute is one of a series of solutions to high-wear problems in the process industries. Other Alanx products include large, complex parts and prototypes for pump, valve, and cyclone components, chute and liner tiles and plates, and nozzles and tubes for process applications.

Nozzles and Pins. The use of ceramic welding nozzles on equipment that is hand-manipulated is common. In these welding operations, the limitations on the nozzle's life are quite often dependent on physical handling, which sometimes results in accidental breakage. However, if the welding operation is automatic and repetitive, the limitation is likely to be a material failure of the ceramic welding nozzle due to a buildup of molten spatter or the stress of thermal shock. Orbital welding of pipe is a typical automatic operation. Syalon 101 has been used successfully in the orbital welding of fuel manifold inlet tubes for the aerospace industry, completing thousands of cycles before replacement.

In one application, a European manufacturer was using hardened steel pins to position nuts being welded to automotive and commercial vehicle chassis. These pins survived 7000 operations before wear and the buildup of spatter caused damage to the threads of the nut. By switching to location pins made with Syalon 101, it was reported that over five million operations were completed successfully.

Metal-Forming Tools. PSZ has been found suitable for dies, punches, rolls, mandrels, and guides. Because of its outstanding wear and corrosion resistance, Mg–PSZ has been used in pump shafts and bearings, in sucker-rod pump balls and valve seats for oil wells, and in hammers for demonstration purposes. Balls of Mg–PSZ, 0.5 to 1 mm in diameter, are used in ball-point pens because of their ability to resist water-based inks, which are more corrosive than traditional inks.[173]

Conclusions and Recommendations

Substantial progress has been made in the use of ceramics in diesel engines. The different levels of diesel engines in which ceramics will be introduced are uncooled, adiabatic, turbocompound, and minimum-friction. The uncooled and adiabatic diesel-engine concepts have already been proved with field tests, but the commercial introduction is expected only in the middle or late 1990s.

Estimates according to analysts and researchers[23] predict that the U.S. markets for advanced structural ceramics in automotive applications could go to $310 million in 1995 and $820 million in 2000. These same experts indicate that demand for ceramics in heat exchangers will go up to $100 million by the year 2000, while the markets for ceramic tool inserts are expected to grow to $500 million in 2000. A major market segment for ceramic bearings will be for automobile applications, and the market will grow to $320 million in 1995 and $720 million in 2000.

The defense-related applications of ceramics for radomes and infrared windows in missiles, aircraft, and spacecraft, and in armor for helicopters as a protection against small-arms armor-piercing projectiles, as well as their expected use in land-based military vehicles could increase to $390 million in 2000.

It is very difficult to predict the market for the other important aerospace applications, beyond the tile fabrication for the Space Shuttle's thermal protection systems. The material currently available for this application is a new generation of rigidized, lightweight HTP. HTP's potential applications are in materials for the thermal protection systems in future vehicles, such as the Mach-25 National Aerospace Plane (NASP), the Shuttle 2, the Advanced Aerospace Vehicle (AAV), the Aeroassisted Orbital Transfer Vehicle (AOTV), and the Orient Express.[23]

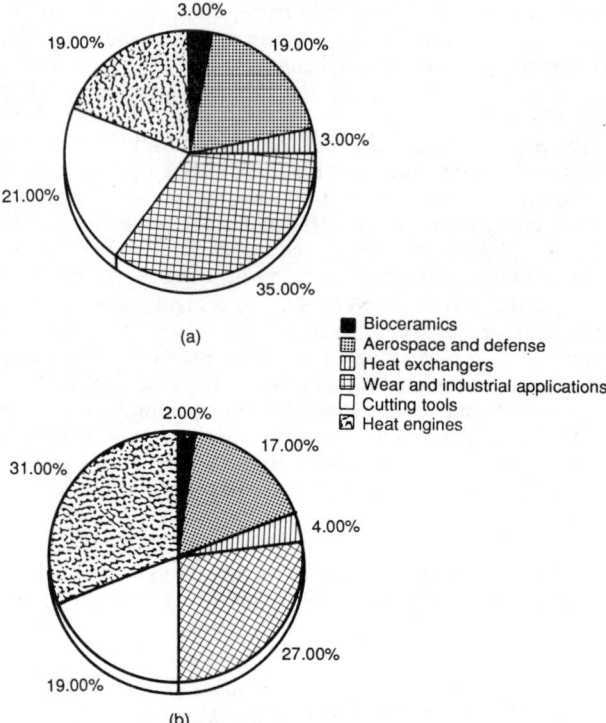

FIGURE 8.26 Segments of advanced structural ceramic market. (*a*) 1990–1995. (*b*) 1995–2000.[23]

The increased use of ceramics as prosthetic devices in the human body and dental implants and crowns will increase the bioceramic market to $60 million in 2000.

Finally, analysts have predicted a 22 percent growth for advanced ceramics and applications for 1990 to 1995 with an expected market of $433 million and an 18 percent growth for 1995 to 2000 with an expected market of $2.6 billion (Fig. 8.26).[23]

8.5 NONDESTRUCTIVE EVALUATION

Much has been said in several previous chapters about improving the reliability of advanced ceramics by nondestructive evaluation, and a wide range of techniques are being developed in industrial and government laboratories to achieve this goal.

Nondestructive evaluation is used both to define flaws and to find them.[174] In some cases, thermomechanical modeling is used to determine stresses in a part over its lifetime, and fracture mechanics is used to define critical flaws as a function of location. These are followed by materials testing in conjunction with nondestructive testing, fabrication, and final nondestructive evaluation of the completed products.

In most cases the wide variation in strength found in advanced ceramics limits their usefulness. Nondestructive testing is used before strength is tested to define the flaw population and residual stress of the material. The testing shows the response of the flaws to stress and the environment. It is also used to determine the flaw population introduced by processing steps and as an aid in the selection of different processing methods. Defects are seeded into the materials to determine the effect of the test environment on flaws.[174]

Flaw Classes

Two classes of flaws are important in establishing the performance capability of structural ceramics. Class I flaws are those large in degree but small in extent (for example, pores, inclusions, foreign matter, and microcracks). Conversely, class II flaws are small in degree and large in extent. Examples of the latter include long-range variations in microstructure (such as phase content, porosity, composition, and grain size). The nature and extent of both classes of flaws are highly dependent on material processing. Both classes of flaws can have a significant effect on material performance. Class I flaws control material strength characteristics directly, in particular the fast-fracture strength. Class II flaws can also affect strength through control of the thermal and elastic properties as well as fracture toughness and oxidation behavior.

Nondestructive Evaluation Methods

The method developed for metals is not appropriate for ceramics since ceramics have much smaller critical flaws. The two best techniques are based on X-rays and ultrasonics. X-rays, which are used both as CAT (computer-assisted tomography) scans and planar microfocus radiography, can detect internal flaws as small as 2 μm. Ultrasonic techniques enable the detection of internal flaws in the 10- to 300-μm range and surface flaws ≥20 μm.

X-Ray Computer Tomography. Engineers and scientists at Argonne National Laboratories are evaluating ceramic engine components with X-ray computer tomography (CT), which is a bulk characterization technique that can display real-time, two-dimensional X-ray sections of complex parts, such as turbocharger rotors and engine valve components. Current X-ray systems take several minutes for the beam to scan a part, but the use of a cone-beam X-ray system to collect all the data at once, coupled with powerful computer processing, will make future systems much faster.

X-ray CT can be used to characterize organic binder compounds in green bodies to within 1 vol % and to determine with extreme sensitivity long- and short-range density gradients within both green and dense bodies. Conventional X-ray can only detect 1 to 2 percent density variations, whereas X-ray CT is 100 times more sensitive and can detect density variations ranging from 0.01 to 0.02 percent. In addition, X-ray CT can be used to determine the quality of a ceramic part from the beginning to the end of processing. In order not to miss defects, it is important to use the optimum X-ray photon energy for the thickness and density of the part and the proper plane of X-ray sectioning. X-ray CT is not as sensitive to cracks as conventional X-rays generated by microfocus X-ray tubes; however, these can only be used for thin parts, and thick parts require the more intense X-rays from linear accelerators.[174]

Ultrasonics. In the past, the standard nondestructive inspection procedure in the ceramics industry has been to use liquid-dye penetrants, which are pulled into surface cracks by capillary action and then bleed out to reveal their exact location and pattern. A severe limitation of this approach is that cracks must break the surface in order to be detected. In the absence of surface cracks, subsurface cracks and voids become active crack-initiation sites under load, making their detection essential also. Ultrasonics answers the need for a nondestructive method that can detect subsurface flaws and also is adaptable to the rapid inspection of large shapes.[175,176]

Ultrasonics is a sensitive technique to detect fracture-controlling flaws of 25 to 200 µm in the structural ceramic materials. The sensitivity for flaw detection is, of course, determined by the wavelength λ, and thus the frequency, of the incident ultrasonic beam and microstructure (such as grain size, porosity, second phase) of the materials. Because of the necessity for the detection of small flaws, frequencies much higher than those in conventional use have to be used. The practical application of high-frequency ultrasonics, however, is limited by transducer technology, specimen geometry, ultrasonic instrumentation, and attenuation.

The ultrasonic inspection method itself consists of numerous variants, many of which must be considered when evaluating a particular component. The major variables relate to the selection of the transducers (size, operating frequency, focal length), configuration (one or two transducers on opposite sides), orientation (normal or angled relative to the surface), and coupling (immersion, water jet, or contact). In determining the use of ultrasonics to evaluate ceramic armor, engineers at Martin-Marietta Corporation used the following criteria in their order of relative importance to select the method:

1. Minimum detectable defect size of cracks and voids in sections larger than 25.4 mm in thickness
2. Insensitivity to defect orientation and depth
3. Scalability to large components
4. Inspection time and cost

The use of through-transmission ultrasonic inspection to nondestructively inspect thick-section Al_2O_3 ceramics for critical defects was successful.[175] In particular, planar subsurface cracks running both parallel and perpendicular to the scanned surface were imaged using this approach.

Subsequent pulse-echo scanning also imaged cracks running parallel to the scanned surface that returned an echo. Measuring the time of arrival of this echo provided the capability for gaging the depth of these cracks.

The through-transmission method was also applied successfully to the detection of simulated cylindrical voids in these ceramics. Their detection threshold, as determined by the fraction of the void's projected area relative to the ultrasonic beam size, was determined to be 2 percent.[175]

Researchers at the National Institute of Standards and Technology are using ultrasonics to monitor the general quality of ceramics during processing and to detect and define flaws. One application is the evaluation of ceramic powder density during compaction by measuring the ultrasonic velocity, which is proportional to the density of the ceramic. Good and defective ceramic parts can be distinguished by the higher densities of the good parts during and after pressing. The second application is the determination of surface roughness using a noncontact air-coupled system, which measures the relative echo amplitude of a reflected 4-MHz ultrasonic wave. The wave is scattered by surface roughness; the greater the roughness, the less the amplitude.[174]

To detect flaws in engine ceramic material,[94] two types of flaw detective ultrasonic equipment were used. The system consisted of an x-y 50.0-MHz scanner, a 75.0-MHz broadband pulse and receiver, and a 50.0-MHz nominal-frequency focused transducer. The reflected radio frequency signals (output of the pulse and receiver) from the specimen were imaged with a radio-frequency peak detector and an Isoscan generator.

Both the peak detector and the Isoscan generator systems were found capable of detecting volume flaws of approximately 100 μm. The advantage of the peak detector over the Isoscan generator is that it is used in the interface gate mode, while the other is not. The advantage of the Isoscan generator over the peak detector is that it provides both the C-scan and the B-scan views. The two systems are therefore complementary.

Scanning Laser Acoustic Microscopy (SLAM). Acoustic microscopy is a technique for imaging localized changes in the elastic properties of materials. The physical properties that govern sound propagation are modulus and density. The variability in these two properties directly affects the acoustic properties of attenuation, velocity, and impedance of the material. The variation in the acoustic properties changes the microstructural insonification behavior of the specimen. Microstructural variations and the presence of flaws (voids, inclusions, and cracks) both in the bulk and on the surface of the specimen are observed as perturbations in the transmitted acoustic amplitude. The sensitivity and the resolution capability of the technique depend on the relative acoustic properties of the parent material and the anomaly (Fig. 8.27).

The scanning laser acoustic microscope consists of a 100-MHz transducer mounted inside a water stage. The specimen is placed on the stage and insonified by an incident bulk wave. This causes dynamic ripples on the specimen back surface. These surface perturbations are continuously scanned by a focused laser beam. The reflected light is received by an optoacoustic detector and processed electronically to provide a signal whose amplitude and phase are replicas of the surface perturbation. The use of a scanning laser beam to detect surface displacements (namely, the sound field) allows images of the transmitted and scattered-mode converted sound fields to be visualized independently.[94,177,178]

This type of acoustic microscope is a through-transmission technique that provides images through the entire thickness of the sample. The resolution and penetration are determined by the frequency. For a typical advanced ceramic, a 10-MHz beam will penetrate a few millimeters of material with resolution of 250 μm, whereas a 500-MHz beam generates a resolution of 5 μm.

Scanning Acoustic Microscopy (SAM) and C-Mode Scanning Acoustic Microscopy (C-SAM). These are ultrasonic surface and subsurface reflection techniques, respectively (Fig. 8.27). SAM and C-SAM are applied for the detection of surface flaws in advanced ceramics and bulk flaws in thin ceramic components, such as electronic substrates and devices.

Scanning Photoacoustic Microscopy (SPAM). Laser-scanned photoacoustic microscopy shows excellent potential for detecting surface and near-surface flaws

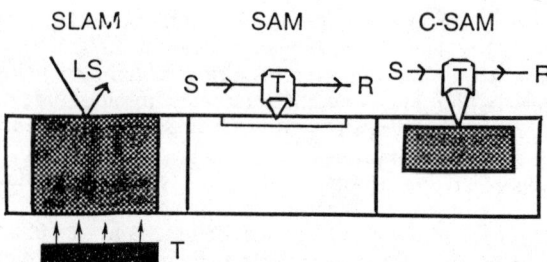

FIGURE 8.27 Three types of acoustic microscopy. Shaded areas are zones of application within samples. T—transducer; LS—laser scanner; S—send pulse; R—receive pulse.[174]

in opaque ceramics. In this technique the material to be examined is placed in a closed cell and scanned with a modulated laser light.[94]

To establish the feasibility of SPAM to detect surface and near-surface flaws, an experimental study was initiated. The two most important conclusions drawn from the study[94] were as follows:

1. SPAM is capable of detecting fracture-controlling flaws in the ceramic turbine-engine program (CATE) in the configuration for ceramic blades. This shows that this technique can be applied successfully to evaluate complex shapes.

2. The highest SPAM signal amplitude was not necessarily from a failure-causing flaw. Flaw characteristics (type, size, shape, and orientation) have a significant effect on the amplitude of the SPAM signal. Substantial development work has been done to determine the empirical relationship, if any, between the flaw characteristics and the SPAM signal.[94]

Ultrasonic Velocity. The ultrasonic properties of solids, including velocity and attenuation, are dependent on microstructure and, as such, are sensitive functions of chemistry, phase content, grain size, and porosity. Characterization of these properties can, therefore, be used as an effective means of monitoring material variability. In particular, ultrasonic velocity can be easily measured with great accuracy. The ultrasonic velocity of waves through a material is related to the density and elastic modulus.[94]

By measuring a part's thickness and the transit time of the acoustic wave through the medium, a relationship is developed that provides a means of determining the ultrasonic velocity of the material.

From studies[94] and experience gained in measuring ultrasonic velocity, blades and test bars have shown that velocity does indeed provide a valuable tool for detecting density variations and characterizing materials. Velocity is a material property that can be quickly and simply measured, nondestructively, on a wide variety of materials and components. Velocity measurement can act both as a research tool and as a quality-control test method.

Impulse Vibration Excitation Technique. This technique displays the dynamic elastic modulus as a characteristic relaxation time. The method can be applied to parts made from any material over a temperature range from 180 to 1500°C, provided the aspect ratio of the parts is greater than 2:1. The absolute modulus can be determined, but a relative modulus is sufficient for evaluation. Since, at room temperature, the method is so simple and fast—for example, only 1 s is required to inspect a part, and this can be decreased if desired—it can be used for 100 percent part inspection at every stage of the manufacturing process, both to detect defective parts and to make the process more uniform. It has been used to optimize grinding wheels for which proof-testing is expensive and dangerous.[174]

Conclusions and Recommendations

At the present time 5-μm flaws are not detectable, 50-μm flaws are always detectable, and flaws of intermediate size may be detectable, such as 10-μm tungsten inclusions in silicon-based ceramics. In addition, surface flaws as small as 1 μm are detectable.

There is no single technique that will detect all flaws. A number of techniques must be used, and these must be carefully optimized for the material, part, and application. X-ray CT is a much more widely useful technique than neutron radiography since the latter requires a nuclear reactor to provide enough flux to examine ceramic parts properly. However, ultrasonic techniques may be preferred over X-ray methods since ultrasonic waves present no hazard to the operator.

Finally general methods that measure the overall quality of ceramic parts are useful for improving the uniformity of the ceramic manufacturing process and for eliminating grossly defective parts, whereas specific methods are required to detect strength-limiting flaws in ceramics used in high-stress applications. Further development of these methods is needed to improve the reliability of advanced ceramics and to remove restrictions on their use.

REFERENCES

1. J. B. Wachtman, Jr., "The Materials Effect in the Manufacturing Revolution: Emphasis on Advanced Ceramics," in M. V. Nevitt and N. D. Peterson (Eds.), ANL 89-3, CONF 8806303, W31-109EN038, Feb. 1989, pp. 52–63.
2. K. Easterling, *Tomorrow's Materials,* Inst. of Metals, Gt. Br., 1988, 109 pp.
3. J. H. Gibbons, "New Structural Materials Technologies," in "Opportunities for Use in Advanced Ceramics and Composites," Congress Office of Tech. Assessment, Tech. Memo OTA TME 32, 1986, 88 pp.
4. *High Technology Ceramics in Japan,* NMAB-418, National Academy Press, 1984.
5. D. E. Niesz et al., "Materials," *Res. Develop.,* pp. 266–272, June 1984.
6. "Trends and Opportunities in Materials Research," Mater. Res. Advisory Comm., NSF, 1984.
7. L. L. Hench and J. Wilson, "Biocompatibility of Silicates for Medical Use," in *Silicon Biochemistry, CIBA Foundation Symp.,* Wiley, Chichester, 1986, pp. 231–246.
8. L. L. Hench, University of Florida, personal communication, Aug. 1986.
9. U.S. Department of Commerce, "A Competitive Assessment of the U.S. Advanced Ceramics Industry," U.S. Govt. Printing Office, Washington, D.C., Mar. 1984.
10. *High-Tech Materials Alert,* vol. 4, pp. 5–6, Aug. 1987.
11. R. H. J. Hannink, M. J. Murray, and H. G. Scott, "Friction and Wear of Partially

Stabilized Zirconia; Basic Science and Practical Applications," in *Wear 100*, Elsevier, Sequoia, Netherlands, 1984, pp. 355–366.

12. "Tribology of Ceramics," Committee on Tribology of Ceramics, Rep. NMAB 435, MDA903-86KO 220, NASA CR 182506, 1988, 117 pp.
13. J. N. Pennington, "Ceramics Add Life to Extrusion Dies," *Mod. Met.*, pp. 82–88, May 1990.
14. M. A. Burke, "Ceramics Enter the Foundry," *Des. News*, pp. 55–64, June 16, 1986.
15. S. C. Lattin, "Tough Valves with Ceramics," *Mach. Des.*, pp. 83–85, Aug. 6, 1987.
16. M. D. Pugh et al., "Fabrication of Si_3N_4 Valve Discs," in D. S. Wilkinson (Ed.), *Proc. Intl. Symp. on Advanced Structural Materials*, Montreal, Que., Aug. 28–31, 1988, Pergamon, New York, pp. 139–147.
17. K. Fukuda et al., "Wear Properties of Ceramics Against Some Metals," in D. S. Wilkinson (Ed.), *Proc. Intl. Symp. on Advanced Structural Materials*, Montreal, Que., Aug. 28–31, 1988, Pergamon, New York, pp. 148–164.
18. "Ceramic Valve Analysis," *Automot. Eng.*, vol. 96, pp. 46–53, May 1988.
19. D. W. Richerson, R. E. King, and J. Beall, "Ceramics; Applications in Manufacturing," SME, 1988, 217 pp.
20. E. P. Rothman et al., "Potential of Ceramic Materials to Replace Cobalt, Chromium, Manganese, and Platinum in Critical Applications," Materials Processing Center, M.I.T., Cambridge, Mass., FR 1/84, Jan. 6, 1984, 292 pp.
21. R. P. Bergstrom, "Here's What Norton Did to Aluminum—And Why," *Production*, pp. 83–85, Oct. 1989.
22. T. B. Troczynski et al., "Advanced Ceramic Materials for Metal Cutting," in D. S. Wilkinson (Ed.), *Proc. Intl. Symp. on Advanced Structural Materials*, Montreal, Que., Aug. 28–31, 1988, Pergamon, New York, pp. 157–168.
23. "Advanced Structural Ceramics: Technologies, Economics and Market Opportunities," GB-107, BCC Inc., Norwalk, Conn., Dec. 1987, 354 pp.
24. P. K. Mehrotra, "Hot Isostatic Pressing of Ceramic Metalcutting Tools," *Met. Prog.*, pp. 506–509, July/Aug., 1987.
25. K. J. A. Brookes, "Guide to the World's Coated Inserts," *Am. Mach.*, pp. 83–96, July 1990.
26. J. B. Pond, "Cutting Tool Buyers to Seek and Find Broader Offerings," *Metalwork. News*, p. 34A, Aug. 15, 1988.
27. L. D. Wedeven and T. A. Harris, "Rolling Element Bearings—Operating at the Extremes," *Mach. Des.*, pp. 72–76, Aug. 6, 1987.
28. J. G. Hannoosh, "Ceramic Bearings Enter the Mainstream," *Des. News*, pp. 224–229, Nov. 21, 1988.
29. J. F. Dill, R. A. Harmon, and E. M. Lenoe, "A Review of the State-of-the-Art in Rolling Element Bearing Technology in Japan," *ONFRE Sci. Bull.*, vol. 12, pp. 57–75, Oct.–Dec. 1987.
30. E. V. Zaretsky, "Ceramic Bearings for Use in Gas Turbine Engines," NASA TM 100288, ASME 88 GT 138, June 1988, 13 pp.
31. D. A. Granov, "Silicon-Nitride Based Materials for Modern Machining," presented at the ASM Intl. Ann. Mtg., Indianapolis, Ind., Oct. 2, 1989.
32. D. L. Deadmore and H. E. Sliney, "Friction and Wear of Monolithic and Fiber Reinforced Silicon-Ceramics Sliding against In-718 Alloy at 25 to 800°C in Atmospheric Air at Ambient Pressure," NASA TM 100294, Feb. 1988, 31 pp.
33. "Ceramics Offer Better Bearings," *Engineering*, pp. 895–896, Dec. 1986.
34. J. W. Boretos, "Ceramics in Clinical Care," *Am. Ceram. Soc. Bull.*, vol. 64, pp. 630–636, Aug. 1985.
35. *Adv. Mater.*, vol. 9, pp. 2–3, Mar. 23, 1987.

36. L. A. Lay, "Corrosion Resistant Ceramics," *Met. Mater.*, pp. 250–253, May 1987.
37. R. D. McIntyre, "Medical Metals; Some Current Uses for Metals in, on, and around Your Body," *Mater. Eng.*, pp. 40–47, Sept. 1982.
38. *New Mater. Jpn.*, vol. 6, p. 10, Nov. 1989.
39. A. K. Tjernlund et al., "Bioimplant Materials of Hot Isostatically Pressed Alumina and Zirconia: Mechanical Properties and Biocompatibility," in *Proc. Science of Ceramics Conf.*, vol. 14, 1988, pp. 799–804.
40. J. L. Drummond, "In Vitro Aging of Yttria-Stabilized Zirconia," *J. Am. Ceram. Soc.*, vol. 72, pp. 675–676, Apr. 1989.
41. R. W. Davidge, "Engineering Performance Prediction for Ceramics," *Mater. Sci. Technol.*, vol. 2, pp. 902–909, Sept. 1986.
42. I. C. Clarke, "Titanium Alloy Alumina Ceramics and UHMWPE Use in Total Joint Replacements," in *Proc. 30th Natl. SAMPE Symp.*, Mar. 19–21, 1985, pp. 1639–1647.
43. S. J. Marz, "High-Tech Hips," *Mach. Des.*, pp. 88–92, Nov. 9, 1989.
44. L. M. Sheppard, "Cure It with Ceramics," *AM&P*, vol. 129, pp. 26–31, May 1986.
45. S. M. Johnson and D. J. Rowcliffe, "Ceramics for Electric Power-Generating Systems," SRI Intl., NASA Tech Briefs, Apr. 1988, pp. 42–43.
46. S. S. Singh, "Design of a High Temperature Gas-Fired Heating System Utilizing Ceramics," *Ind. Heating*, pp. 18–20, Nov. 1988.
47. C. Majani, "Ceramics Materials for Fission and Fusion Nuclear Reactors," Energia Nucleare e Delle Energie Alternative, Italy, ENEA RT 71B, 88-1, Jan. 1988.
48. *High Tech Ceramics News*, vol. 1, pp. 1–4, Jan. 1990.
49. *High Tech Ceramics News*, vol. 1, pp. 1–9, Oct. 1989.
50. L. M. Sheppard, "Ceramics for Future Power Sources," *AM&P*, vol. 131, pp. 46–51, June 1987.
51. M. G. Coombs, "Industrial Compact Ceramic Finned-Plate Recuperator," FR 8/82–3/87, GRI 87-0349, Sept. 1987, 325 pp.
52. T. Stillwagon, "Performance Verification of Industrial Ceramic Materials; Environmental Effects on Performance," GRI 88-0144, Apr. 1988, 157 pp.
53. G. R. Peterson et al., "Evaluation of Advanced Ceramic High Temperature Burner Duct Recuperator," *Ind. Heating*, pp. 34–35, Jan. 1988,
54. S. L. Blum, S. H. Kalos, and J. B. Wachtman, Jr., "Ceramics Gain Leading Edge in Electronics," *Ceram. Ind.*, pp. 40–45, Apr. 1985.
55. "Advanced Ceramic Materials," Charles River Assoc., U.S. Dept. of Commerce, and Natl. Research Council, 1985, 651 pp.
56. "Improved Zirconia Oxyden-Separation Cell," NASA Tech. Briefs, Apr. 1988, pp. 42–43.
57. "High Temperature Fuel Cells Could Cut Power Costs," *Mach. Des.*, pp. 60–61, Nov. 6, 1986.
58. "Survey of Supply/Demand Relationship for Japanese Technical Information in United States; Field of Advanced Ceramics Research and Development," Dept. of Commerce, PB88-210943, Mar. 1988, 139 pp.
59. V. J. Tennery, "Ceramics in Engines—An International Status Report," *Ceram. Bull.*, vol. 68, pp. 362–365, Feb. 1989.
60. "Armor: Lightweight, Ceramic-Faced Composites," Military Spec. Mil-A-46103D, Aug. 1987, 41 pp.
61. D. J. Viechnicki et al., "Lightweight Armor—A Status Report," U.S. Army Mater. Tech. Lab., MTK TR 89-8, Jan. 1989.
62. T. A. Nobbe, "Lightening Armor's Load," *Mach. Des.*, pp. 44–50, Feb. 12, 1987.
63. R. K. Bart and J. C. Lindberg, "Ceramic Bodyguards," *AM&P*, vol. 132, pp. 69–72, Sept. 1987.

64. B. G. Newland, "Ceramics—Materials at the Sharp End of Defence," *Met. Mater.*, vol. 2, pp. 334–339, June 1986.
65. J. R. Bolch and R. J. Drake, "Silicon Carbide Makes Superior Mirrors," *Laser Focus World*, 4 pp., Aug. 1989.
66. G. Campbell et al., "High Strength Ceramic Composite Antenna Window Materials Traces Study," AFWAL TR 88-4247, F33-615-86C5040, Jan. 1989, 145 pp.
67. "Ceramic Parts for Turbines," NASA Tech. Briefs MFS-27081, 1987, 76 pp.
68. W. P. Keith, "Batch Furnace Technology for Space Shuttle Tile Manufacture," *Ceram. Bull.*, vol. 68, pp. 1307–1308, July 1989.
69. C. Covault, "Buran Inspection Shows Soviet Shuttle Details," *Aviation Week Space Technol.*, pp. 46–51, June 19, 1989.
70. R. G. O'Lone, "New Thermal System for Shuttle Urged," *Aviation Week Space Technol.*, pp. 45–50, June 4, 1979.
71. J. F. Creedon, Y. D. Izu, and W. H. Wheeler, "Advanced Composite Insulation; Keeping Cool at 2900°F," *Mater. Eng.*, pp. 57–60, May 1985.
72. W. H. Bennethum and L. T. Sherwood, "Sensors for Ceramic Components in Advanced Propulsion Systems," Task 3, PR4-12/87; "Summary of Literature Survey and Concept Analysis," NASA CR 180900, NAS3-25140, July 1988, 100 pp.
73. R. N. Katz, "Applications of High Performance Ceramics in Heat Engine Design," *Mater. Sci. Eng.*, vol. 71, pp. 227–249, 1985.
74. "Ceramic Technology for Advanced Heat Engines," FR 7/87, NMAB431, July 1987, 73 pp.
75. "Competing in Composites," *Business Tokyo*, pp. 26–31, July 1990.
76. "Technological and Economic Assessment of Advanced Ceramic Materials," vol. 1: "Summary and Conclusions," Charles River Assoc., Boston, Mass., NBS, Gaithersburg, Md., CRA Rep. 684, Aug. 1984.
77. "Slow Growth for Ceramics," *AM&P*, vol. 135, pp. 29–44, Jan. 1989.
78. S. Somiya, E. Kanai, and K. Ando, "Ceramic Components for Engines," Elsevier Appl. Sci., London, 1986, 848 pp.
79. B. J. Busovne and J. P. Pollinger, "Development of Silicon Nitride Engine Components for Advanced Gas Turbine Applications," ASME 90 GT 248, DEN3-335, DEN3-336, July 1990, 7 pp.
80. J. E. Blendell, "NIST's Ultra-Clean Ceramic Processing Laboratory," *J. Met.*, p. 53, Jan. 1989.
81. L. Cartz, "Engineering Materials for Very High Temperatures," presented at the ORNL Workshop, Office of Naval Res., Gt. Br., ORNL 8-016R, Aug. 1988.
82. L. M. Sheppard, "Global Outlook for the Ceramic Heat Engine," *Adv. Ceram. Mater.*, vol. 3, pp. 309–315, 1988.
83. E. P. Rothman, "Advanced Structural Ceramics: Technical and Economic Process Modeling of Production and a Demand Analysis for Cutting Tools and Turbochargers," Materials Systems Lab., M.I.T., Cambridge, Mass., Aug. 1985.
84. R. J. Kobayashi, "Development of Cost-Effective Manufacturing Process for Producing Ceramic Turbocharger Rotors," vol. 2: "Appendix," TACOM TR 13270 V2, DAAE 07-85C-R147, Aug. 1987.
85. H. Droscha, "Ceramic Components for Internal-Combustion Engines," FSTC HT 524-8, ADB086364, July 1984.
86. J. C. Bentz, "Ceramics Manufacturing Technology Development," AR1 12/84, MTL TR 86-13, DAAG 46-83C002, May 1986.
87. D. Fuller, "Little Engines That Can," *High Technol.*, pp. 12–17, June 1986.

88. J. B. Wachtman, Jr., et al., "Japanese Structural Ceramics Research and Development," FASAC, Tech. Assessment Rep., SAIC, La Jolla, Calif., July 1989, 259 pp.
89. S. B. Lasday, "Alpha Silicon Carbide Properties Advantageous for Automotive Water Pump Seal Faces Produced at New Facility in West Germany," *Ind. Heating,* pp. 35–39, Aug. 1990.
90. R. R. Wills and R. E. Southam, "Ceramic Engine Valves," *J. Am. Ceram. Soc.,* vol. 72, pp. 1261–1264, July 1989.
91. D. J. Landini, H. D. Lesher, and J. A. Gesing, "Ceramic Composite Valve Seat Inserts," in V. J. Tennery (Ed.), *Ceramic Materials and Components for Engines, Proc. 3d Intl. Symp.,* Las Vegas, Nev., Nov. 27–30, 1988, 1559 pp.
92. J. Neil, G. Bandyopadhyay, and D. Sordelet, "Development in Injection Molding Silicon Nitride Turbine Components," ASME 90 GT 186, June 1990, 6 pp.
93. R. W. Ohnsorg, J. E. Funk, and H. A. Lawler, "Fabrication of ATTAP Ceramic Turbine Components," ASME 90 GT 185, June 1990, 6 pp.
94. H. E. Helms et al., "Ceramic Applications in Turbine Engines," FR 10/84, NASA CR 174715, DOE/NASA/0017-6, Contract DEN3-17; DE-A101-77CS-51040, EDR-11442, Oct. 1984, 268 pp.
95. I. Stambler, "U.S. and Japan Funding Automotive and Industrial Ceramic Engine R&D Programs," *Gas Turbine World,* pp. 32–35, July–Aug. 1989.
96. M. Taguchi, "Applications of High-Technology Ceramics in Japanese Automobiles," *Adv. Ceram. Mater.,* vol. 2, pp. 754–762, Apr. 1987.
97. H. C. Yeh, "Improved Silicon Nitride for Advanced Heat Engines," AR 2/87, NASA CR 179525, NAS3-24385, Feb. 1987, 113 pp.
98. J. M. Schoenung, "Modeling the Fabrication Costs of Advanced Ceramic Engine Components," presented at the World Materials Cong., Chicago, Ill., Sept. 26, 1988, 23 pp.
99. M. D. Meiser, "Development of Silicon Nitride Components for Advanced Applications," presented at the World Materials Cong., Chicago, Ill., Sept. 26, 1988, 15 pp.
100. G. L. Boyd, "Ceramics for Gas Turbine Applications," presented at the World Materials Cong., Chicago, Ill., Sept. 26, 1988, 12 pp.
101. R. P. Larsen and L. R. Johnson, "Ceramic Turbochargers: A Case Study of a Near-Term Application of High-Strength Ceramics," ANL/CNSV-47, Contract W-31-109-ENG-38, Aug. 1984, 18 pp.
102. T. Sakamoto, H. Horinouchi, and T. Maeda, "Ceramics-to-Metal Joining Technology for Gas Turbine Rotors," ASME 89 GT 302, June 1989, 6 pp.
103. L. M. Sheppard, "Automotive Performance Accelerates with Ceramics," *Ceram. Bull.,* vol. 69, pp. 1012–1021, June 1990.
104. D. C. Larsen and J. W. Adams, "Evaluation of Ceramics and Ceramic Composites for Turbine Engine Applications," IITRI, Tech. Rep. 13, Contract F33615-82-C-5101, June 1983, 97 pp.
105. L. J. Schioler, "Heat Engine Ceramics," *Ceram. Bull.,* pp. 268–294, Feb. 1985.
106. *1987 Proc. Intl. Conf. on PM Aerospace Materials,* Metal Powder Rep., MPR Publ. Services Ltd., Gt. Br., 1988.
107. K. Katayama, M. Sasaki, and T. Itoh, "Development of Ceramic Turbine Rotors," ASME 88 GT 282, June 1988, 8 pp.
108. R. L. Holtman, "Design and Development of Ceramic Components," AIAA-88-3054, presented at the AIAA/ASME/SAE/ASEE 24th Joint Prop. Conf., Boston, Mass., July 11–13, 1988, 6 pp.
109. "Japan's Rising Star," *The Engineer,* pp. 30–31, Mar. 22, 1984.
110. H. E. Helms, "AGT 100 Project Summary," ASME 88 GT 223, June 1988, 11 pp.

111. Y. Hamano et al., "Development of Ceramic Components for High-Temperature Gas Turbines," in *Proc. MRS Intl. Mtg. on Advanced Materials,* May 31–June 3, 1988, Materials Research Soc., Tokyo, Japan, vol. 5, pp. 229–239.
112. T. N. Strom and P. T. Kerwin, "Automotive Ceramic Turbine Research," in *Proc. MRS Intl. Mtg. on Advanced Materials,* May 31–June 3, 1988, Materials Research Soc., Tokyo, Japan.
113. F. L. Boyd and D. M. Kreiner, "AGT 101/ATTAP Ceramic Technology Development," ASME 89 GT 105, June 1989; *J. Eng. Gas Turbines and Power,* vol. 111, pp. 158–167, 1989.
114. M. Patiky, "Ceramic Turbines for Cars Could Wind Up on Planes," *High Technol.,* pp. 56–58, Apr. 1987.
115. "Gas Turbine Combustor Made of Ceramic," 90-04-007-2, *New Technol. Jpn.,* vol. 18, pp. 35–36, 1990.
116. T. Bornemisza and J. Napier, "Comparison of Ceramic vs. Advanced Superalloy Options for a Small Gas Turbine Technology Demonstrator," ASME 88 GT 228, June 1988, 10 pp.
117. "R&D of Automotive Ceramic Gas Turbines in Japan," Special Issue of *New Technol. Jpn.,* 1988.
118. E. M. Lenoe, "Ceramics," *ONFRE Sci. Inform. Bull.,* vol. 14, pp. 59–77, 1989.
119. S. H. Updike, "Internal Combustion Engine Performance Testing of Sialon Structural Ceramic Components," *ONFRE Sci. Inform. Bull.,* vol. 14, 1989.
120. S. Sekiyama and H. Kawamura, "Development Status of Isuzu Ceramic Engine," *ONFRE Sci. Inform. Bull.,* vol. 14, 1989.
121. T. M. Yonushonis and J. C. Bentz, "Ceramic Materials in Diesel Engines," *ONFRE Sci. Inform. Bull.,* vol. 14, 1989.
122. P. Walzer, H. Heinrich, and M. Langer, "Ceramic Components in Passenger-Car Diesel Engines," *SAMPE Quart.,* vol. 17, p. 32, 1986.
123. D. W. Richerson, J. M. Wimmer, and S. M. Wander, "Ceramic Technology Requirements for 1425°C Uncooled Power Generation Applications," in E. M. Lenoe, R. N Katz, and J. J. Burke (Eds.), *Ceramics for High-Performance Applications,* vol. 3: *Reliability,* Plenum, New York, 1983.
124. S. O. Kronogard and L. Malmrup, "Ceramic Engine Research and Development in Sweden," in E. M. Lenoe, R. N Katz, and J. J. Burke (Eds.), *Ceramics for High-Performance Applications,* vol. 3: *Reliability,* Plenum, New York, 1983.
125. R. J. Lumby et al., "Syalon Ceramics for Advanced Engine Components," SAE #850521, presented at the Intl. Cong. and Exposition, Detroit, Mich., Feb. 25–Mar. 1, 1985, 9 pp.
126. W. Bryzik and R. Kamo, "TACOM/Cummins Adiabatic Engine Program," SAE #830314, "The Adiabatic Diesel Engine," Rep. SP-543, 1983.
127. R. Kamo and W. Bryzik, "Ceramics for Adiabatic Turbocompound Diesel Engines," in *Proc. Intl. Symp. on Ceramic Components for Engines,* Japan, 1983, pp. 59–99.
128. M. Marmach and M. V. Swain, "Suitability of Mg-PSZ for Adiabatic and Conventional Diesel Engine Applications," in *Proc. Intl. Symp. on Ceramic Components for Engines,* Japan, 1983, pp. 650–659.
129. I. Oda, M. Matsui, and T. Soma, "Strength and Ductility of PSZ Ceramics," in *Proc. Intl. Symp. on Ceramic Components for Engines,* Japan, 1983, pp. 660–671.
130. L. Sheppard, "Ceramic Engines Are Hot," *Mater. Eng.,* pp. 41–47, Oct. 1984.
131. "Advances in Ceramics Spur Adiabatic Engine Development," *Mech. Eng.,* pp. 64–67, July 1985.
132. R. Kamo, "Ceramics Play Integral Role in Adiabatic's Future," *Ceram. Ind.,* pp. 26–29, July 1984.

133. P. Boch, J. F. Coudert, and C. Gault, "Study of the Thermal Fatigue Resistance of Plasma-Sprayed Zirconia Coatings for Diesel Components," in *Proc. Intl. Symp. on Ceramic Components for Engines,* Japan, 1983, pp. 682–689.

134. J. J. Brennan, "Investigation of Lithium Aluminosilicate (LAS)/SiC Fiber Composites for Naval Gas Turbine Applications," FR 9/82–9/83, UTRC R 83-91623, Oct. 1983.

135. H. Gasthuber et al., "HIPRBSN Prechambers for the Diesel Engine," in V. J. Tennery (Ed.), *Ceramic Materials and Components for Engines, Proc. 3d Intl. Symp.,* Las Vegas, Nev., Nov. 27–30, 1988, pp. 1477–1488.

136. T. Suzuki, M. Tsujita, and Y. Mori, "An Observation of Combustion Phenomenon on Heat Insulated Turbo-Charged and Inter-Cooled D.I. Diesel Engines," SAE#861187, 1986.

137. K. Arakawa et al., "A Two-Piece Articulated Ceramic Piston for Diesel Engines," in V. J. Tennery (Ed.), *Ceramic Materials and Components for Engines, Proc. 3d Intl. Symp.,* Las Vegas, Nev., Nov. 27–30, 1988, pp. 1407–1417.

138. D. L. Hartsock, "A Simplified Structural Ceramic Design Technique," ASME 85 GT 100, 1985.

139. D. L. Hartsock, "Partially Stabilized Zirconia Piston Bowl Reliability," *J. Eng. Gas Turbines and Power,* vol. 109, pp. 367–373, Oct. 1987.

140. *Production,* p. 27, May 1986.

141. S. Goto, Isuzu Motors America, and H. Matsuoka and H. Kawamura, Isuzu Ceramics Res. Institute Ltd., personal communication, 1989.

142. M. Yoshida and A. Kokaji, "Firing Up the Future with Ceramic Engine Parts," *Mach. Des.,* pp. 58–64, Oct. 26, 1989.

143. M. P. Czyz and D. J. Yanko, "Advanced Ceramics for Electronics," *Mach. Des.,* pp. 77–80, June 25, 1987.

144. E. J. Stefanides, "Beryllia Does It Best in Traveling Wave Tubes," *Des. News,* pp. 246–247, Nov. 23, 1987.

145. G. L. S. Buchanan, "SMT on Thick Film Ceramics Ensures Miniaturization, Reliability, Stability," *Surface Mount Technol.,* pp. 13–15, June 1988.

146. T. M. Kuhl, "Passive Suppliers; Active in SMT," *Surface Mount Technol.,* pp. 42–45, Aug. 1988.

147. A. J. Lyke, "CAD Speeds and Simplifies Ceramic IC Package Design," *Microelectron. Manuf. Test.,* pp. 21–23, Nov. 1988.

148. R. L. Keusseyan, "Brazing to Low-Temperature-Fired Thick Films," *IEEE Trans. Comp., Hybrids, Manuf. Tech.,* vol. 13, pp. 219–221, Mar. 1990.

149. J. C. Southern, "Ceramic Aluminas; Manufacturing, Use and Characteristics," EM89-118, presented at Advanced Ceramics '89, Philadelphia, Penn., Feb. 20–22, 1989, 23 pp.

150. W. C. Shumay, Jr., "Copper's Expanding Role in Microelectronics," *AM&P,* vol. 132, pp. 54–60, Dec. 1987.

151. J. Intrater, Adv. Tech. Inc., personal communication, July 1990.

152. J. Intrater, "The Challenge of Bonding Metals to Ceramics," *Mach. Des.,* 6 pp., Nov. 23, 1989.

153. W. C. Shumay, Jr., "Super-Conductor Materials Engineering," *AM&P,* vol. 134, pp. 49–58, Nov. 1988.

154. W. C. Shumay, Jr., "Superconductors to Market?" *AM&P,* vol. 132, pp. 44–49, Sept. 1987.

155. J. Teresko, "Superconductivity—Do All Applications Have To Wait?," *Ind. Week,* pp. 47–49, Sept. 21, 1987.

156. D. Stovicek, "Temperature Is Rising on Superconductivity," *Automation,* pp. 14–15, Sept. 1988.

157. K. Salama et al., "The Influence of Fabricating Technologies on the Structure and Properties of $YBa_2Cu_3O_{7-x}$," *J. Met.*, pp. 6–10, Aug. 1988.
158. B. D. Nordwall, "U.S. Falling behind Japan in Superconductor Research," *Aviation Week Space Technol.*, pp. 57–59, Jan. 16, 1989.
159. T. A. Heppenheimer, *Superconductivity; Research, Applications and Potential Markets,* Pasha Publ., Arlington, Va., 1988, 90 pp.
160. G. M. Robinson, "Superconductors; High-Tech Sweepstakes," *Des. News,* pp. 60–64, June 20, 1988.
161. C. F. Lewis, "Conductive Ceramics," *Mater. Eng.,* pp. 27–30, June 1987.
162. "Superconducting Composite Materials," *MMICIAC,* vol. 8, 3 pp., Mar. 1988.
163. E. T. Smith et al., "Superconductors: The All-Out Pursuit of Zero Resistance," *Business Week,* pp. 56–59, Mar. 14, 1988.
164. G. Kordas, K. Wu, and U. S. Brahme, "High-Temperature Ceramic Superconductors Derived from the Sol-Gel Process," *Mater. Lett.,* vol. 5, pp. 417–419, Oct. 1987.
165. L. E. Murr, A. W. Hare, and N. G. Eror, "Metal-Matrix High Temperature Superconductor," *AM&P,* vol. 132, pp. 37–44, Oct. 1987.
166. D. Hughes, "Pentagon Boosts Research Spending to Develop Practical Superconductors," *Aviation Week Space Technol.,* pp. 57–61, Nov. 16, 1987.
167. D. Hughes, "Aerospace Agencies Foster Research, Application Studies on Superconductors," *Aviation Week Space Technol.,* pp. 89–92, Nov. 23, 1987.
168. H. Schwartz, AREMCO Products, private communication, Nov. 1990.
169. G. Fisher, "Refractory Uses—Practicality of High Technology Ceramics," *Ceram. Bull.,* vol. 66, pp. 1103–1108, July 1987.
170. W. Gazoag, "Ceramic Gage Blocks; Durable," *Am. Mach.,* pp. 46–50, June 1990.
171. E. J. Stefanides, "Beryllia Ceramic Extends Life of Jet Engine Part," *Des. News,* pp. 96–97, Jan. 20, 1986.
172. F. Marc de Piolenc, "Livermore Lab Seeks Partnership with Industry for CHE Development Program," *Gas Turbine World,* 36 pp., Oct. 1988.
173. C. F. Lewis, "Zirconia; The Tough Contender," *Mater. Eng.,* pp. 43–45, May 1989.
174. R. F. Firestone, "NDE; Improving Reliability of Advanced Ceramics," *Ceram. Bull.,* vol. 68, pp. 1177–1186, June 1989.
175. C. L. Friant, "Ultrasonic NDE Approaches for Structural Ceramics," *SAMPE Quart.,* vol. 26, pp. 31–34, Apr. 1990.
176. M. Horino et al., "Strength and Evaluation of Ceramic Joints by the Ultrasonic Method," in M. Doyama and S. Somiya (Eds.), *Proc. MRS Mtg. on Advanced Materials,* Tokyo, June 2–3, 1988, vol. 8, pp. 119–124.
177. D. J. Roth et al., "Reliability of Void Detection in Structural Ceramics Using Scanning Laser Acoustic Microscopy," NASA TM-87035 (N85-32337/NSP), 1985, 54 pp.
178. S. K. Klima and G. Y. Baaklini, "Nondestructive Characterization of Structural Ceramics," *SAMPE Quart.,* vol. 17, no. 3, p. 13, 1986.

BIBLIOGRAPHY

Akimune, Y, Y. Katano, and K. Matoba, "Spherical-Impact Damage and Strength Degradation in Silicon Nitrides for Automobile Turbocharger Rotors," *J. Am. Ceram. Soc.,* vol. 72, pp. 1422–1428, Nov. 1989.

Anderson, N., "Finding Non-Automotive Applications for Silicon Nitride," presented at Ceramtec '90, Dearborn, Mich., June 12–13, 1990.

Ashley, S., "Ceramic-Metal Composites; Bulletproof Strength," *Mech. Eng.*, vol. 112, pp. 46–51, July 1990.

Baker, A. R., D. J. Dawson, and D. C. Evans, "Ceramics and Composite Materials for Precision Engine Components," *Metal. Des.*, vol. 8, p. 315, 1987.

Bandyopadhyay, G., and K. W. French, "Fabrication of Near-Net-Shape Silicon Nitride Parts for Engine Application," ASME 86 GT 11, June 8–12, 1986, 4 pp.

Bandyopadhyay, G., et al., "Fabrication and Development of Axial Silicon Nitride Gas Turbine Rotors," ASME 90 GT 47, June 1990, 8 pp.

Bright, E., "Development of Statistical Process Control Schemes for Ceramic Engine Component Manufacturing," presented at Ceramtec '90, Dearborn, Mich., June 12–13, 1990.

Buchanan, R. C., *Ceramic Materials for Electronics*, Marcel Dekker, New York, 1986, 470 pp.

Buckley, D., and K. Miyoshi, "Tribological Properties of Structural Ceramics," NASA TM 87105 (N86-10341/NSP), LEW-14387, Tech. Briefs, p. 64, Oct. 1987.

Bunshah, R. F., and C. V. Deshpandey, "Processing Science and the Technology of High T_c Films," *R&D*, pp. 65–79, Jan. 1989.

Burnfield, K. E., and D. Schweizer, "The Effects of Organic Additives on Ceramic Body Processing," SME EM 90-136, Feb. 1990, 15 pp.

"Ceramics: More Choices for Designer," *Des. News*, pp. 34–36, July 4, 1988.

Cole, W. E., et al., "Research and Development of a Ceramic CVD Composite Heat Exchanger for Industrial Waste Heat Recovery," Ann. Rep., Phase II, DOE/ID/12544-2(DE88025789), May 1987, 110 pp.

Davis, J. G., and H. N. Murrow, "Enabling Technologies Research and Development Structures; High Temperature Panel-Edge Seal Test Rig," AIAA 89-5011, presented at the AIAA 1st Natl. Aero-Space Plane Conf., Dayton, Ohio, July 20–21, 1989, 26 pp.

DeYoung, H. G., "Marching into the New Stone Age," *High Technol.*, pp. 50–52, Aug. 1985.

DuFrane, K. F., "Wear Performance of Ceramics in Ring/Cylinder Applications," *J. Am. Ceram. Soc.*, vol. 72, pp. 691–695, Apr. 1989.

"Engineering Ceramics in Component Design," *Eng. Mater. Des.*, pp. 47–51, Oct. 1986.

Europ. Conf. on Advanced Materials and Processes, RD 6355 MS02, DAJA45-90M0049, Deut. Ges. f. Metallkunde, Aachen, Germany, Nov. 24, 1989.

Fairbanks, J., "The Enigma of the Adiabatic Engine," presented at Ceramtec '90, Dearborn, Mich., June 12–13, 1990.

Fisher, G., "Zirconia; Ceramic Engineering's Toughness Challenge," *Ceram. Bull.*, vol. 65, pp. 1355–1360, Oct. 1986.

Fox, D. S., N. S. Jacobson, and J. L. Smialek, "Hot Corrosion of Ceramic Engine Materials," FR 10/88, NASA TM 101439, DOE NASA 50111-2, DE-A101-85CE50111, Oct. 1988, 15 pp.

Gee, M. G., "Electrical Ceramics," *Met. Mater.*, vol. D, pp. 769–775, Dec. 1986.

Gruninger, M. F., "Sol-Gel Ceramics as Coatings," *Ceram. Ind. Intl.*, vol. 98, no. 1074, pp. 22–24, 1989.

Hamling, P., "1700°C Rapid Cycle Furnace Insulation Design Part II—Chamber Design," *Ind. Heating*, pp. 20–22, May 1988.

Helms, H. E., and S. R. Thrasher, "Ceramic Applications in Turbine Engines (CATE) Development Testing," ASME 83 GT 179, 1983, 7 pp.

Hirosaki, N., A. Okada, and M. Mitomo, "Effect of Oxide Addition on the Sintering and High-Temperature Strength of Si_3N_4 Containing Y_2O_3," *J. Mater. Sci.*, vol. 25, pp. 1872–1876, Mar. 1990.

Jang, S. D., "Recent Developments in the Advanced Ceramics Industry of Korea," *Ceram. Bull.*, vol. 67, pp. 1531–1536, Sept. 1988.

Jones, R. D., et al., "Ceramic Parts for Turbines," MFS-27 801, NASA Tech. Briefs, Apr. 1987, 2 pp.

Jones, S. L., "Two Defense Units Target Superconductor Applications," *Metalwork. News,* p. 8, Feb. 22, 1988.

Kamo, R., "High Temperature Tribology for Adiabatic Engines," presented at Ceramtec '90, Dearborn, Mich., June 12–13, 1990.

Kubel, E. J., Jr., "Development and Application of Seeded Sol-Gel Achievement," *ASM News,* pp. 4, 14, Oct. 1989.

Langer, M., H. Heinrich, and J. E. Siebels, "Development of Ceramic Components for Car Gas Engines," *High Temp. Technol.,* vol. 2, pp. 225–228, Nov. 1984.

Larsen, R., "Development of Ceramic Engine Components for Wear Application," presented at Ceramtec '90, Dearborn, Mich., June 12–13, 1990.

Lasday, S. B., "New Chemical Deposition Process Speeds Production of Thin-Film Superconductors on Flexible Metal or Ceramic Filaments," *Ind. Heating,* pp. 18–19, Oct. 1990.

Levine, J. B., et al., "Venture Capital's New Gold Rush," *Business Week,* pp. 66–71, Oct. 5, 1987.

Lindberg, L. J., "Elevated Temperature Durability of Ceramic Materials," AIAA-88-3055, presented at the AIAA/ASME/SAE/ASEE 24th Joint Propulsion Conf., Boston, Mass., July 11–13, 1988.

McDermott, J., "Ceramics," *Technology,* pp. 19–30, Nov.–Dec. 1981.

Milberg, M. E., "Ceramics for Cars," *Chemtech,* pp. 552–558, Sept. 1987.

Mitchell, T. E., et al., "Processing Ceramic Superconductors," *J. Met.,* pp. 6–10, Jan. 1989.

Mito, S., "The Ceramics Revolution; An Industrial Army to Bring about the Fourth Industrial Revolution," transl., L. Kanner Assoc., Redwood City, Calif., FSTC-HT-0215-85, Mar. 1986, 196 pp.

Narasimhan, M., "Processing and Applications of High-T_c Superconductors," *J. Met.,* p. 24, Jan. 1989.

"New Era for Advanced Ceramics," *Des. News,* pp. 30–37, Mar. 21, 1988.

Nutter, K. M., "Developments in Ceramic Materials and Ceramic-Based Ballistic Armour," *J. Inst. Refract. Eng.,* pp. 2–5, Autumn, 1989.

Ohnsorg, R., M. Teneyck, and T. Sweeting, "Development of Injection Molded Rotors for Gas Turbine Applications," ASME 86 GT 45, June 1986, 8 pp.

Poeppel, R. B., S. E. Dorris, and C. A. Youngdahl, "Shape Forming High-T_c Superconductors," *J. Met.,* pp. 11–13, Jan. 1989.

Rhys-Jones, T. N., and F. C. Toriz, "Thermal Barrier Coatings for Turbine Applications in Aero Engines," *High Temp. Technol.,* vol. 7, pp. 73–81, 1989.

Robinson, G. M., "Superconductivity Starts to Go Commercial," *Des. News,* pp. 24–25, May 8, 1989.

Sernetz, F., "Production of Orthodontic Parts from Metaland Ceramics by Metal-Injection-Molding and Hot Isostat Presses," *Powder Metall. Intl.,* vol. 22, pp. 69–74, June 1990.

Shannon, J. L., Jr., "Structural Ceramics," NASA CP-2427, May 1986, 229 pp.

"Silicon Nitride Balls for Cryogenic Bearings," MSFC, AL and RI Corp., NASA Tech. Briefs, p. 63, July 1990.

Smidt, F. A., and R. J. Blau, "Engineered Materials for Advanced Friction and Wear Applications," in *Proc. ASM Intl. Conf.,* Gaithersburg, Md., Mar. 1–3, 1988, 262 pp.

Smith, E. T., "The Superconductors' New Bag of Tricks," *Business Week,* pp. 91–92, Oct. 24, 1988.

Stadler, H. L., "Why Ceramic Engines?," in *Proc. 12th Automotive Materials Conf.,* Ann Arbor, Mich., Mar. 14–15, 1984, vol. 5, p. 281.

"Study on Biomechanomimetic Materials (Artificial Bone, Joints and Muscle)," Mech. Eng. Lab. Rep., MITI/AIST, Mar. 1987.

"Superconductivity; NASA Rises to the Challenge," NASA Tech. Briefs, p. 8, Apr. 1988.

"Superconductor Ceramics—An Industry Roundtable," *Des. News,* pp. 22–24, Nov. 21, 1988.

Swab, J. J., "Properties of Yttria-Tetragonal Zirconia Polycrystal (Y-TZP) Materials after Long-Term Exposure to Elevated Temperatures," MTL TR 89-21, AD-A207 064, Contract DE-A105-84OR21411, Mar. 1989, 24 pp.

"Theory Puts Limits on High-Temperature Superconductors," *R&D,* pp. 20–22, Nov. 1988.

Thummler, F., "Structural Ceramics," *Keram. Z.,* vol. 4, pp. 744–745, Oct. 1989.

Torre, J. P., Y. Raynal, and Y. Rouaux, "Selecting Ceramics for Diesel Engine Applications—Methodological Aspects," in *Proc. Intl. Symp. on Ceramic Components for Engines,* Japan, 1983, pp. 120–129.

Tovell, J. F., "Ceramics and the Reciprocating Internal Combustion Engine," *Mater. Des.,* vol. 5, p. 215, 1984.

Woods, M. E., and T. L. Scofield, "Designing Adiabatic Engine Components," in *Proc. Intl. Symp. on Ceramic Components for Engines,* Japan, 1983, pp. 130–145.

Woodward, R. M., "CERH$_x$ Ceramic Heat Exchangers; Five Years of Energy Conservation in the Aluminum Industry," Hague Intl., South Portland, Maine, Feb. 1985, pp. 771–779.

CHAPTER 9
THE FUTURE OF CERAMICS

9.1 INTRODUCTION

Ceramic materials are unglamorous, and therefore their importance is little understood. But as finished products, they are indispensable throughout industry. In the next 25 years, new structural materials will provide a powerful leverage point for the manufacturing sector of the economy. Ceramic and ceramic composite components can not only deliver superior performance, they also enhance the performance and value of the larger systems, such as aircraft and automobiles, in which they are incorporated. Given this multiplier effect, it is likely that the application of advanced structural ceramic materials will have a dramatic impact on gross national product, balance of trade, and employment in the United States. All of the industrialized countries have recognized these opportunities and are competing actively for shares of the large commercial and military markets at stake.

9.2 CERAMICS AND PROCESSES

Sol-Gel[1-3]

There is an ever-increasing demand for engineering ceramics which can resist hostile environments, particularly high temperatures and corrosive atmospheres. Sol-gel processing has proved very effective for producing nuclear ceramics such as thoria, urania, or plutonia, which can successfully withstand the rigors of fast breeder reactors. Having proven that nuclear sol-gel processes produce consistently dense spheres, which are being used in the fabrication of nuclear ceramic fuel elements, the following general example has been serving to prepare nonnuclear ceramics:

$$\text{Starting material} \rightarrow \text{sol} \rightarrow \text{gel} \rightarrow \text{oxide}$$

There are opportunities for exploiting the major advantages of the sol-gel process, namely, free-flowing powders of controlled size, shape, porosity, and composition, produced by plasma-spray equipment. The hope is that this plasma process will contribute to the development of new and innovative coating techniques. The main reason for pursuing the field of thermal spraying and sol-gel-processed ceramic powders is that free-flowing powders, with very small particle sizes (<20 μm), are widely predicted to yield smooth coatings.

One area of ongoing development that is leading to ceramic materials with more predictable mechanical performance is refinement in processing methods through cleaner processing facilities. Experience in the industry is growing with regard to clean-room conditions and the handling of submicrometer ceramic powders.

Sol-gel powders demonstrate flow properties at these small particle sizes which enable them to be sprayed using existing equipment. Material with various chemical compositions can be prepared and supplied in relatively narrow-particle-size distributions at a realistic cost, with the goal of improving deposit efficiencies and allowing a greater degree of control over the plasma-spray process. Fully reacted powders of a fixed composition and similar particle sizes can be supplied with a wide range of bulk densities.

Such materials will enable plasma-spraying techniques to be developed and optimized to extend the range of engine components which can be coated. Both the line-of-sight deposition and reasonably constant coating thickness, with increasing distance from the nozzle, indicate strongly that sol-gel powders may be the key to plasma-spraying components with more complex geometries. The flexibility of sol-gel processing permits a rapid response to any future demands of the advanced aerospace industry for improved powders, such as new chemical compositions or particle-size distributions.

Gelcasting

The gelcasting process is being developed as a possible alternative net-shape-forming method for advanced ceramics. The process offers the potential for manufacturing advantages and improved reliability of products compared with forming by slip casting or injection molding. The gelcasting process disperses ceramic powder in an aqueous solution of acrylamide monomer, and "gelling" is accomplished by thermal polymerization. Initial processing studies have been conducted with Al_2O_3 at Oak Ridge National Laboratory and according to Stiegler and Weir,[4] pieces of various sizes and shapes have been formed and sintered. Future plans include measurement of mechanical properties to evaluate the effects of different process parameters. Potentially significant advantages of the gelcasting process are the high strength and excellent machinability of green parts, even though the process is intended for net-shape forming.

In addition to the processing of monolithic Al_2O_3, the water-based gelcasting process was also evaluated as a technique for fabricating ceramic-fiber-reinforced ceramic-matrix composites. Materials investigated included the forming and densification of continuous and chopped Al_2O_3 (FP) and Al_2O_3–ZrO_2 (PRD 166) fibers in Al_2O_3 and ZrO_2 matrices. Initial results indicated that the degree of uniformity of dispersion of the fibers in a slurry strongly influenced the quality of the composites. As the sintering temperature increased, grain growth and void formation in the fibers and reaction between fiber and matrix became severe. A thin carbon coating on the fibers was not effective in preventing these reactions. PRD 166 fiber was more stable when sintered at high temperatures in an Al_2O_3 matrix than was FP.

Other studies continue in the investigation of the suspension chemistry of gelcasting slurries and complex shapes in silicon powder were gelcast successfully prior to nitridation to produce reaction-bonded Si_3N_4 parts.[4]

Microwave

Microwave processing experiments and studies are continuing to determine the outstanding potential of this heating medium. Studies have examined the grain growth and diffusion on Al_2O_3 and sintering ZrO_2. These studies[4] have evaluated both conventional and microwave (28-GHz) heating of a dense hot-pressed Al_2O_3 ceramic. Microstructural evolution was the same for both heating methods; grain growth followed a typical power-law form. However, the rate of grain growth at a given temperature was much higher with microwave heating.

In the sintering of ZrO_2 it was found that the sintering temperature for ZrO_2–8 mol % Y_2O_3 is lowered by 100 to 150°C by 2.45-GHz microwave heating, compared with conventional firing.

New investigations include the identification of those aspects of microwave processing of dense Si_3N_4 that might (1) accelerate densification, (2) permit sintering to high density with much lower levels of sintering aids, (3) lower the sintering temperature, or (4) produce unique microstructures. Differences have been found in sintering at two different power levels. Therefore testing is being done at 2.45 and 28 GHz.

Testing also is going on whereby reaction-bonded Si_3N_4 is being evaluated versus sintered Si_3N_4. Reaction-bonded Si_3N_4 offers many advantages. Nitridation in the microwave appears to be increased by at least a factor of 3 as compared with conventionally nitrided specimens. Increased nitridation has been observed with higher iron content and smaller-particle-size silicon powders in both microwave-heated and conventionally heated specimens. Nitridation is also enhanced by the use of Y_2O_3 and Al_2O_3 sintering aids. Thus direct nitridation of high-purity silicon followed by a sintering step may yield dense Si_3N_4.

Microwave heating is a novel approach to the processing of finely dispersed oxide powders from the inorganic or organic solutions by microwave techniques and seems to be a very promising way for batchwise or continuous production of pure or mixed oxides. This technique also allows the decomposition of sol-gel combinations with subsequent recovery of the organic precursors and the solvents. Kladnig and Horn[5] believe that microwave decomposition of solution (MDS) could be a powerful method to make ceramics, pigments, or electronic ceramic materials in a continuous and economical way.

Microwave sintering can also be a fairly economical and time-saving process because heat-up and sintering times are much faster and capital costs lower. Researchers at Los Alamos National Laboratory have sintered B_4C to 95 percent of theoretical density without carbon or other sintering aids in less than 12 min using 2.45-GHz radiation. The energy used for this microwave sintering was found to be 18 percent less than that for inductive hot pressing. Others are combining conventional and microwave sintering. It is less expensive to use conventional heating to remove 95 percent of the water or other additives, and microwave energy is faster and more cost-efficient in removing the remaining 5 percent (Fig. 9.1).

Other researchers at the United Technologies Research Center have been looking at using microwaves to join ceramics. Microwaves have the advantage of being an energy that is easily focused. High-quality joints have been produced in Al_2O_3, mullite, and Si_3N_4 by microwave heating the ends of two rods butted together. In some cases, the microwave process proved to be eight times faster than conventional sealing processes and used about 5 percent of the energy.

The foregoing information indicates that considerable progress has been made in processing a wide variety of ceramic materials. The results also show the versatility and adaptability of the microwave process to applications involving liquid, gaseous (plasma), and solid states.

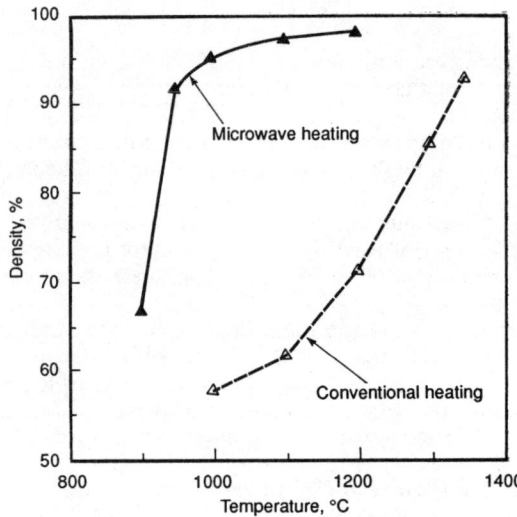

FIGURE 9.1 When 28-GHz microwaves are used, higher densities are achieved in sintered ceramics. Material is AKP-50 Al_2O_3 (0.1 wt % MgO) from Sumitomo.[2]

Since microwave processing is new to most of the ceramics industry, considerable time may be required to develop and implement it as a novel production process. The use of microwaves may be difficult to justify on energy costs alone,[6] since microwave energy is more expensive than fossil fuel energy. However, other benefits from the microwave process, such as much faster throughput of materials, improved quality and yields, less floor space, lower inventory, and new or unique properties, may show that the total economics of the process is not only viable but highly desirable.

Unfortunately many of the benefits are not realized at the initiation of an economic analysis and do not become evident until after the laboratory tests have been completed. In some cases, major savings in processing time, energy, and costs can be achieved by hybridizing the process, that is, by using both conventional and microwave processing,[6] or possibly by using microwaves for more than one process step.

In recent years, much progress has been made in microwave processing technology. Many problems still need to be solved, ranging from basic scientific studies on microwave-material interactions and loss mechanisms to engineering application programs involving both process and equipment design. There is a critical need for a broad database on dielectric properties to high temperatures at various frequencies for a wide range of ceramic materials. Such data are needed to develop process models that will predict the internal fields and the heating patterns and rates, so that optimum processing parameters can be developed to meet material and product requirements.

There is also a need for greater sophistication in processing equipment and applicators designed to meet specific product requirements, and to provide high-power densities to achieve maximum use of energy, heating uniformity, and more efficient heating of low-loss (low tan δ) materials. This includes the need to further develop sensors for continuously monitoring and controlling the process parameters to provide feedback for in-process control.

Another concern is cost, especially that of the microwave power required. Is it really as economical as some claim? Though the cost is high initially, it should decrease significantly as production increases. It was also pointed out that in order to implement the technology, a total systems approach is required; otherwise the application will fail. Microwave processing could be considered a "risk" investment because of its long-term nature, and therefore ceramic manufacturers must be willing to put in the time and money to make it work.

With regard to capital costs, which may seem high initially, according to one expert such costs are actually nominal because the return on investment is 100 percent within months. Depending on the application, capital costs can also be much lower, and other studies have indicated the same for energy costs. Therefore economics may be a major driving force behind this technology. Still other concerns are safety and the potential hazards to employees, which often inhibit companies from investigating this process.

Injection Molding

Injection molding and ceramic injection molding have been used to produce turborotors and fiber-optic ferrules, as well as to predict orientation effects during densification.

When consolidating ceramic composites, optimizing processing conditions becomes even more complicated, especially when injection molding is to be used. Scientists have found that injection-molded samples of Si_3N_4 containing 30 percent SiC whiskers, processed HIP to greater than 99 percent density using glass encapsulation, distort in a systematic manner.[7] Thus whisker orientation and aspect ratio have been measured based on digitized scanning electron micrographs in order to evaluate the microstructure with respect to the injection-molding direction.

A computer program was developed to digitize whisker data for input into the mainframe computer and to provide visual feedback during digitizing. The program also calculates length, width, aspect ratio, orientation angle, and intersected area for discrete whiskers. The whisker data can be archived for future use.

By being able to predict the whisker orientation, parts such as thin airfoil sections could be designed to have a high degree of orientation. These parts could be fabricated with whiskers aligned perpendicular to the likely crack direction caused by foreign object damage. Therefore maximum material toughness could be achieved.

Predictions on the future of ceramic injection molding have been made, based on the actual production rate of 1000 turborotors per month, to include other products producing $50 million in U.S. sales in 1992.

Finally it should be noted that over the last few years, rapid strides have been made in understanding the injection-molding process of complex ceramic parts of high tolerance. The other advantages of the process are high rates of production, low scrap rates, and low cost per part. The most common materials to be injection-molded have been reaction-bonded and sintered Si_3N_4.

Now injection-molded parts can be produced from ZrO_2–12 mol % CeO_2 powders. The material after sintering forms Ce–TZP with 100 percent tetragonal phase content and with reasonable toughness. These materials are expected to show R-curve behavior and consequently are flaw-tolerant.[8]

Tape Casting

In the past, tape casting has been used to produce thin layers of ceramic-loaded polymers which can be stacked and laminated into multilayered structures. Today tape casting is the basic fabrication process which provides the materials which are the backbone of two $1 billion industries: multilayered capacitors (MLC) and multilayered ceramic packages (MLCP).

Since the late 1970s, the environmental and health aspects of the tape-casting process have received a great deal of attention. Many formulations based on water as the solvent have appeared in the literature, but there are problems associated with the use of water-based tape systems.

In spite of the drawbacks, the potential benefits of low toxicity and environmentally safe systems have led to workable systems. As with all water-processing systems, the control of pH at all stages of batching, milling, and forming is essential. The use of a heating system to expedite drying is also necessary for aqueous tape-cast products.[9]

The basis for today's multilayered ceramics and multilayered capacitors is the ability to cofire or cosinter the cast tape and the metallization simultaneously. The process has been described in detail, from the initial patent to the current literature[9,10] on low firing temperatures for the ceramic packages. It is essential that the binder-burnout sinter cycle be controlled precisely, especially for low-firing ceramics where a phase begins to form at temperatures as low as 500°C in an atmosphere of inert gas. Precise burnout and sintering control can produce excellent results for the multilayered ceramic package and the multilayered capacitor, which are ready for a final metallization procedure, and for lead-pin or semiconductor-chip attachment.

The practical application of this new process technology, also called multilayer polyimide (MLP), permits the design of high-density structures using microwavelike construction, or burying signals between ground layers with typical lines less than 25 μm wide and 60-μm pitch. Multilayer polyimide combines new low-dielectric materials (dielectric constant 2.9) that have enhanced stability and a thermal coefficient of expansion of 3, with a photolithography-based process that permits structures to be patterned with fine-line geometries measured in micrometers rather than millimeters.

By alternating several layers of polyimide films onto a ceramic, this process can be applied to the production of state-of-the-art multichip modules. It can also be used as a fundamental technology in high-speed integrated-circuit interconnection and single-die packaging. In addition, the multilayer polyimide process can be used for probing applications, where the options for testing circuits and system functions at operating speed are designed into the system at its conception (Fig. 9.2).

Rabin[11] reports that a simple modified tape-casting procedure has been developed for application to ceramic joining when the joining materials are in powder form. The method involves preparation of a slurry from the powder, solvent, and thermoplastic binder, and then casting directly onto the joining surface using a moving doctor blade. Handling of the tape prior to joining is not necessary. Therefore the binder content is minimized, plasticizers are not required, and the viscosity is controlled by solvent content (Fig. 9.3).

The utility of the modified tape-casting method for producing joints with thin, uniform interlayers has been demonstrated by Rabin with hot-pressed SiC joined to itself with TiC + Ni powder material. The joint thickness was approximately 75 μm.[11]

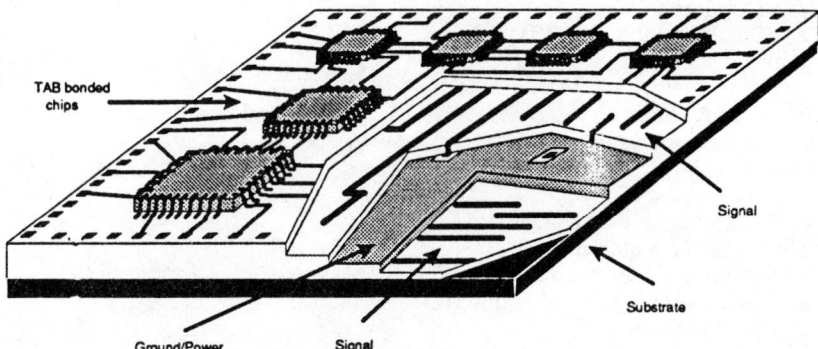

FIGURE 9.2 Multilayer polyimide construction is used in multichip module application. Interconnected structure of multiple high-speed ICs made of several layers of thin-polymer-film dielectric and plated-metal wiring.

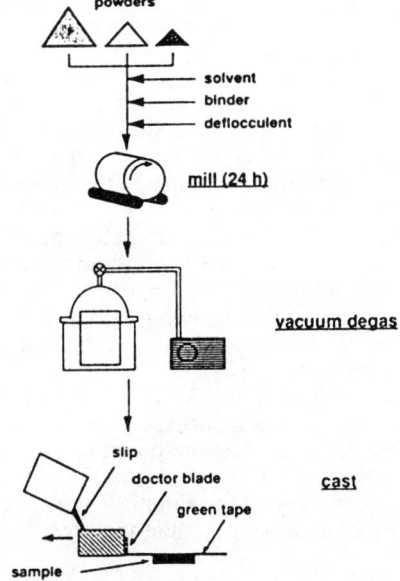

FIGURE 9.3 Schematic of modified tape-casting procedure used in Rabin.[11]

Chemical-Vapor Deposition and Chemical-Vapor Infiltration

These processes and their offshoots appear to present potential improvements in the mechanical properties of ceramic-matrix composite materials. The chemical-vapor infiltration (CVI) process initially involved SiC as the matrix material with both continuous and discontinuous fiber reinforcement in various preform geometries, as follows:

1. Matrix material
   a. Silicon carbide by CVI*
2. Reinforcement types
   a. Ceramic fiber*
   b. Graphite fiber
   c. Ceramic monofilament
3. Fiber length
   a. Continuous*
   b. Discontinuous (felt, mat, etc.)
4. Ceramic-fiber materials
   a. Silicon carbide (Nicalon)*
   b. Silica-alumina-boria (Nextel 312)*
   c. Mullite (Nextel 440)
   d. Alumina (FP, Saffil)
   e. Others
5. Preform types
   a. Textiles
   b. Braided forms*
   c. Unidirectional lay-ups
   d. Laminates*
   e. Three-dimensional preforms

The primary development focused on the areas marked by asterisks, that is, continuous ceramic fibers in braided and laminated preforms.

Chemical-vapor deposition (CVD) SiC involves the thermal decomposition of an organosilane such as methyltrichlorosilane (CH_3Cl_3Si) in the presence of hydrogen on a heated substrate, as illustrated in Fig. 9.4. The process involves the steps of mass transport, diffusion of reactants, absorption, reaction, desorption, and diffusion of gaseous products. A fairly good understanding of most of these steps was presented in the literature.[12-15] Experience has shown that successful infiltration of multi-ply textiles usually requires both low temperature and low pressures. The preform geometry is also important in producing a uniform-density, low-porosity composite by CVI.

The low-temperature, low-pressure CVI process minimizes the chances of fiber degradation or breakage, allows tailoring of the fiber-matrix interface by controlling deposition parameters, and has the potential for eventual low-cost fabrication. By combining those advantages of the CVI process with the use of ceramic textile preforms, the resulting ceramic composite material has the following unique and impressive set of potential features:[12]

High-temperature capability

Oxidation-resistant

Chemically inert

Low density

High strength

Good fracture toughness

Extremely hard

Extremely stiff (high modulus)

High emissivity

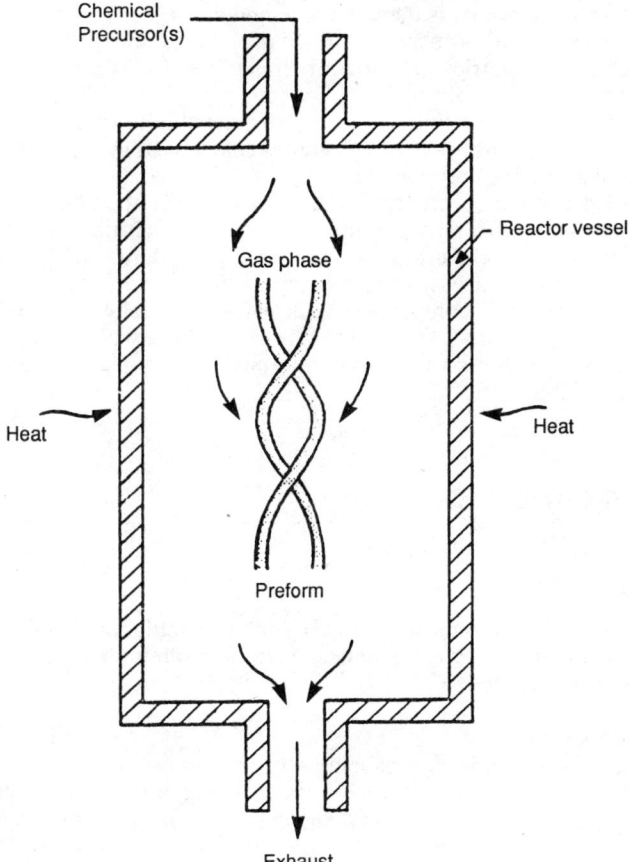

FIGURE 9.4 CVD/CVI process.[12]

Low thermal expansion
Good thermal shock resistance
Impervious to gas diffusion
Semiconductive
Formable to net shape
No strategic materials
Low cost

Continuous, high-strength Nicalon (SiC-based) fibers incorporated into brittle SiC matrices have been fabricated by the forced-chemical-vapor infiltration (FCVI) process. A goal of this program was to infiltrate with Si_3N_4 because of potential improvements in mechanical properties resulting from the close match in the coefficient of thermal expansion of the Si_3N_4 matrix to that of the Nicalon fibers.[4]

Another new program[12] is developing a model of CVI. This model will predict infiltration density, uniformity, and processing time, and will be tested against various preform geometries and fiber architectures. Infiltration experiments will not only include FCVI, but also isothermal chemical-vapor infiltration (ICVI). While particulates and continuous fibers are amenable to the CVI process, other ceramic materials are also being evaluated, namely TiB_2 matrices and techniques using thermal and pressure gradients.

The ICVI process incorporates continuous ceramic fibers in a ceramic matrix without damaging the fibers. It simultaneously uses a thermal gradient and a pressure gradient to reduce the infiltration time from the many weeks required in conventional isothermal processes to less than 24 h.

Fiber-reinforced SiC composites made with this technology have exhibited typical room-temperature flexure strengths of 345 MPa with 10 percent open porosity. Fracture toughness values as determined by various techniques vary from 9.8 to over 24.1 MPa · $m^{1/2}$.

9.3 FABRICATION PROCESSES

Machining Ceramics

Improvements in traditional and nontraditional machining processes and techniques must take place if the advances made in ceramic manufacturing and fabrication are to be utilized.

Traditional Processes
  Grinding. Grinding is the machining process most widely used in finishing ceramic materials. Material removal in the grinding of ceramics is due in part to brittle fracture and in part to the dislodging of individual grains. Past experience shows that effective grinding of ceramics requires abrasives with hardnesses greater than that of the workpiece, such as diamond or cubic BN.

We know the abrasive hardness, cleavage, and chemical reactivity, which are necessary for good grinding performance. We know from experience the desired wheel systems, and the machine stiffness required for a few industrially important ceramics. But we do not have sufficient general knowledge of these parameters to be able to predict them for the new ceramics that are being developed, nor have we collected our existing knowledge into readily accessible form. Despite the many applications of grinding, we still do not have a good method to predict optimum grinding conditions for a new ceramic material.

Another major problem is caused by the lack of a rapid nondestructive technique to determine the nature of strength-limiting flaws introduced by grinding. Grinding damage can be evaluated in a research environment using transmission-electron microscopy using the newest ion-beam or atom-beam thinning techniques, but these are not yet practical for use in industry. Techniques are available for determining general values of grinding-induced residual stresses by X-ray diffraction. It also seems possible to improve ultrasonic techniques to detect strength-limiting flaws.[16]

The areas of grinding research requiring attention include improved dimensional tolerance, finer finish, complex geometries, and part reproducibility in production quantities.[16] It is further important to define the relationships among the grinding parameters and between these parameters and the desired grinding properties.

Polishing. While grinding requires that the grit penetrate the surface of the workpiece, polishing requires no grit penetration. Grinding removes material primarily by brittle fracture; polishing removes material plastically.

The polishing of a surface can be either mechanical, such as lapping with an abrasive, or nonmechanical, such as flame polishing or ion-beam machining. The specific application determines the polishing technique.

One major research need for polishing systems is to determine optimum combinations of machines and polishing materials, and the conditions of their use. No single source is presently available which can be consulted to determine polishing grits, machines, speeds, and feeds for an arbitrary ceramic material and finish. Another research need is to study the fundamental mechanisms of polishing, including electrochemical interactions between slurry, grit, and the workpiece. A third significant research need is for surface and subsurface characterization to be applied to polishing.

Nontraditional Processes. Ceramics have traditionally been shaped by abrasive grinding techniques. Increasing interest in these materials has led to experimentation in adapting metal-removal techniques such as ultrasonic machining, electric-discharge machining, and water-jet machining to ceramics. In addition, the laser has been applied successfully to ceramic machining.

These nontraditional machining methods have inherent problems which are commonly associated with the traditional machining methods. Although the material-removal mechanism of each method is different, the machined surface and subsurface may be damaged. Any machining process will introduce flaws, microcracks, residual stresses, and possibly thermal distortion. Since these machining mechanisms are completely different from those of grinding and other traditional machining mechanisms, it is necessary to investigate the effects of nontraditional machining methods on the surface and subsurface and relate these effects to other design properties. Models are also needed to predict for each method the process parameters such as material-removal rates, roughness, accuracy, and flaw population from machine parameters such as power and speed.[16]

Superplasticity and Ceramics

It has been demonstrated that superplastic ceramics offer the potential benefit of forming net-shape or near-net-shape parts. This could be particularly useful for forming shapes, such as freestanding sheets or foils, which are difficult to achieve using conventional techniques. An additional advantage of developing superplastic ceramics is the fact that these ceramics exhibit a high-temperature stress-strain curve identical to that of metals. This feature would assist structural engineers in accepting structural ceramics into their designs. From a design viewpoint, for sophisticated aerospace applications it is far better to design from a conventional stress-strain curve rather than to use classical Weibull plots based on the probability of brittle failure.[17]

Superplastic structural ceramics (Y–TZP, Al_2O_3, Si_3N_4, and their composites) that can withstand biaxial stretching to large strains have been developed recently. The microstructural design of these ceramics first requires an ultrafine grain size that is stable against coarsening during sintering and deformation. A low sintering temperature is a necessary, but not a sufficient, condition for achieving the required microstructure. In many cases, the selection of an appropriate phase, such as the tetragonal phase in ZrO_2 or the α-phase in Si_3N_4, which is resistant to grain growth, is crucial. The use of sintering aids and grain-growth inhibitors, particularly those that segregate to the grain boundaries, can be beneficial. Second-phase particles are especially effective in suppressing static and dynamic grain growth. Another major concern is to maintain an adequate grain-boundary cohesive strength relative to the flow stress, to mitigate cavitation or grain-boundary cracking during large strain deformation.

Since the discovery of ceramic superplasticity in 1986, considerable progress in its application to ZrO_2 and other superplastic ceramics has been achieved. The interrelated material considerations for various structure-property-processing relationships have been summarized by Chen and Xue[18] in a block diagram (Fig. 9.5). Although the implementation and specifics may vary from ceramic to ceramic and further basic and applied research is still desirable, it is felt that Fig. 9.5 nevertheless provides a sketchy but useful "road map" to guide future superplastic ceramic development.

Despite the fact that superplasticity in metals is relatively well-developed, a fundamental understanding of superplasticity in ceramics is still in its infancy. A number of issues such as the precise microstructural requirements, deformation mechanisms, and constitutive equations are poorly understood. These questions will certainly impact the future development and processing of ceramics for structural applications by superplastic forming technology.

Looking forward, it is Chen and Xue's opinion that most, if not all, ceramics can be rendered superplastic at reasonable forming temperatures and forming times no higher or longer than those currently used in commercial practice for ceramic sintering. From their experience with ZrO_2, Al_2O_3, Si_3N_4, and their composites, their optimism has been heightened.

A number of research groups and teams have explored the superplasticity phe-

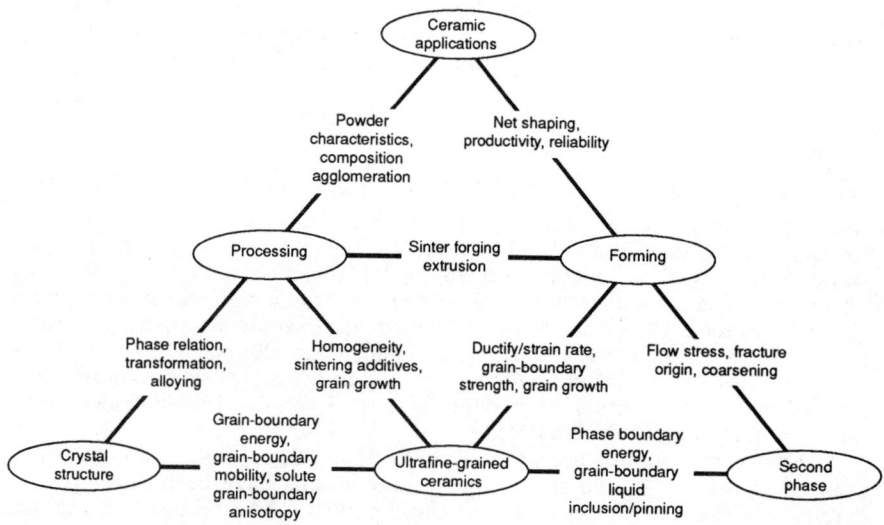

FIGURE 9.5 Major issues in structure-property-processing relationships in superplastic ceramics for structural applications.[18]

nomenon and reached the point of selling nozzles, which are among the first commercial products made from superplastic ceramic. A group of aerospace engineers are developing a turbine blade of superplastic fine-grained Y–TZP. The hope is to make a reliable blade able to withstand temperatures higher than the limits for present superalloys. The Y–TZP is being slip-cast to near-net shape.[19–22]

9.4 NONDESTRUCTIVE AND DESTRUCTIVE TESTING

Nondestructive Evaluation

Nondestructive evaluation techniques were discussed earlier. However, various improvements and new approaches are being investigated. Some of these approaches involve the use of analytical ultrasonics to characterize monolithic ceramic microstructures, acoustoultrasonics for characterizing ceramic-matrix composites, damage monitoring in impact specimens by microfocus X-ray radiography and scanning ultrasonics, and high-resolution computer tomography to identify structural features in fiber-reinforced ceramics.[23]

Analytical Ultrasonics. Ultrasonic methodology for determining microstructure and mechanical properties of ceramic materials is of high interest. Empirical relationships between ultrasonic velocity, attenuation, bulk density, grain size, and strength properties have been observed and conceptual models for explaining and predicting empirical correlations have been advanced according to Klima and Kautz.[23] The use of ultrasonic velocity is a viable quality-control option for the qualitative ranking of specimens of a given ceramic material on the basis of material density.

Acoustoultrasonics. The analysis of ultrasonic signals recovered from composites by conventional methods is often impossible because the heterogeneous structure of the laminate creates a tortuous path for sound waves, resulting in multimodal transmission and superposition of multiple signals arriving at the receiver at about the same time. The result is a complex "noisy" waveform that is difficult to interpret. A new technique termed acoustoultrasonics has emerged, which offers a unique way of evaluating composite panels by measuring bulk material effects on ultrasonic transmission rather than detecting discrete defects.[23]

Stress-wave factor measurements were made on SiC_f–Si_3N_4 tensile specimens exposed to an oxidizing environment at temperatures ranging from 700 to 1200°C. The data showed that an empirical relationship exists between the normalized stress-wave factor and the elastic modulus measured in tension. The relationship suggests that the stress-wave factor may be sensitive to a number of factors that contribute to the stiffness of the material, including the integrity of the interfacial bond between fibers and matrix. Thus acoustoultrasonic measurements may be useful for estimating the overall quality of composite materials.[23]

Damage Monitoring. There is increasing interest in utilizing nondestructive examination procedures for monitoring damage accumulation in test specimens and actual components of advanced ceramic materials to help identify failure modes. Although fractographic analysis is an extremely useful technique for determining fracture origins in monolithic ceramics after failure, it has limited applicability to fiber-reinforced composites because these materials are less sensitive to discrete defects. Thus the need for evaluating specimens by nondestructive techniques that complement one another is evident.[23]

Computer Tomography. Only in the last 3 years has this technique been evaluated for material characterization, and results of initial studies of high-resolution computer tomography for the evaluation of ceramic composites are very promising.[23]

Destructive Testing

Although there appears to be no immediate prospect in the United States for an ASTM fracture toughness test method for ceramics, two methods are pending approval as Japanese Industrial Standards (JIS).[24]

Single-Edge Precracked Beam. This test is used for measuring K_{Ic}. It uses three-point bending, a loading that is common for fracture toughness tests of both metals and ceramics. For ceramics, the problem with this specimen is introducing a sufficiently sharp starter notch.

The Japanese solution is to use the bridge-indentation method. The bend bar is provided with a microhardness impression and is placed face down in the precracking fixture.[25] When the fixture is loaded in compression, a sharp through-crack is formed. The specimen is then transferred to a three-point loading fixture and broken.

The Japanese[24] claim that the benefits of the single-edge precracked beam are reproducibility and conservatism (that is, the reported toughness values are both low and consistent).

Indentation Fracture. This is an auxiliary test method, which is simpler but less reliable. The method involves using a microhardness indentor with a load sufficiently high to produce cracking at the tips of the hardness impression. Toughness is calculated from crack length, indentation length, indentation load, and Young's modulus. The method has drawbacks: there is dispute as to specimen calibration and the indentation cracks are sometimes difficult to measure.[24,25]

9.5 CERAMIC APPLICATIONS

Ceramic applications were discussed in Chap. 8. However, there are several areas of development that will open up new markets.

Bearings

Joint ventures of several companies are attacking this fruitful application. One foresees a Si_3N_4 bearing that is glowing white hot, running at 10,000 rev/min with a solid lubricant at 649°C. This is impossible today with a steel bearing. However, bearing developers are looking only at the extreme-performance markets because that is all that can be afforded since ceramic bearings are still very expensive.

Although the life of a ceramic bearing is 5 to 50 times that of steel, the ceramic will replace steel only where steel cannot perform because of temperature, corrosion, lubrication, or other limitations. Ceramic applications in gyroscopes, extremely high-speed machine-tool spindles, and commercial uses such as hydraulic and pneumatic activator systems and nozzles for sandblasters and processing equipment are currently foreseen for Si_3N_4 and SiC bearings.

Engines

The gas-turbine engine, said to be a next-generation power plant for ocean-going vessels, will have an overall thermal efficiency of more than 60 percent and an output power of approximately 250 kg/cm². The superfreighter's speed will be 93 km/h or faster, or 2.5 times the average speed of ordinary container ships now in service. The engine will use new ceramic bearing materials and adopt ceramic and other advanced materials in pistons, cylinders, and other parts in the combustion chamber. This will make the chamber highly resistant to heat and pressure as well as wear.

Engineers designing future aircraft engines see metal- and ceramic-matrix composites growing rapidly, and by the year 2010, perhaps close to 60 percent of the weight of high-performance jet engines will be constituted by the two composite materials (Fig. 9.6). During the development of these high-performance materials there will be a parallel need for radical changes in the way components are designed and manufactured. Metal- and ceramic-matrix composites could not be directly substituted for current metal alloys. Engineers in the future will not merely choose the material best suited to the temperature and stress levels of a particular component, but will decide the temperature and stress levels needed by a component for a new engine, and will then design a material to meet those requirements.[26-36]

Requirements for compressors for the start of the 21st century will lead to the use of fiber-reinforced titanium. Glass, however, offers an alternative matrix to titanium, with better compatibility with SiC fibers needed for higher temperature

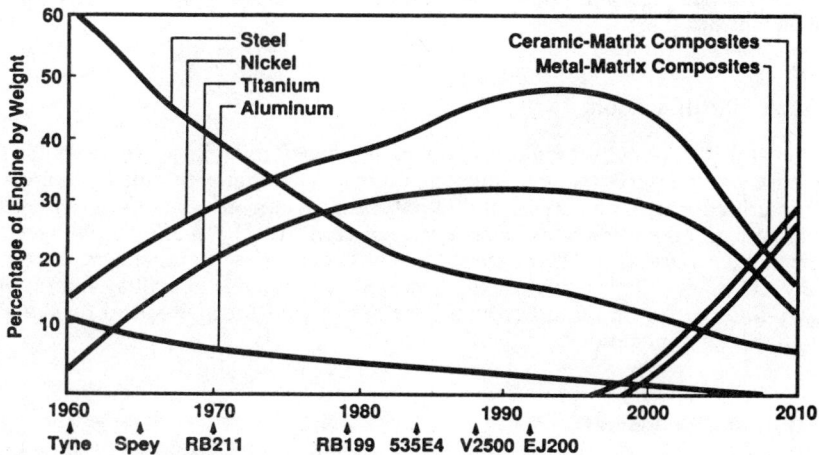

FIGURE 9.6 Use of metal- and ceramic-matrix composites in future Rolls-Royce aircraft-engine design.

performance. Brittleness can be offset by designing a tough composite structure. In the future, turbine disks will need higher-strength materials than are used today and the ability to operate at 800°C. Likely materials are nickel aluminides, nickel matrix composites, and, in the longer term, reinforced ceramics.

Two U.S. government-sponsored high-temperature engine development programs are under way[37,38] in which the goals are 21st-century civil propulsion systems with greatly increased fuel economy, improved reliability, extended life, and reduced operating costs. The first is a high-speed civil transport (HSCT) utilizing ceramic-, metal-, and polymer-matrix composition, while the second is an ultrahigh-bypass engine (UBE).

Automotive gas-turbine engine designers and developers are progressing slowly with government and industry initiatives and cooperative programs in the United States and Europe and are concentrating heavily on fuel-efficient reciprocating and gas-turbine engines, primarily for automotive use in the civil and military markets. These programs include ACT[39,40] and ATTAP.[41–43]

Polymer Ceramics

Electronic packaging engineers have longed to apply the low dielectric constant of easily processed polymers to high-speed circuit substrates. Anisotropic ceramic-fiber laminates now offer that opportunity.

Electronics industry trends toward denser and faster circuits are placing demands on microelectronic packaging, encouraging the development of new dielectric materials for improved heat removal, lower dielectric constant, and improved reliability. Recent development work[44] indicates that a composite of ceramic fibers and polymers can achieve an excellent property balance for electronics: the high thermal conductivity and stability of ceramics, coupled with the low dielectric constant and ease of processing of polymers. In combining these materials, the orientation and morphology of the ceramic filler can be manipulated to provide a favorable anisotropy, a directionality for key properties, in the microelectronic package—a primary consideration in the composites approach utilizing ceramic-fiber composites with Al_2O_3 spherical particles in polyimide thin films.

Ceramic Cutting Tools

Sialon and Al_2O_3–SiC_w ceramic tools, with enhanced reliability, are being used to increase productivity in machining superalloys. The choice of cutting-tool material is critical for maximizing productivity improvements. Sialons provide high toughness and thermal-shock resistance, whereas the Al_2O_3–SiC_w tools have enhanced wear resistance. These basic material properties have been exploited by applying sialons for rough turning, interrupted cutting, and milling operations. Al_2O_3–SiC_w tools, on the other hand, are best suited for high-speed finishing or semifinishing applications.[45]

Aircraft and Spacecraft

Programs to develop the technology for manned, reusable, single-stage-to-orbit flight vehicles are now under way, and the vehicles will reach hypersonic speeds (over 6500 km/h) to Mach 25 through the atmosphere, be propelled into earth orbit by hydrogen-fuel air-breathing engines, maneuver in orbit, deorbit, and return. That scenario will pose formidable demands on engine and airframe materials.

Ceramic-matrix composites offer potential use as uncooled lightweight structures. Besides also being suitable for use at temperatures in excess of 1371°C, they are inherently much more oxidation-resistant than the carbon-carbon composites, and the use of fiber reinforcements can compensate for their inherent brittleness by serving to blunt propagation.[46]

For intermediate temperatures, to about 816°C, glass-ceramic-matrix composites are relatively well characterized. Current technology allows glass-ceramic-matrix materials to be fabricated into many forms, including honeycomb-core panels, truss-core panels, and other complex shapes.[47]

Ceramic Gloves and Mittens

A fabric is woven from fibers coated with ZrC, which enables the fabric to absorb the sun's ultraviolet rays and produce heat—as much as 5°C more than conventional outerware. It is now used in skiwear ranging from gloves and mittens to one-piece suits.

9.6 NEW CERAMIC AND METAL MATERIALS

Reaction-Based Processing

Under this umbrella of processing a variety of multiphase reactions are grouped, such as gas-solid, gas-gas, and gas-liquid. Lanxide Corporation has developed a unique method to produce ceramic composites. The process forms a ceramic-metal composite in situ during the formation of an oxide from a molten-metal solution in a refractory container.[48,49] The reaction is controlled by forming the ceramic-matrix product with a gaseous oxidant, with the resultant reaction product growing outward from the metal surface. By adding filler materials (in the form of platelets, particles, fibers, and other reinforcement shapes) adjacent to the surface of the molten metal, various composites can be produced.

Transport of the metal occurs by wicking of the liquid metal along microscopic channels in the reaction product. The resulting matrix material is a ceramic-metal composite composed of an interconnected ceramic with some interconnected metal, ranging from typically 5 to more than 10 percent, depending on the process parameters. The mechanical characteristics tend to be dominated by the ceramic.

The growth rate of the matrix is rapid and approximately constant; for instance, Al_2O_3 grows to 2.5 to 3.8 cm in 24 h in an air atmosphere. Large pieces can be produced. So far, Al_2O_3 bodies to 20 cm thick have been made. The microstructure and properties of the matrix depend on dopants, temperature, and time.[7]

Numerous ceramic-metal matrix and ceramic-ceramic systems have been demonstrated including Al–Al_2O_3, Al–AlN, Ti–TiN, and Zr–ZrN. The composites that can be fabricated from the filler materials include an Al–Al_2O_3 matrix with Al_2O_3, SiC, ZrO_2, and $BaTiO_3$; an Al–AlN matrix infiltrated with Al_2O_3, AlN, TiB_2, and B_4C; a Ti–TiN matrix infiltrated with TiC, TiN, TiB_2, and Al_2O_3; and Zr–ZrN with ZrN and ZrB fillers. One of the key elements in producing these composites is that the rapid oxidation reaction is promoted by introducing minor elemental or oxide powders as dopants, such as magnesium in combination with silicon, germanium, tin, or lead for the Al–Al$_2$O$_3$ system.[30]

Researchers at Lanxide Corporation believe that there are as many as three toughening mechanisms operative in their ceramic-matrix composites, depending on the type of reinforcement: crack clamping by ductile-metal ligaments, crack-particle interactions, and fiber toughening. During crack clamping, cracks propagate around an adherent metal phase. These metal-phase inclusions remain as ligaments that bridge crack faces. The unfractured ductile metal ligaments that remain behind the crack front apply closure forces.

For the second type of mechanism that occurs with particle reinforcements, thermal-expansion differences between matrix and filler produce internal stresses that can lead to crack deflections. The crack path is affected by the strength of the matrix-filler bond.

To achieve fiber toughening, the fibers are usually coated to induce low fiber-matrix bond strength. Cracks propagate through the matrix, around the fibers. The fibers remain intact behind the crack front and bridge the crack opening. Fiber pullout also contributes to toughening.

Lanxide has already demonstrated the fabrication of near-net shapes for applications such as piston engine components (camshafts, valve seats, and exhaust-port liners), ballistic tiles, and wear rings and plates. Other shape-forming techniques are under development. Work also continues in developing process kinetics and microstructure control methods and techniques to verify the three toughening mechanisms listed.[7]

In Situ Precipitation Techniques

As opposed to mechanical mixing, the in situ growth of dispersoids, such as TiB$_2$, ensures clean adherent bonds between dispersoid particles and matrix. Also, mixtures of similar or dissimilar dispersoids, from fine spherical particles to whiskers, can be combined in the matrix. In many cases, the dispersoids are quite stable so that ingots can be remelted and cast into shapes as well as reduced into mill forms. Thus the resulting composites can serve as a matrix for continuous-fiber-reinforced composites.

This promising technology (XD process) has been applied to making titanium-aluminide composites by Martin Marietta Laboratories,[50] where ingots of 113 to 136 kg have been produced. Having gained success with the γ-titanium aluminide, work has progressed to aluminum-matrix composites and will continue to examine and evaluate other materials in the future.

Monolithic Si$_3$N$_4$

A new Si$_3$N$_4$ ceramic whose fracture stress of 11.3 MPa · m$^{1/2}$ is twice that of conventional ceramics has been developed in Japan. The scientists at NKK Corporation[51] researched ways for controlling the microstructure of Si$_3$N$_4$ to reduce brittleness and ensure reliability. In NKK's production process, needle-shaped crystals in the microstructure have been enlarged, thus suppressing crack propagation. These needle-shaped crystals are much larger than the whiskers used for ceramic-ceramic composites, and as they increase in size, they increase in strength.

In general, if the fracture stress of a ceramic is increased, its bending strength decreases. The new ceramic retains a strength greater than 100 kg/mm^2 and a Weibull modulus of 26.1, both comparable to those of ceramics currently in use. The high Weibull modulus means a homogeneous structure, which ensures the reliability of ceramic products. The new ceramic does not contain foreign elements such as whiskers. This permits easy production of complicated high-precision near-net shapes by injection molding, especially for the manufacture of engine parts.

Iron Aluminide and Ceramic Reinforcement

Aluminide-based intermetallic matrix composites are currently being considered as potential high-temperature materials. One of the key factors in the selection of a reinforcement material is its chemical stability in the matrix. Chemical interactions between iron aluminides and several potential reinforcement materials, which include carbides, oxides, borides, and nitrides, have been analyzed from thermodynamic considerations. Several chemically compatible reinforcement materials were identified for the iron aluminides with aluminum concentrations ranging from 40 to 50 wt %.[52] These were HfC, TiC, ZrC, HfB$_2$, ScB$_2$, TiB$_2$, ZrB$_2$, Al$_2$O$_3$, BeO, La$_2$O$_3$, Sc$_2$O$_3$, Y$_2$O$_3$, and HfN. Of all these materials, Al$_2$O$_3$ is probably the most stable reinforcement material in the iron aluminide matrices.

Because the reinforcement materials for the iron aluminide matrices are expected to carry a major portion of the load at high temperatures, a strong bond between matrix and reinforcement is required for effective load transfer. Thus a limited extent of chemical interactions at the matrix-reinforcement interface would be beneficial in this regard. The reactions that lead to the formation of new compounds at the interface are probably not desirable because of the adverse effects of these compounds, which are normally brittle, on the mechanical properties of the composites. On the other hand, reactions in which a small amount of the reinforcement material is dissolved in the matrix might be beneficial for creating a strong bond between the matrix and the reinforcement material. Another approach would be to apply a thin layer of interfacial coating to promote such a strong bond.[52]

Ion Implantation

Ion-implanted single-crystal Y$_2$O$_3$-stabilized cubic ZrO$_2$ was investigated.[53] Inert gas ions (neon, argon, xenon) and nitrogen, silicon, titanium, and tungsten were implanted. As a result various microhardness values of ZrO$_2$ implanted with various species over a range of fluences showed that the principal variable causing hardness changes is the damage energy and not ion fluence or ion species. For all implants studied, the hardness versus damage energy gives a unified plot. At low doses the hardness rises with increasing deposited damage energy to a value of 15 percent higher than that of unimplanted ZrO$_2$. With additional damage the hardness drops to a value 15 percent lower than that of the unimplanted ZrO$_2$.[53]

Additional investigations and research continue since the range of attainable hardnesses, both increases and decreases, and the alteration of the friction mechanism show that surface modification by ion beams can help shape the mechanical surface properties of ZrO$_2$ for a wide variety of functions and applications.

Self-Propagating High-Temperature Synthesis

This self-sustaining reaction (SHS) utilizes applied pressure which produces dense compacts (Table 9.1). The attractiveness of the process, which needs more research and development, is the energy savings associated with the use of self-sustaining reactions, the simplicity of the process, the relative purity of the products, and the possibility of simultaneous formation and densification of the product. In addition, in some cases it has been demonstrated that products prepared by the SHS process are superior to those prepared by conventional means. Another advantage of the SHS process is its suitability for the preparation of solid solutions, composites, and metastable phases.

The products of synthesis by combustion can be in powder form. Powders produced by this method are used as end products (for example, TiC as abrasive material) or in subsequent forming operations to prepare desired shapes and densities. The products are quite often porous bodies. These have been used as fibers or as crucible materials. To obtain highly dense bodies, two approaches have been utilized. The first involves the application of pressure during the combustion process, and the second takes advantage of the molten state of the products to form cast bodies.

Ceramic-Matrix Composites

Monolithic ceramics have yet to be applied to structural components in aircraft engines, chiefly because of poor structural reliability and reproducibility. It is difficult to control the number and size of flaws during conventional ceramic processing or to prevent more from forming during operation in aggressive environments.[54]

Monolithic versus Composite Ceramics. The requirement for future combustor airfoils and reheat and exhaust components can only be satisfied with ceramic-type materials. Of the possible monolithic ceramics, Si_3N_4 and SiC have the greatest potential for application in gas-turbine engines in components operating at temperatures to about 1400 and 1600°C, respectively. These materials are stronger than nickel superalloys above 1000°C, have superior creep strength and oxidation resistance, and are potentially much cheaper. In addition, their density is less than half that of the superalloys (typically 3.2 g/cm^3 compared to 7.9 g/cm^3).

Before being accepted in the gas-turbine engine, ceramic components must

TABLE 9.1 Examples of Materials Prepared by SHS Method

Borides	CrB, HfB_2, NbB_2, TaB_2, TiB_2, LaB_6, MoB_2
Carbides	TiC, ZrC, HfC, NbC, SiC, Cr_3C_2, B_4C, WC
Carbonitrides	TiC–TiN, NbC–NbN, TaC–TaN
Cemented carbides	TiC–Ni, TiC–(Ni,Mo), WC–Co, Cr_3C_2–(Ni,Mo)
Chalcogenides	MoS_2, $TaSe_2$, NbS_2, WSe_2
Composites	TiC–TiB_2, TiB_2–Al_2O_3, B_4C–Al_2O_3, TiN–Al_2O_3
Hydrides	TiH_2, ZrH_2, NbH_2
Intermetallics	NiAl, FeAl, NbGe, TiNi, CoTi, CuAl
Nitrides	TiN, ZrN, BN, AlN, Si_3N_4, TaN (cubic and hexagonal)
Silicides	$MoSi_2$, $TaSi_2$, Ti_5Si_3, $ZrSi_2$

demonstrate a reliability in operation at least as good as the metal components they replace. This can be achieved in one of three ways.

The first approach is to learn to live with the brittleness (low K_c) and develop a fundamental understanding of the micromechanics of failure (that is, flaws and their relation to strength). In this way statistical methods, nondestructive evaluation, or proof test methodologies can be used to specify design parameters such as strength or component life.[55]

The second approach is to identify the sources of strength-degrading flaws and then develop improved processing methods to eliminate these strength-limiting defects totally (that is, reduce the size of the largest flaw). As the critical defect size for ceramics is less than 100 μm, that is, some two orders of magnitude less than that for metallic materials at typical operating stresses, this solution is not as easy as it may first appear. Substantial advances have, however, been made over the last 10 years with both approaches, with proof of concept and component demonstration having been carried out successfully by several companies. Rolls-Royce has bench tested monolithic ceramic blades, turbine shroud rings, and air bearings in an experimental helicopter engine (Fig. 9.7). Garrett, Ford, and Allison have run ceramic radial turbine rotors, stators, combustors, regenerators, and seals in the AGT101 and AGT100 programs aimed at automobile applications, and work in Japan and Germany on monolithic ceramic components for vehicular gas turbines has been ongoing at an intense pace for several years.

The two approaches discussed so far rely, however, heavily on the design stress being kept below the failure stress at all times. If for any reason the failure stress is exceeded, catastrophic failure will ensue. In meeting current and future application requirements, a basic change in failure mechanism from this flaw-sensitive, brittle, catastrophic failure to a more forgiving noncatastrophic failure is seen to be essential for aircraft-engine operation. This can only be achieved by the third approach: designing ceramic microstructures with improved resistance to fracture, and hence some defect tolerance.

Fiber-reinforced ceramic-matrix composites have this potential and consequently have received a great deal of attention for use in high-temperature structural applications. Adding a reinforcing or toughening phase can improve the functional reliability of ceramics. A reinforcing phase having the proper mechanical, physical, and chemical properties can introduce internal deformation and

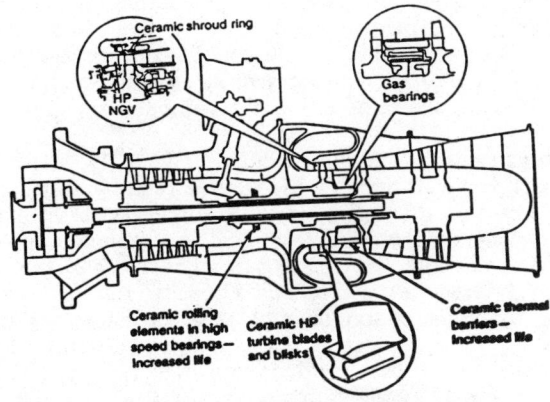

FIGURE 9.7 Ceramics in small helicopter engines.[55]

fracture mechanisms that decrease the sensitivity of the matrix to flaws and increase the fracture toughness of the ceramic. As with other composite systems, the reinforcing ceramic second phase can have a variety of shapes, ranging from nearly spherical particles through whiskers and chopped fibers to effectively continuous fibers. Compact particles and whiskers lend themselves to conventional ceramic processing methods. In addition to toughening parts in all directions, this reinforcement approach is relatively inexpensive, and thus it is well suited to mass production.[56-63]

Continuous fibers have structural advantages over particulates and whiskers. Stiff fibers with diameters less than 10 μm and well aligned in the direction of the principal load significantly raise both the stress and the strain at which flaws begin to grow. In addition, if the fibers are considerably stronger than the matrix and are properly bonded to it, unstable matrix cracks can pass around fibers and not through them. Fibers remain intact and can prevent catastrophic failure. Glass matrices reinforced with continuous SiC and graphite fibers behave this way.[1,27,56,64-67]

Dramatic improvements in the fracture properties of ceramics have been obtained by reinforcing with continuous high-strength fibers. Optimum microstructures result in composites that do not fail catastrophically, and therefore have mechanical properties that are very different from those of monolithic ceramics.

As a result, ceramic-matrix composites present an opportunity to design composites for specific engineering applications. This will require a detailed understanding of the micromechanics of failure and explicit quantitative relations between mechanical properties and microstructural characteristics. The most important breakthrough in ceramic composites will come with the development of new high-temperature fibers that can be processed with a wider range of matrix materials. New chemical methods to form fibers and matrices together could yield even more revolutionary advances.[68-78]

Ceramic-matrix materials are divided into two categories based on failure mode and processability. Whisker- and particulate-reinforced ceramic-matrix composites are relatively easy to process but can fail catastrophically, with failure generally controlled by the properties of the matrix. Densification of such composites can be achieved by sintering techniques similar to those developed for monolithic ceramics. Continuous-fiber-reinforced ceramics (FRCs) fail noncatastrophically if fiber-lay-up architecture and fiber-matrix interfacial bonding are optimized so that mechanical behavior is fiber-dominated. Continuous-fiber-reinforced ceramics usually require special consolidation techniques to incorporate the matrix into the fiber network, and full composite densification frequently is difficult to attain, especially without damaging the fibers (Fig. 9.8).[30,56,60,61]

Numerous organizations and laboratories[4,13,14,27,28,43,54,56,58,62,66,67,72-76,79,80] have been homing in on ceramic matrices reinforced by stiff continuous fibers, such as SiC. On the theoretical side, researchers have been conducting microstructural design studies using analytical and computer approaches. On the practical side, they have been trying to surmount the two main obstacles to attaining quality high-temperature ceramic composites reinforced with long fibers: lack of a strong, stiff, thin, continuous fiber with mechanical properties not drastically degraded by composite processing or use, and absence of net-shape composite processing methods that yield uniform microstructures of nondegraded aligned fibers surrounded by low-porosity matrices.

Fiber development has also continued at many companies in the development

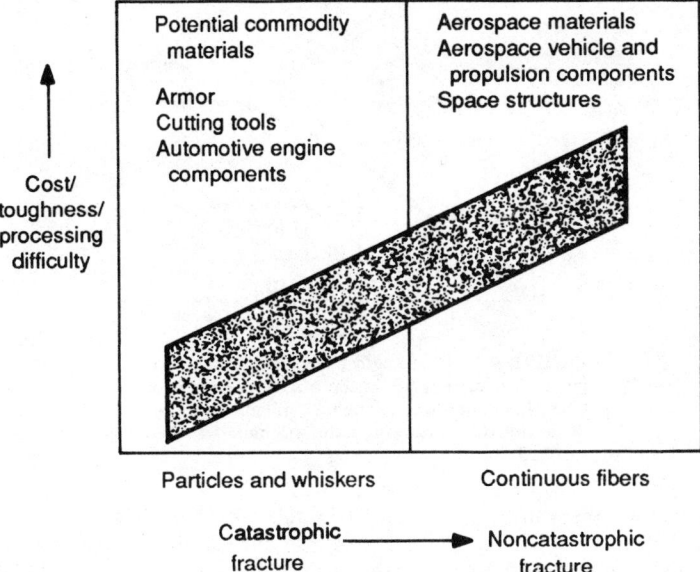

FIGURE 9.8 Ceramic-matrix composites are divided into two categories based on failure mode and processability.[56,80]

of commercially available high-performance fibers, with emphasis on Al_2O_3, SiC, Nextel, Nicalon, and others, that is, SiC fibers produced by CVD. Such fibers have significantly greater tensile strength and stiffness than continuous strands produced by other methods, such as conversion of ceramic precursor polymers to fibers followed by pyrolysis to accomplish conversion to a ceramic.[68,70] Their resistance to stress and combustion gases at high temperatures and compatibility with various matrix compositions are being investigated. Reactions with the gases will be blocked by fiber coatings. Besides acting as diffusion barriers, coatings can help tailor the strength of the fiber-matrix bond, which is of prime importance for avoiding catastrophic failure.

A variety of processing approaches were investigated and discussed in Chaps. 4 and 8, in which fibers are first made into shapes, matrix precursor is then added, and finally the combination is heated to a high enough temperature to turn the precursor into a ceramic matrix, but low enough to preserve the fibers. Three-dimensional fiber shapes can be formed by weaving, two-dimensional shapes by tape lay-up, and bodies of revolution by either method.

The matrix is made by reaction bonding, polymer pyrolysis, CVD, or sol-gel processing.[1–4,27,30] In reaction bonding,[32,42] alternate plies of fiber tape and silicon sheet are laid up, then heated in nitrogen to convert the silicon to Si_3N_4 and bond the layers together. CVD slowly builds up the matrix on the surfaces of the fibers until the spaces between them fill up. The remaining two methods, polymer pyrolysis and sol-gel processing, consist essentially of adding liquid and heat treating.[13–15,56,68,70,72,73,75,76,78–84]

An approach recently developed[54] shows promise of yielding strong and tough high-temperature composites: formation of reaction-bonded Si_3N_4 matrices reinforced by continuous CVD SiC fibers.[85] Commercially available large-diameter

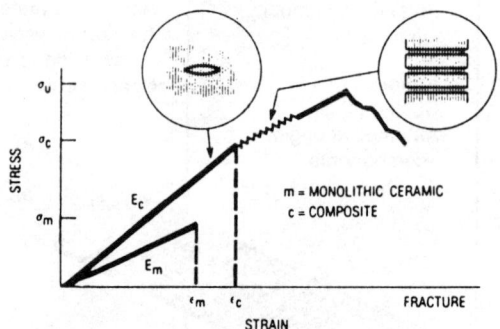

FIGURE 9.9 In strength test of ceramic composite, jagged line signals formation of cracks in matrix; but when fibers are long enough to bridge cracks, sample does not fail until ultimate strength of fibers is reached.[54]

(~140-μm) SiC fibers are uniformly distributed within a Si_3N_4 matrix with fiber content of up to 40 vol %. Jagged stress-strain curves for this material reveal the development of multiple matrix cracks before the composite ultimately fails when the fibers break (Fig. 9.9). Because of the large diameter of the fibers, they do not bridge the much smaller defects in the ceramic and serve as crack stoppers. Thus first matrix cracking appears at strains no greater than those at which the unreinforced matrix fractures. But since the 3893-GPa modulus of elasticity of the fiber is higher than the 900-GPa modulus of the matrix, the matrix cracks at a higher stress than when unreinforced, and the difference increases with the fiber content. Compared to unreinforced matrices, high-fiber-content composites of this type have significantly greater ultimate strength (Fig. 9.10).

Experience gained over the past 20 years with polymer-, metal-, and carbon-matrix composites, and the growing acceptance of these materials by designers, have opened the way for expanded ceramic-matrix composite research. However, these composites require major improvements in design and application approaches over other composite systems, so their development is still only in its very early stages.

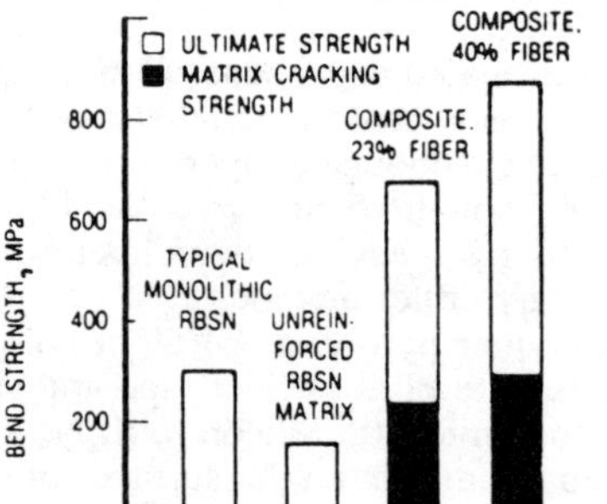

The concept of whisker reinforcement of ceramics to enhance toughness and strength has been demonstrated in SiC-whisker-reinforced Al_2O_3 for cutting tools.[45,56] The current use of monolithic ceramic turbochargers in automotive engines also paves the way for similar uses of whisker-reinforced ceramics to provide lighter weight and faster engine response. Initially, however, such parts will be subjected primarily only to low thermal stresses at relatively low temperatures.[33] Ceramic-whisker com-

FIGURE 9.10 Alone, reaction-bonded Si_3N_4 has strength shown by first two bars and solid portions of second two bars. Si_3N_4 fibers more than double strength.[54]

posites, because of their hardness, also can provide increased wear life in a variety of applications such as valves and bushings.

In satellite structures, continuous-fiber-reinforced ceramic composites could provide high dimensional stability over a broad temperature range, resistance to attack by atomic oxygen, damage tolerance to particle bombardment, and perhaps high structural damping capacity. For example, carbon-reinforced glass has been considered for use in high-performance mirrors for laser systems and precision antennas.

Heat exchangers and catalysts that operate at higher temperatures and pressures than those permitted by current metallic materials could provide higher heat transfer or chemical reaction rates, and at the same time reduce energy consumption. However, the use of ceramic-matrix composites in heat exchangers requires better strength capability in instances where thin cross sections, long tubes, or high differential pressures are involved. In such applications, these composites must be able to withstand loads due to thermal stresses and thermal shock, as well as providing resistance to corrosive environments.

An even more rigorous but extremely high-payoff application for ceramic-matrix composites would be as hot, highly loaded structural components in automotive, aircraft, and rocket engines. Here they would help increase engine efficiency by raising cycle temperatures, reducing cooling losses, and decreasing weight. In automotive applications, however, very low cost is an additional important factor, and must be attained if ceramic-matrix composites are to be considered. The feasibility for ceramic-matrix composites in automotive and unmanned aircraft gas-turbine engine applications should be established in the early 1990s.[56,86]

Processing of Ceramic-Matrix Composites. The interest in ceramic-matrix composites is not only confined to the military. These newly developed material combinations of matrices and fibers, whiskers, and particulates have also commercial value. Of course, the interest goes beyond the curiosity of how to make the material economically and, therefore, sell the product.

The Ti-Al alloys and aluminide-SiC composites are new products which some consider to be metal-matrix composites.[87–89] In previous studies, the composites were formed by plasma spraying,[87] hot pressing,[88] and melting.[89] The work by Rawers et al.[87] reports the synthesizing of titanium, aluminum, and SiC powders into intermetallic composites by: (1) hot-press reactive sintering (RESHOP), (2) self propagating, high-temperature synthesis (SHS), and (3) a differential scanning calorimeter (DSC). The titanium and aluminum powders reacted exothermically to form intermetallics, and the resulting mixture of phases depended on the fabrication technique used and the amount of SiC present. Addition of small quantities of SiC improved the transverse rupture stress of the composite.

Ball and roller bearings are too often viewed as commodity products, purchased and installed as single items. This ignores the fact that bearings are a highly complex system of interactive components, consisting of raceways, rolling elements, cages, seals, and lubricants. In some applications, the system strains the limits of current technology. This is especially true of bearings used at very low and very high temperatures and at high speeds. Lubrication is also an important consideration in extreme environments.[90]

Ceramic (monolithic) materials of the future are Si_3N_4, SiC, TiC, and sialon. Si_3N_4 is favored because it has good high-temperature strength and hardness, favorable strength-to-weight properties, and outstanding resistance to rolling-contact fatigue when properly lubricated. However, for bearing operation at ultrahigh temperature, above 538°C, the search is on for better solid lubricants than graphite and MoS_2.

If all conditions of manufacture and operation were perfect, balls or rollers might space themselves naturally, eliminating the need for separators or cages. However, such perfection is not possible. Cage loading comes from the rolling elements and from centrifugal and thermal effects. Stellites and René 41 can operate at up to 538°C; ceramic composite materials are under consideration to exceed that temperature. Currently under serious consideration are hot-pressed Si_3N_4 containing SiC whiskers, and SiC with SiC continuous fibers.[91] Each of these materials has high bending strength (340 to 544 MPa) and excellent fracture toughness (37.9 to 55.9 MPa · $m^{1/2}$). A common aircraft bearing-cage steel, AISI 4340, has a fracture toughness of 98 MPa · $m^{1/2}$.

The problems and limitations associated with the pressurized sintering method are well recognized. The sol-gel technique shows potential in overcoming the disadvantages. The advantages of the sol-gel technique as a forming method for ceramic composites stem from: (1) a small particle size, which can lower the sintering temperature, (2) a greater homogeneity and purity in the sol-gel-derived matrix, and (3) the possibility of forming large, intricate shapes and near-net shapes by casting.[92]

Since CVD techniques can produce high-density and high-purity materials, they have been used extensively for forming fibers and coatings. The CVD process or, more specifically, CVI is particularly useful when yarns, made up of fine filaments, are the reinforcing agents.[93] Methyltrichlorosilane was used as the most common precursor in the formation of SiC matrix composites.[94] The CVI process has an advantage over some other composite-forming techniques which require pressure that tends to break fibers during fabrication. The CVI process has been demonstrated in the formation of carbon matrices with glass fibers.[95]

In this work, composites of SiC (Nicalon) yarn–CVD SiC matrix and mullite (Nextel 440)–CVD SiC matrix were prepared using methyl-dichlorosilane. The SiC-reinforced composite had a higher strength (450 versus 120 MPa) and exhibited more fiber pullout than the mullite-reinforced composite. This difference can be attributed, in part, to a relatively weak fiber-matrix interface for Nicalon-SiC compared with Nextel-SiC composites.[92]

A novel process for ceramic composites involves the growth of ceramic matrices through shaped preforms using directed oxidation reactions of molten metals. The preforms may consist of reinforcing fibers, whiskers, platelets, or particles, as needed to produce the desired properties in the finished component. This new technology is being developed by Lanxide Corporation and is being applied to gas-turbine engine components.[96]

This process with its unusual technology is illustrated schematically in Fig. 9.11. The first step involves forming a shaped preform of the filler material. For example, this could be a lay-up of reinforcing fibers or whiskers. Figure 9.11 shows the preparation of a particle preform using uniaxial pressing to form the shape. Next, the preform is brought into contact with the parent metal alloy (selected to contain the necessary constituents to enable the growth process and to obtain the desired characteristics of the grown matrix), and a gas-permeable growth barrier is applied to the surfaces. In the growth step, the assembly is heated to the process temperature (typically in the range of 900 to 1150°C for aluminum-based systems) in a suitable refractory container, and the parent metal reacts with the surrounding gas atmosphere to grow the ceramic reaction product through and around the filler or reinforcing material to form a ceramic-matrix composite. For example, the growth step might involve aluminum reacting with

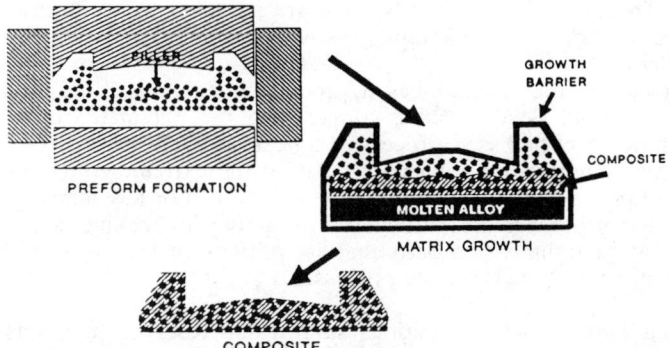

FIGURE 9.11 Simplified process schematic illustrating: (1) filler preform formation, (2) growth of oxidation reaction product matrix into preform, and (3) recovery of composite part.[96]

oxygen in ambient air to form Al_2O_3 or with nitrogen to form AlN. The reaction continues until the growing matrix reaches the barrier, at which point it stops, thus preventing overgrowth of the matrix and preserving the desired shape of the composite part. Finally, after cooling to ambient, the growth barrier and any adherent residual parent metal are removed to recover the part.[96]

Reinforced ceramic composites present some of the most difficult processing problems known for ceramics. The combination of the requirements for highly specified matrix-reinforcement interfaces for optimized mechanical properties, and the relative instability of many reinforcements with respect to temperature, atmosphere, and chemical reaction with the matrix, place stringent limitations on processing temperatures and times. These concerns are reflected in increasing attention to coatings that can protect the reinforcements and preserve interfacial properties. Furthermore, composite systems are inherently difficult to densify without applied pressure because of two distinct physical effects.

According to Chiang et al.[97] one physical effect is the reduced sintering stress that results when an internal hydrostatic tensile stress occurs upon matrix shrinkage in the presence of nondensifying inclusions.[98–100] The second physical effect is the ultimate limit to densification presented by the bridging of inclusions or fibers at the percolation limit.

Therefore, composites containing desirable volume fractions (more than about 20 percent) of fibers, whiskers, or particles will not be easily sinterable. This is certainly consistent with the overwhelming bias toward hot pressing as the preferred consolidation method for both glass- and crystalline-matrix composites. However, the limitations of hot pressing (and, to a lesser degree, sintering) in preparing complex shapes having near-final dimensions are equally clear. As a result, reaction-bonding techniques capable of forming nonshrinking matrices in situ are believed to be key areas of opportunity for fiber-, whisker-, and particulate-reinforced ceramic composites.

The various approaches can be distinguished (somewhat arbitrarily) as condensed-phase–gas-phase reactions (typified by gas-phase reaction-bonding and directed oxidation) and condensed-phase–condensed-phase reactions (such as liquid-phase reaction bonding, solid-solid exchange reactions, and SHS processes). Some of these processes have been applied to the fabrication of ceramic-matrix composites, whereas others appear to have promise, but have not been explored. The examples previously used are intended to be illustrative rather than encyclopedic, and in this rapidly moving field, the omission of some new results is inevitable.[101]

In conclusion, the reaction-based processing methods offer promise even though they are not as advanced as conventional ceramic-matrix composite processes. These reaction-based processes appear uniquely capable of fabricating parts for high-temperature applications due to the extremely high-purity levels and unusual phase chemistries that can be achieved. For low-temperature applications, these processes are particularly attractive for making large, complex, net-shape, and net-dimension parts that are difficult or expensive with conventional densification-based ceramic processes.

Current and Future Studies. Various studies and research programs covering ceramic-matrix composites reveal new findings about variations in thermal expansion coefficients. In a program where three fibers, namely, FP alumina (Al_2O_3), Nextel (mullite), and Nicalon (Si–C–O–N), were studied,[79] it was found that for those ceramic-ceramic composites where the coefficient of thermal expansion of the fibers is greater in magnitude than that of the matrix, and the composite was initially processed at relatively high temperatures, the thermal expansion during heating may vary from that of the lowest coefficient of thermal expansion (the SiC matrix in this case) to the level predicted by the rule of mixtures. Similarly, for ceramic-ceramic composites whose constituents do have a relatively large mismatch of thermal expansion, such as Al_2O_3–SiC, evidence suggests that the magnitude of the thermal expansion hysteresis may be related to the integrity of the fiber-matrix bond.[79]

Other studies reveal different microstructures dependent on the fabrication methods such as sintering [79,84,101] where an Al_2O_3–TiC ceramic composite was fabricated by high-pressure self-combustion sintering (HPCS). This material had a unique microstructure, which was quite different from that of commercial Al_2O_3–TiC made by conventional processing. The HPCS material had a higher K_{Ic} value than the commercial Al_2O_3–TiC.[101]

Modeling techniques to evaluate testing[64] and crack-growth resistance, CVD/CVI techniques to produce both strength and toughness in SiC yarns in ceramic-matrix composites,[72,93] combustion synthesis and dynamic consolidation of TiC–Al_2O_3 composites,[82] and erosion of SiC fiber–SiC matrix composites[81] as well as the effects of solid-particle erosion of gas-pressure sintered Si_3N_4, PSZ, sintered SiC, and soda-lime silica glass[98] are being studied.

Applications. A fast, uniform method is now available for fabricating ceramic-matrix composites by CVI, 12.7 to 25.4 mm thick, in about 24 h, compared with 1 month or longer for conventional processing.[102]

Most processes hold the fiber preform at constant temperature, which results in rapid surface buildup that blocks further diffusion. However, at the Oak Ridge National Laboratory[102] the vapors are introduced under a pressure gradient from one side of the preform, with water cooling on that side. This allows the vapors to pass to the far, hot side of the preform before dissociating to deposit matrix material. The preform thus remains porous until the infiltration process is complete, with deposition rates up to 1 mm/h.

A variety of reinforcements can be used, including Nippon Carbon Nicalon SiC, Dupont FP Al_2O_3, 3M Nextel 440 aluminoborosilicate, and UBE Tyranno SiC. Matrices include SiC and Si_3N_4. Densities up to 90 percent are the most frequently used composition. Tests have shown that the mechanical properties of the composites are not highly dependent on density in the 75 to 90 percent range. Recently developed process models are expected to improve the ability to make complex shapes.

In the next 3 years the following ceramic parts will be commonplace items in most automobiles[103]:

Piston caps

Cylinder and exhaust manifold liners

Valve heads

Turbocharger rotors

The same holds for the top-performing cutting characteristics of ceramic tools. Hot-pressed Si_3N_4 containing 8 percent Y_2O_3 doubles the productivity of conventional cutting tools in machining cast iron auto parts such as wheel drums and clutch components. Si_3N_4-based ceramic containing Y_2O_3, Al_2O_3, and 30 percent TiC as a dispersed phase shows a tenfold increase in the number of brake drums produced per tool. Sialon is another top tool cutter.

Finally, look for thrust-vectoring nozzles, hollow blades, and magnetic bearings in tomorrow's gas turbines being made from ceramic composite materials. Reversing, pitch, and yaw-vectoring nozzles, built of carbon-carbon composites, will give future fighters supermaneuverability and shorter take-off capabilities.[104]

The main advantage of C–SiC composite nozzles for large, liquid-fueled engines is their capability to retain their mechanical characteristics up to 1800°C, eliminating the need for a nozzle cooling system. As a result, the nozzle is lighter than metallic units that require cooling. Other expected advantages of C–SiC nozzles include their low erosion rates, high resistance to oxidation, very high shear strength, and resistance to thermal shock.

Other hypersonic vehicle propulsion systems are being tested with ceramic-matrix composites. These include SiC-reinforced carbon compressor blades, ceramic-matrix composite ramjet combustion chamber, rocket exhaust nozzles,[105] and controls for aerospace planes (Fig. 9.12).[106]

Finally the flight of a Mirage 2000 fighter aircraft has validated the use of a ceramic-matrix composite from the Société Européenne de Propulsion (SEP).[107] SEP made up half of the "hot" and "cold" flaps for the engine using CVI technology, whereby SiC fibers in an SiC matrix were incorporated into the engine flaps. Ceramic-matrix composite on one side and metal on the other allowed for a head-to-head comparison. Now that the initial flight has proved the performance of the ceramic-matrix composite part, its life cycle benefits versus metal are being investigated.[108,109]

Fibers and Whiskers

Low-aspect (length-to-diameter)-ratio particulate and whisker reinforcements lend themselves to conventional ceramics processing methods which use powder blending to form green compacts prior to firing. Thus, besides offering three-dimensional toughening, these reinforcements also offer economic advantages and therefore are well suited for large-volume ceramic-matrix composite production such as for automotive engine applications.

In terms of performance, however, high-aspect-ratio second phases, such as continuous fibers, have structural advantages over particulates and whiskers. For example, if the fibers possess a high modulus, a small diameter (<10 μm), and are

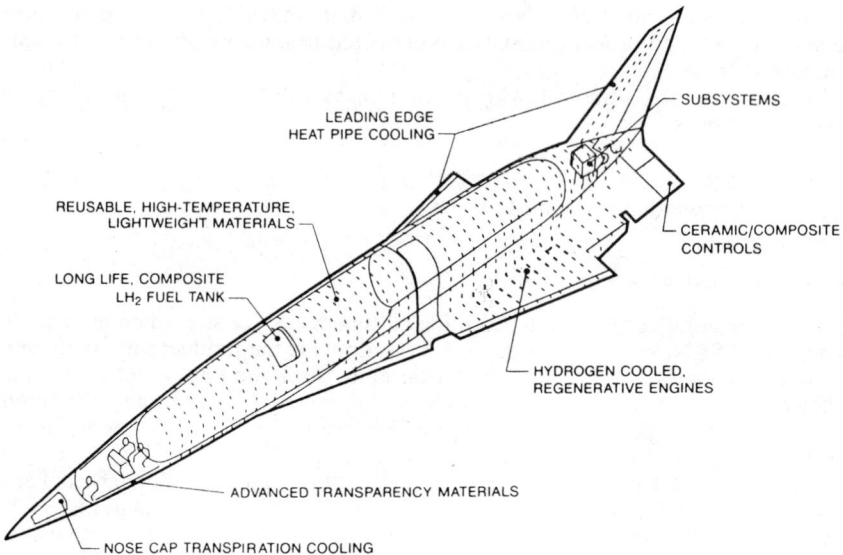

FIGURE 9.12 Advanced materials for future space plane.[106]

aligned with high-volume fraction in the principal load direction, both the stress and strain levels required for matrix flaw propagation can be increased significantly.[110–112] In addition, if the fibers are considerably stronger than the matrix, and are properly bonded to the matrix, then unstable matrix cracks can pass around the fiber and not through them, so that after a matrix fracture, the reinforcing fibers remain intact and prevent catastrophic failure of the composite. Further loading would allow continued matrix cracking, but without composite failure, until the fiber-bundle strength is reached at a much greater fracture strain than that of the monolithic ceramic matrix. This type of improved microscopic and macroscopic toughening behavior has been observed in ceramic-matrix composites consisting of glass matrices reinforced with continuous SiC and graphite fibers and should also be effective for other continuous-fiber-reinforced ceramic composite systems.[113]

Developments to obtain quality high-temperature ceramic composites reinforced with long fibers include (1) high-strength, high-modulus, small-diameter, continuous-length fibers whose mechanical properties are not drastically degraded by composite processing or use conditions, and (2) net-shape composite processing methods which result in uniform microstructures of nondegraded, aligned fibers surrounded by low-porosity matrices. Fiber and matrix properties are dictated on the basis of making optimum use of available toughening mechanisms.[112]

Whiskers. The understanding and further development of whisker-reinforced ceramics have been advanced by the analysis of the toughening mechanisms and the development of a comprehensive theoretical analysis of the toughening behavior. The current results show that elastic bridging, frictional bridging, and pull-out bridging can contribute to the toughness of various whisker-reinforced ceramics. Key features in determining which process dominates the toughening process are the whisker strength and diameter, as well as the whisker-matrix interface properties. The contributions of these parameters to the toughness have now been demonstrated experimentally. Thus the theoretical descriptions of whisker reinforcement by whisker bridging provide a means to design other whisker-matrix composite systems.[114]

In the development of such composites it is important to keep in mind that whisker reinforcement can be combined with other toughening mechanisms to obtain rather dramatic increases in both toughness and fracture strength. Thus matrix microcracking and grain bridging associated with grain-size effects in anisotropic materials and toughening associated with additional dispersed phases, such as transformation toughening with ZrO_2 additions, can be combined with whisker reinforcement. Observations of the mullite–SiC–ZrO_2 system show that at least fivefold increases in toughness and fracture strengths greater than 750 MPa can be achieved with such approaches.[1,2,4,28,30,54,56,58,65,83,84,115,116]

Finally, a wide variety of mechanical properties can be enhanced with whisker reinforcement. In the Al_2O_3–SiC whisker composites, high fracture strengths (650 to 800 MPa for surfaces prepared by 180 grit diamond surface grinding) have been obtained along with the increased toughness. These same Al_2O_3-based composites also exhibit much greater resistances to slow crack growth and thermal shock, retain their high flexure strengths and toughness to temperatures of at least 1000°C, and exhibit much lower creep rates at elevated temperatures.

The reinforcement of Al_2O_3 by short, discontinuous SiC whiskers (~0.5-μm diameter by 30-μm length) greatly improves strength and fracture toughness, and both increase with the whisker content. Typical properties of both whisker-reinforced and ZrO_2-toughened Al_2O_3 are listed in Table 9.2.[117] The primary toughening mechanism appears to be crack deflection by the SiC whiskers, although there is some contribution by whisker pullout.

Ceramics with SiC-whisker contents of over 50 vol % have been produced. The major problem in producing ceramics with large whisker loadings is the uniform dispersion of the whiskers. The use of ultrasonic homogenization or ball milling improves the dispersion of the whiskers.[117] These ceramics can be produced by cold pressing followed by sintering around 1550°C. An inert environment should be used to avoid oxidation of the SiC. However, most whisker-reinforced Al_2O_3 are hot-pressed using graphite dies at temperatures anywhere between 1500 and 1850°C. This results in texturing of the whiskers and some anisotropy in properties. Tiegs and Becher[117] have reported achieving 98 percent theoretical density with whisker loadings of up to 50 percent.

With careful consideration of the design criteria with regard to the microstructural (compositional) and material properties, there is considerable po-

TABLE 9.2 Typical Properties of Toughened Aluminas

	Bend strength (4-point), MPa		K_{Ic}, MPa·m$^{1/2}$	Coefficient of thermal expansion, 10^{-6}/°C	Thermal conductivity, W/m·K
	RT	800°C			
Al_2O_3, 15%ZrO_2	400	350	5.0	8.2	7.0
SiC whisker, toughened Al_2O_3*	690	590†	8.0	—	—

*Greenleaf Technical Ceramics WG 300 (~50% whisker loading)
†Extrapolated from another material
Source: From Becher and Tiegs.[114,117]

tential for expanded development and application of whisker-reinforced ceramics.[114–118]

Campbell et al.[119] have studied two whisker-toughened materials with the objective of identifying the mechanisms that provide the major contribution to toughness. They concluded that, for composites with randomly oriented whiskers, bending failure of the whiskers obviates pullout, whereupon the major toughening mechanisms are the fracture energy consumed in creating the debonded interface and the stored strain energy in the whiskers, at failure, which is dissipated as acoustic waves. The toughening potential is thus limited. High toughness requires extensive pullout and, hence, aligned whiskers with low-fracture energy interfaces.

Another series of whisker studies was reported by Karasek et al.[120] in which whiskers from six manufacturers were characterized by bulk chemical techniques, X-ray photoelectron spectroscopy, X-ray diffraction, and scanning transmission electron microscopy. From a morphological viewpoint, significant differences in diameter, debris level, straightness, and types and quantities of defects were observed from one manufacturer to another, which showed that industry must continue to improve processing.

Other researchers have been examining SiC-whisker–Al_2O_3[121] and Si_3N_4[122] matrices for impact and fracture-strength properties, while Iio et al.[123] studied the mechanical properties of Al_2O_3–SiC-whisker composites with emphasis on the effect of whisker content and of the hot-pressing temperature. The fracture toughness of Al_2O_3 was markedly improved with increasing whisker content up to 40 wt %. In the case of the high SiC-whisker content of 40 wt %, the fracture toughness of the sample hot-pressed at 1900°C decreased significantly, in spite of densification, compared with one hot-pressed at 1850°C. Fracture toughness strongly depended on the microstructure, especially the distribution of SiC whiskers, rather than on the grain size of the Al_2O_3 matrix.

Akimune[124] examined the mechanical properties of SiC-whisker–sialon composites, which exhibited higher strength at elevated temperatures than the sialon matrix. This was attributed to the fact that the SiC whiskers were tightly compacted in the microstructure of α- and β-sialon matrices. The SiC whiskers in the sialon matrix sustain the imposed load and enable the composites to avoid strength degradation.

Fibers. Ceramic-matrix composites incorporate fibers primarily as a means to bridge matrix microcracks, thereby extending the strain capability of the brittle matrix material. Fibers also provide toughening by functioning as crack deflectors. Theoretically, matrix-strain enhancement is expected to be inversely proportional to the fiber diameter. Therefore, small-diameter fibers would reduce interfiber distances in the composite, better limiting the size of matrix cracks. In addition, small-diameter fibers are more suitable for weaving and complex-shape fabrication. Thus, their development is a key future requirement[56] (Table 9.3).

Many of the small-diameter fibers are polymer or sol-gel derived, having compositions that deviate from stoichiometry and, therefore, show thermodynamic instabilities at higher temperatures. Fibers forming glassy phases are subject to creep, as are fine-grain-oxide fibers. The SCS-6 SiC fiber in Table 9.3, produced by CVD, has a large (143-μm) diameter and high stiffness, which prevents weaving it to fabricate complex shapes. While CVD has been used successfully to produce large monofilaments, fabrication of small-diameter fibers using this approach is more difficult to control. Problems include bridging between fibers and

TABLE 9.3 Commercially Available Ceramic Fibers*

Manufacturer	Trade name	Composition	Phases	Diameter, μm	Strength, MPa	Modulus, GPa	Density, g/cm³
			Domestic				
Alpha Associates	Alphaquartz	SiO_2	Amorphous	1–15	886	70	—
Babcock & Wilcox	Kaowool XE	SiO_2–Al_2O_3	—	2.8	1310	116	—
Dow Corning/Celanese	MPDZ	47Si–30C–15N–8O	—	10–15	1750–2100	175–210	2.3
	HPZ	59Si–10C–28N–3O	—	10	2100–2450	140–175	2.35
	MPS	60Si–30C–10O	—	10–15	1050–1400	175–210	2.6–2.7
DuPont	FP	>99 α-Al_2O_3	α-Al_2O_3	20	>1400	996	3.9
	PRD-166	Al_2O_3, 15–25ZrO_2	α-Al_2O_3, t-ZrO_2	20	2100–2450	385	4.2
Fiber Materials	—	SiO_2	Amorphous	9–15	3400	—	—
ICI Americas	Saffil	Al_2O_3–SiO_2	α-Al_2O_3, mullite	3	2000	300	—
Owens Corning	E glass	SiO_2–Al_2O_3–B_2O_3	Amorphous	10	3450	72	—
	S-2 glass	SiO_2–Al_2O_3–B_2O_3	Amorphous	10	4580	87	—
Textron (Avco)	SCS-0	SiC on C core	SiC, C	140	2200	415	—
	SCS-2	SiC on C core	SiC, C	140	3450	415	—
	SCS-6	SiC on C core	SiC, C	143	3920	406	3.0
	—	Si, C		6–10	>2800	280–315	—
	CVD boron	B on C core	B, C	100, 140	3585	422	2.7
3M	Nextel 312	$62Al_2O_3$–$14B_2O_3$–$24SiO_2$	9A-2B, glass	11	1750	154	
	Nextel 440	$70Al_2O_3$–$28SiO_2$–$2B_2O_3$	γ-A, mullite, glass	10–12	2100	189	3.05
	Nextel 480	$70Al_2O_3$–$28SiO_2$–$2B_2O_3$	Mullite	10–12	2275	224	3.05
Zircar Products	Nextel Z-11	ZrO_2–SiO_2	$ZrSiO_4$, glass	11	1335	77	—
	ZYT	ZrO_2	$cZrO_2$	3–6	—	—	—

TABLE 9.3 Commercially Available Ceramic Fibers (*Continued*)*

Manufacturer	Trade name	Composition	Phases	Diameter, μm	Strength, MPa	Modulus, GPa	Density, g/cm³
International							
Akzo Poly. & Fib. (Neth.)	HT silica	SiO_2	Amorphous	10–15	800	66	—
British Petroleum (U.K.)	Sigma	SiC on W core	SiC, W	100	3600	420	—
Electrofuel (Canada)	BNF-HP	BN	Hex-BN	2–6	880	42–70	—
Quartz and Silice (France)	Astroquartz II	SiO_2	Amorphous	9	3500	—	—
Mitsui Mining (Japan)	—	99.5%Al_2O_3	—	10–13	1950	342	—
Nippon Carbon	Nicalon	59Si–31C–10O	SiC	10–20	2520–3290	182–210	2.55
Sumitomo	Sumica	85Al_2O_3–15SiO_2	γAl_2O_3, glass	9–12	1800–2600	210–250	3.2
Ube	Tyranno	Si–Ti–C–O	Amorphous	8–10	>2970	>200	2.3–2.5
Shimadzu	—	Silicon oxynitride glass + Mg,Ca,Al (up to 7%N; metal oxide permits more N to be added with no crystallization)	Glass fiber	7–20	4800	195	3.03
Toa Nenryo Kogyo K.K.	(Tonen Si_3N_4)	Si_3N_4 + O + C	—	10	2500	300	2.5

*All data were provided by fiber manufacturers and their distributors. Mechanical property data are for room temperature. Fiber gauge lengths tested may differ, making direct comparison of tensile strength data questionable.
Source: Synterials Co., Herndon, Va.

substrate nonuniformities. Therefore research is required to produce small-diameter, thermally and environmentally stable, and creep-resistant fibers.[56,125]

Fibers are continually being improved and developed domestically and overseas. At Rhone-Poulenc, a silicon carbonitride ceramic fiber with good thermomechanical behavior at 1300 to 1400°C is being developed.

Fiberamic is a continuous textile yarn of 250 or 500 filaments obtained through pyrolysis of a polysilazane precursor. This synthesis permits the production of continuous ceramic filaments with small diameters. Fiberamic has demonstrated good high-temperature durability. It retains 80 percent of room-temperature strength at 1400°C and exhibits high-temperature stability after aging in an oxidizing atmosphere.

The typical composition is 57 wt % Si, 22 wt % N, 13 wt % C, and 8 wt % O, and the fiber exhibits typical brittle failure with an average tensile strength in the range of 1500 to 2200 MPa. The fiber also retains more than 85 percent of the initial stress after 100 h at 1000°C or after 10 h at 1200°C.[126]

Ube Industries, Ltd.,[127] has developed a high-performance composite fiber, called Tyranno fiber, made of titanium, carbon, and silicon. A high-tenacity ceramic composite material was subsequently developed, called Tyrannohex, which can withstand 1400°C in air. It has a fracture toughness of 15 MPa · $m^{1/2}$ at room temperature, which does not change even at 1400°C and is ideal for use as a superhigh thermal-resistant material for next-generation aircraft and space rockets.

Tyrannohex is produced by bundling Tyranno fiber and sintering the bundle at above 1800°C at a pressure of 300 to 700 kg/cm^2. The fiber's cross-sectional area is transformed into a densely packed hexagonal shape, and it is a fiber-cohesion-type ceramic composite material requiring little matrix material.

The new material's detailed structure has not been clarified, but it consists of elements such as silicon, carbon, and titanium and is a sintered body consisting of regularly arranged fibrous raw materials.

The material's strength differs according to the state of the fiber arrangement, but has a three-point bending strength of 200 to 500 MPa that deteriorates slightly in air at about 1400°C. This is a big improvement over conventional high-tenacity ceramic composite materials which can withstand only up to 1200°C.

A SiC fiber tow has been developed which is a reinforcing ultrahigh-temperature, high-purity, small-diameter SiC fiber that is chemically and crystallographically stable to 1800°C in air or inert atmosphere. The material has a high modulus (4000 GPa) and low density (3.2 g/cm^3). It can be made in diameters of 5 to 12 μm in a tow size of 1000 to 12,000 fibers per tow, which makes it useful for reinforcing ceramics, intermetallics, or metals. Applications include the manufacture of high-temperature components in gas-turbine engines, hypersonic vehicles, integrated-circuit engines, and gun barrels.

An Al_2O_3-enhanced thermal-barrier (AETB), rigid, fibrous ceramic tile material now extends the high-temperature capability of insulating materials. Although it is intended primarily for use in the heat shield of the Space Shuttle orbiter, the new material has obvious potential for terrestrial use in kilns, furnaces, heat engines, and other applications in which light weight and high operating temperature are specified.

The lightweight aluminoborosilicate (ABS) silica material (Fig. 9.13) was developed to replace much of the denser all-SiO_2 high-temperature reusable surface insulation originally used on the Space Shuttle. AETB is made from high-purity SiO_2 fibers 1 to 3 μm in diameter, Al_2O_3 fibers 2 to 4 μm in diameter, and ABS fibers 2 to 4 μm in diameter with a nominal composition of 62 percent Al_2O_3, 14 percent B_2O_3, and 24 percent SiO_2 by weight.[128]

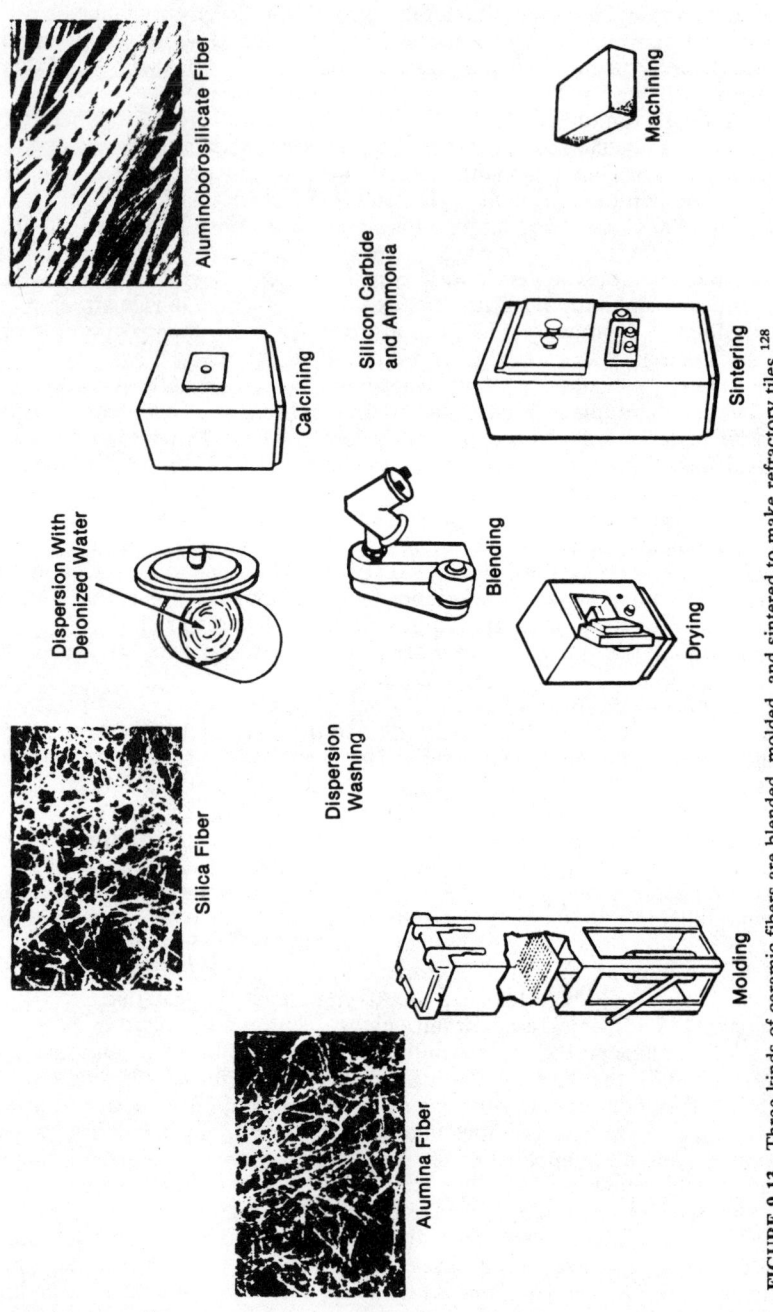

FIGURE 9.13 Three kinds of ceramic fibers are blended, molded, and sintered to make refractory tiles.[128]

High-performance fibers have been successfully incorporated into glasses to form composite materials which have higher strength and toughness. This was carried out as a direct extension of metal and resin matrix composite efforts. In each case high-elastic-modulus fibers have been incorporated into a lower-elastic-modulus matrix to achieve structural reinforcement. Glass and glass-ceramic-matrix elastic moduli are in the range of 60 to 85 GPa, while the reinforcing fibers are generally characterized by elastic moduli in excess of 210 GPa and in some cases as high as 700 GPa. Another important aspect of the work was that the composites were fabricated in a manner totally analogous to that used for resin-matrix composites. This is due to the fact that the glass matrix can be readily deformed and flowed in its low-viscosity state at elevated temperatures. Not only were glasses used in this process, but also glass-ceramics, which provided the greatest potential for high-temperature applications. Glass-ceramics provide the unique capability to densify a composite in the glassy state and subsequently crystallize the matrix to achieve high-temperature stability. Most, if not all, of the techniques used to fabricate resin-matrix composites were adapted to glass and glass-ceramic-matrix composites.[129]

The resultant glass or glass-ceramic composites have high strength, stiffness, toughness, and can withstand temperatures as high as 1200°C. In addition, these composites are not expected to be susceptible to environmental degradation due to moisture, oils, or fuels.

The United Technologies Research Center[27] has been actively involved in the development of the fiber-reinforced glasses and glass-ceramics. Table 9.4 shows the types of fibers and their properties most frequently used at UTRC to reinforce glasses and glass-ceramics. The variety of the fiber forms and the compositions makes available different levels of tensile strength, elastic modulus, chemical reactivity, electrical conductivity, and density. Of all the reinforcements listed, the carbon yarns offer the greatest range of mechanical properties and also potential for the lowest cost. A wide range of carbon fibers has been found to be compatible with glasses and glass-ceramics for composite fabrication. However,

TABLE 9.4 Materials Used to Reinforce Glass and Glass-Ceramic Matrices

Material	Diameter, μm	Density, g/cm^3	E modulus, GPa	Ultimate tensile strength, GPa	Thermal expansion coefficient, 10^{-6}/°C
Boron monofilament	100–200	2.5	400	2.75	4.7
Silicon carbide monofilament	140	3.3	425	3.45	4.4
Carbon yarn	7–10	1.7–2.0	200–700	1.4–5.5	−0.4 to −1.8
Silicon carbide yarn*	10–15	2.55	190	2.4	3.1
FP alumina yarn†	20	3.9	380	1.4	5.7
Alumino-borosilicate yarn‡	10	2.5	150	1.7	—
VLS-SiC whisker§	6	3.3	580	8.4	—

*Nicalon, Nippon Carbon Co., Tokyo, Japan.
†E. I. Du Pont de Nemours & Co., Inc., Wilmington, Del.
‡Nextel 312, 3M Co., St. Paul, Minn.
§Los Alamos National Laboratory, Los Alamos, N. Mex.
Source: United Technologies Research Center, East Hartford, Conn.

carbon fibers are susceptible to oxidation at elevated temperatures, hence SiC fibers have been used in high-temperature oxidizing applications.[55]

While fibers used for improving the properties of the composites are important, the matrix glass and glass-ceramic materials are equally important to tailor a successful material. The matrix compositions not only contribute to such composite properties as elastic modulus, coefficient of thermal expansion, and high-temperature creep resistance, but, more importantly, matrix composition provides a means of controlling fiber-matrix reaction during the composite densification step. This reactivity, or lack thereof, in large measure determines the nature of the fiber-matrix interface, which subsequently determines the composite fracture mode. Table 9.5 gives a partial list of the matrix materials being used.

Glass-matrix composites can be easily fabricated. The following processes are in current use or under study[34,43,55,130,131]:

1. Hot pressing of infiltrated unitape and fabric lay-ups
2. Hot matrix transfer into woven preforms
3. Hot injection molding of chopped fiber compounds or preforms[130,131]

9.7 RELIABILITY AND DESIGN

Research has focused on improving ceramic material processing and properties, as well as on establishing a sound design methodology. Because of the variable severity of inherent flaws, the nature of ceramic failure is probabilistic, and optimization of design requires the ability to accurately determine the reliability of a loaded component. Methods of quantifying this reliability and the corresponding failure probability have been investigated and refined. The result of this effort is a public-domain computer program called CARES (ceramics analysis and reliability evaluation of structures). CARES calculates the fast-fracture reliability of macroscopically isotropic ceramic components.[132,133]

The design methodology used by CARES combines three major elements: (1) linear elastic fracture mechanics (LEFM) theory, which relates the strength of ceramics to the size, shape, and orientation of critical flaws; (2) extreme-value statistics to obtain the characteristic flaw-size distribution function, which is a material property; and (3) material microstructure.

Probabilistic component design requires the determination of a fracture strength distribution from simple-geometry flexural or tensile test specimens. The statistical material parameters are estimated as a function of temperature, specimen loading, and geometry. From these data the reliability for a complex-component geometry and loading is then predicted. Appropriate design changes are made until an acceptable probability of failure has been reached.

In conclusion, the potential use of structural ceramics for high-temperature applications depends on the strength, toughness, and reliability of these materials. Components using ceramics can be designed for high reliability in service if the contributing factors that cause material failure are accounted for. The basis of this design methodology must combine the statistical nature of strength-controlling flaws with fracture mechanics to allow for multiaxial stress states and concurrent flaw populations. This is accomplished using the NASA/CARES public domain computer program for predicting the fast-fracture reliability of structural ceramic components.[133-137]

TABLE 9.5 Glass and Glass-Ceramic Matrices of Interest

Matrix type	Major constituents	Minor constituents	Major crystalline phases	Maximum use temperature in composite form, °C
Glasses:				
7740 borosilicate	B_2O_3, SiO_2	Na_2O, Al_2O_3		600
1723 aluminosilicate	Al_2O_3, MgO, CaO, SiO_2	B_2O_3, BaO		700
7930 high silica	SiO_2	B_2O_3		1150
Glass-ceramics:				
LAS-I	Li_2O, Al_2O_3, MgO, SiO_2	ZnO, ZrO_2, BaO	B-spodumene	1000
LAS-II	Li_2O, Al_2O_3, MgO, SiO_2, Nb_2O_5	ZnO, ZrO_2, BaO	B-spodumene	1100
LAS-III	Li_2O, Al_2O_3, MgO, SiO_2, Nb_2O_5	ZrO_2	B-spodumene	1200
MAS	MgO, Al_2O_3, SiO_2	BaO	Cordierite	1200
BMAS	BaO, MgO, Al_2O_3, SiO_2		Barium osumilite	1250
Ternary mullite	BaO, Al_2O_3, SiO_2		Mullite	~1500
Hexacelsian	BaO, Al_2O_3, SiO_2		Hexacelsian	~1700

Source: United Technologies Research Center, East Hartford, Conn.

9.8 FUTURE TRENDS

Fiber-reinforced glass and glass-ceramic matrix composites have provided an excellent demonstration of the development of high-performance ceramic-matrix composite materials. High levels of strength and toughness are clearly attainable, but only if the fiber-matrix interfacial region is controlled in a manner that prevents crack propagation across this boundary. Thus, even more than in polymer- and metal-matrix composite systems, the composite must be considered to consist of three equally important and tailorable regions: fiber, interface, and matrix.

The current rate of development of ceramic-matrix composites is encouraging. It is anticipated that their true potential will soon be fully recognized and that they will take their place, along with polymer- and metal-matrix composites, as a major class of materials for structural and nonstructural applications. It is also to be expected that many different routes to composite fabrication, along with the broad range of fiber-matrix combinations available, will combine to provide many unique composite systems.[138]

For the future the challenges are clear. First is the need to improve the understanding of composite mechanical and physical properties and their relationship to fiber, matrix, and interface characteristics. There are many aspects to this issue, including the need to actually make and test engineering components. Without this knowledge, future guidelines for composite developments will be inadequate. Second, and equally important, the many factors controlling fiber-matrix and environmental-chemical interaction at elevated temperatures must be explored more fully. Ceramic-matrix composites are intended for use in environments beyond the capabilities of metals and polymers and will be fabricated and used at high temperatures, where extensive reactions can take place. Finally, the traditional thinking of ceramists that the most important applications exist at very high temperatures and under high stress should be constructively amended to include a much broader range of possibilities, including lower-temperature, lower-stress structural and nonstructural applications.[138]

A key challenge and question for ceramic-fiber composites is the extent of their utility at high temperatures. While many have viewed their high fracture toughness and resultant greater mechanical reliability as the answer to using ceramics in a variety of high-temperature applications, the majority of composites to date have shown serious high-temperature limitations. A key problem has been the embrittlement that has occurred with most of these composites after high-temperature exposure. The toughness depends on limited bonding between fibers and matrix, at least in continuous-fiber composites. Exposure to high temperatures can result in strong fiber-matrix bonding due either to sintering between fiber and matrix, or to their bonding because of reactions, such as oxidation. Environmental access to the interior of the composite, which accelerates such reactions, can be enhanced significantly by porosity in the composite, and hence can be related to processing methods. However, some environmental access to the composite will generally be obtained in even dense composites under sufficient loading due to microcracking that characteristically occurs because of the lower strain of failure of the matrices vis-à-vis the fibers. Continuous-fiber composites made by CVI have in general shown the best retention of toughness at high temperatures, apparently due to fiber encapsulation in the CVI process. Whisker composites have generally been next in high-temperature performance.

Another important need is to better identify, document, and understand the mechanical behavior of ceramic-fiber composites. Thus much more needs to be known about the effects of preexisting microcracks and microcrack generation during stressing, their origin and effects. One important aspect of this is mechanical fatigue, which clearly occurs, but has only had limited study. Even less has been done to document, or study, thermal fatigue and mechanical impact.[139]

Broader needs can generally be classified in terms of materials, processes, understanding, and applications. The key need relative to fibers is obtaining greater fiber temperature capability to provide greater compatibility with both processing and use conditions. A wider range of fibers to provide a wider range of compatibility with different matrices would also be useful. However, much of this function may also be served by suitable fiber coatings. Thus, a critical need is identified as further development of ceramic coatings not only to increase the array of coating-fiber combinations, but also to continue to look at more sophisticated coatings, such as two- and three-layer coatings.

Many necessary improvements in processing and matrix development can be identified from the mostly exploratory studies. An overall need is to further increase the homogeneity of matrix distribution among the fibers. Reducing both the amount and the inhomogeneity in porosity distribution is also important. While this is in part related to the homogeneity of the matrix distribution within the fibers, it also depends on the processing method, as well as in many cases on both the size and the shape of the composite being fabricated. Because of their greater versatility, chemical methods of processing need to be further explored. Further development of potential matrices, which may themselves be (particulate) composites or more complex compounds, such as indicated by the ZrO_2–SiO_2 and ZrO_2–TiO_2 developments, should be beneficial.

The critical need for continuous-fiber composites is to identify specific applications and demonstrate that they satisfy such applications physically and economically. This is essential to establish both production processing as well as practical performance as a baseline for further developing composites and their applications.

There is a significant increase in the diversity of composites based on different fibers, fiber coatings, matrices, and processing methods, at least in the study and development phases. It remains to be determined whether subsequent applications may ultimately increase or decrease this diversity. It is likely to depend, at least in part, on how extensive the applications are. Finally, the promise of ceramic whisker composites is sufficiently high that they will find more applications beyond the two that have thus far been established (cutting tools and heat-treating fixtures, utilizing whisker composites). Identifying and developing more specific applications will be a critical factor in determining both the extent and the course of future development, along with whisker costs.

Finally, two other basic trends are suggested. The first is in process approaches. There is a significant shift in the mix of processing technologies utilized for making ceramic composites relative to those utilized for making monolithic ceramics. Second, the most basic trend that can be seen is for substantial growth and diversification of these composite materials because of the significant improvements they offer in mechanical performance.[139]

REFERENCES

1. J. D. Buckley, in *Proc. 11th Ann. Mtg. Metal Matrix, Carbon, and Ceramic Matrix Composites 1987,* NASA CP 2482, Oct. 1987.

2. L. Kempfer, "Forming the Pieces of the Ceramic Puzzle," *Mater. Eng.*, pp. 23–26, June 1990.
3. L. L. Hench and W. Vasconcelos, "Gel-Silica Science," *Ann. Rev. Mater. Sci.*, vol. 20, pp. 269–298, 1990.
4. J. O. Steigler and J. R. Weir, Jr., "Metals and Ceramics Division," ORNL 6601, DE-AC 05-84OR21400, Apr. 1990, 166 pp.
5. W. F. Kladnig and J. E. Horn, "Submicron Oxide Powder Preparation by Microwave Processing," *Ceram. Intl.*, vol. 16, pp. 99–106, Feb. 1990.
6. W. H. Sutton, "Microwave Processing of Ceramic Materials," *Ceram. Bull.*, vol. 68, pp. 376–386, Feb. 1989.
7. L. M. Sheppard, "Advances in Processing of Ceramics," *Ceram. Bull.*, vol. 67, pp. 1649–1653, Oct. 1988.
8. S. B. Bhaduri, A. Chakraborty, and J. J. Reddy, "Injection Moulding of Ceria-Stabilized Tetragonal Zirconia Polycrystals (CE-TZP)," *J. Mater. Sci. Lett.*, vol. 9, pp. 209–210, Feb. 1990.
9. R. E. Mistler, "Tape Casting: The Basic Process for Meeting the Needs of the Electronics Industry," *Ceram. Bull.*, vol. 69, pp. 1022–1026, June 1990.
10. H. W. Stetson, "Method of Making Multilayer Circuits," US Patent 3189978, 1965.
11. B. H. Rabin, "Modified Tape Casting Method for Ceramic Joining; Application to Joining of Silicon Carbide," *J. Am. Ceram. Soc.*, vol. 73, pp. 2757–2759, Sept. 1990.
12. R. E. Fisher, C. V. Burkland, and W. E. Bustamante, "Ceramic Composites Based on Chemical Vapor Infiltration," in *Proc. Metal and Ceramic Composite Processing Conf.*, Columbus, Ohio, Nov. 13–15, 1984, vol. 2, pp. 115–143.
13. C. V. Burkland et al., "CVI-Processed Ceramic Matrix Composites," in *Proc. Metal and Ceramic Composite Processing Conf.*, Columbus, Ohio, Nov. 13–15, 1984.
14. P. J. Geoghegan, "Ceramic Matrix Composites by Chemical Vapor Infiltration," in *Proc. Metal and Ceramic Composite Processing Conf.*, Columbus, Ohio, Nov. 13–15, 1984.
15. D. Miller, "Preparation of Advanced Composite Ceramics Via Novel Processing Methods," presented at the 5th Ann. ASM/ESD Advanced Composites Conf. and Exposition, Dearborn, Mich., Sept. 25–28, 1989.
16. W. J. Kennedy and E. C. Skaar, "Improving the Machining of Ceramics," in *Nontraditional Machining*, Orlando, Fla., SME-MS 89-813, Oct. 1989, 16 pp.
17. T. G. Nieh and J. Wadsworth, "Superplastic Ceramics," *Ann. Rev. Mater. Sci.*, vol. 20, pp. 117–140, 1990.
18. I. W. Chen and L. A. Xue, "Development of Superplastic Structural Ceramics," *J. Am. Ceram. Soc.*, vol. 73, pp. 2585–2609, Sept. 1990.
19. *World Report on Advanced Ceramics*, vol. 1, pp. 4–5, Aug. 1989.
20. A. S. Brown, "Ceramics Learn to Bend Instead of Break," *Aerospace Am.*, pp. 24–28, Aug. 1990.
21. T. G. Langdon, "Superplastic Ceramics—They're Not a Stretch of the Imagination Anymore," *J. Met.*, vol. 42, pp. 8–13, July 1990.
22. *Adv. Ceram. Mater.*, vol. 1, p. 7, 1986.
23. S. J. Klima and H. E. Kautz, "Nondestructive Evaluation of Advanced Ceramics," NASA TM 101489, Oct. 1988, 11 pp.
24. T. Fujii and T. Nose, "Evaluation of Fracture Toughness for Ceramic Materials," *ISIJ Intl.*, vol. 29, pp. 717–725, 1989.
25. A. R. Rosenfield and W. H. Duckworth, "Japanese Standard Fracture Toughness Test," CAB, no. 199, pp. 3–4, Mar. 1990.

26. *Aerospace Eng.,* p. 40, Feb. 1989.
27. K. M. Prewo, "Ceramic and Carbon Fiber Reinforced Glasses," in *Proc. Metal and Ceramic Composite Processing Conf.,* Columbus, Ohio, Nov. 13–15, 1984, vol. 1, pp. 139–152.
28. F. D. Lemkey et al. (Eds.), in *High Temperature/High Performance Composites,* Materials Research Soc., Pittsburgh, Penn., 1989, 383 pp.
29. J. W. Holmes, "Elevated Temperature Mechanical Behavior of SiC Fiber-Reinforced Si_3N_4 Composites," in F. D. Lemkey et al. (Eds.), *High Temperature/High Performance Composites,* Materials Research Soc., Pittsburgh, Penn., 1989.
30. J. J. Mecholsky, Jr., "Engineering Research Needs of Advanced Ceramics and Ceramic-Matrix Composites," *Ceram. Bull.,* vol. 68, pp. 367–375, Feb. 1989.
31. F. I. Hurwitz, "Ceramic Matrix and Resin Matrix Composites Comparison," NASA TM 89830, 1987, 14 pp.
32. D. M. Dawson, R. F. Preston, and A. Briggs, "Ceramic Fibre/Glass Matrix Composites with Outstanding Mechanical Performance," AERE, Harwell, Gt. Br., AERE R120 25, Oct. 1986.
33. D. C. Larsen, "Evaluation of Ceramics and Ceramic Composites for Turbine Engine Applications," AFWAL TR 88-4202, F33615-82C5101, FR9/82-6/86, Dec. 1988.
34. G. Schnittgrund, "Ceramic Matrix Composites for Advanced Turbomachinery," presented at Aeromat '90, Advanced Aerospace M/P Conf. and Exposition, Long Beach, Calif., May 21–24, 1990.
35. R. Vaidya and K. N. Subramanian, "Metallic Glass Reinforcement of Glass-Ceramics," presented at Aeromat '90, Advanced Aerospace M/P Conf. and Exposition, Long Beach, Calif., May 21–24, 1990.
36. M. Meiser, "Development of Silicon Nitride Components for Advanced Application," presented at Aeromat '90, Advanced Aerospace M/P Conf. and Exposition, Long Beach, Calif., May 21–24, 1990.
37. J. R. Stephens, "Composites Boost 21st Century Aircraft Engines," *AM&P,* vol. 137, pp. 35–38, Apr. 1990.
38. *Mater. Eng.,* p. 51, Jan. 1990.
39. A. Bennett, "Overview of the Advanced Ceramics for Turbines (ACT) Programme," *Adv. Mater. J.,* vol. 1, pp. 21–26, Sept. 1990.
40. D. C. Phillips et al., "Behavior of Ceramic Matrix, Fiber Composites under Combined Impact and Tensile Stresses," AERE, Harwell, Gt. Br., AFWAL TR 88-4016, AERE R12941, F33615-86C5144, May 1988, 87 pp.
41. S. L. Jones, "DOE Wants Ceramic R&D to Continue," *Metalwork. News,* pp. 1, 39, June 29, 1987; S. L. Jones, "Ceramic Components Yet to Meet Engine Test Levels," ibid., p. 11, July 13, 1987.
42. J. W. McCauley and D. J. Viechnicki, "Ceramics R&D at the U.S. Army Materials Technology Lab," *Ceram. Bull.,* vol. 67, pp. 1340–1344, Aug. 1988.
43. A. P. Majidi and T. W. Chou, "Elevated Temperature Studies of Continuous and Discontinuous Fiber Reinforced Ceramic Matrix Composites," 89-GT-124, June 1989, 8 pp.
44. J. D. Bolt and R. H. French, "Polymer/Ceramic Composites; New Properties for Electronics," *AM&P,* vol. 134, pp. 32–35, July 1988.
45. P. K. Mehrotra, "Advanced Ceramic Tools for High Productivity Machining of Superalloys," *Adv. Ceram. '90,* SME-EM90-140, Feb. 1990, 15 pp.
46. J. A. Vaccari, "The Challenges of the 'Orient Express'," *Amer. Machinist,* pp. 55–57, Jan. 1990.
47. L. Kempfer, "Materials Take Hypersonic Leap Into Space," *Mater. Eng.,* pp. 19–22, Aug. 1990.

48. M. S. Newkirk, A. W. Urquhart, and H. R. Zwicker, "Formation of Lanxide® Ceramic Composite Materials," *J. Mater. Res.,* vol. 1, no. 1–2, 1986.
49. J. E. Garnier, S. Keck, and A. S. Fareed, "Structural Ceramic and Metal Matrix Composites Fabricated by the Directed Metal Oxidation and Pressureless Metal Infiltration Processes," presented at Aeromat '90, Advanced Aerospace M/P Conf. and Exposition, Long Beach, Calif, May 21–24, 1990.
50. *Amer. Machinist,* p. 63, Nov. 1988.
51. "High-Strength Silicon Nitride Ceramic," *Jetho,* 90-04-001-5, p. 19, 1990.
52. A. K. Misra, "Identification of Thermodynamically Stable Ceramic Reinforcement Materials for Iron Aluminide Matrices," *Met. Trans. A.,* vol. 21A, pp. 441–446, Feb. 1990.
53. E. L. Fleischer et al., "The Effect of Ion Induced Damage on the Hardness, Wear, and Friction of Zirconia," *J. Mater. Res.,* vol. 5, pp. 385–391, Feb. 1990.
54. S. R. Levine, "Ceramics for Engines," *Aerospace Am.,* pp. 22–27, May 1987.
55. *Proc. 1987 Intl. Conf. on PM Aerospace Materials,* Metal Powder Rep., MPR Publ. Services Ltd., U.K., 1988, pp. 43.2–43.18.
56. S. J. Grisaffe, "Ceramic Matrix Composites," *AM&P,* vol. 137, pp. 43–44, 93–94, Jan. 1990.
57. M. D. Meiser, "Development of Monolithic and Composite Ceramics at Allied-Signal Aerospace Company," in M. Doyama, S. Sōmiya, and R. P. H. Chang (Eds.), *Structural Ceramics—Fracture Mechanics, Proc. MRS Intl. Mtg. on Advanced Materials,* Tokyo, Japan, May 31–June 3, 1988, vol. 5, pp. 187–200.
58. R. M. McMeeking, "Toughening Mechanisms in Brittle Matrix Composites," in M. Doyama, S. Sōmiya, and R. P. H. Chang (Eds.), *Structural Ceramics—Fracture Mechanics, Proc. MRS Intl. Mtg. on Advanced Materials,* Tokyo, Japan, May 31–June 3, 1988, vol. 5, pp. 277–294.
59. M. J. Koczak et al., "Inorganic Composite Materials in Japan; Status and Trends," ONRFEM7, Nov. 1989, 53 pp.
60. A. G. Evans and D. B. Marshall, "The Mechanical Behavior of Ceramic Matrix Composites," *Acta Metall.,* vol. 37, pp. 2567–2583, Oct. 1989.
61. R. J. Kerans et al., "The Role of the Fiber-Matrix Interface in Ceramic Composites," *Ceram. Bull.,* vol. 68, pp. 429–442, Feb. 1989.
62. A. S. Brown, "Taming Ceramic Fiber," *Aerospace Am.,* pp. 14–22, May 1989.
63. R. Warren, "Ceramic-Matrix Composites," *Composites,* vol. 18, pp. 86–88, Apr. 1987.
64. H. Cao and M. D. Thouless, "Tensile Tests of Ceramic-Matrix Composites: Theory and Experiment," *J. Am. Ceram. Soc.,* vol. 73, pp. 2091–2094, July 1990.
65. M. Yoshimura et al., "Fabrication of Whisker/Glass Composites from Hydrothermally Oxidized Si_3N_4 Whisker," in M. Doyama, S. Sōmiya, and R. P. H. Chang (Eds.), *Structural Ceramics—Fracture Mechanics, Proc. MRS Intl. Mtg. on Advanced Materials,* Tokyo, Japan, May 31–June 3, 1988, vol. 5, pp. 497–502.
66. L. M. Butkus, L. P. Zawada, and G. A. Hartman, "Room Temperature Tensile and Fatigue Properties of Silicon-Carbide Fiber-Reinforced Ceramic Matrix Composites," in M. Doyama, S. Sōmiya, and R. P. H. Chang (Eds.), *Structural Ceramics—Fracture Mechanics, Proc. MRS Intl. Mtg. on Advanced Materials,* Tokyo, Japan, May 31–June 3, 1988, vol. 5.
67. F. S. Galasso, *Advanced Fibers and Composites,* Gordon and Breach, New York, 1988, 145 pp.
68. R. M. Laine, "Silicon Nitride Ceramic Fibers from Preceramic Polymers," TR 8 (SRI), N00014-84C0392, June 1987.

69. M. U. Islam, "Artificial Composites for High Temperature Applications; Review," Natl. Research Council of Canada, DME007, Jan. 1987, 84 pp.
70. W. H. Atwell et al., "Polymer Processing," AFWAL-TR-85-4099, F33615-83-C-5006, Dec. 1987, vol. 1: "Fiber Technology," 394 pp., vol. 2: "Composites Technology," 622 pp.
71. R. G. Munro and C. R. Hubbard, "Property Database for Gas-Fired Applications of Ceramics," *Ceram. Bull,* vol. 68, pp. 2084–2090, Dec. 1989.
72. R. D. Veltri and F. S. Galasso, "Chemical-Vapor-Infiltrated Silicon Nitride, Boron Nitride and Silicon Carbide Matrix Composites," *J. Am. Ceram. Soc.,* vol. 73, pp. 2137–2140, July 1990.
73. N. H. Tai and T. W. Chou, "Modeling of Chemical Vapor Infiltration (CVI) in Al_2O_3/ SiC Composites Processing," in J. D. Buckley (Ed.), *Proc. 12th Ann. Mtg., Metal Matrix, Carbon, and Ceramic Matrix Composites 1988*, NASA Conf., Cocoa Beach, Fla., Jan. 20–22, 1988, pp. 237–246.
74. P. Reagan and F. N. Huffman, "Chemical Vapor Composite Deposition," in J. D. Buckley (Ed.), *Proc. 12th Ann. Mtg., Metal Matrix, Carbon, and Ceramic Matrix Composites 1988* NASA Conf., Cocoa Beach, Fla., Jan. 20–22, 1988, pp. 247–258.
75. M. I. Mendelson, "Characterization of Nextel® 3-D Woven Fiber Structures, in J. D. Buckley (Ed.), *Proc. 12th Ann. Mtg., Metal Matrix, Carbon, and Ceramic Matrix Composites 1988* NASA Conf., Cocoa Beach, Fla., Jan. 20–22, 1988, pp. 259–270.
76. C. V. Burkland and J. M. Yang, "Development of 3-D Braided Nicalon/Silicon Carbide Composite by Chemical Vapor Infiltration," in J. D. Buckley (Ed.), *Proc. 12th Ann. Mtg., Metal Matrix, Carbon, and Ceramic Matrix Composites 1988*NASA Conf., Cocoa Beach, Fla., Jan. 20–22, 1988, pp. 271–280.
77. S. Majumdar, D. Kupperman, and J. Singh, "Analytical and Experimental Determinations of Residual Thermal Stresses in a Ceramic-Ceramic Composite," in J. D. Buckley (Ed.), *Proc. 12th Ann. Mtg., Metal Matrix, Carbon, and Ceramic Matrix Composites 1988* NASA Conf.,.Cocoa Beach, Fla., Jan. 20–22, 1988, pp. 281–301.
78. R. M. Salinger et al., "Progress in the Formation of Si/N-C Advanced Ceramic Fibers from Polymer Precursors," in J. D. Buckley (Ed.), *Proc. 12th Ann. Mtg., Metal Matrix, Carbon, and Ceramic Matrix Composites 1988* NASA Conf., Cocoa Beach, Fla., Jan. 20–22, 1988, pp. 21–28.
79. A. J. Eckel and R. C. Bradt, "Thermal Expansion of Laminated, Woven, Continuous Ceramic Fiber/Chemical-Vapor-Infiltrated Silicon Carbide Matrix Composites," *J. Am. Ceram. Soc.,* vol. 73, pp. 1334–1338, May 1990.
80. L. Leonard, "Ceramic-Matrix Composites; Mettle for the Nasty Jobs," *Adv. Composites,* pp. 37–43, July/Aug. 1990.
81. B. Q. Wang and A. V. Levy, "Erosion Behavior of SiC Fiber-SiC Matrix Composites," *Wear,* vol. 138, no. 1–2, pp. 125–136, 1990.
82. B. H. Rabin, G. E. Korth, and R. L. Williamson, "Fabrication of Titanium Carbide-Alumina Composites by Combustion, Synthesis and Subsequent Dynamic Consolidation," *J. Am. Ceram. Soc.,* vol. 73, pp. 2156–2157, July 1990.
83. H. W. Lee and M. D. Sacks, "Pressureless Sintering of SiC-Whisker-Reinforced Al_2O_3 Composites; I: Effect of Matrix Powder Surface Area," *J. Am. Ceram. Soc.,* vol. 73, pp. 1884–1893, July 1990.
84. H. W. Lee and M. D. Sacks, "Pressureless Sintering of SiC-Whisker-Reinforced Al_2O_3 Composites; II: Effects of Sintering Additives and Green Body Infiltration," *J. Am. Ceram. Soc.,* vol. 73, pp. 1894–1900, July 1990.
85. R. H. Bhatt et al., "Thermal Effects on the Mechanical Properties of SiC Reinforced Reaction Bonded Silicon Nitride Matrix (SiC/RBSN) Composites," NASA TM-101348, AVSCOM Tech Rep. 88-C-028, NASA-E-4375, Oct. 1988, 16 pp.
86. P. Nagy and J. Kalady, "Status of Ceramic Composites for Cruise Missile Propulsion

Applications," presented at Aeromat '90, Advanced Aerospace M/P Conf. and Exposition, Long Beach, Calif., May 21–24, 1990.
87. J. C. Rawers et al., "Ti-SiC Composites Prepared by High Temperature Synthesis," *Mater. Sci. Technol.*, vol. 6, pp. 187–191, Feb. 1990.
88. P. K. Brindley, in C. C. Koch et al. (Eds.), *Proc. Symp. High Temperature Ordered Intermetallic Alloys*, Materials Research Soc., Pittsburgh, Penn., 1987, vol. 81.
89. J. E. Heany, *Aviation Week. Space Technol.*, vol. 100, Oct. 12, 1987.
90. L. D. Wedeven and T. A. Harris, "Rolling Element Bearings Operating at the Extremes," *Machine Des.*, pp. 72–76, Aug. 6, 1987.
91. J. F. Dill, R. A. Harmon, and E. M. Lenoe, "A Review of the State-of-the-Art in Rolling Element Bearing Technology in Japan," *ONRFE*, vol. 12, pp. 57–70, Oct.–Dec. 1987.
92. B. I. Lee and S. Y. Park, "Sol-Gel Processing of SiC-Whisker-Reinforced Silica-Based Ceramic Composites," *J. Am. Ceram. Soc.*, vol. 72, pp. 2381–2385, Dec. 1989.
93. R.O. Veltri, D. A. Condit, and F. S. Galasso, "Chemical Vapor Deposited SiC Matrix Composites," *J. Am. Ceram. Soc.*, vol. 72, pp. 478–480, Mar. 1989.
94. A. J. Caputo et al., "Fiber-Reinforced SiC Composites with Improved Mechanical Properties," *Am. Ceram. Soc. Bull.*, vol. 66, pp. 368–372, Feb. 1987.
95. D. A. Condit, R. D. Veltri, and F. S. Galasso, "Low-Density Composites of Oriented Carbon on Fused Quartz Wool or Reticulated Vitreous Carbon," *Am. Ceram. Soc. Bull.*, vol. 66, pp. 359–362, Feb. 1987.
96. G. H. Shiroky, A. W. Urquhart, and B. W. Sorenson, "Ceramic Composites for Gas Turbine Engines via a New Process," ASME 89-GT-316, June 1989, 6 pp.
97. Y. M. Chiang et al., "Reaction-Based Processing Methods for Ceramic-Matrix Composites," *Ceram. Bull.*, vol. 68, pp. 420–428, Feb. 1989.
98. R. Raj and R. K. Bordia, "Sintering Behavior of Bi-Modal Powder Compacts," *Acta Metall.*, vol. 32, pp. 1003–1019, July 1984.
99. C. H. Hsueh et al., "Viscoelastic Stresses and Sintering Damage in Heterogeneous Powder Compacts," *Acta Metall.*, vol. 34, pp. 927–936, May 1986.
100. C. H. Hsueh, A. G. Evans, and R. M. McMeeking, "The Influence of Multiple Heterogeneities on Sintering Rates," *J. Am. Ceram. Soc.*, vol. 69, pp. C-64–C-66, Apr. 1986.
101. S. Adachi, T. Wada, and T. Mihara, "High-Pressure Self-Combustion Sintering of Alumina-Titanium Carbide Ceramic Composite," *J. Am. Ceram. Soc.*, vol. 73, pp. 1451–1452, May 1990.
102. M. Hunt, "Future Federal Materials Here Today," *Mater. Eng.*, pp. 35–38, Feb. 1990.
103. "High Interest in Ceramics at Toyota," *AM&P*, vol. 136, pp. 21–25, Sept. 1989.
104. D. J. Bak, "Turbine Engines 2001," *Des. News*, pp. 113–118, Apr. 9, 1990.
105. S. W. Kandebo, "Spaceplane Conference Highlights International Hypersonic Programs," *Aviation Week. Space Technol.*, pp. 66–68, Nov. 12, 1990.
106. F. W. Tortolano, "Birth of a Space-Age Plane," *Des. News*, pp. 52–57, Apr. 4, 1988.
107. R. F. Gilby, "Ceramic Matrix Earns its Wings on French Fighter," *Des. News*, pp. 104–106, Nov. 20, 1989.
108. G. Marsh, "Engineering Ceramics—Part I," *Aerosp. Composites Mater.*, vol. 1, pp. 34–37, Nov.–Dec. 1990.
109. J. Constance, "Industry Turns to Ceramic Composites," *Aerospace Am.*, pp. 22–26, Mar. 1990.
110. M. Taya et al., "Toughening of a Particulate-Reinforced Ceramic Matrix Composite by Thermal Residual Stress," *J. Am. Ceram. Soc.*, vol. 73, pp. 1382–1391, May 1990.

111. T. C. Lei and Y. Zhou, "Effect of Sintering Process on Microstructure and Properties of Al_2O_3-ZrO_2 Ceramics," in M. Doyama, S. Somiya, and R. P. H. Chang (Eds.), *Structural Ceramics—Fracture Mechanics, Proc. MRS Intl. Mtg. on Advanced Materials*, Tokyo, Japan, May 31–June 3, 1988, vol. 5, pp. 325–330.

112. J. A. DiCarlo, *J. Met.*, vol. 37, pp. 44–49, 1985.

113. E. R. Thompson and K. M. Prewo, "Structures," in *Structural Dynamics and Materials Conf. Tech. Papers*, AIAA, New York, 1984, pp. 539–543.

114. P. F. Becher, "Recent Advances in Whisker-Reinforced Ceramics," *Ann. Rev. Mater. Sci.*, vol. 20, pp. 179–195, 1990.

115. S. A. Bradley et al., "Silicon Carbide Whisker Stability during Processing of Silicon Nitride Matrix Composites," *J. Am. Ceram. Soc.*, vol. 72, pp. 628–636, Apr. 1989.

116. C. F. Lewis, "Ceramic Matrix Composites; The Ultimate Materials Dream," *Mater. Eng.*, pp. 41–45, Sept. 1988.

117. T. N. Tiegs and P. F. Becher, in *Proc. 23d Auto. Tech. Dev. Contractors' Coord. Mtg.*, SAE, Warminster, Penn., 1985, vol. P-165, pp. 209–213.

118. E. M. Lenoe, *ONRFE Sci. Inform. Bull.*, vol. 14, Mar. 1989, 72 pp.

119. G. H. Campbell et al., "Whisker Toughening; A Comparison between Aluminum Oxide and Silicon Nitride Toughened with Silicon Carbide," *J. Am. Ceram. Soc.*, vol. 73, pp. 521–530, Mar. 1990.

120. K. R. Karasek et al., "Characterization of Silicon Carbide Whiskers," *J. Am. Ceram. Soc.*, vol. 72, pp. 1907–1913, Oct. 1989.

121. R. K. Govila, "Fracture of Hot-Pressed Alumina and SiC-Whisker-Reinforced Alumina Composite," *J. Mater. Sci.*, pp. 3782–3791, Oct. 1988.

122. Y. Akimune, Y. Katano, and K. Matoba, "Spherical-Impact Damage and Strength Degradation in Silicon Carbide Whisker/Silicon Nitride Composites," *J. Am. Ceram. Soc.*, vol. 72, pp. 791–798, May 1989.

123. S. Iio et al., "Mechanical Properties of Alumina/Silicon Carbide Whisker Composites," *J. Am. Ceram. Soc.*, vol. 72, pp. 1880–1884, Oct. 1989.

124. Y. Akimune, "High-Temperature Strength of SiC Whisker-Sialon Composites," *J. Mater. Sci. Lett.*, vol. 9, pp. 816–817, July 1990.

125. H. Kodama, H. Sakamoto, and T. Miyoshi, "Silicon Carbide Monofilament-Reinforced Silicon Nitride or Silicon Carbide Matrix Composites," *J. Am. Ceram. Soc.*, vol. 72, pp. 551–558, Apr. 1989.

126. "Ceramic Fibers," *Mater. Eng.*, p. 14, June 1990.

127. "Heat-Resistant High-Tenacity Ceramic Composite Material," *Jetho*, 90-3-002-478, p. 23, Mar. 1990.

128. "Alumina-Enhanced Thermal Barrier," *NASA Tech. Briefs*, p. 78, Apr. 1989.

129. "Advanced Structural Ceramics: Technologies, Economics and Market Opportunities," GB-107, BCC Inc., Norwalk, Conn., Dec. 1987, 354 pp.

130. R. Chaim and A. H. Heuer, "The Interface between (Nicalon) SiC Fibers and a Glass-Ceramic Matrix," *Adv. Ceram. Mater.*, vol. 2, pp. 154–158, Feb. 1987.

131. R. U. Vaidya and K. N. Subramanian, "Metallic Glass Ribbon Reinforced Glass-Ceramic Matrix Composites," *J. Mater. Sci.*, vol. 25, pp. 3291–3296, July 1990.

132. N. N. Nemeth, J. M. Manderscheid, and J. P. Gyekenyesi, "Ceramic Analysis and Reliability Evaluation of Structures (CARES), User's and Programmer's Manual," NASA TP-2916, Washington, D.C., 1989.

133. N. N. Nemeth, J. M. Manderscheid, and J. P. Gyekenyesi, "Designing Ceramic Components with the CARES Computer Program," *Ceram. Bull.*, vol. 68, pp. 2064–2072, Dec. 1989.

134. J. A. M. Boulet, "Assessment of State of Art in Predicting Failure of Ceramics," ORNL Sub 86-57598-1, DE-AC05-840R21400, Mar. 1988, 39 pp.

135. I. A. Aksay, "Microdesigning of Lightweight/High Strength Ceramic Materials," AFOSR TR 87-1595, AFOSR83-0375, Aug. 1987.
136. S. F. Duffy, J. M. Manderscheid, and J. L. Palko, "Analysis of Whisker-Toughened Ceramic Components—A Design Engineer's Viewpoint," *Ceram. Bull.*, vol. 68, pp. 2078–2083, Dec. 1989.
137. E. J. Kubel, Jr., "Structural Ceramics; Materials of the Future," *AM&P*, vol. 134, pp. 25–33, Aug. 1988.
138. K. M. Prewo and J. J Brennan, "Fiber Reinforced Glasses and Glass Ceramics for High Performance Applications," in S. M. Lee (Ed.), *Reference Book for Composites Technology*, Technomic Publ., Lancaster, Penn., 1989, pp. 97–115.
139. R. W. Rice and D. Lewis, III, "Ceramic Fiber Composites Based upon Refractory Polycrystalline Ceramic Matrices," in S. M. Lee (Ed.), *Reference Book for Composites Technology*, Technomic Publ., Lancaster, Penn., 1989, pp. 117–141.

BIBLIOGRAPHY

Abkowitz, S. M., et al., "Ductile Alloy Encapsulated Ceramic Armor Development," FR 6/87-11/89, Dynamet Tech. Inc., MTL-TR-90-2, DAAL04-87-C-0029, Jan. 1990.

Aikin, R. M., Jr., "The Mechanisms of Dispersion Strengthening and Fracture in Al-Based XD® Alloys," NASI-18531, NASA CR 4276, MML TR 89-54C, Feb. 1990, 81 pp.

Arya, P. V., et al., "A Plasma Assist CVD Process for Producing TiB_2 and Other Ceramics," Materials and Elec. Research Corp., Tucson, Ariz., DAAH01-89-C-0707, Dec. 5, 1989.

Banerjee, P. K., "Development of BEM for Ceramic Composites," NASA CR183313, NAG3-888, PR 3-12/88, 1988, 76 pp.

Buckley, J. D., "Metal Matrix, Carbon, and Ceramic Matrix Composites 1986," in *Proc. 10th Ann. Mtg., NASA/DOD Joint Conf.*, Cocoa Beach, Fla., Jan. 21–24, 1986, NASA Conf. Publ. 2445, 345 pp.

Carlsson, L., "Cyclic Fatigue of Hot-Isostatic-Pressed Silicon Nitride," Statens Provningsanstalt, Boras, Sweden, 1989, 27 pp.

Cartz, L., "Ceramic-Ceramic Composites Meeting in Belgium," Office of Naval Res. Gt.Br., ONRL 7-020R, Aug. 1987.

Coblenz, W. S., and D. Lewis, III, "In Situ Reaction of B_2O_3 with AlN and/or Si_3N_4 to Form BN-Toughened Composites," *J. Am. Ceram. Soc.*, vol. 71, pp. 1080–1085, Dec. 1988.

DiCarlo, J. A., "CMCs for the Long Run," *AM&P*, vol. 135, pp. 41–44, June 1989.

Easterling, K., *Tomorrow's Materials*, Inst. of Metals, London, 1988, 109 pp.

Eckel, A. J., and R. C. Bradt, "Strength Distribution of Reinforcing Fibers in a Nicalon Fiber/Chemically Vapor Infiltrated Silicon Carbide Matrix Composite," *J. Am. Ceram. Soc.*, vol. 72, pp. 455–458, Mar. 1989.

Evans, A. G., and R. Mehrabian, "Processing and Mechanical Properties of High Temperature/High Performance Composites," AR 9/88-9/89, Univ. of Calif., Santa Barbara, N0001 4-86K0753; book 5, sec. 4: "Processing: Matrices and Composites, Part 1"; book 2, sec. 2: "Strength and Fracture Resistance, Part 1"; book 1, sec. 1: "Coatings and Interfaces"; book 3, sec. 2: "Strength and Fracture Resistance, Part 2"; book 6, sec. 4: "Processing: Matrices and Composites, Part 2"; Oct. 1989.

Greil, P., "Opportunities and Limits in Engineering Ceramics," *P/M Intl.*, vol. 21, pp. 40–45, Apr. 1989.

Habib, F. A., R. G. Cooke, and B. Harris, "Cracking in Brittle Matrix Composites," *Brit. Ceram. Trans. J.*, vol. 89, pp. 115–124, Apr. 1990.

Halverson, D. C., et al., "Processing of Boron Carbide-Aluminum Composites," *J. Am. Ceram. Soc.,* vol. 72, pp. 775–780, May 1989.

Inst. for Mater. Sci. & Engineering, "Ceramics: Technical Activities 1987," NBSIR 87-3612, Natl. Bureau of Stds., Gaithersburg, Md., Nov. 1987, 86 pp.

Jenkins, M. G., "Fracture Resistance of TiB_2 Particle/SiC Matrix Composite at Elevated Temperature," NASA TM 100967, June 1988, 18 pp.

Johnson, D. R., and J. O. Stiegler, "Structural Ceramics R&D," *AM&P,* vol. 138, pp. 55–61, Sept. 1990.

Kellett, B. J., et al., "Processing Science to Increase the Reliability of Ceramics," Rockwell International, N00014-84-C-0298, SC 5410 FR, Sept. 1986.

Kempfer, L., "Glass Goes Ceramic," *Mater. Eng.,* pp. 21–24, Sept. 1990.

Kumar, K. S., "Nickel Aluminide/Titanium Diboride Composites via XD**® Synthesis," FR 6/85-4/89, MML TR 89-102(C), N00014-85C0639, Nov. 1989.

Lawn, B. R., "Strength and Microstructure of Ceramics," NIST, AFOSR TR 90-0013, 2306 A2, FR 1987-89, AFOSR-ISSA87-0034, AFOSR-ISSA88-0005, Nov. 1989.

Layden, G. K., K. Prewo, and E. Thompson, "Study of SiC Whisker Reinforced Glass and Glass-Ceramic Matrix Composites," UTRC/R85-916943-1, N00014-84-C-0396, June 1985, 59 pp.

Lee, J. D., "Ultra-High Temperature Composites Concepts Evaluation," FR 10/86-12/87, AFWAL TR88-4114, F33615-86C5113, July 1988.

Luthra, K. L., "Chemical Interactions in High-Temperature Ceramic Composites," *J. Am. Ceram. Soc.,* vol. 71, pp. 1114–1120, Dec. 1988.

Ma, Z. Y., and C. K. Yao, "Microstructure and Properties of SiC_W/6061Al Composite," *Mater. Chem. Phys.,* vol. 25, pp. 463–474, May 1990.

Maehara, Y., and T. G. Langdon, "Review Superplasticity in Ceramics," *J. Mater. Sci.,* vol. 25, pp. 2275–2286, 1990.

Maloney, L. D., "Make Way for 'Engineered Ceramics'," *Des. News,* pp. 64–74, Mar. 13, 1989.

Marsh, G., "Engineering Ceramics—Part II," *Aerosp. Composites Mater.,* pp. 24–26, Mar.–Apr. 1990.

Petrovic, J. J., and R. E. Honnell, "$MoSi_2$ Particle Reinforced-SiC and Si_3N_4 Matrix Composites," *J. Mater. Sci. Lett.,* vol. 9, pp. 1083–1084, Sept. 1990.

Reynolds, W. N., "Nondestructive Testing Techniques for Metal Matrix Composites," AERE, Harwell, Gt. Br., AERE R13040, June 1988, 16 pp.

Salem, J. A., "Strength and Toughness of Monolithic and Composite Silicon Nitrides," NASA TM 102423, E5188, DE-AI05-870R21749, Jan. 1990, 20 pp.

Sheppard, L. M., "A Global Perspective of Advanced Ceramics," *Ceram. Bull.,* vol. 68, pp. 1624–1633, Sept. 1989.

Swab, J. J., "Performance of Y–TZP Materials between 800°C and 1200°C," Army Lab. Command, FR 1/90, MTL TR90-3, DE-AI05-840R21411, Jan. 1990.

Sykes, M. T., R. O. Scattergood, and J. L. Routbort, "Erosion of SiC-Reinforced Alumina Ceramic Composites," *J. Am. Ceram. Soc.,* vol. 18, pp. 153–163, Apr. 1987.

Tan, T. M., C. M. Pastore, and F. K. Ko, "Engineering Design of Tough Ceramic Matrix Composites for Turbine Components," 89-GT-294, June 1989, 10 pp.

Tebbe, F. N., et al., in *Proc. Symp. on Chemical Precursors to Ceramics,* Am. Chem. Soc., Miami Beach, Fla., AFOSR-TR-89-1355, AFOSR-89-0448, Sept. 12, 1989.

World Report on Advanced Ceramics, Tech. Insights, Inc., vol. 1, no. 6, pp. 1–2, Apr. 1989.

Yoshimura, M., "Hydrothermal Processing in 21st Century," *Ceram. Jpn.,* vol. 25, no. 1, pp. 11–12, 1990.

INDEX

Abrasive-jet machining (AJM), **7.14** to **7.17**
  cutting, **7.15**
  drilling, **7.17**
  equipment, **7.14**, **7.15**
  method comparison, **7.17**
  milling, **7.17**
  turning, **7.16**
Abrasive machining (AM), **7.7** to **7.14**
  creep-feed grinding, **7.10**
  cubic boron nitride (CBN), **7.10**
  diamond abrasives, **7.7**
  diamond tools, **7.12**
  drilling, **7.11** to **7.13**
  grinding, **7.8** to **7.10**
  lapping, **7.8**
  milling, **7.13**
  particle size, **7.7**
  polishing, **7.8**
Abrasives, **7.19**
Acheson process, **4.14**
  SiC, **4.41**
Active-metal brazing, **7.72** to **7.74**, **7.79**
Additives, **5.3** to **5.5**, **6.5** to **6.7**
  bactericide or fungicide, **5.3**
  chemical flaws, **4.13**
  coagulant, **5.3**, **5.4**
  deflocculant, **5.3**, **5.4**
  flocculant or binder, **5.3** to **5.5**
  liquid or solvent medium, **5.3**
  lubricant, **5.3**, **5.5**
  surfactant (wetting agent), **5.3**, **5.4**
Adhesive bonding (*see* Adhesives)
Adhesives, **7.56** to **7.58**, **7.60** to **7.66**
  applications of, **7.62**, **7.63**
  ceramic-based, **7.60**, **7.62**, **7.63**
  ceramic design factors, **7.60**
    ceramic property limits, **7.61**
    clearance of mating parts, **7.61**
    coefficient of thermal expansion, **7.60**
    stress analysis, **7.60**

Adhesives (*cont.*):
  design with, **7.58**
  inorganic, **7.56**
  organic, **7.56**
  processing, **7.56**
  properties, **7.57**
  sealants, **7.66**
Advanced turbine technology
  applications project (*see* ATTAP
  engine)
Aerosol decomposition, **4.16**
Agglomerates, **4.3**, **4.13**, **6.5**
  silicon nitride (Si_3N_4), **3.30**
  zirconia (ZrO_2), **3.35**
AGT (automotive gas-turbine) engine,
  8.46 to **8.50**, **9.16**, **9.21**
  other engine components and materials,
    8.49, **8.50**
  turbocharger, **8.46** to **8.49**
    materials, **8.49**
      Si_3N_4, **8.49**
    processing, **8.47**
      brazing, **8.47**
      injection molding, **8.47**
      slip casting, **8.47**
Al_2O_3 (*see* Alumina; Aluminum oxide)
Alumina, **3.11**
  alumina-matrix ceramic composites,
    3.11, **3.13**, **3.18** to **3.20**, **8.10**, **8.16**,
    9.2, **9.16**
  Al_2O_3–TiC, **9.28**
  β-Al_2O_3, **8.26**, **8.29**
  β''-Al_2O_3, **8.29**
  fibers, **4.55**, **9.2**, **9.28**
    Al_2O_3 aluminosilicate, **4.55**
    FP fibers, **9.2**, **9.28**
  mechanical properties, **3.11** to **3.13**,
    9.31
  processing, **4.48**
    Bayer, **4.48**, **4.49**

Alumina (*Cont.*):
  transformation toughening, **3.11, 3.17, 8.21**
    zirconia-toughened alumina (ZTA), **4.48**
  whiskers and particulates, **4.73, 4.74**
Aluminide, **9.19, 9.25**
  iron, **9.19**
  titanium, **9.25**
Aluminoborosilicate, **9.28, 9.35**
  fiber, **9.28**
  3M Nextel 440, **9.35**
Aluminum nitride (AlN), **1.15, 4.45, 8.67, 8.70, 9.17**
Aluminum oxide, **1.13, 8.11, 8.14, 8.22, 8.23, 8.32, 8.67, 8.68, 9.2, 9.11, 9.17, 9.24, 9.31, 9.35**
Amorphous covalent ceramics, **4.76**
Analysis, **2.9 to 2.11**
  strength–critical flaw, **2.11**
  stress, **2.19**
  Weibull, **2.9**
Anisotropy, **2.27**
Antenna windows, **8.34**
  boron nitride (BN), **8.34**
  Si_3N_4, **8.34**
  silicon oxide (SiO_2), **8.34**
Applications of adhesives, **7.62, 7.63**
Aqueous process (*see* Hydrothermal processes)
Armor, **8.30 to 8.32**
  boron carbide (B_4C), **8.30**
  Compglass, **8.32**
  Lanxide, **8.32**
  SiC, **8.31**
  titanium boride (TiB_2), **8.30**
Artificial hip joints, **8.22**
ATTAP engine, **8.50 to 8.52, 9.16, 9.21**
  other components, **8.51**
    disks, **8.51**
    ducts, **8.51**
    scroll, **8.51**
    springs, **8.51**
    stator, **8.51**
  turbine rotor, **8.50**
    materials, **8.50**
      α-SiC, **8.50, 8.51**
      aluminosilicate, **8.51**
      SiC toughened with TiB_2, **8.51**
      Si_3N_4 with SiC-whisker reinforcement, **8.50**
    NDT, **8.51**
      X-ray computer tomography, **8.51**

ATTAP engine, turbine rotor (*Cont.*):
  processing, **8.51**
    extrusion, **8.51**
    injection molding, **8.51**
    slip casting, **8.51**
Attrition (*see* Turbomilling process)
Automotive gas-turbine engine (*see* AGT)

Barriers for heat engines, **8.40 to 8.42**
  measurement- and standards-related, **8.41**
  technological, **8.41**
Bayer process, **4.48, 4.49**
  Al_2O_3, **4.48**
Bearings, **8.17 to 8.20, 9.14, 9.15**
  advantages of ceramics, **8.18**
  applications, **8.18 to 8.20**
    engines, **8.20**
    helicopter mast, **8.19**
    instruments, **8.20**
    pump, **8.18, 8.19**
    tool spindle, **8.19**
  materials, **8.20**
    ruby sapphire, **8.20**
    Si_3N_4, **8.20**
    ZrO_2, **8.20**
  properties, **8.17 to 8.19**
Behavior, **2.1 to 2.3, 2.28, 3.2, 3.3**
  ceramics, **2.1**
    brittleness, **2.1 to 2.3**
  composites, **2.28**
  mechanical, **3.2, 3.3**
  oxidation, **8.77**
Beryllium oxide, **1.13, 8.67**
Bioceramics, **8.21 to 8.23**
  applications, **8.22, 8.23**
    artificial hip joints, **8.22**
      Al_2O_3, **8.22**
    dental components, **8.22**
      Al_2O_3, **8.32**
      ZrO_2, **8.22**
    ear-bone and jawbone replacement and reconstruction, **8.23**
    heart pacemakers, **8.23**
      Al_2O_3, **8.23**
  categories, **8.21**
    nearly inert, **8.21**
      Si_3N_4-based, **8.21**
      transformation-toughened Al_2O_3, **8.21**
      transformation-toughened ZrO_2, **8.21**

Bioceramics, categories (*Cont.*):
  resorbable, 8.21
    calcium phosphate salts, 8.21
    polylactic acid–carbon composite, 8.21
    trisodium phosphate, 8.21
    surface active, 8.21
      glass-ceramics, 8.21
      hydroxyapatite, 8.21
      surface-active composites, 8.21
      surface-active glass, 8.21
Bonding, 1.3
  covalent, 1.3, 1.4
  ionic, 1.3, 1.4
Borides, 1.14, 1.15
Boron-based fibers, 4.56, 4.57
Boron carbide (B_4C), 1.14, 4.20, 4.22, 8.9
Boron nitride (BN), 1.15, 4.28, 4.45, 8.34, 8.67
Braze filler metals, 7.75, 7.76, 7.79, 7.81
Brittle failure, 2.2
Brittleness, 1.3, 2.1, 4.3, 9.21
  index, 2.2

Carbides, 1.14, 4.15, 4.20, 4.22, 8.9, 8.11, 8.24, 8.25, 8.27, 8.33, 8.36, 8.43, 8.67, 9.14 to 9.16, 9.22, 9.26, 9.28, 9.35
  boron, 1.14, 4.20, 4.22, 8.9
  silicon, 1.14, 8.11, 8.24, 8.25, 8.27, 8.33, 8.36, 8.43, 8.67, 9.14 to 9.16, 9.22, 9.26, 9.28, 9.35
Carbothermal synthesis, 4.9, 4.15, 4.17, 4.22
  processes, 4.10, 4.22 to 4.24, 4.74
    electric arc, 4.10
    self-propagating, high-temperature (SHS), 4.10, 4.22, 9.27
    thermite, 4.10
  processing, 4.22 to 4.24
    combustion synthesis, 4.22 to 4.24, 4.74
    silica reduction method, 4.43, 4.44
      Si_3N_4, 4.43, 4.44
    thermal decomposition, 4.22
CARES (ceramics analysis and reliability evaluation of structures), 9.38
Casting, 5.17 to 5.19, 5.38 to 5.45
  compocast, 5.18, 5.19
  gel, 5.46
  pressure, 5.39 to 5.42
  slip, 5.1, 5.15 to 5.17, 5.37, 5.38
  squeeze, 5.42 to 5.45
  tape, 5.1, 5.17, 5.38, 5.39

Cement, 1.9 1.11
Ceramic-based adhesives, 7.60, 7.62, 7.63
Ceramic-ceramic composites (CCC), 3.10, 4.69 to 4.71, 5.31, 5.47, 9.2, 9.9, 9.15, 9.17, 9.18, 9.28
  processing techniques, 4.69 to 4.77
    gas-pressure sintering, 4.69
    HIP, 4.69
    injection molding, 4.70, 4.72, 8.51
    squeeze casting, 4.70, 4.72, 5.45
  properties, 5.32
Ceramic composites (*see* Ceramics)
Ceramic cutting tools, 7.38 to 7.48, 9.16, 9.29
  barriers, 7.39, 7.40
    institutional, 7.40
    measurement-related, 7.39
    technical, 7.39
  future use, 7.47, 9.29
  material properties, 7.42 to 7.44, 9.16, 9.29
    Al_2O_3–SiC_w, 9.16
    carbides, 7.42 to 7.44
    carboxides, 7.42 to 7.44
    nitrides, 7.42 to 7.45, 9.29
    oxides, 7.42 to 7.44
    sialons, 7.45, 9.16, 9.29
    whiskers, 7.46, 9.16
Ceramic fibers, 4.27
  (*See also* Ceramic reinforcements)
Ceramic liners and scrappers, 8.10
Ceramic materials, 1.1, 1.7, 1.8, 1.19, 4.41
  classes of, 1.9
  development of, 1.7, 1.20
  fracture of, 2.10
  usage, 1.2
Ceramic-matrix composites (CMC), 3.10, 4.53, 5.31, 9.17, 9.18, 9.20 to 9.30, 9.40, 9.41
  composite ceramics, 9.20
    SiC-whisker-reinforced Al_2O_3, 9.24
  monolithic ceramics, 9.20
    sialon, 9.25
    SiC, 9.20, 9.25
    Si_3N_4, 9.20, 9.25
    TiC, 9.25
  processability and failure mode, 9.22
    continuous-fiber-reinforced ceramics (FRC), 9.22, 9.25
    particulate-reinforced, 9.22
    whisker-reinforced, 9.22

Ceramic-matrix composites (*Cont.*):
  processing, **9.23**, **9.24**
    CVD, **9.23**, **9.24**, **9.26**
    hot-press reactive sintering (RESHOP), **9.25**
    polymer pyrolysis, **9.23**
    SHS, **9.25**
    sol-gel, **9.23**
Ceramic military, **8.29** to **8.34**, **9.16**
  applications, **8.29** to **8.34**
  armor, **8.30** to **8.32**
    Compglass, **8.32**
    Lanxide, **8.32**
    SiC, **8.31**
    TiB_2, **8.30**
  bearings for missiles, **8.33**
    Si_3N_4, **8.33**
  engines, **8.30**
    diesel, **8.30**
    turbine, **8.30**, **9.16**
  gun barrel liners, **8.34**
    α-SiC, **8.34**
  infrared windows, **8.32**, **8.33**
    Al_2O_3, **8.32**
    fused SiO_2, **8.32**
    MgO_2, **8.32**
  laser mirrors, **8.33**, **8.34**
    RBSC (siliconized SiC), **8.34**
    SiC, **8.33**
  radomes, **8.32**, **8.33**
    sintered fused SiO_2, **8.33**
Ceramic reinforcements, **4.52** to **4.77**, **9.16**, **9.19**
  foams, **4.76**, **4.77**
    processes, **4.76**
      CVD, **4.77**
  glass matrixes, **4.52**
    SiC/LAS, **4.52**
  inorganic fibers, **4.54**
    carbides, **4.57**, **4.58**, **9.19**
    nitrides, **4.57** to **4.59**, **9.19**
    oxides, **4.55** to **4.59**, **9.19**
  particulates, **4.61**, **4.62**
  spheres, **4.76**, **4.77**, **9.16**
  whiskers, **4.60** to **4.63**, **5.31**
    toughening mechanisms, **4.63**
Ceramic seals, **8.10**
Ceramic space, **8.34** to **8.38**
  applications, **8.34**
  antenna windows, **8.34**
    BN, **8.34**
    Si_3N_4, **8.34**
    SiO_2, **8.34**

Ceramic space (*cont.*):
  diesel and turbine engine components, **8.34** to **8.37**
    blades, **8.36**, **8.37**
    caps, **8.36**, **8.37**
    ducts, **8.36**, **8.37**
    materials, **8.36**
      SiC, **8.36**
      Si_3N_4, **8.36**
    seals, **8.36**, **8.37**
    vanes, **8.36**, **8.37**
  radiators, **8.38**, **8.39**
  reentry vehicles, **8.36**, **8.38**
    HRSI (black tile), **8.38**
    HTP, **8.38**
    TPS, **8.36**, **8.38**
    fused SiO_2, **8.38**
Ceramics, **1.2** to **1.4**, **1.8**, **1.9**, **1.11** to **1.16**, **2.1** to **2.6**, **2.8** to **2.10**, **2.16** to **2.24**, **3.2**, **3.4**, **3.5**, **3.8** to **3.10**, **4.2**, **4.12**, **4.27**, **5.2**, **5.3**, **5.7** to **5.22**, **7.1** to **7.48**, **8.1** to **8.5**
  amorphous covalent, **4.27**
  behavior of, **2.1**
  brittle failure of, **2.2**
  brittleness, **2.2**
  chemically bonding, **1.15**
  classes, **1.9**
    cements and concrete, **1.9**, **1.11**
    composites, **1.9**, **1.12**
    glasses, **1.9**, **1.10**
    natural ceramics (rocks and minerals), **1.9**, **1.11**
    refractories, **1.9**, **1.11**, **1.13** to **1.15**
    vitreous ceramics or clays, **1.9**
  conductive, **7.22**
  database, **3.8** to **3.10**
  design of, **2.16** to **2.24**
  forming processes, **5.7** to **5.22**
  fracture of, **2.10**
  machining of, **7.1** to **7.48**
  nondestructive evaluation of, **2.29**
    powders, **4.2**
    pores, **4.12**
  processing of, **1.4**
    processing aids, **5.2**, **5.3**
  properties of, **1.3**, **1.4**, **1.16**, **3.2**, **8.1** to **8.5**
    chemical, **8.1** to **8.3**
    creep, **2.22**, **3.5**
    electromechanical, **8.2**, **8.3**
    electronic, **8.2**, **8.3**
    fatigue, **3.5**

Ceramics, properties of (*Cont.*):
  fracture toughness, **2.2**
  hardness, **2.2**
  magnetic, **8.2, 8.3**
  mechanical, **8.3, 8.4**
  optical, **8.2, 8.3**
  strain tolerance, **2.3**
  thermal shock, **2.22**
  tribological, **8.3, 8.4**
 standardization, **3.6** to **3.10**
 structural, **1.2**
 testing, **2.4, 3.4**
  biaxial, **2.5**
  C-ring, **2.6**
  compression, **2.8**
  diametral compression, **2.6**
  expanded ring, **2.6**
  flexure, **2.4, 3.5**
  tensile, **2.4, 2.6, 3.5**
 traditional, **1.8**
 Weibull analysis, **2.9**
Chemical bonding, **7.48**
 joining methods, **7.48**
  atomic contact interface, **7.48**
   pressure, **7.48**
   wetting, **7.48**
  chemical equilibrium interface, **7.48**
   ceramic-ceramic, **7.48**
   ceramic-metal, **7.48, 7.52**
   glass-metal, **7.48, 7.52**
   metal-metal, **7.48**
Chemical families, **3.1, 3.10**
 structural ceramics, **3.10** to **3.36**
  Al_2O_3, **3.10** to **3.14**
  sialon, **3.10, 3.35**
  SiC, **3.10, 3.14, 3.15, 3.20** to **3.24**
  Si_3N_4, **3.10, 3.15, 3.16, 3.20** to **3.26, 3.29** to **3.34**
  ZrO_2, **3.10, 3.16, 3.17, 3.35, 3.36**
Chemical/processing functions, **8.1, 8.3**
 microporous ceramic materials, **8.1, 8.3**
 chemical sensors, **8.1, 8.3, 8.6**
 coal gas filters for removal of sulfur, **8.1, 8.3**
 liquid metals, **8.1**
 membranes for chemical processing, **8.1, 8.3**
Chemical vapor deposition (CVD), **4.10, 4.36** to **4.38, 9.8**
 organometallic precursors (OMCVD), **4.37, 9.28, 9.32**
 process, **4.36, 4.45, 4.46**

Chemical vapor infiltration (CVI), **5.47, 9.7** to **9.10, 9.28, 9.29**
 ceramic-ceramic composite, **5.47, 9.7, 9.8, 9.10, 9.28, 9.29**
 fibers, **9.8**
  continuous, **9.8**
  discontinuous, **9.8**
 matrix, **9.8**
  SiC, **9.8, 9.9**
 processing, **9.8**
 advantages, **9.8**
Chemical vapor precipitation (CVP), **4.21**
 laser activation, **4.21**
 thermal, **4.21**
Coatings, **7.84** to **7.94, 8.20, 8.21**
 applications, **7.86, 8.20, 8.21**
 benefits, **8.20**
 materials used, **7.84**
  carbides, **7.84**
   B_4C, **7.87**
   TiC, **7.87**
  oxides, **7.85, 7.86**
   Al_2O_3, **7.87**
  nitrides, **7.87**
   TiN, **7.87**
   Ti carbonitride, **7.87**
 NDT tests, **7.94**
 process types, **7.84** to **7.96**
  combination, **7.86**
  CVD, **7.86**
   applications, **7.88**
   disadvantages, **7.88**
   other CVD processes, **7.88**
    laser-assisted CVD, **7.89**
    plasma-activated or plasma-assisted CVD, **7.88, 7.89**
    sputtering, **7.89**
   process definition, **7.87, 7.88**
  ion-assist, **7.94, 7.96**
  PVD, **7.86**
   applications, **7.90, 7.92**
   materials used, **7.90**
    carbides, B_4C, **7.90**
    nitrides, TiN, **7.91**
    oxides, ZrO_2 (stabilized with Y_2O_3 and Yb_2O_3), **7.92, 7.93**
   other PVD processes, **7.90** to **7.94**
    cathodic arc deposition, **7.90**
    evaporative electron-beam melting, **7.92, 7.93**
    laser, **7.93, 7.94**
    sputter ion plating, **7.90, 7.91**
    sputtering, **7.90**

Coatings, process types, PVD (*Cont.*):
   tests, **7.**93
   powder, **7.**84 to **7.**86
   sol-gel, **7.**94
   special RS (rapidly solidified), **7.**86
   thermal spraying, **7.**84
   wet processes, **7.**84
Cold pressing, **6.**8
Combustion sintering, **4.**38
   high-pressure (HPCS), **4.**38, **6.**17
Compocasting, **5.**18, **5.**19
Composite machining, **7.**36 to **7.**38
   laser and ultrasonics, **7.**36
   machining effects, **7.**37
Composites:
   behavior of, **2.**28
   classes, **4.**52
   design, **2.**26
   interface properties of, **2.**27
   multiphase, **1.**6
   (*See also* Ceramic-Ceramic composites; Ceramic-matrix composites)
Computer-assisted tomography (CAT), **8.**77, **8.**78
Computer modeling, **1.**7
Concrete, **1.**9, **1.**11
Condensed-phase—condensed-phase reactions, **9.**27, **9.**28
   liquid-phase reaction bonding, **9.**27
   SHS, **9.**27
   solid-solid exchange reactions, **9.**27
Condensed-phase–gas-phase reactions, **9.**27, **9.**28
   directed oxidation, **9.**27
   gas-phase reaction bonding, **9.**27
Conductive ceramics, **7.**22
Continuous fibers, **4.**59, **4.**65, **9.**22, **9.**29, **9.**30
Controlled firing (*see* Sintering)
Covalent or ionic bonding, **1.**3, **1.**17, **6.**3, **8.**2
Crack growth, **2.**23
Crack propagation, **3.**4
Creep, **3.**5
Creep-feed grinding, **7.**10
Cryochemical processing, **4.**39
   salts, **4.**39
Cubic boron nitride (CBN), **7.**10
Cutting, **7.**15, **7.**16
Cutting and machining, **7.**31, **7.**32
Cutting tools, **8.**13 to **8.**17, **9.**29
   classes, **8.**13, **8.**14
      Al_2O_3, **8.**14

Cutting tools, classes (*Cont.*):
   sialon, **8.**14, **9.**29
   Si_3N_4, **8.**14, **9.**29
   material types, **8.**15 to **8.**17
      Al_2O_3 base, **8.**16
      Al_2O_3–SiC-whisker-reinforced, **8.**16
      Al_2O_3–TiC, **8.**16, **9.**28
      ceramic-coated carbide, **8.**16
      others, **8.**17
      Si_3N_4, **8.**16, **9.**29
   properties, **8.**14 to **8.**16
CVD coating processes, **7.**86 to **7.**89
   laser, **7.**89
   plasma, **7.**88, **7.**89
   sputter, **7.**89

Database, **3.**8 to **3.**10
   NIST National Standard Data Reference Database Series, **3.**9
   Oak Ridge National Laboratory (ORNL), **3.**9
Dental components, **8.**23
Design considerations, **2.**16 to **2.**24, **2.**26
   anisotropic, **2.**27
   brittle design, **2.**17
   compliant layer approach, **2.**18
   composite, **2.**26
   deterministic, **2.**21
   probabilistic, **2.**21
Diamond
   abrasives, **7.**7, **7.**10
   tools, **7.**11 to **7.**13
Die pressing, **5.**9
Die sinking machines, **7.**22 to **7.**24
Dies, **8.**9
Diesel engines, **8.**56 to **8.**66
   ceramic challenges, **8.**64
   ceramic components, **8.**61 to **8.**63, **9.**29
      level 1 ceramic turbocharger and ceramic components, **8.**56, **8.**57
      level 2 ceramic cylinder and piston, **8.**56 to **8.**58
      thermal insulation plus exhaust train, **8.**56 to **8.**58
      advanced minimum-friction engine (MFE), **8.**56, **8.**57, **8.**60
   market, **8.**60, **8.**62 to **8.**66
   potential materials, **8.**56, **8.**58 to **8.**65
      coatings, **8.**59, **8.**60
      Mg–PSZ, **8.**59
      PSZ, **8.**59
      SSC, **8.**58, **8.**60

Diesel engines, potential materials (*Cont.*):
   SSN, **8.58**, **8.59**
   ZrO_2, **8.59**, **8.60**
Diffusion bonding, **7.66 to 7.82**
   applications, **7.68**, **7.69**, **7.80**
   brazing and metallizing techniques, **7.69 to 7.81**
      braze filler materials, **7.75**, **7.76**, **7.79**, **7.81**
      keys to brazing, **7.78**, **7.79**
      metallizing processes, **7.70 to 7.81**
         active hydride, **7.80**
         active-metal brazing, **7.72 to 7.74**, **7.79**
         copper-copper compound, **7.72**
         direct metallization, **7.72**
         Intragene, **7.74**, **7.77**
         sintered metal powder (moly-manganese), **7.70**
         vapor deposition, **7.77**, **7.78**
   electrostatic bonding method, **7.69**
   methods with interlayers, **7.66 to 7.68**
      aluminum, **7.67**
      precious metal, **7.67**
      transition metal, **7.67**, **7.68**
   new method, **7.81**, **7.82**
      microwave heating, **7.81**, **7.82**
Direct joining, **7.82 to 7.84**
   fusion welding, **7.82**
      arc, **7.82**
      electromagnetic laser beam, **7.83**
      electron beam, **7.83**
      friction, **7.83**, **7.84**
      ultrasonic, **7.83**
Direct nitriding method, **4.43**, **4.44**
   Si_3N_4, **4.43**, **4.44**
Discontinuous reinforcement (*see* Whiskers)
Dopants, **4.14**, **9.18**
Double-cantilever test method, **2.19**
Drilling, **7.11**, **7.13**, **7.17**, **7.34**, **7.35**
Dry pressing, **5.1**, **5.8**, **5.9**, **5.25**, **5.39**
   axial, **5.25**
   isostatic dry, **5.26**
   slurry, **5.27**, **5.28**
Drying, **5.22 to 5.24**
   methods, **5.24**
   conduction, **5.24**
   controlled humidity, **5.24**
   convection, **5.24**
   infrared, **5.24**
   microwave, **5.24**
   slurry, **5.24**
   spray, **5.24**

Drying (*Cont.*):
   supercritical, **5.24**
   vacuum-assisted, **5.24**
Ductility, **1.3**, **1.17**
Dynamic compaction (DCT), **4.1**, **5.19**
   applications, **5.21**, **5.22**
   key factors, **5.20**
   properties, **5.21**

Education, **1.19**
Electric-discharge grinding, **7.27**, **7.28**
Electric-discharge machining (EDM), **7.21 to 7.28**
   conductive ceramics, **7.22**
   grinding machines, **7.27**, **7.28**
      die sinking, **7.22 to 7.24**
      wire machining, **7.22 to 7.24**
   materials, **7.24**
   methods comparison, **7.24**
   process parameters, **7.25**, **7.26**
   properties, **7.25 to 7.27**
Electroacoustic dewatering (EAD), **5.46**
Electrochemical devices, **8.25 to 8.29**
   applications, **8.26 to 8.29**
      batteries, **8.29**
      fuel cells, **8.27**, **8.29**
      oxygen concentration cells, **8.27**
      oxygen sensors, **8.26**
   materials, **8.26**
      β-Al_2O_3, **8.26**, **8.29**
      β''-Al_2O_3, **8.29**
      ZrO_2, **8.26**, **8.27**
Electromechanical functions, **8.2**, **8.3**
   micropositioners, **8.2**, **8.3**
   pickups, **8.2**, **8.3**
   transducers, **8.2**, **8.3**
Electronic ceramics, **8.66 to 8.71**, **9.16**
   applications, **8.66**
   integrated-circuit packaging, **8.67**, **9.16**
      AlN, **8.67**
      BeO, **8.67**
      BN, **8.67**
      glass-bonded Al_2O_3, **8.67**
      SiC, **8.67**
   laser tubes, **8.67**
   power transistors, **8.67**
   printed-circuit boards, **8.68**
      Al_2O_3, **8.68**
   resistor cores, **8.67**
   superconductivity, **8.70**, **8.71**
   surface-mount technology, **8.68**
   thin films and metallization, **8.67**, **8.69**, **8.70**
      AlN, **8.67**, **8.70**

Electronic functions, **8.2, 8.3**
  absorbers of radio frequency, **8.2, 8.3**
  dielectrics for capacitors, **8.2, 8.3**
  electrodes translucent, **8.2, 8.3**
  electron emitters, **8.2, 8.3**
  electronic substrates, **8.2, 8.3**
  filters, **8.2, 8.3**
  insulators, **8.2, 8.3**
  overvoltage protection, **8.2, 8.3**
  semiconductors, **8.2, 8.3**
  solid electrolytes, **8.2, 8.3**
  superconductors, **8.3, 8.4**
Engineering ceramics, **1.11, 2.2**
Engines, **8.38 to 8.66**
  categories for ceramics, **8.38**
  advanced diesels, **8.38, 8.56 to 8.66**
    coatings, **8.38**
    monolithic hot sections, **8.38**
  metal reciprocating, **8.38**
    turbochargers, **8.38, 8.44, 8.46, 9.29**
  turbine, **8.38, 8.45, 8.46, 8.48, 8.52, 9.15, 9.16**
    all-ceramic, **8.38, 8.52**
Etching, **7.36**
Evaporation decomposition (*see* Aerosol decomposition)
Extrusion, **5.1, 5.29, 8.51**
  binders, **5.12**
  jiggering, **5.11, 5.12**
  process steps, **5.12**
  wet pressing, **5.11**

Fabrication, **2.17**
Fatigue, **3.5**
Fibers, **9.32 to 9.38, 9.8 to 9.10**
  ceramic, **4.27, 9.8, 9.9, 9.10**
    hollow, **4.77**
  inorganic, **4.54**
  properties, **4.68, 9.33, 9.34**
Finite-element analysis (FEA), **4.52**
Flame synthesis, **4.39**
Flaws, **2.4, 4.12, 8.77**
  nondestructive method, **4.13**
Flexure strength, **3.27, 3.31**
  SiC, **3.27**
  Si_3N_4, **3.31**
Forced chemical-vapor infiltration (FCVI), **9.9, 9.10**
Forging, **6.28, 6.29**
  sinter, **6.28**
  slow, **6.29**

Forming processes, **5.7 to 5.22**
  casting, **5.14 to 5.19**
  extrusion, **5.11, 5.12, 5.29**
  injection molding, **5.13, 5.14, 5.29**
  pressing, **5.8 to 5.13**
    die, **5.9**
    dry, **5.8, 5.9, 5.25**
    isostatic, **5.9, 5.10, 5.26**
    roll, **5.8**
    slurry, **5.27, 5.28**
  warm molding, **5.7**
Fractography, **2.11**
Fracture, **2.10**
Fracture mechanics, **2.12**
  linear elastic, **2.13**
Fracture toughness, **2.2, 3.4, 8.77**
  tests, **2.13**
    double-cantilever, **2.14**
    hertzian, **2.15**
    indentation, **2.15**
    notched beam, **2.14**
    straight-through notch, **2.14**
    surface-flaw bend, **2.15**
Fuel cells, **8.27, 8.29**
Functionally gradient materials (FGM), **4.37, 4.38**
Fusion welding, **7.82 to 7.84**
  arc, **7.82**
  electromagnetic laser beam, **7.83**
    laser-activated brazing, **7.84**
  electron beam, **7.83**
  friction, **7.84**
  ultrasonic, **7.83**
Future considerations, **9.40, 9.41**
  applications, **9.41**
  approaches to processing, **9.40**
  ceramic whisker composites, **9.41**
  coatings, **9.41**
  mechanical performance improvements, **9.41**

Gas-phase reactions **4.6, 4.21**
  chemical vapor precipitation (CVP), **4.21**
    laser activation, **4.21**
    thermal activation, **4.21**
  flame synthesis, **4.39**
Gas-pressure-assisted sintering (GPS), Si_3N_4, **3.30**
  (*See also* Overpressure sintering)
Gasoline engines, **8.42, 8.43, 8.45**
Gel casting, **5.46, 9.2**
  ceramic-fiber-reinforced CCC, **9.2**

Gel casting (*Cont.*):
  fibers, **9.2**
    Al_2O_3 (FP), **9.2**
    Al_2O_3 –ZrO_2 (PRD 166), **9.2**
  materials, **9.2**
    Si_3N_4, **9.2**
  matrixes, **9.9**
    Al_2O_3, **9.2**
    ZrO_2, **9.2**
Glass ceramics, **1.15**, **4.71**, **9.37**, **9.39**
  organometallic compounds, **4.15**
Glass matrixes, **4.52**, **9.15**, **9.17**, **9.38**, **9.39**
  processes, **9.38**
  hot injection molding, **9.38**
  hot matrix transfer, **9.38**
  hot pressing, **9.38**
Glasses, **1.9**, **1.10**, **4.54**, **4.71**, **9.37**, **9.38**
Glow-discharge powder synthesis, **4.16**, **4.19**
Grain size, **1.6**
  control, **4.4**
Granulation, **5.6**
Grinding, **7.8** to **7.10**, **9.10**, **9.11**
Grinding media, **8.9**, **8.10**

Hardness, **2.2**
Heart pacemakers, **8.23**
Heat engines, **8.38** to **8.56**
  AGT, **8.46**
  ATTAP, **8.50** to **8.52**
  barriers, **8.40**
    measurement- and standards-related, **8.41**, **8.42**
      chemical analysis, **8.41**
      instrumentation , **8.41**
      modeling, **8.41**
      NDT, **8.41**
      testing, **8.41**
      technological, **8.41**
    reliability improvement, **8.41**, **8.52**
  benefits, **8.39**, **8.40**
  diesel, **8.43** to **8.45**
    components, **8.43**
  gas turbines and components, **8.52** to **8.56**
    auxiliary power unit (APU), **8.53**, **8.54**
    combustors, **8.52**
  gasoline, **8.42**, **8.43**, **8.45**
    cam followers, **8.42**
    exhaust port liners, **8.42**
    turbocharger components, **8.42**, **8.44**, **8.46**, **9.29**
  market, **8.42**

Heat engines (*Cont.*):
  materials, **8.43**
    Lanxide Al_2O_3 matrix composite, **8.44**
    SiC, **8.43**
    Si_3N_4, **8.43**
Heat exchangers, **8.23** to **8.25**
  economics, **8.24**, **8.25**
  fabrication methods, **8.24**
    extrusion, **8.24**
    high-pressure, **8.24**
    low-pressure, **8.24**
  materials, **8.24**
    SiC, **8.24**
Hertzian fracture method, **2.15**
High-pressure self-combustion sintering (HPCS), **9.28**
HIP maps, **6.22**
Hot isostatic pressing (HIP), **6.17** to **6.26**
  advantages, **6.18**
  atmospheres, **6.22**
  encapsulation methods, **6.18**, **6.21**
    coating, **6.18**
    cladding, **6.21**
    glass bath, **6.21**
    glass particle, **6.21**
    glass tubes, **6.18**
    sinter canning, **6.21**
    welded containers, **6.18**
  materials, **6.23** to **6.26**
  processes, **6.18** to **6.26**
    pressure-assisted sintering (PAS), **6.18**, **6.27**
    pulsed sinter/HIP (PUSH), **6.26**
    sinter/HIP, **6.18**, **6.22** to **6.26**
    sinter-plus HIP, **6.17**
  properties, **6.23** to **6.26**
    SiC, **3.25**
    Si_3N_4, **3.30**
Hot pressing, **6.8** to **6.10**
Hydrodynamic machining (HDM) (*see* Abrasive-jet machining)
Hydrothermal synthesis, **4.24**
  methods, **4.1**, **4.25**
    crystallization, **4.25**
    hydrolysis, **4.25**
    oxidation, **4.25**
    precipitation, **4.25**
  processes, **4.9**
Impact grinding, **7.18**, **7.20**
Impulse vibration excitation technique, **8.80**, **8.81**

In situ precipitation technique, 9.18
  XD process, 9.18
  TiB$_2$ dispersoid, 9.18
Indentation method, 2.15
Indirect joining, 7.49, 7.56 to 7.58, 7.60 to 7.82
  adhesives, 7.56 to 7.58, 7.60 to 7.66
  diffusion bonding, 7.66 to 7.82
Infrared windows, 8.32, 8.33
  Al$_2$O$_3$, 8.32
  fused SiO$_2$, 8.32
  MgO, 8.32
Injection molding, 5.1, 5.25, 5.29 to 5.33, 8.47, 8.51
  applications, 5.29 to 5.33
    high-pressure, 5.33
    low-pressure, 5.33
  materials, 9.5
    CE-TZP (ZrO$_2$–12%ECeO), 9.5
    RBSN, 9.5
    SSN, 9.5
  process steps, 5.13, 5.14
  properties, 5.32
    Si$_3$N$_4$–SiC, 5.32
  whiskers, 4.70, 4.72, 5.31, 5.32, 8.50
    SiC, 4.70, 5.31, 5.32, 9.5
    Si$_3$N$_4$, 4.72
Inorganic adhesives, 7.56
Inorganic fibers, 4.54 to 4.60
  Al$_2$O$_3$ aluminosilicate, 4.55
    slurry and solution processes, 4.55
  alumina-silica-boria, 4.56
  B$_4$C, 4.57
  BN, 4.57
  fabrication methods, 4.55
  glass, 4.54
  polycrystalline α-alumina, 4.56
  SiC, 4.57, 4.58
  Si$_3$N$_4$, 4.58, 4.59
  SiO$_2$, 4.56
  ZrO$_2$, 4.56, 4.59
Integrated-circuit packaging, 8.67, 9.16
  AlN, 8.67
  BeO, 8.67
  BN, 8.67
  glass-bonded Al$_2$O$_3$, 8.67
  SiC, 8.67
Interlayer brazing methods, 7.66 to 7.68
  aluminum, 7.67
  precious metal, 7.67
  transition metal, 7.67, 7.68

Intermetallics, 1.15
  aluminides, 1.15
    nickel aluminum, 1.15
    titanium aluminum, 1.15
  beryllides, 1.15
    columbium, 1.15
    tantalum, 1.15
    zirconium, 1.15
  silicides, 1.15
Ion implantation, 9.19
  materials, 9.19
    Y$_2$O$_3$-stabilized cubic ZrO$_2$, 9.19
Isoscan generator system, 8.78
Isostatic pressing, 5.9, 5.10
  dry bag, 5.9
  wet bag, 5.9
Isothermal chemical vapor infiltration (ICVI), 9.10
  fibers, 9.10
    continuous, 9.10
  matrixes, 9.10
    TiB$_2$, 9.10
    SiC composites, 9.10

Joining, 7.48 to 7.84
  fundamental requirements, 7.48
    chemical bonding, 7.48
    minimal stress levels, 7.48
  key parameters, 7.49 to 7.50
    interfaces, 7.50
    spreading, 7.50
    surface properties, 7.50, 7.54
    wetting, 7.50, 7.54
  major techniques, 7.49
    direct, 7.49
    indirect, 7.49
    mechanical, 7.49
  methods, 7.55 to 7.66
    adhesive bonding, 7.55 to 7.58, 7.60, 7.66
    diffusion bonding, 7.66
    mechanical interlocking, 7.58 to 7.60
      adhesives and glazing, 7.60
      bolts and rivets, 7.58
      clamping, 7.60
      crimping, 7.59
      threads, 7.59
    special processes, 7.63 to 7.66
      pressing, 7.63
      slip/CIP, 7.64
      solders, 7.64 to 7.66
        high-melting glass, 7.66
        low-melting glass, 7.64, 7.65

Joining, methods, special processes,
    solders (*Cont.*):
        metal, **7**.64
    types of materials, **7**.62 to **7**.84
        AlN, **7**.77
        Al_2O_3, **7**.62, **7**.69, **7**.72, **7**.77, **7**.81 to **7**.84
        SiC, **7**.64, **7**.69
        Si_3N_4, **7**.62, **7**.74, **7**.77, **7**.78, **7**.81, **7**.82
        ZrO_2, **7**.70, **7**.78, **7**.84
            Mg-PSZ, **7**.70
            PSZ, **7**.78, **7**.84
Joining parameters, **7**.50 to **7**.54
    interfaces, **7**.50
    spreading, **7**.50
        surface energies, **7**.51
    surface properties, **7**.50, **7**.54
        hydriding, **7**.52, **7**.53
        intermediate-layer foils, **7**.52
        metallizing, **7**.51 to **7**.54
        plating, **7**.51
        vapor deposition, **7**.51
    wetting, **7**.50, **7**.54
        active metal braze filler, **7**.50 to **7**.54
        contact angles, **7**.51, **7**.54
Joint stress, **2**.23
    ceramic-ceramic, **2**.23
    ceramic-metal, **2**.23
        interlayers, **2**.23

Lanxides, **1**.5, **3**.20, **5**.48 to **5**.51, **8**.44, **9**.17, **9**.18, **9**.26, **9**.27
    applications, **5**.49, **9**.18
    ceramic-metal and ceramic-ceramic
        systems, **9**.17, **9**.18, **9**.26, **9**.27
        Al–Al_2O_3 matrix/Al_2O_3, **9**.17
        Al–Al_2O_3 matrix/$BaTiO_3$, **9**.17
        Al–Al_2O_3 matrix/SiC, **9**.17
        Al–Al_2O_3 matrix/ZrO_2, **9**.17
        Al–AlN matrix/AlN, **9**.17
        Al–AlN matrix/Al_2O_3, **9**.17
        Al–AlN matrix/B_4C, **9**.17
        Al–AlN matrix/TiB_2, **9**.17
        Ti–TiN matrix/Al_2O_3, **9**.17
        Ti–TiN matrix/TiB, **9**.17
        Ti–TiN matrix/TiC, **9**.17
        Ti–TiN matrix/TiN, **9**.17
        Zr–ZrN matrix/ZrB, **9**.17
        Zr–ZrN matrix/ZrN, **9**.17
    processes, **5**.48 to **5**.51, **9**.17, **9**.18
        DIMOX, **9**.17, **9**.18
        PRIMEX, **9**.17, **9**.18
    properties, **5**.50

Lapping, **7**.8
Laser-beam machining (LBM), **7**.28 to **7**.36
    cutting, **7**.31
    drilling, **7**.34
    etching, **7**.36
    machining, **7**.31
    material properties, **7**.33
    method comparisons, **7**.32, **7**.33
    process principles, **7**.29
    shaping, **7**.35
    types of lasers, **7**.29 to **7**.32
        CO_2, **7**.29 to **7**.32
        excimer, **7**.31, **7**.32
        Nd:YAG, **7**.29 to **7**.32
Laser mirrors, **8**.33, **8**.34
    RBSC (siliconized SiC), **8**.34
    SiC, **8**.33
Laser processing, **7**.28 to **7**.31
Laser sintering, **6**.36
Laser synthesis, **4**.8, **4**.17, **4**.18
    Si_3N_4, **4**.20
Laser types, **7**.30, **7**.32
Lasers, **4**.1
Liquid-phase sintering, **3**.30
    β-sialon, **4**.50
Lithium aluminosilicate, **8**.51

Machining energy, **7**.3 to **7**.36
    chemical, **7**.4 to **7**.6
        chemical machining (CM), **7**.6
        electrochemical machining (ECM), **7**.4 to **7**.6
    mechanical, **7**.7 to **7**.21, **9**.11
        abrasive-jet machining (AJM), **7**.14 to **7**.17
        traditional abrasive machining (AM), **7**.7 to **7**.14
        ultrasonic abrasive machining (UAM), **7**.17 to **7**.21
    thermoelectric, **7**.21 to **7**.36
        electric-discharge machining (EDM), **7**.21 to **7**.28
        electron-beam machining (EBM), **7**.6
        ion-beam machining (IBM), **7**.7
        laser-beam machining (LBM), **7**.28 to **7**.36
Machining methods, **7**.3 to **7**.7, **9**.10, **9**.11
    advantages, **7**.4 to **7**.7
    disadvantages, **7**.4 to **7**.7
    generation, **7**.3
    process definitions, **7**.4 to **7**.7
    transfer, **7**.3

Magnesium oxide, **1.**13, **8.**32
Magnetic functions, **8.**2 to **8.**4
  hard magnets, **8.**3, **8.**4
  motor parts, **8.**3, **8.**4
  soft magnets, recording heads, magnetic tape, **8.**3, **8.**4
Markets, **1.**18
  ceramic composite, **1.**18
Materials, **1.**1, **1.**2
  ceramic, **1.**1, **1.**2
  metallic, **1.**2
Matrixes, **4.**69 to **4.**77
Mechanical behavior, **3.**2
  data, **3.**4
Mechanical functions, **8.**3 to **8.**5
  ceramic composites, **8.**3
  cutting tools, **8.**3, **8.**4, **8.**6
  dies, **8.**3, **8.**5
  engine parts, **8.**3, **8.**5, **8.**6
  pump parts, **8.**3, **8.**5
  valves, **8.**3, **8.**5
Mechanical joining, **7.**49
  clamping, **7.**49
  press fit, **7.**49
  shrink fit, **7.**49
  tie down with hooks, **7.**49
Mechanical properties, **3.**2, **3.**9, **3.**11
  Al_2O_3, **3.**11 to **3.**13
  grain size, **3.**2
  microstructure, **3.**2
  porosity, **3.**2
  SiC, **3.**14, **3.**20 to **3.**25
  Si_3N_4, **3.**15, **3.**21 to **3.**34
  TiC, **3.**16
  ZrO_2, **3.**16 to **3.**18, **3.**36 to **3.**39
Melting temperature, **1.**16
Metal-matrix composites (MMC), **3.**10, **4.**74, **5.**1, **5.**44, **9.**15
  primary processes, **5.**1
    combining, **5.**1
      liquid-metal infiltration, **5.**2
      modified coating, **5.**2
      plasma spraying, **5.**1
    consolidating, **5.**1
      diffusion bonding, **5.**1
      hot pressing, **5.**1
  secondary processes, **5.**1
    joining, **5.**1
      diffusion bonding, **5.**1
    shaping, **5.**1
      diffusion bonding, **5.**1
      hot pressing, **5.**1

Metal nitridation, **4.**10
Metallizing processes, **7.**70 to **7.**81
  active metal or active-hydride, **7.**80
  copper-copper compound, **7.**72
  direct metallization, **7.**72
  Intragene, **7.**74, **7.**77
  moly-manganese, **7.**70
  vapor deposition, **7.**77, **7.**78
Metals, **1.**2
Microwave heating, **7.**81, **7.**82, **9.**3 to **9.**5
  costs, **9.**5
  nitridation, **9.**3
  principles, **9.**4
  processing, **9.**3
Microwave sintering, **6.**30 to **6.**35, **9.**3
Milling, **7.**13, **7.**14, **7.**17
Military engines, **8.**30, **9.**15, **9.**16
  diesel, **8.**30
  turbine, **8.**30, **9.**16
Miscellaneous ceramic applications, **8.**71 to **8.**75, **9.**17
  barrel liners, **8.**73
  castable tooling, **8.**73
  chutes, **8.**74
  cutting blades, **8.**74
  flame lance, **8.**74
  gage blocks, **8.**73
  honeycomb panels, **8.**71, **8.**72
  hydroclones and pumps, **8.**74
  metal-forming tools, **8.**75
  mirrors, **8.**72
  mitts, **8.**73, **9.**17
  nozzles and pins, **8.**75
  spark igniter, **8.**73
  sports equipment, **9.**17
  spray equipment, **8.**74
  wafer seals, **8.**71
  wrist watches, **8.**73
Mixed oxide, **1.**4
Molecular composites (*see* Particulate composites)
Mullite, **4.**70, **9.**28

Natural ceramics (rocks and minerals), **1.**9, **1.**11
Nicalon (Si–C–O–N), **9.**28
Nitrides, **1.**14, **1.**15
  aluminum, **1.**15
  boron, **1.**15
  sialon, **1.**14, **8.**14
  silicon, **1.**14
Nitriding, **6.**14

Nondestructive evaluation, **4.10**, **8.51**, **8.76** to **8.81**
  ceramics, **2.29**
  methods, **8.77** to **8.81**
    C-mode scanning acoustic microscopy (C-SAM), **8.79**
    impulse vibration excitation technique, **8.80**, **8.81**
    Isoscan generator system, **8.78**
    peak detector system, **8.79**
    scanning acoustic microscopy (SAM), **8.79**
    scanning laser acoustic microscopy (SLAM), **8.79**
    scanning photoacoustic microscopy (SPAM), **8.79**, **8.80**
    ultrasonic velocity, **8.80**
    ultrasonics, **8.77** to **8.81**, **9.13**
      acoustoultrasonics, **9.13**
      process definition, **8.78**
      through-transmission, **8.78**
    X-rays, **8.77**, **8.78**, **8.81**
      CAT (computer-assisted tomography), **8.77**, **8.78**
      planar microfocus radiography, **8.77**
  sintered parts, **4.10**
    acoustic microscopy, **4.10**
    high-frequency ultrasonics, **4.10**
    holographic interferometry, **4.10**
    infrared scanning, **4.10**
    radiography, **4.10**
  types of flaws, **8.77**
    class I flaws, **8.77**
      material strength, **8.77**
    class II flaws, **8.77**
      fracture toughness, **8.77**
      oxidation behavior, **8.77**
  unfired parts, **4.10**
    low-frequency acoustic method, **4.10**
    microradiography, **4.10**
Notched-beam test method, **2.14**

Optical functions, **8.2** to **8.4**
  frequency doublers, **8.3**, **8.4**
  light detectors, **8.3**, **8.4**
  light guides, **8.3**, **8.4**
  light sources, **8.3**, **8.4**
  memory systems, **8.3**, **8.4**
  reflectors, **8.3**, **8.4**
  shutters, **8.3**, **8.4**

Organic adhesives, **7.56**
Organometallic compounds, **4.15**
  glass ceramics, **4.15**
  glass-ceramic composites, **4.15**
Organometallic polymers, **4.27**, **4.28**
  aminoborazine, **4.27**
  polycarbosilane, **4.27**, **4.28**
Organometallic synthesis, **4.7**, **4.15**
  CVD, **4.37**
  sol-gel, **4.7**, **4.8**
Osprey process (*see* Spray casting)
Overpressure sintering, **6.16**, **6.17**
  combustion sintering by gas pressure (HPCS), **6.17**
Oxides, **1.13**, **1.14**
  Al_2O_3, **1.13**
  BeO, **1.13**
  MgO, **1.13**
  mixed, **1.14**
  ThO_2, **1.13**
  TiO_2, **1.14**
  ZrO_2, **1.13**
  synthesis, **4.7**, **4.15**, **4.16**
    aerosol, **4.7**, **4.15**, **4.16**
    aqueous precipitation, **4.7**
    CVD, **4.7**, **4.10**
    flame hydrolysis, **4.7**, **4.9**
    hydrothermal, **4.7**, **4.9**
    metal nitridation, **4.7**, **4.10**
    organometallic, **4.7**
    plasma, **4.7**, **4.9**
    sol-gel, **4.7**, **4.8**
    thermite, **4.7**

Particle packing, **5.6**
Particle sizes, **4.12**, **7.7**
Particles (*see* Particulate composites; Particulates)
Particulate composites, **4.75**, **9.22**
  dual toughening, **4.75**
  sol-gel, **4.75**
Particulates, **4.61** to **4.63**, **4.66**, **4.74**, **9.22**, **9.29** to **9.32**
  Al_2O_3, **4.74**
  ceramic types, **4.74**
  MMC, **5.2**
  Si_3N_4 matrix, **3.32**
Peak detector system, **8.79**
Physical vapor deposition (PVD), **4.10**
Planar microfocus radiography, **8.77**
Plasma, **4.1**
Plasma sintering, **6.27**

Plasma synthesis, **4.9, 4.16** to **4.18**
   reactive electrode submerged arc (RESA), **4.16, 4.19**
Polishing, **7.8, 9.11**
Polymer-matrix composites (PMC) (*see* Resin-matrix composites)
Polymer pyrolysis, **4.25**
   precursors, **4.27** to **4.29**
Powder processes, **4.14**
Powder processing, problem of, **1.5**
Powders, **4.1** to **4.12**
   agglomerates, **4.3**
   inorganic inclusions, **4.4**
   organic inclusions, **4.4**
   production, **4.14**
   properties, **4.5, 4.6, 4.12**
     purity, **4.6**
     size, **4.5, 4.12**
     size distribution, **4.5, 4.12**
     shape, **4.5**
     state of aggregation, **4.5**
   synthetic processing, **4.6** to **4.39**
   types of, **4.45**
     AlN, **4.45**
     BN, **4.45**
Preceramic polymer chemistry, **4.27**
Precursors, **4.27, 4.28**
   polycarbosilane, **4.28**
   polysilazane, **4.27**
Pressure-assisted sintering (PAS), **6.27**
Pressure sintering (*see* Hot pressing)
Pressureless sintering, **6.10** to **6.13**
   properties, **6.11, 6.12**
Printed-circuit boards, **8.68**
   Al_2O_3, **8.68**
Processing, **1.4, 4.1, 4.2**
   advances in, **1.4**
   powder, **4.1, 4.2**
     consolidation to engineering shape, **4.2**
     densification, **4.2**
     manufacture, **4.2**
     preparation for consolidation, **4.2**
PVD coating processes, **7.86** to **7.93**
   electron beam, **7.92, 7.93**
   sputter, **7.90**
   sputter ion plate, **7.90, 7.91**

Radiators, **8.38, 8.39**
Radomes, **8.32, 8.33**
   sintered fused SiO_2, **8.33**
Rapid omnidirectional consolidation (ROC), **6.30**

Rapid solidification technology (RST), **4.1**
Reaction bonding, **3.30, 4.20, 9.23, 9.27, 9.28**
   Si_3N_4, **3.30, 4.20**
Reaction hot pressing (RHP), **4.24**
Reaction sintering, **6.13** to **6.16**
   RBSC or SSC, **3.25, 6.14**
   RBSN, **6.14**
Recuperators, **8.25** to **8.27**
   constraints, **8.25**
   materials, **8.25**
     SiC, **8.25, 8.27**
Reentry vehicles, **8.36, 8.38**
   fused SiO_2, **8.38**
   HRSI (black tile), **8.38**
   HTP, **8.38, 8.75**
   LRSI (white tile), **8.38**
   TPS, **8.36, 8.38**
Refractories, **1.9, 1.11, 1.13**
   borides, **1.14**
   carbides, **1.14**
     B_4C, **1.14**
     SiC, **1.14**
   nitrides, **1.14**
Reliability and design, **9.38**
   CARES, **9.38**
Resin-matrix composites (RMC), **3.10**
Rheocasting, **5.45**
   processing, **5.45**
Rheology, **5.6**

Scanning acoustic microscopy (SAM), **8.79**
Scanning laser acoustic microscopy (SLAM), **8.79**
Scanning photoacoustic microscopy (SPAM), **8.79, 8.80**
Sealant adhesives, **7.66**
Self-propagating high-temperature synthesis (SHS), **4.1, 4.10, 6.35, 9.20, 9.27**
   combustion synthesis, **4.22** to **4.24, 4.38**
     control reactions of, **4.24**
       chemical activators, **4.24**
       chemical furnace, **4.24**
       kinetic braking, **4.24**
       thermal explosion, **4.24**
   exothermic filtration combustion, **4.38**
   nitrides, **4.23**
   reaction hot pressing (RHP), **4.24**
Semiconductors, **8.2, 8.3**
Sensors, **6.37**
Shaping, **7.35**

Shock-wave processing (*see* Dynamic compaction)
Sialon (Si$_3$Al$_3$O$_3$N$_5$), **1.14**, **3.35**, **4.48**, **9.16**
  forms, **4.48**
  β-sialon, **4.48**, **4.50**
  processing, **4.50**
    liquid-phase sintering, **4.50**
    self-combustion synthesis (sintering), **6.35**, **6.36**
    slip casting, **5.38**
  whiskers, **4.74**
SiC (*see* Silicon carbide)
Si$_3$N$_4$ (*see* Silicon nitride)
SiC/LAS (silicon carbide fibers in lithium aluminosilicate), **8.51**
  glass matrixes, **4.52**
Silica, **8.33**, **8.34**, **8.38**
  fused and sintered, **8.33**, **8.34**, **8.38**
Silica reduction method, **4.43**, **4.44**
  Si$_3$N$_4$, **4.43**, **4.44**
Silicon carbide, **1.14**, **8.11**, **8.24**, **8.25**, **8.27**, **8.33**, **8.36**, **8.43**, **8.67**, **9.14** to **9.16**, **9.22**, **9.26**, **9.28**, **9.35**
  α-phase, **4.41**, **8.50**, **8.51**
  β-phase, **4.41**
  composite, **8.10**, **8.51**, **9.10**
    SiC/Al$_2$O$_3$, **8.10**, **8.51**
    SiC/TiB$_2$, **9.10**
  creep, **3.29**
  fibers, **4.57**, **4.58**, **4.68**, **5.45**, **9.8**, **9.9**, **9.28**
  mechanical properties, **3.14**, **3.20** to **3.24**
  powders, **4.16**, **4.18**, **4.22**, **4.41**
  processing, **3.25**, **4.41**, **4.42**, **6.13**
    Acheson, **4.41**
    hot-pressed, **3.25**, **6.9**
    pressureless or self-sintered, **4.41**, **6.13**
    reaction-sintered, **3.25**, **4.42**, **6.13**
    sintered, **3.25**, **4.42**, **6.14**, **6.23** to **6.26**
  whiskers, **4.64** to **4.71**, **5.38**, **5.45**, **8.50**, **9.5**, **9.16**, **9.24**, **9.32**
    β-type properties, **4.68**, **5.31**, **5.32**
    sol-gel, **4.34**, **9.26**
    VLS process, **4.67**, **4.69**
Silicon carbonitride, **9.35**
Silicon nitride, **1.14**, **8.10** to **8.12**, **8.16**, **8.20**, **8.21**, **8.30**, **8.34**, **8.43**, **8.47**, **8.50**, **9.2**, **9.5**, **9.9**, **9.11**, **9.14**, **9.15**, **9.18**, **9.19**, **9.26**, **9.29**
  α-phase, **4.43**
  β-phase, **4.43**, **4.71**
  fibers, **4.58**, **4.59**

Silicon nitride (*Cont.*):
  matrixes, **4.69**, **9.9**
  mechanical properties, **3.15**, **3.31** to **3.34**
  monolithic controlled microstructure, **9.18**, **9.19**
  powders, **4.16**, **4.18**, **4.20** to **4.23**, **4.43**
  processing, **4.20**, **4.43**, **6.14**, **6.16**, **6.23**, **6.26**, **9.5**
    hot pressing and hot isostatic pressing (HIP), **3.30**, **4.43**, **6.23** to **6.26**, **9.26**
    injection molding, **9.5**
    RBSN, **9.5**
    SSN, **9.5**
    reaction bonding, **4.20**, **4.21**, **4.43**, **9.23**, **9.27**
    sintering and gas-pressure-assisted sintering, **3.30**, **4.43**, **4.44**, **6.4**, **6.7**, **6.14**, **6.16**
    slip casting, **5.37**, **5.38**
    squeeze casting, **4.72**, **5.45**
  whiskers, **3.32**, **3.34**, **5.31**, **5.32**
    Si$_3$N$_4$ matrix, **3.32**, **5.31**, **5.32**, **9.28**, **9.32**
    flexural strength, **3.33**, **3.34**
    flexural toughness, **3.33**, **3.34**
Silicon oxynitride (Si$_2$ON$_2$), **4.28**, **4.74**
Siliconizing, **6.14**, **8.34**
Sinter/HIP (S/H), **6.22** to **6.26**
  materials, **6.23** to **6.26**
    oxides, **6.23**
    sialon, **6.24**
    SiC, **6.25**
    Si$_3$N$_4$, **6.26**
  properties, **6.23** to **6.26**
Sintering, **6.1** to **6.36**
  atmospheres, **6.7**
  green-forming aids, **6.2**
    lubricants, **6.2**
    plasticizers, **6.2**
  key process parameters, **6.3**, **6.4**
  mechanisms, **6.2**, **6.5**
    liquid-phase, **3.30**, **6.2**, **6.13**
    solid-state, **6.2**
    vitrification, **6.2**
  processes, **6.16** to **6.35**
    cold pressing, **6.8**
    high-frequency sintering, **6.28**
    hot isostatic pressing (HIP), **6.17** to **6.26**
    hot pressing, **6.8** to **6.10**
    laser sintering, **6.36**

Sintering, processes (*Cont.*):
   microwave sintering, **6.30** to **6.35**
      advantages, **6.33**
      applications, **6.34**
      parameters, **6.30**, **6.31**
   overpressure sintering, **6.16**, **6.17**
   plasma sintering, **6.27**, **6.28**
      hollow-cathode discharge (HCD), **6.27**
      induction-coupled plasma (ICP), **6.27**
      microwave-induced plasma (MIP), **6.27**
   pressureless sintering, **6.10** to **6.13**
   reaction sintering, **6.14**, **6.28**, **6.29**, **6.35**
      nitriding (RBSN), **6.14**
      siliconizing (RBSC or SSC), **6.14**
   self-combustion synthesis, **6.35**
   sinter forge, **6.28**
   slow forging, **6.29**
   properties, **3.26**
      SiC, **3.25**, **4.42**, **6.13**, **6.14**, **6.23**
      Si_3N_4, **3.30**, **6.4**, **6.5**, **6.9**, **6.14**, **6.23**
Slip casting, **5.1**, **5.15** to **5.17**, **5.37**, **5.38**, **8.47**, **8.51**
   method, **5.14** to **5.16**
   process casting variations, **5.14** to **5.17**
   process future, **5.16**, **5.17**
   process limitations, **5.37**
      centrifugal, **5.37**
      compocasting, **5.18**
      pressure, **5.18**, **5.39**
      tape, **5.17**
      thixotropic, **5.18**
      vacuum-assisted, **5.18**
Slurry pressure casting, **5.39** to **5.42**
   advantages, **5.42**
   differences in casting processes, **5.41**
Slurry pumps, **8.10**
Sol-gel, **4.1**, **4.8**
   definition, **4.29**
      gel, **4.29**
      sol, **4.29**
   future developments, **9.1**, **9.2**
   particular composites, **4.75**
   process selection, **4.34** to **4.36**
   processes, **4.30** to **4.36**
      external gelations, **4.31**, **4.35**
      internal gelations, **4.31**, **4.35**
      precursors, **4.29**
         metal alkoxides, **4.29**
         metal salts, **4.29**

Solution techniques, **4.15**
   precipitation filtration, **4.6**
   solvent combustion, **4.7**
   solvent extraction filtration, **4.6**, **4.14**
   sol-gel, **4.7**
   solvent vaporization, **4.6**, **4.14**
Spray casting (Osprey process), **5.47**
Spray drying, **4.16**, **4.21**
   (*See also* Vapor-phase techniques)
Spray pyrolysis or roasting (*see* Aerosol decomposition)
Squeeze casting, **5.42** to **5.45**
   applications, **5.42**
   MMC, **5.44**
   process steps, **5.42**, **5.43**
   whiskers, **4.70**
      SiC, **4.70**
      Si_3N_4, **4.72**
Standardization, **3.6** to **3.8**, **3.36**
   International Energy Agency, **3.6**
   Japan Fine Ceramics Association (JFCA), **3.6**
   U.S. Advanced Ceramics Association (USACA), **3.10**
Straight-through notch method, **2.14**
Stress analysis, **2.19**
Stress intensity factor, **3.4**
Stress rupture, **3.5**
Structural ceramics, **1.2**, **3.9**, **3.10**
   Al_2O_3, **3.11**
   sialon, **3.10**
   SiC, **3.9**
   Si_3N_4, **3.9**
   ZrO_2, **3.10**
Superconductors, **8.2**, **8.31**, **8.70**, **8.71**
   magnets, **8.71**
   microelectronics, **8.71**
   transformers, **8.71**
Superplasticity, **9.11** to **9.13**
   materials, **9.11** to **9.13**
      AlO, **9.11**
      Si_3N_4, **9.11**
      Y–TZP, **9.11**
   process theories, **9.12**
Surface effects, **2.24** to **2.26**, **3.5**
   chemical deposits, **2.25**
   forming residual stress, **2.24**
   machining, **2.24**
   removal, **2.25**
Surface-flaw bend method, **2.15**
Surface-mount technology, **8.68**, **9.16**
Surface roughness, **7.2**

Synthetic ceramic powder, **4.5 to 4.7**
  economics, **4.7**
  processing, **4.6, 4.7 to 4.12, 4.14, 4.15**
    gas-phase reactions, **4.6, 4.43**
    salt decomposition, **4.6, 4.7, 4.14, 4.29, 4.39**
    solution techniques, **4.6, 4.7, 4.14, 4.15**
    vapor-phase techniques, **4.6, 4.7, 4.14, 4.15, 4.44**

Tape casting, **5.1, 5.17, 5.38, 5.39, 9.6, 9.7**
  applications, **5.39, 9.5**
    multilayered capacitors (MLC), **9.6**
    multilayered ceramic packages (MLCP), **9.6**
  carriers, **5.39**
  process technology, **9.6**
  multilayer polyimide (MLP), **9.6, 9.7**
Tests, **2.4, 4.10, 8.51, 9.13**
  destructive, **2.4, 9.14**
    biaxial, **2.5**
    C-ring, **2.6**
    compression, **2.8**
    diametral, **2.6**
    expanded ring, **2.6**
    flexure, **2.4**
    indentation fracture, **9.14**
    single-edge precracked beam, **9.14**
    tensile, **2.4, 2.6**
  nondestructive, **4.10, 8.51, 9.13**
Thermal conductivity, **1.16**
Thermal expansion, **1.16**
Thin films and metallization, **8.67, 8.69, 8.70**
  AlN, **8.67, 8.70**
Thorium oxide (ThO_2), **1.13**
Titanium boride (TiB_2), **4.22, 8.51, 9.10, 9.18**
  applications, **4.47**
Titanium carbide (TiC), **3.16, 9.28**
Titanium nitride (TiN), **9.17**
Titanium oxide (TiO_2), **1.14**
Toughness, **1.16**
Transformation toughening, **3.5, 6.36**
  sol-gel, **4.31**
Tribological functions, **8.3, 8.4**
  bearings, **8.3, 8.5, 8.6**
  guides, **8.3, 8.5, 8.6**
  seals, **8.3, 8.5, 8.6**
Turbine rotors, **8.50, 8.51**
  α-SiC, **8.50, 8.51**

Turbine rotors (*Cont.*):
  SiC/LAS, **8.51**
  SiC toughened with TiB_2, **8.51**
  Si_3N_4 with SiC_w, **8.50**
  ZrO_2, **3.35**
Turbocharger rotors, **5.35**
  injection molding, **5.35**
  overpressure sintering, **6.16**
  Si_3N_4, **6.16**
Turbochargers, **8.46 to 8.49**
Turbomilling process, **4.40**
Turning, **7.16**
Tyranno fiber, **9.35**

Ultrasonic abrasive machining (UAM), **7.17 to 7.21**
  abrasives, **7.19**
  advantages, **7.20**
  equipment, **7.19**
  impact grinding, **7.18, 7.20**
  vibration grinding, **7.20, 7.21**
Ultrasonic velocity system, **8.80**
Ultrasonics, **8.77 to 8.81, 9.13**
  acoustoultrasonics, **9.13**
  process definition, **8.78**
  through-transmission, **8.78**

Valves, **8.12, 9.29**
Vapor-phase techniques, **4.15 to 4.22, 4.44**
  vapor condensation, **4.7, 4.15**
    aerosol, **4.7**
  vapor decomposition, **4.7, 4.15**
  vapor-solid reactions, **4.7, 4.15**
  vapor-vapor reactions, **4.7, 4.15**
    laser, **4.17, 4.18, 4.21**
    plasma, **4.16 to 4.18**
  spray drying, **4.15, 4.21**
Vibration grinding, **7.20, 7.21**
Vitreous ceramics, **1.10**

Wear and abrasive parts, **8.8 to 8.13**
  ceramic seals, **8.10**
    Al_2O_3, **8.11**
    SiC, **8.11**
    Si_3N_4, **8.11**
  ceramic liners and scrappers, **8.10**
  dies, **8.9**
    PSZ, **8.9**
    TTZ, **8.9**
  grinding media, **8.9, 8.10**
    B_4C, **8.9**
    PSZ, **8.9**

Wear and abrasive parts (*Cont.*):
  slurry pumps, 8.10
    SiC/Al_2O_3 composite, 8.10
    Si_3N_4, 8.10
    TTZ, 8.10
    ZrO_2, 8.10
  valves, 8.12
    Si_3N_4, 8.12
  wire and cable use, 8.11, 8.12
    Al_2O_3, 8.11
    PSZ, 8.12
    TTZ, 8.12
Weibull analysis, 2.9
Weibull distribution, 3.3, 3.4
Whiskers, 4.54, 4.60, 4.74, 5.31, 5.32, 9.16, 9.26, 9.32
  Al_2O_3, 4.73, 4.74, 9.31
  cutting tools, 7.46
  fabrication, 5.2
  methods, 4.54, 4.60
    forming, 4.54
      basal growth, 4.54
      tip growth, 4.60
    techniques, 4.64
      liquid metals, 4.64
      other methods, 4.64
      solid-state methods, 4.64
  miscellaneous ceramics, 4.73
    Si–Al–O–N, 4.74, 9.32
    SiO_2–N_2, 4.74
    TiB_2, 4.73
  MMC, 5.2
  properties, 4.62, 4.68
    fracture toughness and strengthening, 9.32
  SiC, 4.64, 4.66, 4.69 to 4.71, 5.31, 5.32, 5.38, 5.45, 9.16, 9.26, 9.31, 9.32
  Si_3N_4, 4.71, 5.45, 9.32

Whiskers (*Cont.*):
  toughening mechanisms, 4.63, 9.30 to 9.32
  ZrO_2, 4.70
Wire and cable use, 8.11, 8.12
Wire machining, 7.22 to 7.24

X-ray computer tomography, 8.51, 9.14
X-rays, 8.77, 8.78, 8.81

Zirconia fibers, 4.56, 4.59, 9.2, 9.17
  ion implantation, 9.19
    Y_2O_3-stabilized cubic ZrO_2, 9.19
  mechanical properties, 3.16 to 3.18, 3.36 to 3.39,
  microstructure classification, 4.51
    conventional PSZ, 4.51
    fine-grained PSZ, 4.51
    fine-grained monoclinic PSZ, 4.51
    overaged conventional PSZ, 4.51
    single-crystal PSZ, 4.51
    TZP, 4.51
  phase transformation, 4.34
    sol-gel, 4.34
  processing, 4.50
    chemical hydrolysis, 4.50
    sintering CE–TZP, 9.5
    sintering TZP, 6.17
  superplasticity, 9.11
    Y–TZP, 5.38, 9.11 to 9.13
  transformation toughening, 3.5, 8.21
  types, 3.35
    partially stabilized PSZ, 3.35 to 3.38, 4.25, 4.50, 8.9, 8.12
    tetragonal TZP, 3.35 to 3.38
    TTZ, 8.9, 8.10, 8.12
Zirconia-toughened alumina (ZTA), 4.48
  zirconium carbide (ZrC), 9.17
  zirconium nitride (ZrN), 9.17
ZrO_2 (*see* Zirconia fibers)